Electromagnetics, Control and Robotics

A Problems & Solutions Approach

Electromagnetics, Control and Robotics

A Problems & Solutions Approach

Harish Parthasarathy
Professor
Electronics & Communication Engineering
Netaji Subhas Institute of Technology (NSIT)
New Delhi, Delhi-110078

CRC Press
Taylor & Francis Group
Boca Raton London New York

CRC Press is an imprint of the
Taylor & Francis Group, an **informa** business

Manakin
PRESS

First published 2023
by CRC Press
4 Park Square, Milton Park, Abingdon, Oxon, OX14 4RN

and by CRC Press
6000 Broken Sound Parkway NW, Suite 300, Boca Raton, FL 33487-2742

CRC Press is an imprint of Informa UK Limited

British Library Cataloguing-in-Publication Data
A catalogue record for this book is available from the British Library

ISBN: 9781032384337 (hbk)
ISBN: 9781032384351 (pbk)
ISBN: 9781003345046 (ebk)

DOI: 10.1201/9781003345046

Typeset in Arial, Minion Pro, Symbol, Calis MT Bold, Times New Roman, Rupee Foradian, Wingdings, Zap Dingbats, Euclid, MT-Extra
by Manakin Press, Delhi

Manakin
PRESS

Preface

This book is a research monograph covering a variety of problems in statistical state and parameter estimation in nonlinear stochastic dynamical system in both the classical and quantum scenario, propagation of electromagnetic waves in a plasma described by the Boltzmann Kinetic Transport Equation, Classical and Quantum General Relativity etc. It will be of use to Engineering undergraduate students interested in analysing the motion of robots subject to random perturbation. It will also be very useful for research scientists working in Quantum Filtering and studying problems of inhomogeneous perturbation to in a homogeneous and isotropic universe leading to the formation of galaxies in an expanding universe.

The mathematical prerequisite for studying this book include Ordinary and Partial Differential Equations, Riemannian Geometry, Stochastic Calculus of Brownian Motion and Poisson Processes, Perturbation Theory of Dynamical System, Group Theory and Group Representation Theory and Basic Operator Theory Especially in the Boson Fock Space as formulated by R.L. Hudson & K.R. Parthasarathy.

Author

Table of Contents

Master-slave Robots, Optimal Control, Quantum Mechanics and Information

[1] Basic Markov chain theory:

Let $X(t)$, $t \geq$ be a stochastic process on a probability space taking values in $\mathbb{Z}$ with the property

$$P(X(t) \in B \mid X(\tau) : \tau \leq s) = P(X(t) \in \mathrm{B} \mid X(s))$$

for all $B \subset \mathbb{Z}$ and $0 \leq s \leq t$. Then $X(t)$ is called a Markov chain. It is completely characterized by the infinitesimal generator $\Lambda = ((\lambda_{ab}))$ where

$$P(X(t+dt) = b \mid X(t) = a) = \delta_{a,b} + dt.\lambda_{ab} + 0(dt)$$

Let $\qquad P_a(t) = P(X(t) = a)$

for some initial probability distribution of $X(0)$ on $\mathbb{Z}$. Then, the Chapman - Kolmogorov equation

$$P_a'(t) = \sum_b P_b(t)\lambda_{ba}$$

or equivalently, in terms of matrices.

$$\mathbf{P}'(t)^T = \mathbf{P}(t)^T \Lambda$$

which gives

$$\mathbf{P}(t)^T = \mathbf{P}(0)^{\mathrm{T}} \exp(t\Lambda)$$

Let $0 \leq t_1 < t_2 < \ldots < t_p$. Then for $f : \mathbb{Z} \to \mathbb{R}$, we have

$$\mathbb{E}(f(X(t_1))\ldots f(X(t_p))) = P(0)^T \exp(t_1\Lambda)D_f.$$
$$\exp((t_2 - t_1)\Lambda)D_{\mathrm{f}}\ldots\exp((t_p - t_{p-1})\Lambda)Dfu$$

where

$D_f = diag(f(a) : a \in \mathbb{Z})$

and

$$\mathbf{u} = [1,1,\ldots]^T$$

Let $T_1 < T_2 < \ldots$be the jump times of the Markov chain. Assume that $X(0)=i_0$. Then for $i_1 \neq i_0$, $P(T_1(t, t+dt], X(T_1+) = i_1\ P(T_1 > \mathrm{t})\ \lambda(i_0, i_1)dt$

Now,

$$P(T_1 > t + dt) = P(T_1 > t)\,(1+\lambda(i_0, i_0)\,dt)$$

and hence

$$P(T_1 > t) = \exp(t\lambda(i_0, i_0))$$

We are using the notation $\lambda(a, b)$ in place of λ_{ab}. Thus, we get the joint distribution of $(T_1, X(T_1+))$:

$$P(T_1 \in (t, t+dt), X(T_1+) = i_1) = \lambda(i_0, i_1).\exp(t\lambda(i_0, i_0))dt$$

Likewise, we get using the Markov property

$$P(T_n \in (t, t+dt), X(T_n+) = i_n \mid T_{n-1}, (X(T_{n-1}+) = i_{n-1})$$

$$= \lambda(i_{n-1}, i_n) \exp(t\lambda(i_{n-1}, i_{n-1}))\, dt$$

It is clear that $(T_n, X(T_n+)), n \geq 0$ where T_0 = is a discrete time Markov process with state space $\mathbb{R}_+ \times \mathbb{Z}$.

Now suppose we wish to compute the expectation

$$u(t, x) = \mathbb{E}\left[\exp\left(\int_0^t f(X(s))\,ds\right) g(X(t)) \middle| X(0) = x\right], t \geq 0, x \in \mathbb{Z}$$

We have by the Markov property and homogeneity (stationary transition probabilities) on neglecting $0\,(dt)$ terms,

$$u(t+dt, x) = (1 + f(X)dt\, \mathbb{E}[u(t, X(dt)) | X(0) = x]$$

$$= (1 + f(x)dt)(u(t, x) + dt)\sum_u u(t, y)\lambda(x, y)$$

so that

$$\frac{\partial u(t, x)}{\partial t} = f(x)\, u(t, x) + \sum_y \lambda(x, y) u(t, y)$$

or equivalently, defining the column vector $\mathbf{u}(t) = (u(t, x))_{x \in \mathbb{Z}}$

$$\mathbf{u}'(t) = D_f \mathbf{u}(t) + \Lambda \mathbf{u}(t)$$

The initial condition is

$$u(0, x) = g(x)$$

or equivalently, in terms of column vectors,

$$\mathbf{u}(0) = \mathbf{g}$$

Thus,

$$\mathbf{u}(t) = \exp(t(Df + \Lambda))\, \mathbf{g}$$

Assume that the eigenvalues of $D_f + \Lambda$ are λ_1, λ_2,... arranged in decreasing order or real parts , *i.e.*, $\text{Re}\lambda_k \geq \text{Re}\lambda_{k+1} \forall k$.

Let υ_k, $k = 1, 2,$...be the corresponding right eigenvectors and w_k, $k = 1, 2,$... the corresponding left eigenvectors. Then, we have

$$Df + \Lambda = \sum_k \lambda_k v_k w_k^T$$

Note that

$$w_k^T v_i = \delta_{kj}$$

Thus,
$$\mathbf{u}(t) = \sum_k \exp(\lambda_k t) v_k w_k^T$$

Suppose λ_1 is real and $\lambda_1 > \text{Re}(\lambda_2)$. Then, assuming $w_1^T \mathbf{g} \neq 0$ and $v_k(x) \neq 0$ for some fixed x, we get

$$\lim_{x\to\infty} t^{-1}\log\left(u(t,x)\right) = \lambda_1$$

❖❖❖❖❖

[2] Basic quantum mechanics and quantum field theory:

Quantum mechanics is based on the following postulates : (1) The state of the system at any time is described by a density operator ρ in a Hilbert space $\mathcal{H}$. By a density operator, we mean that it is positive semidefinite and its trace is unity. The state ρ is said to be pure if it has rank one. In that case, we can write $\rho = |\psi\rangle\langle\psi|$ where $|\psi\rangle$ is a unit vector in $\mathcal{H}$. The state ρ plays the role of a probability measures in classical probability theory while the Hibert space $\mathcal{H}$ plays the role of the sample space. (2) An event is an orthogonal projection operator P in $\mathcal{H}$. The probability of the event P occurring when the system is in the state ρ is defined as $Tr(\rho P)$. The Boolean σ algebra of events in classical probability is replaced by $\mathcal{P}(\mathcal{H})$, the lattice of orthogonal projections in $\mathcal{H}$. Just as in classical probability theory, we talk of a set of mutually exclusive and collectively exhaustive events (*i.e.*, pairwise disjoint and whose union is the entire sample space), the analog in quantum probability is a spectral resolution of the identity, *i.e.*, a countable set $\{P_i\}$ of orthogonal projections in $\mathcal{H}$ such that $P_iP_j = P_i\delta_{ij}$ and $\Sigma_i P_i = I$. The analogue of an observable in the quantum theory is a Hermitian operator X in the Hilbert space. Suppose we take a classical mechanical system described by generalized coordinates $(q_1,\ldots,q_n)$ and generalized momenta $(p_1\ldots p_n)$, then we have the Poisson bracket relation

$$\{q_i, p_j\} = \delta_{ij}$$

An observable for this system is real valued function of $(t, q_1,\ldots,q_n, p_1,\ldots, p_n)$. We denote it by $f(t, q_1,\ldots, q_n, p_1,\ldots, p_n)$.

The classical system evolves in time in accord with a Hamiltonian.

$$H(t, \mathbf{q}, \mathbf{p})$$

with the equations of motion being

$$\mathbf{q}'(t) = \nabla_p H(t,\mathbf{q},\mathbf{p}), \mathbf{p}'(t) = -\nabla_q H(t,\mathbf{q},\mathbf{p})$$

The rate of change of the observe f is then

$$\begin{aligned}\frac{df}{dt} &= \frac{\partial f}{\partial t} + \sum_i q'(t)\frac{\partial f}{\partial q_i} + p_i'(t)\frac{\partial f}{\partial p_i} \\ &= \frac{\partial f}{\partial t} + \sum_i \frac{\partial H}{\partial pi}\frac{\partial f}{\partial qi} - \frac{\partial H}{\partial qi}\frac{\partial f}{\partial pi} \\ &= \frac{\partial f}{\partial t} + \{f,H\}\end{aligned}$$

The Poisson bracket in quantum mechanics is the replaced by $-i$ times the commutator for the following reason: The properties of the classical Poisson bracket imply that for observables u_1, u_2, v_1, v_2 (maintaining the order in which the observables appear in the Poisson bracket)

$$\{u_1u_2, v_1v_2\} = v_1\{u_1u_2, v_2\} + \{u_1u_2, v_1\}\, v_2$$
$$= v_1\{u_1, v_2\}u_2 + v_1u_1\{v_2, v_2\} + \{u_1, v_1\}u_2v_2 + u_1\{u_2, v_1\}v_2$$

on the one hand and on the other,

$$\{u_1u_2\,, v_1v_2\} = \{u_1, v_1v_2\}\, u_2 + u_1\, \{u_2, v_1v_2\}$$
$$= \{u_1, v_1\}\, v_2u_2 + v_1\, \{u_1, v_2\}\, u_2 + u_1\{u_2, v_1\}\, v_2 + \{u_2, v_2\}$$

Equating these two expressions gives us

$$\{u_1, v_1\}\,(u_2v_2 - v_2u_2) = (u_1v_1 - v_1u_1)\{u_2, v_2\}$$

and hence $\{u_1, v_1\}$ must be replaced by a constant times $u_1v_1 - v_1u_1 = [u_1, v_1]$ in the quantum, theory.

Phase volume and Liouville's theorem: Assume that the classical system is described by the Hamiltonian $H\,(t, \mathbf{q}, \mathbf{p})$.

Let the solution to the Hamiltonian equations be $T_t(\mathbf{q}(0),\mathbf{p}(0))$ Let $(\mathbf{q}(0), \mathbf{p}(0))$ be a random vector with probability density $\rho_0\,(\mathbf{q}, \mathbf{p})$. Let $\rho_t\,(\mathbf{q}, \mathbf{p})$ be the probability density of $(\mathbf{q}(t), \mathbf{p}(t)$. Then we have by the standard change of variable formula.

$$\rho_t\,(T_t\,(\mathbf{q}, \mathbf{p})\; \mathbf{J}_t(\mathbf{q}, \mathbf{p}) = \rho_0\,(\mathbf{q}, \mathbf{p})$$

where J_t is the Jacobian determinant of T_t. The Jacobian of the infinitesimal transformation $(\mathbf{q}(t), \mathbf{p}(t)) \to (\mathbf{q}(t+dt), \mathbf{p}(t+dt)$ is one since the divergence of the Hamiltonian vector field $(\nabla_p H, -\nabla_q H)$ vanishes:

$$\nabla_q . \nabla_p H - \nabla_p . \nabla_q H = 0$$

Hence $J_t = 1$. Thus we get

$$\rho_t(T_t(\mathbf{q}, \mathbf{p}) = \rho_0\,(\mathbf{q}, \mathbf{p})$$

and hence

$$\frac{d}{dt}\rho_t\,(T_t\,(\mathbf{q},\mathbf{p}) = 0$$

and this can equivalently be expressed as

$$\frac{d}{dt}\rho_t(\mathbf{q}(t),\mathbf{p}(t)) = 0$$

or making use of the Hamiltonian equations of motion,

$$\frac{\partial p_t\,(\mathbf{q}, \mathbf{p})}{\partial t} + \{p_t(\mathbf{q},\mathbf{p}), H(t,\mathbf{q},\mathbf{p})\} = 0$$

This is called Liouville's equation.

If there is Brownian noise in the Hamiltonian system , then Liouville's acquires an Ito second order correction term. The equation of motion of the density ρ_t and observable f are dual to each other. The average value of the observable f at time t is

$$\langle f \rangle\,(t) = \int f(t,\mathbf{q},\mathbf{p})\rho_t(\mathbf{q},\mathbf{p})d\mathbf{q}d\mathbf{p}$$
$$= \int f(t,\mathbf{q},\mathbf{p})\rho_0(T_t^{-1}(\mathbf{q},\mathbf{p}))d\mathbf{q}d\mathbf{p}$$

$$= \int f(t, T_t(\mathbf{q},\mathbf{p}))\rho_0(\mathbf{q},\mathbf{p})d\mathbf{q}d\mathbf{p}$$

Hence, the rate of change of this average can be expressed in two different ways, one as

$$\frac{d<f>(t)}{dt} = \int\left(\frac{\partial f}{\partial f} + \{f, H\}\right)\rho_0(\mathbf{q},\mathbf{p})d\mathbf{q}d\mathbf{p}$$

and two as

$$\frac{d<f>}{dt} = \int\left[\frac{\partial f}{\partial t}\rho_t + f\{H, \rho_t\}\right]d\mathbf{q}d\mathbf{p}$$

One can be derived from the other by using integration by parts. So we have a duality in classical mechanics, namely in computing the rate of change of the average value of an observable, we can either keep the density fixed and calculate w.r.t. this density the average of the rate of change of the observable or keep the observable fixed w.r.t. its phase space variables and calculate the average value of its partial derivative w.r.t. time using a density that evolves in accord with Liouville's equation. Likewise in the quantum theory, the state of the system evolves according to Schrodinger' equation

$$\rho'(t) = -i[H(t), \rho(t)]$$

in the Schrodinger picture and in this picture, observables X do not evolve with time. The average value of the observable at time t is thus

$$\langle X\rangle(t) = Tr(\rho(t)X)$$

and its rate of change is

$$\langle X\rangle'(t) = Tr(p'(t)X) = -Tr(i[H(t), \rho(t)]X) = i.Tr(\rho(t)[H(t), X])$$

so we can alternately assume that ρ does not evolve in time but X evolves in time in accord with

$$X'(t) = i[H(t), X(t)]$$

This is the Heisenberg picture or matrix mechanics. We write

$$T_t = I + \sum_{n\geq 1}(-i)^n \int_{0<t_n<\ldots<t_1<t} ad(H(t_1))..ad(H(t_n))dt_1\ldots dt_n$$

and then the solution to Schrodinger's equation is

$$\rho(t) = T_t(\rho(0))$$

We note that

$$\frac{dT_t}{dt} = -i.ad(H(t))_0 T_t$$

so

$$\frac{dT_t(p(0))}{dt} = -iad(H(t))_0(T_t(\rho(0))) = i[H(t), T_t(\rho(0))]$$

i.e.

$$T_t(\rho(0)) = p(t)$$

in the Schrodinger picture. We then have

$$\langle X\rangle(t) = Tr(T_t(\rho(0))X) = Tr(\rho(0)\, T_t^*(X))$$

where

$$T_t^* = I + \sum_{n\geq 1}(i)^n \int_{0<t_n<...<t_1<t} ad(H(t_n))...ad(H(t_1))dt_{1...}dt_n$$

and hence

$$\frac{dT_t^*}{dt} = iT_t^* o ad(H(t))$$

so

$$\frac{dT_t^*(X)}{dt} = iT_t^*([H(t), X])$$

If the $H(t)$, s_n at different times commute, then this reduces to

$$\frac{dT_t^*(X)}{dt} = i[H(t), T_t^*(X)]$$

which is the Heisenberg equation of motion.

Optimal control in the Heisenberg picture: Assume that the Hamiltonian H depends on a parameter vector θ that can be made to vary with time, *i.e.*, $\theta = \theta(t)$. We write the Hamiltonian as $H(\theta(t))$. The observable X evolves in accord with

$$X'(t) = i[H(\theta(t)), X(t)]$$

We are given an initial state ρ and wish to select the parameter evolution $\theta(t)$ over $[0, T]$ so that $Tr(\rho X(t))$ follows as closely as possible a given function $f(t)$, *i.e*, we wish to minimize

$$\int_0^T (Tr(pX(t)) - f(t))^2\, dt$$

After incorporating the constraints corresponding to the equations of motion using Large multiplier operator $\Lambda(t)$,

the functional to be minimized is

$$S[\theta, X, \Lambda] = \int_0^T (Tr(\rho X(t)) - f(t))^2 dt - \mathrm{Re}\int_0^T Tr(\Lambda(t)(X'(t)\ 0 - i[H(\theta(t)), X(t)]))dt$$

$\theta(t)$ is assume to be a real parameter vector.

$$\delta_\theta S = 0$$

gives

$$\mathrm{Re}\, Tr(\Lambda(t)[\nabla_\theta H(\theta(t)), X(t)]) = 0$$

$$\delta_\chi S = 0$$

gives

$$-\Lambda'(t) - i[\Lambda(t), H(\theta(t))] + 2(Tr(\rho X(t)) - f(t))\rho = 0$$

Finally,

$$\delta_\Lambda S = 0$$

gives the Heisenberg equations of motion

$$X'(t) - i[H(\theta(t)), X(t)] = 0$$

Quantum stochastic differential equations in the Hudson-Parthasarathy setting: $\mathcal{H}$ is a Hilbert space and χ_t, $t \geq 0$ is an increasing family of right-continuous orthogonal projections in $\mathcal{H}$. For $u \in \mathcal{H}$, $e(u)$ is the exponential vector in the Boson Fock space $\Gamma_s(\mathcal{H})$. For $u \in \mathcal{H}$, we define $u_t = \chi_t u$ for $a(u)$, $a^*(u)$ are the annihilation and creation operator fields in $\Gamma_s(H)$. The family of operators $A_t(u) = a(\chi_t u)$, $t \geq 0$ is called the annihilation process and the family $A_t^*(u) = a^*(\chi_t u)$, $t \geq 0$ the creation process. They satisfy the commutation rules

$$\left[a(u), a(v)^*\right] = \langle u, v\rangle, \left[A_t(u), A_s(v)^*\right] = \langle u_t, v_s\rangle = \left\langle u, \chi\left[0, \min(t,s)\right]^v\right\rangle$$

For an operator H in $\mathcal{H}$ that commutes with the spectral family $\{\chi_t\}$, we define the conservation process

$$\Lambda_t(H) = \lambda(H_t)$$

where

$$H_t = H_{\chi t} = {}_{\chi t}H$$

and $\lambda(H)$ is the operator in $\Gamma_s(H)$ defined by

$$\exp(it\lambda(H)) = W(0, \exp(itH)), t \in \mathbb{R}$$

where $W(u,U), u \in \mathcal{H}, U \in \pounds(\Gamma_s(\mathcal{H}))$ are the Weyl operators. They are unitary when U is unitary. We have

$$\langle e(w)|\lambda(H)a(u)|e(v)\rangle = \langle u,v\rangle\langle w|H|v\rangle\langle e(w)|e(v)\rangle$$

Note that

$$\langle e(w)|\exp(it\lambda(H))|e(v)\rangle = \langle e(w)|W(0,\exp(itH))|e(v)\rangle$$

$$= \langle e(w)|e(\exp(itH)v)\rangle = \exp\left(\langle w|\exp(itH)|v\rangle\right)$$

so

$$\langle e(w)|\lambda(H)|e(v)\rangle = \langle w|H|v\rangle \exp(\langle w,v\rangle)$$

$$= \langle w|H|v\rangle\langle e(w)|e(v)\rangle$$

Also

$$\langle e(w)|a(u)\lambda(H)|e(v)\rangle = \left\langle a(u)^* e(w)|\lambda(H)|e(v)\right\rangle$$

$$= \frac{d}{dt}\langle e(w+tu)|\lambda(H)|e(v)\rangle\Big|_{t=0}$$

$$= \frac{d}{dt}\langle w+tu|H|v\rangle \exp\left(\langle w+tu|v\rangle\right)\Big|_{t=0}$$

$$= \left(\langle u|H|v\rangle + \langle w|H|v\rangle\langle u|v\rangle\right)\langle e(w)|e(v)\rangle$$

Thus,

$$\langle e(w)|[\lambda(H), a(u)]|e(v)\rangle = \langle u|H|v\rangle\langle e(w)|e(v)\rangle$$

On the other hand,

$$\langle e(w)|a(H^*u)|e(v)\rangle = \langle H^*u|v\rangle\langle e(w)|e(v)\rangle = \langle u|H|v\rangle\langle e(w)|e(v)\rangle$$

Thus, $[\lambda(H), a(u)] = a(H^*u)$

Taking the adjoint of this equation gives

$$[a(u)^*, \lambda(H)] = a(H^*u)^*$$

and replacing H by H^*, we get

$$[a(u)^*, \lambda(H)] = a(Hu)^*$$

Thus,

$$[\Lambda_t(H), A_s(u)] = A_{\min(t,s)}((H^*u))$$

and

$$[A_t(u)^*, \Lambda_s(H)] = A_{\min(t,s)}(Hu)^*$$

Finally,

$$\langle e(w)|\lambda(H_1)\lambda(H_2)|e(v)\rangle$$

$$= -\frac{\partial^2}{\partial t \partial s}\langle e(w)|e(\exp(itH_1).\exp(isH_2)v)\rangle\Big|_{t=s=0}$$

$$-\frac{\partial^2}{\partial t \partial s}\exp\left(\langle w|(1+itH_1)(1+isH_2)|v\rangle\right)\Big|_{t=s=0}$$

$$= -i\frac{\partial}{\partial s}\langle w|H_1(1+isH_2)|v\rangle \exp\left(\langle w|1+isH_2|v\rangle\right)\Big|_{s=0}$$

$$= \left[\langle w|H_1H_2|v\rangle + \langle w|H_1|v\rangle\langle w|H_2|v\rangle\right]\langle e(w)|e(v)\rangle$$

Thus,

$$\langle e(w)|[\lambda(H_1), \lambda(H_2)]|e(v)\rangle = \langle w|H_1, H_2|v\rangle\langle e(w)|e(v)\rangle$$

and therefore,

$$[\lambda(H_1), \lambda(H_2)] = \lambda([H_1, H_2])$$

from which, it follows that

$$[\Lambda_t(H_1),\Lambda_s(H_2)] = \Lambda_{\min(t,s)}([H_1,H_2])$$

Quantum filtering theory: Let A_t, $t \geq 0$ be an increasing family of Non-Neumann algebras, for example $\mathcal{H}_t = \mathcal{X}_t\mathcal{H}$ is an increasing family of Hilbert spaces and $A_t = (\pounds_s(\mathcal{H}_t))$ is the family of linear operators on the Boson Fock space $\Gamma_s(\mathcal{H}_t)$. We can regard (via a unitary isomorphism),

$$\Gamma_s(\mathcal{H}_{t+s}) = \Gamma_s(\mathcal{H}_t)\otimes\Gamma_s(\mathcal{H}_{(t,t+s]})$$

where

$$\mathcal{H}_{(t,\,t+s]} = (\mathcal{X}_{t+s}-\mathcal{X}_t)\mathcal{H}$$

❖❖❖❖❖

[3] Solving RDRA problem with arbitrary boundary using orthogonal curvilinear co-ordinate:

An orthogonal curvilinear system of coordinates in the *xy* plane can be easily constructed using the real and imaginary parts of an analytic function $q = q_1 = jq_2$ of the complex variable $Z = x+jy$. We can select the analytic function $q = F(Z)$ such that q_1 = constant corresponds to the boundary curve of the cross-section of the waveguide or cavity resonator or microstrip antenna. We can evaluate the Lame's coefficients for this orthogonal curvilinear system easily from the original analytic function F. We assume that F is invertible and then express $Z = G(q)$ where G is the inverse of F. We also assume that G is an analytic function of a complex variable in the domain of interest. The Lame's coefficients are

$$H_1(q_1, q_2) = \left(\left(\frac{\partial x}{\partial q_1}\right)^2 + \left(\frac{\partial y}{\partial q_1}\right)^2\right)^{1/2} = |G'(q)|^2$$

and likewise

$$H_2(q_1, q_2) = \left(\left(\frac{\partial x}{\partial q_2}\right)^2 + \left(\frac{\partial y}{\partial q_2}\right)^2\right)^{1/2} = |G'(q)|^2$$

i.e., $H_1 = H_2$. We note that since

$$G(q) = G(F(Z)) = Z,$$

we have

$$G'(q)F'(Z) = 1$$

and hence

$$H_1 = H_2 = |F'(Z)|^{-1} = \left|F'\left(F^{-1}(q)\right)\right|^{-1}$$

The Laplacian in the *xy* plane can be expressed in terms of q:

$$\nabla_\perp^2 = \frac{1}{H_1H_2}\left(\frac{\partial}{\partial q_1}(H_2/H_1)\frac{\partial}{\partial q_1} + \frac{\partial}{\partial q_2}(H_1/H_2)\frac{\partial}{\partial q_2}\right.$$

$$= |G'(q)|^{-2}\left(\frac{\partial^2}{\partial q_1^2}+\frac{\partial^2}{\partial q_2^2}\right)$$

Consider now a *TM* mode in the microstrip antenna patch of height *d*. The top and bottom surfaces of this patch are perfect electric conductors while the sidewalls are perfect magnetic conductors. The Maxwell equations imply the boundary conditions $E_x = E_y = 0$ at $z = 0.\ d$ or equivalently, $E_{q1}(q_1, q_2, z) = E_{q2}(q_1, q_2, z) = 0$ when $z = 0, d$ and $H_{q1}(c, q_2) = 0$. We note that $\hat{q}_1$ is normal to the boundary curve and $\hat{q}_2$ is tangential to it. For the *TM* mode, the boundary condition on H_{q1} can easily be show on to translate into $E_{z,q_1}(c, q_2) = 0$ where $E_{z,q_1} = \frac{\partial E_z}{\partial q_1}$. Thus E_z consists of a linear combination of functions of the form $\psi(q_1, q_2)\sin(\pi pz/d)$ where

$$\nabla_\perp^2\psi(q_1,q_2)+h^2\psi(q_1,q_2) = 0$$

where

$$h^2 = \omega^2\mu\varepsilon-(\pi p/d)^2$$

or

$$\omega = (\mu\,\varepsilon)^{-1/2}\left(h^2+(\pi p/d)^2\right)^{1/2}$$

The possible values of h^2 are determined from the Neumann boundary condition

$$\psi_{,q_1}(c, q_2) = 0 \quad ..(1)$$

We define

$$K(q_1, q_2) = |G'(q)|^2$$

Then the 2-*D* Laplace equation for ψ can be expressed as

$$\left[\left(\partial_1^2+\partial_2^2\right)+h^2K(q_1,q_2)\right]\psi(q_1,q_2) = 0 \quad ...(2)$$

where

$$\partial_k = \frac{\partial}{\partial qk},\ k=1,2$$

Remarks: (*a*) Here for any vector *f*, f_{q_1} denotes its components along the q_1 direction *i.e.*,

$$f_{q_1} = (f,\nabla q_1)/|\nabla q_1| \quad (1)$$

and likewise,

$$f_{q_2} = (f,\nabla q_2)/|\nabla q_2| \quad (1)$$

We note that in view of the orthogonality of the curvilinear system, ∇q_k is proportional to $\frac{\partial x}{\partial qk}\hat{x}+\frac{\partial y}{\partial qk}\hat{y},\ k = 1, 2$.

(*b*) Suppose that we are given a closed curve Γ in the *xy* plane. We wish to find a function $\Gamma(q)$ of a complex variable such that when $q_1 = c$, the curve $q_2 \to \Gamma(c + jq_2)$ approximates the curve as closely as possible. For this, we sample the curve Γ at points (x_k, y_k), $k = 1, 2..., N$ and choose points q_{2k}, $G_k = 1,2,... N$. We then select analytic test functions $\Gamma k(q)$, $k = 1, 2,...,p$ (for example, $G_k(q) = q_{k-1}$ or $G_k(q) = exp(\lambda_k q)$ and complex numbers $c_1,..., c_p$ so that

$$E(c_1,..., c_p) = \sum_{k=1}^{N} \left| \sum_{m=1}^{p} c_m G_m(c + jq_{2k}) - x_k - jy_k \right|^2$$

is a minimum. This optimization amounts to solving a set of linear equations for c_m, $m = 1, 2,...,p$.

Example: Let

$$Z = G(q) = \exp(q)\,(1 + \delta\chi(q))$$

Thus,

$$q = \log(Z)\,(1-\delta_\chi(\log(Z))) + O(\delta^2)$$

δ is a small perturbation parameter. Let

$$\phi(Z) = \chi(\log(Z))$$

Thus,

$$q = \log(Z)\,(1-\delta\phi(Z))$$

Then writing $\phi_1 = \text{Re}\phi$, $\phi_2 = Im(\phi)$, we have

$$\begin{aligned} q_1 &= \log(r)(1-\delta\phi_1(r, \theta)) + \delta\theta\phi_2(r, \theta)) + O(\delta_2) \\ &= \log(r) + \delta(q\phi_2 - \log(r)\phi_1) + O(\delta_2) \end{aligned}$$

and

$$q_2 = \theta(1-\delta\phi_1(r, \theta)) - \delta.\log(r).\,\phi_2(r, \theta)$$

The curve $q_1 = c$ is a small $O(\delta)$ deviation of a circular curve of radius exp (c).

Solution using Fourier series: The function $\psi(q_1, q_2)$ satisfies the generalized Helmholtz equation (2) with the Neumann boundary condition (1). We express the solution as a Fourier series in q_2 in analogy with the above example where q_2 equals the polar angular coordinate θ plus a small perturbation:

$$\psi(q_1, q_2) = \sum_n \psi_n(q_1)\exp(inq_2) \qquad ...(3)$$

Let

$$K(q_1, q_2) = \sum_n K_n(q_1)\exp(inq_2) \qquad ...(4)$$

Then we get from (2), (3) and (4),

$$\psi_n''(q_1) - n^2\psi_n(q_1) + \sum_r K_{n-r}(q_1)\psi_r(q_1) = 0 \qquad ...(5)$$

This is a sequence of linear ordinary differential equations for the sequence of functions ψ_n, $n \in \mathbb{Z}$. The Neumann boundary condition implies that

$$\psi'_n(c) = 0 \forall n$$

Thus, we can expand ψ_n as

$$\psi_n(q_1) = \sum_{k=1}^{p} c[n,k](q_1 - c)^k \qquad ...(6)$$

and determine the coefficients $c[n, k]$ approximately by applying the method of moments. Substituting (6) into (5) gives

$$\sum_k c[n,k]k(k-1)(q_1-c)^{k-2}$$

$$-n^2\sum_k c[n,k](q_1-c)^k + \sum_{r,k} K_{n-r}(q_1)c[r,k](q_1-c)^k = 0 \qquad ...(7)$$

We multiply both sides by $(q_1 - c)^s$, $s \geq 1$ and integrate from $q_1 = 0$ to $q_1 = c$ to get a set of linear equations for the coefficients $\{c[n,k]\}$:

$$-\sum_{k\geq 2} k(k-1)(k-1+s)^{-1}(-c)^{s+k-1}c[n,k]$$

$$+n^2\sum_k (s+k+1)^{-1}(-c)^{s+k+1}c[n,k]$$

$$+\sum_{r,k} c[r,k]\int_0^c K_{n-r}(q_1)(q_1-c)^{s+k}dq_1 = 0, s \geq 1, n \in \mathbb{Z} \qquad ...(8)$$

This set linear equations can be truncated and solved for $\{c[n,k]\}$.

Small perturbations from a circular boundary: We consider now the above example

$$Z = \exp(q)(1 + \delta\chi(q))$$

Then,

$$K(q_1, q_2) = |G'(q)|^2 = \left|\exp(q)\left(1+\delta\left(\chi(q)+\chi'(q)\right)\right)\right|^2$$

$$= \exp(2q_1)(1+2\delta\,\mathrm{Re}(\chi(q)+\chi'(q))) + O(\delta^2)$$

Writing

$$\eta(q_1, q_2) = 2\exp(2q_1).\mathrm{Re}\left(\chi(q)+\chi'(q)\right)$$

we have

$$K(q_1, q_2) = \exp(2q_1) + \delta.\eta(q_1, q_2)$$

The Helmholtz equation to be solved is

$$\left[\partial_1^2 + \partial_2^2 + h^2\exp(2q_1) + \delta h^2\eta(q_1, q_2)\right]\psi(q_1, q_2) = 0 \qquad ...(9)$$

with the boundary condition

$$\delta_1\psi(c, q_2) = 0 \qquad ...(10)$$

We solve using perturbation theory:

$$\psi(q_1, q_2) = \psi_0(q_1, q_2) + \delta.\psi_1(q_1, q_2) + O(\delta^2) \quad ...(11)$$

$$h^2 = h_0^2 + \delta.h_1^2 + O(\delta^2) \quad ...(12)$$

Substituting (11) and (12) into (9) equating coefficients of δ_0 and δ_1 gives us respectively

$$\left[\partial_1^2 + \partial_2^2 + h_0^2 \exp(2q_1)\right]\psi_0(q_1, q_2) = 0 \quad ...[13(a)],$$

with the boundary condition

$$\partial_1\psi_0(c, q_2) = 0 \quad ... [13(b)]$$

and

$$\left[\partial_1^2 + \partial_2^2 + h_0^2 \exp(2q_1)\right]\psi_1(q_1, q_2) + h_1^2 \exp(2q_1)\psi_0(q_1, q_2)$$

$$+ h_0^2\, \eta(q_1, q_2)\psi_0(q_1, q_2) = 0 \quad ...[14(a)]$$

with the boundary condition

$$\partial_1\psi_1(c, q_2) = 0$$

The unperturbed equation [(13*a*)] with the boundary condition [(13*b*)] is same as the usual Helmholtz equation inside a circular boundary with Neumann boundary conditions as can be seen by making the change of variables

$$\exp(q_1) = r,\ q_2 = \phi$$

in [13(*a*)] which results in

$$\left[r^{-1}\partial_r r\partial_r + r^{-2}\partial_\phi^2 + h_0^2\right]\psi_0 = 0$$

The general solution to this equation with the above boundary condition which translate to $\frac{\partial\psi_0(r,\phi)}{\partial r}\Big|_{r=\exp(c)} = 0$ is known using separation of variable and is given by

$$\psi_0(r, \phi) = \sum_{m,n} J_m\left(\alpha_{mn}r/R\right)\left(C(m,n)\cos(m\phi) + D(m,n)\sin(m\phi)\right)$$

where J_m is the Bessel function of order *m*, *i.e.*, it satisfies

$$x^2 J_m''(x) + xJ_m'(x) + (x^2 - m^2)J_m(x) = 0$$

and α_{mn} are the roots of $J_m'(x)$. Here $R = \exp(c)$.

The solutions are superpositions of $J_m\left(\alpha_{m,n}r/R\right)\cos(m\phi)$, $J_m\left(\alpha_{m,n}r/R\right)\sin(m\phi)$ where $\alpha_{m,n}$ are the roots of $J_m'(x)$.

The corresponding unperturbed modal eigenvalue h_0^2 is $h_0 = \alpha_{m,n}/R = h_0[m, n]$ or equivalently,

$$\omega = (\mu\varepsilon)^{-1/2}(h_0[m,n]^2 + (\pi p/d)^2)^{1/2}$$

$$= (\mu\, \varepsilon)^{-1/2} (\alpha^2_{m,n}/R^2 + (\pi p/d^2)^{1/2}$$

$$= \omega_0[m, n, p].$$

These are the oscillation frequencies inside the cavity assuming a perfectly circular boundary, *i.e.*, no perturbation to the boundary. Now, we denote the above normalized eigenfunctions by $u_{m,n}(q_1, q_2)$ and $v_{m,n}(q_1, q_2)$. We note that there is a two-fold degeneracy of the modal eigenvalue. We denote the unperturbed area measure by

$$dS_0(q_1, q_2) = rdrd\phi = \exp(2q_1)\, dq_1 dq_2$$

Then the unperturbed eigenfunctions satisfy the orthogonality relations:

$$\langle u_{m,n}, u_{p,q}\rangle = \delta_{m,p}\delta_{n,q},$$

$$\langle v_{m,n}, v_{p,q}\rangle = \delta_{m,p}\delta_{n,q},$$

$$\langle u_{m,n}, v_{p,q}\rangle = 0$$

where the inner product is taken w.r.t. the unperturbed area measure. Writing

$$\psi_0(q_1, q_2) = C_1 u_{m,n}(q_1, q_2) + C_2 v_{m,n}(q_1, q_2),$$

substituting into [14(*a*)] with $h_0 = h_0\,[m, n]$, and taking inner products with $v_{m,n}$ and $v_{m,n}$ respectively gives us the secular matrix system for the perturbed eigenmode h_1^2:

$$h_1^2 \langle u_{m,n}, \exp(2q_1)(C_1 u_{m,n} + C_2 v_{m,n})\rangle$$

$$+ h_0[m,n]^2 \langle u_{m,n}, \eta(q_1, q_2)(C_1 u_{m,n} + C_2 v_{m,n}\rangle = 0$$

$$h_1^2 \langle u_{m,n}, \exp(2q_1)(C_1 u_{m,n} + C_2 v_{m,n})\rangle$$

$$+ h_0[m,n]^2 \langle v_{m,n}, \eta(q_1, q_2)(C_1 u_{m,n} + C_2 v_{m,n}\rangle = 0$$

These two equations can be expressed in matrix form as

$$\begin{pmatrix} h_0^2[m,n] < u_{m,n}, \eta u_{m,n} > + h_1^2 < u_{m,n}, \exp(2q_1) u_{m,n} > & h_0^2[m,n] < u_{m,n}, \eta v_{m,n} > + h_1^2 < u_{m,n}, \exp(2q_1) v_{m,n} > \\ h_0^2[m,n] < v_{m,n}, \eta u_{m,n} > + h_1^2 < v_{m,n}, \exp(2q_1) u_{m,n} > & h_0^2[m,n] < v_{m,n}, \eta v_{m,n} > + h_1^2 < v_{m,n}, \exp(2q_1) v_{m,n} > \end{pmatrix} \begin{pmatrix} C_1 \\ C_2 \end{pmatrix}$$

Setting the determinant of the above secular matrix to zero gives us two values of h_1^2 which represent the perturbation of the modal eigenvalue $h_0^2\ [m, n]$ into two parts lifting the two-fold degeneracy of the $(m,n)^{\text{th}}$ mode.

[10] Optimal control of the wave function and density matrix of a quantum system by modulation of the perturbing potential: The wave function $\psi(t)$ takes values in a Hilbert space $\mathcal{H}$ and satisfies the Schrodinger equation

$$\psi'(t) = -i\left(H_0 + \varepsilon\sum_{k=1}^{m} f_k(t)V_k\right)\psi(t) \quad ...(1)$$

The objective is to choose the functions *fk*(.) over the interval [*O*, *T*] so that

$$\int_0^T \left\|\psi(t) - \psi_d(t)\right\|^2 dt$$

is a minimum where ψ_d is a given wave function. This optimization problem can be solved by using a complex Lagrange multiplier function $\lambda(t) \in \mathcal{H}$ with the objective then becoming to minimize

$$E(\psi, \{f_k\},\lambda) = \int_0^T \left\|\psi(t) - \psi_d(t)\right\|^2 dt$$

$$-2\,\mathrm{Re}\left(\int_0^T < \lambda(t), i\psi'(t) - \left(H_0 + \varepsilon\sum_{k=1}^{m} fk(t)V_k\right)\psi(t) > dt\right)$$

We have that

$$\delta E/\delta f_k(t) = 0$$

gives

$$\psi m\,(< \lambda(t), V_k\psi\,(t) >) = 0,\ 0 \le t \le T \quad ...(2)$$

Regarding $\psi\,(t)$ and $\bar{\psi}(t)$ as independent variables,

$$\delta E/\delta f_k(t) = 0$$

gives

$$\psi(t) - \psi_d(t) + (H_0 + \varepsilon\sum_k f_k(t)V_k)\lambda(t) - i\lambda'(t) = 0 \quad ...(3)$$

Finally, regarding $\lambda(t)$ and $\bar{\lambda}(t)$ as independent variables,

$$\delta E/\delta\bar{\lambda}(t) = 0$$

gives the Schrodinger equation (1). Further, $\delta E/\delta\psi(T) = 0$ gives

$$\lambda(\mathrm{T}) = 0$$

(1) and (3) (*i.e.*, the state and co – state equations) can be combined into the equation

$$i\frac{d}{dt}\begin{pmatrix}\psi(t)\\ \lambda(t)\end{pmatrix}$$

$$= \begin{pmatrix} H_0 + \varepsilon\sum_k f_k(t)V_k & 0 \\ I & (H_0 + \varepsilon\sum_k f_k(t)V_k)\end{pmatrix}\begin{pmatrix}\psi(t)\\ \lambda(t)\end{pmatrix}$$

$$-\begin{pmatrix}0\\ \psi_d(t)\end{pmatrix}$$

❖❖❖❖❖

[4] Optimal control of the average values of a set of observables in a quantum system.

We now consider a quantum system described buy the Schrodinger equation with controllable classical signals $fk(t)$, $k = 1, 2,...,n$ to be selected so that if $X_1,...,X_r$ is a set of observable, we wish that $\langle\psi(t), X_m\psi(t)\rangle$, $0 \le t \le t\, T$ tracks a given function $\phi_m(t)$ for each $m = 1, 2,...,r$. Thus, we solve the Schrodinger equation

$$i\psi'(t) = \left(H_0 + \varepsilon\sum_{k=1}^{n} f k(t)V_k\right)\psi(t)$$

with the $f_k's$ chosen over the time interval [0, T] so that

$$\sum_{m=1}^{r}\int_0^T \left(\langle\psi(t), X_m\psi(t)\rangle - \phi_m(t)\right)^2 dt$$

is a minimum,

This problem can be modified to take into account the collapse postulate. Suppose we that start the system in an initial state $|\psi(0)>$ and allow it to evolve upto time t_1 under the Schrodingr dynamics. The error energy upto this time is $\int_0^{t_1}\left\|\psi_d(t) - U(t,0)\psi(0)\right\|^2 dt$. Note that $U(t, s)$ is a functional of $f_k(\cdot)$. Specifically,

$$i\frac{\partial U(t,s)}{\partial t} = \left(H_0 + \varepsilon\sum_k f_k(t)V_k\right)U(t,s), t \ge s,$$

$$U(s, s) = I$$

Then at time t_1, we measure an observable X and note the outcome, say α_1 with the corresponding eigenstate $|\alpha_1>$ of X. The system then evolves from the state $|\alpha_1>$ to the state $U(t_2,t_1)\,|\alpha_1>$ at time t_2 and the error energy in this time interval is

$$\int_{t_1}^{t_2}\left\|\psi_d(t) - U(t\,|\,t_1)\,|\,\alpha_1>\right\|^2 dt$$

At time t_2 we measure X and note the outcome a_2. The wave function then evolves from $|\alpha_2>$ at time t_2 to the state $U(t_3, t_2)\,|\alpha_2>$ at time t_3. In this way we take repeated measurements at times $t_1 < t_2 < ... < t_N$ taking measurements of X at these times with out comes $\alpha_1, \alpha_2, ...,\alpha_N$ respectively. The wave-function tracking error energy is then

$$E_T = \sum_{l=0}^{N_1}\int_{t_k}^{t_{k+1}}\left\|\psi_d(t) - U(t,t_k)\middle|\alpha_k>\right\|^2 dt$$

Now suppose we take measurements of X but do not note the outcomes. Then if at time $t\_$, the state is $\rho\_$, after the measurement at time t_+, the state is

$$\rho_+ = \sum_{k=1}^{d}|e_k\rangle\langle e_k|\rho - |e_k\rangle\langle e_k|$$

where

$$X = \sum_k |e_k\rangle c(k)\langle e_k|$$

is the spectral resolution of X. Moreover, the outcome of measurement is

$$Tr(\rho\_X) = \sum_k \langle e_k \rho - X|e_k\rangle = \sum_k c(k)\langle e_k|\rho - |e_k\rangle$$

We note that

$$Tr\,(\rho\_X) = d = Tr\,(\rho_+ X)$$

Apart from considering the state tracking error energy

$$E_T = \sum_{k=0}^{N-1} \int_{t_k}^{t_{k+1}} \|\rho_d(t) - \rho(t)\|^2$$

we consider in addition the error in the mean value of the observable at each time, *i.e.*,

$$E_0 = \sum_{K=1}^{N} \left|\phi_m[k] - Tr(\rho(t_k -)X)\right|^2$$

where

$$\rho(t) = U(t, t_k)\rho(t_k +)U(t, t_k)^* t_k \langle t \langle t_{k+1}$$

with

$$\rho(t_k +) = \sum_m |e_m\rangle\langle e_m|\rho(t_k -)|e_m\rangle\langle e_m|$$

Optimal control of the wave function of a quantum system using external electromagnetic fields.

Mathematical prerequisites for robotics:

❖❖❖❖❖

EXERCISES

[1] Motion of a finite system of particles under mutual interaction external forces.

[2] D' Alembert's principle of virtual work and the derivation of Lagrange's equations of motion from this principle.

Let $q_1 ,..., q_k$ denote the generalized coordinates.

[3] Motion of a rigid bodies: Angular momentum equations, Euler's equations for a rigid body, Motion of a spinning top.

[4] Motion of a d link Robot arm under machine torque and external forces.

[5] Motion of a master-slave d link Robot system with matrix PID feedback applied to slave Robot.

[6] Maximum likelihood estimation of the PID parameters of a master slave system.

[7] Maximum likelihood estimation of parameters in the presence of non-Gaussian white noise and coloured Gaussianj noise.

[8] The Cramer-Rao lower bound (CRLB) for vector valued parameters with bias taken into account.

[9] Least squares and Least mean squares based determination of PID feed back parameters of a master-slave Robot system for optimal master-slave tracking.

[10] Non-random optimum control of time varying PID parameters of a master-slave system based on variational calculus: (*a*) Continuous time case, (*b*) Discrete time case.

[11] Stochastic optimal control theory and its application to PID parameters control for optimal master-slave tracking.

[12] Quantum Robotics:

[13] The tensor product and its applications in the analysis of nonlinear dynamical systems.

[14] Markov chains as a model for spike noise in Robots: Estimating the chain parameters from master-slave position measurements.

[15] Stochastic processes with independent increments: Brownian motion, Poisson process and other Levy processes.

[16] Basics of electromagnetics with applications to Robotics.

[17] Belavkin's theory of quantum nonlinear filtering in the Husdon-Parthasarathy quantum Ito framework.

[18] Quantum gate design using fractional delay filters.

[19] Non-Gaussian master and slave Tremor estimation in master-slave Robot dynamics.

[20] Least squares problems in Robotics with nonlinear parameter dependence.

[21] Statistical analysis of slave Robot dynamics

[22] Electromagnetic fields generated in inhomogeneous wave-guides and cavity resonators used for controlling a slave Robot.

[12] Quantum Robotics

(*a*) Quantization of the angular variables and momenta using the Schrodinger equation.

Consider a single link of length L, with centre of mass located at $(x(t),y(t))$ and angle $\theta(t)$ made by the link relative to the x-axis. The kinetic energy of the link is

$$T = \frac{m}{2}(x'^2 + y'^2) + I\theta'^2/2$$

and the potential energy

$$V(x, y, \theta) = mgy$$

where

$$I = mL^2/12$$

The Hamiltonian obtained after a Legendre transformation is then

$$H(x, y, \theta, p_x, p_y, p_\theta) = (2m)^{-1}(p_x^2 + p_y^2) + p_\theta^2/2I + mgy$$

The energy eigenstates of this Hamiltonian are

$$|k_x, n, k_\theta\rangle = exp(i(k_x x + k_\theta \theta))u_n(y)$$

in the position-angle representation. where k_x, $k_\theta \in \mathbb{R}$ and u_n satisfies the eigen-equation

$$-u_n''(y)/2m + mgyu_n(y) = E_{y,n}u_n(y)$$

where $E_{y,n}, n = 1, 2,...$ are the quantized energy levels of the link associated with the y-direction motion. Motion along the x and θ directions is free. The energy of this eigenstate is

$$E(k_x, n, k_\theta) = E_{y,n} + k_x^2/2m + k_\theta^2/2I$$

In the case machine torque $\tau(t)$ around the CM, the Hamiltonian gets perturbed by $-\tau(t)\theta$ and we would be interested in calculating transition probabilities.

(*b*) Quantization of a single Robot using the Dirac relativistic wave equation.

(*c*) Action of the quantum Klein-Gordon field on a rigid body carrying current.

(*d*) Action of the quantum electromagnetic field on a rigid Robot arm carrying current.

(*e*) Joint quantization of Robot angles and Klein-Gordon field acting on the Robot current

Let $\phi(x)$ be the KG field and $J(x|\theta(t))$ be the current density in the rigid Robot arm with $\theta(t)$ as the d-component link angels. $x = (t, \mathbf{r})$ is the space-time variable. The action for the field and the Robot is

$$S[\phi,\theta] = \frac{1}{2}\int \partial_\mu \phi(x)\partial^\mu \phi(x)d^4x$$

$$-\frac{m^2}{2}\int \phi^2(x)d^4x + \frac{1}{2}\int \theta'^T(t)J_0(\theta(t))\theta'(t)dt$$

$$-\int V(\theta(t))dt + \int J(x|\theta(t))\phi(x)d^4x$$

The classical equations of the KG field $\phi(x)$ and the link angles $\theta(t)$ are obtained from the variational principle

$$\delta_\theta s = 0, \delta_\phi S = 0$$

These result in

$$(J_0(\theta(t))\theta'(t))' - \frac{1}{2}\theta'(t)^T J_0'(\theta(t))\theta'(t)$$

$$+ V'(\theta(t)) - \int \frac{\partial J(t,r|\theta(t))}{\partial\theta}\phi(t,\mathbf{r})d^3r = 0$$

$$\partial_\mu\partial^\mu\phi(x) + m^2\phi(x) - J(x|\theta(t)) = 0$$

We quantize the field and the Robot link motion using the Feynman path integral: The propagator for the field in the presence of the interaction with the Robot is given by

$$\Lambda_\phi(x, y) = \langle vac|T\{\phi(x)\phi(y)|vac\rangle$$

$$= C.\int \exp(iS[\phi,\theta])\phi(x)\phi(y)\Pi_{\xi\in\mathbb{R}^4} d\phi(\xi)\Pi_{t\in\mathbb{R}} d\theta(t)$$

Likewise, we can construct the propagator for the Robot link system:

$$\Delta_{\theta,a,b}(t_1, t_2) = C.\int \exp(iS[\phi,\theta])\theta_a(t_1)\theta_b(t_2)\Pi_{\xi\in\mathbb{R}^4} d\phi(\xi)\Pi_{t\in\mathbb{R}} d\theta(t)$$

and the joint propagator of the field and link system:

$$\Delta_{\phi\theta,a}(x,t) = C.\int \exp(iS[\phi,\theta])\phi(x)\theta_a(t)\Pi_{\xi\in\mathbb{R}^4} d\phi(\xi)\Pi_{s\in\mathbb{R}} d\theta(s)$$

where

$$C^{-1} = \int \exp(iS[\phi,\theta])\Pi d\phi(x)\Pi d\theta(t)$$

In the absence of the interaction, these propagators are Gaussian integrals and can be easily evaluated. In the presence of the perturbation, we expand the current $J(x|\theta)$ as a Taylor series in θ and use formulae for the higher order moments of Gaussian random vectors to calculate perturbative expansions for the propagator.

❖❖❖❖❖

[5] Optimal control of observable averages in quantum mechanics.
Stochastic optimal control (Stochastic Bellman-Hamilton-Jacobi equation) Consider the sde:

$$dx(t) = f(t, x(t), u(t))\,dt + g(t, x(t), u(t))dB(t)$$

where $x(t) \in \mathbb{R}^N$, $f : \mathbb{R}_+ \times \mathbb{R}^N \times \mathbb{R}^q \to \mathbb{R}^N$ and $g : \mathbb{R}_+ \times \mathbb{R}^N \times \mathbb{R}^q \to \mathbb{R}^{N\times p}$ and $B(\cdot)$ is standard p-dimensional Brownian motion. We wish to minimize

$$V(u, x_0) = \mathbb{E}\left[\int_0^T L(x(t), u(t), t)dt | x(0) = x_0\right]$$

with respect to all control inputs $u(\cdot)$ that can be expressed as $u(t) = \phi(t, x(t))$ where ϕ is a non-random function. We define

$$V(t, x(t)) = \min_{u(s),(t)\le s\le T} \mathbb{E}\left[\int_t^T L(x(s), u(s), s)\,ds | x(t)\right]$$

Then we have by the Markov property (the assumption on $u(\cdot)$ implies that (x) is a Markov process),

$$V(t, x(t)) = \min_{u(t),}(L(x(t),u(t),t)dt + \mathbb{E}(V(t+dt, x(t+dt)|x(t)))$$

and this lead immediately to the Stochastic Bellman- Hamilton-Jacobi equation

$$\min_u (L(x,u,t) + V_{,t}(t,x) + T_t(u)V(t,x)) = 0 \quad ...(1)$$

where $T_t\ (u)$ is the generator of the Markov process $x(.)$ at time t assuming $u(t) = u \in \mathbb{R}^q$ is fixed *i.e.*,

$$T_t(u)\phi(x) = f(t,x,u)^T \nabla_x \phi(x) + \frac{1}{2} Tr(g(t,x,u)g(t,x,u)^T \nabla_x \nabla_x^T \phi(x))$$

❖❖❖❖❖

[6] Single link robot

Recursive least squares algorithm for estimating PID feedback coefficients in a Master-slave Robot system: The discretized slave dynamics is

$$\theta_s[n+2] = \Phi_1(\theta_s[n+1], \theta_s[n]) + F_2(\theta_s[n+1], \theta_s[n])(K_0\psi_1(\theta_m[n], \theta_s[n]) + K_1\psi_2(\theta_m[n], \theta_s[n], \theta_m[n-1], \theta_s[n-1])$$

$$+K_2 \sum_{k=0}^{n} \psi_3(\theta_m[k], \theta_s[k])) + W_s[n]$$

Using the delay operator Z^{-1}, this difference equation can be expressed as (on applying the operator $1-Z^{-1}$ to both sides),

$$\begin{aligned}\theta_s[n+2] - \theta_s[n+1] = {} & F_1(\theta_s[n+1], \theta_s[n]) - F_1(\theta_s[n] - \theta_s[n-1]) \\ & + F_2(\theta_s[n+1]), \theta_s[n])(K_0\psi_1(\theta_m[n], \theta_s[n]) \\ & + K_1\psi_2(\theta_m[n], \theta_s[n], \theta_m[n-1], \theta_s[n-1]) \\ & + K_2 \sum_{k=0}^{n} \psi_3(\theta_m[k], \theta_s[k])) \\ & - F_2(\theta_s[n], \theta_s[n-1])(K_0\psi_1(\theta_m[n-1], \theta_s[n-1]) + K_1\psi_2 \\ & (\theta_m[n-1], \theta_s[n-1], \theta_m[n-2], \theta_s[n-2]) \\ & + K_2\psi_3(\theta_m[n], \theta_s[n])) \\ & + W_s[n] - W_s[n-1]\end{aligned}$$

Here K_0, K_1, K_2 are matrices.

A remark: The position of the tip of the slave Robot system is of the form $x_s = h_s(\theta_s) \in \mathbb{R}^3$ where $\theta_s = (\theta_{sk})_{k=1}^d$ are the slave link angles and likewise, the position of the master Robot system if of the form $x_m = h_m(\theta_m) \in \mathbb{R}^3$. The slave tip moves with the velocity $x_s'(t) = h_s'(\theta_s(t))\theta_s'(t)$ and the master tip moves with the velocity $x_m'(t) = h_m'(\theta_m(t))\theta_m'(t)$. The Jacobian matrices $h_s'(\theta)$ and $h_m'(\theta)$ are $3 \times d$ matrices. The difference between the master and slave tip positions,
velocities and position integral are respectively $x_m(t) - x_s(t), x_m'(t) - x_s'(t)$ and

$\int_0^t (x_m(t') - x_s(t'))dt'$. The PID controller takes these as input and produces force

$$F_s(t) = K_0(x_m(t) - x_s(t)) + K_1(x'_m(t) - x'_s(t)) + K_2 \int_0^t (x_m(t') - x_s(t'))dt'$$

that acts at the slave tip. This feedback force after discretization can be expressed as

$$\begin{aligned} F_s[n] &= (K_0\psi_1(\theta_m[n], \theta_s[n]) \\ &\quad + K_1\psi_2(\theta_m[n], \theta_s[n], \theta_m[n-1]), \theta_s[n-1] \\ &\quad + K_2 \sum_{k=0}^{n} \psi_3(\theta_m[k], \theta_s[k])) \end{aligned}$$

Now if this force F_s acts at the tip of the slave link system, it generate an equivalent torque τ_s defined by the principle of virtual work

$$\tau_s^T \delta\theta_s = F_s^T \delta x_s = F_s^T h'_s(\theta_s)\delta\theta_s$$

so that

$$\tau_s = h'_s(\theta_s{}^T F_s$$

This justifies the above difference equation for the slave dynamics.
Now the above difference equation with the matrices K_0, K_1, K_2 as parameters can be cast in the form

$$f[n] = A_0[n]K_0g_0[n] + A_1[n]K_1g_1[n] + A_2[n]K_2g_2[n]$$

where $f[n]$, $g_k[n]$, $k = 1, 2, 3$ are vectors and $A_k[n]$ K_k, $k = 0, 1, 2$ are matrices possibly of different size but such that just left and right sides are vectors of the same dimension. Now if K is a matrix with columns $k_1,\ldots,k_p$, A is a matrix having the same number of column m as the number of rows of K and $g = (g_r)$ is a column vector of dimension equal to p, then

$$\begin{aligned} AK_g &= A\sum_{r=1}^{p} k_r g_r = A\left[g_1 I_m, \ldots, g_p I_m\right] Vec(K) \\ &= A(g^T \otimes I_m)Vec(K) \end{aligned}$$

where

$$Vec\ (K) = \begin{pmatrix} k_1 \\ k_2 \\ \ldots \\ k_p \end{pmatrix} \in \mathbb{R}^{mp}$$

The problem of estimating the parameters K_0, K_1, K_2, from measurements on the master and slave angles after discretization can thus be cast in the form of estimating the vector parameter $\xi \in \mathbb{R}^p$ from the dynamical model

$$y[n] = F[n]\xi + G[n]w[n], n = 0, 1, 2, ..., N$$

where $y[n] \in \mathbb{R}^q, F[n] \in \mathbb{R}^{q\times p}, G[n] \in \mathbb{R}^{q\times s}$ are functions of the data $\theta_s[n]$, $\theta_m[n]$ and their delayed and advanced versions by finite amounts and $ww[n] \in \mathbb{R}_s$ is noise, it can be either coloured Gaussian or non-Gaussian. In the case of white non-Gaussian noise $w[n]$, with pdf $p_w(w)$ assuming that $s = q$ and that $G[n]$ is non-singular, the joint pdf of the observations is given by

$$\Pi_{n=0}^{N} \left|G[n]\right|^{-1} p_w(G[n]^{-1}(y[n] - F[n]\xi))$$

and the log-likelihood function after deleting additive terms that do not depend on the parameter vector ξ is given by

$$L_N(\xi) = \sum_{n=0}^{N} \log(p_w(G[n]^{-1}(y[n] - F[n]\xi))$$

The *ML* estimate of ξ based on observations upto time N is thus

$$\hat{\xi}[N] = \text{argmax}_\xi \sum_{n=0}^{N} \log(p_w(G[n]^{-1}(y[n] - F[n]\xi))$$

In case when w[n] is white with zero mean and covariance R_w, we get

$$\hat{\xi}[N] = \arg\min_\xi \sum_{n=0}^{N} (y[n] - F[n])_\xi^T G[n]^{-T} R_w^{-1} G[n]^{-1}(y[n] - F[n]\xi)$$

$$= \left[\sum_{n=0}^{M} F[n]^T (G[n]R_w G[n]^T)^{-1} F[n]^{-1}\right]$$

$$\left[\sum_{n=0}^{N} F[n]^T (G[n]R_w G[n]^T)^{-1} y[n]\right]$$

and this can be implemented in recursive form (RLS algorithm) using the matrix inversion lemma.

❖❖❖❖❖

[7] Klein-garden field interacting with a robot classical and quantum cases: Kronecker tensor product, its properties and generalization to infinite dimensional Hilbert spaces. Given two vector spaces U, V, consider a third vector space W and a bilinear map $\phi: U \times V \rightarrow W$ such that (1) span (Range (ϕ)) = W and (2) X is another vector space and $T: U \times V \rightarrow X$ is any bilinear map, then there exists a linear map $H: W \rightarrow X$ such that $HO\phi = T$, *i.e.*, $H\,\phi(f(u, v)) = T(u,v)\ \forall u \in U, v \in V$. Then, we say that (ϕ, W) is a tensor product of U and V. As an example, let $U = \mathbb{C}^n$, $V = \mathbb{C}^m$ and $W = \mathbb{C}^{cm}$. Let ϕ be the Kronecker tensor product of U and V, *i.e.*, for $x = (x_i) \in \mathbb{C}^n$ and $y = (y_i) \in \mathbb{C}^m$, we define

$$\phi\,(x, y) = (x_i y)_{i=1}^{n} \in \mathbb{C}^{nm}$$

Then, it is clear that span (Range(ϕ)) = W. To see this simply consider $x = e_i$, $y = f_j$ where e_i is the $n \times 1$ vector with a one in its ith position and zero at all the other positions and likewise, f_j is the $m \times 1$ vector with a one in its jth position and zero at all the other positions. Then, $\phi(e_i, f_j)$ is the $nm \times 1$ vector with a one in its $(m(i-1) + j)^{\text{th}}$ position and a zero at all the other positions. Further $m(i-1) + j$ varies over the entire set $\{1, 2, ..., nm\}$ as i varies over $1, 2, ..., n$ and j over $1, 2, ..., m$ and hence any $z \in W$ can be expressed as a linear combination of the vectors $\{\phi(e_i, f_j): 1 \le i \le n, 1 \le j \le m\}$. In fact, this argument shows that $\{\phi(e_i, f_j): 1 \le i \le n, 1 \le j \le m\}$ is a permutation of the standard ordered basis for $W = \mathbb{C}^{nm}$. The second property is verified as follows: Let X be any vector space and $T: U \times V \to W$ be a bilinear map. Then,

$$T(x, y) = T\left(\sum x_i e_i, \sum y_j f_j\right) = \sum_{i,j} x_i y_j T(e_i, f_j)$$

Now define a linear map S_{ij} from $W \to X$ so that $S_{ij}\,(\phi(e_a, f_b)) = \delta_{a,\,i}\delta_{b\,,j}T(e_i, f_j)$. This is possible since as we saw above $\phi(e_a, e_b)$, $1 \le a \le n$, $1 \le b \le m$ is a basis for $W = \mathbb{C}^{nm}$. Then we have

$$S_{ij}\,(\phi(x, y)) = S_{ij}\left(\phi\left(\sum_a x_a e_a, \sum_b y_b f_b\right)\right)$$

$$= \sum_{a,b} x_a y_b S_{ij}(\phi(e_a, f_b)) = \sum_{a,b} x_a y_b \delta_{a,i}\delta_{b,j} T(e_i, f_j)$$

$$= x_i y_j T(e_i, f_j)$$

and hence

$$\left(\sum_{ij} S_{ij}\,(\phi(x, y))\right) = T\,(x, y)$$

Thus, $S = \sum_{ij} S_{ij}$ is a linear transformation (unique) from $W \to X$ that satisfies $So\phi = T$.

Some of the useful properties of the tensor product are as follows. We denote by $\otimes$ the Kronecker tensor product of two or more vector spaces relative to specified ordered bases. Suppose U, V are two vector spaces with bases $\{e_i\}$ and $\{f_j\}$ respectively. We assume that they are finite dimensional spaces. Then if $x \in U$, $y \in V$, we have

$$(x \otimes y) = \sum_{i,j} x_i y_j e_i \otimes f_j$$

where $\{e_i \otimes f_j)$ denotes an ordered basis for $W = U \otimes V$. On $U\,(V)$, we define inner product as

$$\langle x, y\rangle = \sum \bar{x}_i y_i$$

relative to bases $\{e_i\}$ $(\{f_j\})$. Correspondingly, on W we define the inner product so that

$$\langle e_i \otimes f_j, e_a \otimes f_b \rangle = \delta_{a,i}\delta_{b,j}$$

Then, we have for $x, u \in U, y, v \in V$,

$$\langle x \otimes y, u \otimes v \rangle = \sum_{i,j} \bar{x}_i \bar{y}_j u_i v_j$$

$$= (\sum_i \bar{x}_i u_i).(\sum_j \bar{y}_j v_j) = \langle x, u \rangle \langle y, v \rangle$$

❖❖❖❖❖

[8] Stochastic optimal control.
Markov chains as a model for spikes in robots: The Master-slave Robot system with position - velocity feedback through the *pd* coefficient vector θ and in the presence of Markov chain noise $X(t)$ can be expressed as

$$\mathbf{q}''(t) = F_1(q(t), q'(t), t) + F_2(q(t), q'(t))\theta + \varepsilon F_3(q(t), q'(t))X'(t)$$

Suppose $X(t)$ has state space the integers $\mathbb{Z}$ and infinitesimal generator matrix $(\lambda(a, b))$, then with T_k, $k = 1, 2,...$ denoting the jump times of the process $X(t)$, we have the spike noise $X'(t) = dX(t)/dt$ given by

$$X'(t) = \sum_{k \geq 1} (X(T_k+) - X(T_k-))\delta(t - T_k)$$

We assume that the noise is small (terror in master operator and slave environment) and that the non- random components $f_h(t)$ $f_e(t)$ in the master operator and environment forces are taken into account by making F_1 depend explicitly on time t.

❖❖❖❖❖

[9] RLS algorithm for P.D. feedback coefficient estimation in a robot system.
Basics of electromagnetics with application to Robotics:

1. A flat plate of radius R carries a surface charge density $\sigma(p)$ where p is the distance from the centre in the xy plane and the plate centre coincides with the origin and is located in the xy plane. Derive integral expressions for the potential and electric at any point on the z-axis.
2. Write down the Maxwell equations in the absence of external charge and current density and derive the wave equations for the electric and magnetic fields. Prove that for plane wave solutions at a given frequency, the electric field vector, the magnetic field vector and the direction of propagation are mutually prependicular.
3. A perfectly conducting sphere of radius R is placed in a constant electric field $E_0\hat{z}$. Calculate the modified electric field in space and the surface charge density induced on the spherical surface using the poisson equation.
4. Assume that the dielectric constant of a medium ε fluctuates slowly with time but is constant in space. Derive the modified wave equation for the electric and magnetic field.

5. Assume that the plane $z = 0$ separates the two dielectric regions $z > 0$ and $z < 0$ having parameters (ε_1, μ_1) and (ε_1, μ_2) . A plane electromagnetic wave in incident on this plane at an angle θ_i relative to the z axis in the region $z > 0$. Assuming that the electric field vector in this incident wave makes an angle α with the plane of incidence (*i.e.*, the plane formed by the incident wave vector and the z axis). Then calculate the ratios E_r/E_i, E_t/E_i, H_r/H_i, H_t/H_i where E_i, E_r, E_t are respectively the amplitudes of the incident, reflected and transmitted electric fields and likewise for the magnetic fields H_i, H_r, H_t.

6. (*a*) Give two applications of Faraday's law of induction.

(*b*) If the magnetic field through the surface spanned by a planar wire loop changes as $B(t)\hat{z}$ and the area of the loop changes as $A(t)$, then calculate the emf induced in the wire.

(*c*) If a wire loop moves in space such that a point $\mathbf{r}(s)$ parameterized by the length parameter s along the wire from a given point on it has a velocity $\mathbf{v}(s)$ and if the magnetic field in space is $\mathbf{B}(\mathbf{r})$, then show that the emf induced in the wire loop is

$$\int_0^l (\mathbf{v}(s) \times \mathbf{B}(s), dx(s))$$

What is the name given to this kind of emf?

7. (*a*) Derive the expression for the power flux and electromagnetic energy density in space (Poynting's theorem) from first principles.

(*b*) Calculate the electrostatic energy in the field produced by a spherical ball of charge carrying total charge Q, having radius R and having a uniform charges density within it. Your calculation must take into account both the field inside and outside the ball.

8. The plane $lx + my + nz = d$ carries a surface current density $\mathbf{K}_s$ along the a the tangential direction (a, b, c) where $la + mb + nc = 0$. The region (I): $lx + my + nz > d$ of space has permeability μ_1 and the region (II): $lx + my + nz < d$ has permeability μ_2. Let $\mathbf{H}_1 = (H_{1x}, H_{1y}, H_{1z})$ denote the magnetic field in the region (1) and $H_2 = (H_{2x}, H_{2y}, H_{2z})$ the magnetic field in the region (II). Derive formulae for $\mathbf{H}_2$ in terms of $\mathbf{H}_1$ and $\mathbf{K}_s$.

9. Nonlinear waveguides: Taking into account the hysterisis effect in the inductance part of a waveguide and likewise nonlinearity and memory effects in the capacitance part, owing to nonlinear field polarization relations for large field amplitudes, the weveguide equations can be cast in the form

$$-v_{,z}(t, z) = R_0(z)i(t, z) + (L_0(z)i(t, z))_{,t} + \frac{\partial}{\partial t}\int h_L(\tau_1, \tau_2, z)i(t - \tau_2, z)i(t - \tau_2, z)d\tau_1 d\tau_2$$

and

$$-i_{,z}(t, z) = G_0(z)\, v\,(t, z) + C_0(z)v_{,t}(t, z)\frac{\partial}{\partial t}\int h_c(\tau_1, \tau_2, z)v(t - \tau_1, z)v(t - \tau_2, z)d\tau_1 d\tau_2$$

Second order Volterra systems have been used here to model the hysteresis and nonlinear capacitive component of the induced voltage and current in the line over an infinitesimal element. In the temproral Fourier domain, these translate to

$$-V_{,z}(\omega, z) = Z(\omega, z)I(\omega, z) + \delta\int H_L(\omega_1, \omega - \omega_1, z)I(\omega_1, z)I(\omega - \omega_2, z)d\omega_1\,,$$

$$-I_{,z}(\omega, z) = Y(\omega,z)V(\omega,z)+\delta\int H_C(\omega_1,\omega-\omega_1,z)V(\omega_1,z)V(\omega-\omega_1,z)d\omega_1$$

where

$$V(\omega, z) = \int_{\mathbb{R}} v(t,z)\exp(-j\omega t)dt, I(\omega,z) = \int_{\mathbb{R}} i(t,z)\exp(-j\omega t)dt$$

$$\delta.H_L(\omega_1, \omega_2, z) = \frac{j(\omega_1+\omega_2)}{2\pi}\int_{\mathbb{R}^2} h_L(\tau_1,\tau_2,z)\exp(-j(\omega_1\tau_1+\omega_2\tau_2))d\tau_1 d\tau_2,$$

$$\delta.H_C(\omega_1, \omega_2, z) = \frac{j(\omega_1+\omega_2)}{2\pi}\int_{\mathbb{R}^2} h_c(\tau_1,\tau_2,z)\exp(-j(\omega_1\tau_1+\omega_2\tau_2))d\tau_1 d\tau_2$$

Here

$$Z(\omega, z) = R_0(z) + j\omega L_0(z),\ Y(\omega, z) = G_0(z) + j\omega C_0(z)$$

δ is a small perturbation parameter that signifies that the nonlinear terms are small compared with the linear terms. The line length is d and we make a spatial Fourier series expansion of the different quantities:

$$Z(\omega, z) = \sum_n Z[\omega,n]\exp(j2\pi nz/d), Y(\omega,z) = \sum_n Y[\omega,n]\exp(j2\pi nz/d)$$

where

$$Z[\omega, n] = R_0[n] + j\omega L_0[n],\ Y[\omega, n] = G_0[n] + j\omega C_0[n],$$

$$R_0(z) = \sum_n R_0[n]\exp(j2\pi nz/d, L_0(z) = \sum_n L_0[n]\exp(j2\pi nz/d),$$

$$G_0(z) = \sum_n G_0[n]\exp(j2\pi nz/d), C_0(z) = \sum_n C_0[n]\exp(j2\pi nz/d),$$

$$H_L(\omega_1, \omega_2, z) = \sum_n H_L[\omega_1,\omega_2,n]\exp(j2\pi nz/d),$$

$$H_C(\omega_1, \omega_2, z) = \sum_n H_C[\omega_1,\omega_2,n]\exp(j2\pi nz/d),$$

$$V(\omega, z) = \exp(-\gamma(\omega)z)\sum_n V[\omega,n]\exp(j2\pi nz/d),$$

$$I(\omega, z) = \exp(-\gamma(\omega)z)\sum_n I[\omega,n]\exp(j2\pi nz/d)$$

Then the time equation become

$$(\gamma - j2\pi n/d)\, V[\omega,n] = \sum_k Z[\omega,n-k] + I[\omega,k] + \delta.\sum_{k_1,k_2}\int H_L[\omega_1,\omega-\omega_1,n-k_1-k_2]$$
$$I[\omega_1, k_1]\, I[\omega-\omega_1, k_2]d\omega_1,$$

$$(\gamma - j2\pi nd)\, I[\omega, n] = \sum_k Y[\omega,n-k]V[\omega,k] + \delta.\sum_{k_1,k_2}\int H_C[\omega_1,\omega-\omega_1,n-k_1-k_2]$$
$$V[\omega_1, k_1]\, V[\omega-\omega, k_2]d\omega_1.$$

This is a nonlinear eigenvalue problem. We now solve this system of equations upto $O(\delta)$ using first order perturbation theory:

$$V[\omega, n] = V_0[\omega, n] + \delta.V_1[\omega, n] + O(\delta^2),$$

$$I[\omega, n] = I_0[\omega,n] + \delta.I_1[\omega,n] + O(\delta^2)$$

$$\gamma(\omega) = \gamma_0(\omega) + \delta\gamma_1(\omega) + O(\delta^2$$

Then the $O(\delta_0)$ equations are

$$(\gamma_0 - j2\pi n/d)V_0[\omega,n] = \sum_k Z[\omega,n-k]I_0[\omega,k],$$

$$(\gamma_0 - j2\pi n/d)V_0[\omega,n] = \sum_k Z[\omega,n-k]I_0[\omega,k],$$

By truncating the Fourier series so that n, k vary over $\{-N, -N+1,..., N-1, N\}$ and defining the matrices

$$\mathbf{Z}(w) = ((Z[\omega,n-k]))_{|n|,|k|\le N} \in \mathbb{C}^{2N+1\times 2N+1}$$

$$\mathbf{Y}(\omega) = ((Y[\omega,n-k]))_{|n|,|k|\le N} \in \mathbb{C}^{2N+1\times 2N+1}$$

and defining the voltage and current fourier series vectors

$$\mathbf{V}_0(\omega) = ((V_0[\omega,n]))_{|n|\le N} \in \mathbb{C}^{2N+1},$$
$$\mathbf{I}_0(\omega) = ((I_0[\omega,n]))_{|n|\le N} \in \mathbb{C}^{2N+1},$$

and defining the diagonal matrix

$$\mathbf{D} = \text{diag}[j2\pi n / d\text{: } |n| \le N]$$

we can write the above zeroth order equations as an eign–problem:

$$\begin{pmatrix} \mathbf{D}-\gamma_0\mathbf{I}_{2N+1} & \mathbf{Z}(\omega) \\ \mathbf{Y}(\omega) & \mathbf{D}-\gamma_0\mathbf{I}_{2N+1} \end{pmatrix} \begin{pmatrix} \mathbf{V}_0(\omega) \\ \mathbf{I}_0(\omega) \end{pmatrix} = 0 \quad ...(1)$$

Denote the eigenvalues γ_0 (obtained by setting the determinant of the above $4N+2 \times 4N+2$ matrix to zero by $\gamma_{0k}(\omega)$, $k = 1, 2,..., 4N+2$ and the corresponding eigenvectors by

$$[\mathbf{V}_{0k}(\omega)^T, \mathbf{I}_{0k}(\omega)^T]^T = \xi_{0k}[\omega] \in \mathbb{C}^{4N+2}$$

Also let $\mu_{0k}[\omega]$, $k = 1, 2,...,4N+2$ denote the corresponding right eigenvectors of the matrix in (1) normalized so that

$$\xi_{0k}[\omega]^T \mu_{0m}[\omega] = \delta_{k,m,}, 1 \le k,m \le 4N+2$$

The general solution to the zeroth order equations is a superposition

$$\mathbf{V}_0(\omega, z) = \sum_{k=1}^{4N+2} c[k]\mathbf{e}(z)^T \mathbf{V}_{0k}(\omega)\exp(-\gamma_{0k}(\omega)z),$$

$$\mathbf{I}_0(\omega, z) = \sum_{k=1}^{4N+2} c[k]\mathbf{e}(z)^T \mathbf{I}_{0k}(\omega)\exp(-\gamma_{0k}(\omega)z)$$

where $\{c[k]\}$ are arbitrary complex numbers and

$$\mathbf{e}(z) = ((\exp(j2\pi nz/d)))_{|n|\leq N} \in \mathbb{C}^{2N+1}$$

The counting in the column vector arrangement begins from $n = -N$ and increases to $n = N$. The $O(\delta)$ equations are now given by

$$(\gamma_0 - j2\pi n/d)V_1[\omega,n] + \gamma_1 V_0[\omega,n]$$

$$= \sum_k Z[\omega,n-k]I_1[\omega,k] + \sum_{k_1,k_2} \int H_L[\omega_1,\omega-\omega_1,n-k_1-k_2]$$

$$I_0[\omega, k_1]\, I_0[\omega - \omega_1, k_2]\delta\omega_1,$$

$$(\gamma_0 - j2\pi n/d)I_1[\omega,n] + \gamma_1 V_0[\omega,n]$$

$$= \sum_k Y[\omega,n-k]V_1[\omega,k] + \sum_{k_1,k_2} \int H_C[\omega_1,\omega-\omega_1,n-k_1-k_2]$$

$$V_0[\omega_1, k_1]V_0[w-1, k_2]d_1$$

Let

$$\mathbf{A}(w) = \begin{pmatrix} \mathbf{D} & \mathbf{Z}(\omega) \\ \mathbf{Y}(\omega) & \mathbf{D} \end{pmatrix}$$

We can write the above equations as

$$(\gamma_0(\omega)\mathbf{I}_{4N+2} - \mathbf{A}(\omega))\xi_1[\omega] + \gamma_1(\omega)\xi_0[\omega]\ N[\omega, \xi 0[.]] \quad ...(2)$$

where N is a equation nonlinear operator acting on the vector valued function $\xi_0[.]$. This equation does not tell us how a general solution to the zeroth order equation gets perturbed but rather how a given eigenmode (corresponding to a given value of $\gamma_0(\omega)$ gets perturbed). Since the matrix $\gamma_0(\omega)\,\mathbf{I}_{4N+2} - \mathbf{A}(\omega)$ is singular, the above equation may not have any solution, However, formally we get from (2) by setting $\gamma_0 = \gamma_{0k}$, $\xi_0 = \xi_{0k}$ and taking the inner product on both sides with μ_{0m},

$$(\gamma_{0k}(\omega) - \gamma_{0m}(\omega))\eta_{0m}(\omega)^T \xi_1(\omega)$$

$$+\gamma_1(\omega)\delta_{k,m} = \eta_{0m}(\omega)^T N[\omega, \xi_{0k}[\cdot]]$$

Taking $m = k$ gives us

$$\gamma_1(\omega) = \gamma_{1k}(\omega) = \eta_{0k}(\omega)^T N[\omega, \xi_{0k}[\cdot]]$$

i.e., we have determined the perturbation $\delta.\gamma_{1k}(\omega)$ to the kth modal eigenvalue $\gamma_{0k}(\omega)$ of the linear system due to the non-linearity. We also obtain the corresponding perturbed values of the voltage and current Fourier series vector, For $K \neq m$.

$$\eta_{0m}(\omega)^T \xi_1(\omega) = \eta_{0m}(\omega)^T \xi_{1k}(\omega)$$

$$= \frac{\mu_{0k}(\omega)^T N(\omega, \xi_{0k}[\cdot]]}{\gamma_{0k}(\omega) - \gamma_{0m}(\omega)}$$

We combine this with the equation

$$\xi_{1k}(\omega) = \sum_{m\neq k} (\eta_{0m}(\omega)^T \xi_{1k}(\omega)\varepsilon_{om}(\omega)/(\eta_{0m}(\omega)^T \xi_{0m}(\omega))$$

to obtain the perturbed voltage–current Fourier series vector.

❖❖❖❖❖

[10] Quantum gate design using fractional delay filters:

$$f(t) = \sum_n I[n]h(t-\tau_n)$$

Sampling time interval is Δ. Discretized signal is

$$f(n\Delta) = f[n] = \sum_n I[n]h(n\Delta-\tau_n)$$

$\tau_n \in [0, \Delta)$ and $\{I[n]\}$ are to be chosen so that

$$\|U_d - U(T)\|$$

is a minimum where $U(t)$ satisfies the Schrodinger equation

$$iU'(t) = (H_0 + \varepsilon f(t)V)U(t)$$

To simulate the evolution in discrete time, we use the Cayley transform:

$$U\,U[n+1] = \frac{I - i(\nabla/2)(H_0\varepsilon f[n]V)}{I + i(\Delta/2)(H_0 + \varepsilon g[n]V)} U[n] = T(f[n])\,U[n]$$

The Cayley transform $\frac{I - i\varepsilon X}{I + i\varepsilon X}$ for X Hermitian is a unitary operator and for small ε, it provides a good approximation to the non-unitary operator $I - 2i\varepsilon X$. We have

$$f[n] = \sum_k I[k]h[n-\alpha_k]$$

where

$$h[x] = h(\Delta x), \alpha_n \in [0,1]$$

The $I[n]'s$ and $\alpha'_n s$ are to be chosen so that $\|U[N] - U_d\|$ is a minimum. For simulation purpose, we can choose $U_d = M^{-1/2}((\exp(-j2\pi kn/M)))_{0\leq k,n\leq M-1}$, the DFT matrix and H_0 so that

$$N\Delta H_0 = i\log(U_d) + X$$

where X is a small random Hermitian matrix. If $X = 0$, then the evolution under H_0 gives $U(T) = U_d$ after time $N\Delta$. Suppose $X \neq 0$. Then

$$\exp(-iN\Delta H_0) = \exp(\log(U_d) - iX) \neq U_d$$

However, if X is small, we can use the Campbell = Baker = Hausdorf formula to expand $\exp(\log(U_d) - X)$ as $U_d Y$ where Y is expressed in terms of commutators of U_d with X (see for example the book by V.S.Varadarajan, "Lie Groups, Lie Algebras and their Representations"). The optimization problem for the gate design in then

$$\{\hat{I}[k], \hat{\alpha}_k\} = \text{Arg}\min_{\{I[k],\hat{\alpha}k} \left\|U_d - \Pi_{n=0}^{N-1} T(f[n])\right\|^2$$

$$= \text{Arg}\min_{\{I[k],\hat{\alpha}k} \left\| U_d - \Pi_{n=0}^{N-1} T(\sum_k I[k]h[n-\alpha_k]) \right\|$$

This is a highly non-linear optimization problem and can be accomplished using a gravitational search optimization algorithm (GSO).

❖❖❖❖❖

[11] Non-Gaussian tremor estimation in master slake dynamics: $q[n] = [q_M[n]T, q_s[n]T]T \in \mathbb{R}_{zd}$ where the master and slave Robots are d-link systems These angles satisfy the difference equation

$$q[n+2] - 2q[n+1] + q[n] = F_1[n, q[n], q[n+1]] + F_2[q[n], q[n+1]]W[n+2]$$

$$+F_3[q[n], q[n+1]]\theta$$

where $q \in \mathbb{R}^p$ are the master-slave pd controller coefficient feedback parameters, the dependence of F_1 in n indicates the presence of external master operator and slave environment forces that are known functions of time $F_3\theta$ is the feedback term involving feeding back the error between master and slave positions and velocities as a force to the master and slave Robot tips and the consequent conversion of these forces into torques using the transpose of the Jacobian matrix that connects the master and slave angles and the position of their tips. F_2 is assumed to be a $2d \times 2d$ non-singular matrix and $W[n]$ is a $2d \times 1$ iid non-Gaussian random vector sequence with weakly non-Gaussian pdf:

$$p_w(w)(2\pi)^{-d} \exp(-\|w\|^2/2(1+\varepsilon\phi(w))$$

Define

$$\xi[n] = q[n+2] - 2q[n+1] + q[n] - F_1[n, q[n], q[n+1]],$$

$$F_3[n] = F[q[n], q[n+1]], F_2[n] = F_2[q[n], q[n+1]]$$

Then the negative of the logarithm of the joint pdf of $\{q[n]\}$ given θ with neglect of an terms which do not depend on θ (but depend on $q[.]$ in the form of logarithm of determinants is given by (upto $O(\varepsilon)$)

$$L(q[.]|\theta] = \frac{1}{2}\sum_n \left\| F_2[n]^{-1}(\xi[n] - F_3[n]\theta \right\|^2 - \varepsilon\sum_n \phi(F_2[n]^{-1}(\xi[n] - F_3[n]\theta))$$

The optimal maximizing equation for θ (maximum likelihood estimate) obtained as

$$\nabla_\theta L\left(q[.]|\hat{\theta}[N]\right) = 0$$

and this gives

$$\sum_{n=0}^{N} F_3[n]^T (F_2[n]F_2[n]^T)^{-1}(\xi[n] - F_3[n]\hat{\theta}[N])$$

$$-\varepsilon\sum_{n=0}^{N}(F_2[n]^{-1}f_3[n])^T\phi'(F_2[n])^{-1}(\xi[n]-F_3[n]\check{\theta}[N]))=0$$

Writing

$$\hat{\theta}[N] = \hat{\theta}_0[N]+\varepsilon\hat{\theta}_1[N]+O(\varepsilon^2)$$

we get upto $O(\varepsilon^0)$, the standard least squares solution

$$\hat{\theta}_0[N] = (\sum_{n=0}^{N}F_3[n]^T(F_2[n]F_2[n]^T)^{-1}F_3[n])^{-1}(\sum_{n=0}^{N}F_3[n]^T\ (F_2[n]F_2[n]^T)^{-1}\xi[n])$$

This solution can be cast in recursive form, *i.e.*, recursive least squares as

$$\hat{\theta}_0[N+1] = \hat{\theta}_0[n]+K[N+1](\xi[N+1]-F_3[N+1]\hat{\theta}_0[N])$$

where $K[N+1]$ is the Kalman gain. The computation of $K[N+1]$ is based on the matrix inversion lemma applied to

$$X[N+1] = X[N]+F_3[N+1]^T(F_2[N+1]F_2[N+1]^T)^{-1}F_3[N+1]$$

and

$$y[N+1] = y[N]+F_3[N+1](F_2[N+1]F_2[N+1]^T)^{-1}\xi[N+1]$$

The O (ε) gives

$$\sum_{n=0}^{N}F_3[n]^T(F_2[n]F_2[n]^T)^{-1}F_3[n]\hat{\theta}_0[N]))$$

$$+\sum_{n=0}^{N}(F_2[n]^{-1}F_3[n])^T\phi'(F_2[n]^{-1}(\xi[n]-F_3[n]\hat{\theta}_0[N]))=0$$

and this equation can be immediately inverted for $\hat{\theta}_1[n]$. By regarding the term $K[N+1](\xi[N+1]-F_3[N+1]\hat{\theta}_0[n])$ in $\hat{\theta}_0[N+1]$ as being small, *i.e.*, of $O(\varepsilon)$, we can be obtain are recursive algorithm for $\hat{\theta}_1[N]$ too.

❖❖❖❖❖

[12] To see how the decomposition of the dynamics explicitly into the master and slave components occurs, we write down the equations of motion in continuous time:

$$F_m(q_m,q_m',q_m'') = J_m(q_m)q_m''+N_m(q_m,q_m') = \tau_m \quad ...(1)$$

$$F_s(q_s,q_s',q_s'') = J_s(q_s)q_s''+N_s(q_s,q_s') = \tau s \quad ...(2)$$

Here, $J_m(q_m)$ is the master Robot moment of inertia matrix $(d\times d)$ and so is the slave Robot moment of inertia matrix $J_s(q_s).N_m(q_m,q_m')$ takes into account the moment of inertia contribution on the centrifugal term $J_m'(q_m)(q_m'\otimes q_m')$, the damping frictional terms proportional to the angular velocity $q_m'\in\mathbb{R}^d$ and the gravitational forces, τ_m is the torque applied to the master and τ_s is the torque applied to the slave. The master torque process $\tau_m(t)$ is assumed to be known and the slave torque is derived as follows. We consider the master-slave position - velocity errors $q_m - q_s$ and $q'_m - q'_s$ and pass these through a pd controller resulting in an "error process"

$$e(t) = K_p(q_m(t) - q_s(t)) + K_d(q'_m(t) - q'_s(t)) \quad ..(3)$$

where K_p, K_d are d matrices. This error process is added to the master angular acceleration q''_m giving an interface torque output

$$\tilde{\tau}_m = F_m(\tilde{q}_m, \tilde{q}'_m, q''_m + e(t)) \quad ...(4)$$

that is used to drive the slave, *i.e.*,

$$\tau_m = \tilde{\tau}_m \quad ...(5)$$

The interface master angles $\tilde{q}_m$, angular velocities $\tilde{q}'_m$ and angular accelerations $\tilde{q}''_m$ are directly connected to those of the original master outputs. In other words, the interface Robot is just a program with

$$\tilde{q}_m = q_m, \tilde{q}'_m = q'_m \quad ...(6)$$

We then find that

$$\tau_s = \tilde{\tau}_m = J_m(q_m)(q''_m + e(t)) + N_m(q_m, q'_m)$$
$$= \tau_m + J_m(q_m)\, e_m(t) \quad ...(7)$$

where we have made use of (1), (4) and (5). Substituting (7) into (2) gives

$$J_s(q_s)q''_s + N_s(q_s, q'_s) = J_m(q_m)e(t) + \tau_m(t)$$

or equivalently,

$$q''_s = \psi_1(q_s, q'_s) + L(t, q_s)e(t) + \psi_2(q_s)\tau_m(t) \quad ...(8)$$

where

$$\psi_1(q_s, q'_s) = -J_s(q_s)^{-1}N_s(q_s, q'_s) \in \mathbb{R}^d,$$

$$\psi_2(q_s) = J_s(q_s)^{-1} \in \mathbb{R}^{d\times d}, L(t, q_s(t)) = J_s(q_s(t))^{-1}J_m(q_m(t)) \in \mathbb{R}^{d\times d}$$

We can also consider the situation when there is environmental noise $W(t)$ in the slave torque. The appear when the slave hits a vibrating wall or moves in a fluctuating environment. In that case we have to replace $\tau_s(t)$ in (7) by an additional term $W(t)$ which results in (8) getting modified to

$$q''_s = \psi_1(q_s, q'_s) + L(t, q_s)e(t) + \psi_2(q_s)\tau_m(t) + W(t) \quad ...(9)$$

(9) is our fundamental equation for the slave dynamics. We note that the slave motion does not influence the master dynamics in this model since no error feedback is provided to the master. However, the master motion influences the slave in that feedback is provided to the slave motion via the error process $e(t)$. We now assume that $q_s(t)$ is close to $q_m(t)$ and linearize (9) about $q_m(t)$. Thus writing

$$q_s(t) = q_m(t) + \delta q_s(t)$$

we find that upto linear orders in $\delta q_s(t)$, (9) becomes

$$\delta q_s''(t) = \psi_{1,1}(t)\delta q_s(t) + \psi_{1,2}(t)\delta q_s'(t)$$

$$-(K_p\delta q_s(t) + K_d\delta q_s'(t)) + \psi_{2,1}(t)(I_d \otimes \tau_m(t))\delta q_s(t) + \psi_2(t)W(t) \quad ...(10)$$

We are assuming here that $W(t)$ is of the same order of magnitude as $\delta q_s(t)$ and the following notations are used

$$\psi_{1,1}(t) = \frac{\partial\psi_1(q_m(t), q_m'(t))}{\partial q_s}$$

$$= \left[\frac{\partial\psi_1(q_m(t), q_m'(t))}{\partial q_{s1}} \Big| \ldots \Big| \frac{\partial\psi_1(q_m(t), q_m'(t))}{\partial q_{sd}}\right] \in \mathbb{R}^{d\times d},$$

$$\psi_{1,2}(t) = \frac{\partial\psi_1(q_m(t), q_m'(t))}{\partial q_s'}$$

$$= \left[\frac{\partial\psi_1(q_m(t), q_m'(t))}{\partial q_{s1}'} \Big| \ldots \Big| \frac{\partial\psi_1(q_m(t), q_m'(t))}{\partial q_{sd}'}\right] \in \mathbb{R}^{d\times d},$$

$$\psi_{2,1}(t) = \frac{\psi_2(q_m(t))}{\partial q_s}$$

$$= \left[\frac{\partial\psi_2(q_m(t))}{\partial q_{s1}} \Big| \ldots \Big| \frac{\partial\psi_2(q_m(t))}{\partial q_{sd}}\right] \in \mathbb{R}^{d\times d},$$

$$\psi_2(t) = \psi_2(q_m(t))$$

We can use the discretized version of either (9) or its linearized version (10) for constructing the maximum likelihood estimate of the pd parameter matrices K_p, K_d from discrete measurements of $q_m(t)$, $q_s(t)$ assuming that the noise $W(t)$ is white Gaussian. Using these estimates, we can reconstruct the sample trajectory of the environmental tremor $\{W(t)\}$. The advantage in this scheme is that apriori, the pd controller K_p, K_d has been designed to make the slave track the master. This design may be based on a deterministic optimal control algorithm and is carried out in the absence of environmental noise $W(t)$. The values of these parameters are fixed but are unknown to the new user of the system who is aiming at using the system in a noisy environment. Thus, he first needs to estimate the parameters K_p, K_d from master slave angular position measurements and then decode the environmental noise process $W(t)$ using these parameters and the angular position measurements. This idea can also applied to estimating the master operator tremor, *i.e.*, when there is white noise in the master torque $\tau_m(t)$.

Consider first the nonlinear equation (9). We discretize it with a time step of Δ to get the stochastic difference equation

$$q_s[n+2] - 2q_s[n+1] + q_s[n]$$

$$= \Delta^2\psi_1(q_s[n], (q_s[n+1] - q_s[n])/\delta) + \Delta^2 L[n, q_s[n]]$$

$$(K_p(q_m[n]-q_s[n])$$

$$+\Delta^{-1}K_d(q_m[n+1]-q_m[n]-q_s[n+1]+q_s[n]))+\Delta^2\psi_2(q_s[n])(\tau_m(n+2))$$

$$+\varepsilon W[n+2]/\sqrt{\Delta}) \quad ...(10)$$

where $\{W[n]\}$ is an $\mathbb{R}^d$ valued iid Gaussian sequence with zero mean and identity covariance. Note that if $W(t)$ is white Gaussian with zero mean and auto correlation $I_d\delta(t-t')$, then the random variable $W(t)dt$ are independent Gaussian with zero mean and variable dt. Hence $W(n\Delta)$ should be modelled as an iid Gaussian sequence with zero mean and covariance I_d/Δ. Thus, we may set $W(n\Delta) = W[n]/\sqrt{\Delta}$ where $W[n]$ is an iid Gaussian sequence in $\mathbb{R}^d$ with zero mean and covariance I_d. We define the $\mathbb{R}^d$ valued vectors

$$\xi[n] = q_s[n+2]-2q_s[n+1]+q_s[n]-\Delta^2\psi_1(q_s[n])\nabla^2\psi_2(q_s[n]\tau_m[n+2]$$

Also noting that

$$L[n,q_s[n]]K_p(q_m[n]-q_s[n]) = L[n,q_s[n]]((q_m[n]-q_s[n])^T \otimes I_d)Vec(K_p)$$

and

$$L[n,q_s[n]]K_d(q_m[n+1]-q_m[n]-q_s[n+1]+q_s[n])$$

$$= L[n,q_s[n]]((q_m[n+1]-q_m[n]-q_s[n+1]+q_s[n])^T \otimes I_d)Vec(K_d)$$

we have

$$\Delta^2 L[n,q_s[n]](K_p(q_m[n]-q_s[n])$$

$$+\Delta^{-1}K_d(q_m[n+1]-q_m[n]q_s[n+1]+q_s[n])) = X[n]\theta$$

where

$$\theta = \begin{pmatrix} Vec(K_p) \\ Vec(K_d) \end{pmatrix} \in \mathbb{R}^{2d^2}$$

and

$$X[n] = [\Delta^2 L[n,q_s[n]](q_m[n]-q_s[n])^T \otimes I_d | \Delta L[n,q_s[n]](q_m[n+2]$$

$$-q_m[n]-q_s[n+1]+q_s[n])^T \otimes I_d] \in \mathbb{R}^{d\times 2d^2}$$

(10) can then be expressed as

$$\xi[n] = X[n]\theta + \Delta^{3/2}\psi_2(q_s[n])W[n+2]$$

and hence the negative logarithm of the joint *pdf* of $\{q_s[n]\}$ given θ (negative log-likelihood function) can be expressed as

$$\pounds(q_s[\cdot]|\theta) = \left(\Delta^{-3}/2\right)\sum_n (\xi[n]-X[n]\theta)^T(\psi_2[n]\psi_2[n]^T)^{-1}(\xi[n]-X[n]\theta)$$

Setting the gradient of this w.r.t. θ to zero gives us the maximum likelihood estimate of θ:

$$\hat{\theta} = (\sum_n X[n]^T (\psi_2[n]\psi_2[n]^T)^{-1} X[n])^{-1}$$

$$(\sum X[n]^T (\psi_2[n]\psi_2[n]^T)^{-1}\xi[n])$$

where we have used the abbreviated notation $\psi_2[n]$ in place of $\psi_2(q_s[n])$.

❖❖❖❖❖

[13] Consider the measurement model:

$$X = F(\theta) + W = F_0\theta + \varepsilon F_1(\theta) + W$$

Where F_0 is a constant $N \times d$ matrix, $F_1 : \mathbb{R}^d \to \mathbb{R}^N$ is a nonlinear function and W is Gaussian noise in $\mathbb{R}^N$ with mean zero and covariance $\sigma^2 I_N$. The MLE of θ is also the least squares estimate of θ and is given by

$$\hat{\theta}_0 = \hat{\theta}(X) = argmin_\theta \|X - F(\theta)\|^2$$

Now,

$$E(\theta) = \|X - F(\theta)\|^2 = \|X - F_0\theta - \varepsilon F_1(\theta)\|^2$$

$$= \|X - F_0\theta\|^2 - 2\varepsilon F_1(\theta)^T (X - F_0\theta) + \varepsilon^2 \|F_1(\theta)\|^2$$

Setting the gradient of $E(\theta)$ at $\hat{\theta}$ to zero gives

$$F'(\hat{\theta})^T (X - F(\hat{\theta})) = 0$$

or

$$(F_0^T + \varepsilon F_1'(\hat{\theta})^T)(X - F_0\hat{\theta} - \varepsilon F_1(\hat{\theta})) = 0$$

or

$$F_0^T (X - F_0\hat{\theta}) = \varepsilon(F_0^T F_1(\hat{\theta}) - F_1'(\hat{\theta})^T (X - F_0\hat{\theta})) + O(\varepsilon^2) \quad ...(1)$$

We write

$$\hat{\theta}_0 = \hat{\theta}_0 + \varepsilon\,\hat{\theta}_1 + O(\varepsilon^2)$$

Then the coefficient of ε^0 in (1) gives

$$\hat{\theta}_0 = (F_0^T F_0)^{-1} F_0^T X \quad ...(2)$$

and the coefficient of $\in^1$ in (1) gives

$$F_0^T F_0 \hat{\theta}_1 = F_0^T F_1(\hat{\theta}_0) - F_1'(\hat{\theta}_0)^T X \quad ..(3)$$

From (2) and the measurement model, we get

$$\hat{\theta}_0 = (F_0^T F_0)^{-1} F_0^T (F_0\theta + \varepsilon F_1(\theta) + W)$$

$$= \theta + \varepsilon PF_1(\theta) + PW$$

where

$$P = (F_0^T F_0)^{-1} F_0^T$$

Thus,

$$\mathbb{E}(\hat{\theta}_0) = \theta + \varepsilon PF_1(\theta)$$

and

$$\text{Cov}(\hat{\theta}_0) = sigma^2 PP^T = \sigma^2 (F_0^T F_0)^{-1}$$

The coefficient of ε in (1) gives

$$-F_0^T F_0 \hat{\theta}_1 = (F_0^T F_1(\hat{\theta}_0) - F_1'(\hat{\theta}_0)^T (X - F_0 \hat{\theta}_0)) + O(\varepsilon) \quad ..(4)$$

In this equation we can substitute for $\hat{\theta}_0$ and X respectively from (2) and the measurement model and neglect $O(\varepsilon)$ terms to get

$$-F_0^T F_0 \hat{\theta}_1 = F_0^T F_1(\theta + PW) - F_1'(\theta + PW)^T W$$

or

$$\hat{\theta}_1 = -PF_1(\theta + PW) + PF_1'(\theta + PW)W$$

We now have the complete expression

$$\hat{\theta} = \hat{\theta}_0 + \varepsilon \hat{\theta}_1 + O(\varepsilon^2)$$

$$= \theta + \varepsilon PF_1(\theta) + PW - \varepsilon PF_1'(\theta + PW) + \varepsilon PF_1'(\theta + PW)W + O(\varepsilon^2)$$

From this expression, we can obtain integral expressions for the estimator bias $B(\theta)$ $= \mathbb{E}(\hat{\theta}) - \theta$ upto $O(\varepsilon)$ and the estimator mean square error $\sigma_e^2(\theta) = \mathbb{E}\left(\left\|\hat{\theta} - \theta\right\|^2\right)$ upto $O(\varepsilon^2)$.

❖❖❖❖❖

[14] General statistical analysis of slave Robot dynamics. The slave state $q[n] = [\theta_s[n]^T, \omega_s[n]^T]^T$ satisfies the difference equation

$$q[n+1] = F_1[n, q[n]] + F_2[q[n]]K + F_3[q[n]]W[n+1]$$

where $q[n] \in \mathbb{R}^{2d}, K \in \mathbb{R}^p, W[n+1] \in \mathbb{R}^d$. $\{W[n]\}$ is a zero mean white Gaussian sequence with covariance $\sigma^2 I_d$. K is the position-differentiation (pd) parameter vector F_1 takes values in $\mathbb{R}^{2d}$. Its explicit dependence on n shows the presence of the external and force. $F_2[q[n]]$ is an $\mathbb{R}^{2d \times p}$ matrix and $F_3[q[n]]$ is an $\mathbb{R}^{2d \times d}$ matrix. Wherever the inverse does not exist, we replace it with the Moore – Penrose pseudo-inverse. We wish to compute the maximum likelihood estimate (mle) of K based on $q[n], 0 \le n \le N$. The negative log likelihood function is apart from an additive constant,

$$-\log(p(q[.]|K)) = \frac{1}{2\sigma^2} \sum_n (q[n+1] - F_1[n, q[n]] - F_2[q[n]]K)^T (F_3[q[n]]$$

$$F_3[q[n]]^T)^{-1}(q[n+1] - F_1[n, q[n]] - F_2[q[n]]K)$$

and setting the gradient of this w.r.t. K to zero gives the mle of K:

$$\widehat{K}[N] = (\sum_{n=0}^{N-1} F_2[q[n]]^T (F_3[q[n]]F_3[q[n]]^T)^{-1} F_2[q[n]])^{-1} (\sum_{n=0}^{N-1} F_2[q[n]]^T (F_3[q[n]]]$$

$$F_3[q[n]]^T)^{-1}[q[n+1] - F_1[n, q[n]]]$$

Define the $\mathbb{R}_{p\times p}$ matrix valued function H on $\mathbb{R}_{2d}$ by

$$H[q] = F_2[q]^T (F_3[q]F_3[q]^T)^{-1} F_2[q]$$

We get on substituting for $q[n+1]$ from the slave dynamics

$$\widehat{K}[N] = K + \left(\sum_{n=0}^{N-1} H[q[n]]\right) \cdot \left(\sum_{n=0}^{N-1} P[q[n]]W[n+1]\right)$$

where

$$P[q] = F_2[q]^T (F_3[q]F_3[q]^T)^{-1} F_2[q]$$

Let $q_0[n]$ denote the solution to the state equations in the absence of noise, *i.e.*, we may assume that $q_0[n]$ is the master Robot trajectory while $q[.]$ is the slave trajectory. In the absence of environmental noise W[.], the slave follows the master. The linearized dynamics for the deviation $\delta q[n] = q[n] - q_0[n]$ is

$$\delta q[n{+}1] = F_1'[n]\delta_q[n] + F_2'[n](I_{2d} \otimes K)\delta_q[n]) + F_3[n][n+1]$$

where

$$F_1'[n] = \left[\frac{\partial F_1[n, q_0[n]]}{\partial q_1} \middle| \ldots \frac{\partial F_1[n, q_0[n]]}{\partial q_{2d}} \right] \in \mathbb{R}^{2d\times 2d}$$

$$F_2'[n] = \left[\frac{\partial F_2[q_0[n]]}{\partial q_1} \middle| \ldots \frac{\partial F_2[q_0[n]]}{\partial q_{2d}} \right] \in \mathbb{R}^{2d\times 2d\, p},$$

$$F_3[n] = F_3[q_0[n]]$$

We define

$$A[n,K] = F_1'[n] + F_2'[n](I_{2d} \otimes K) \in \mathbb{R}^{2d\times 2d},$$

Then,

$$\delta q[n+1] = A[n,K]\delta q\,[n] + F_3[n]W[\mathrm{n}+1]$$

and hence defining the error correlation matrix

$$R_q[n] = \mathbb{E}(\delta q[n]\delta q[n]^T),$$

we get

$$R_q[n+1] = A[n,K]R_q[n]A[n,K]^T + G[n]$$

where

$$G[n] = \sigma^2 F_3[n]F_3[n]^T$$

Let A, X, B be matrices of compatible sizes, $B = [b_1,..., b_n]$, b_j being the jth column of B. Then

$$Vec(XB) = \begin{pmatrix} Xb_1 \\ Xb_2 \\ ... \\ Xb_n \end{pmatrix} = (I \otimes X)Vec(B)$$

Now let

$$X = [x_1,...,x_m],$$

x_j being j^{th} column of X. Then,

$$A\,X\,B = [Ax_1,..., Ax_m]B$$

Let $c_1^T,...,c_m^T$ be the rows of B. Then

$$A\,X\,B = \sum_j Ax_j c_j^T \,,$$

Now suppose u, v are column vectors. Then

$$Vec(uv^T) = \begin{pmatrix} uv_1 \\ uv_2 \\ ... \\ uv_n \end{pmatrix} = v \otimes u$$

Thus,

$$Vec\,(AXB) = \sum_j Vec(Ax_j c_j^T) = \sum_j c_j \otimes Ax_j$$

$$= (I \otimes A)\sum_j c_j \otimes r_j$$

On the other hand,

$$(B^T \otimes I)Vec(X) = [c_1 \otimes I,..., c_m \otimes]Vec(X) = \sum_j c_j \otimes x_j$$

and hence, we get

$$Vec(AXB) = (B^T \otimes A)Vec(X)$$

Thus our equation of evolution of $R_q[n]$ can be expressed as

$$Vec(R_q[n]) = (A[n,K] \otimes A[n,K])Vec(R_q[n]) + Vec(G[n])$$

and this equation can be easily solved by recursion using the state transition matrix

$$\Phi\,[n, m] = (A[m,K]A[m-1,K]...A[n+1,K])^{\otimes 2}$$

We leave the details to the reader, Upto linear orders in the noise correlation matrix, we have

$$\mathbb{E}[(\widehat{K}[N]-K)(\widehat{K}[N]-K)^T]$$

$$= \left(\sum_{n=0}^{N=0} H[q_0]]\right)\cdot\left(\sum_{n=0}^{N-1} P[q_0[n]])P[q_0[n]^T\right)\cdot\left(\sum_{n=0}^{N-1} H[q_0[n]]\right)^T$$

To get a better approximation, we must expand $H[q[n]]$ and $P[q[n]]$ about $q_0[n]$:

$$H[q[n]] \approx H[q_0[n]] + H'[q_0[n]](\delta q[n]\otimes I),$$

$$P[q[n]] \approx P[q_0[n]] + P'[q_0[n]](\delta q[n]\otimes I),$$

Then, upto quardratic orders in $\delta q[n]$ (we assume that $W[n]$ is of the same order of magnitude as $\delta q[n]$), we have

$$\widehat{K}[N]-K \approx (\sum_{n=0}^{N-1}(H[q_0[n]] + H'[q_0[n]](\delta q[n]\otimes I)).(\sum_{n=0}^{N-1}(P[q_0[n]] + P'[q_0[n]]$$

$$(\delta q[n]\otimes I))W[n+1])$$

$$= (\sum_{n=0}^{N-1} H[q_0[n]])(\sum_{n=0}^{N-1} P[q_0[n]]W[n+1])$$

$$+(\sum_{n=0}^{N-1} H[q_0[n]])(\sum_{n=0} N-1P'[q_0[n]](\delta q[n]\otimes W[n+1]))$$

$$+(\sum_{n=0}^{N-1} H'[q_0[n]](\delta q[n]\otimes I)).(\sum_{n=0}^{N-1} P[q_0[n]]W[n+1])$$

The evaluate the expectation of the estimation error $e_k[N] = \widehat{K}[N]-K$ upto linear orders in the correlation R_q from this expression, we need to compute $\mathbb{E}(\delta_q[n+1]W[m]^T)$ for all *n*, *m*. By causality of the system, this correlation is zero for $m > n$ and for $m \le n$, we can calculate it using

$$\mathbb{E}(\delta_q[n+1]W[m]^T) = A[n,K]\mathbb{E}(\delta q[n]W[m]^T) + \sigma^2 F_3[n]\delta[n+1-m], n \ge m$$

and

$$E(\delta_q[n]W[n]^T) = \sigma^2 F_3[n]$$

The error correlation $E(e_K[N]e_K[N]^T)$ can also be evaluated using the above expression upto quadratic orders in the correlation $\{R_q[n]\}$. We leave it as an exercise to the reader.

❖❖❖❖❖

[15] Electromagnetic fields generated in inhomogeneous wave=guides and cavity resonators for controlling a slave Robot. A slave Robot consisting of just a single link of length d carries a current $i(t)$ with Fourier transform $I(\omega)$.

An external em field ($\mathbf{E}(t, \mathbf{r}), \mathbf{B}(t, \mathbf{r})$) is incident on this link. This field is generated by a cavity resonator filled with a dielectric having inhomogeneous permittivity (may be even a meta–material having inhomogeneous negative permittivity). The aim is to control the current source at the probe used to excite the resonator so that the Robot link follows a given trajectory.

❖❖❖❖❖

[16] Generalized Sudarshan–Lindblad equation: Let $\mathfrak{h}$ be the system Hilbert space and $\mathcal{H} = L^2(\mathbb{R}_+)$ the noise Hilbert space. $\Gamma_s(\mathcal{H}) = \Gamma_s(L^2(\mathbb{R}_+))$ is the noise Boson Fock space. The evolution $U(t)$ of the system plus bath in the system plus noise Hilbert space $\mathfrak{h} \otimes \Gamma s(\mathcal{H})$ satisfies the Husdon–Parthasarathy noisy Schrodinger equation

$$dU(t) = (-iH(t)dt + L_1(t)dA(t) - L_2(t)dA^*(t) + S(t)d\Lambda(t))U(t)$$

where $L_1(t)$, $L_2(t)$, $S(t)$, $H(t)$ are system operators *i.e.* in $L(\mathfrak{h})$ and $A(t)$, $A^*(t)$ $\Lambda(t)$ act in Boson Fock space $\Gamma_s(\mathcal{H})$. More precisely, $A(t)$, $A^*(t)$ and $\Lambda(t)$ act in $\Gamma_s(\mathcal{H}_{t]})$ and act as identity operators in $\Gamma_s(\mathcal{H}_{(t})$. Here, we identity the Hilbert space $\Gamma_s(\mathcal{H}) = \Gamma_s(\mathcal{H}_{t]} \otimes \mathcal{H}_{(t})$ with $\Gamma_s(\mathcal{H}_{t]}) \otimes \Gamma_s(\mathcal{H}_{(t})$ via the canonical isomorphism defined using exponential vectors:

$$e(u) \approx e(u\chi_{[}0,t]) \otimes e(u\chi_{(t,\infty)})$$

Note that for any Borel subset B of $\mathbb{R}_+$, $(u\chi_B)(t)$ equals $u(t)$ for $t \in B$ and zero for $t \notin B$. The quantum Ito formula is

$$dA(t)dA^*(t) = dt, dA^*(t)dA(t) = 0, d\Lambda(t)dA^*(t) = dA^*(t), d\Lambda.d\Lambda = d\Lambda$$

$$dA(t)d\Lambda(t) = dA(t), d\Lambda(t)dA(t) = 0, dA^*(t)d\Lambda(t) = 0$$

Let $X \in L(\mathfrak{h})$ be Hermitian. Define $X(t) \in L(\mathfrak{h} \otimes \Gamma_s(\mathcal{H}))$ by

$$X(t) = U^*(t)XU(t) = U^*(t)(X \otimes I)U(t)$$

By the quantum Ito formula, we have

$$\begin{aligned} d(U^*(t)U(t)) &= dU^*dU + U^*dU + dU^*dU \\ &= U^*(i(H^* - H)dt + (L_1^* - L_2)dA^* + (L_1 - L_2^*)dA + (S^* + S)d\Lambda)U \\ &\quad + U^*(L_1 * dA^* - L_2^*dA + S^*d\Lambda)(L_1dA - L_2dA^* + Sd\Lambda)U \\ &= U^*[(i(H^* - H) - L_2^*L_2)dt + (S^* + S + S^*S)d\Lambda \\ &\quad + (L_1^* - L_2 - S^*L_2)dA^* + (L_1 - L_2^* - L_2^*S)dA]U \end{aligned}$$

For $U(t)$ to be unitary for all t, we require that $d(U^*U) = 0$ and this happens iff

$$i(H^* - H) + L_2^*L_2 = 0, S + S^* + S^*S = 0, L_1 = L_2^*(1 + S)$$

We assume that these conditions are satisfied. Another application of the quantum *Ito* formula yield the Heisenberg equation for $X(t)$:

$$
\begin{aligned}
dX(t) &= dU^*.X.U + U^*XdU + dU^*XdU \\
&= U^*(i(H^*X - XH - L_2^*XL_2)dt + (L_1^*X - XL_2 - S^*XL_2)dA^* \\
&\quad +(XL_1 - L_2^*X - L_2^*XS)dA + (XS^* + SX + S^*XS)d\Lambda)U \qquad ...(1)
\end{aligned}
$$

We get for $\rho_s(0) \in \mathcal{L}(\hbar)$ being the initial system state and $u \in \mathcal{H}$,

$$
Tr\left(\rho_s(0) \otimes |e(u)\rangle\langle e(u)| dX(t)\right) = Tr_1\left(\rho_s(0) Tr_2\left(|e(u)\rangle\langle e(u)| dX(t)\right)\right)
$$

Now from (1),

$$
\begin{aligned}
&Tr\left(\rho_s(0) \otimes |e(u)\rangle\langle e(u)| dX(t)\right) \\
&= i.Tr\left[U(t)\rho_s(0) \otimes |e(u)\rangle\langle e(u)| U(t)^*)(H^*X - XH - L_2^*XL_2\right]dt \\
&\quad + Tr\left(U(t)(\rho_s(0) \otimes |e(u)\rangle\langle e(u)|)U(t)^*(L_1^*X - XL_2 - S^*XL_2)\right)\bar{u}(t)dt \\
&\quad + Tr\left(U(t)(\rho_s(0) \otimes |e(u)\rangle\langle e(u)|)U(t)^*(XL_1 - L_2^*X - L_2^*XS)\right)u(t)dt \\
&\quad + Tr\left(U(t)(\rho_s(0) \otimes |e(u)\rangle\langle e(u)|)U(t)^*(XS^* - SX - S^*XS)\right)|u(t)|^2 dt
\end{aligned}
$$

It follows that if we define the system state at time t as

$$
\rho_s(t) = \rho_s(t, u) = Tr_2\left(U(t)\left(\rho_s(0) \otimes |e(u)\rangle\langle e(u)|\right)U(t)^*\right)
$$

then $\rho_s(t)$ satisfies the generalized Sudarshan – Lindblad equation

$$
\begin{aligned}
\rho_s'(t) &= -i\left(H(t)\rho_s(t) - \rho_s(t)H(t)^* - L_2(t)\rho_s(t)L_2(t)^*\right) \\
&\quad + \bar{u}(t)\left(\rho_s(t)L_1(t)^* - L_2(t)\rho_s(t) - L_2(t)\rho_s(t)S(t)^*\right) \\
&\quad + u(t)\left(L_1(t)\rho_s(t) - \rho_s(t)L_2(t)^* - S(t)\rho_s(t)L_2(t)^*\right) \\
&\quad + |u(t)|^2\left(S(t)^*\rho_s(t) - \rho_s(t)S(t)^* - S(t)\rho_s(t)S(t)^*\right)
\end{aligned}
$$

Problem: Given a state evolution $\rho_d(t), 0 \le t \le T$, assuming L_1, L_2, H, S to be constant operator in $\hbar$ (*i,e*, independent of time), (*a*) determine the optimum function $u(t), 0 \le t \le T (u \in L^2(\mathbb{R}_+))$ so that $\int_0^T \|p_s(t,u) - p_d(t)\|^2 dt$ is a minimum, (*b*) Assuming the function $u(t), 0 \le t \le T$ and H given, determin the operators L_1, L_2, S, so that $\int_0^T \|p_s(t,u) - p_d(t)\|^2 dt$ is a minimum.

❖❖❖❖❖

[17] Quantum image processing:

First suppose that the image signal field x is specified. We start a quantum system in the state $\rho(0)$ and allow it to evolve under the Schrodinger dynamics corresponding to a Hamiltonian $H_x(t)$ dependent on the image signal field x (Note that x may be a time dependent field). Let $U_x(t,s)$ denote the corresponding Schrodinger evolution operator.

$$i\frac{\partial U_x(t,s)}{\partial t} = H_x(t)U_x(t,s), t \geq s,$$
$$U_x(s, s) = I$$

At time t_1 we make a POV (Positive operator valued) measurement $\{\Pi_a : a = 1,2,...,N\}(\Pi_a \geq 0, \sum_a \Pi_a = I)$. If Π_a is true, then we assign a value α_a to the measurement. We again allow the system to evolve under the same dynamics till time t_2 and make the same POVM measurement. In this way, we get by applying the standard collapse postulate of quantum quantum mechanics. that the probability of observing the values $\alpha_{a(k)}$, $k = 1, 2,...,M$ at times $t_1 < t_2 < ... < t_M$ is

$$P(\alpha_{a(1)},...,\alpha_{a(M)}, t_1,...,t_m | x)$$

$$= Tr[\Pi_{\alpha_{a(M)}} U_x(t_M, t_{M-1})\Pi_{\alpha_{a(M-1)}} U_x(t_{M-1}, t_{M-2})...\Pi_{\alpha_{a(1)}} U_x(t_1, 0)p(0)$$

$$U_x(t_1,0)^* \Pi_{\alpha_{a(1)}} U_x(t_1,t_2)^* \Pi_{\alpha_{a(2)}}... U_x(t_M,t_{M-1})^* \Pi_{\alpha_{a(M)}}]$$

By maximizing this probability over the permissible image fields x, we obtain a quantum mechanical image estimator,

Problem: Assume that the image consists of pixels at $\mathbf{R}_{nm} = (n\Delta, m\Delta, 0), 0 \leq n,m \leq K-1$ in the $x - y$ plane. The pixel at (n,m) as an RGB complex vector magnetic field amplitudes of $\mathbf{H}_k(n, m)$, $k = 1, 2, 3$ respectively. Let ω_k, $k = 1, 2, 3$ denote the RGB frequencies. Then the surface current density phasors corresponding to these RGB amplitudes are respectively $\mathbf{J}_{sk}(n,m) = \hat{z} \times \mathbf{H}_k(n, m), k = 1, 2, 3$. Clearly these surface current densities do not have any z component, *i.e.*, we can express these as

$$\mathbf{J}_{sk}[n, m] = J_{sxk}[n,m]\hat{x} + J_{syk}[n,m]\hat{y}, k = 1, 2, 3$$

where

$$J_{sxk}[n, m] = -H_{yk}[n,m], J_{syk}[n,m] = H_{xk}[n, m], K = 1, 2, 3$$

The magnetic vector potential at $\mathbf{r} = (x, y, z)$ is then

$$\mathbf{A}(t, \mathbf{r}) = \left(\mu\Delta^2/4\pi\right) \sum_{nm,k} \mathrm{Re}\left[\mathbf{J}_{sk}[n,m]\exp(j\omega_k\left(t - |\mathbf{r} - \mathbf{R}_{nm}|/c\right)\right]/|r - \mathbf{R}_{nm}| \quad ...(1)$$

The electric scalar potential Φ $(t, \mathbf{r})$ can then be determined using the Lorentz gauge:

$$\Phi(t, \mathbf{r}) = -C^2 \int_0^t div\mathbf{A}(s, \mathbf{r})ds$$

Clearly $\mathbf{A}(t, \mathbf{r})$ does not have a z–component. Assume that we excite a two dimensional quantum harmonic oscillator having unperturbed Hamiltonian

$$H_0 = (p_1^2 + \alpha_1^2 q_1^2)/2 + (p_2^2 + \alpha_2^2 q_2^2)/2$$

with this radiation field coming from the image. The resulting Hamiltonian assuming a charge $-e$ on the oscillator particle is given by

$$H(t) = (p_1 + eA_x)^2/2 + (p_2 + eA_y)^2/2 + (\alpha_1^1 q_1^2 + \alpha_2^2 q_2^2)/2 - e\Phi$$

$$= H_0 - e(\Phi + i(A_x \partial_1 + A_y \partial_2 + A_{x,1} + A_{y,2}) + (e^2/2)(A_x^2 + A_y^2)$$

The image signal field can be taken as $\{(J_{sx}[n,m], J_y[n,m]) : 0 \le n,m \le K-1\}$ and we can express (1) in "Kernel form"

$$A_x(t, \mathbf{r}) = \sum_{n,m,k} \text{Re}(K(t,\mathbf{r}|n,m,k) J_{sxk}[n,m]),$$

$$A_y(t, \mathbf{r}) = \sum_{n,m,k} \text{Re}(K(t,\mathbf{r}|n,m,k) J_{syk}[n,m]))$$

where

$$K(t, \mathbf{r}|n, m, k) = (\mu/4\pi)\exp(i\omega_k(t - |\mathbf{r} - \mathbf{R}_{nm}|/c))/|\mathbf{r} - \mathbf{R}_{nm}|$$

❖❖❖❖❖

[18] Computation of the information transmitted by a classical source through a quantum channel defined by Schrodinger's equation

Abstract: The Von-Neumann entropy of a state p is defined as $S(\rho)- = Tr(\rho.\log(\rho))$. Suppose we send a random classical alphabet x with probability $p[x]$ and resulting state is ρx. Then the conditional entropy of the output given the input is given by $S(Y \mid X) = \sum_x p(x) S(\rho_x)$ and the entropy of ouytput is given by $S(Y) = S(\sum_x p(x)_{px})$. The information transmitted from the input to the input output is then

$$I(Y, X) = S(Y) - S(X \mid Y)$$

and it is also called the mutual information. We evaluate this mutual information approximately for the case when the output state px is obtained as the solution of Schrodinger's equation with the input "alphabet" being a random information bearing classical sequence $\{I[n]\}$. This classical random sequence modulates a set of quantum potentials via time varying pulses and this modulated potential perturbs the Hamiltonian of a quantum system causing the state to evolve in time. After time T the output state is forced. the advantage of evaluating this mutual information is that we can maximize it over all probability distributions of the input alphabet $p(\{I[n]\}$ and hence derive the Holevo capacity of the Schrodinger channel.

The information sequence to be transmitted is $I = \{I[n] : 1 \le n \le N\}$. This sequence modulates a sequence of n pulses $p_n(t)$, $n = 1, 2,...,N$ and if $V_1,..., V_N$ are potentials, then the Hamiltonian of the perturbed quantum system is

$$H(t) = H_0 + \varepsilon \sum_{n=1}^{N} I[n] p_n(t) V_n$$

The unitary quantum evolution operator $U(t)$ satisfies Schrodinger's equation

$$U'(t) = -iH(t)U(t), t \geq 0$$

The approximate solution to this equation upto $O(\varepsilon^2)$ is obtained by well known methods as

$$U(t) = U_0(t)W(t), U_0(t) = \exp(-itH_0),$$

$$W(T) = I - i\varepsilon\sum_n I[n]\int_0^T p_n(t)\tilde{V}_n(t)dt - \varepsilon^2\sum_{n,m} I[n]I[m]\int_{0<l_2<t_1<T} p_n(t_1)$$

$$p_m(t_2)\tilde{V}_n(t_1)\tilde{V}_m(t_2)dt_1dt_2$$

$$= I - i\varepsilon\sum_n I[n]W_n^{(1)} - \varepsilon^2\sum_{n,m} I[n]I[m]W_{nm}^{(2)}$$

with $W_n^{(1)}$ and $W_{nm}^{(2)}$ having the obvious meanings.

We wite $U(t, I)$ and $W(t, I)$ for $U(t)$ and $W(t)$ to emphasize that this corresponds to the classical information sequence I being transmitted. The state (output) at time T is

$$\rho(T, I) = U(T, I)\rho(0)U(T, I)^*$$

and hence the entropy of the quantum output Y given the classical input $X = I$ is given by

$$S(Y \mid X) = S(\rho(T, I)) = S(\rho(0))$$

where $S(\rho) = Tr(\rho.\log(\rho))$ is the Von-Neumann entropy of ρ. Secondly, the output state is obtained by averaging $\rho(T, I)$ over all possible classical inputs. I *i.e.*, the output state is

$$\rho(T) = \sum_I \rho(I)\rho(T, I) = \sum_I \rho(I)U(T, I)^*\rho(0)U(T, I)$$

$$= U_0(T)\sum_I p(I)W(T, I)^*\rho(0)W(T, I)U0(T)^*$$

and hence, the entropy of the output state is

$$S(Y) = S\left(\sum_I p(I)W(T, I)\rho(0)W(T, I)\right)$$

We have

$$W(T, I)^*\rho(0)W(T, I) = \rho(0) - i\varepsilon\sum_n I[n]\left[W_n^{(1)}, \rho(0)\right]$$

$$-\varepsilon^2\sum_{nm} I[n]I[m]\left(W_{nm}^{(2)}\rho(0) + \rho(0)W_{nm}^{(2)*}\right)$$

$$+\varepsilon^2\sum_{nm} I[m]I[n]W_n^{(1)}\rho(0)W_m^{(1)}$$

$$O(\varepsilon^3)$$

Thus.

$$\sum_I p(I)W(T,I)^*\rho(0)W(T,I)$$

$$\rho(0)-i\varepsilon\sum_n\langle I[n]\rangle\left[W_n^{(1)},\rho(0)\right]$$

$$+\varepsilon^2\sum_{nm}\langle I[n]I[m]\rangle\left(W_n^{(1)}\rho(0)W_m^{(1)}-W_{nm}^{(2)}\rho(0)W_{nm}^{(2)*}\right)$$

$$+O(\varepsilon^3)$$

$$=\rho(0)+\varepsilon\rho_1+\varepsilon^2\rho_2+O(\varepsilon^3)$$

To compute the entropy of this state which we denote by $\tilde{\rho}$, we shall first determine using standard using standard perturbation theory, the eigenvalues of $\tilde{\rho}$ upto $O(\varepsilon^2)$. Let $\lambda_a^{(0)}$, $v_a^{(0)}$, a = 1, 2, ..., M denote the eigenvalues and corresponding eigenvectors of ρ_0. Assume for simplicity that all the eigenvalues of $\rho(0)$ are distinct. Let $v_a^{(0)}+\varepsilon v_a^{(1)}+\varepsilon^2 v_a^{(2)},\lambda_a^{(0)}+\varepsilon\lambda_a^{(1)}+\varepsilon^2\lambda_a^{(2)}$ denote the corresponding eigenvalues and eigenvalues and eigenvectors of $\tilde{\rho}$ upto $O(\varepsilon^2)$. Then we have by standard perturbation theory,

$$\left(\rho(0)-\lambda_a^{(0)}\right)v_a^{(1)}\rho_1 v_a^{(0)}-\lambda_a^{(1)}v_a^{(0)}=0$$

$$\left(\rho(0)-\lambda_a^{(0)}\right)v_a^{(2)}+\left(\rho_1-\lambda_a^{(1)}\right)v_a^{(1)}+\left(\rho_2-\lambda_a^{(2)}\right)v_a^{(0)}=0$$

These give on taking inner products with $v_b^{(0)}$ and using their eigenproperty for $\rho(0)$ and orthonormality,

$$\left(\lambda_b^{(0)}-\lambda_a^{(0)}\right)\left\langle v_b^{(0)},v_a^{(1)}\right\rangle+\left\langle v_b^{(0)},\rho_1 v_a^{(0)}\right\rangle-\lambda_a^{(1)}\delta_{b,a}=0$$

$$\left(\lambda_b^{(0)}-\lambda_a^{(0)}\right)\left\langle v_b^{(0)},v_a^{(2)}\right\rangle+\left\langle v_b^{(0)},\rho_1 v_a^{(1)}\right\rangle-\lambda_a^{(1)}\left\langle v_b^{(0)},v_a^{(1)}\right\rangle$$

$$+\left\langle v_b^{(0)},\rho_2 v_a^{(0)}\right\rangle-\lambda_a^{(2)}\delta_{b,a}=0$$

setting $b = a$, in the first equation gives

$$\lambda_a^{(1)}=\left\langle v_a^{(0)},\rho_2 v_a^{(0)}\right\rangle$$

and for $b \neq a$, we get

$$\left\langle v_b^{(0)}, v_a^{(1)} \right\rangle = \frac{\left\langle v_b^{(0)}, \rho_1 v_a^{(0)} \right\rangle}{\lambda_a^{(0)} - \lambda_b^{(0)}}$$

and

$$v_a^{(1)} = \sum_{b \neq a} v_b^{(0)} \left\langle v_b^{(0)}, v_a^{(1)} \right\rangle$$

The other method for computing the rate of change of quantum entropy for a state $\rho(t)$ that evolves according to the Sudarshan-Lindblad equation is based on the Baker-Campbell-Hausdorff formula in Lie groups and Lie groups and Lie algebras:

$$S(t) = Tr(\rho(t)).\log(\rho(t))$$

Thus

$$S'(t) = -Tr\left(\rho'(t).\log(\rho(t)) - Tr\left(\rho(t)\left(\log(\rho(t))\right)\right)'\right)$$

Now let

$$\log(\rho) = Z$$

Thus,

$$\rho = \exp(Z)$$

Using the result that for two matrices A, B

$$\frac{d}{dt}\exp(A + tB)\Big|_{t=0} = \exp(A).\frac{(I\exp(-adA))}{ad(A)}(B)$$

$$= \exp(A)\sum_{n=0}^{\infty}\frac{(-1)^n ad(A)^n}{n!}(B)$$

To prove this, we consider

$$F(t, s) = \exp(s(A + tB))$$

Then

$$F_{,s}(t, s) = (A + tB)F(t, s)$$

and hence writing

$$F(t, s) = \exp(sA)G(t, s)$$

we have

$$G_{,s}(t,s) = t.\exp(-sadA)(B)G(t,s)$$

Differentiating both sides w.r.t. at $t = 0$, gives

$$G_{,ts}(0,s) = \exp(-sadA)(B)\,G(0,s) = \exp.(-\,\mathrm{sad}A)(B)$$

and thus

$$G_{,t}(0,1) = \int_0^1 \exp(-sad\,A)(B)ds = \frac{I-\exp(-ad\,A)}{ad\,A}(B)$$

proving the claim. It thus follows that

$$\rho' = (\exp(Z))' = \exp(Z)\frac{(I-\exp(-ad\,Z))}{ad\,Z}(Z')$$

$$= \rho\sum_{n=1}^{\infty}(-1)^n\frac{(ad\,Z)^n}{(n+1)!}(Z')$$

Inverting this equation gives

$$Z' = \frac{ad\,Z}{I-\exp(-ad\,Z)}(\rho^{-1}\rho')$$

Consider now the function

$$f(z) = \frac{z}{1-\exp(-z)}, z \in \mathbb{C}, z \neq 0$$

Clearly,

$$\lim_{z\to 0} f(z) = 1$$

Hence $f(z)$ has a Taylor expansion

$$f(z) = 1+\sum_{n=1}^{\infty} c_n z^n$$

which converges absolutely for $|\exp(-z)| < 1$, or equivalently for $Re z > 0$. If follows that for the operator $Z = \log\rho$, we have

$$f(adZ) = I+\sum_{n\geq 1} c_n (adZ)^n$$

Convergence issue needs to be discussed. Let p_k, $k \geq 1$ be the eigenvalues of ρ with corresponding eigenvectors v_k, $k = 1, 2,\ldots$. We have $\langle v_k, v_m\rangle = \delta_{k,m}p_k > 0$, $\sum_k p_k = 1$. It follows that

$$ad(Z)(|v_k\rangle\langle v_m|) = Z|v_k\rangle\langle v_m| - |v_k\rangle\langle u_m|Z$$

$$= (\log(p_k)-\log(p_m))|v_k\rangle\langle v_m|$$

and hence, the eigenvalues of $ad(Z)$ and log (p_k/p_m) k, m = 1, 2.... So the above infinite series for $f(adZ)$ will converge provided that for all k, m, the series

$$\sum_{n\geq 1} c_n \left(\log\left(p_k/p_m\right)\right)^n$$

converges, For absolute convergence of the series, we require that

$$\sum_{n\geq 1} |c_n| \left|\log\left(p_{\max}/p_{\min}\right)\right| < \infty$$

Assuming this, we have,

$$Tr\left(\rho(\log\rho)'\right) = Tr(\rho Z') = Tr\left(\rho\left(\rho^{-1}\rho' + \sum_{n\geq 1} c_n\rho(adZ)^n\left(\rho^{-1}\rho'\right)\right)\right) = 0$$

since

$$Tr(\rho') = Tr(\rho)' = 0$$

and $n \geq 1$,

$$Tr\left(\rho(ad\,Z)^n\left(\rho^{-1}\rho'\right)\right) = Tr\left(\rho(ad\,Z)X_n\right) = Tr\left(\rho[Z, X_n]\right) = Tr([Z, pX_n]) = 0$$

since $[Z, \rho] = 0$ Here, we have defined

$$X_n = (adZ)^{n-1}\left(\rho^{-1}\rho'\right)$$

Thus,

$$S'(t) = Tr(\rho'.\log\rho)$$

We write $H(t, I)$ for the Hamiltonian $H(t)$ defined above and then $U(t, I)$ denotes the Schrodinger evolution unitary family of operators corresponding to $H(t, I)$. Then,

$$\rho(t) = \sum_i p(I)U(t,I)\rho(0)U(t,I)^*$$

so that

$$q'(t) = i\sum_i p(I)\left[H(t,I), \rho(t,I)\right]$$

where

$$\rho(t, I) = U(t,I)\rho(0)U(t,I)^*$$

Thus,

$$S'(t) = i\sum_I p(I)Tr\left(\left[H(t,I), \rho(t,I)\right]\right)\log(\rho(t))$$

$$= i\sum_I p(I)Tr\left(\left[H(t,I), \rho(t,I)\right]\right)\log\left(\sum_j p(J)\rho(t,J)\right)$$

❖❖❖❖❖

[19] Optimal control of a dynamical system using pd controllers: The dynamical system has the form

$$q'(t) = A(t,\theta)q(t)+F(q(t),t) \qquad ...(1)$$

where $q(t) \in \mathbb{R}^n$, $A(t, \theta) \in \mathbb{R}^{n\times n}$ and $F(q, t) \in \mathbb{R}^n$. The aim is to select the parameter vector θ on which the matrix A depends so that

$$\int_0^T \|q(t)-q_d(t)\|^2 dt$$

is a minimum. An example of such a dynamical system is in robotics where $q(t) \varepsilon \mathbb{R}^d$ satisfies the differential equation

$$M(q(t))q''(t)+N(q(t),q'(t)) = F(q(t),q'(t))(K_p(q_d(t)-q(t)))$$
$$+K_d(q_d'(t)-q_d(t))+\tau(t) \qquad ...(2)$$

Here $q_d(t)$ is the desired signal which the robot angular process $q(.)$ is intended to track and K_p, K_d are the pd control coefficients. $\tau(t)$ is the externally applied torque. Ideally this control problem must be solved by minimizing the energy functional

$$E(q(.),\lambda(.),\theta) = \int_0^T \|q(t)-q_d(t)\|^2 dt$$
$$-\int_0^T \lambda(t)^T (q'(t)-A(t,\theta)q(t)-F(q(t),t))dt \qquad ...(3)$$

The optimal equations

$$\delta E/\delta q = 0, \delta E/\delta\lambda = 0, \frac{\partial E}{\partial\theta} = 0 \qquad ...(4)$$

give

$$2(q(t)-q_d(t))+\lambda'(t)+A(t,\theta)^T\lambda(t)+F'(q(t),t)^T\lambda(t) = 0 \qquad ...(5)$$

the equations of motion (1) and

$$\int_0^T \lambda(t)^T \frac{\partial A(t,\theta)}{\partial\theta}(I_p \otimes q(t))dt = 0 \qquad ...(6)$$

❖❖❖❖❖

[20] Control of Robot with updating of parametric uncertainties based on the Lyapunov energy method. $M(q) = M(q,\theta_0)$, $\widehat{M}(q) = M(q, \theta_0+\delta\theta(t))$. θ_0 is the true value of the parameter vector and $\theta(t)$ is the estimate of the parameter vector at time t. $\delta\theta(t) = \theta(t) - \theta_0$ is the discrepancy in the estimate of the parameter vector at time t. Likewise the term $N(q, q', \theta)$ comprising the centrifugal and centripetal torque and frictional and gravitational torque also depend on the parameter θ. We write $\widehat{N}(q, q') = N(q, q', \theta(t))$ and $N(q, q') = N(q, q', \theta_0)$. The desired trajectory is $q_d(t)$. The trajectory error is $e(t) = q_d(t) - q(t)$. K_p, K_d are the pd controller coeffcients. The computed torque at time t (based on the estimate $\theta(t) = \theta_0 + \delta\theta(t)$) of the parameter vector is given by

$$\tau(t) = \widehat{M}(q)\left(q_d'' + K_p e + K_d e'\right) + \widehat{N}(q, q')$$

$$= \widehat{M}(q)\left(q'' + e'' + K_p e + K_d e'\right) + \widehat{N}(q, q') \quad \text{...(1)}$$

We write

$$\delta M = \delta M(q) = \widehat{M}(q) - M(q)$$

$$= M(q, \theta_0 + \delta\theta) - M(q, \theta_0) \approx \left(\frac{\partial M(q, \theta_0)}{\partial\theta}\right)(\delta\theta \otimes I_d)$$

$$\delta N = \delta N(q, q') = N(q, q', \theta_0 + \delta\theta(t)) - N(q, q', \theta_0)$$

$$\approx \frac{\partial N(q, q', \theta_0)}{\partial\theta}\delta\theta$$

The equations of motion

$$M(q)q'' + N(q, q') = \varepsilon(t) = (M + \delta M)\left(q'' + e'' + K_p e + K_d e'\right) + (N + \delta N) \quad \text{...(2)}$$

Now assume the form $\left(\widehat{M} = M + \delta M\right)$

$$e'' + K_p e + K_d e' = -\widehat{M}(q)^{-1}(\delta M q'' + \delta N)$$

$$= -M(q)^{-1}(\delta M q'' + \delta N) = W(q, q', q'')\delta\theta(t) \quad \text{...(3)}$$

where

$$W(q, q', q'') = -M(q)^{-1}\left(\frac{\partial M(q, \theta_0)}{\partial\theta}\left(I_p \otimes q''\right) + \frac{\partial N(q, q', \theta_0)}{\partial\theta}\right)$$

We can replace q'' by $q_d'' + e''$ and neglect the product term $e'' \otimes \delta\theta$ to get the approximate equation

$$W = W(t, e, e')$$

where

$$W(t,e,e') = -M(q)^{-1}\left(\frac{\partial M(q,\theta_0)}{\partial \theta}\left(I_p \otimes q_d''(t)\right) + \frac{\partial N(q,q',\theta_0)}{\partial \theta}\right)$$

with $q = q_d - e, q' = q_d' - e'$. (3) can be expressed as

$$\frac{d}{dt}\begin{pmatrix} e \\ e' \end{pmatrix} = \begin{pmatrix} 0 & I_d \\ -K_p & -K_d \end{pmatrix}\begin{pmatrix} e \\ e' \end{pmatrix}$$

$$+\begin{pmatrix} 0 \\ W(t,e,e') \end{pmatrix}\delta\theta(t)$$

We write this as

$$\frac{d}{dt}\tilde{e}(t) = A\tilde{e}(t) + B(t,\tilde{e})\delta\theta(t)$$

To obtain the update equation for $dq(t)$, we construct a Lyapunov function

$$V(\tilde{e},\delta\theta) = \frac{1}{2}\tilde{e}^T Q\tilde{e} + \frac{1}{2}\delta\theta^T P\delta\theta$$

where Q, P are respectively $d \times d$ and $p \times p$ positive definite matrices. Here, d is the number of Robot links and p is the number of parameters. For stability of the algorithm (*i.e.*, tracking error $\tilde{e}(t)$ as well as parametric error $\delta\theta(t)$ both should converge to zero), we require that

$$\frac{dV}{dt} \le 0$$

Thus,

$$\tilde{e}^T Q\frac{d\tilde{e}}{dt} + \delta\theta^T P\frac{d\delta\theta}{dt} \le 0$$

i.e., $\tilde{e}^T Q(A\tilde{e} + B\delta\theta) + \delta\theta^T P\delta\theta' \le 0$

or

$$\frac{1}{2}\tilde{e}^T\left(QA + A^T Q\right)\tilde{e} + \delta\theta^T\left(P\delta\theta' + B^T Q\tilde{e}\right) \le 0$$

This can be achieved if

$$QA + A^T Q \leq 0,$$

and

$$\delta\theta' = -P^{-1}B^T Q\tilde{e} - P^{-1}R\delta\theta$$

where $R \leq 0$. We then get

$$\frac{dV}{dt} = \tilde{e}^T \left(QA + A^T Q\right)\tilde{e} + \delta\theta^T R\delta\theta \leq 0$$

This completes the derivation for the pd control of the robot with parametric uncertainty updates.

❖❖❖❖❖

Quantum State Estimation Robot System Estimation, Rotation Matrix Estimation, Quantum Filtering

[21] Estimating the initial state of a quantum system from continuous measurements of the average value of an observable. The state follows the Sudarshan-Lindblad equation

$$\rho'(t) = -i[H(t), \rho(t)] + \varepsilon^2 \theta(\rho(t)) \qquad ...(1)$$

where

$$\theta(\rho) = -\frac{1}{2}\sum_{k=1}^{p}\left(L_k^* L_k \rho + \rho L_k^{*-2} L_k \rho L_k^*\right) \qquad ...(2)$$

$$H(t) = H_0 + \varepsilon V(t) \qquad ...(3)$$

Here $H_0, V(t)$, L are operators in the system Hilbert space $\hbar$ and $\rho(t)$ is the quantum state in $\hbar$ at time t. We write the solution as perturbation series upto $O(\varepsilon^2)$:

$$\rho(t) = \rho_0(t) + \varepsilon r_1(t) + \varepsilon^2 \rho_2(t) + O(\varepsilon^3) \qquad ...(4)$$

Substituting this into the differential equation (1) and equating coefficients of ε^m, $m = 0, 1, 2$ successively gives us

$$\rho_0'(t) = -i[H_0, \rho_0(t)]$$

$$\rho_1'(t) = -i[H_0, \rho_1(t)] - i[V(t), \rho_0(t)]$$

$$\rho_2'(t) = -i[H_0, \rho_2(t)] - i[V(t), \rho_1(t)] + \theta(\rho_0(t))$$

The solutions are given by

$$\rho_0(t) = T_{0t}(\rho(0)), T_{0t} = \exp(-itad(H_0))$$

$$\rho_1(t) = -i\int_0^t T_{0,t-s} ad\left(V(s)\right)\left(\rho_0(s)\right)ds$$

$$= -i\int_0^t T_{0,t-s} o ad\left(V(s)\right) o T_{0,s}\left(\rho(0)\right)ds \; T_{1,t}\left(\rho(0)\right)$$

where

$$T_{1,t} = -i\int_0^t T_{0,t-s} o ad\left(V(s)\right) o T_{o,s} ds$$

Finally,

$$\rho_2(t) = -i\int_0^t T_{0,t-s} o ad\left(V(s)\right) o \rho_1(s)ds + \int_0^t T_{0,t-s} o\theta\left(\rho_0(s)\right)ds$$

$$= -i\int_0^t T_{0,t-s} o ad\left(V(s)\right) o T_{1,s}\left(\rho(0)\right)ds + \int_0^t T_{0,t-s} o\theta o T_{0,s}\left(\rho(0)\right)ds$$

$$= T_{2,t}\left(\rho(0)\right)$$

where

$$T_{2,t} = -i\int_0^t T_{0,t-s} o ad\left(V(s)\right) o T_{1,s} ds + \int_0^t T_{0,t-s} o\theta o T_{0,s} ds$$

We write

$$T_t = T_{0,t} + \varepsilon T_{1,t} + \varepsilon^2 T_{2,t}$$

with $O(\varepsilon^3)$ terms neglected. In short, we have

$$\rho(t) = T_t\left(\rho(0)\right), t \geq 0$$

where T_t is a linear operator on the Banach algebra $\mathcal{B}(\mathcal{H})$. We can express this relationship as

$$\rho(t) = T_t(\rho(0)) = \int_0^t K_1(t,s)\rho(0)K_2(t,s)ds$$

where $K_1(t, s)$ and $K_2(t, s)$ are linear operators in $\mathcal{H}$. Since $\rho(t)$ is always Hermitian for $\rho(0)$ Hermitian, it follows that

$$K_2(t, s) = K_1(t, s)^*$$

So defining

$$K(t, s) = K_1(t, s)$$

it follows that

$$\rho(t) = \int_0^t K(t,s)\rho(0)K(t,s)^* ds$$

The average value of an Heisenberg observable *X* at time *t* is then

$$\langle X\rangle(t) = Tr(\rho(t)X) = \int_0^t Tr\left(\rho(0)K(t,s)^* XK(t,s)\right)ds$$

$$= Tr(\rho(0)\,X(t))$$

where

$$X(t) = \int_0^t K(t,s)^* XK(t,s)\,ds$$

Suppose we have noisy measurements

$$z(t) = \langle X\rangle(t) + w(t)$$

Then, $\rho(0)$ can be estimated by minimizing

$$E(\rho(0)) = \int_0^T \left(z(t) - Tr(\rho(0)X(t))\right)^2 dt$$

subject to the constraint

$$Tr(\rho(0)) = 1$$

Incorporating this constraint using a Lagrange multiplier λ and setting the variational derivative of E ($\rho(0)$) with constraint ot zero gives

$$-2\int_0^T X(t)\left(z(t) - Tr(\rho(0)X(t))\right)dt - \lambda I = 0$$

With respect to an o.n.b. $\{e_n : n \geq 1\}$ for $\mathcal{H}$, this becomes

$$\lambda\delta_{mn}/2 = -\int_0^T z(t)\langle m|X(t)|n\rangle dt$$

$$+\sum_{p,q}\left(\int_0^T \langle q|X(t)|p\rangle\langle m|X(t)|n\rangle dt\right)\langle p|\rho(0)|q\rangle$$

This is a system of linear equations for the matrix elements $\langle p|\rho(0)|q\rangle$, $p, q = 1, 2, \ldots$

❖❖❖❖❖

[22] Estimating the configuration of two link Robot system using group representation theory. The two links are modelled as three dimensional rigid bodies. The bottom link is attached to the origin at its base point while the top link is attached at its base point to a fixed point p_0 on the bottom link. The final configuration of this robot system is obtained by first applying a rotation $R_1 \in$ SO to the bottom link with the top link being held rigidly. This means that R_1 also (3) acts on the top link. After that,we apply another rotation R_2 to the top link relative to the bottom link. At time $t = 0$ we assume that the region of space occupied by the bottom link is $B_1 \subset \mathbb{R}^3$ and that by the top link is $B_2 \subset \mathbb{R}^3$. After the application of the two rotations, a point r_0 in the bottom link moves to $r = R_1 r_0$ and a point s_0 in the top link moves to $s = R_1 p_0 + R_2 R_1 (s_0 - p_0)$.

Note that $r_0 \in B_1$ and $s_0 \in B_2$. The group action can thus be described by

$$\begin{pmatrix} r_0 \\ s_0 \end{pmatrix} \rightarrow \begin{pmatrix} R_1 & 0 \\ R_2 R_1 & \end{pmatrix} \begin{pmatrix} r_0 \\ s_0 \end{pmatrix} + \begin{pmatrix} 0 \\ (R_1 - R_2 R_1) p_0 \end{pmatrix}$$

Equivalently,

$$\begin{pmatrix} r_0 \\ s_0 \\ p_0 \end{pmatrix} \rightarrow \begin{pmatrix} R_1 & 0 & 0 \\ 0 & R_2 R_1 & R_1 - R_2 R_1 \\ 0 & 0 & I \end{pmatrix} \begin{pmatrix} r_0 \\ s_0 \\ p_0 \end{pmatrix}$$

We denote this above 9×9 matrix by $T(R_1, R_2)$ In the following exercise, we discuss the configuration estimation problem,

Exercise: Let R_1, R_2, S_1, S_2 $SO(3)$. Expree $T(S_1, S_2)\,T(R_1, R_2)$ in the form $T(S, R)$ where S, R $SO(3)$. Now let $J_1(r)$ be the current density in the bottom link at time $t = 0$ and $J_2(r)$ that in the top link. Thus $J_1(r) = 0$ for $r \notin B_1$

$$J_0(r) = J_1(r) + J_2(r)$$

After time t the current density field is obviously

$$J(r) = J_1\left(R_1^{-1} r\right) + J_2\left(p_0 + (R_2 R_1)^{-1}(R_1 p_0)\right)$$

The far field magnetic vector potential is then

$$A(r) = \left(\mu.\exp(-jk|r|)/4\pi|r|\right) P(\hat{r})$$

where

$$P(\hat{r}) = \int J(r')\exp\left(jk(\hat{r}, r')\right) d^3 r'$$

$$= P_1(\hat{r}) + P_2(\hat{r})$$

where

$$P_1(\hat{r}) = \int J_1(R_1^{-1}r')\exp(jk\hat{r}.r')d^3r'$$

$$= \int J_1(r')\exp(jk(R_1^{-1}\hat{r}.r'))d^3r' = Q_1(R_1^{-1}\hat{r})$$

and

$$P_2(\hat{r}) = \int J_2(p_0 + (R_2R_1)^{-1}(r' - R_1p_0))\exp(jk\hat{r}.r')d^3r'$$

$$= \int J(r')\exp(jk(\hat{r}.R_1p_0 + R_2R_1(r' - p_0)))d^3r'$$

$$= Q_2(R_1^{-1}R_2^{-1}\hat{r}).\exp(jk((R_1^{-1} - R_1^{-1}R_2^{-1})\hat{r}, p_0))$$

Writing $R = R_1 S = R_2 R_1$, we get

$$P(\hat{r}) = Q_1(R^{-1}\hat{r}) + \exp((R^{-1} - S^{-1})\hat{r}, p_0)Q_2(S^{-1}\hat{r})$$

where

$$Q_1(\hat{r}) = \int J_1(r')\exp(jk\hat{r}.r')d^3r'$$

$$Q_2(\hat{r}) = \int J_2(r')\exp(jk\hat{r}.r')d^3r'$$

The Problem is to estimate the rotations R, S from measurements of the field pattern $P(\hat{r})$ at different points $\hat{r}$.

❖❖❖❖❖

[23] Evans-Hudson diffusions, control using quantum pd controllers: In the format of the usual quantum stochastic calculus setting, we B_0 is the initial algebra and B_t is the algebra after time t, j_t: $B_0 \to B_t$ is a star unital homomorphism, *i.e.*, (*a*) $j_t(aX + bY) = aj_t(X) + bj_t(Y) a, b \in \mathbb{C}$ and $X, Y \in B_0, j_t(1) = 1\ j_t(XY) = j_t(X)\ j_t(X)\ j_t(Y)\ \forall\ X, Y\ B_0$ and if B_t has involution *, then $j_t(X^*) = j_t(X)^*$. Then j_t is called a quantum flow if it satisfies the following qsde:

$$dj_t(X) = j_t(\theta_j^i(X))d\Lambda_i^j$$

the summation being over all i, j. Here $\Lambda_0^0(t) = t, \Lambda_0^j(t) = A_j(t), \Lambda_0^j = A_j^*(t), j \geq 1$, For $j, k \geq 1, \Lambda_k^j(t) = \Lambda_{|k\rangle\langle j|}(t)$ where $\{|k\rangle : k = 1, 2, ...\}$ is an orthonormal basis for a Hilbert space V. The bath Hibert space $\mathcal{H} = V \otimes L^2(\mathbb{R}_+)$. We have

$$d\Lambda_k^j . d\Lambda_q^p = \delta_q^j d\Lambda_k^p, j \geq, q \geq 1,$$

otherwise the product is zero. As an example, we consider the following classical setup : $\xi(t)$ is a classical diffusion satisfying the sde

$$d\xi(t) = \mu(\xi(t))dt + \sigma(\xi(t))dB(t)$$

where

$$\xi(t) \in \mathbb{R}^n, B(t) \in \mathbb{R}^d$$

Then for any function $f \in C^\infty(\mathbb{R}^n)$, we have by the Ito rule

$$\begin{aligned} df(\xi(t)) &= \left[\mu^T \nabla f(\xi(t)) + \frac{1}{2} Tr\left(\left(\sigma\sigma^T \nabla\nabla^T f\right)\right)\varepsilon(t)\right] dt \\ &\quad + dB^T \sigma(\xi(t))^T \nabla f(\xi(t)) \\ &= Lf(\xi(t))dt + \sum_{k=1}^{d} dB_k(t)\left(\sigma^T \nabla f\right)_k (\xi(t)) \end{aligned}$$

where L is the generator of the process. We define the homomorphism $f \rightarrow f(\xi(t))$ from $C^\infty(\mathbb{R})$ into $B\left(L^2\left(\Omega, \mathcal{F}_t^\xi, P\right)\right)$ where $f(\xi(t))$ acts by multiplication. Denote this homomorphism by j_t. Thus, $j_t(f) = f(\xi(t))$. Then, another way to write down the above sde in qsde form is

$$df_t(f) = j_t(Lf)dt + \sum_{k=1}^{d} jt(L_k f)dB_k$$

where

$$Lf = \mu^T \nabla f + \frac{1}{2} Tr\left(\sigma\sigma^T \nabla\nabla^T f\right)$$

and

$$L_k f = \left(\sigma^T \nabla f\right)_k = \sum_{j=1}^{n} \sigma_{jk}(\xi) \frac{\partial f(\xi)}{\partial \xi_j}$$

Another example comes directly from Hamiltonian mechanice with quantum noise. The system is an atom or a collection of atoms. The bath is the electromagnetic field consisting of photon, The bath operators are A, A^*, Λ. A describes the annihilation of a photon, A^* the creation of a photon and Λ is a counting process describing the number of photons. The system Hilbert space is $\mathfrak{h}$ and the noise Boson Fock space is $\Gamma_s(\mathcal{H})$ where $H = L^2(\mathbb{R}_+)$. We consider the evolution operator $U(t)$ satisfying the qsde

$$dU(t) = \left(-iHdt + L_1 dA(t) + L_2 dA^*(t) + Sd\Lambda(t)\right)U(t)$$

where H, L_1, L_2, S are in $\mathcal{L}(\text{h})$. The condition for $U(t)$ to be unitary is that $d(U(t)^* U(t)) = 0 \forall t$. In other words,

$$dU(t)^*.U(t) + U(t)^*.dU(t) + dU(t)^*.dU(t) = 0$$

Applying the quantum Ito formulae $dA.\ dA^* = dt$, $dA^* dA = 0$, $dA.\ d\Lambda = dA$, $d\Lambda.\ dA^* = dA^*$, $dA^*.\ d\Lambda = 0$, $d\Lambda\ dA = 0$, $d\Lambda.\ d\Lambda = d\Lambda$, we get as the unitarity condition

$$i\left(H^* - H\right) + L_2^* L_2 = 0$$

$$L_2^* + L_1 + L_2^* S = 0,\ S^* + S + S^* S = 0$$

Writing $H = P + iQ$ where P, Q are Hermitian and $S = F - I$, These conditions translate to

$$Q = -L_2^* L_2/2,\ L_1 = -L_2^* F,\ F^* F = I$$

Now we come to the pd controller part. In classical mechanics and control theory, if the position $q(t)$ deviates from a reference trajectory $q_d(t)$, then we can give a position error feedback as a force $K_p\left(q_d(t) - q(t)\right)$. This corresponds to adding a Harmonic oscillator potential $K_p\left(q_d - q\right)^2/2$ to a Hamiltonian. However, we also give a velocity error feedback force $K_{dp}\left(q_d' - q'\right)$. This force is essentially a velocity damping and can only be realized in the quantum theory by appropriate choice of the Sudarshan-Lindblad operators, which here is L_2. So our pd controlled quantum system is governed by the qsde

$$dU(t) = \left(-\left(iP + L_2^* L_2/2\right)dt - L_2^* F dA + L_2 dA^* + F d\Lambda\right)U(t)$$

where

$$P = P_0 + K_1 P_1 L_2 = L_{20} + K_2 L_{21}$$

with K_1, K_2 real and P_0, P_1 Hermitian.

Let X be a system observable and put

$$j_t(X) = U(t)^* X U(t)$$

Then

$$dj_t(X) = j_t\left(i[P, X] - \frac{1}{2}\left(L_2^* L_2 X + X L_2^* L_2 - 2L_2^* X L_2\right)\right)dt$$

$$+ j_t\left(-X L_2^* F + L_2^* X + L_2^* X F\right)dA$$

$$+j_t\left(XL_2 - F^*L_2X + F^*XL_2\right)dA^*$$

$$+j_t\left(F^*XF + F^*X + XF\right)d\Lambda$$

or equivalently defining

$$\theta_0(X) = i[P, X] - \frac{1}{2}\left(L_2^*L_2X + XL_2^*L_2 - 2L_2^*XL_2\right)$$

$$\theta_1(X) = \left[L_2^*, X\right]F + L_2^*X, \theta_2(X) = F^*[X, L_2] + XL_2$$

$$\theta_3(X) = F^*XF + F^*X + XF$$

Then we have the Evans-Hudson flow

$$dj_t(X) = j_t(\theta_0(X))dt + j_t(\theta_1(X))dA(t)$$

$$+j_t(\theta_2(X))dA^*(t) + j_t(\theta_3(X))d\Lambda(t)$$

We note that the operator $\theta_0(.)$ is a linear function of K_1 and a linear quadratic function of K_2. Also $\theta_1(.)$ is a linear function of K_2 and finally,. $\theta_3(.)$ is independent of K_1, K_2. The control problem is to design K_1, K_2 so that

$$E_T(K_1, K_2) = \int_0^T \langle fe(u), j_t(X)fe(u)\rangle dt$$

is a minimum. In the context of classical tracking $j_t(f)(q(t)) = f(t, q(t))$ can be taken as $\|q_d(t) - q(t)\|^2$ and the above optimization problem then amounts to minimizing the time avergage mean square error. Here, $f \in h$ is such that $\|f\| = \exp\left(-\|u\|^2/2\right)$. We can design an approximate adaptive algorithm for updating K_1, K_2 as follows. Suppose $K(t-dt) = (K_1(t-dt), K_2(t-dt))$ has been chosen so that $E_t(K)$ is a minimum. Then we chooes $K(t)$ so that $\langle fe(u), dj_t(X)fe(u)\rangle$ is a minimum . Such a minimization may be carried out subject to a constraint on $K(t)$. We note that

$$\langle fe(u), dj_t(X)fe(u)\rangle = (\langle fe(u), j_t(\theta_0(X))fe(u)\rangle dt + 2\text{Re}(u)(t)$$

$$+\langle fe(u), j_t(\theta_3(X))fe(u)\rangle |u(t)|^2 dt$$

$$+\langle fe(u), j_t(\theta_3(X))fe(u)\rangle |u(t)|^2 dt$$

Since θ_3 is independent of K, this amounts to minimizing

$$\langle fe(u), j_t(\theta_0(X))fe(u)\rangle + 2\,\text{Re}\left(u(t)\langle fe(u), jt(\theta_1(X))fe(u)\rangle\right)$$

Now,

$$\theta_0(X) = i[P, X] - \frac{1}{2}\left(L_2^* L_2 X + X L_2^* L_2 - 2L_2^* X L_2\right)$$

$$= i[P_0, X] + iK_1[P_1, X] - \frac{1}{2}\left(L_{20}^* L_{20} X + X L_{20}^* L_{20} - 2L_{20}^* X L_{20}\right)$$

$$-\frac{K_1^2}{2}\left(L_{21}^* L_{21} X + X L_{21}^* L_{21} - 2L_{21}^* X L_{21}\right)$$

$$-\frac{1}{2}\left(\left(L_{20}^* L_{21} + L_{21}^* L_{20}\right)(X)\right.$$

$$\left. + X\left(L_{20}^* L_{21} + L_{21}^* L_{20}\right) - 2L_{20}^* X L_{21} - 2L_{21}^* X L_{20}\right)$$

This can be expressed as

$$\theta_{00}(X) = \theta_{00}(X) + K_1\theta_{01}(X) + K_2\theta_{02}(X) + K_2^2\theta_{03}(X)$$

Likewise,

$$\theta_1(X) = \left[L_2^*, X\right]F + L_2^* X$$

$$= \left[L_{20}^*, X\right]F + L_{20}^* X + K_2\left(\left[L_{21}^*, X\right]F + L_{21}^* X\right)$$

$$= \theta_{10}(X) + K_2\theta_{11}(X)$$

Thus the quantity to be minimized for determining $K(t)$ is

$$2K_1 \operatorname{Re}\left(\left\langle fe(u), j_t\left(\theta_{01}(X)\right) fe(u)\right\rangle\right)$$

$$2K_2 \operatorname{Re}\left(\left\langle fe(u), j_t\left(\theta_{02}(X)\right) fe(u)\right\rangle\right)$$

$$+2K_2^2 \operatorname{Re}\left(\left\langle fe(u), j_t\left(\theta_{03}(X)\right) fe(u)\right\rangle\right)$$

$$+2K_2 \operatorname{Re}\left(\left\langle fe(u), j_t\left(\theta_{11}(X)\right) fe(u)\right\rangle\right)$$

Taking dt an finite, we get on discretization an adaptive algorithm for the tracking problem.

❖❖❖❖❖

[24] Belavkin quantum filter:

The state process is $j_t(X)$ defined by the qsde

$$j_t(X) = V_t^* X V_t, X \in \mathcal{L}(h)$$

$$dV_t = \left(-iHdt + L_1 dA + L_2 dA^* + Sd\Lambda\right)V_t$$

where H, L_1, L_2, S are in l

$\mathcal{L}(h)$ (system operators) and are chosen so that V_t is unitary for all t. The measurement process

$$Y_{out}(t) = V_t^* Y_{in}(t) V_t, Y_{in}(t)$$

$$= \int_0^t \left(c_1(s)dA(s) + \bar{c}_1(s)dA^*(s) + c_2(s)d\Lambda(s)\right)$$

with c_1 a complex valued function of time and c_2 a real valued function of time. It is clear that the process $Y_{in}(t)$ takes values in $\mathcal{L}(\Gamma_s(\mathcal{H}))$ where $\mathcal{H}=L^2(\mathbb{R}_+)$ and therefore it commutes with system operators ie1 with $\mathcal{L}(h)$. Now by unitarity of V_t and the fact the $Y_{in}(s)$ commutes with H, L_1, L_2, S, it follows that

$$d_t\left(V_t^* Y_{in}(s) V_t\right) = 0, t > s$$

Hence,

$$Y_{out}(s) = V_t^* Y_{in}(s) V_t, t \geq s$$

Remark: The following abbreviations are used:

$$V_t^* Y_{in}(s) V_t = V_t^*(I \otimes Y_{in}(s)) V_t, V_t^* X V_t = V_t^*(X \otimes I) V_t$$

where X is a system operator.

It follows that for $t \geq s$

$$[j_t(X), Y_{out}(s)] = \left[V_t^* X V_t, V_t^* Y_{in}(s) V_t\right] = V_t^*[X, Y_{in}(s)] V_t = 0$$

Now

$$dY_{out}(t) = d\left(V_t^* Y_{in}(t) V_t\right)$$

$$= dY_{in}(t) + dV_t^* dY_{in}(t) V_t + V_t^* dY_{in}(t) dV_t + dV_t^* dY_{in}(t) dV_t$$

$$= dY_{in}(t) + \bar{c}_1(t) V_t^* L_2^* V_t dt + c_2(t) V_t^* L_2^* V_t dA(t)$$
$$+ \bar{c}_1(t) V_t^* S^* V_t dA^*(t) + c_1(t) V_t^* L_2 V_t dt$$
$$+ c_1(t) V_t^* S V_t dA(t) + c_2(t) V_t^* L_2 V_t dA^*(t)$$
$$+ c_2(t) V_t^* S^* V_t d\Lambda(t) + c_2(t) V_t^* S V_t d\Lambda(t)$$
$$+ c_2(t) V_t^* S^* S V_t d\Lambda(t) + c_2(t) V_t^* L_2^* S V_t dA(t)$$
$$+ c_2(t) V_t^* S^* L_2 V_t dA^*(t) + c_2(t) V_t^* L_2^* L_2 V_t dt$$

$$= dY_{in}(t) + jt\left(\bar{c}_1(t) L_2^* + c_1(t) L_2 + c_2(t) L_2^* L_2\right) dt$$

$$+j_t\left(c_2(t)L_2^* + c_1(t)S + c_2(t)L_2^*S\right)dA(t)$$

$$+j_t\left(\overline{c_1}(t)S^* + c_2(t)L_2 + c_2(t)S^*L_2\right)dA^*(t)$$

$$+j_t\left(c_2(t)\left(S + S^* + S^*S\right)\right)d\Lambda(t)$$

To proceed further in the filter computation, we need the notion of quantum conditional expectation. Let $\mathcal{M}$ be an Abelian Von-Neumann algebra and let $\mathcal{A}$ be an algebra containing $\mathcal{M}$. For $X \in \mathcal{A}$ we may define $\varepsilon(X)$ by for examply choosing a pure or mixed state ρ in $\mathcal{A}$ and defining $\mathbb{E}(X) = Tr\,(\rho X)$. We then define $\mathbb{E}(X|M)$ as the projection of X on $\mathcal{M}$ in the sense that $\mathbb{E}(X|M) \in \mathcal{M}$ is uniquely determined by the condition that for any $Y \in \mathcal{M}$,

$$\mathbb{E}(XY) = \mathbb{E}(\mathbb{E}(X \mid \mathcal{M})Y)$$

It is then easy to see to see that $Y \in \mathcal{M}$, then for all $X \in \mathcal{A}$, we have

$$\mathbb{E}(XY \mid \mathcal{M}) = \mathbb{E}(X \mid \mathcal{M})Y = Y\mathbb{E}(X \mid \mathcal{M})$$

the last equality following form the face the $\mathcal{M}$ is Abelian. Now suppose $\mathcal{A} = \mathcal{B}\,(\mathcal{H})$ where $\mathcal{H}$ is Hilbert space and $\mathcal{M}$ is an Abelian subalgebra of $\mathcal{A}$. Let V be a unitary operator in $\mathcal{H}$ and define the Abelian algebra

$$\mathcal{M}_0 = V^*\mathcal{M}V = \left\{V^*YV : Y \in \mathcal{M}\right\}$$

Then consider the following operators:

$$\mathbb{E}(V^* XV \mid \mathcal{M}_0) \qquad ...(1)$$

and

$$\mathbb{E}_0(X|\mathcal{M}) \qquad ...(2)$$

where E_0 is the expectation defined by

$$\mathbb{E}_0(X) = \mathbb{E}\left(V^*XV\right)$$

It then follows that for $Y \in \mathcal{M}$, $V^*YV \in \mathcal{M}_0$ and therefore,

$$\mathbb{E}\left(\mathbb{E}\left(V^*XV \mid \mathcal{M}_0\right)V^*YV\right) = \mathbb{E}\left(V^*XVV^*YV\right)$$

$$= \mathbb{E}\left(V^*XYV\right) = \mathbb{E}_0(XY)$$

$$= \mathbb{E}_0\left(\mathbb{E}_0(X \mid \mathcal{M})Y\right)$$

$$= \mathbb{E}\left(V^*\mathbb{E}_0(X \mid \mathcal{M})YV\right) = \mathbb{E}\left(V^*\mathbb{E}_0(X \mid \mathcal{M})VV^*YV\right)$$

and hence

$$E\left(V^{*}XV \mid \mathcal{M}_0\right) = V^{*}\mathbb{E}_0\left(X \mid \mathcal{M}\right)V$$

In other, words, the operators defined in (1) and (2) are unitarily similar. Now let $\mathcal{M}'$ denote the commutant of $\mathcal{M}$, *i.e.*,

$$\mathcal{M}' = \left\{X \in \mathcal{A} : [X,Y] = 0 \forall Y \in \mathcal{M}\right\}$$

Suppose $F \in \mathcal{M}'$ is given and we define the expectation $\mathbb{E}_F$ by

$$\mathbb{E}_F(X) = \mathbb{E}\left(F^{*}XF\right), X \in \mathcal{M}'$$

Then we claim that the conditional expectation given $\mathcal{M}$ associated with the expectation $\mathbb{E}_F$ is given by

$$\mathbb{E}_F(X|\mathcal{M}) = \frac{\mathbb{E}\left(F^{*}XF \mid \mathcal{M}\right)}{\mathbb{E}\left(F^{*}F \mid \mathcal{M}\right)}, X \in M'$$

Proving this is equivalent to proving that

$$\mathbb{E}_F\left(X \mid \mathcal{M}\right)\mathbb{E}\left(F^{*}F \mid \mathcal{M}\right) = \mathbb{E}\left(F^{*}FX \mid \mathcal{M}\right)$$

Note that $\mathcal{M}$ is an Abelian algebra. We have by the definition of conditional expectation, for any $Y \in \mathcal{M}$,

$$\mathbb{E}\left[\mathbb{E}_F\left(X \mid \mathcal{M}\right)\mathbb{E}\left(F^{*}F \mid \mathcal{M}\right)Y\right] = \mathbb{E}\left[F^{*}F\mathbb{E}_F\left(X \mid \mathcal{M}\right)Y \mid \mathcal{M}\right]$$

$$\mathbb{E}_F\left[\mathbb{E}_F\left(X \mid \mathcal{M}\right)Y\right] = \mathbb{E}_F[XY] = \mathbb{E}\left[F^{*}FXY\right]$$

$$= \mathbb{E}\left[\mathbb{E}\left[F^{*}F\,X \mid \mathcal{M}\right]Y\right]$$

This proves the claim. Note that we have used the fact that $Y \in \mathcal{M}$, $\mathcal{M}$ is Abelian and that F^{*} and F commute with $\mathcal{M}$.

Reference: Gough *et. al.* "Quantum filtering in coherent states".

Exercise: Let $\mathcal{A}$ be a Von-Neumann algebra and $\mathcal{M}$ a subalgebra of $\mathcal{A}$. Suppose an expectation $\mathbb{E}$ defined by a density operator ρ in $\mathcal{A}$ is given, *i.e.* $\mathbb{E}(X) = Tr(\rho X)$, $X \in \mathcal{A}$. Let $\mathbb{E}(.|M) : \mathcal{A} \to \mathcal{M}$ be a linear operator such that $\mathbb{E}(\mathbb{E}(X|\mathcal{M})Y) = \mathbb{E}(XY)$ for all $X \in \mathcal{A}$ and $Y \in \mathcal{M}$ and such that the square of the operator $\mathbb{E}(.|\mathcal{M})$ is itself, *i.e.*, this operator is idempotent. Then show that $\mathcal{A} = \mathcal{M}'$ and $\mathcal{M} \subset \mathcal{M}'$, *i.e.*, $\mathcal{M}$ is Abelian.

❖❖❖❖❖

[25] Estimating parameters in nonlinear systems driven by *iid* Gaussian noise. The signal model is

$X(n) = H(X(n-1: n-p))\,\theta + G(X(n-1:n-p))\,V(n), n \geq p$ ...(1)

where $\{V(n)\}$ is an iid $N(0,\Sigma_v)$ sequence in $\mathbb{R}^d$ and $X(n) \in \mathbb{R}^d$ $X(n-1:n-p) = [X(n-1)^T,\ldots,X(n-p)^T]^T \in \mathbb{R}^{pd}$.

(1) desribes a pth order Markov model and *It* implies assuming that $X(p-1:0)$ is independent of $(V(n) : n \geq p)$ that $X(n : 0)$ is independent of V(n), $n \geq p + 1$. Equivalently, $X(n)$ is a Borel function of $(X(p-1 : 0))$, $(V(n : 0))$. We assume that G is nonsingular for all data $X(n-1:n-p)$. Then the pdt of $(X(n): p \leq n \leq n)$ given $(X(n): 0 \leq n \leq p-1)$ and θ is given by

$$P\left(\left(X(n): p \leq n \leq N\right)/\theta\right) = (2\pi)^{-d(N-p+1)/2}$$

$$\left[\prod_{n=p}^{N} \det\left[G\left(X(n-1:n-p)\right)\sum_v G\left(X(n-1:n-p)\right)^T\right]^{-1/2}\right]$$

$$\exp-\frac{1}{2}\sum_{n=p}^{N}\left(X(n)-H_n\left(X(n-1:n-p)\right)\theta\right)^T\left(G_n\left(X(n-1:n-p)\right)\right.$$

$$\left.\sum_v G_n\left(X(n-1:n-p)\right)^T\right)^{-1}\left(X(n)-Hn\left(X(n-1:n-p)\right)\theta\right).$$

It follows that the ML estimator of θ given $X(N : 0)$ given by maximizing p above w.r.t. θ equals

$$\hat{\theta}(N) = F_N^{-1}\sum_{N=p}^{N} L_n X_n$$

where

$$F_N = \sum_{N=p}^{N} H_n^T\left(G_n \sum_v G_n\right)^{-1} H_n$$

and

$$L_n = H_n^T\left(G_n \sum_v G_n\right)^{-1}$$

We can express this estimator as

$$\hat{\theta}(N) = \theta + F_N^{-1}\sum_{n=p}^{N} L_n G_n V_n$$

We note that $G_n = G_n\left(X(n-1:n-p)\right)$ is independent of V_n. Now we use perturbation theory: Let

$$H_n\left(X(n-1:n-p)\right) = H_{n0} + \varepsilon H_{n1}\left(X(n-1:n-p)\right),$$

$$G_n\left(X(n-1:n-p)\right) = G_{n0} + \varepsilon G_{n1}\left(X(n-1:n-p)\right)$$

where H_{n0}, G_{n0} are data independent constant matrices. Then,

$$K_n = G_n \sum{}_v G_n^T = K_{n0} + \varepsilon K_{n1} + O(\varepsilon^2)$$

$$K_{n0} = G_{n0} \sum{}_v G_{n0}^T$$

$$K_{n1} = G_{n0} \sum v G_{n1}^T + G_{n1}^T \sum v G_{n0}^T$$

K_{n0} is data independent while K_{n1} is data dependent. Now,

$$K_n^{-1} = K_n^{-1} - \varepsilon K_{n0}^{-1} K_{n1} K_{n0}^{-1} + O(\varepsilon^2)$$

$$L_n = H_n^T K_n^{-1} = L_{n0} + \varepsilon L_{n1} + O(\varepsilon^2)$$

where

$$L_{n0} = H_{n0}^T K_{n0}^{-1}$$

$$L_{n1} = -H_{n1}^T K_{n0}^{-1} K_{n1} K_{n0}^{-1} + H_{n1}^T K_{n0}^{-1}$$

$$F_N = \sum_{n=p}^{N} L_n H_n = F_{N0} + \varepsilon F_{N1} + O(\varepsilon^2)$$

where

$$F_{N0} = \sum_{n=p}^{N} L_{n0} H_{n0} = \sum_{n=p}^{N} H_{n0}^T K_{n0}^{-1} H_{n0}$$

$$= \sum_{n=p}^{N} H_{n0}^T \left(G_{n0} \sum{}_v G_{n0}^T\right)^{-1} H_{n0}$$

and

$$F_{N1} = \sum_{n=p}^{N} (L_{n0} H_{n1} + L_{n1} H_{n0})$$

$$= \sum_n \left(H_{n0}^T K_{n0}^{-1} H_{n1} + H_{n1}^T K_{n0}^{-1} H_{n0} - H_{n1}^T K_{n0}^{-1} K_{n1} K_{n0}^{-1} H_{n1}\right)$$

We then find that

$$e(N) = \hat{\theta}(N) - \theta = F_N^{-1} \sum_{n=p}^{N} L_n G_n V_n$$

$$= (F_{N0} + \varepsilon F_{N1})^{-1} \sum_n (L_{n0} + \varepsilon L_{n1})(G_{n0} + \varepsilon G_{n1}) V_n$$

$$= e_0(N) + \varepsilon e_1(N) + O(\varepsilon^2)$$

where

$$e_0(N) = F_{N0}^{-1}\sum_n L_{n0}G_{n0}V_n$$

$$e_1(N) = F_{N0}^{-1}\sum_n \left(L_{n0}G_{n0} + L_{n1}G_{n0}\right)V_n$$

$$-F_{N0}^{-1}F_{N1}F_{N0}^{-1}\sum_n L_{n0}G_{n0}V_n$$

We note that

$$\langle e_0(N)\rangle = 0, Cov\left(e_0(N)\right) = F_{N0}^{-1}$$

$$\langle e_1(N)\rangle = -F_{N0}^{-1}\left\langle F_{N1}F_{N0}^{-1}\sum_{n=p}^{N} L_{n0}G_{n0}V_n\right\rangle$$

Note that $\langle G_{n1}V_n\rangle = 0$ but F_{N1} and $V_{n,P}p \leq n \leq N$ are correlated.

Approximate statistics. We write

$$X_n = X_n^{(0)} + \varepsilon X_n^{(1)} + O(\varepsilon^2)$$

Then from the signal model,

$$X_n^{(0)} = H_{n0}\theta + G_{n0}V_n$$

implying that $X_n^{(1)}$, $n \geq 0$ is a Gaussian process. further the coefficient of ε^1 gives

$$X_n^{(1)} = H_{n1}\left(X_{n-1:n-p}^{(0)}\right)\theta + G_{n1}\left(X_{n-1:n-p}^{(0)}\right)V_n$$

showing that in general $X_n^{(1)}$ is a non-Gausisian process. We note that $X_n^{(1)}$ can be expressed entirely in terms of $\{V_n\}$as

$$X_n^{(1)} = H_{n1}\left(H_{n-1:n-p0}\theta + D_{G,n-1:n-p}V_{n-1:n-p}\right)\theta$$

$$+G_{n1}\left(H_{H_{n-1:n-p},0}\theta + D_{G,n-1:n-p}V_{n-1:n-p}\right)V_n$$

where

$$D_{G,n-1:n-p} = diag\left[G_{n-1,0},\ldots,G_{n-p,0}\right]$$

These equations express $X_n^{(1)}$ explicitly as a nonlinear function of the noise $\{V_n\}$ and their mean and covariances can at least formally be expressed as multivariate Gaussian integrals.

❖❖❖❖❖

[26] Lyapunov functions in control:

Consider the dynamical system

$$X'(t) = f(X(t)), t \geq 0, f : \mathbb{R}^n \to \mathbb{R}^n$$

We construct a Lyapunov function of the form

$$V(X) = X^T J(X, \theta) X$$

where $J(X, \theta)$ is a positive definite matrix for all X belonging to the range in which $X(t)$ takes values. We wish that the parameter θ be selected so that

$$\frac{dV(X(t))}{dt} < 0$$

For this we find that

❖❖❖❖❖

[27] Estimating the configuration of a two link 3-D Robot carrying current from the electromagnetic field pattern abstract: A two link 3-D robot can be modelled as a system of two tops (rigid bodies), with the bottom link pivoted to the ground at the origin and the top link pivoted to the bottom link at some point p on it. The two links are assumed to be carrying a d.c. current density produced by a voltage or current source within them. The configuration of such a robot at a given time instant can be modelled by two rotations R, $S \in SO(3)$ as follows. First a rotation R_1 is applied to both the links about the origin O. Then a rotation R_2 is applied to the top link about its pivot p (on the lower link). We take $R = R_1$, $S = R_2 R_1$. By applying the Biot-Savart law, we derive expressions for the magnetic field produced by the robot in its final configuration in terms of the rotation matrices R, S. We then apply the theory of representations of the group $SO(3)$ to express the spatial Fourier transform of the magnetic field produced by the two links in terms of the matrices representing the irreducible representations of $SO(3)$. Finally, MATLAB programs for calculating R, S using truncated approximations of the Fourier transform are written down and a simulation example is presented.

Problem formulation: At time $t = 0$, the bottom link which is modelled as a 3–D top occupies a volume $B_1 \subset \mathbb{R}^3$ and the top which is modelled as another 3–D top occupies a volume $B_2 \subset \mathbb{R}^3$. The pivot of the bottom link is attached to the origin O and that of the top link is attached to a point p on the top surface of the bottom link. At time t, the configuration of this system is described as follows: A rotation R_1 has been applied to both the links about O followed by a rotation R_2 to the top link about $R_1(p)$. Thus, a point r_1 in the botom link goes over to $R_1(r_1)$ and a point r_2 in the top link goes over to $R_1(p)+R_2R_1(r_2-p)$. This entire transformation group action can be expressed as

$$\begin{pmatrix} r_1 \\ r_2 \\ p \end{pmatrix} \to \begin{pmatrix} R_1 & 0 & 0 \\ 0 & R_2R_1 & R_1 - R_2R_1 \\ 0 & 0 & P \end{pmatrix} \begin{pmatrix} r_1 \\ r_2 \\ p \end{pmatrix}$$

Let $J_1(r)$ be the current density in the bottom link and $J_2(r)$ that in the top link at time $t = 0$. Thus, $J_1(r) = 0$ for $r \notin B_1$ and $J_2(r) = 0$ for $r \notin B_2$. We note that $B_1 \cap B_2 = \phi$. Then at time t, the current density field due to both the links is given by

$$J(r) = J_1\left(R_1^{-1}r\right) + J_2\left(p + (R_2R_1)^{-1}\left(r - R_1(P)\right)\right)$$

or equivalently, writing R in place of R_1 and S in place of R_2R_1, we have

$$J(r) = J_1\left(R_1^{-1}r\right) + J_2\left(p + S^{-1}\left(r - R(P)\right)\right)$$

The (non-relativistic) magnetic field produced by this d.c. current density is given by the Biot-Savart law:

$$B(r) = \int_{\mathbb{R}^3} \frac{J(r') \times (r - r')}{|r - r'|^3} d^3r' + \int G(r - r')J(r')d^3r'$$

where $G(r)$ is the 3×3 skew symmetric matrix given by

$$g(r) = G(x, y, z) = |r|^{-3} \begin{pmatrix} 0 & z & -y \\ -z & 0 & x \\ y & -x & 0 \end{pmatrix}$$

We can write after transforming the integration variables,

$$B(r) = \int_{B_1} G(r - Rr')J_1(r')d^3r' + \int_{B_2} G\left(r - R(p) - S(r' - p)\right)J_2(r')d^3r'$$

The spatial Fourier transform of the magnetic field is then

$$\hat{B}(k) = \int B(r)\exp(ik.t)d^3r = \hat{G}(k)\left[\int \exp\left(i(k, Rr')\right)J_1(r')d^3r'\right.$$
$$\left. + \int \exp\left(i\left(k, R(p) + S(r' - p)\right)\right)J_2(r')d^3r'\right.$$

or equivalently,

$$F(k) = \hat{G}(k)^{-1}\hat{B}(k) = \int \exp\left(i(k, Rr')\right)J_1(r')d^3r'$$
$$+ \int \exp\left(i\left(k, R(p) + S(r' - p)\right)\right)J_2(r')d^3r'$$

We now consider the following spherical harmonic expansion

$$exp(i(k,r)) = \exp\left(i|r|\left(k,\hat{r}\right)\right) = \sum_{1\geq 0,|m|\leq l} C\left(l,m,k|r|\right)Y_{lm}\left(\hat{r}\right)$$

where

$$C\left(l,m,k,|r|\right) = \int_{S^2} \exp\left(i|r|\left(k,\hat{r}\right)\right)\bar{Y}_{lm}\left(\hat{r}\right)d\,S\left(\hat{r}\right)$$

We note that $\left\{Y_{lm} : |m| \leq l\right\}$ is an orthonormal basis for an irreducible representation π_l of the rotation group $SO(3)$. More precisely, with S^2 denoting the unit sphere in $\mathbb{R}^3$, we have the decomposition

$$L^2\left(S^2, \sin(\theta)\,d\theta d\phi\right) = \overset{\infty}{\underset{l=0}{\oplus}} \mathcal{H}_l$$

where $\mathcal{H}_1$ is a subspace of $L^2(S^2)$ invariant under SO (3) and for any $R \in SO(3)$, we have

$$Y_{lm}\left(R^{-1}\hat{r}\right) = \sum_{|m'|\leq l} \left[\pi_1(R)\right]_{m'm} Y_{lm'}\left(\hat{r}\right)$$

with

$$\pi_l\text{: } SO(3)U(\mathbb{C}^{2l+1})$$

being an irreducible representation of $SO(3)$. Moreover, $\pi_l : l \geq 0$ exhaust all the in equivalent irreducible representations of $SO(3)$. We then have

$$F(k) = \int \exp\left(i\left(k, Rr\right)\right)J_1(r)d^3r + \int \exp\left(i\left(k, R(p) + S(r-p)\right)\right)J_2(r)d^3r$$

$$= \sum_{l,m}\int C\left(l,m,k,|r|\right)Y_{lm}\left(R\hat{r}\right)J_1(r)d^3r + \exp\left(i\left(k,(R-S)(p)\right)\right)$$

$$\sum_{l,m}\int C\left(l,m,k,|r|\right)Y_{lm}\left(S\hat{r}\right)J_2(r)d^3r$$

$$= \sum_{l,m,m'} \left[\bar{\pi}_l(R)\right]_{mm'} \int C\left(l,m,k,|r|\right)Y_{lm}\left(\hat{r}\right)J_1(r)d^3r$$

$$+ \exp\left(i\left(k,(R-S)(p)\right)\right)$$

$$\sum_{l,m,m'} \left[\bar{\pi}_l(S)\right]_{mm'} \int C\left(l,m,k,|r|\right)Y_{lm}\left(\hat{r}\right)J_2(r)d^3r$$

Define

$$J_1\left[l,m,m',k\right] = \int C\left(l,m,k,|r|\right)Y_{lm'}\left(\hat{r}\right)J_1(r)d^3r$$

$$J_2\left[l,m,m',k\right] = \int C\left(l,m,k,|r|\right)Y_{lm'}\left(\hat{r}\right)J_2(r)d^3r$$

Then we can express the above equation as

$$F(k) = \sum_{l,m,m'} [\bar{\pi}_l(R)]_{mm'} J_1[l,m,m',k] + \exp(i(k,(R-S)(p)))$$

$$\sum_{l,m,m'} [\bar{\pi}_l(S)]_{mm'} J_2[l,m,m',k]$$

Noting that

$\exp(i(k,(R-S)(p)) = \exp(i(k, Rp)).\ \exp(-i(k, Sp))$

$$= \sum_{l_1,m_1,l_2,m_2} C(l_1,m_1,k,|p|)\bar{C}(l_2,m_2,k,|p|)Y_{l1,m1}(R\hat{p})\bar{Y}_{l2,m2}(S\hat{p})$$

$$= \sum_{l_1,m_1,m_1' l_2,m_2,m_2'} C(l_1,m_1,k,|p|)\bar{C}(l_2,m_2,k,|p|)Y_{l_1,m_1'}(\hat{p})\bar{Y}_{l_2,m_2'}$$

$$(\hat{p})[\bar{\pi}_{l_1}(R)]_{m_1,m_1'}[\pi_{l_2}(S)]m_2,m_2'$$

$$= \sum_{l_1,l_2 m_1,m_2,m_1',m_2'} A[l_1,l_2,m_1,m_2,m_1',m_2' k,p]$$

$$[\bar{\pi}_{l_1}(R)]_{m_1,m_1'}[\pi_{l_2}(S)]_{m_2,m_2'}$$

where

$$A[l_1,l_2,m_1,m_2,m_1',m_2' k,p]$$

$$= C(l_1,m_1,k,|p|)\bar{C}(l_2,m_2,k,|p|)$$

$$Y_{l_1},m_1'(\hat{p})\bar{Y}_{l_2},m_2'(\hat{p})$$

Our final expression for the spatial Fourier transform of the magnetic field is then given by

$$F(k) = \hat{G}(k)^{-1}\hat{B}(k)$$

$$= \sum_{l,m,m'} [\bar{\pi}_l(R)]_{mm'} J_1[l,m,m',k]$$

$$+\Big(\sum_{l_1,l_2 m_1,m_2,m_1',m_2'} A[l_1,l_2,m_1,m_2,m_1',m_2' k,p]$$

$$[\bar{\pi}_{l_1}(R)]_{m_1,m_1'}[\pi_{l_2}(S)]_{m_2,m_2'}\Big)$$

$$\Big(\sum_{l,m,m'} [\bar{\pi}_l(S)]_{mm'} J_2[l,m,m',k]\Big).$$

❖❖❖❖❖

[28] Tunnelling problems in quantum mechanics: Assume first a rectangular potential barrier of height $V_0 > E$ where E is the particle's energy. The barrier extends from $x = 0$ to $x = L$ solving the stationary state Schrodinger equation over the three regions $x < 0$, $0 < x < L$, $x > L$ respectively after noting that in the region $x > L$ there is no reflected wave gives the solutions

$$\psi(x) = C_1.\exp(ikx) + C_2.exp(-ikx), x < 0,$$
$$\psi(x) = B_1.\exp(\alpha x) + B_2.exp(-\alpha x), 0 < x < L,$$
$$\psi(x) = A.\exp(ikx), x > L$$

where

$$k = (2mE)^{1/2}, \alpha = (2m(V_0-E))^{1/2}$$

The boundary conditions are that $\psi(x)$ and $\psi(x)$ should be continuous everywhere. Thus

$$\psi(0-) = \psi(0+), \psi'(0-) = \psi'(0+), \psi(L-) = \psi(L+), \psi'(L-) = \psi'(L+)$$

These yield four homogeneous linear equations using which the ratios C_2/C_1, B_1/C_1, B_2/C_1, A/C_1 are easily calculated. Exercise: Calculate the current in the three regions.

Now consider a barrier $V(x)$ of arbitrary shape. The stationary Schrodinger equation in $\mathbb{R}$ is solved approximately using the quasi-classical approximation.

Hint: Write the wave function as $\psi(x) = A(x).\exp(iS(x)/h)$ (where h is Planck's constant divided by 2π) We have

$$\psi' = (A' + iAS'/h)\exp(iS/h),$$

$$\psi'' = \left(A'' + 2iA'S'/h + iAS''/h - AS'^2/h^2\right)\exp(iS/h)$$

Substituting this into the Stationary Schrodinger equation

$$\psi''(x) + 2m(E - V(x))\psi(x)/h^2 = 0$$

and equating the real and imaginary parts repectively gives

$$A''(x) - A(x)S'^2(x)/h^2 + 2m(E - V(x))A(x)/h^2 = 0$$

$$2A'(x)S'(x) + A(x)S''(x)/h = 0$$

The second gives

$$\left(A^2S'\right)' = 0$$

so that

$$C/\sqrt{S'(x)}$$

where C is a constant. Substituting this into the first equation gives a nonlinear differential equation for $S(x)$. This equation can be solved by perturbation theory, *i.e.*, by expanding $S(x)$ in powers of h:

$$S(x) = \sum_{m=0}^{\infty} h^m S_m(x)$$

and equating equal powers of h on both sides. We leave this an an exercise to the reader. Note that if $A(x)$ is assumed to tbe slowly varying so that $A''(x)$ is approximately zero, then the first equation gives

$$S'(x) \neq \left(2m(E - V(x))\right)^{1/2}$$

i.e., $S'(x)$ is approximately the momentum $p(x)$ of a classical particle at the point x when its energy is E. We then find that approximately

$$A(x) = C/\sqrt{S'(X)} = C/\sqrt{p'(x)}$$

and hence

$$\psi(x) \approx C.\left(p'(x)\right)^{-1/2} \exp\left((i/h)\int_0^x p(y)dy\right)$$

❖❖❖❖❖

[29] K-variate generalization of Bernoulli filters:

The state space of the signal (state vector) is $E = \bigcup_{k=0}^{K-1} \{k\} \times \mathbb{R}^k$ where $\{0\} \times \mathbb{R}^0 \{k\}$. At time n, let the state be $(k(n), X(n))$. Thus, $X(n) \in \mathbb{R}^{k(n)}$ and $k(n) \in \{0,1, ..., K-1\}$. The discrete time process $\{k(n), X(n)\}_{n\geq 0}$ is a Markov process with transition with transition probability density

$$\phi(k, X \mid r, Y), 0 \leq k, r \leq K\text{–}1, X \in \mathbb{R}^k, Y \in \mathbb{R}^r$$

When $k = 0$, we read (k, X) as simply 0. Thus we put

$$\phi(0, X|r, Y) = P(r, Y), 1 \leq r \leq K-1, \phi(0X \mid 0, Y) = P(0),$$

$$\phi(k, X \mid 0, Y) = Q(k, X), 1 \leq k \leq K-1$$

We have by the law of total probabilities

$$\sum_{k=0}^{k-1} \int_{\mathbb{R}^k} \phi(k, X \mid r, Y) d^k X = 1$$

by which we mean that

$$P(r, Y) + \sum_{k=0}^{\mathrm{k}-1} \int \phi(k, X \mid r, Y) d^k X = 0$$

For $r = 0$ this reads

$$P(0)+\sum_{k=0}^{k-1}\int_{\mathbb{R}^k} Q(k,X)d^kX = 1$$

$\phi(k, X \mid r, Y)d^kX$ can be interpreted for $k, r \geq 1$ as the probability that at time $n + 1$, there are k targets and the signals emitted by them fall in the k dimensional infinitesimal volume element d^kX given that at time n there are r targets located at Y. For $k = 0, r \geq 1$, $\phi(k, X|r, Y) = \phi(0, X|r, Y) = P(r, Y)$ is the probability that at time $n + 1$ there are zero targets given that at time n, there are r targets located at Y. For $k \geq 1, r = 0$, $\phi(k, X|r, Y)\, d^kX = \phi(k, X|0, Y)d^kX = Q(k, X)\, d^kX$ is the probability that at time $n+1$ there are k targets located within the volume element d^kX given that at time n there are zero targets and finally for $k = r = 0$, $\phi(k, X|r, Y) = f(0, X|0, Y) = P(0)$ is the probability that at time $n + 1$ there are zero targets given that at time n there are zero targets. Here the state space of one target is $\mathbb{R}$ and the state space of $k \geq 1$ targets is therefore $\mathbb{R}^k$. We can generalize this to the case when the state space of one targets is a measurable space $(F. \mathcal{F})$ and then the state space of the Markov process $(k(n), X(n))$, $n \geq 0$ is $\bigcup_{k=0}^{K-1}\{k\}\times F^k$. This state model can be used to model situations when targets disappear from the field or new targets appear into the field.

The measurement model is given by

$$z(n) = h_n(k(n), X(n)) + v(n), n \geq 0$$

where

$$h_n : E \rightarrow \mathbb{R}^p$$

and $v(n) \in \mathbb{R}^p$ is a sequence of iid random vectors independent of the state process $\{(k(n), X(n))\}$. Let $p_v(v)$ denote the density of $v(n)$. The observations upto time n are defined by

$$Z(n) = \{z(k) : 0 \leq k \leq n\}$$

The posterior density of the state at time n given observations upto time n is

$$P_{n|n}(k(n), X(n) \mid Z(n))$$

and we also have the one "step ahead posterior density"

$$P_{n+1|n}(k(n+1), X(n+1) \mid Z(n))$$

From basic Bayes' theory of conditioning, we have

$$P_{n+1|n}\left(k(n+1), X(n+1) \mid Z(n)\right)$$

$$= \sum_{k(n)=0}^{K-1} \int \phi\left(k(n+1), X(n+1) \mid k(n), X(n)\right) p_{n|n}$$

$$\left(k(n), X(n) \mid Z(n)\right) d^{k(n)} X(n) \quad ...(1)$$

For $k(n + 1) > 0$, this equation is to interpreted as

$$p_{n+1|n}\left(k(n+1), X(n+1) \mid Z(n)\right)$$

$$= Q\left(k(n+1), X(n+1)\right) p_{n+1}\left(k(n+1), X(n+1)\right)$$

$$+ \sum_{k(n)>0} \int \phi\left(k(n+1), X(n+1) \mid k(n), X(n)\right) p_{n|n}(k)$$

and for $k\,(k + 1) = 0$ as

$$p_{n+1|n}\left(0 \mid Z(n)\right) = P(0)\, p_{n|n}(0 \mid 0)$$

$$+ \sum_{k(n)>0} \int P\left(k(n), X(n)\right) p_{n|n}\left(k(n), X(n) \middle| Z(n)\right) d^{k(n)} X(n) \quad ...(2b)$$

(1) for equivalently (2) is called the state update equation to the posterior density because it tells us how the evolution of the state according to its Markov generator ϕ enables us to construct the one step ahead posterior density from the posterior density (*i.e*, the predictor density $p_{n+1|n}$ form the filtered density $p_{n|n}$.) The measurement law defined by h_n and p_v do not play any role in this update formula. We now determine the measurement update equation:

$$p_{n+1|n+1}\left(k(n+1), X(n+1) \mid Z(n+1)\right)$$

$$= p\left(z(n+1), Z(n)\left(k(n+1), X(n+1)\right) / p\left(z(n+1), (n)\right)\right)$$

$$= p\left(z(n+1) \mid k(n+1), X(n+1)\right)\Big) p_{n+1|n}$$

$$\left(k(n+1), X(n+1) \mid Z(n)\right) / p\left(z(n+1), Z(n)\right)$$

$$= p_v\left(z(n+1) - h_{n+1}\left(k(n+1), X(n+1)\right)\right) p_{n+1|n}$$

$$\left(k(n+1), X(n+1) \mid Z(n)\right)$$

$$= \sum_{k(n+1)=1}^{k-1} \int p_v\left(z(n+1) - h_{n+1}\left(k(n+1), X(n+1)\right)\right) p_{n+1|n}$$

$$\left(k(n+1), X(n+1) \middle| Z(n)\right) d^{k(n+1)} X(n+1)$$

Note that $h_{n+1}(0, X(n+1)), = h_{n+1}(0)$. The rest is just expanding the expressious.

❖❖❖❖❖

[30] Disturbance observer and parametric uncertainty equation in Robotics:
θ is the true parameter vector of the robot. It consists of the masses and lengths of the links. θ_0 is the true value this parameter vector and $\hat{\theta}(t)$ is its estimate at time t. $\delta\theta(t) = \hat{\theta}(t) - \theta_0$ is the parameter estimation error at time t. Let $M(q, \theta)$ be the moment of inertia matrix of the robot link system and $N(q, q', \theta)$ the term arising from centrifugal, coriolis, gravitational and frictional forces. All these forces except the frictional component are derived from the Lagrangian:

$$\frac{1}{2}q'^T M(q,\theta)q' - m_1 g L_1 \sin(q_1)|2 - m_2 g\left(L_1 \sin(q_1) + L_2 \sin(q_2)|2\right)$$

We writer

$$M_0(q) = M(q, \theta_0), \widehat{M}(q) = M\left(q, \hat{\theta}(t)\right)$$

$$N_0(q, q') = N(q, q', \theta_0), \widehat{N}(q, q') = N\left(q, q', \hat{\theta}(t)\right)$$

Thus,

$$\widehat{M}(q) = M_0(q) + \frac{\partial M(q, \theta_0)}{\partial \theta}(\delta\theta \otimes l_2)$$

with neglect of $O\left(|\delta\theta|^2\right)$ terms. Likewise,

$$\widehat{N}(q, q') = N_0(q, q') + \frac{\partial N(q, q', \theta_0)}{\partial \theta}\delta\theta(t)$$

Let $q_d(t)$ be the reference angular vector for the Robot. Then incorporating the computed torque $\tau_e(t)$ with PD feedback into the equations of motions, the dynamicl equations of the robot are

$$M_0(q)q'' + N_0(q, q')$$

$$= \widehat{M}(q)\left(q_d'' + K_p(q_d - q) + K_d(q_d' - q')\right) + \widehat{N}(q, q') + d(t) - \hat{d}(t)$$

$$= \tau_e(t) + d(t)$$

where $d(t)$ is teh disturbance and $\hat{d}(t)$ is the disturbance estimate. The disturbance estimate equations are

$$\hat{d}(t) = p(q'(t)) + z(t)$$

$$z'(t) = L(q,q')\big(N_0(q,q') - \tau_0(t) - p(q'(t)) - z(t)\big)$$

The whole point is that the disturbance observer equations should be based only on measurements of (q, q') and not on q''. Indeed, if q'' can be measured, then we can directly solve for the disturbance *d*(*t*) from the dynamical equation (without the lerm $\hat{d}(t)$). We select the function $L : \mathbb{R}^2 \times \mathbb{R}^2 \to \mathbb{R}^{2\times 2}$ so that

$$L(q,q') = p'(q')M_0(q)^{-1}$$

where $p'(q') \in R^{2\times 2}$ is the 2 × 2 Jacobian matrix of p. If we do so, then we get

$$\widehat{d'} = p'(q')q'' + z' = LM_0q'' + z'$$

$$= L\left(M_0q'' + N_0 - \tau_e - \hat{d}\right) = L\left(d - \hat{d}\right)$$

so defining the disturbance estimate error

$$f(t) = d(t) - \hat{d}(t)$$

our disturbance observer dynamics assumes the form

$$f' = d' - \widehat{d'} = d' - Lf \qquad \text{...(1)}$$

We define the angular position error

$$e(t) = q_d(t) - q(t)$$

and then the state dynamics can be cast as

$$\left(e'' + K_pe + K_dr'\right) + f = \widehat{M}(q)^{-1}\left(\widehat{M}(q) - M(q)\right)q'' + \left(\widehat{N}(q,q') - N(q,q')\right)$$

$$\approx W(t,\tilde{e})\delta\theta$$

where

$$\tilde{e} = \left[e^T, e'^T\right]^T$$

and we neglect $O(|\bar{e}||\delta\theta|)$ terms. Here

$$W(t,\tilde{e}) = -M_0(q)^{-1}\left[\frac{\partial M(q,\theta_0)}{\partial\theta}(I \otimes q''d(t)) + \frac{\partial N(q,q',\theta_0)}{\partial\theta}\right]$$

We note that on the rhs of this equation, q, q' are to be replaced by $q_d - e$ and $q'_d - e$ respectively and then it becomes a function of $(t, \tilde{e})$ rather than of (t, q, q'). $q_d(t)$ is known $\mathbb{R}^2$ valued signal.

❖❖❖❖❖

[31] Performance analysis of the disturbance observer in robots.

The dynamical equation of the robot in the presence of disturbance $d(t)$ and disturbance estimte $\hat{d}(t)$ with disturbance canceller incorporated is given by

$$M(q)q'' + N(q,q') = \tau(t) + d(t) - \hat{d}(t)$$

where

$$\tau(t) = M(q_d)\left(q_d'' + K_p e + K_d e'\right) + N(q_d, q_d')$$

with $q_d(t)$ as the desired trajectory. $\tau(t)$ is the computed torque with pd controller incorporated in order to make the robot trajectory track the desired trajectory. The disturbance observer for computing $\hat{d}(t)$ has the form

$$\begin{aligned}\hat{d}(t) &= p(q'(t)) + z(t)\\ \hat{z}'(t) &= L(q(t),q'(t))\left(N(q(t),q'(t)) - \left(\tau(t) - \hat{d}(t)\right) - p(q(t)) - z(t)\right)\\ &= L(q(t),q'(t))\left(N(q(t),q'(t)) - \tau(t)\right)\end{aligned}$$

Here, p (.) is a vector valued function of q' and is to be chosen apropriately. Simple computations show that that if we select the matrix valued function L such that

$$L(q,q') = p'(q')M(q)^{-1}$$

then the disturbance estimate $\hat{d}(t)$ satisfies the differential equation

$$\hat{d}'(t) = L(t)\left(d(t) - \hat{d}(t)\right)$$

where

$$L(t) = L(q(t)\cdot q'(t)) = p'(q'(t))M(q(t))^{-1}$$

Note that $p'(q')$ is the Jacobian matrix of the mapping $q' \to p(q')$. For a d-link robot, q, q' are $\mathbb{R}^d$-vectors $p(q')$ is also an $\mathbb{R}^d$ -vector and $p'(q')$ in an $\mathbb{R}^{d\times d}$ matrix. $M(q)$ is a positive definite $d \times d$ matrix for all $q \in \mathbb{R}^d$. Since the components of q are the angles of the robot links relative to the x-axis, we may assume that $q \in [0, 2\pi]^d$. Writing $f(t) = d(t)\,\hat{d}(t)$ the disturbance estimation error, we have

$$f'(t) = d'(t) - L(t)f(t)$$

and hence if $\Phi(t, \tau)$ is the state transition matrix associated with $L(t)$, *i.e.*,

$$\Phi_{,t}(t, \tau) = -L(t)\Phi(t, \tau),\ t \geq \tau,$$

$$\Phi(\tau, \tau) = I$$

then $$f(t) = \int_0^t \Phi(t,\tau)d'(\tau)dr = d(t) - \int_0^t \Phi_{,\tau}(t,\tau)d(\tau)dr$$

we note that

$$\Phi(t,\tau) = I + \sum_{n=1}^{\infty}(-1)^n \int_{r<t_1<\ldots<t_n<t} L(t_1)\ldots.L(t_n)dt_1\ldots dt_n$$

and hence

$$\phi_{,r}(t,\tau) = -\Phi(t,\tau)L(\tau)$$

We have

$$f(t) = \int_0^t \Phi(t,\tau)d'(\tau)d\tau$$

$$d(t) = \int_0^t \Phi_{,\tau}(t,\tau)d(\tau)d\tau$$

Let

$$R_{ff}(t,t) = \mathbb{E}\left(f(t)f(\tau)^T\right)$$

We have approximately

$$f(t) \approx d(t) - \int_0^t L(\tau)d(\tau)dr$$

and hence with this approximation

$$R_{ff}(t_1,t_2) = R_{dd}(t_1,t_2) + \int_{[0,t]^2} L(\tau_1)R_{dd}(\tau_1,\tau_2)L(\tau_2)^T d\tau_1 d\tau_2$$

With any approximation, we have

$$f(t) = d(t) + \int_0^t \Phi(t,\tau)L(\tau)d(\tau)d\tau$$

and hence we have exactly,

$$R_{ff}(t_1,t_2) = R_{dd}(t_1,t_2)$$

$$+ \int_{[0,t_1]\times[0,t_2]} \Phi(t_1,\tau_1)L(\tau_1)R_{dd}(\tau_1,\tau_2)L(\tau_2)^T \Phi(t_1,\tau_2)Td\tau_1 d\tau_2$$

$$+\int_0^t \Phi(t,\tau)R_{dd}(\tau,t)dr + \int_0^t R_{dd}(t,\tau)\Phi(t,\tau)^T d\tau$$

If $d(t)$=0, $\hat{d}(t) = 0$ then $f(t) = 0$ and hence in this case, $q_d(t)$ is an exact solution to the dynamical equation. This desired trajectory can thus be called as the as the noiseless output. If $d(t) \neq 0$, then $\hat{d}(t) \neq 0$ and hence $f(t) \neq 0$ which implies that the solution to the dynamical equation can be expressed approximately as

$$q(t) = q_d(t) + \delta q(t)$$

where $\delta q(t)$ is the linearized response to the dynamical equation:

$$M(q_d)\delta q'' + N_{,1}(q_d, q_d')\delta q + N_{,2}(q_d, q_d')\delta q' = -M(q_d)\left(K_p\delta q + K_d\delta q'\right) + f$$

We can rearrange this equation as

$$M(q_d)\delta q'' + \left(N_{,2}(q_d, q_d') + K_d M(q_d)\right)\delta q' + N_{,1}(q_d, q_d') + K_p M(q_d)\delta q = f$$

or equivalently,

$$\delta q''(\mathrm{t}) = A(t)\delta q'(t) + B(t)\,\delta q(t) + C(t)f(t)$$

where

$$A(t) = -M(q_d(t))^{-1}N_{,2}(q_d(t), q'_d(t)) - K_d I,$$

$$B(t) = -M(q_d(t))^{-1}N_{,1}(q_d(t), q'_d(t)) - K_p I,$$

$$C(t) = M(q_d(t))^{-1}$$

The above linearized dynamical equation for $\delta q(t)$ can be expressed in state variable form as

$$\frac{d}{dt}\begin{pmatrix}\delta q(t)\\ \delta q'(t)\end{pmatrix} = \begin{pmatrix}0 & I\\ B(t) & -A(t)\end{pmatrix}\begin{pmatrix}\delta q(t)\\ \delta q'(t)\end{pmatrix} + \begin{pmatrix}0\\ C(t)\end{pmatrix}f(t)$$

Let $\psi(t, \tau)$ be the state transition matrix ($2d \times 2d$) associated with this linearized state equation. Then with $\psi_{12}(t, \tau)$ denoting the top right hand corner block, we have

$$\delta q(t) = \int_0^t \psi_{12}(t, \tau)C(\tau)f(r)d\tau$$

Letting

$$F(t, r) = \psi_{12}(t, \tau)C(\tau)$$

we can write this equation as

$$\delta q(t) = \int_0^t F(t, \tau)f(\tau)d\tau$$

and hence

$$R_{qq}(t_1, t_2) = \mathbb{E}\left(\delta q(t_1)\delta q(t_2)^T\right)$$

$$= \int_{[0,t_1]\times[0,t_2]} F(t_1,\tau_1)R_{ff}(\tau_1,\tau_2)F(t_2,\tau_2)d\tau_1 d\tau_2$$

The mean square fluctuation error in the position process due to disturbance with disturbance canceller incorporated is

$$E\left(\|\delta q(t)\|^2\right) = Tr\left(R_{qq}(t,t)\right)$$

The performance of the robot after incorporating the disturbance canceller over the duration [0, *T*] may then be defined as

$$NSR_T = \frac{\int_0^T \mathbb{E}\|\delta q(t)\|^2 dt}{\int_0^T \|q_d(t)\|^2 dt}$$

Largen the value of NSR_T, worse is the performance.

Now coming to the structure of the disturbance observer defined completely by the function $p : \mathbb{R}^d \to \mathbb{R}^d$, we usually choose p to be a lingerer function of q', *i.e.*,

$$p(q') = \left(\left(p_k(q')\right)\right)_{k=1}^{d}$$

where

$$p_k(q') = \sum_{j=1}^{d} c(k,j)q'_j$$

Then, the Jacobian of this transformation is the constant matrix

$$p'(q') = C = \left(\left(c(k,j)\right)\right) \in \mathbb{R}^{d\times d}$$

and $$L(t) = CM(q(t))^{-1}$$

Now to determine C, we construct the Lyapunov function

$$V(t) = V(f,q) = \frac{1}{2}f(t)^T J\left(q(t)\right)f(t)$$

where $J(q)$ is a positive definite matrix valued function opf $q \in \mathbb{R}^d$. J must be selected so that $dV/dt < 0$ for that will guarantee that $f(t) \to 0$ as $t \to \infty$, Assuming that $d(t) \to$ constant as $t \to \infty$, it follows that asymptotically, f satisfies

$f'(t) - L(t)f(t)$

Now,

$$dV/dt = f^T J(q)f' + \frac{1}{2}\sum_k f^T J_{,k}(q) f q'_k$$

$$= -f^T JLf + \frac{1}{2}\sum_k f^T J_{,k} f q'_k$$

$$= -\frac{1}{2} f^T \left(JL + L^T J - \sum_k q'_k J_{,k} \right) f$$

we require that the matrix

$$X = JL + L^T J - \sum_k q'_k J_{,k}$$

be positive definite at all times, Substitution for L, we get

$$X = JCM^{-1} + M^{-1}C^T J - \sum_k q'_k J_{,k}$$

To simplify the construction of the Lyapunov matrix $J(q)$, we shall choose it so that

$$A = J(q)CM^{-1}(q)$$

is a constant $d \times d$ matrix. Equivalently, we shall require that

$$J(q) = AM(q)C^{-1}$$

where A is constant matrix. The choice

$$A = aC^{-T}$$

where $\alpha > 0$ guarantees that $J(q)$ will be positive definite. We then have

$$J(q) = \alpha C^{-T} M(q) C^{-1}$$

and hence

$$X = \alpha\left(C^{-1} + C^{-1}\right) - C^{-T} \sum_k q'_k M_{,k}(q) C^{-1}$$

$$= \alpha C^{-1} \left(C + C^T - \sum_k q'_k M_{,k}(q) \right) C^{-T}$$

We thus require to select the matrix C so that for all values of q in $[0, 2\pi]^d$ and for all value of the angular velocity vector $q' = (q'_k)$ in a prescribed range, the matrix

$$C + C^T - \sum_k q'_k M_{,k}(q)$$

be positive definite. This gives us a relation between the matrix C and the prescribed range D of q'.

❖❖❖❖❖

[32] Some aspects of quantum robotics:

$$H_0 = \frac{1}{2}p^T M(q)^{-1} p + V(q)$$

where
$$M(q) = \begin{pmatrix} a + b\cos(q_2) & c + d.\cos(q_2) \\ c + d.\cos(q_2) & f \end{pmatrix}$$

where a, b, c, d, f are real numbers chosen to make $M(q)$ positive definite for all $q \in [0, 2\pi]^2$. In other words, we require that (a)

$$a + f + b.\cos(q_2) > 0$$

and (b)

$$(a + b.\cos(q_2))f > (c + d.\cos(q_2))^2$$

$\forall q_2 \in [0, 2\pi]$. In the presence of disturbance $d(t)$, the Hamiltonian becomes

$$H(t) = H_0 - (\tau(t) + d(t))^T q$$

The disturbance observer is constructed as

$$\hat{d}(t) = \langle P(q')\rangle(t) + z(t),$$

where $z(t)$ satisfies

$$z'(t) = L(\langle q(t)\rangle, \langle q'(t)\rangle)(\tau(t) - N(\langle q(t)\rangle, \langle q'(t)\rangle) - \hat{d}(t))$$

where if $|\psi_0 1>$ is the initial state of the system,

$$\langle q'(t)\rangle = i\langle\psi(t)|[H_{0,q}]|\psi(t)\rangle$$

and more generally,

$$\langle\phi(q, q')\rangle(t) = \langle\psi(t)|\phi(q, i[H_{0,q}])|\psi(t)\rangle$$

with
$$\psi(t) = U(t)\,\psi_0,$$
$$iU'(t) = H(t)U(t)$$

Equivalently,

$$i\psi'(t) = H(t)\psi(t),\ t \geq 0$$

These equations are the quantum analogue of the classical disturbance observer equations. Other formulations of the distubance observer exist, for example,

$$z'(t) = \langle L(q, q')\rangle(t)(\tau(t) - \langle N(q, q')\rangle(t) - \hat{d}(t))$$

$L(q, q')$ can be chosen by the classical algorithm

$$L(q, q') = CM(q)^{-1}$$

where C is a constant matrix. Note that at time $t = 0$,

$$q'_k = i[H_{0,}q_k] = i\left[\frac{1}{2}p^T M(q)^{-1} p, q_k\right]$$

$$= \frac{1}{2}\left(\left[p^T M(q)^{-1}\right]_k + \left[M(q)^{-1} p\right]_k\right)$$

Equivalently,

$$q' = \frac{1}{2}\left(M(q)^{-1} p + \left[p^T M(q)^{-1}\right]^T\right)$$

Sometimes, it is more accurate to model the disturbance $d(t)$ as a quantum stochastic process in terms of creation, annihilation and conservation operators. The disturbance dynamics will then also be described in terms of quantum stochastic differential equations. We thus start with the qsde's

$$dU(t) = \left(-i\widehat{H}(t)dt + L_1 dA(t) + L_2 dA^*(t) + Sd\Lambda(t)\right)U(t)$$

where $\tilde{H}(t)$ is $H(t)$ plus some terms involving L_1, L_2 and

$$H(t) = H_0 - \tau(t)^T q, H_0 = \frac{1}{2}p^T M(q)^{-1} p$$

is a system operator, and the system operators L_1, L_2, S are chosen so that $\{U(t)\}$ is a unitary family. A, A^*, Λ are the noise process in quantum stochastic calculus that model the distrubance. They satisfy the quantum Ito rules

$$dA.\, dA^* = dt,\ dA.d\Lambda = dA,\ d\Lambda.\, dA^* = dA^*, (d\Lambda)^2 = d\Lambda$$

with all the other differential products zero. Take an obsevable X in the system Hilbert space like q, p and compute $dX(t)$ where $X(t) = U(t)^* XU(t) = U(t)^*(X \otimes I)$ $U(t) = j_t(X)$. The result is

$$dX(t) = j_t\left(i\left[H(t), X\right]\right)dt + \text{noise term}$$

where the noise term is given by

$$j_t\Big(\left(L_1^* X + XL_2\right)dA^* + \left(L_2^* X + XL_1\right)dA$$

$$+\left(S^* X + XS^*\right)d\Lambda S^* XSd\Lambda + L_2^* XSdA + S^* XL_2 dA^*\Big)$$

$$= j_t\left(L_1^* X + XL_2 + S^* XL_2\right)dA^*$$

$$+j_t\left(L_2^* X + XL_1\right)dA + j_t\left(S^* X + XS^* + S^* XS\right)d\Lambda$$

$$= j_t\left(\theta_1(X)\right)dA^* + j_t\left(\theta_2(X)\right)dA + j_t\left(\theta_3(X)\right)d\Lambda$$

We denote this differential by $dN(t, X)$ of simple $dN_t(X)$, N standing for noise. Specifically, $X = q$ and $X = p$ gives

$$dq_k(t) = j_t\left(i\left[\frac{1}{2}p^T M(q)^{-1} p, q_k\right]\right)dt + j_t\left(\theta_1(q_k)\right)dA^*$$

$$+j_t\left(\theta_2(q_k)\right)dA + jt\left(\theta_3(q_k)\right)d\Lambda$$

$$= j_t\left(i\left[\frac{1}{2}p^T M(q)^{-1} p, q_k\right]\right)dt + dN_t(q_k)$$

and likewise,

$$dp_k(t) = j_t\left(i\left[\frac{1}{2}p^T M(q) p, q_k\right]\right)$$

$$-i\tau(t)^T\left[q, p_k\right]\Big)dt + dN_t(p_k)$$

Now,

$$\left[\frac{1}{2}p^T M(q)^{-1} p, q_k\right] = \frac{1}{2}\left(\left(p^T M(q)^{-1}\right)_k + M(q)^{-1} p\right)_k$$

We denote this by $[M(q)^{-1}p]_k$, *i.e.*, [.] denotes symmetrization: $[AB] = \frac{1}{2}(AB + BA)$. We write

$$K(q) = M(q)^{-1}$$

Further

$$\left[\frac{i}{2}pTM(q)^{-1} p, p_k\right] = -\frac{1}{2}p^T K_{,k}(q)p$$

where

$$K_{,K}(q) = \frac{\partial K(q)}{\partial q_k}$$

In this way, we get

$$dq_k(t) = j_j\left(\left[K(q)p\right]_k\right) + dN_t(q_k), dp_k$$

$$= -\ j_t\left(\frac{1}{2}p^T K_{,k}(q)p\right)dt + dN_t(p_k)$$

In the presence of a gravitational potential potential $V(q)$, H_0 gets replaced by

$\frac{1}{2}p^T M(q)^{-1} p + V(q)$ and the above equations become

$$dp_k(t) = j_t\left(\left[K(q)p\right]_k dN_t(q_k), dp_k(t)\right)$$

$$= -j_t\left(\frac{1}{2}p^T K_{,k}(q)p\right)dt - j_t\left(V_{,k}(q)\right) + dN_t\left(p_k\right)$$

Using the homomorphism property of j_t, these equations can be expressed as

$$dp(t) = \left[K(q(t))p(t)\right] + dN_t(q), dp(t)$$

$$= -p^T(t)\nabla K(q(t))p(t) - \nabla V(q(t)) + dN_t(p)$$

These two qsde's describe the dynamics of the robot system in the presence of quantum noise (disturbance process). Now we discuss the constuction of the quantum disturbance observer. Classically, it has the form

$$d\hat{d}(t) = L(q(t), q'(t))\left(d(t)dt - \hat{d}(t)dt\right)$$

Quantum mechanically, we expect therefore that the disturbance observer will be described by the qsde

$$d\hat{d}(t) = L(t)\left(L_1 dA(t) + L_2 dA^*(t) + Sd\Lambda(t) - \hat{d}(t)dt\right)$$

where $L(t)$ is an adapted process w. r. t. the Hilbert space filtration $\left\langle \otimes\Gamma_s\left(L^2([0,t])\right)\right\rangle$, $t \geq 0$. As in the classical case, we shall take

$$L(t) = C.M(q(t))^{-1}$$

where C is a constant matrix. We define the quantum distubance estimate error process by

$$f(t) = d(t) - \hat{d}(t)$$

Then

$$df(t) = dd(t) - L(t)f(t)dt$$

$$= L_1 dA(t) + L_2 dA^*(t) + Sd\Lambda(t) - L(t)f(t)dt$$

❖❖❖❖❖

[33] Consider the following problem:

A dynamical system is specified as

$$x'(t) = f(t, x(t), \theta) \qquad ...(1)$$

where $x(t) \in \mathbb{R}^n$, $\theta \in \mathbb{R}^p$. θ is a parameter vector chosen to minimize the cost function

$$x[\theta] = \int_0^T L(t, x(t))dt \qquad ...(2)$$

Now suppose the system gets corrupted by noise $d(t)$. The new dynamics is given by

$$x'(t) = f(t, x(t), \theta) + \varepsilon d(t) \qquad ...(3)$$

where is a small perturbation parameter that signifies that the noise amplitude is small. Let $x_0(t)$ denote the dynamics withe zero noise and θ_0 the corresponding optimal parameter vector that minimizes $C(\theta)$. After noise corruption, we seek to determine the first order correction $\varepsilon\delta\theta$ to the parameter vector θ_0 so that

$$\bar{C}(\theta_0 + \varepsilon\delta\theta) = \mathbb{E}\int_0^T L(t, x(t))dt \qquad ...(4)$$

is a minimum. To this end, we write

$$x(t) = x_0(t) + \varepsilon\delta x(t) \qquad ...(5)$$

Then linearization of the dynamical system gives

$$dx'(t) = f_{,x}(t, x_0(t), \theta_0)\delta x(t) + f_{,\theta}(t, x_0(t), \theta_0)\delta\theta + d(t) \qquad ...(6)$$

Defining

$$A(t) = f_{,x}(t, x_0(t), \theta_0) \qquad ...(7a)$$

$$B(t) = f_{,\theta}(t, x_0(t), \theta_0) \qquad ...(7b)$$

we have

$$dx'(t) = A(t)\delta x(t) + B(t)\delta\theta + d(t) \qquad ...(8)$$

and hence if $\Phi(t, \tau)$ denotes the state transition matrix corresponding to $A(t)$, we have

$$\delta x(t) = \int_0^t \Phi(t, \tau)(B(\tau)\delta\theta + d(\tau))d\tau \qquad ...(9)$$

Note the $\delta\theta$ is a non-random parameter vector perturbation. We have

$$C_1(\delta\theta) = \bar{C}(\theta_0 + \varepsilon\delta\theta) - C_0(\theta_0)$$

$$\varepsilon\int_0^T E\left(L_{,x}(t, x_0(t))\mathbb{E}(\delta x(t))dt + (\varepsilon^2/2)\right)$$

$$\int_0^T Tr\left(L_{,xx}(t, x_0(t))\mathbb{E}\left(\delta x(t)\delta x(t)^T\right)\right)dt + O(\varepsilon^3) \qquad ...(10)$$

Note that we have retained upto $O(\varepsilon)$ terms in the perturbed dynamics while upto $O(\varepsilon^2)$ terms have been retained in the cost function. This is because typically, the cost is a quadratic functional of the state vector. We have

$$\mathbb{E}(\delta x(t)) = \int_0^t \Phi(t,\tau)\left(B(\tau)\delta\theta + \mu_d(t)\right)d\tau \quad ...(11)$$

where

$$\mu_d(t) = \mathbb{E}(d(t)) \quad ....(12)$$

and likewise,

$$\begin{aligned}\mathbb{E}\left(\delta x(t)\delta x(t)^T\right) = & \int_{[0,t]^2} \Phi(t,\tau)B(\tau)\delta\theta\delta\theta^T B^T(\tau')\Phi(t,\tau')^T d\tau d\tau' \\ & + \int \Phi(t,\tau)B(\tau)\delta\theta_{\mu d}(\tau')\Phi(t,\tau')^T d\tau d\tau' \\ & + \int \Phi(t,\tau)_{\mu d}(\tau)\delta\theta^T B(\tau')^T \Phi(t,\tau')^T d\tau d\tau' \\ & + \int \Phi(t,\tau)R_{dd}(\tau,\tau')\Phi(t,\tau')^T d\tau d\tau' \quad ...(13)\end{aligned}$$

where $$R_{dd}(t_1, t_2) = \mathbb{E}\left(d(t_1)d(t_2)^T\right) \quad ...(14)$$

Substituting (11) and (13) into (10) gives us an expression for $C_1(\delta\theta)$ as a linear-quadratic function of $\delta\theta$ and this expression can be minimized by setting its gradient w.r.t. $\delta\theta$ to zero yielding a linear matrix equation for $\delta\theta$. We note that the final expression for the optimal correction $\delta\theta$ to the parameter vector does not depend on the noise correlation R_{dd} although it depends on the noise mean μ_d. The minimum value of the average cost change $C_1(\delta\theta)$ however, depends on R_{dd} also.

❖❖❖❖❖

[34] Quantum channel estimation:

Let $U_1, ..., U_n$ be n unitary matrices and $p_1, ..., p_n$ be non-negative numbers adding to unity. Then

$$\rho \to T(\rho) = \sum_{k=1}^{n} p_k U_k \rho U_k^*$$

defines a quantum channel, *i.e.*, T is a completely positive trace preserving (CPTP) map. Suppose ρ_1 and ρ_2 are two mixed states and we wish to determine unitary matrices $U_1, ..., U_n$ such that

$$\sum_{m=1}^{n} \left(Tr(\rho_2 Y_m) - Tr(T(\rho_1)Y_m)\right)^2$$

is a minimum for given observables Y_m, $m = 1, 2, ..., N$ If δH is an infinitesimal

Hermitian operator and U a unitary operator, then $(1 + i\delta H)U$ us unitary upto first order. Using, this we deduce that the above error energy is a minimum where

$$\sum_m \left(Tr(\rho_2 Y_m) - Tr(T(\rho_1)Y_m)\right)Tr\left(Y_m \delta T(\rho_1)\right) = 0$$

where

$$\delta T(\rho) = i\sum_k p_k\left[\delta H_k, U_k \rho U_k^*\right]$$

with δH_k being arbitrary infinitesimal Hermitian operators. Equating to zero the coefficient of δH_k gives

$$\sum_m \left(Tr(\rho_2 Y_m) - Tr(T(\rho_1)Y_m)\right)p_k Tr\left[Y_m\left(\delta H_k U_k \rho_1 U_k^* - U_{k\rho 1} U_k^* \delta H_k\right)\right] = 0$$

from which we deduce using $Tr(AB) = Tr(BA)$ that

$$\sum_{m,k} p_k\left(Tr(\rho_2 Y_m) - Tr(T(\rho_1)Y_m)\right)\left[U_k \rho_1 U_k^*, Y_m\right] = 0$$

This is a fourth degree polynomial equation in $U_k, U_k^*, k = 1, 2, ..., n$ and can be solved only by numerical methods. We assume now that the $U'_k s$ are known and they suffer small multiplicative perturbations $1 + i\delta H_k$. We wish to estimate the perturbations δH_k subject to knowledge about the perturbed output density matrix $\rho_2 + \delta\rho_2$. A straightforward linearization argument gives

$$\delta\rho_2 = \sum_k p_k\left[\delta U_{k\rho 1} U_k^* + U_{k\rho 1}\delta U_k^*\right]$$

$$= i\sum_k p_k\left[\delta H_k, U_{k\rho 1} U_k^*\right]$$

and we estimate δH_k so that

$$E(\{\delta H_k\}) = \sum_m \left(Tr\left(\delta_{\rho 2} Y_m\right)\right) - i\sum_k p_k Tr\left(\left[\delta H_k, U_{k\rho 1} U_k^*\right]Y_m\right)^2$$

is a minimum. For convenience of notation, we denote δH_k by H_k and then δH_k will denote the perturbation in H_k (*i.e*, the perturbation in the original δH_k). Thus, we have to minimize

$$E(\{H_k\}) = \sum_m \left(b[m] - i\sum_k p_k Tr\left(\left[H_k, F_k\right]\right)Y_m\right)^2$$

where

$$b[m] = Tr(\delta\rho_2 Y_m),\ F_k = U_k \rho_1 U^*_k$$

Setting the variational derivative of E w.r.t. H_r to zero gives

$$\sum_m (b[m] - i\sum_k p_k Tr([H_k, F_k]Y_m))^2 Tr([\delta H_r, F_r]Y_m) = 0$$

which gives on using the arbitrariness of the Hermitian perturbation δH_r,

$$\sum_m (b[m] - i\sum_k p_k Tr([H_k, F_k]Y_m))[F_r, Y_m] = 0$$

or equivalently,

$$\sum_m b[m][F_r, Y_m] = i\sum_k Tr([H_k, F_k]Y_{m)})[F_r, Y_m], 1 \le r \le$$

These are linear equations for the optimum perturbations $\{H_r\}$ of the channel.

❖❖❖❖❖

[35] Estimation of the initial quantum state based on measurements with state collapse incorporated.

❖❖❖❖❖

[36] Statistical problems in robotic vision: The tip of a two link robot has its coordinates given by

$$\mathbf{r}_2(t) = \mathbf{r}_1 + ((l_1 \cos(q_1) + l_2 \cos(q_2))\cos(\phi), (l_1 \cos(q_1) + l_2 \cos(q_2))\sin(\phi), l_1 \sin(q_1) + l_2 \sin(q_2))$$

where

$$\mathbf{r}_1 = (x_1, y_1, z_1)$$

is fixed, *i.e.*, the base point of the robot and q_1, q_2, ϕ are functions of time. The kinetic energy of this robot can be expressed as

$$T(t) = \frac{1}{2}\mathbf{q}'^T M(\mathbf{q})\mathbf{q}' + \frac{1}{2}M_3(\mathbf{q})\phi'^2$$

where

$$\mathbf{M}(\mathbf{q}) = \begin{pmatrix} M_{11}(\mathbf{q}) & M_{12}(\mathbf{q}) \\ M_{12}(\mathbf{q}) & M_{22}(\mathbf{q}) \end{pmatrix}$$

is a positive definite 2×2 matrix valued function of $\mathbf{q} = (q_1, q_2)^T$, the link angles and ϕ is angle which specifies how much the plane spanned by the two links has turned relative to the $x - z$ plane. $M_3(\mathbf{q})$ is a positive scalar function of $\mathbf{q}$. The gravitational potential energy of the robot is given by

$$\mathbf{V}(\mathbf{q}) = m_1 g l_1 \sin(q_1)/2 + m_2 g(l_1 \sin(q_1) + l_2 \sin(q_2)/2)$$

❖❖❖❖❖

[37] Simulink model for filter design with em field inputs that guarantee optimum tracking of a 3-D quantum harmonic oscillator unitary evolution matrix with a given unitary evolution. The Hamiltonian of the oscillator in an em field that is constant in space but varies with time is approximately given by the following scalar and magnetic vector potentials:

$$\Phi(t, \mathbf{r}) = -(\mathbf{E}(t), \mathbf{r}) = -(E_1(t)x + E_2(t)y + E_3(t)z),$$

$$\begin{aligned}\mathbf{A}(t, \mathbf{r}) &= \tfrac{1}{2}\mathbf{B}(t)\times\mathbf{r}\\ &= \tfrac{1}{2}(B_2(t)z - B_3(t)y, B_3(t)x - B_1(t)z, B_1(t)y - B_2(t)x)\end{aligned}$$

The exact em field corresponding to these potentials are

$$\begin{aligned}\mathbf{E}(t, \mathbf{r}) &= -\nabla\phi - \mathbf{A}_{,t}\\ &= \mathbf{E}(t) - \frac{1}{2}\mathbf{B}'(t)\times\mathbf{r}\\ \mathbf{B}(t, \mathbf{r}) &= \nabla\times\mathbf{A}(t,\mathbf{r}) = \mathbf{B}(t)\end{aligned}$$

We assume a slow magnetic field variation so that $|\mathbf{B}'|d << |\mathbf{E}|$ where d is the typical dimension of the oscillator. Then, the expression for the electric field is approximately

$$\mathbf{E}(t,\mathbf{r}) \approx \mathbf{E}(t)$$

The oscillator Hamiltonian in this em field is given by

$$\begin{aligned}H(t) &= (\mathbf{P} + e\mathbf{A})^2/2 + \mathbf{q}^2/2 - e\Phi\\ &= H_0 + (e/2)(\mathbf{p},\mathbf{A}) + (\mathbf{A},\mathbf{p}) + e^2 A^2/2 + e(\mathbf{E}(t),\mathbf{q})\\ &= H_0 + (e/2)(\mathbf{B}(t),\mathbf{L}) + e(\mathbf{E}(t),\mathbf{q}) + e^2(\mathbf{B}(t)\times\mathbf{q})^2/8\end{aligned}$$

where

$$\begin{aligned}\mathbf{q} &= \mathbf{r}, \mathbf{L} = \mathbf{r}\times\mathbf{p},\\ H_0 &= (\mathbf{P}^2 + \mathbf{q}^2)/2\end{aligned}$$

is the unperturbed harmonic oscillator Hamiltonian. We now assume that the em field is generated by passing known signals (vector valued) through matrix FIR filter as follows:

$$\begin{aligned}\mathbf{E}(t) &= \sum_{n=0}^{p}\mathbf{H}[n]\mathbf{S}_1(t - n\Delta)\\ \mathbf{B}(t) &= \sum_{n=0}^{p}\mathbf{G}[n]\mathbf{S}_2(t - n\Delta)\end{aligned}$$

where $\mathbf{S}_1(t)$ and $\mathbf{S}_2(t)$ are 3×1 vector valued non-random function of time and $\mathbf{H}[n]$, $\mathbf{G}[n]$, $n = 0, 1,..., p$ are 3×3 matrix valued FIR filters of length $p + 1$.

❖❖❖❖❖

[38] Cavity resonator in an inhomogeneous medium. The basic Maxwell equations at frequency ω are

$$\nabla \times \mathbf{E} = -j\omega\mu\mathbf{H}$$

$$\nabla \times \mathbf{H} = j\omega\varepsilon\mathbf{E}$$

where

$$\varepsilon = \varepsilon(\omega, \mathbf{r}), \mu = \mu(\omega, \mathbf{r})$$

Write

$$\mathbf{E} = E(x, y)\exp(-\gamma z), H = H(x, y) \exp(-\gamma z)$$

Then,

$$(\nabla_\perp - \gamma\hat{z}) \times (E_\perp + E_z\hat{z}) = -j\omega\mu(H_z\hat{z} + H_\perp)$$

the other curl equation can be derived from duality. The above gives

$$\nabla_\perp E_z \times \hat{z} - \gamma\hat{z} \times E_\perp = -j\omega\mu H_\perp. \quad ...(1)$$

and by duality,

$$\nabla_\perp H_z \times \hat{z} - \gamma\hat{z} \times H_\perp = j\omega\varepsilon E_\perp. \quad ...(2)$$

From (1), it follows on taking $\hat{z}\times$ that

$$\nabla_\perp E_z + \gamma E_\perp = -j\omega\mu\hat{z} \times H_\perp \quad ...(3)$$

Elimination $\hat{z} \times H_\perp$ between (2) and (3) gives us

$$\nabla_\perp H_z \times \hat{z} + (\gamma/j\omega\mu)(\nabla_\perp E_z + \gamma E_1) = j\omega\varepsilon E_\perp$$

so that

$$(j\omega\varepsilon - \gamma^2/j\omega\mu)E_\perp = \nabla_\perp H_z \times \hat{z} + (\gamma/j\omega\mu)\nabla_\perp E_z$$

or defining

$$h^2 = h^2(\omega, x, y, z) = h^2(w, \mathbf{r}) = \gamma^2 + \omega^2\mu(\omega,\mathbf{r}) \in (\omega,\mathbf{r})$$

we get

$$E_\perp = (-j\omega\mu/h^2)\nabla_\perp H_z \times \hat{z} - (\gamma/h^2)\nabla_\perp E_z \quad ...(4)$$

and by duality

$$H_\perp = (j\omega\varepsilon\mu/h^2)\nabla_\perp E_z \times \hat{z} - (\gamma/h^2)\nabla_\perp H_z \quad ...(5)$$

We substitute (4) and (5) into the z component of the Maxwell curl equations to get

$$\nabla_\perp \times [(j\omega\epsilon/h^2)\nabla_\perp E_z \times \hat{z} - (\gamma/h^2)\nabla_\perp H_z] = j\omega\varepsilon E_z \quad ...(6)$$

$$\nabla_\perp \times [(j\omega\mu/h^2)\nabla_\perp H_z \times \hat{z} - (\gamma/h^2)\nabla_\perp E_z] = j\omega u H_z \quad ...(7)$$

(6) can be expanded to give

$$(-j\omega\varepsilon/h^2)\nabla_\perp^2 E_z - (\nabla_\perp(j\omega\varepsilon/h^2), \nabla_\perp E_z)$$

$$-(\hat{z}, \nabla(\gamma/h^2)\nabla_\perp H_z) - j\omega\varepsilon E_z = 0$$

or equivalently,

$$[\nabla_\perp^2 + h^2]E_z + \left(\nabla_\perp\left(\log\left(\varepsilon/h^2\right)\right), \nabla_\perp E_z\right)$$

$$-(\gamma/j\omega\varepsilon)\left(\hat{z}, \nabla\left(\log\left(h^2\right)\right) \times \nabla_\perp H_z\right) = 0 \quad ...(8)$$

and by duality, we get (7) in the form

$$[\nabla_\perp^2 + h^2]H_z + (\nabla_\perp(\log(\mu/h^2)), \nabla_\perp H_z)$$

$$+(\gamma/j\omega\mu)(\hat{z}, \nabla(\log(h^2)) \times \nabla_\perp E_z) \quad ...(9)$$

We now observe that

$$h^2 = \gamma^2 + \omega^2\mu\varepsilon = \gamma^2 + k^2(1 + \delta(\chi_e + \chi_m)) + O(\delta^2)$$

where we are assuming that

$$\varepsilon = \varepsilon_0(1 + \delta\chi_e(\omega, \mathbf{r})),$$

$$\mu = \mu_0(1 + \delta\chi_m(\omega, \mathbf{r}))$$

with δ as a small perturbation parameter and

$$k^2 = \omega^2\mu_0\varepsilon_0$$

is a constant. Thus, defining

$$h_0^2 = \gamma^2 + k^2$$

This generalized eigenvalue problem is of the form

$$(\nabla_\perp^2 + h_0^2)\psi(x, y) + \delta.L(\gamma, \omega)\psi(x, y) = 0$$

with $\psi(x, y) = (E_z(x, y), H_z(x, y))^T$ and ψ satisfies the boundary condition that H_z vanishes at $x = 0, a$ and at $y = 0, b$ and likewise, $E_{z,x}$ vanishes when $x = 0, a$ and $E_{z,y}$ vanishes when $= 0, b$. F. Further, H_z, E_x, E_y vanish when $z = 0, d$. We are assuming here that the sidewalls $x = 0, a$ and $y = 0, b$ are perfect magnetic conductors while the top and bottom surfaces $z = 0, d$ are perfect electric conducting surfaces. This implies that the tangential components of the magnetic field and normal component of the electric field vanish on the sidewalls while the normal component of the magnetic field and tangential components of the electric field vanish on the top and bottom surfaces. In the case of the cavity resonator, we have to replace γ by the operator $-\frac{\partial}{\partial z}$ and hence, we end up with the 3-D eigenvalue problem

$$\left(\nabla^2 + \omega^2/c^2\,\psi(x,y,z) + \delta.L\left(-\frac{\partial}{\partial z},\omega\right)\psi(x,y,z)\right) = 0$$

owing to the boundary conditions at $z = 0, d$. These conditions cannot be satisfied by assuming a z dependence of exp $(-\gamma z)$. We must instead consider linear combinations of exp $(\pm\gamma z)$ and the boundary conditions on z imply that the z dependence is either $\cos(p\pi z/d)$ or $\sin(p\pi z/d)$ depending on whether we require the field derivative w.r.t. z to vanish at $z = 0, d$ or whether we require the field itself to vanish at $z = 0, d$. It follows from these considerations that E_z is a linear combination of

$$u_{nmp}(x, y, z) = (2\sqrt{2}/\sqrt{abd})\cos(n\pi x/a)\cos(mgpy/b)\cos(p\pi z/d)$$

while H_z is linear combination of

$$u_{nmp}(x, y, z) = (2\sqrt{2}/\sqrt{abd})\sin(n\pi x/a)\sin(n\pi y/b)\sin(p\pi z/d)$$

with n, m, p taking values 1, 2, ... The boundary condition imply that ω can take only discrete values. This boundary value problem is a special case of the following general boundary value problem: Let L_0, $L_1(\gamma)$ be linear differential operators in a Hilbert space with L_1 depending on the a complex parameter γ. We than wish to solve the generalized eigenvalue problem

$$[(L_0 - \lambda I) + \delta.L_1(\lambda)]\psi(x) = 0, x \in D$$

In our situation, L_0 is a second order partial differential operator and L_1 is also a first order partial differential operator and the boundary conditions on ψ can be expressed as $M\psi(x) = 0$ for $x \in \partial D$ where M is a first order partial differential operator. In our problem, L_0 is the three dimensional Laplacian and L_1 is linear in $\frac{\partial}{\partial x}$ and $\frac{\partial}{\partial y}$ and also linear in $\frac{\partial}{\partial z}$. More precisely, L_1 is built as linear combination of the operators $\frac{\partial}{\partial x}$, $\frac{\partial}{\partial y}$, $\frac{\partial^2}{\partial x \partial z}$ and $\frac{\partial^2}{\partial y \partial z}$.

❖❖❖❖❖

[39] Radon transform based estimation of rotations of an object in the presence of noise.

The 3-D object is $f(x)$ $x \in \mathbb{R}^3$. Its Randon transform is

$$Rf(\hat{n}, p) = \int f(x)\delta(p - (\hat{n}, x))d^3x$$

Suppose that while taking the Radon transform of f w.r.t. a plane $(\hat{n}, p)$, zero mean Gaussian noise gets added. Then the noise corrupted Radon transform is given by

$$F(\hat{n}, p) = Rf(\hat{n}, p) + w(\hat{n}, p)$$

We assume ω to be white, *i.e.*,

$$\mathbb{E}(w(\hat{n}, p)w(\hat{n}', p')) = \sigma^2\delta(\hat{n} - \hat{n}')\delta(p - p')$$

When a rotation $K \in SO(3)$ is applied to the object f, the object becomes $f(K^{-1}x)$ and its Radon transform after adding noise is

$$F_K(\hat{n}, p) = \int f(K^{-1}x)\delta(p - (\hat{n}, x))d^3x + w_K(\hat{n}, p)$$

which can be expressed as

$$F_K(\hat{n}, p) = \int f(x)\delta(p - (\hat{n}, Kx))d^3x + w_K(\hat{n}, p)$$

$$= Rf(K^{-1}\hat{n}, p) + w_K(\hat{n}, p)$$

we assume that w, w_K are iid noise fields. We take the spherical harmonic transform of both of these Randon transforms. For doing this, we require measuring $F(\hat{n}, p)$ and $F_K(\hat{n}, p)$ for all $\hat{n} \in S^2$ and for a finite set of p's, *i.e.*, we take the radon transform on all planes that have distances $p_1, \ldots, p_m$ where m is a finite integer. The result is

$$TF(l, m, p) = \int F(\hat{n}, p)Y_{lm}(\hat{n})d\Omega(\hat{n})$$

$$= TRf(l, m, p) + Tw(l, m, p),$$

$$TF_K(l, m, p) = \sum_{m'} [\pi_l(K^{-1})]_{m',m} TRf(l, m', p) + Tw_K(l, m, p)$$

We observe that Tw and Tw_K are iid Gaussian random field with mean zero and covariance

$$\mathbb{E}(Tw(l, m, p)\bar{T}w(l', m', p'))$$

$$= \int E(w(\hat{n}, p)\bar{w}(\hat{n}', p'))Y_{lm}(\hat{n})\bar{Y}_{l'm'}(\hat{n}')d\Omega(\hat{n})d\Omega(\hat{n}')$$

$$= \sigma^2\delta(p - p')\int Y_{lm}(\hat{n})Y_{l'm'}(\hat{n})d\Omega(\hat{n})$$

$$= \sigma^2 \delta u' \delta_{mm'} \delta(p - p')$$

We can write our noisy model for the spherical harmonic transform of the Radon transform of the original and rotated object as

$$\mathbf{TF}(l, p) = \mathbf{TRf}(l, p) + \mathbf{Tw}(l, p) \quad ...(1)$$

$$\mathbf{TF}_K(l, p) = \bar{\pi}(K)\mathbf{TRf}(l, p) + \mathbf{Tw}_K(l, p) \quad ...(2)$$

where for $g(l, m, p)$, we define

$$\mathbf{g}(l, p) = \left(\left(g(l, m, p)\right)\right)_{|m| \le l} \in \mathbb{C}^{2l+1}$$

The negative log likelihood function of K, $\mathbf{TRf}(l, p)$ is thus given by

$$\sigma^2 E(K, \mathbf{TRf})$$

$$= \sum_{l, p} \left\| \mathbf{TF}(l, p) - \mathbf{TRf}(l, p) \right\|^2$$

$$+ \sum_{l, p} \left\| \mathbf{TF}_K(l, p) - \bar{\pi}(K)\mathbf{TRf}(l, p) \right\|^2.$$

Setting the derivative of E w. r. t. $\overline{TRf}(l, p)$ at $[TRf(l, p)]_{ML}$ to zero gives

$$TF(l, p) - [TRf(l, p)]_{ML} - \pi(K)^T \left(TF_K(l, p) - \bar{\pi}(K)[TRf(l, p)]\right)_{ML} = 0$$

❖❖❖❖❖

[40] Recursive least squares lattice algorithm.

[*a*] Time updates: Let $X[n] \in \mathbb{R}^{M \times p}$, $y[n] \in \mathbb{R}^M$ and $h[N] \in \mathbb{R}^p$ be such that

$$\hat{h}[N] = \operatorname{argmin}_h \sum_{n=0}^{N} \lambda^{N-n} \left\| y[n] - X[n]h \right\|^2$$

It is clear that

$$\hat{h}[N] = \left(\sum_{n=0}^{N} \lambda^{N-n} X[n]^T X[n]\right)^{-1} \left(\sum_{n=0}^{N} \lambda^{N-n} X[n]^T y[n]\right)$$

To cast this in recursive form define

$$R[N] = \sum_{n=0}^{N} \lambda^{N-n} X[n]^T X[n]$$

$$r[N] = \sum_{n=0}^{N} \lambda^{N-n} X[n]^T y[n]$$

Then,

$$R[N+1] = \lambda R[N] + X[N+1]^T X[N+1]$$

from which it follows that

$$R[N+1]^{-1} = \lambda^{-1}R[N]^{-1} - \lambda^{-2}R[N]^{-1}X[N+1]^T$$

$$\left(I + \lambda^{-1}X[N+1]R[N]^{-1}X[N+1]^T\right)^{-1} X[N+1]R[N]^{-1}$$

Further,

$$r[N+1] = \lambda r[N] + X[N+1]^T y[N+1]$$

It follows that

$$h[N+1] = h[N] + \lambda^{-1}R[N]^{-1}X[N+1]^T y[N+1]$$

$$-\lambda^{-1}R[N]^{-1}X[N+1]^T(I + \lambda^{-1}X[N+1]R[N]^{-1}X[N+1]^T)^{-1}$$

$$X[N+1]h[N]$$

$$-\lambda^{-2}R[N]^{-1}X[N+1]^T(I + \lambda^{-1}X[N+1]R[N]^{-1}X[N+1]^T)^{-1}$$

$$X[N+1]R[N]^{-1}X[N+1]^T y[N+1]$$

$$= h[N] + \lambda^{-1}R[N]^{-1}X[N+1]^T(I+\lambda^{-1}X[N+1]R[N]^{-1}X[N+1]^T)^{-1}$$

$$(y[N+1]-X[N+1]h[N])$$

This is the fundamental time update equation.

[*b*] Order updates: Let $x[n], y[n] \in \mathbb{R}^M$ and let $H[n, k]$ be $M \times M$ matrix valued filter coefficients determined so that

$$= \sum_{n=p}^{N} \lambda^{N-1} \left\| y[n] - \sum_{k=0}^{p} H[N,k]x[n-k] \right\|^2$$

is a minimum. Thus from the orthogonality principle,

$$\sum_{k=0}^{p} H[N,k]\left(\sum_{n=p}^{N} \lambda^{N-n} x[n-k]x[n-m]^T \right)$$

$$= \sum_{n=p}^{N} \lambda^{N-n} y[n]x[n-m]^T, \; 0 \le m \le p$$

Define the matrices

$$R_x[k, m, N] = \sum_{n=p}^{N} \lambda^{N-n} x[n-k]x[n-m]^T \in \mathbb{R}^{M \times M}$$

and let

$$R_x[N] = \left(\left(R_x[k, m, N]\right)\right)_{0\leq k,\, m\leq p} \in \mathbb{R}^{M(p+1)\times M(p+1)},$$

$$r_{yx}[m, N] = \sum_{n=p}^{N} \lambda^{N-n} y[n] x[n-m]^T \in \mathbb{R}^{M\times M}, 0 \leq m \leq p.$$

$$r_{yx}[N] = \left(\left(r_{yx}[m, N]\right)\right)_{m=0}^{p} \in \mathbb{R}^{M(p+1)\times M}$$

$$H[N] = \left(\left(H[N, k]^T\right)\right)_{k=0}^{p^T}$$

$$= \left[H[N, 0], H[N, 1], \ldots, H[N, p]\right] \in \mathbb{R}^{M\times M(p+1)}$$

Then, the optimal filter is given by

$$H[N] = y_m[N] R_x[N]^{-1}$$

To cast these equations in the form of orthogonal projections, we define the Hilbert space $\mathcal{H}[n, p, M]$ to be all matrices in $\mathbb{R}^{M\times N-p+1}$ of the form

$$\sum_{k=0}^{p} A[k]\left[x[p-k], x[p+1-k], \ldots, x[N-k]\right]$$

with norm given by

$$\left\|\xi[p], \ldots, \xi[N]\right\|^2 = \sum_{n=p}^{N} \left\|\xi[n]\right\|^2$$

❖❖❖❖❖

Quantum Field Theory, Gravitational and Electromagnetic Waves in Cosmology, Electromagnetics and Fluids, Quantum Gate Design

[41] Propagator computation for the Klein-Gordon field, electromagnetic field and Dirac field. The Dirac equation is given by

$$= \left(i\gamma^{\mu}\partial_{\mu} - m\right)\psi = 0$$

In the presence of an electromagnetic field, it gets modified to

$$(\gamma^{\mu}(i\partial_{\mu} + eA_{\mu}) - m)\psi = 0$$

In Dirac γ matrices (γ^{μ}) satisfy the anti-commutation relations

$$\gamma^{\mu}\gamma^{\nu} + \gamma^{\nu}\gamma^{\mu} = 2\eta^{\mu\nu}$$

where

$$((\eta^{\mu\nu})) = diag[1, -1, -1, -1]$$

We take

$$\gamma^{r} = \begin{pmatrix} 0 & \sigma_r \\ -\sigma_r & 0 \end{pmatrix}, r = 1, 2, 3$$

where the $\sigma'_r s$ are the Pauli spin matrices. Clearly since

$$\sigma_r\sigma_s + \sigma_s\sigma_r = 2\delta_{rs}$$

and

$$\sigma_r^{*} = \sigma_r$$

it follows that

$$\gamma^{r}\gamma^{s} + \gamma^{s}\gamma^{r} = -2\delta_{rs}$$

and

$$\gamma^{r*} = -\gamma^{r},$$

In our notation, Greek symbols like μ, ν, σ, ρ, α, β take values 0, 1, 2, 3, while Roman characters like r, s, n, m, l take values 1, 2, 3. Let

$$\gamma^{0} = \begin{pmatrix} I & 0 \\ 0 & -I \end{pmatrix}$$

Then,

$$\gamma^{0*} = \gamma^{0}, \gamma^{0}\gamma^{r} + \gamma^{r}\gamma^{0} = 0$$

Thus our construction of the Dirac gamma matrices satisfies all the required conditions.

Consider now the dirac equation in an external ex field. We derive from this equation, the equation

$$\left(\gamma^{\mu}\left(i\partial_{\mu}+eA_{\mu}\right)+m\right).\left(\gamma^{v}\left(i\partial_{v}+eA_{v}\right)-m\right)\psi=0$$

which expands to

$$\left(\gamma^{\mu}\gamma^{v}\left(i\partial_{\mu}+eA_{\mu}\right)\left(i\partial_{v}+eA_{v}\right)-m^{2}\right)\psi=0$$

Now,

$$\gamma^{\mu}\gamma^{\nu}=\frac{1}{2}\left\{\gamma^{\mu},\gamma^{v}\right\}+\frac{1}{2}\left[\gamma^{\mu},\gamma^{v}\right]$$

$$=\eta^{\mu\nu}+\sum{}^{muv}$$

where

$$\Sigma^{\mu\nu}=\frac{1}{2}\left[\gamma^{\mu},\gamma^{\nu}\right]$$

and we thus get

$$\left[\left(i\partial_{\mu}+eA_{\mu}\right)\left(i\partial^{\mu}+eA^{\mu}\right)-m^{2}\right]\psi+\sum{}^{\mu v}\left(i\partial_{\mu}+eA_{\mu}\right)\left(i\partial_{v}+eA_{v}\right)\psi=0$$

Since $= \Sigma^{\mu\nu *} - \Sigma^{\nu\mu}$, it follows that

$$\sum{}^{\mu v}\left(i\partial_{\mu}+eA_{\mu}\right)\left(i\partial_{v}+eA_{v}\right)\psi$$

$$=\frac{1}{2}\sum{}^{\mu v}\left[i\partial_{\mu}+eA_{\mu},i\partial_{v}+eA_{v}\right]\psi$$

$$=(ie/2)\sum{}^{\mu v}\left(A_{\nu,\mu}-A_{\mu,nu}\right)\psi=(ie/2)\sum{}^{\mu v}F_{\mu\nu}\psi$$

Thus as a consequence of the Dirac equation in an external em field, we get the modified Klein-Gordon equation

$$\left[\left(i\partial_{\mu}+eA_{\mu}\right)\left(i\partial^{\mu}+eA^{\mu}\right)-m^{2}\right]\psi+ie\sum{}^{\mu v}F_{muv}=0$$

Thus last term in this expression corresponds to interaction of the spin of the electron with the em field.

Derivation of the Dirac equation from a Lagrangian density: Let $\bar{\psi}=\psi^{*}\gamma^{0}$ and consider

$$L\left(\psi,\bar{\psi},\psi_{,\mu}\right)=\bar{\psi}\gamma^{\mu}\left(i\partial_{\mu}+eA_{\mu}\right)\psi-m^{2}\bar{\psi}\psi$$

It is easy to see using

$$\gamma^{0*} = \gamma^0, \gamma^{r*} = \gamma^r, \gamma^0\gamma^r = -\gamma^r\gamma^0$$

and integration by parts that

$$S[\psi, \bar{\psi}] = \int L(\psi, \bar{\psi})d^4x$$

is real. The equations of motion for ψ that follow from the variational principal

$$\delta S = 0$$

and firstly

$$\left[\gamma^\mu \left(i\partial_\mu + eA_\mu\right) - m\right]\psi = 0$$

i.e., the Dirac equation (this is a consequence of $\delta_{\bar{\psi}} S = 0$ and secondly,

$$-i\gamma^\mu \partial_\mu \bar{\psi}^T + eA_\mu \gamma^{\mu T} \bar{\psi}^T - m^2 \bar{\psi}^T = 0$$

this is a consequence of $\delta_\psi S = 0$. The second equation is equivalent to the Dirac equation. We leave it as an exercise to the reader to verify this. Consider now the Dirac current

$$J^\mu = -e\bar{\psi}\gamma^\mu\psi = -e\psi^*\gamma^0\gamma^\mu\psi$$

We leave it as an exercise to the reader to show using the Dirac equation in an em field that the charge is conserved, *i.e.*,

$$\partial_\mu J^\mu = 0$$

❖❖❖❖❖

[42] Cosmological metric (Robertson-Walker metric) and the analysis of gravitationsl wave propagation and Maxwell's equation in this metric.

[*a*] Spatial line element for a spherical universe:

$$dl^2 = dx^2 + dy^2 + dz^2 + du^2$$

where

$$x^2 + y^2 + z^2 + u^2 = S^2$$

Writing

$$r^2 = x^2 + y^2 + z^2$$

we get

$$u = \sqrt{S^2 - r^2}$$

and hence

$$du = -\frac{rdr}{\sqrt{S^2 - r^2}}$$

Thus,

$$dl^2 = dr^2 + r^2\left(d\theta^2 + \sin^2(\theta)d\phi^2\right) + \frac{r^2dr^2}{S^2 - r^2}$$

$$= \frac{S^2dr^2}{S^2 - r^2} + r^2\left(d\theta^2 + \sin^2(\theta)d\phi^2\right)$$

Change to commoving coordinates: $S = S(t)$, $r = S(t)r1$. Thus

$$dl^2 = \frac{S^2dr_1^2}{S^2 - r_1^2} + S^2dr_1^2\left(d\theta^2 + \sin^2(\theta)d\phi^2\right)$$

We write r in place of the comoving radial coordinate r_1 so that the entire space-time metric assumes the form

$$d\tau^2 = dt^2 - \frac{S^2(t)}{1 - kr^2} - S^2(t)r^2\left(d\theta^2 + \sin^2(\theta)d\theta^2\right) \qquad ...(1)$$

Here the curvature factor k has been incorporated. This corresponds to the spatial part of the universe defined by the 3 – D surface

$$kr^2 + u^2 = S^2, r^2 = x^2 + y^2 + z^2$$

where $k = 0$ for a flat universe, $k = 1$ for a spherical universe and $k = -1$ for a hyperbolic universe.

Exercise: Verify that the metric defined in (1) gives (r, θ, ϕ) as comoving coordinates in the sense that the world line $t \rightarrow (t, r_0, q_0, \phi_0)$ is a geodesic for any constant r_0, θ_0, ϕ_0.

The Einstein field equations for the universe (1): Matter moves along geodesics and since our coordinate system is comoving, we have the velocity field of the matter given by

$$v^\mu = (1, 0, 0, 0)$$

Then the energy momentum tensor of our matter distribution given by

$$T^{\mu\nu} = \left(\rho(t) + p(t)\right)v^\mu v^\nu - p(t)g^{\mu\nu}$$

has the form

$$\left(\left(T^{\mu\nu}\right)\right) = diag\left[\rho, -pg^{11}, -pg^{22}, -pg^{33}\right]$$

where

$$g_{11}=1/g_{11} = -1/\left(S^2(t)f(r)\right), g^{22} = -1/\left(S^2(t)r^2\right), g^{33}$$
$$= -1/\left(S^2(t)r^2\sin^2(\theta)\right)$$

where

$$f(r) = \frac{1}{1-kr^2}$$

❖❖❖❖❖

[43] Filtering theory applied to robot trajectory measurement from noisy measurements of the position of an object by a camera on the robot with the image recorded on a flat curved screen.

The robot camera position is given by

$$\mathbf{r}_1(t) = \left(a + l_1.\cos(q_1(t)) + l_2.\cos(q_2(t))\right)\hat{x}$$
$$+\left(l_1.\sin(q_1(t)) + l_2.\sin(q_2(t))\right)\hat{z}$$

where the angles $q(t) = (q_1(t), q_2(t))^T$ of the two links satisy the dynamical equations

$$dq(t) = (t)dt,$$
$$d\omega(t) = F(t, q(t), \omega(t))dt + \sigma G(q(t))dB(t)$$

with $$F(t, q, \omega) = M(q)^{-1}(-N(q, \omega) + \tau(t))$$

$\tau(t)$ is the known applied machine torque and $N(q, \omega)$ is the contribution to the torque coming from centrifugal, coriolis, gravitational and frictional forces. Here,

$$G(q) = M(q)^{-1}$$

and $B(t)$ is standard $\mathbb{R}^2$ valued Brownian motion. σ is a positive constant that determines the intensity of the noisy torque. To see how the above equation is derived, we start with the standard Lagrangian for the two link robot:

$$L(q, q', t) = \frac{1}{2}q'^T M(q)^{-1} q' - V(q) + \tau(t)^T q$$

where $M(q) = M(q_2)$, the mass moment of inertia matrix has the form

$$M(q) = \begin{pmatrix} a + b.\cos(q_2) & c + d.\cos(q_2) \\ c + d.\cos(q_2) & f \end{pmatrix}$$

The constants a, b, c, d, f are expressed in terms of the masses and lengths of the two links. These constants satisfy the conditions $Tr(M(q_2)) > 0$ and $det\,(M(q_2)) > 0 \forall q_2\ [0, 2\pi]$. $\tau(t)$ is the applied torque and $V(q)$ the gravitational potential energy of the two links system. The Euler-Lagrange equations

$$\frac{d}{dt}\frac{\partial L}{\partial q'} = \frac{\partial L}{\partial q}$$

give

$$\frac{d}{dt}\left(M(q)q'\right)+V'(q)-\tau(t)\ -\frac{1}{2}\left(q'^T\otimes I_2\right)M'(q)^T q' = 0$$

which further expands to

$$M(q)q''+M'(q)(q'\otimes q')+V'(q)-\tau(t)\ -\frac{1}{2}\left(q'^T\otimes I_2\right)M'(q)^T q' = 0$$

Here,

$$M'(q) = \left[M_{,q_1}\middle|M_{,q_2}\right] = \left[0_2\middle|M_{,q_2}\right]\in\mathbb{R}^{2\times 4}$$

Thus, we can identify

$$N(q,q') = V'(q)+M'(q)(q'\otimes q')-\frac{1}{2}\left(q'^T\otimes I_2\right)M'(q)^T q'$$

Here friction has been neglected. The screen is assuumed to be the $y-z$ plane. Thus, if the position of the object is $r_0=(x_0, y_0, z_0)$, the position of its image on the screen is given by $(0,\xi_1,\xi_2)$ where

$$\begin{aligned}(0,\xi_1,\xi_2) &= r_0+\lambda(r_1-r_0)\\ &= (x_0,y_0,z_0)+\lambda(x_1-x_0,-y_0,z_1-z_0)\\ &= \left(x_0+\lambda(x_1-x_0),(1-\lambda)y_0,z_0+\lambda(z_1-z_0)\right)\end{aligned}$$

and hence,

$$\lambda = \frac{x_0}{x_0-x_1}$$

so that

$$\xi_1 = (1-\lambda)y_0 = \frac{x_1y_0}{x_1-x_0}$$

$$\begin{aligned}\xi_2 &= z_0+\lambda(z_1-z_0) = z_0-\frac{x_0(z_1-z_0)}{x_1-x_0}\\ &= \frac{x_1z_0-x_0z_1}{x_1-x_0}\end{aligned}$$

x_1, z_1 are functions of (q_1,q_2):

$$\begin{aligned}x_1 &= l_1.\sin(q_1)+l_2.\sin(q_2)\\ z_1 &= l_1.\sin(q_1)+l_2.\sin(q_2)\end{aligned}$$

and hence ξ_1, ξ_2 are also expressible as functions of (q_1, q_2). We denote these functions by $h_1(q_1, q_2)$ and $h_2(q_1, q_2)$ respectively. In the presence of measurement noise, these, measurements are given by

$$dz_1(t) = h_1(q)dt + \sigma_1 dW_1(t), dz_2(t) = h_2(q)dt + \sigma_2 dB_2(t)$$

where $W_1(.)$, $W_2(.)$ are standard one dimensional independent Brownian motions independent of $B(.)$. The problem is to obtain the estimates $\hat{q}(t) = E(q(t)|Z_t)$ and $\hat{\omega}(t) = \mathbb{E}(\omega(t)Z_t)$ where

$$Z_t = \sigma(z_1(s), z_2(s): s \le t)$$

is the σ field generated by the measurements upto time t.

The general theory of Kushner filtering: Consider the stochastic dynamical system (in the Ito sense)

$$dx(t) = \mu(t, x(t))dt + g(t, x(t))dB(t)$$

where $x(t) \in \mathbb{R}^n$, $B(t) \in \mathbb{R}^d$ and B is standard d-dimensional Brownian motion. The measurement model is

$$dz(t) = h(t, x(t))\,dt + dV(t)$$

where $V(.)$ is p-dimensional Brownian motion with

$$dV(t)dV(t)^T = R(t)dt$$

and V is assumed to be independent of B. Let

$$Z_t = (z(s): s \le t)$$

we have

$$p(x(t+dt)|Z_{t+dt}) = \int p(x(t+dt)|x(t)p(x(t)|Z_{t+dt}))dx(t)$$

This is because conditioned on $x(t)$, $x(t + dt)$ is independent of Z_{t+dt} Thus for any function f on $\mathbb{R}^n$, we have

$$\begin{aligned}\mathbb{E}(f(x(t+dt)Z_{t+dt}) &= \int \mathbb{E}(f(x(t+dt)|x(t)))p(x(t)|Z_{t+dt})dx(t) \\ &= \int [f(x(t)) + dt.L_t f(x(t))]p(x(t)|Z_{t+dt})\,dx(t)\end{aligned}$$

where

$$L_t = -\mu(t, x)^T \nabla_x + \frac{1}{2}Tr(b(t, x)\nabla_x \nabla_x^T(.))$$

is the generator of the diffusion process $x(t)$. Now

$$\begin{aligned} p\left(x(t)\middle|Z(t+dt)\right) &= p\left(x(t), Z_t, dz(t)\right)/p\left(dz(t), Z_t\right) \\ &= p\left(dz(t)\middle|x(t)\right)p\left(x(t)\middle|Z_t\right)/p\left(dz(t)\middle|Z_t\right) \\ &= C.\exp\left(-\frac{1}{2}\left(dz(t)-h(t,x(t))dt\right)^T \right. \\ &\quad \left. \left(R(t)dt\right)^{-1}\left(dz(t)-h(t,x(t))dt\right)/p\left(dz(t)\middle|Z_t\right)\right) \end{aligned}$$

and hence we deduce

$$\mathbb{E}\left(f\left(x(t+dt)\right)\middle|Z_{t+dt}\right)$$

$$= \frac{\int\left[f(x(t))+dt.L_t f(x(t))\right]\exp\begin{pmatrix}-\frac{1}{2}\left(dz(t)-h(t,x(t))\right)^T \\ \left(R(t)dt\right)^{-1}\left(dz(t)-h(t,x(t))\right)\end{pmatrix}p\left(x(t)\middle|Z_t\right)dx(t)}{\exp\left(-\frac{1}{2}\left(dt(t)-h(h,x(t))\right)^T\left(R(t)dt\right)^{-1}\left(dz(t)-h(t,x(t))dt\right)\right)p\left(x(t)\middle|Z_t\right)dx(t)}$$

Equivalently, defining $\pi_t(f) = \mathbb{E}\left(f(x(t))\middle|Z_t\right)$

we can write $\quad \pi_{t+dt}(f) = \dfrac{\sigma_{t+dt}(f)}{\sigma_{t+dt}(1)}$

where

$$\begin{aligned} \sigma_{t+dt}(f) &= \int\left[f(x(t))+dt.L_t f(x(t))\right]\exp\left(dz(t)^T R(t)^{-1}h(t,x(t))\right. \\ &\quad \left.-\frac{1}{2}h(h,x(t))^T R(t)^{-1}h(t,x(t)dt)\,p\left(x(t)\middle|Z_t\right)dx(t)\right. \\ &= \int\left[f(x(t))+dt.L_t f(x(t))\right] \\ &\quad \left[1+dz(t)^T R(t)^{-1}h(t,x(t))\right]p\left(x(t)\middle|Z_t\right)d_x(t) \\ &= \pi_t(f)+dt.\pi_t(L_t f)+dz(t)^T R(t)^{-1}\pi_t(fh_t) \end{aligned}$$

where we have made use of Ito's formula in the form

$$dz(t)dz\,(t)^T = R\,(t)dt$$

and the notation $h_t(x) = h(t, x)$

has been used. Since $\pi_t\,(t) = 1$, it follows that

$$\pi_{t+dt}(f) = \frac{\pi_t(f) + dt\pi_t(L_t f) + dz^T R^{-1}\pi_t(fh_t)}{1 + fz^T R^{-1}\pi_t(h_t)}$$

$$= \left(\pi_t(f) + dt\pi_t(L_t f) + dz^T R^{-1}\pi_t(fh_t)\right)$$

$$\left(1 - dz^T R^{-1}\pi_t(h_t) + \pi_t(ht)^T R^{-1}\pi_t(ht)\right)dt\Big)$$

$$= \pi_t(f) + dt.\pi_t(L_t f) + dz^T R^{-1}\pi_t(fh_t)$$

$$-dt.\pi_t(fh_t)^T R^{-1}\pi_t(h_t) + \pi_t(ht)^T R^{-1}\pi_t(h_t)\pi_t(f)dt$$

$$= +\left(dz - \pi_t(h_t)dt\right)^T R^{-1}\left(\pi_t(fh_t) - \pi_t(f)\pi_t(h_t)\right)$$

Thus, we obtain Kushner's equation for the evolution of the conditional mean of an arbitrary function of the state:

$$d\pi_t(f) = \pi_t(L_t f) + \left(dz(t) - \pi_t(h_t)dt\right)^T \left(\pi_t(fh_t) - \pi_t(f)\pi_t(h_t)\right)$$

❖❖❖❖❖

[44] Propagation of em waves in an inhomogeneous and anisotropic medium. The permittivity tensor has the form

$$\varepsilon(x, y, w) = \begin{pmatrix} \varepsilon_{11}(x,y,w) & \varepsilon_{12}(x,y,w) & \varepsilon_{13}(x,y,w) \\ \varepsilon_{21}(x,y,w) & \varepsilon_{22}(x,y,w) & \varepsilon_{23}(x,y,w) \\ \varepsilon_{31}(x,y,w) & \varepsilon_{32}(x,y,w) & \varepsilon_{3}(x,y,w) \end{pmatrix}$$

and likewise, the permeability tensor has the form

$$\mu(x, y, w) = \begin{pmatrix} \mu_{11}(x,y,w) & \mu_{12}(x,y,w) & \mu_{13}(x,y,w) \\ \mu_{21}(x,y,w) & \mu_{22}(x,y,w) & mu_{23}(x,y,w) \\ \mu_{31}(x,y,w) & \mu_{32}(x,y,w) & \mu_{3}(x,y,w) \end{pmatrix}$$

We can abbreviate these expressions to

$$= \begin{pmatrix} \varepsilon_{\perp} & \varepsilon_{1:2,3} \\ \varepsilon_{3,1:2} & \varepsilon_{3} \end{pmatrix}$$

and likewise for μ. The Maxwell curl equations with a z-dependence of $\exp(-\gamma z)$ assume the form

$$\left(\nabla_{\perp}-\gamma\hat{z}\right)\times\left(E_{\perp}+E_{z}\hat{z}\right) = j\omega\left(\mu_{\perp}H_{\perp}+\mu_{1:2,3}H_{z}+\left(\mu_{3,1:2}H_{\perp}+\mu_{3}H_{z}\right)\hat{z}\right) \quad ...(1)$$

and likewise for H. The equation for H can be obtained by applying duality to (1) From (1). We get on equation the transverse and longitudinal components,

$$\nabla_{\perp}E_{z}\times\hat{z}-\gamma\hat{z}\times E_{\perp} = -j\omega\left(\mu_{\perp}H_{\perp}+\mu_{1:2,3}H_{z}\right) \quad ...(2a)$$

$$\nabla_{\perp}\times E_{\perp} = -j\omega\left(\mu_{3}H_{z}+\mu_{3,1:2}H_{\perp}\right)\hat{z} \quad ...(2b)$$

and by duality,

$$\nabla_{\perp}H_{z}\times\hat{z}-\gamma\hat{z}\times H_{\perp} = j\omega\left(\varepsilon_{\perp}E_{\perp}+\varepsilon_{1:2,3}E_{z}\right) \quad ...(3a)$$

$$\nabla_{\perp}\times \mathrm{E}_{\perp} = j\omega\left(\varepsilon_{3}H_{z}+\varepsilon_{3,1:2}H_{\perp}\right)\hat{z} \quad ...(3b)$$

For any transverse vector $F = \left(F_x, F_y\right)^T$, we can writer the vector $\hat{z}\times F = \left(-F_y, F_x\right)^T$ as JF where

$$J = \begin{pmatrix} 0 & -1 \\ 1 & 0 \end{pmatrix}$$

Thus ($2a$) and ($3a$) can be expressed as

$$J\nabla_{\perp}E_{z}+\gamma JE_{\perp} = j\omega\mu_{\perp}H_{\perp}+j\omega\mu_{1:2,3}H_{z}, \quad ...(4)$$

$$J\nabla_{\perp}H_{z}+\gamma JH_{\perp} = -j\omega\varepsilon_{\perp}E_{\perp}-j\omega\mu\varepsilon_{1:2,3}E_{z} \quad .....(5)$$

(4) and (5) can be further arranged as

$$\begin{pmatrix} \gamma J & -j\omega\mu \\ j\omega\varepsilon_{\perp}\gamma J & \end{pmatrix}\begin{pmatrix} E_{\perp} \\ H_{\perp} \end{pmatrix}$$

$$\begin{pmatrix} j\omega\mu_{1:2,3}H_{z} & -J\nabla_{\perp}E_{z} \\ -j\omega\varepsilon_{1:2,3}E_{z} & -J\nabla_{\perp}H_{z} \end{pmatrix}$$

❖❖❖❖❖

[45] MHD equations after spatial discretization: The continuous time MHD equations are the Navier-Stokes equation, the equation of continuity and the Maxwell curl equations:

$$v_{,t}+\left(v,\nabla\right)v = -\nabla p/\rho+(\eta/\rho)\nabla^{2}v+(\sigma/\rho)(E+v\times B)\times B \quad ...(1)$$

$$\nabla.v = 0, \quad ...(2)$$

$$\nabla\times E = -\mathrm{B}_{,t}, \nabla\times B = \mu\sigma\,(E+v\times B)+c^{-2}E_{,t} \quad ...(3)$$

The continuity equation means that we can write

$$v = \nabla \times \psi \quad ...(4)$$

where ψ is the stream vector field. Without loss of generality, we may assume that

$$\nabla.\psi = 0 \quad ...(5)$$

Then, the vorticity is given by

$$\Omega = \nabla \times v = -\nabla^2 \psi \quad ...(6)$$

The Navier-Stokes equation can be expressed as

$$v_{,t} + \Omega \times v + \nabla v^2/2 = -\nabla p/\rho + (\eta/\rho)\nabla^2 v + (\sigma/\rho)(E + v \times B) \times B \quad ...(7)$$

Taking the curl of this equation gives

$$\Omega_{,t} + \nabla \times (\Omega \times v) = (\eta/\rho)\nabla^2 \Omega + (\sigma/\rho)\big(\nabla(E \times B) + \nabla \times ((v \times B) \times B)\big) \quad ...(8)$$

After substituting for Ω from (6) into (8), we

$$\nabla^2_{\psi,t} + \nabla \times \left(\nabla^2 \psi \times (\nabla \times \psi)\right) - (\eta/\rho)\left(\nabla^2\right)^2 \psi$$

$$+(\sigma/\rho)\nabla \times (E \times B) + (\sigma/\rho)\nabla \times \left((\nabla \times \psi) \times B\right) \times B\Big) = 0 \quad ...(9)$$

After spatial discretization, the fields $\psi(t, \mathbf{r})$, $\mathbf{E}(t, \mathbf{r})$ and $\mathbf{B}(t, \mathbf{r})$ become $\mathbb{R}^N$ values functions of time *i.e.*, $\psi(t)$, $E(t)$, $B(t) \in \mathbb{R}^N$ where N is the number of spatial pixels. A linear differential or integral operator L acting on the spatial part of a vector field $F(t, \mathbf{r})$, becomes on spatial discretization a vector $D(L)\ F(t) \in \mathbb{R}^M$ where $M = N$ if LF is also a vector field and LF is a scalar field, then $M = N/3$. Here, $D(L)$ is a matrix. Thus, (9) can be expressed as after applying $(\nabla^2)^{-1}$ as

$$\psi'(t) = A_1(\psi(t) \otimes \psi(t)) + (\eta/\rho)A_2\psi(t) + (\sigma/\rho)A_3(E(t) \otimes B(t))$$
$$+(\sigma/\rho)A_4(B(t) \otimes B(t) \otimes \psi(t)) \quad ...(10)$$

Likewise, the Maxwell curl equations on discretization have the form

$$B'(t) = A_5 E(t), ...(11) E'(t) = c^2 A_6 B(t) + (\sigma/\varepsilon)A_\tau(B(t) \otimes \psi(t)) \quad ...(12)$$

Eqns. (10), (11), (12) are the fundamental discretized MHD equations when the electric field cannot be ignored. Here A_k, $1 \le k \le 7$ are matrices of the appropriate size. These equations can be solved approximately using perturbation theory assuming the nonlinear terms in the state vector

$$\xi(t) = \left(\psi(t)^T, E(t)^T, B(t)^T\right) \in \mathbb{R}^{3N}$$

to be small

❖❖❖❖❖

[46] Problems in signal processing and control:

[1] Derive using the Bellman-Hamilton-Jacobe dynamic programming method the optimal control input $u(t)$ as a function of $(t, x(t))$ for which

$$\mathbb{E}\int_0^T L(t, x(t), u(t))dt$$

is a minimum where $x(t)$ is a Markov process with infinitesimal generator $K(t, u(t))$, *i.e.*,

$$\mathbb{E}(\phi(x(t+h))|x(t),u(t)) = \phi(x(t)) + h(K(t, u(t))\phi)(x(t)) + o(h)$$

[2] Consider a quantum channel which takes as input the state ρ(0) and outputs the state

$$\rho(T) = \sum_{k=1}^{r}\int_0^T \left(F_k(t)\rho(0)G_k(t)^* + G_k(t)\rho(0)F_k(t)^*\right)dt$$

where $F_k(t)$, $G_k(t)$ are linear operators in the Hilbert space of pure states satisfying the CPTP (Completely positive trace preserving) condition $\rho(0) \geq 0$ implies $\rho(t) \geq 0$ and $Tr(\rho(T)) = 1$ when $Tr(\rho(0)) = 1$. The latter condition is the same as

$$\sum_{k=1}^{r}\int_0^T \left(G_k(t)^* F_k(t) + F_k(t)^* G_k(t)\right)dt = I$$

Suppose now that $F_k(t) = Fk(t, \theta)$, $G_k(t) = G_k(t, \theta)$ depend on an unknown parameter vector θ to be estimated from noisy measurements ξ_k of the averages $Tr(\rho(T)X_k)$ of observables X_k, $k = 1,2,...N$. Then assuming N sufficiently large, first compute the optimum input state $\rho(0) = \rho(0, \theta)$ for which

$$\sum_{k=1}^{r}\left(\xi_k - \sum_{k=1}^{r}\int_0^T\left(Tr\left(F_m(t,\theta)\rho(0)G_m(t,\theta)^* X_k\right)\right)\right.$$
$$\left.+Tr\left(G_m(t,\theta)\rho(0)F_m(t,\theta)^* X_k\right)dt^2\right)$$

is a minimum. Then substitute the optimum ρ(0, θ) insto this error energy and design a gradient descent algorithm for minimizing this energy w.r.t. θ. This problem thus solves the joint initial state and channel parameter estimation problem in the quantum theory.

Nonlinear systems, example:

❖❖❖❖❖

[47] Consider a nonlinear system defined by the difference equation

$$y(n) = \sum_{k=1}^{M} a(k)\psi_k\left(y(n-1),...,y(n-p),x(n),...,x(n-q)\right)+\delta.v(n)$$

where ψ_k: $\mathbb{R}^{q+q+1} \rightarrow \mathbb{R}$ are given functions and $a(k)$ are the parameters to be estimated based on measurements of the non-random input $x(n)$ and the random output $y(n)|v(n)$ is white Gaussian noise with zero mean and variance σ^2. For estimating the parameters $a(k)$, introduce a forgetting factor $0 < \lambda \leq 1$ and compute the estimate of $a(k)$ as

$$\hat{a}_N = \arg\min_a \sum_{n=0}^{N} \lambda^{N-n}\left(y(n)-\psi\left(y_p(n), x_q(n)\right)^T a\right)^2$$

where

$$y_p(n) = (y(n-1),...,y(n-p))^T, x_q(n) = (x(n),....,x(n-q))^T$$

$$\psi\left(y_p(n), x_q(n)\right) = \left(\left(\psi_k\left(y_p(n), x_q(n)\right)\right)\right)_{k=1}^{M} \in \mathbb{R}^M$$

After computing this estimate, develop a recursive least square method for computing $\hat{a}_{N+1}$ in terms of $\hat{a}_N$ and the newly arrived data $y(N+1)$, $x(N+1)$. Finally obtain approximate values of $E\left(\left(\hat{a}_N - a\right)\left(\hat{a}_N - a\right)^T\right)$ by calculating $y(n)$ as $y_0(n) = y_0(n)+\delta y_1(n)+O\left(\delta^2\right)$ so that $y_0(n)$ satisfies the noiseless difference equation

$$y_0(n) = \psi\left(y_{0p}(n), x_q(n)\right)^T a$$

and $y_1(n)$ is the perturbation to $y_0(n)$ caused by noise. It satisfies the linear difference equation

$$y_1(n) = \sum_{k=1}^{M} a_k \frac{\partial\psi_k\left(y_{0p}(n), x(n)\right)}{\partial y_p} y_{1p}(n)+v(n)$$

where

$$y_{0p}(n) = \left(y_0(n-1),..., y_0(n-p)\right)^T, y_{1p}(n)$$

$$= \left(y_1(n-1),..., y_1(n-p)\right)^T$$

Defining the state vector at time $n-1$

$$\xi[n-1] = (y_1(n-1),...,y_1(n-p)^T \qquad ...(1)$$

if follows that (1) can expressed as

$$\xi[n] = A[n]\,\xi\,[n-1] + bv[n] \quad ...(2)$$

where $A[n]$ is a $p \times p$ matrix expressible in terms of the parameter vector a and $y_0\,(n-1),...y_0(n-p), x(n),...,x(n-q)$. From (2), we easily derive

$$A[0]^{-1}...A[n]^{-1}\xi[n] = A[0]^{-1}...A[n-1]^{-1}[n-1] + A[0]^{-1}...A[n]^{-1}bv[n]$$

and summing this equation from $n = 1$ to $n = m$, we get

$$\xi[m] = \Phi[m,1]\xi[0]\sum_{k=1}^{m}\Phi[m,k+1]bv[k]$$

where

$$\Phi[n, m] = A[n]...A[m],\ n \geq m,\ \Phi[n, n+1] = I$$

Then we find that assuming $\xi[0]$ to be independent of $v[k]$, $k \geq 1$

$$\mathbb{E}\left(\xi[m]\xi[r]^T\right) = \Phi[m,1]\mathbb{E}\left(\xi[0]\xi[0]^T\right)\Phi[r,1]^T$$

$$+\sigma^2\sum_{k=1}^{\min(m,r)}\Phi[m,k+1]bb^T\Phi[r,k+1]$$

In particular, $\mathbb{E}\left(\xi[m]\xi[m]^T\right)$ increases with increasing m which means that the entropy of $\xi\,[m]$ also increases with increasing m (If X is a $N(\mu, R_X)$ random vector of size N, then the entropy of X is $-\mathbb{E}(\log(f(X)) = (N/2)\log(2\pi e) + \log(det(R_x))/2$. Here $f(X) = (2\pi)^{-N/2}\,det\,(R_x)^{-1/2}.\ \exp\left(-X^T R_X^{-1} X/2\right)$ is the density of X. The least squares estimate of a based on input output data collected upto time N is given by

$$a_N = \left[\sum_{n=0}^{N}\lambda^{N-n}\psi\left(y_p(n), x_q(n)\right)\psi\left(y_p(n), x_q(n)\right)^T\right]^{-1}$$

$$\left[\sum_{n=0}^{N}\lambda^{N-n}y[n]\psi\left(y_p(n), x_q(n)\right)\right]$$

We define

$$R_\psi[N] = \sum_{n=0}^{N}\lambda^{N-n}\psi\left(y_p(n), x_q(n)\right)\psi\left(n\right), x_q(n))^T,$$

$$R_{\psi y}[N]\sum_{n=0}^{N}\lambda^{N-n}\psi\left(y_p(n), x_q(n)\right)\psi(n)$$

We have

$$\psi(\psi_p(n), x_q(n)) = \psi\left(y_{0p}(n), x_q(n)\right) + \delta.A(n)y_{1p}(n) + O\left(\delta^2\right)$$

and hence, with the notation

$$\psi_0[n] = \psi(y_{0p}(n), x_q(n)),$$

we get

$$R_\psi[N] = R_0[N] + \delta.R_1[N] + O\left(\delta^2\right)$$

$$R_{\psi y}[N] = S_0[N] + \delta.S_1[N] + O\left(\delta^2\right)$$

where
$$R_0[N] = \sum_{n=0}^{N} \lambda^{N-n}\psi_0[n]\psi_0[n]^T$$

$$S_0[N] = \sum_{n=0}^{N} \lambda^{N-n}\psi_0[n]y_0[n]$$

and
$$R_1[N] = \sum_{n=0}^{N} \lambda^{N-n}\left(A[n]\xi[n]\psi_0[n]^T + \psi_0[n]\xi[n]^T A[n]^T\right)$$

and

$$S_1[N] = \sum_{n=0}^{N} \lambda^{N-n}y_0[n]A[n]\xi[n-1] + y_1[n]\psi_0[n]$$

We get

$$a_N = \left(R_0[N] + \delta.R_1[n]^{-1}\left(S_0[N] + \delta S_1[N]\right)\right) + O\left(\delta^2\right)$$

We leave it as an exercise to express a_N upto $O(\delta)$ terms and hence calculate its covariance matrix upto $O(\delta^2)$ terms.

❖❖❖❖❖

[48] Wavelet based parameter estimation of nonlinear system:
Let the nonlinear system be described by a potential

$$U(x, q) = U_0(x) + \delta \cdot \sum_{k=1}^{p} \theta_k U_k(x)$$

where δ is a small perturbation parameter. Here $x \in \mathbb{R}^N$ and this potential along with a velocity damping term $-\Gamma x'$ describes the motion of N particles moving in an externally applied force $f(t) \in \mathbb{R}^N$. The equations of motion are then

$$X_0''(t)\Gamma x'(t)U_0'\left(x(t)\right) + \delta\sum_{k=1}^{p} \theta_k U_k'(x) = f(t)$$

$\theta = (\theta_1, \ldots \theta_p)$ is an unknown parameter vector to be estimated from measurements on the system trajectory $x(t)$ over a given time interval. We write the solution to the above differential equation as

$$x(t) = x_0(t) + \delta.y(t) + O(\delta^2)$$

On substituting this into the equation of motion and equating coefficients of δ^θ and δ' respectively, we get

$$x_0''(t)\Gamma x_0'(t) + U_0'(x(t)) = f(t) \qquad ...(1)$$

$$y''(t) + \Gamma y'(t) + U_0'(x_0(t))y(t)$$

$$+\sum_k \theta_k U_k'(x_0(t)) = 0 \qquad ...(2)$$

Assume that (1) has been solved for $x_0(t)$. Then (2) can be expressed in state variable form as

$$\frac{d}{dt}\begin{pmatrix} y(t) \\ y'(t) \end{pmatrix} = \begin{pmatrix} 0 & I \\ -U_0'(x_0(t)) & -\Gamma \end{pmatrix} - \sum_k \theta_k \begin{pmatrix} 0 \\ U_k'(x_0(t)) \end{pmatrix}$$

It follows that with $\Phi(t, \tau) \in \mathbb{R}^{2N\times 2N}$ denoting the state transition matrix corresponding to the force matrix

$$A(t) = \begin{pmatrix} 0 & I \\ -U_0'(x_0(t)) & -\Gamma \end{pmatrix}$$

we have

$$\begin{pmatrix} y(t) \\ y'(t) \end{pmatrix} = -\sum_k \theta_k \int_0^t \Phi(t, \tau) \begin{pmatrix} 0 \\ U_k'(x_0(t)) \end{pmatrix}$$

from which we get

$$y(t) = -\sum_k \theta_k \int_0^t \left(Phi_{12}(t, \tau) U_k'(x_0(\tau))\right) d\tau$$

Talking the wavelet transform on both sides gives

$$W_{a,b}(y) = a^{-1/2} \int_{\mathbb{R}} y(t)\psi((t-b)/a)dt = -\sum_K \theta_k F(k, a, b)$$

where $F_k(a, b)$ is the wavelet transform of the process $-\int_0^T \phi_{12}(T, \tau) U_k'(x_0(\tau))d\tau$. We estimate the wavelet coefficients $W_{a,\,b}(y)$ for the scaling and translation variable (a, b) falling in a certain domain D chosen to get a good approximation of the signals involved with sufficient data compression (*i.e.*, D must not be too large in cardinality). Then, with $\tilde{W}_{a,b}$ denoting the estimated wavelet coefficient of $W_{a,\,b}(y)$ the estimates of the parameters $\{\theta_k\}$ are given by

$$\hat{\theta} = \text{argmin}_{\theta} \sum_{(a,b)\in D}\left(\widehat{W}_{a,b} - \sum_k \theta_k F(k,a,b)\right)^2$$

We can improve our accuracy by taking into account higher order perturbation terms in the solution. In general, we can then express our wavelet coefficients of the measured signal approximately as a multivariate polynomial in θ of degree $M \geq I$ and base our estimates of θ by matching this polynomial to the wavelet coefficients estimated from the measured data.

❖❖❖❖❖

[49] Large deviation methods in cosmological models:

We first consider Newtonian cosmology: Let $\rho(t)$ be the density of a sphere of radius $R(t)$ at time. A point mass on the spherical surface satisfies the equation of motion

$$R''(t) = -4\pi G\rho(t)R(t)/3$$

in the absence of pressure. We have with M denoting the total mass of the sphere,

$$M = 4\pi\rho(t)R^3(t)/3$$

and hence

$$R''(t) = -GM/R^2(t)$$

or equivalently

$$R^2(t)R''(t) = -K$$

where K is a positive constant. A small noisy perturbation to this equation arising from random pressure fluctuations can be studied as follows. Let $p(t)$ denote the pressure at time t within the sphere of radius $R(t)$. Then the total outward radial pressure force on a shell of radius $[r, r + dr]$ within the sphere is given by $-\left(4\pi r^2 p(t)\right)_{,r}dr = -8\pi r dr.p(t)$ and the total outward radial gravitational force on this shell is given by

$$-G.4\pi r^2 dr\rho(t)\left(4\pi r^3\rho(t)/3r^2\right) = -16\pi r^2 r^3 \rho^2(t)dr/3$$

$$= -4\pi G M r^3\rho(t)dr/R^3(t)$$

Taking $r = R(t)$ gives the equation of motion of the surface radius:

$$4\pi R^2(t)\rho(t)R''(t) = -4GM\pi\rho(t) - 8\pi R(t)p(t)$$

or after making appropriate cancellations,

$$R^2(t)R''(t) = -GM - 2R(t)p(t)/\rho(t)$$

Assuming that the matter within the sphere obeys an equation of state $p(t) = f(\rho(t))$, we get

$$R^2(t)R''(t) = -GM - 2R(t)f(\rho(t))/\rho(t)$$

Equivalently,

$$R^2(t)R''(t) = -GM - 2R(t)g(3M/4\pi P^3(t))$$

where

$$g(x) = f(x)/x$$

To apply large deviation theory to this problem, we assume that there is a small randomly fluctuating component $w(t)$ in the pressure. Then $p(t)$ is replaced by $f(\rho(t)) + w(t)$ and the equation of motion assumes the form

$$R^2(t)R''(t) = -GM - 2R(t)(g(K/R^3(t))R^3(t)w(t)/K)$$

$$-GM - 2R(t)g\left(K/R^3(t)\right) - 2R^4(t)w(t)/K$$

❖❖❖❖❖

[50] Nonlinear system identification using fractional delay Volterra filters. Assume that $x(t)$ is the input to a nonlinear system and $y(t)$ is the output. Two two are related by a nonlinear functional equation of the form

$$y(t) = f\left(t, x(s), s \le t\right) + \varepsilon\omega(t)$$

where $f(t..)$ is a nonlinear functional and $w(.)$ is Gaussian noise with zero mean. ε is a perturbation parameter assumed to be small. We approximate this system by a system of the form

$$y(t) = \sum_{n=1}^{p} h[n]x(t-\tau_n) + \sum_{k,m=1}^{p} g[k,m]x(t-\tau_k)x(t-\tau_m)$$

In the frequency domain, this gives

$$Y(\omega) = \sum_{n=1}^{p} h[n]\exp(-j\omega\tau_n)X(\omega)$$

$$+\frac{1}{2\pi}\sum_{k,m=1}^{p} g[k,m]\int \exp\left(-j\left(\omega_1\tau_k + (\omega-\omega_1)\tau_m\right)\right)$$

$$X(\omega_1)X(\omega-\omega_1)d\omega_1$$

We choose a frequency discretization step size as Δ and approximate the above equation by

$$Y(n\Delta) = X(n\Delta)\sum_{k} h[k]\exp(-jn\Delta\tau_k) + (\Delta/2\pi)$$

$$\sum_{k,m,r} g[k,m]\exp(-j\Delta((n-r)\tau_m + r\tau_k))X(r\Delta)X((n-r)\Delta)$$

We write $Y[n]$ in place of $Y(n\Delta)$ and $X[n]$ in place of $X(n\Delta)$. Then the frequency domain adaptive algorithm for estimating the filter coefficients $h[k]$, $g[k, m]$ and the fractional delay τ_κ is given by

$$h[k, n+1] = h[k,n] - \mu.\frac{\partial e^2[n]}{\partial h[k,n]}$$

$$g[k, m, n+1] = g[k,m,n] - \mu.\frac{\partial e^2[n]}{\partial g[k,m,n]}$$

$$\tau_k[n+1] = \tau_k[n] - \mu.\frac{\partial e^2[n]}{\partial \tau_k[n]}$$

where

$$e[n] = Y[n] - X[n]\sum_k h[k,n]\exp(-jn\Delta\tau_k[n])$$

$$+(\Delta/2\pi)\sum_{k,m,r} g[k,m,n]\exp(-j\Delta((n-r)\tau_m[n] + r\tau_k[n]))X[r]X[n-r]$$

We note that

$$\frac{\partial e[n]}{\partial g[k,n]} = -X[n].\exp(-jn\Delta\tau_k[n])$$

$$\frac{\partial e[n]}{\partial g[k,m,n]} = \Delta\sum_r \exp(-j\Delta((n-r)\tau_m[n] + r\tau_k(n)))X[r]X[n-r]$$

$$\frac{\partial e[n]}{\partial \tau_k[n]}$$

$$= jnX[n]\Delta.\exp(-jn\Delta\tau_k[n])$$

$$-(j\Delta^2/\pi)\sum_{r,m} g[k,m\,n]rX[r]X[n-r]$$

$$\exp(-j\Delta((n-r)\tau_m[n] + r\tau k[n]))$$

We are assuming that $g[k, m, n] = g[m, k, n]$ without sot of generality which means that the update equations for $g[k, m, n]$ need to be carried out only for $k \leq m$.

Experiments in Astronomy

1. write down the equation of an ellipse in $\mathbb{R}^3$ when ellipse is specified by its major axis a, its minor axis b, the unit normal **n** to the plane of the ellipse and the position r_0 of its centre.

2. Prove the following Kepler's laws about two body motion in an inverse square attractive law:

(*a*) the trajectory is an ellipse

(*b*) Equal areas are swept in equal intervals of time of equivalently, the angular momentum is conserved.

(*c*) If T is the time of an orbit and a is the major axis of the ellipse then T^2/a^3 is a constant dependent only on the mass M of the attracting body and the gravitational constant G.

3. Explain using the previous problem how by estimating the major axis of the ellipse and the time period of motion, the mass of the attracting body can be calculated.

4. Calculate the general relativistic correction to the orbit in the two body problem using the Schwarzchild metric.

5. Derive formulas for the major axis. minor axis and eccentricity of an elliptic orbit in terms of the energy and angular momentum of the orbiting mass. Explain how by measuring the posing of the orbiting particle at different time and using a least square fit. we can estimate the energy and angular momentum of the orbiting mass.

6. Explain the triangulation method, *i.e.* how by measuring the angles of elevation of a body from two different points on the earth and knowing the distance between the two points, the distance of the body from the earth can calculated.

7. If two convex lenses of focal lengths f_1, f_2 are respectively the objective and eyepiece of a telescope with $f_1 < f_2$, then calculate the magnification produced by this telescope from first principles.

8. If a parabolic concave mirror is specified by the equation $x^2 + y^2 = 4az$, then calculate the focal length of the mirror form first principles ie using Snell's laws of reflection.

9. Write down the equations of the evolution of the radius of the universe $R(t)$ in terms of the mass density $\rho(t)$ and pressure $p(t)$ using (*a*) Newtonian cosmology in which the universe in modeled as sphere of radius $R(t)$ With the Einstein field equations.

10. It the moon is assumed to be moving in the periodic gravitational field generated by the earth and the sun, with the earth moving around the sun in an orbit of radius R and time period $T = 365$ days, then write down the equation of motion of the moon and derive approximate solutions to the orbit.

11. Suppose the location of the sunspots at the same time on different days are measured using the refracting telescope with a screen. Explain how a least squares algorithm can be obtained for modeling the position of these sunspots using an auto regressive model.

12. If the distance of an object from three distinct points in space is computed, then derive formulas for the position of the object in space relative to a fixed set of coordinate axes when the positions of the three distinct points relative to these axes in knows. How many solutions do you get.

13. Write down the equations of motion in the two body problem in the presence of gravitating dust when the effect of the gravitational force exerted by the dust is modeled using stochastic differential equations. Write down the mean and covariance propagation equations for this sde and derive algorithms for estimating the drift and diffusion parameters from trajectory measurements.

14. Derive formulas for the Doppler shift of the frequency of radiation emitted by a reeding body. Explain how using this formula, Hubble's constant *H*(*t* = *R*'(*T*)/*R*(*t*)) can be calculated knowing the original frequency of the spectral line of a given element in the receding body.

❖❖❖❖❖

[51] Quantum gravitational search algorithm (QGSO) Let a system of N quantum mechanical particles evolve according to the inverse square law of gravitation. The joint wave function of this system at time The potential energy of this system is

$$V\left(r_1, \ldots r_N / M_1, \ldots, M_N\right) = -\sum_{1 \leq i < j \leq N} \frac{GM_i M_j}{\left|r_i - r_j\right|}$$

Let us associate the error energy $E(i)$ with the ith particle. At time t, the masses are computed as $M_i(t) = \frac{E(\max, t) - E(i, t)}{E(\max, t) - E(\min, t)}$ where $E(max, t) = max_i E(i, t)$, $E(min, t) = minE(i, t)$. The updated potential energy is then

$$V\left(r_1, \ldots r_N / M_1\left(t + dt\right), \ldots, M_N\left(t + dt\right)\right)$$

The wave function at time $t + dt$ is then

$$\psi_{t+dt}\left(\tau_1, \ldots, r_N\right) = \psi_t\left(\tau_1, \ldots, r_N\right) - idt.H\left(t, r_1, \ldots, r_N\right)\psi_t\left(r_1, \ldots, r_N\right)$$

where

$$H(t, r_1,...,r_N) = -\sum_{i=1}^{N} \nabla_{ri}^2 / 2M_i(t) + V(r_1,...r_N / M_1(t),..., M_N(t))$$

is the Hamilonian differential operator at time t. The energies at time $t + dt$ is computed from the average positions of the masses and their position fluctuations at time $t + dt$, *i.e*,

$$E(i, t + dt) = F\Big(\int r_i \psi(t + dt, r_1,..., r_N) d^3r_1...d^3r_N$$
$$\int r_i^2 \psi(t + dt, r_1,..., r_N) d^3r_1...d^3r_N\Big)$$

This iteration continues until the energy difference E (min, t) is smaller than a given threshold and the value of i at which $E(min, t) = E(i, t)$ is the desired solution. If r_i is interpreted as the ith guess for the filer coefficient vector, then its mean and variance must be such that $E(i) = F(\langle ri \rangle, \langle r_i^2 \rangle)$ is small. The function $F(\mu, \sigma^2)$ must be chosen so that F increases with increasing σ^2. For example. suppose we wish to design a filter coeffcient vector $h \in \mathbb{R}^p$ such that $h^TQh + h^Tq$ is a minimum. This error energy comes from an expression of the form $\mathbb{E}(h^T X - Y)^2$ where X is an $\mathbb{R}^p$ valued random vector and Y is a scalar random variable. Q is then the positive definite matrix $\mathbb{E}(XX^T)$ and $q = -\mathbb{E}(XY) \in \mathbb{R}^p$. This error energy does not take into account fluctuations in h. We therefore incorporate an additional term $\mathbb{E}(\|\delta h\|^2)$ where δh is the filter vector dispersion. We encode the position r of a particle to a choice of the filter weight vector.

Specifically, $\langle r \rangle$ is the filter weight vector and $\langle r^2 \rangle$ is the filter vector dispersion. Our QGSO takes both of these into account by specifying the joint wave function $\psi_1(r_1,...,r_N)$ of the N particles rather than just the mean filter coefficients.

❖❖❖❖❖

[52] Tele-opration for linear control systems. The response signal on side 1 is $x_1(t)$ and the response signal on side 2 is $x_2(t)$. If $f_1(t)$ is the total force applied 0 side 1, then the response signal x_1 satisfies

$$\left(s^2 m_1 + sb_1 + sZ_1(s)\right) X_1(s) = F_1(s)$$

and likewise if $f_2(t)$ is the total force applied on side 2, then the response signal x_2 satisfies

$$\left(s^2 m_2 + sb_2 + sZ_2(s)\right) X_2(s) = F_2(s)$$

The total force on side 1 is the sum of an applied component f_{01} (t) and an error feedback component $c_1(t)*(x_2(t)-x_1(t))$ and likewise, the total force on side 2 is the sum of an applied component $f_{02}(t)$ and an error feedback component $c_2(t)*(x_1(t)-x_2(t))$. In the Laplace domain, it follows that denoting

$$G_k(s) = \left(s^2 m_k + sb_k + sZ_k(s)\right)^{-1}, k = 1, 2$$

we have

$$G_1(s)^{-1} X_1(s) = F_{01}(s) + C_1(s)(X_2(s) - X_1(s))$$

$$G_2(s)^{-1} X_2(s) = F_{02}(s) + C_2(s)(X_1(s) - X_2(s))$$

Or equivalently, in matrix notation,

$$\begin{pmatrix} 1+G_1C_1 & -G_1C_1 \\ -G_2C_2 & 1+G_2C_2 \end{pmatrix}\begin{pmatrix} X_1 \\ X_2 \end{pmatrix} = \begin{pmatrix} G_1F_{01} \\ G_2F_{02} \end{pmatrix}$$

and hence defining the matrix

$$\mathrm{A(s)} = \begin{pmatrix} A_{11} & A_{12} \\ A_{21} & A_{22} \end{pmatrix} = \begin{pmatrix} 1+G_1C_1 & -G_1C_1 \\ -G_2C_2 & 1+G_2C2 \end{pmatrix}^{-1}\begin{pmatrix} G_1 & 0 \\ 0 & G_2 \end{pmatrix}$$

we get

$$X_1 = A_{11}F_{01} + A_{12}F_{02}, X_2 = A_{21}F_{01} + A_{22}F_{02}$$

Now suppose we perturb the feedback transfer functions $C_1(s)$, $C_2(s)$ by $\delta C_1(s)$, $\delta C_2(s)$ respectively. Then defining

$$K(s) = \begin{pmatrix} 1+G_1C_1 & -G_1C_1 \\ -G_2C_2 & 1+G_2C_2 \end{pmatrix}$$

$$\delta k(s) = -\begin{pmatrix} 1+G_1C_1 & -G_1C_1 \\ -G_2C_2 & 1+G_2C2 \end{pmatrix}^{-1}\begin{pmatrix} G_1\delta C_1 & -G_1\delta C_1 \\ -G_2\delta C_2 & G_2\delta C_2 \end{pmatrix}$$

$$\begin{pmatrix} 1+G_1C_1 & -G_1C_1 \\ -G_2C_2 & 1+G_2C2 \end{pmatrix}^{-1}$$

$$= P_1(s)\delta C_1(s) + P_2(s)\delta C_2(2)$$

say, where $P_1(s)$ $P_2(s)$ are 2×2 matrices. We leave it as an exercise to obtain explicit formulae for P_1, P_2 in terms of G_1, G_2, C_1, C_2.

❖❖❖❖❖

[53] Tele-operation for nonlinear robotic systems:
The basic dynamical equations for the master and slave robot with tele-operation of delay T taken into account in the error feedback are

$$M(q_m)q_m'' + N(q_m, q_m') = \tau_m(t) + \sum_{k\geq 0} c_m[k](q_s(t-kT) - q_m(t-kT))$$

$$M(q_s)q_s'' + N(q_s, q_s') = \tau_s(t) + \sum_{k\geq 0} c_s[k](q_m(t-kT) - q_s(t-kT))$$

We study the stochastic dynamics in the presence of shot noise, *i.e.*,

$$\tau m(t) = \varepsilon \int h_m(t-t_1)dN_m(t_1)$$

$$\tau s(t) = \varepsilon \int h_s(t-t_1)dN_s(t_1)$$

where N_m and N_s are independent Poisson processes with arrival rates of λ_m and λ_s respectively. Equivalently, if $t_{m,k}, t_{s,k}, k \geq 0$ are the jump times of the processes N_m and N_s respectively, then $t_{m,k+1} - t_{m,k}$ $k \geq 0$ and $t_{s,k+1} - t_{s,k}$ $k \geq 0$ are independent sequences of iid exponential random variables with means of $1/\lambda_m$ and $1/\lambda_s$ respectively. The above stochastic integrals for the shot noise τ_m and τ_s can be expressed as

$$\tau_m(t) = \varepsilon \sum_{k\geq 0} h_m(t - t_{m,k}), \tau_s(t) = \varepsilon \sum_{k\geq 0} h_s(t - t_{s,k})$$

Note that formally, we have

$$N'm(t) = \varepsilon \sum_{k\geq 0} \delta(t - t_{m,k}), N_s'(t) = \sum_{k\geq 0} \delta(t - t_{s,k})$$

The linearized teleoperation system satisfies the dynamical equations

$$M(q_m(t))\delta q_m''(t) + M'(q_m(t))(\delta q_m'(t) \otimes q_m''(t))$$
$$+N_{,1}(q_m(t), q_m'(t))\delta q_m(t) + N_{,2}(q_m(t), q_m'(t))\delta q_m'(t)$$
$$= \tau_m(t) + \sum_k c_m[k](\delta q_s(t-kT) - \delta qm(t-kT))$$

$$M(q_s(t))\delta q_s''(t) + M'(q_s(t))(\delta q_s'(t) \otimes q_s''(t))$$
$$+N_{,1}(q_s(t), q_s'(t))\delta q_s(t) + N_{,2}(q_s(t), q_s'(t))\delta q_s'(t)$$
$$= \tau_s(t) + \sum_k c_s[k](\delta q_m(t-kT) - \delta q_s(t-kT)).$$

❖❖❖❖❖

[54] Fluid dynamics in a pipe of non-uniform time varying cross section and nonlinear viscosity.

The z direction is along the length of the pipe and the r direction is along the radial coordinate of the pipe. the cross section radius is $R(t, z)$ Let $u(t, r, z)$ and $\omega(t, r, z)$ denote respectively the velocity along the radial and longitudinal directions. The non-zero components of the stress tensor of the fluid are $\tau_{rr}, \tau_{rz}, \tau_{zz}$ and these are nonlinear functions of $(u_{,r}, u_{,z}, w_{,r} w_{,z}.$ $p(t, r, z))$ is the pressure field. The Navier-Stokes equations for the fluid are

$$u_{,t} + uu_{,r} + wu_{,z} = (r\tau_{rr})_{,r} + \tau_{rz,z} - p,_r \qquad ...(1)$$

$$w_{,t} + uw_{,r} + ww_{,z} = (r\tau_{rz}), r + t_{zz,z} - p_{,z} \quad ...(2)$$

Finally, the mass conservation equation (*i.e.* the equation of continuity) assuming incompressibility of the fluid is

$$r^{-1}(ru)_{,r} + w_{,z} = 0 \quad ...(3)$$

The boundary conditions are that both u and w vanish when $r = R(t, z)$. We define the normalized radial coordinate as

$$x = r/R(t, z)$$

and regard (x, z) as the independent variables rather than (r, z). Then

$$\frac{\partial}{\partial r} = R^{-1}\frac{\partial}{\partial x}$$

and the above three equation become

$$u_{,t} + \left(u/R^2\right)u_{,x} + wu_{,z} = \left(x\tau_{rr}\right)_{,x} + \tau_{rz,z} - R^{-1}p_{,x} \quad ...(4)$$

$$w_{,t} + R^{-1}uw_{,x} + ww_{,z} = \left(x\tau_{rz}\right)_{,x} + \tau_{zz,z} - p_{,z} \quad ...(5)$$

$$R^{-1}x^{-1}\left(xu\right)_{,x} + w_{,z} \quad ...(6)$$

Eliminatint the pressure between (4) and (5) by applying the operator $R\frac{\partial}{\partial z}$ to (4) and the operator $\frac{\partial}{\partial x}$ to (5) and subtracting gives

$$Ru_{,tz} - w_{,tx} + \left(\left(u/R\right)u_{,x}\right)_{,z} + R\left(wu_{,z}\right)_{,z} - R^{-1}\left(uw_{,x}\right), x - R\left(ww_{,z}\right), x$$

$$= R(x\tau_{rr})_{,xz} + R\tau_{rz,zz} - (X\tau_{rz})_{,xx} - \tau_{zz,zx} \quad ...(7)$$

(6) and (7) are now our fundamental equations and they are valid for $0 \le x \le 1$, $0 \le z \le L$ where L is the length of the pipe.

[55] Gravitational N-body problem in general relativity. Let the rest masses of the N bodies be $m_1 ... m_N$. Let the world line of the Kth body be $x_k = \left(x_k^{\mu}\right) 0 \le \mu \le 3$. The differential proper time interval measured by a clock attached to the Kth body is $d\tau_k = \left(g_{\mu\nu}\left(x_k\right)dx_k^{\mu}dx_k^{\nu}\right)^{1/2}$ and the energy momentum tensor this system is

$$T^{\mu\nu}(x) = \sum_k \left(-g\left(x_k\right)\right)^{-1/2} m_k \left(d\tau_k/dt\right) v_k^{\mu} v_k^{\nu} \delta^3\left(x - x_k\right)$$

where

$$v_k^{\mu} = dx_k^{\mu}/d\tau_k$$

is the four velocity of the kth particle. It is easy to see that $T^{\mu\nu}$ is a tensor field since $\sqrt{-g}d^4x$ is a differential four volume invariant and

$$\int T^{\mu\nu}\sqrt{-g}d^4x = \int\left(\sum_k m_k v_k^\mu v_k^\nu\right)d\tau_k$$

is a tensor. Note that $m_1,...,m_N$ are scalars. The Eintein field equations are

$$R^{\mu\nu} - \frac{1}{2}Rg^{\mu\nu} = KT^{\mu\nu} = K\sum_k mk\delta^3(x-x_k)(-g(x_k))^{-1/2}(d\tau_k/dt)v_k^\mu(t)v_k^\mu(t)$$

Note that $t = x^0$ is the coordinate time (measured by a clock at rest located infinitely far from the system of N bodies.

We define

$$u_k^\mu(t) = dx_k^\mu(t)/dt$$

and also

$$\gamma k(t) = \left(g_{\mu\nu}(x_k)u_k^\mu u_k^\nu\right)^{-1/2} = (d\tau_k/dt)^{-1} = dt/d\tau k$$

Then

$$u_k^0 = 1, v_k^r = (dt/d\tau_k)(dx_k^r/dt) = \gamma k u_k^r$$

or equivalently,

$$v_k^\mu = \gamma k u_k^\mu$$

The Einstein field equations imply the geodesic equations

$$dv_k^\mu/d\tau_k + \Gamma_{\alpha\beta}^\mu(x_k)v_k^\alpha v_k^\beta = 0$$

or equivalently,

$$\frac{d}{dt}\left(\gamma k u_k^\mu\right) + \gamma k\Gamma_{\alpha\beta}^\mu(x_k)u_k^\alpha u_k^\beta = 0$$

We note that

$$\gamma_k^{-2} = g_{00}(x_k) + 2g_{0r}(x_k)u_k^r + g_{rs}(x_k)u_k^\tau u_k^s$$

which follows immediately from

$$g_{\mu\nu}(x_k)v_k^\mu v_k^\nu = 1$$

❖❖❖❖❖

[56] Let $f(t, r)$ be random force field (input field) applied to a fluid dynamical system whose output is the velocity field $v(t, r)$. The fluid is assumed to be incompressible and hence the equations of motion and continuity are

$$(v, \nabla)v + v_{,i.t} = -p/s + v\,\nabla_2 v + f/s$$
$$\nabla, v = 0$$

From first of these equation we derive

$$\Omega_{,t} + \nabla \times (\Omega \times v) = v\nabla^2\Omega + g$$

where

$$\Omega = \nabla \times v, g = \nabla \times f/\rho$$

The second implies the existence of a stream function ψ such that

$$v = \nabla \times \psi, \; div\psi = 0$$

Thus,

$$\Omega = -ndbla^2\psi$$

and hence ψ satisfies

$$\nabla^2_{\psi,t} + \nabla \times ((\nabla^2\psi)\times(\nabla\times\psi)) = v(\nabla^2)^{\psi} + g$$

and hence denoting ∇^2 by Δ we can write

$$\psi_{,t} + \Delta^{-1}(\nabla \times ((\Delta p) \times \nabla \times \psi)) \; L(psi \otimes \psi)$$
$$= v\Delta\psi + \Delta^{-1}g$$

where Δ^{-1} is the inverse of the Laplacian operator taking into account the boundary conditions. We can after discretization of the spatial indices, express the above equation as

$$\psi_{,t} + L(psi \otimes \psi) = v\,\Delta\,\psi + h$$

where L is linear operator and $h = \Delta^{-1}g$. The solution to this nonlinear equation can be expressed as a perturbation series by introducing a pertubation paranmeter δ into the nonlinear term, expanding ψ in powers of δ and equating coeffcients of same powers of δ on both the sides

❖❖❖❖❖

[57] Gravitatinal swarm optimization for the design of quantum gates:
The perturbed quantum system has Hamiltonian

$$H(t) = H_0 + \varepsilon \sum_{k=1}^{n} I_k f_k(t) V$$

where $I_1 \ldots I_n$ are real numbers and $f_1(t) \ldots f_n(t)$ are real signals. The evolution operator for this system upto $O(\varepsilon^3)$ is given by

$$U(T) = U_0(T)W(T), U_0(T) = \exp(-iTH_0)$$

$$W(T) = I - i\varepsilon \sum_k I_k \int_0^T f_k(t)\tilde{V}(t)dt$$

$$+(-i\varepsilon)^2 \sum_{k,m} I_k I_m \int_{0<t_2<t_1<T} f_k(t_1) f_m(t_2)\tilde{V}(t_1)\tilde{V}(t_2)dt_1 dt_2$$

$$+(-i\varepsilon)^3 \sum_{k,m,p} I_k \mathrm{I_m} I_p \int_{0<t_3<t_2<t_1<T} f_k(t_1) f_m(t_2)$$

$$f_p(t_3)\tilde{V}(t_1)\tilde{V}(t_3)\,dt_1 dt_2 dt_3$$

$$= I - i\varepsilon \sum_k I_k F_1(k) + (-i\varepsilon)^2 \sum_k I_k I_m F_2(k,m) + (-i\varepsilon)^3$$

$$\sum_{k,m,p} I_k I_m I_p F_3(k,m,p)$$

where

$$\tilde{V}(t) = \exp(-itH_0)V.\exp(itH_0)$$

$$F_1(k) = \int_0^T f_k(t)\tilde{V}(t)\,dt,\ F_2(k,m)$$

$$= \int_{0<t_2<t_1<T} f_k(t_1) f_m(t_2)\tilde{V}(t_1)\tilde{V}(t_2)\,dt_1 dt_2$$

$$F_3(k,m,p) = \int_{0<t_3<t_2<t_3<T} f_k(t_1) f_m(t_2) f_p(t_3)\tilde{V}(t_1)\tilde{V}(t_2)\tilde{V}(t_3)\,dt_1 dt_2 dt_3$$

The gate $U(T)$ is to be designed by selecting $I_1,\ldots I_n$ so that $E(I) = \left\| U_g - U(T) \right\|^2$ is a minimum where U_g is a given unitary gate. The minimization is to be calculated upto $O(\varepsilon^3)$. We observe that

$$E(I) = \Bigg\| U_0(-T)U_g - I + i\varepsilon \sum_k I_k I_n F_1(k) i\varepsilon^2$$

$$\sum_{k,m} I_k I_m F_2(k,m) - i\varepsilon^3 \sum_{k,m,p} I_k I_m I_p F^3(k,m,p^2) \Bigg\|$$

$$\left\| W_g \right\|^2 - 2\varepsilon \sum_k I_k I_m \left(Tr\left(W_g^* F_1(k) \right) \right)$$

$$+\varepsilon^2 \left(\sum_k I_k I_m \left(\mathrm{Re}\, Tr\left(F_1(k) * F_1(m) \right) + 2\,\mathrm{Re}\left(W_g^* F_2(k,m) \right) \right) \right)$$

$$+ \epsilon^2 \sum_{k,m,p} I_k I_m I_p \left(2 I_m \left(Tr\left(W_g^* F_3(k,m,p) \right) \right) \right)$$

$$+ 2 I_m Tr\left(F_1(p)^* F_2(k,m) \right) \Big) \Big) + O(\varepsilon^4)$$

where

$$W_g = U_0(-T)\, U_g - I$$

We observe that $E(I)$ is, in this approximation, a cubic polynomial in $I = (I_1,....I_n)$. We can use the GSO to minimize this cube polynormial.

❖❖❖❖❖

[58] Nono-robotice using quantum mechanices:

The nono-robot is of a very small size, of the order of cellular dimensions. Its motion takes place inside the living body and it may for example be sensitive to chemical secretions. The motion of a nono-robot inside the fluid may be modelled using the Langevin equation and if the dimensions are very small then we must use the quantum Langevin equation which is a special case of the Evans-Hudson flows modeled using the quantum stochastic calculus of Hudson and Parthasarathy. Consider the unitary evolution operator $U(t)$ satisfying the quantum stochastic differential equation

$$dU(t) = \left(-iH.dt + S_b^a d\Lambda_a^b(t)\right)U(t)$$

where H is a self-adjonit operator and $\Lambda_a^b(t)$ are creation, conservation and annihilation processes satisfying the quantum Ito formula

$$d\Lambda_b^a.\Lambda_d^c = \varepsilon_b^a d\Lambda_d^c, a, b = 0,1,2....$$

where

$$\begin{aligned}
\Lambda_0^0 &= t, \Lambda_0^a = A_a(t), a \geq 1, \Lambda_a^0 = A_a(t)^*, a \geq 1 \\
dA_a(t)dA_b(t)^* &= \delta_{ab}dt, a, b \geq 1 \\
\varepsilon_0^0 &= 0, dA_a(t)d\Lambda_d^c(t) = \delta_d^a dA_c(t), a, c, d \geq 1 \\
d\Lambda_b^a(t)dA_c(t)^* &= \delta_{ac}dA_b(t)^*, \ a, b\ c \geq 1, \\
\varepsilon_a^0 &= \varepsilon_0^a = 0, a \geq 1, S_b^a = \delta_{ab}, a,b \geq 1
\end{aligned}$$

We find the constraints on the system operator H, S_b^a, $a, b \geq 0$ to be unitary for all t. The condition obtained using quantum Ito's formula is

$$\begin{aligned}
0 &= d\left(U^*(t)U(t)\right) = dU^*.U + U^*.dU + dU^*dU \\
&= U^*\left(S_b^{a*}d\Lambda_b^a + S_b^a d\Lambda_a^b + S_b^{a*}S_d^c d\Lambda_a^b d\Lambda_a^d\right)U
\end{aligned}$$

and hence,

$$\left(S_b^{a*} + S_a^b\right)d\Lambda_b^a + S_d^{b*}S_a^c \varepsilon_c^d d\Lambda_b^a = 0$$

or equivalently

$$S_b^{a*} + S_a^b + \varepsilon_c^d S_d^{b*}S_a^c = 0$$

Note that the Einstein summation convention has been adopted. The above is equivalent to

$$S_b^{a*} + S_a^b + \sum_{j \geq 1} S_d^{b*}S_a^j = 0$$

Let X be a system operator and define

$$X(t) = U(t)^* XU(t)$$

Then by quantum Ito's formula,

$$\begin{aligned} dX(t) &= U(t)^*\left(i(HX - XH)dt + \left(S_b^{a*}X + XS_a^b + S_b^{d*}XS_a^c\varepsilon_c^d\right)d\Lambda_b^a\right)U(t) \\ &= U(t)^*\left(i(HX - XH)dt + \left(S_b^{a*}X + XS_a^b + S_b^{d*}XS_a^c\varepsilon_c^d\right)d\Lambda_b^a\right)U(t) \\ & U(t)^*\left(\left(i[H,X] + S_0^{0*}X + XS_0^0 + S_0^{d*}XS_0^c\varepsilon_c^d\right)\right)dt \\ & +S_0^{a*}X + XS_a^a + \sum_{k\geq 1}\left(S_0^{k*}X + XS_k^0 + \sum_{j\geq 1} S_0^{j*}XS_k^j\right)dA_k(t) \\ & +\sum_{k\geq 1}\left(S_k^{0*}X + XS_0^k + \sum_{j\geq 1} S_k^{j*}XS_0^j\right)dA_k(t)^* \\ & +S_k^{j*}X + XS_j^k + \sum_{m,j,k\geq 1} S_k^{m*}XS_j^m d\Lambda_k^j(t)U(t) \end{aligned}$$

Formally, this equation can be expressed in Evans-Hudson form as

$$d_{jt}(x) = j_t\left(\theta_b^a(X)\right)d\Lambda_a^b(t)$$

where

$$\begin{aligned} \theta_0^0(X) &= i[H,X] + S_0^{0*}X + XS_0^0 + \sum_{m\geq 1} S_0^{m*}XS_0^m \\ \theta_k^0(X) &= S_0^{k*}X + XS_k^0 + \sum_{j\geq 1} S_0^{j*}XS_k^j \end{aligned}$$

❖❖❖❖❖

[59] Three dimensional quantum harmonic oscillator in an electromagnetic field. The Hamiltonian of the system is given by

$$H(t) = \left(p + eA(t,q)\right)^2/2m + q^2/2 - e\Phi(t,q)$$

where $A(t, q)$ is the magnetic vector potential and is the electric scalar potential. Suppose we take

$$A(t,q) = \frac{1}{2}B(t)\times q,\ \Phi(t,q) = -(E(t), q)$$

where $B(t)\ E(t) \in \mathbb{R}^3$ are vector valued functions of time only, then we get

$$\nabla\times A(t,q) = B(t), -\nabla\Phi(t,q) - A_{,t}(t,q) = E(t) - \frac{1}{2}B'(t)\times q$$

It follows that if the oscillator characteristic dimension d is such that $|B'(t)| << |E(t)|$, then the electromagnetic field corresponding to the four potential (A, Φ) is approximately $(E(t), B(t))$. We shall assume that this is the case. Then

$$H(t) = H_0 + eV_1(t) + e^2V_2(t)$$

where

$$H_0 = \left(p^2 + q^2\right)/2$$

is the unperturbed oscillator potential,

$$V_1(t) = -\Phi(t, q) + ((p, A) + (A, p))$$

$$= (E(t), q) + \left[(p, B(t)\times q) + (B(t)\times q, p)\right]/2$$

$$= (E(t), q) + \frac{1}{2}(B(t), q\times p - p\times q)$$

$$= (E(t), q) + (B(t), L)$$

where

$$L = q \times p$$

is the angular momentum of the 3-D oscillator about the origin. Note that $p \times p = -q \times p$ even though the vector operator q and p do not commute. Further,

$$V_2(t) = \frac{1}{2}(B(t)\times q)^2 = \frac{1}{2}B^2(t)q^2 - (B(t), q)^2$$

We can now apply time dependent second order perturbation theory to calculate the unitary evolution operator $U(t)$ upto. Specifically,

$$U'(t) = -iH(t)U(t)$$

Writing

$$U(t) = U_0(t)W(t), U_0(t) = \exp(-itH_0)$$

we find that

$$W'(t) = -i\left(e\tilde{V}_1(t) + e^2\tilde{V}_2(t)\right)W(t)$$

where

$$\tilde{V}_1(t) = U_0(-t)V_1(t)U_0(t), \tilde{V}_2(t) = U_0(-t)V_2(t)U_0(t)$$

Then,

$$W(t) = I - ie\int_0^T \tilde{V}_1(t_1)dt_1 + e^2\left(\int_{0<t_2<t_1<t} \tilde{V}_1(t_1)\tilde{V}_1(t_2)dt_1dt_2\right.$$

$$\left. -i\int_0^t \tilde{V}_2(t_1)dt_1\right) + O\left(e^3\right)$$

$$= I + eW_1(t) + e^2W_2(t) + O\left(e^3\right)$$

say, Here,

$$W_1(t) = -i\int_0^t \tilde{V}_1(t_1)dt_1$$

$$W_2(t) = \int_{0<t_2<t_1<t} \tilde{V}_1(t_1)\tilde{V}_1(t_2)dt_1dt_2$$

$$-i\int_0^t \tilde{V}_2(t_1)dt_1$$

The normalized eigen functions of the unperturbed oscillator Hamitonian H_0 are expressed as $|n_1\, n_2\, n_3\rangle, n_1, n_2, n_3 = 0,\ 1,...$where

$$H_0|n_1\, n_2\, n_3\rangle = (n_1 + n_2 + n_3 + 3/2)|n_1\, n_2\, n_3\rangle$$

We have

$$\langle n_1\, n_2\, n_3|\tilde{V}_1(t)|m_1\, m_2\, m_3\rangle$$

$$= \exp\left(i(n_1 - m_1 + n_2 - m_2 + n_3 - m_3)t\right)\left\langle n_1n_2n_3\left|(E(t), q) + B(t), L\right|m_1m_2m_3\right\rangle$$

This can be evaluated using the matrix elements of the position and momentum of a I-D oscillator. Specifically

$$\langle n_1n_2n_3|q|m_1m_2m_3\rangle = (\langle n_1|q_1|m_1\rangle\delta[n_2 - m_2]\delta[n_3 - m_3], \langle n_2|q_2|m_2\rangle,$$

$$\delta[n_1 - m_1]\delta[n_3 - m_3], \langle n_3|q_3|m_3\rangle\delta[n_1 - m_1]\delta[n_2$$

and

$$\langle n_1n_2n_3|L|m_1m_2m_3\rangle$$

$$= (\langle n_2|q_2|m_2\rangle(\langle n_3|q_3|m_3\rangle - \langle n_3|q_3|m_3\rangle\langle n_2|q_2|m_2\rangle,$$

$$\langle n_3|q_3|m_3\rangle\langle n_1|p_1|m_1\rangle - \langle n_1|q_1|m_1\rangle\langle n_3|p_3|m_3\rangle$$

$$\langle n_1|q_1|m_1\rangle\langle n_2|p_2|m_2\rangle - \langle n_2|q_2|m_2\rangle\langle n_1|p_1|m_1\rangle)$$

To write this in a more convenient form define the matrix elements of the position and momentum of a dimensional oscillator

$$\langle a|q_1|b\rangle = Q(a, b), \langle a|p_1|b\rangle = P(a, b)$$

Then

$$\langle n_1n_2n_3|L|m_1m_2m_3\rangle$$

$$= \big(Q(n_2,m_2)P(n_3,m_3)-Q(n_3,m_3)P(n_2,m_2)$$

$$Q(n_3,m_3)P(n_1,m_1)-Q(n_1,m_1)P(n_3,m_3)$$

$$Q(n_1,m_1)P(n_2,m_2)-Q(n_2,m_2)P(n_1,m_1)\big)$$

This is same as the formula

$$\langle n_1n_2n_3|L|m_1m_2m_3\rangle = \langle n_1n_2n_3|q|m_1m_2m_3\rangle \times \langle n_1n_2n_3|p|m_1m_2m_3\rangle$$

❖❖❖❖❖

[60] Problems in control and signal processing:

[1] The charge density in space is given in terms of the current density by the charge conservation law:

$$\rho(t,r) = \rho(0,r) - \int_0^t divJ(t',r)dt'$$

Determine differential/integral equations satisfied by the functions $\rho(0, r)$ and $J(t, r)$ so that the electromagnetic field $E(t, r)$. $H(t, r)$ follows a given pattern $E_d(t, r)$, $H_d(t, r)$ over the space-time region $t \in [0, T)$, $r \in B \subset \mathbb{R}^3$ in the closest possible way in the sense that

$$w_1 \int_{[0,T]\times B} |E(t,r)-E_d(t,r)|^2 dtd^3r + w_2 \int_{[0,T]\times B} |B(t,r)-B_d(t,r)|^2 dtd^3r$$

is a minimum subject to the constraints dictated by the Maxwell equations:

$$\text{curl } H(t,r) = \varepsilon E_{,t}(t,r) + J(t,r)$$

$$\text{curl } E(t,r) = -\mu H_{,t}(t,r)$$

Use may be made of the vector potential approach in solving this problem.

[2] A source voltage $v_s(t)$ with internal impedance $Z_s(w)$ is connected to a transmission line having propagation constant

$$\gamma(w) = [(R+jwL)(G+jwC)]^{1/2}$$

and characteristic impedance

$$Z_0(w) = \left[\frac{R+jwL}{G+jwC}\right]^{1/2}$$

The load impedance of the line is $Z_L(w)$ and the length of the line is d. Calculate the optimal source voltage $V_s(w)$ in the frequency domain so that line voltage $V(w, z)$ and line current $I(w, z)$ track as closely as possible a given pattern $V_d(w, z)$, $I_d(w, z)$ over the frequency range $w \in [-\sigma, \sigma]$ in the sense that

$$w_1 \int_{[0,d]\times\mathbb{R}} |V(z,w)-V_d(z,w)|^2 dzdw + w_2 \int_{[0,d]\times\mathbb{R}} |I(z,w)-I_d(z,w)|^2 dzdw$$

is a minimum. Repeat this problem in the time domain, *i.e.* design the optimal source voltage pattern $v_s(t)$ so that

$$w_1 \int_{[0,d]\times[0,T]} \left|v(z,t)-v_d(z,t)\right|^2 dzdt + w_2 \int_{[0,d]\times[0,T]} \left|i(z,t)-i_d(z,t)\right|^2 dzdt$$

is a minimum. In solving this problem, you may make use of the equations

$$V(z,w) = V_+(w)\exp(-\gamma(w)z)+V_-(w)\exp(\gamma(w)z),$$

$$I(z,w) = Z_0(w)^{-1}\left(V_+(w).\exp(-\gamma(w)z)-V_-(w).\exp(\gamma(w)z)\right),$$

$$V_s(w)-I(0,w)Z_s(w) = V(0,w),$$

$$V(d,w) = I(d,w)Z_L(w)$$

[3] Solve the quantum optimal control problem: The state vector $\psi(t) \in \mathcal{H}$ evolves according to the Schrodinger equation

$$\psi'(t) = \left(H_0+\sum_{k=1}^{p} f_k(t)V_k\right)\psi(t), t \geq 0$$

Determine the optimal control inputs $f_k(t)$, $0 \leq t \leq T$, $k = 1, 2, \ldots, p$ so that

$$\int_0^T \left\|\psi_d(t)-\psi(t)\right\|^2 dt$$

is a minimum subject to a quadratic energy constraint:

$$\sum_{k,m=1}^{p} \int_0^T \sum q(k,m) f_k(t) f_m(t) dt = E$$

❖❖❖❖❖

Brownian Motion and Levy Processes, Nonlinear Filtering Plasma Waves, Quantum Stochastics and Parameter Estimation

[61] Nonlinear filtering in discrete time, approximate algorithms:
Let $x[n]$, $n = 0, 1, 2, \ldots$ be a Markov process with transition generator P_n, *i.e.*

$$\mathbb{E}(\phi(x[n+1])|x[n]) = (P_n\phi)(x[n])$$

Let the measurement process be given by

$$z[n] = H_n(x[n]) + v[n]$$

where $v[n]$, $n \geq 0$ is iid with pdf $p_v(v)$ and this sequence is independent of the sequence $x[n]$, $n \geq 0$. Let

$$Z_n = \{z[k]P : 0 \leq k \leq n\}$$

and consider the conditional probability density of $x[n]$ given Z_n:

$$p(x[n]|Z_n)$$

Also consider the conditional expectation

$$\pi_n(\phi(x)) = E(\phi(x[n])|Z_n)$$

Derive an evolution equation for $\pi_n(\phi(x))$.

hints: $$p(x[n+1]|Z_{n+1}) = \frac{p(x[n+1], Z_n, z[n+1])}{p(z_n, z[n+1])}$$

$$= \frac{\int p(z[n+1]|x[n+1])\, p(x[n+1]|x[n])\, p(x[n]|Zn)\, dx[n]}{\int \text{numerator } dx[n+1]}$$

so noting that

$$p(z[n+1]|x[n+1]) = p_v(z[n+1] - H_{n+1}(x[n+1])),$$

we get

$$\pi_{n+1}(\phi(x)) = \int \phi(x[n+1])\, p(x[n+1]|Z_{n+1})\, dx[n+1]$$

$$= \frac{\sigma_{n+1}(\phi(x))}{\sigma_{n+1}(1)}$$

where $$\sigma_{n+1}(\phi(x)) = \pi_n\left(P_n\left[p_v\left(z[n+1]-H_{n+1}(x)\right)\phi(x)\right]\right)$$

We finally get

$$\pi_{n+1}(\phi(x)) = \frac{\pi_n o P_n\left[p_v\left(z[n+1]-H_{n+1}(x)\right)\phi(x)\right]}{\pi_n o P_n\left[p_v\left(z[n+1]-H_{n+1}(x)\right)\right]}$$

In the special case when

$$x[n+1] = F_n(x[n]) + w[n+1]$$

where $w[n]$ is iid with pdf $p_w(w)$ and the sequences $w[.]$ and $v[.]$ are mutually independent, the above formula becomes

$$\pi_{n+1}(\phi(x)) = \pi_n \frac{\left(\int p_v\left(z[n+1]-H_{n+1}\left(F_n(x)+w\right)\right)\phi\left(F_n(x)+w\right)p_w(w)dw\right)}{\pi_n\left(\int p_v\left(z[n+1]-H_{n+1}\left(F_n(x)+w\right)\right)p_w(w)dw\right)}$$

The numerator of this expression is $\sigma_{n+1}(\phi(x))$ and can be approximated (upto the second moments of w) by (assuming that w has mean zero)

$$\sigma_{n+1}(\phi(x)) \approx \pi_n\left(\int p_v z[n+1] - H_{n+1}\left(F_n(x)\right)\right.$$

$$\left. -H'_{n+1}\left(F_n(x)\right)w + \frac{1}{2}H''_{n+1}\left(F_n(x)\right)(w\otimes w)\right)$$

$$\left(\phi\left(F_n(x)\right) + \phi'\left(F_n(x)\right)w + \frac{1}{2}\phi''\left(F_n(x)\right)(w\otimes w)\right)p_w(w)dt$$

$$\approx \pi_n\left(\int\left[p_v\left(z - H_{n+1}\left(F_n(x)\right)\right)\right.\right.$$

$$\left.\left. -p'_v\left(z-H_{n+1}(Fn(x))\right)\left(H'_{n+1}\left(F_n(x)\right)w + \frac{1}{2}H''_{n+1}(Fn(x))(w\otimes w)\right]\right)$$

$$+\frac{1}{2}p''_v\left(z - H_{n+1}\left(F_n(x)\right)\right)$$

$$\left(H'_{n+1}\left(F_n\left[\phi\left(F_n(x) + \phi'\left(F_n(x)\right)w + \frac{1}{2}\phi''\left(F_n(x)\right)(w\otimes w)\right)\right]p_w(w)\right)dw\right)$$

$$\approx \pi_n\left(p_v\left(z - H_{n+1}\left(F_n(x)\right)\right)\left(\phi\left(F_n(x)\right) + \frac{1}{2}\phi''\left(F_n(x)\right)Vec\left(R_w\right)\right)\right)$$

$$-p'_v\left(z - H_{n+1}\left(F_n(x)\right)\right)\left(H'_{n+1}\left(F_n(x)\right)\otimes\phi'\left(F_n(x)\right)\right)Vec\left(R_w\right)$$

$$+\frac{1}{2}\left(H''_{n+1}\left(F_n(x)\right)\otimes\phi\left(F_n(x)\right)\right)Vec\left(R_w\right)$$

$$+\frac{1}{2}\left(p_v''\left(z-H_{n+1}\left(F_n(x)\right)\right)\otimes\phi\left(F_n(x)\right)\right)Vec(Rw)$$

❖❖❖❖❖

[62] Problems in number theory and infinite series:

$$\sum_{r=0}^{n}\binom{n}{r}x^r = (1+x)^n$$

This implies

$$(1+x)^{pn} = \sum_{0\le r_1,\dots,r_p\le n}\left(\Pi_{j=1}^{p}\binom{n}{r_j}\right)x^{r_1+\dots+r_p}$$

and hence

$$\binom{pn}{k} = \sum_{r+1+\dots+r_p=k}\Pi_{j=1}^{p}\binom{n}{r_j}$$

[2] More generally

$$f(x) = \sum_{n=0}^{\infty}f^{(n)}(a)(x-a)^n/n!$$

So

$$(m!)^{-1}\frac{d^m}{dx^m}\left(f(x)^N\right)\Big|_{x=a} = \sum_{r_1+\dots+r_N=m}\left(\Pi_{k=1}^{N}\left(f^{(r_k)}\right)(a)/r_k!\right)$$

❖❖❖❖❖

[63] Propagation of em waves in nonlinear media. The motion of an electron in the electrostatic field of the nucleus and in addition subject to an external electric field $E(t, r)$ is given approximately by

$$M(r)\xi''(t)+\Gamma(r)\xi'(t)+K(r)\xi(t)+\delta.C(r)\left(\xi(t)\otimes\xi(t)\right) = -eE(t,r)$$

where the an-harmonic term $C(t)\xi\otimes\xi$ of the nucleas-electron interaction is accounted for. Here, r is the location of the nucleus and ξ is the location of the electron relative to the nucleus. $M(r)$ is the effective mass matrix of the electron, $\Gamma(r)$ is the velocity damping matrix and $K(r)$ is the harmonic spring constant matrix component of the nucleus-electron interaction. Ideally higher an-harmonic terms like $\sum_{m=3}^{\infty}C_m(t)(\xi(t))^{\otimes m}$ should be taken into account. We solve the above equation approximately using first order perturbation theory:

$$\xi(t) = \xi_0(t)+\delta.\xi_1(t)+O\left(\delta^2\right)$$

Then, the $O(d^0)$ component of this equation gives

$$M(r)\xi_0''(t) + \Gamma(r)\xi_0' + K(r)\xi_0 = -eE(t, r)$$

giving

$$\xi_0(t) = -e\int_0^t h_1(t-\tau, r)E(\tau, r)d\tau$$

where
$$H_1(s, r) = \int_0^\infty h_1(t, r)\exp(-st)\,dt$$
$$= \left(M(r)s^2 + \Gamma(r)s + K(r)\right)^{-1}$$

is a 3 × 3 matrix. The $O(\delta^1)$ component of this equation gives

$$M(r)\xi_1''(t) + \Gamma(r)\xi_1'(t) + K(r)\xi_1(t) + C(r)\left(\xi_0(t)\otimes\xi_0(t)\right) = 0$$

and hence,

$$\xi_1(t) = -\int h_1(t-\tau, r)C(r)\left(\xi_0(\tau)\otimes\xi_0(\tau)\right)d\tau$$
$$= e^2\int h_1(t-\tau, r)C(r)\left(h_1(\tau-\tau_1, r)\otimes h_1(\tau-\tau_2, r)\right)$$
$$\left(E(\tau_1, r)\otimes E(\tau_2, r)\right)d\tau d\tau_1 d\tau_2$$
$$= e^2\int h_2(t-\tau_1, t-\tau_2, r)\left(E(\tau_1, r)\otimes E(\tau_2, r)\right)d\tau_1 d\tau_2$$

where
$$h_2(t_1, t_2, r) = \int h_1(\tau, r)C(r)\left(h_1(t_1-\tau, r)\otimes h_1(t_2-\tau, r)\right)d\tau$$

Note that $h_1(t, r) \in \mathbb{R}^{3\times3}$ and $C(r) \in \mathbb{R}^{3\times9}$. Thus, $h_2(t_1, t_2\ r) \in \mathbb{R}^{3\times4}$ the dipole moment of the atom is given by $p(t) = -e\xi(t) = -e\xi_0(t) - e\delta\xi_1(t) + O(\delta^2)$ and the polarization vector field (dipole moment per unit volume) is given by

$$P(t, r) = N(r)p(t) = -N(r)e\xi_0 t - N(r)e\delta\xi_1(t) + O(e^4)$$

where $N(r)$ is the number of atoms per unit volume. We denote $N(r)e^2h_1(t, r)/\varepsilon_0$ by $h_1(t, r)$ and $N(r)e^3h_2(t_1, t_2, r)/\varepsilon_0$ by $h_2(t_1, t_2, r)$. Then, the polarization field is given by

$$P(t, r) = \varepsilon_0\int h_1(t-\tau, r)E(\tau, r)d\tau$$
$$+\varepsilon\,\delta\int h_2(t-\tau_1, t-\tau_2, r)\left(E(\tau_1, r)\otimes E(\tau_2, r)\right)d\tau_1 d\tau_2 + O(\delta^2)$$

or equivalently, in the frequency domain,

$$P(w, r) = \varepsilon_0 H_1(w,r)E(w,r) + (2\pi)^{-1}$$
$$\varepsilon_0 \delta \int H_2(w_1, w-w_1)\left(E(w_1,r)\otimes E(w-w_1,r)\right)dw_1$$

Symbolically, we can denote this relationship as

$$P = \varepsilon_0\left(\chi_1 \cdot E + \delta\chi_2 \cdot (E\otimes E)\right))$$

where $\chi_1 = H_1$, $\chi_2 = (2\pi)^{-1}H_2$. The electric displacement vector field is then

$$D = \varepsilon_0 E + P = \varepsilon_0\left(E + \chi_1 \cdot E + \delta\chi_2 \cdot (E\otimes E)\right))$$

We now assume that $|\chi_1| << 1$ and that the $\delta|\chi_2.(E\otimes E, | << |\chi_1.E|$. So we write in place of the above expression,

$$D = \varepsilon_0\left(E + \chi_1 \cdot E + \delta^2\chi_2 \cdot (E\otimes E)\right)$$

where $\chi_1.E = \chi_1(w, r)E(w, r)$

with $\chi_1(w, r) \in \mathbb{C}^{3\times 3}$ and

$$\chi_2.E = \int \chi_2(w_1, w-w_1, r)\left(E(w_1,r)\otimes E(w-w_1,r)\right)dw_1$$

with $\chi_2(w_1, w_2, r) \in \mathbb{C}^{3\times 9}$

The Maxwell equation $divD = 0$ can be expressed as

$$\text{div}\left(E + \delta\chi_1 \cdot E + \delta^2\chi_2(E\otimes E)\right) = 0$$

and the Maxwell curl equations are

$$\text{curl } E = -jw\mu H, \text{ curl } H = jw\,\varepsilon_0\left(E + \delta\chi_1 E + \delta^2\chi_2 \cdot (E\otimes E)\right)$$

We finally have div $H = 0$

From these equations, we derive

$$\begin{aligned}\nabla(\text{div}E) - \nabla^2 E &= -jw\mu\,\text{curl}\,H \\ &= w^2\mu\varepsilon_0\left(E + \delta\chi_1 E + \delta^2\chi_2 \cdot (E\otimes E)\right)\end{aligned}$$

or writing $k^2 = w^2\mu\varepsilon_0$,

we get

$$(\nabla^2 + k^2)E + \delta k^2\chi_1 E + \delta^2 k^2 \chi_2 \cdot (E \otimes E) - \nabla(\text{div}E) = 0 \quad ...(1)$$

Now write $E = E_0 + \delta.E_1 + \delta^2 E_2 + O(\delta^3)$

Then, the equation div$D$ = 0 gives

$$\text{div}E_0 = 0,\ \text{div}E_1 + \text{div}(\chi_1 E_0) = 0,$$

$$\text{div}E_2 + \text{div}(\chi_2 \cdot (E_0 \otimes E_0)) + \text{div}(\chi_1 E_1) = 0$$

Thus, $\text{div}E_1 = -(\nabla\chi_1, E_0)$

$$\text{div}E_2 = -\text{div}(\chi_2 \cdot (E_0 \otimes E_0)) - (\nabla\chi_1, E_1) + \chi_1(\nabla\chi_1, E_0)$$

Further, (1) gives successively on equating the coefficients of δ^m, m = 0, 1, 2 the following differential equations:

$$(\nabla^2 + k^2)E_0 = 0,$$

$$(\nabla^2 + k^2)E_1$$

❖❖❖❖❖

[64] Discrete time nonlinear filtering equations, approximations.

$$\eta[n+1] = F(\eta[n], \chi[n]) + w[n+1] \cdot \chi[n]$$

$$\eta[n] = \begin{pmatrix} \xi[n] \\ a[n] \end{pmatrix}$$

$$\xi[n] = [y[n-1], ..., y[n-p]]^T \cdot \chi[n]$$

$$= [x[n], x[n-1], ..., x[n-q]]^T$$

$$F(\eta, \chi) = \begin{pmatrix} a^T \psi(\xi, \chi) \\ C\xi \\ a \end{pmatrix}$$

where

$$\eta = [\xi^T, a^T]^T, \xi \in \mathbb{R}^p, a \in \mathbb{R}^M, \chi \in \mathbb{R}^{q+1}$$

$$C = [I_{p-1} | 0_{p-1\times 1}] \in \mathbb{R}^{p-1\times p}$$

$$\eta, F(\eta, \chi) \in \mathbb{R}^{(M+p)\times 1}$$

$$z[n] = h^T \eta[n] + v[n]$$

$$Z_n = \{z[k] : k \le n\}$$

$$p(\eta[n+1]|Z_{n+1}) = p(\eta[n+1], Z_n, z[n+1]) / p(Z_n, z[n+1])$$

$$= \frac{\int p(z[n+1]|\eta[n+1])\, p(\eta[n+1]|\eta[n])\, p(\eta[n]|Z_n)\, d\eta[n]}{\int \text{numerator } d\eta[n+1]}$$

$$\pi_n(\phi(\eta)) = \mathbb{E}(\phi(\eta[n])|Z_n)$$

Then,

$$\pi_{n+1}(\phi(\eta)) = \mathbb{E}(\phi(\eta[n+1])|Z_{n+1})$$

$$= \int \phi(\eta[n+1])\, p(\eta[n+1]|Z_{n+1})\, d\eta[n+1]$$

$$= \frac{\sigma_{n+1}(\phi(\eta))}{\sigma_{n+1}(1)}$$

where

$$\sigma_{n+1}(\phi(\eta)) = \int \phi(\eta[n+1])\, p(z[n+1]|\eta[n+1])\, p(\eta[n+1]|\eta[n])$$

$$p(\eta[n]|Z_n)\, d\eta[n] d\eta[n+1]$$

$$= \int \phi(\eta[n+1]) p_w(\eta[n+1] - F(\eta[n], \chi[n]))$$

$$p_v(z[n+1] - h^T \eta[n+1])\, p(\eta[n]|Z_n)\, d\eta[n] d\eta[n+1]$$

$$= \int \phi(F_n(\eta[n]) + w)\, p_w(w) p_v(z[n+1] - h^T(F_n(\eta[n]) + w))$$

$$p(\eta[n]|Z_n d\eta[n] dw$$

where $\quad F_n(\eta) = F(\eta, \chi[n])$

We expand in powers of w retaining only upto quadratic terms in w. Then

$\sigma_{n+1}(\phi(\eta))$

$$= \int p_w(w) \Big(\phi(F_n(\eta[n])) + \phi'(F_n(\eta[n]))^T w + \frac{1}{2}\phi''(F_n(\eta[n]))(w \otimes w)\Big)$$

$$\times \Big(p_v(z[n+1] - h^T F_n(\eta[n])) - p_v'(z[n+1] - h^T F_n(\eta[n])) h^T w$$

$$+ 0.5 p_v''(z[n+1] - h^T F_n(\eta[n]))(h^T w)^2\Big) \times p(\eta[n]|Zn)\, d\eta[n] dw$$

$$= \pi_n\Big\{\phi(F_n(\eta)) p_v(z - h^T F_n(\eta)) - p_v'(z - h^T F_n(\eta)\phi'(F_n(\eta))^T R_w h$$

$$+0.5p_v\left(z-h^TF_n(\eta)\right)Tr\left(\phi''(F_n)R_w\right)+0.5\phi(Fn(\eta))\,p_v''\left(z-h^TF_n(\eta)\right)h^TR_wh\Big\}$$

where $z = z[n+1]$

We use the notation

$$\hat{\eta} = \pi_n(\eta) = \mathbb{E}\left(\eta[n]\|Z_n\right) = \hat{\eta}[n\,|\,n]$$

$$\delta\eta = \eta[n]-\pi_n(\eta) = \delta\eta[n]$$

$$P_\eta[n] = P_\eta[n|n] = \mathbb{E}\left(\delta\eta[n]\cdot\delta\eta[\eta]^T|Z_n\right)$$

$$= \pi_n\left\{\phi(F_n(\eta))p_v\left(z-h^TF_n(\eta)\right)\right\}-p_v'\left(z-h^T(\hat{n})F_n(\hat{\eta})\right)^TR_wh$$

P_η and R_w are assumed to be of the same order of magnitude. Hence, retaining terms only upto linear in these covariances, we get

$$\sigma_{n+1}(\phi(\eta))$$

$$= \pi_n\left\{\phi(F_n(\eta))p_v\left(z-h^TF_n(\eta)\right)\right\}-p_v'\left(z-h^T(\hat{n})F_n(\hat{\eta})\right)^TR_wh$$

$$+0.5p_v\left(z-h^TF_n(\hat{\eta})\right)Tr\left(\phi''\left(F_n(\hat{\eta})\right)R_w\right)+\left(\sigma_w^2/2\right)p_v''\left(z-h^TF_n(\hat{\eta})\right)$$

We use the following notations:

$$\widehat{F} = F_n(\hat{\eta}) = F_n\left(\hat{\eta}[n|n]\right)\in\mathbb{R}^{M+p},$$

$$\widehat{F'} = F'n\left(\hat{\eta}[n|n]\right)\in\mathbb{R}^{(M+p)\times(M+p)},$$

$$\widehat{F''} = F_n''\left(\hat{\eta}[n|n]\right)\in\mathbb{R}^{(M+p)\times(M+p)^2},$$

$$P\eta = P_\eta[n|n]$$

$$\hat{p}_v = p_v\left(z-h^T\widehat{F}\right),$$

$$\widehat{p'}_v = p_v'\left(z-h^T\widehat{F}\right)$$

$$\widehat{p''}_v = p_v''\left(z-h^T\widehat{F}\right)$$

Then we have upto linear orders in R_w, P_η,

$$\pi_n\left\{\phi(F_n(\eta))p_v\left(z-h^TF_n(\eta)\right)\right\}$$

$$= \pi_n\left\{\left[\phi(\widehat{F})+\phi(\widehat{F})^T\left(\widehat{F'}\delta\eta+0.5\widehat{F''}(\delta\eta\otimes\delta\eta)\right)+0.5Tr\left(\phi''(\widehat{F})\widehat{F'}\delta\eta\delta\eta\widehat{F'}^T\right)\right]\right.$$

$$\left.\times\left[\hat{p}_v-\widehat{p'}_v\left(h^T\widehat{F'}\delta\eta+0.5h^T\widehat{F''}(\delta\eta\otimes\delta\eta)+0.5\widehat{p''}_vh^T\widehat{F'}\delta\eta\delta\eta^T\widehat{F'}^Th\right]\right\}$$

$$= \phi(\widehat{F})\widehat{p}_v - \widehat{p'}_v\left(h^T \widehat{F'} P_\eta \widehat{F'} T \phi'(\widehat{F})\right) + 0.5\phi(\widehat{F}) h^T \widehat{F''} Vec(P_\eta)$$

$$+ 0.5\widehat{p}_v \phi'(\widehat{F})^T \widehat{F''} Vec(P_\eta)$$

$$+ 0.5\widehat{p}_v Tr\left(\phi''(\widehat{F})\widehat{F'} P_\eta \widehat{F'}^T\right)$$

$$+ 0.5\widehat{p''}_v \phi(\widehat{F}) h^T \widehat{F'} P_\eta \widehat{F'}^T h$$

$$= \widehat{p}_v\left(\phi(\widehat{F}) + 0.5Tr\left(\phi''(\widehat{F})\widehat{F'} P_\eta \widehat{F'}^T\right) + 0.5\phi'(\widehat{F})\widehat{F''} Vec(P_\eta)\right)$$

$$-\widehat{p'}_v\left(h^T \widehat{F'} P_\eta \widehat{F'}^T \phi'(\widehat{F}) + 0.5\phi(\widehat{F}) h^T F'' Vec(P_\eta)\right)$$

$$+ 0.5\widehat{p''}_v \phi(\widehat{F}) h^T \widehat{F'} P_\eta \widehat{F'}^T h$$

We write

$$\sigma_{n+1}(\phi(\eta)) = \rho_{n+1}(\phi(\eta)) + \tilde{\rho}_{n+1}(\phi(\eta))$$

$$\rho_{n+1}(\phi(\eta)) = \pi_n\left\{\phi(F_n(\eta))\, p_v\left(z - h^T F_n(\eta)\right)\right\}$$

$$= \widehat{p}_v\left(\phi(\widehat{F}) + 0.5Tr\left(\phi''(\widehat{F})\widehat{F'} P_\eta \widehat{F'}^T\right) + 0.5\phi'(\widehat{F})^T \widehat{F''} Vec(P_\eta)\right)$$

$$-\widehat{p'}_v\left(h^T \widehat{F'} P_\eta \widehat{F'} T \phi'(\widehat{F}) + 0.5\phi(\widehat{F}) h^T \widehat{F''} Vec(P_\eta)\right)$$

$$+ 0.5\widehat{p''}_v \phi(\widehat{F}) h^T \widehat{F'} P_\eta \widehat{F'}^T h$$

and $\tilde{\rho}_{n+1}(\phi(\eta)) = -p'_v\left(z - h^T F_n(\hat{\eta})\right)\phi'\left(F_n(\hat{\eta})\right)^T R_w h$

$$+ 0.5 p_v\left(z - h^T F_n(\hat{\eta})\right) Tr\left(\phi''\left(F_n(\hat{\eta})\right) R_w\right)$$

$$+\left(\sigma_w^2/2\right) p''_v\left(z - h^T F_n(\hat{\eta})\right)\phi\left(F_n(\hat{\eta})\right)$$

Taking $\phi(\eta) = \eta_\alpha$

$$\rho_{n+1}(\eta_\alpha) = -\widehat{p'}_v\left(h^T \widehat{F'} P_\eta \widehat{F'} T e_\alpha + 0.5\widehat{F}\alpha h^T \widehat{F''} Vec(P_\eta)\right)$$

$$+ 0.5\widehat{p''}_v \widehat{F}\alpha h^T \widehat{F'} P_\eta \widehat{F'}^T h + \widehat{pv}\left(\widehat{F'}\alpha + 0.5 e_\alpha^T \widehat{F''} Vec(P_\eta)\right)$$

and hence

$$\rho_{n+1}(\eta) = \left(\left(\rho_{n+1}(\eta_\alpha)\right)\right)$$

$$= \widehat{p}_v\left(\widehat{F} + 0.5\widehat{F''} Vec(P_\eta)\right) - p'_v\left(\widehat{F'} P_\eta \widehat{F'}^T h + 0.5\widehat{F} h^T \widehat{F''} Vec(P_\eta)\right)$$

$$+0.5\widehat{p''}_v \widehat{F} h^T \widehat{F'} P_\eta \widehat{F'}^T h$$

and $$\tilde{\rho}_{n+1}(\eta) = -\widehat{p'}_v \sigma_w^2 h + (\sigma_w^2/2)\widehat{p''}_v \widehat{F}$$

Combining these, we get

$$\sigma_{n+1}(\eta) = \widehat{p}_v\left(\widehat{F} + 0.5\widehat{F''} Vec(P_\eta)\right) - \widehat{p'}_v\left(\sigma_w^2 h + \widehat{F'} P_\eta \widehat{F'}^T h\right.$$

$$\left. + 0.5\widehat{F} h^T \widehat{F''} Vec(P_\eta)\right) + 0.5\sigma_w^2 p_v''\left(\widehat{F} + \widehat{F} h^T \widehat{F'} P_\eta \widehat{F'}^T h\right)$$

Now consider $\phi(\eta) = \eta_\alpha \eta_\beta$

We get

$$\rho_{n+1}(\eta_\alpha \eta_\beta) = \widehat{p}_v\left(\widehat{F}_\alpha \widehat{F}_\beta + 0.5Tr\left(\left(e_\alpha e_\beta^T + e_\beta e_\alpha^T\right)\widehat{F'} P_\eta \widehat{F'}^T\right)\right.$$

$$\left. + 0.5\left(\widehat{F}_\alpha e_\beta^T + \widehat{F}_\beta e_\alpha^T\right)\widehat{F''} Vec(P_\eta)\right)$$

$$-\widehat{p'}_v h^T \widehat{F'} P_\eta \widehat{F'}^T \left(\widehat{F'}_\alpha e_\beta + \widehat{F'}_\beta e_\alpha\right) + 0.5\widehat{F}_\alpha \widehat{F}_\beta h^T \widehat{F''} Vec(P_\eta)$$

$$+0.5\widehat{p''}_v \widehat{F}_\alpha \widehat{F}_\beta^T \left(h^T \widehat{F'} P_\eta \widehat{F'}^T h\right)$$

and therefore,

$$\rho_{n+1}(\eta\eta^T) = \left(\left(\rho_{n+1}(\eta_\alpha \eta_\beta)\right)\right)$$

$$= \widehat{p}_v\left(\widehat{F}\widehat{F}^T + \widehat{F'} P_\eta \widehat{F'}^T + 0.5\left(\widehat{F} Vec(P_\eta)^T F''^J + \widehat{F''} Vec(P_\eta)\widehat{F}^T\right)\right)$$

$$-\widehat{p'}_v\left(\left(\widehat{F} h^T \widehat{F'} P_\eta \widehat{F'}^T + \widehat{F'} P_\eta \widehat{F'}^T h \widehat{F}^T\right) + 0.5\left(\widehat{F}\widehat{F}^T\right) h^T \widehat{F''} Vec(P_\eta)\right)$$

$$+0.5\widehat{p''}_v \widehat{F}\widehat{F}^T \left(h^T \widehat{F'} P_\eta \widehat{F'}^T h\right)$$

❖❖❖❖❖

[65] Dispersion relations for gravitational waves in the general theory of relativity:

$$g_{\mu\nu}(x) = \eta_{\mu\nu} + \varepsilon h_{mu\nu}(x)$$

$((\eta_{\mu\nu}))$ is the Minkowskian metric of flat space-time. The Einstein tensor

$$G_{\mu\nu} = R_{\mu\nu} - \frac{1}{2}R g_{\mu\nu} = \varepsilon G_{\mu\nu}^{(1)} + \varepsilon^2 G_{\mu\nu}^{(2)} + O(\varepsilon^3)$$

where

$$G_{\mu\nu}^{(1)} = C_1(\mu\nu\rho\sigma\alpha\beta) h_{\rho\sigma,\alpha\beta} + C_2(\mu\nu\rho\sigma\alpha) h_{\rho\sigma,\alpha} \quad ...(1)$$

$$G^{(2)}_{\mu\nu} = D_1(\mu\nu\rho_1\sigma_1\alpha_1\rho_2\sigma_2\alpha_2)h_{\rho_1\sigma_1,\alpha_1}h_{\rho_2\sigma_2,\alpha_2}$$
$$+D_2(\mu\nu\rho_1\sigma_1\alpha_1\rho_2\sigma_2)h_{\rho_1\sigma_1,\alpha_1}h_{\rho_2\sigma_2} \quad ...(2)$$

Where C_1, C_2, D_1, D_2 are constants defined entirely in terms of the Minkowskian metric and the Kronecker delta symbol. The energy-momentum tensor of matter is

$$T_{\mu\nu} = \varepsilon\left[(\rho+p)v_\mu v_\nu - pg_{\mu\nu}\right] \quad ...(3)$$

We have from the Einstein field equations

$$G^{\mu\nu} = KT^{muv} \quad ...(4)$$

and the Bianchi identity $G^{\mu\nu}_{:\nu} = 0$

that $T^{\mu\nu}_{:\nu} = 0$

and hence $\quad ((\rho+p)v^\nu)_{:\nu}v^\mu + (\rho+p)v^\nu v^\mu_{:\nu} - p^{,\mu} = 0$

Since $v_\mu v^\mu$=1, we have $v_\mu v^\mu_{:\nu}$=0 and hence

$$((\rho+p)v^\nu)_{:\nu} = v_\mu p^{,\mu} = v^\mu p_{,\mu} \quad ...(5)$$

which is the equation of continuity in curved space-time. Thus we get the Navier-Stokes equation in curved space-time:

$$(\rho+p)v^\nu v^\mu_{:\nu} + p_{,\nu}v^\nu v^\mu - p^{,\mu} = 0 \quad ...(6)$$

We now write

$$h_{\mu\nu} = h^{(1)}_{\mu\nu} + \varepsilon h^{(2)}_{\mu\nu} + O(\varepsilon^2) \quad ...(7),$$

$$v^\mu = V^\mu + \varepsilon\delta v^\mu = \rho_0 + \varepsilon\delta\rho + O(\varepsilon^2)$$

$$p = p_0 + \varepsilon\delta p + O(\varepsilon^2) \quad ...(8)$$

Then (5) and (6) give upto $O(\varepsilon^0)$,

$$((\rho_0+p_0)V^\mu)_{,\mu} + p_{0,\mu}V^\mu \quad ...(9),$$

$$(\rho_0+p_0)V^\nu V^\mu_{,\nu} - p_0^{,\mu} + p_{0,\nu}V^\nu V^\mu = 0 \quad ...(10)$$

We note that defining

$$U^r = dx^r/dt,\ r = 1, 2, 3,$$

$$\gamma = (1-U^2)^{-1/2},$$

we have $V^\mu = \gamma U^\mu$, U^0=1

Thus, (9) and (10) can be expressed entirely in terms of U^r, $U^{r,0}$, $U^r_{,s}$ using

$$\gamma_{,0} = \gamma^3 U^r U^r_{,0} = \frac{1}{2}\gamma^3(U^2)_{,0}, U^2 = U^r U^r$$

(summation over $r = 1, 2, 3$) and

$$\gamma_{,r} = \gamma^3 U^s U^s_{,r} = \frac{1}{2}\gamma^3 \left(U^2\right)_{,r}$$

(summation over $s = 1, 2, 3$).

❖❖❖❖❖

[66] Ricci tensor components for the Robertson-Walker metric.

$$g_{00} = 1, g_{11} = -S^2(t)/\left(1 - kr^2\right), g_{22}$$

$$= -S^2(t)r^2, g_{33} = -S^2(t)r^2\sin^2(\theta)$$

$$R_{00} = \Gamma^{\alpha}_{0\alpha,0} - \Gamma^{\alpha}_{00,\alpha} - \Gamma^{\alpha}_{00}\Gamma^{\beta}_{\alpha\beta} + \Gamma^{\alpha}_{0,\beta}\Gamma^{\beta}_{0\alpha}$$

$$\Gamma^{\alpha}_{0\alpha,0} = \left(\Gamma^1_{01} + \Gamma^2_{02} + \Gamma^3_{03}\right)_{,0}$$

$$= 3\left(S'(t)/S(t)\right)'$$

$$\Gamma^{\alpha}_{00} = 0,$$

$$\Gamma^{\alpha}_{0\beta}\Gamma^{\beta}_{0\alpha} = \sum_{r=1}^{3}\left(\Gamma^r_{0r}\right)^2 = 3\left(S'/S\right)^2$$

So

$$R_{00} = 3S''/S$$

$$R_{11} = \Gamma^{\alpha}_{1\alpha,1} - \Gamma^{\alpha}_{11,\alpha} - \Gamma^{\alpha}_{11}\Gamma^{\beta}_{\alpha\beta} + \Gamma^{\alpha}_{1\beta}\Gamma^{\beta}_{1\alpha}$$

$$\Gamma^{\alpha}_{1\alpha} = \Gamma^1_{11} + \Gamma^2_{12} + \Gamma^3_{13}$$

$$0.5\left(\log g_{11}\right)_{,1} + \left(\log g_{22}\right)_{,1} + \left(\log g_{33}\right)_{,1}$$

$$= kr/\left(1 - kr^2\right)^2 + 2/r$$

$$\Gamma^{\alpha}_{11,\alpha}\ \Gamma^{0}_{11,0} + T^{1}_{11,1}$$

$$= -0.5\left(g^{00}g_{11}\right)_{,0} + 0.5\left(\log g_{11}\right)_{,1}$$

$$= -SS'/\left(1 - kr^2\right) + kr/\left(1 - kr^2\right).$$

❖❖❖❖❖

[67] Image processing applied to the study of fluid dynamical perturbations and galactic evolution perturbations in cosmology: We shall be developing algorithms for estimating the parameters of the fluid like the fluid viscosity, density, pressure and the velocity field *via* the stream function from noisy measurements of the velocity field of the fluid at different space-time points.

By estimating these fields we can get a more accurate picture of the fluid field. The algorithms we shall develop shall be based on (1) expressing the Navier-Stokes equation as a partial differential equation in terms of the stream function vector field. The pressure is then eliminated from this equation by a curl operation and the result is a pde from the stream function $\psi(t, \mathbf{r})$, $\mathbf{r} = (x, y, z)$ of the form

$$\frac{\partial \psi}{\partial t} + \Delta^{-1}\left(\Delta\psi \times (\nabla \times \psi)\right) = v\Delta\psi - \Delta^{-1}(\nabla \times f) \qquad ...(1)$$

where $\delta = \nabla^2$ is the three dimensional Laplacian and $f(t, \mathbf{r})$ is the external force field (random) and $v = \eta/\rho$ with η the fluid viscosity and ρ the density. We then represent $\psi(t, \mathbf{r})$ by a vector $\psi(t) \in \mathbb{R}^{3N^3}$ where N^3 is the number of spatial pixels and then (1) can be put in the form of finite dimensional state variable system (this is an approximation to the infinite dimensional problem):

$$\frac{d\psi(t)}{dt} = A\psi(t) + B(\psi(t) \otimes \psi(t)) + g(t) \qquad ...(2)$$

where $\mathbf{A}$ is an $3N^3 \times 3N^3$ matrix and $\mathbf{B}$ is a $3N^3 \times 9N^6$ matrix. $g(t) \in \mathbb{R}^{3N^3}$ is a random process. $g(t)$ can be modeled as a white Gaussian random process with an autocorrelation

$$\mathbb{R}\left(g(t_1)g(t_2)^T\right) = \mathbf{R}_g\delta(t_1 - t_2)$$

where $\mathbf{R}_g$ is a $3N^3 \times 3N^3$ real positive definite matrix. Let $\mathbf{W}(t)$ be a $\mathbb{R}^{3N^3}$ valued standard Brownian motion process. Then the above dynamics of $\psi(t)$ can be represented by a stochastic differential equation

$$d\psi(t) = \left[A\psi(t) + B(\psi(t) \otimes \psi(t))\right]dt + \sqrt{\mathbf{R}_g}\,d\mathbf{W}(t) \qquad ...(3)$$

From this equation, it is easy to write down the Fokker-Planck equation for $p(t, \psi)$, the probability density of $\psi(t)$. We take measurement on the velocity field $\mathbf{v}(t, \mathbf{r}) = \nabla \times \psi(t, \mathbf{r})$ at different times and at different spatial pixels. If measurements are taken in continuous time, then the measurement model for the velocity field is of the form

$$d\xi(t) = \mathbf{C}_1\psi(t)dt + d\,\varepsilon(t) \qquad ...(4)$$

where

$$\xi(t) = \int_0^t \mathbf{v}(\mathbf{s})\mathbf{ds}$$

and C_1 represents a sub matrix of the $3N^3 \times 3N^3$ matrix approximation of the curl operator. The measurement noise $d\varepsilon(t)/dt$ is white Gaussian or equivalently, $R_\varepsilon^{-1/2}\varepsilon(t)$ is an $\mathbb{R}^{3N^3}$ valued vector Brownian motion independent of the fluid state noise $\mathbf{W}(t)$. The problem is then to construct the estimate

$$\mathbb{E}\left(\psi(t)\middle|\xi(s), s \le t\right) = \hat{\psi}(t|t) \qquad ...(5)$$

of the stream function based on velocity measurement upto time t by applying the Kushner filter to (3) and (4). The Kushner filter also gives us the evolution of the error covariance matrix

$$P_{\psi}(t|t) = \text{Cov}(\psi(t)|\xi(s) < s \leq t)$$

$$= \mathbb{E}\left[(\psi(t) - \widehat{\psi}(t|t))(\psi(t) - \widehat{\psi}(t|t))|\xi(s), s \leq t\right]$$

We can then construct the velocity field estimate

$$\hat{v}(t|t) = \mathbf{C}\psi(t|t)$$

We note that the Kushner filter has the form

$$d\pi_t(\phi(\psi))$$

$$= \pi_t(L_t(\phi(\psi)))dt + (\pi_t(\phi(\psi)C_1\psi) - \pi_t(\phi(\psi))\pi_t(C_1\psi))^T$$

$$R_{\in}^{-1}(d\xi(t) - C_1(t)pi_t(\psi)dt) \quad \text{...(6)}$$

where π_t is the conditional expectation operator and $\phi(\psi)$ is a function of the stream function vector. We have

$$\phi_t(\phi(\psi)) = \mathbb{E}(\phi(\psi(t))|\xi(s), s \leq t)$$

and the above formula can be applied successively to $\phi(\psi) = \psi_{\alpha}$ and to $\phi(\psi) = \psi_{\alpha}\psi_{\beta}$ to obtain approximate conditional mean and conditional error covariance propagation. We note that L_t is the backward Kolmogorov operator:

$$L_t(\phi(\psi)) = (\mathbf{A}\psi + \mathbf{B}(\psi \otimes \psi))^T \nabla_{\psi}\phi(\psi) + \frac{1}{2}Tr(\mathbf{R}_g \nabla\nabla\psi^T\phi(\psi)) \quad \text{...(7)}$$

This idea can be extended to study the motion of the galactic fluid *i.e.*, large scale fluid dynamics. Specifically, let $g_{\mu\nu}^{(0)}(x)$ denote the background homogeneous isotropic Robertson-Walker metric and $g_{\mu\nu}^{(0)}(x) + \delta g_{\mu\nu}(x)$ its perturbation caused by small random forces having an energy momentum tensor $\delta\mathcal{F}^{\mu\nu}$. The Einstein field equations

$$R^{\mu\nu} - \frac{1}{2}Rg^{\mu\nu} = K(T^{\mu\nu} + \delta\mathcal{F}^{\mu\nu}) \quad \text{...(8)}$$

with

$$T^{\mu\nu} = (\rho + p)v^{\mu}v^{\nu} - pg^{\mu\nu} \quad \text{...(9)}$$

and the Bianchi identity $\left(R^{\mu\nu} - \frac{1}{2}Rg^{\mu\nu}\right)_{:\nu} = 0$

imply that $\quad (T^{\mu\nu} + \delta\mathcal{F}^{\mu\nu})_{:\nu} = 0$

or

$$((\rho + p)v^{\nu})_{:\nu} v^{\mu} + (\rho + p)v^{\nu}v^{\mu}_{:\nu} - p^{,mu} = \delta\mathcal{F}^{\mu} \quad \text{...(10)}$$

where $\delta\mathcal{F}^{\mu} = -\delta F^{\mu\nu}_{:\nu}$

is the random external four force density field acting on them matter. (10) implies using $v_\mu v^\mu = 1,\ v_\mu v^\mu_{:v} = 0$ that

$$\left((\rho + p)v^v\right)_{:v} = v^\mu\left(p_{,\mu} + \delta\mathcal{F}_\mu\right) \quad ...(11)$$

or equivalently, $\left(\rho + p)v^v\sqrt{-g}\right)_{,v} = \sqrt{-g}v^\mu\left(p_{,\mu} + \delta F_\mu\right)$

This is the generalized form of the mass conservation equation. The Einstein field equations with $\delta\mathcal{F}^{\mu v} = 0$ have an unperturbed solution

$$g_{00} = 1, g_{11} = -S^2(t)/\left(1 - kr^2\right), g_{22}$$
$$= -S^2(t)r^2, g_{33} = -S^2(t)r^2\sin^2(\theta)$$

and $p = p_0(t)$, $\rho = \rho_0(t)$, $v^\mu = V^\mu = (1, 0, 0, 0)$

These are called comoving coordinates since the spatial components of the velocity field to zero. The functions $p_0(t)$, $\rho_0(t)$, $S(t)$ satisfy in view of the unperturbed Einstein field equations, two ordinary differential equations and these lead to the Friedman models for expanding universe, contracting universe and expanding and contracting universes according to a cycloid curve. The unperturbed field equation

$$R^{(0)}_{00} = \frac{K}{2}\left((p_0 + \rho_0) - p_0 - 0.5(\rho_0 - 3p_0)\right)$$
$$= (K/2)(3p_0 + rho_0)$$

gives $$3S''(t)/S(t) = (K/2)\left(3p_0(t) + \rho_0(t)\right) \quad ...(12)$$

The unperturbed mass conservation equation

$$\left((\rho_0 + p_0)V^\mu\sqrt{-g^{(0)}}\right)_{,\mu} = V^\mu p_{0,\mu}\sqrt{-g^{(0)}}$$

gives

$$\left((\rho_0(t) + p_0(t))S^3(t)\right)' = p_0'(t)S^3(t)$$

or equivalently

$$\left(\rho_0 S^3\right)' = -3p_0 S^2 S'$$

or $$\left((4\pi/3)\rho_0 S^3\right)' = -4\pi S^2 S'(t)p_0 \quad ...(13)$$

which is the energy conservation equation. $(4\pi/3)\rho_0 S^3$ is the energy stored inside a sphere of radius $S(t)$ having mass density $\rho_0(t)$. The rate of its increase equations $-4\pi S^2(t)S'(t)p_0(t)$ which is the external pressure force multiplied by the velocity of expansion, namely, the total power input the universe. Thus (13) corresponds to energy balance. (12) and (13) are the basic cosmological equations for the unperturbed homogeneous isotropic universe. Now we look at small perturbations of this universe caused by the random forces described by the energy momentum tensor $\delta\mathcal{F}^{\mu\nu}$. The mass conservation equation

$$\left((\rho+p)\sqrt{-g}v^{\nu}\right)_{,\nu} = v^{\nu}\left(p_{,\nu}+\delta\mathcal{F}^{\nu}\right)$$

gives as the first order perturbed component

$$\left((\rho_0+p_0)\sqrt{-g^{(0)}}\delta v^{\nu}\right)_{,\nu}+\left((\delta\rho+\delta p)\sqrt{-g^{(0)}}\right)_{,0}$$

$$= \left(\delta p_{,0}+\delta\mathcal{F}_0\right)\sqrt{-g^{(0)}}+\delta v^{0}p_{0,0}\sqrt{-g^{(0)}}+\sqrt{-g^{(0)}}\delta p_{,0}+\sqrt{-g^{(0)}}\delta\mathcal{F}_0$$

Here we are assuming that the metric is not perturbed. If in addition, metric perturbations are accounted for, then we would get the extra term

$$\left((\rho_0+p_0)\delta\sqrt{-g}\right)_{,0}$$

on the left and the term $p_{0,0}\delta\sqrt{-g}$

on the right, where $\delta\sqrt{-g} = -\delta g/2\sqrt{-g^{(0)}} = \sqrt{-g^{(0)}}g^{(0)\mu\nu}\delta g_{\mu\nu}$
Finally, substituting the mass conservation equation (11) into the momentum equation (1) gives the cosmological Navier-Stokes equation

$$\left((\rho+p)v^{\nu}v^{\mu}_{:\nu}+v^{\mu}v^{\nu}\left(p_{,\nu}+\delta\mathcal{F}_{\nu}\right)\right) = \delta\mathcal{F}^{\mu}+p^{,\mu}$$

This can be expressed as for $\mu = k = 1, 2, 3$

$$(\rho+p)v^{\nu}\left(v^{k}_{,\nu}+\Gamma^{k}_{\nu\alpha}v^{\alpha}\right)+v^{k}v^{\nu}\left(p_{,\nu}+\delta F_{\nu}\right) = \delta\mathcal{F}^{k}+g^{k\nu}\delta p_{,\nu}$$

The perturbed form of this equation is in the case when metric perturbations are not taken into account,

$$(\rho_0+p_0)\left(\delta v^{k}_{,0}+\Gamma^{(0)k}_{0k}\delta v^{k}\right)+p_{0,0}\delta v^{k} = \delta\mathcal{F}^{k}+g^{(0)kk}p_{,k}$$

in case metric perturbations are accounted for, the left side acquires an additional term $(\rho_0+p_0)\delta\Gamma^{k}_{00}$

and the right side an additional term $\delta g^{k0}p_{0,0}$

The metric perturbations are determined by considering the perturbed Einstein field equations

$$\delta R^{\mu\nu}-\frac{1}{2}\delta R.g^{(0)\mu\nu}-\frac{1}{2}R^{(0)}\delta g^{\mu\nu} = K\left(\delta T^{\mu\nu}+\delta F^{\mu\nu}\right)$$

where $\delta T^{\mu\nu} = (\delta\rho+\delta p)V^{\mu}v^{\nu}+(\rho_0+p_0)\left(V^{\mu}\delta v^{\nu}+V^{\nu}\delta v^{\mu}\right)-\delta p g^{(0)\mu\nu}-p_0\delta g^{\mu\nu}$

where it should be noted that

$$\delta g^{\mu\nu} = -g^{(0)\mu\alpha}g^{(0)\nu\beta}\delta g_{\alpha\beta}$$

By solving these perturbed field equations, we obtain the perturbed density, pressure, velocity and gravitational fields which when plotted show perturbations to galactic evolution.

❖❖❖❖❖

[68] Compound Poisson processes. Let $X(t)$ be a Brownian motion process and Y_n, $n = 1, 2, \ldots$ be iid random variables independent of the process $X(.)$ and let F be the distribution of Y_1. Let $N(t)$ be a Poisson process with rate λ and let the process $N(.)$ be independent of $\{Y_n\}$ and $\{X(t)\}$. Consider the Levy process

$$Z(t) = X(t) + \sum_{k=1}^{N(t)} Y_k$$

$Z(.)$ has independent increments and we see that its characteristic function is given by

$$\psi_Z(\alpha, t) = \mathbb{E}(\exp(i\alpha Z(t))) = \exp(t\Phi(\alpha))$$

where

$$\Phi(\alpha) = -\alpha^2/2 + \lambda \int_{\mathbb{R}} (\exp(i\alpha x) - 1)\, dF(x)$$

We have

$$\psi z(\alpha, t) = \exp\left(-t\alpha^2/2\right) \cdot \sum_{n=0}^{\infty} \exp(-\lambda t)\left((\lambda t)^n/n!\right)\left(\int \exp(i\alpha x)\, dF(x)\right)^n$$

Suppose therefore that $f_0(x, t) = (2\pi t)^{-1/2} \exp\left(-x^2/2t\right)$

and $f_1(x) = F'(x)$

is the density of Y_1, then the density of $Z(t)$ is given by

$$fz(x, t) = f_0(x, t) + \sum_{n \ge 1} \exp(-\lambda t)\left((\lambda t)^n/n!\right) \cdot f_0(x, t) * f_1^{n*}(x)$$

Let for example,

$$f_1(x) = \left(\sigma\sqrt{2\pi}\right)^{-1} \exp\left(-x^2/2\sigma^2\right)$$

Then, the above gives

$$fz(x, t) = f_0(x, t) + \sum_{n \ge 1} \exp(-\lambda t)\left((\lambda t)^n/n!\right)\left(t + n\sigma^2\right)^{-1/2} \phi\left(x/\left(t + n\sigma^2\right)^{1/2}\right)$$

where $\phi(x)$ is the density of a $N(0, 1)$ random variable. Thus, $fz\ (x, t)$ is a superposition of the normal densities $\left(t + n\sigma^2\right)^{-1/2} \phi\left(x/\left(t + n\sigma^2\right)^{1/2}\right)$, $n = 0, 1, 2, \ldots$, *i.e.*, $N(0, n\sigma^2 + t)$ densities with weights $\exp(-\lambda t)(\lambda t)^n/n!, n = 0, 1, 2, \ldots$. The independent increment property of the $Z(.)$ process implies that

$$fz(x, t) * fz(x, s) = fz(x, t + s)$$

We are now given a dynamical system defined by a differential equation

$$F\left(t, q(t), q'(t), q''(t) + G(q(t), q'(t), q''(t)\right)\theta = Z'(t)$$

and wish to estimate θ based on observations of $q(.)$ collected the discrete time points $t = n\Delta$, $n = 0, 1, 2, \ldots$. The discretized version of the above equation is

$$F(n, q[n], q'[n], , q''[n]) + G(q[n], q'[n], q''[n])\theta = Z[n+1] - Z[n]$$

where n is understood to be $n\Delta$ etc. The density of $W[n+1] = Z[n+1] - Z[n]$ is given by

$$f_W(w) = f_0(w, \Delta) + \Delta g(w, \Delta)$$

where $f_0(w, \Delta) = (2\pi\Delta^2)^{-1/2} \cdot \exp(-w^2/2\Delta)$

and $g(w, \Delta) = \sum_{n \geq 1} \exp(-\lambda\Delta)(\lambda^n \Delta^{n-1}/n!)(\Delta + n\sigma^2)^{-1/2} \phi\left(w/(\Delta + n\sigma^2)^{1/2}\right)$

❖❖❖❖❖

[69] Designing an equalizer for a classical electromagnetic channel modelled by a linear frequency dependent susceptibility and a quadratically nonlinear frequency dependent susceptibility. The relation between $D(\omega, r)$ and $E(\omega, r)$ is given by

$$D(\omega, r) = \varepsilon_0 \Big(E(\omega, r) + \delta\chi_1(\omega, r) E(\omega, r) + \delta^2 \int \chi_2(\omega_1, \omega - \omega_1, r)\big(E(\omega_1, r) \otimes E(\omega - \omega_1, r)\big) d\omega_1 \Big)$$

where $\chi_1(\omega, r) \in \mathbb{C}^{3\times 3}$ and $\chi_2(\omega_1, \omega_2, r) \in \mathbb{C}^{3\times 9}$. Here, δ is a small perturbation parameter. Assume that

$$\chi_1(\omega, r) = \sum_n \mathbf{C}_n(\omega)\psi_n(r),$$

$$\chi_2(\omega_1, \omega_2, r) = \sum_n \mathbf{K}_n(\omega_1, \omega_2)\psi_n(r)$$

where $\{\psi_n\}$ are basis functions and $\mathbf{C}_n(\omega)$, $\mathbf{K}_n(\omega_1, \omega_2)$ are matrix valued functions of frequency and frequency pairs to be determined from knowledge of the input em field and the output em field (*i.e.* incident and scattered fields). For solving this electromagnetic system identification problem, we must expand the electric field upto $O(\delta^2)$ terms and use perturbation theory. Let

$$E(\omega, r) = E_0(\omega, r) + \delta.E_1(\omega, r) + \delta^2 E_2(\omega, r) + O(\delta^3)$$

Then, $$\operatorname{div}\left(E + \delta\chi_1 \cdot E + \delta^2\chi_2 \cdot (E \otimes E)\right) = 0$$

gives on equating coefficients of δ^m, $m = 0, 1, 2$ successively,

$$\operatorname{div}(E_0) = 0, \quad \operatorname{div}(E_1 + \chi_1.E_0) = 0,$$

$$\operatorname{div}\left(E_2 + \chi_1.E_1 + \chi_2.(E_0 \otimes E_0)\right) = 0$$

Also curl $E = -j\omega\mu H$, curl $H = j\omega\,\varepsilon_0\left(E + \delta\chi_1.E + \delta^2\chi_2.(E \otimes E)\right)$

give $$\nabla(\text{div}\,E)-\nabla^2 E = \omega^2\mu\,\varepsilon_0\left(E+\delta.\chi_1.E+\delta^2.\chi_2.(E\otimes E)\right)$$

Equating coefficients of δ^m, $m = 0, 1, 2$ successively in this equation, we get (since div $E_0 = 0$, div $(E_1) = -$ div$(\chi_1.E_0)$)

$$\nabla^2 E_0 - \omega^2\mu\,\varepsilon_0 E_0 = 0$$

$$\text{div}(\chi_1.E_0)+\nabla^2 E_1 - \omega^2\mu\,\varepsilon_0(E_1+\chi_1.E_0) = 0$$

$$\nabla(\text{div}E_2)-\nabla^2 E_2 = -\omega^2\mu\varepsilon_0\left(E_2+\chi_1.E_1+\chi_2.(E_0\otimes E_0)\right)$$

These pde's can be successively solved for E_0 (the incident field) and the scattered field $\delta E_1 + \delta^2 E_2$.

❖❖❖❖❖

[70] Coherent states: Let a, a^* denote the annihilation and creation operators of a harmonic oscillator. Consider the operator

$$W(\alpha, \beta) = \exp(\alpha a+\beta a^*), \alpha,\beta\in\mathbb{C}$$

We have the commutation relation

$$[a, a^*] = 1$$

Consider $$\frac{d}{dt}W(t\alpha,t\beta) = (\alpha a+\beta a^*)W(t\alpha,t\beta)$$

Write $$W(t\alpha, t\beta) = \exp(t\alpha a)F(t)$$

Then, $$\frac{dF(t)}{dt} = \beta.\exp(-t\alpha a)a^*.\exp(t\alpha a)F(t)$$

Let $$G(t) = \exp(-t\alpha a)a^*\exp(t\alpha a)$$

Then, $$G'(t) = -\alpha.\exp(t\alpha a)[a,a^*]\exp(t\alpha a) = -\alpha$$

Thus, $$G(t) = a^* - t\alpha$$

This gives $F'(t)=\beta.(a^*-t\alpha)F(t)$

whence, $$F(t) = \exp\left(t\beta a^* - t^2\alpha\beta/2\right)$$

Thus, $$W(\alpha, \beta) = \exp(\alpha a)\exp(\beta a^*)\exp(-\alpha\beta/2)$$

Equivalently, $$W(\alpha, \beta) = \exp(\beta a^*).\exp(\alpha a)\exp(\alpha\beta/2)$$

Thus, $$W(\alpha,\beta)|0\rangle = \exp(\alpha\beta/2)\exp(\beta a^*)|0\rangle$$

Noting that $$|n\rangle = a^{*n}|0\rangle/\sqrt{n!}, n = 0, 1, 2, \ldots$$

it follows that $$W(\alpha,\beta)|0\rangle = \exp(\alpha\beta/2)\sum_{n=0}^{\infty}\frac{\beta^n}{\sqrt{n!}}|n\rangle$$

In particular,

$$W(-\bar{\alpha}a, \alpha a^*)|0\rangle = \exp(\alpha a^* - \bar{\alpha}a)|0\rangle$$

$$= \exp\left(-|\alpha|^2/2\right)\sum_{n\geq 0}\frac{\alpha^n}{\sqrt{n!}}|n\rangle$$

Now, $$a|n\rangle = \sqrt{n}|n-1\rangle$$

so $$aW(-\bar{\alpha}a, \alpha a^*)|0\rangle = \exp\left(-|\alpha|^2/2\right)\sum_{n\geq 1}\frac{\alpha^n}{\sqrt{(n-1)!}}|n-1\rangle$$

$$= \alpha W(-\bar{\alpha}a, \alpha a^*)|0\rangle$$

Thus, denoting $$|\alpha\rangle = \alpha W(-\bar{\alpha}a, \alpha a^*)|0\rangle$$

$$|\alpha\rangle = W(-\bar{\alpha}a, \alpha a^*)|0\rangle$$

We have that $|\alpha\rangle$ is a normalized eigenstate of a with eigenvalue α for each complex number α.

Resolution of the identity: We have

$$\langle n|\alpha\rangle\langle\alpha|m\rangle = (n!m!)^{-1/2}\exp\left(-|\alpha|^2\right)\alpha^n\bar{\alpha}^m$$

Writing $$\alpha = x + iy$$

We have $$\int_{\mathbb{R}^2}\exp\left(-|\alpha|^2\right)\alpha^n\bar{\alpha}^m dxdy$$

$$= \int_0^\infty\int_0^{2\pi}\exp\left(-r^2\right)r^{n+m+1}\exp(i(n-m)\theta)\,drd\theta$$

This is zero if $n \neq m$ and for $n = m$ it equals

$$2\pi\int_0^\infty\exp\left(-r^2\right)r^{2n+1}dr = 2\pi\int_0^\infty\exp(-u)u^n du = 2\pi n!$$

Thus, $$\int\langle n|\alpha\rangle\langle m|\alpha\rangle d^2\alpha = 2\pi\delta[n-m]$$

where by $d^2\alpha$, we mean $dxdy$. This shows that

$$\int|\alpha\rangle d^2\alpha\langle\alpha| = 2\pi I$$

i.e., $\left\{(2\pi)^{-1/2}\,\alpha\rangle\langle\alpha| : \alpha\in\mathbb{C}\right\}$ is a resolution of the identity.

❖❖❖❖❖

[71] GSA (Gravitational search algorithm) for estimating the parameters of a fractional delay based linear time invariant system.

$$y(t) = \sum_{k=1}^{p}h[k]x(t-\tau(k)) + w(t)$$

We wish to estimate the vector $\theta = [h[k], \tau[k], k = 1, 2, ..., p]$ so that

$$E(\theta) = \left(\int_0^T (y(t) - yd(t))^2 dt \right.$$

In the frequency domain, we would consider

$$Y(\omega) = \sum_{k=1}^{p} h[k]\exp(-j\omega\tau[k])X(\omega) + W(\omega)$$

define $H_d(\omega) = Y(\omega)/X(\omega)$ and choose $\{h[k], \tau[k]\}$ so that

$$E[h, \tau] = \int \left| H_d(\omega) - \sum_{k=1}^{p} h[k]\exp(-j\omega\tau[k]) \right|^2 d\omega$$

is a minimum. Let $\mathbf{h} = [h[1], ..., h[p]]^T$

and $$e(\omega, \tau) = [\exp(-j\omega\tau[k]) : k = 1, 2, ..., p]^T$$

Then, $$E[\mathbf{h}, \tau] = \int |H_d(\omega) - h^T e(\omega, \tau)|^2 d\omega$$

Minimizing this w.r.t. **h** first gives for the optimal **h**

$$\mathbf{h} = \left[\int e(\omega, \tau)e(\omega, \tau)^* d\omega\right]^{-1} \left[\int \bar{H}_d(\omega)e(\omega, \tau)d\omega\right]$$

and substituting this expression back into the expression for E gives

$$E_0[\tau] = \int H_d(\omega)\left(\bar{H}_d(\omega) - h^T \bar{e}(\omega, \tau)\right)d\omega$$

Minimizing this w.r.t. τ is equivalent to maximizing

$$F_0(t) = \text{int}\, H_d(\omega)e(\omega, \tau)^* h d\omega$$

$$= \left[\int H_d(\omega)e(\omega, \tau)^* d\omega\right]\left[\int e(\omega, \tau)e(\omega, \tau)^* d\omega\right]^{-1}$$

$$\left[\int \bar{H}_d(\omega)e(\omega, \tau)d\omega\right]$$

This function is computed using frequency discrimination. For example, if the frequency band over which the filter $H_d(\omega)$ assumes prominent values is $[-\sigma, \sigma]$ and we discretize this interval into N equal parts of which Δ so that $\Delta = 2\sigma / N$.

We have the approximations,

$$= \int H_d(\omega)e(\omega, \tau)^* d\omega$$

$$\approx \Delta \sum_{r=-N/2}^{N/2-1} H_d(r\Delta)e(r\Delta, \tau)^* \int e(\omega, \tau)e(\omega, \tau)^* d\omega$$

$$\approx \Delta \sum_{r=-N/2}^{N/2-1} e(r\Delta,\tau)e(r\Delta,\tau)^*, \int \bar{H}_d(\omega)e(\omega,\tau)d\omega$$

$$\approx \Delta \sum_{r=-N/2}^{N/2} H_d(r\Delta)e(r\Delta,\tau)$$

bigskip.

❖❖❖❖❖

[72] Parameter estimation in dynamical models involving Levy noise. Let $B(t)$ be Brownian motion, Y_k, $k = 1, 2, \ldots$ be iid random variables each having distribution F_Y and density f_Y and let $N(t)$ be a Poisson process with rate λ. Assume that the processes $B(.)$, $N(.)$ and $\{Y_k\}$ are all statistically independent of each other and define the independent increment (Levy) process

$$Z(t) = \sigma B(t) + \sum_{k=1}^{N(t)} Y_k$$

Then, the characteristic function of $Z(t)$ is given by

$$\log \mathbb{E}[\exp(i\alpha Z(t))] = -\sigma^2\sigma^2 t/2 + \lambda t\int(\exp(i\alpha x)-1)dF_Y(x)$$

If Y_k is in addition an $N(0,\sigma_Y^2)$ random variable, we thus have that the density of $Z(t)$ is given by

$$f_{Z(t)}(z) = \sum_{n=0}^{\infty} p_n(t)\left(2\pi\left(\sigma^2 t + n\sigma_Y^2\right)\right)^{-1/2} \exp\left(-z^2/2\left(\sigma^2 t + n\sigma_Y^2\right)\right)$$

where $p_n(t)=P(N(t) = n)$

$$= \exp(-\lambda t)(\lambda t)^n/n!, n = 0.\ 1, 2, \ldots$$

In other words, the density of $Z(t)$ is a weighted superposition of $N\left(0,\sigma^2 t + n\sigma_Y^2\right)$ random variables with $n = 0, 1, 2, \ldots$ with weights given by the Poisson distribution. We now consider the stochastic differential equation

$$dX(t) = g\left(X(t),\theta\right)dt + dZ(t), t \in [0,T]$$

where θ is an unknown parameter to be estimated. Formally, the likelihood function for θ is given by the limit of

$$p_T(X|\theta) = \prod_{t\in[0,t]} f_{\delta Z(t)}\left(\delta X(t) - g\left(X(t),\theta\right)\delta t\right)$$

where $\delta X(t) = X(t+\delta) - X(t)$, $\delta t = \delta$, $\delta Z(t) = Z(t+\delta) - Z(t)$

as $\delta \to 0$. The product is interpreted as

$$p_T(X|\theta) = \prod_{n=0}^{N-1} f_{\delta Z(t)}\left(X((n+1)\delta) - X(n\delta) - g\left(X(n\delta),\theta\right)\delta\right)$$

The approximate density of $\delta Z(t)$ is given by

$$f_{\delta Z(t)}(z) = (1-\lambda\delta)\left(2\pi\sigma^2\delta\right)^{-1/2}\exp\left(-z^2/2\sigma^2\delta\right)$$
$$+\lambda\delta\left(2\pi\left(\sigma^2\delta+\sigma_Y^2\right)\exp\left(-z^2/2\left(\sigma^2\delta+\sigma_Y^2\right)\right)\right)$$

where $O(\delta^2)$ terms have been neglected. It follows that

$$-\log f_{\delta Z(t)}(z) = -\log\left((1-\lambda\delta)\left(2\pi\sigma^2\delta\right)^{-1/2}\right)+z^2/2\sigma^2\delta$$
$$+\left(2\pi\sigma^2\right)^{1/2}\lambda\delta^{3/2}\exp\left(\left(z^2/2\right)\left(1/\sigma^2\delta-1/\sigma^2\delta+\sigma_Y^2\right)\right)$$

[73] Classical and quantum nonlinear filtering for parameter estimation in physical models. An LIP nonlinear system in the terminology of Sicuranza *et.al.*[] is defined as a recursive finite memory nonlinear system which is linear in the unknown parameters to be estimated. Here we consider two classes of LIP systems, one using the Fourier sine functions and two a two link robot system and by taking noisy measurements on the output in the second case, the deviation of the robot trajectory from the noiseless one, we apply discrete time nonlinear filtering theory to estimate the model parameters. We compare our estimates with the standard ones present in the literature and show that our filtering based approach is more accurate. Finally, we indicate, how the LIP approach can also be used in identifying a nonlinear, *i.e.* field dependent channel in electromagnetic wave propagation theory by applying a pilot input em field and measuring the output. Simulation studies are carried out using MATLAB.

Let $\psi_1, \ldots, \psi_M$ be given basis functions, all mapping $\mathbb{R}^p \times \mathbb{R}^{q+1}$ into $\mathbb{R}$ and assume that all these functions are bounded. Consider a recursive nonlinear system in discrete time defined by the nonlinear difference equation

$$y[n] = \sum_{k=1}^{M} a[k]\psi_k\left(y[n-1:n-p],x[n:n-q]\right)+w_1[n] \quad \ldots(1)$$

where $w_1[n]$ is white noise with pdf $p_{w1}(w)$. Noisy measurements are taken on $y[n]$:

$$z[n] = y[n] + v[n] \quad \ldots(2)$$

where $v[n]$ is white noise with pdf $p_v(v)$, $x[n]$ is a known nonrandom input. In the nonlinear filtering literature [], the problem is to estimate $y[n]$ based on the observations $Z_n = \{z[k]:k\le n\}$ as

$$\hat{y}[n|n] = \mathbb{E}\left(y[n]|Z_n\right) \quad \ldots(3)$$

along with the error variance

$$V_y(n|n) = \mathbb{E}\left(\left(y[n]-\hat{y}[n|n]\right)^2|Z_n\right) \quad \ldots(4)$$

recursively. Suppose in addition, we are interested in estimating the parameters

$a[k]$ from the noisy output measurements. We may, as in the EKF (Extended Kalman filter) setting assume that these parameters fluctuate according to the dynamics

$$a[k,n+1] = a[k,n]+\varepsilon[k,n] \qquad ...(5)$$

where $(\varepsilon[k,n], k=1,2,...,M)$ forms a vector of white noise vectors. The complete extended state model is then

$$y[n] = a[n]^T \psi(y[n-1:n-p], x[n:n-q]) + w[n] \qquad ...(6)$$

$$a[n+1] = a[n] + \varepsilon[n] \qquad ...(7)$$

$$z[n] = y[n] + v[n] \qquad ...(8)$$

where
$$a[n] = [a[k,n], k=1,2,...,M]^T \qquad ...(9)$$

is the parameter vector at time n. In state variable form, the above equations can be cast as

$$\xi[n+1] = F[\xi[n],n] + w[n] \qquad ...(10)$$

this being the state model and the measurement model is

$$z[n] = h^T \xi[n] + v[n] \qquad ...(11)$$

where

$$\xi[n] = \left[y[n-1:n-p]^T, a[n]^T\right]^T \in \mathbb{R}^{p+M} \qquad ...(12)$$

is the state vector,

$$F[\xi[n],n] = \Big[a[n]^T \psi[y[n-1:n-p], x[n:n-q]],$$
$$(Cy[n-1:n-p])^T, a[n]^T\Big]^T \in \mathbb{R}^{p+M} \qquad ...(13)$$

where
$$C = \left(I_{p-1} \middle| 0_{p-1\times 1}\right) \in \mathbb{R}^{p-1\times p} \qquad ...(14)$$

The standard filtering equations for the above state and measurement model are then derived as follows:

$$p(\xi[n+1]|Z_{n+1}) = p(\xi[n+1], Z_n, z[n+1]) / p(Z_n, z[n+1])$$

$$= \frac{\int p(z[n+1]|\xi[n+1])\, p(\xi[n+1]|\xi[n])\, p(\xi[n]|Z_n)\, d\xi[n]}{\int \text{numerator } d\xi[n+1]}$$

For any smooth function

$$\phi : \mathbb{R}^{p+M} \rightarrow \mathbb{R}$$

we define the conditional expectation

$$\pi_n(\phi(\xi)) = \mathbb{E}(\phi(\xi)|Z_n)$$

of the observable $\phi(.)$ at time n given the observations upto time n. Then,

$$\begin{aligned}\pi_{n+1}(\phi(\xi)) &= \mathbb{E}(\phi(\xi[n+1])|Z_{n+1}) \\ &= \int \phi(\xi[n+1])\, p(\xi[n+1]|Z_{n+1})\, d\xi[n+1] \\ &= \frac{\sigma_{n+1}(\phi(\xi))}{\sigma_{n+1}(1)}\end{aligned}$$

where

$$\begin{aligned}\sigma_{n+1}(\phi(\xi)) &= \int \phi(\xi[n+1])\, p(z[n+1]|\xi[n+1])\, p(\xi[n+1]|\xi[n])\, p(\xi[n]|Z_n) \\ &\quad d\xi[n]d\xi[n+1] \\ &= \int \phi(\xi[n+1])\, p_w(\xi[n+1]-F(\xi[n],n))\, p_v \\ &\quad (z[n+1]-h^T\xi[n+1]) \\ &\quad p(\xi[n]|Z_n)\, d\xi[n]d\xi[n+1] \\ &= \int \phi(F(\xi[n],n)+w)\, p_w(w)\, p_v(z[n+1]-h^T(F(\xi[n],n)+w)) \\ &\quad p(\xi[n]|Z_n)\, d\xi[n]dw\end{aligned}$$

Define $Fn(\xi) = F(\xi, n)$

We expand in powers of w retaining only upto quadratic terms in w. Then

$$\begin{aligned}\sigma_{n+1}(\phi(\xi)) = \int p_w(w)\Big(&\phi(F_n(\xi[n]))+\phi'(F_n(\xi[n]))^T w+\frac{1}{2}\phi''(F_n(\xi[n]))(w\otimes w)\Big) \\ &\times\Big(p_v(z[n+1]-h^T F_n(\xi[n]))-p_v'(z[n+1]-h^T F_n(\xi[n]))h^T w \\ &+0.5p_v''(z[n+1]-h^T F_n(\xi[n]))(h^T w)^2\Big)\times p(\xi[n]|Z_n)\, d\xi[n]dw \\ = \pi_n\Big\{&\phi(F_n(\xi))\, p_v(z-h^T F_n(\xi)-p_v'(z-h^T F_n(\xi))\phi'(F_n(\xi))^T R_w h \\ &+0.5p_v(z-h^T F_n(\xi))Tr(\phi''(F_n)R_w) \\ &+0.5\phi(F_n(\xi))\, p_v''(z-h^T F_n(\xi))h^T R_w h\Big\}\end{aligned}$$

where $z = z[n+1]$

We use the notation

$$\begin{aligned}\hat{\xi} &= \pi_n(\xi) = \mathbb{E}(\xi[n]|Z_n) = \hat{\xi}[n|n] \\ \delta\xi &= \xi[n]-\pi_n(\xi) = \delta\xi[n] \\ P_\xi[n] &= P_\xi[n|n] = \mathbb{E}(\delta\xi[n].\delta\xi[n]^T|Z_n)\end{aligned}$$

$$= \mathbb{E}\left(\xi[n]\xi[n]^T \middle| Z_n\right) - \hat{\xi}[n|n]\hat{\xi}[n|n]^T$$

P_ξ and R_w are assumed to be of the same order of magnitude. Hence, retaining terms only upto linear in these covariances, we get

$$\begin{aligned}
\sigma_{n+1}(\phi(\xi)) &= \pi_n\left\{\phi(F_n(\xi))p_v\left(z - h^T F_n(\xi)\right)\right\} \\
&\quad - p_v'\left(z - h^T F_n(\hat{\xi})\right)^T F_n(\hat{\xi})\phi F_n'(\hat{\xi})\Big) \\
&\quad + 0.5 p_v\left(z - h^T F_n(\hat{\xi})\right) Tr\left(\phi''\left(F_n(\hat{\xi})\right)R_w\right) \\
&\quad + \left(\sigma_w^2/2\right)p_v''\left(z - h^T F_n(\hat{\xi})\right)\phi\left(Fn(\hat{\xi})\right)
\end{aligned}$$

We use the following notations:

$$\begin{aligned}
\widehat{F} &= F_n(\hat{\xi}) = F_n\left(\hat{\xi}[n|n]\right) \in \mathbb{R}^{M+p}, \\
\widehat{F'} &= F_n'\left(\hat{\xi}[n|n]\right) \in \mathbb{R}^{(M+p)\times(M+p)}, \\
\widehat{F''} &= F_n''\left(\hat{\xi}[n|n]\right) \in \mathbb{R}^{(M+p)\times(M+p)^2}, \\
P_\xi &= P_\xi[n|n], \\
\widehat{p}_v &= p_v\left(z - h^T \widehat{F}\right),\ \widehat{p'}_v = p_v'\left(z - h^T \widehat{F}\right), \\
\widehat{p''}_v &= p_v''\left(z - h^T \widehat{F}\right)
\end{aligned}$$

Then we have upto linear orders in R_w, $P\xi$,

$$\pi_n\left\{\phi(F_n(\xi))p_v\left(z - h^T F_n(\xi)\right)\right\}$$

$$\begin{aligned}
&= \pi_n\Big\{\Big[\phi(\widehat{F}) + \phi'(\widehat{F})^T\left(\widehat{F'}\delta\xi + 0.5\widehat{F''}(\delta\xi\otimes\delta\xi)\right) + 0.5 Tr\left(\phi''(\widehat{F})\widehat{F'}\delta\xi\delta\xi\widehat{F'}^T\right)\Big] \\
&\quad \times\Big[\widehat{p}_v - \widehat{p'}_v\left(h^T \widehat{F'}\delta\xi + 0.5 h^T \widehat{F''}(\delta\xi\otimes\delta\xi) + 0.5\widehat{p''}_v h^T \widehat{F'}\delta\xi\delta^T \widehat{F'}^T h\right]\Big\} \\
&= \phi(\widehat{F})\widehat{p}_v - \widehat{p'}_v\left(h^T \widehat{F'}P_\xi \widehat{F'}T\phi'(\widehat{F}) + 0.5\phi(\widehat{F})h^T F'' Vec(P_\xi)\right) \\
&\quad + 0.5\widehat{p}_v\phi'(\widehat{F})^T \widehat{F''}Vec(P_\xi) + 0.5\widehat{p}_v Tr\left(\phi''(\widehat{F})\widehat{F'}P_\xi\widehat{F'}^T\right) \\
&\quad + 0.5\widehat{p''}_v\phi(\widehat{F})h^T \widehat{F'}P_\xi\widehat{F'}^T h \\
&= \widehat{p}_v\left(\phi(\widehat{F}) + 0.5 Tr\left(\phi''(\widehat{F})\widehat{F'}P_\xi\widehat{F'}^T\right) + 0.5\phi'(\widehat{F})^T \widehat{F''}Vec(P_\xi)\right) \\
&\quad - \widehat{p'}_v\left(h^T \widehat{F'}P_\xi\widehat{F'}^T \phi'(\widehat{F}) + 0.5\phi(\widehat{F})h^T \widehat{F''}Vec(P_\xi)\right) \\
&\quad + 0.5\widehat{p''}_v\phi(\widehat{F})h^T \widehat{F'}P_\xi\widehat{F'}^T h
\end{aligned}$$

We write

$$\sigma_{n+1}(\phi(\xi)) = \rho_{n+1}(\phi(\xi)) + \tilde{\rho}_{n+1}(\phi(\xi))$$

where

$$\rho_{n+1}(\phi(\xi)) = \pi_n\left\{\phi(F_n(\xi))p_v\left(z - h^T F_n(\xi)\right)\right\}$$

$$= \hat{p}_v\left(\phi(\hat{F}) + 0.5Tr\left(\phi''(\hat{F})\widehat{F'}P_\xi\widehat{F'}^T\right) + 0.5\phi'(\hat{F})^T\widehat{F''}Vec(P_\xi)\right)$$

$$-\widehat{p'}_v\left(h^T\widehat{F'}P_\xi\widehat{F'}^T\phi'(\hat{F}) + 0.5\phi(\hat{F})h^T\widehat{F''}Vec(P_\xi)\right)$$

$$+0.5\widehat{p''}_v\phi(\hat{F})h^T\widehat{F'}P_\xi\widehat{F'}^T h$$

and $\tilde{\rho}_{n+1}(\phi(\xi)) = -p_v'\left(z - h^T F_n(\hat{\xi})\right)\phi'\left(F_n(\hat{\xi})\right)^T R_w h$

$$+0.5p_v\left(z - h^T F_n(\hat{\xi})\right)Tr\left(\phi''\left(F_n(\hat{\xi})\right)R_w\right) + \left(\sigma_w^2/2\right)p_v''\left(z - h^T F_n(\hat{\xi})\right)\phi\left(F_n(\hat{\xi})\right)$$

Taking $\phi(\xi) = \xi_\alpha$

we get $\rho_{n+1}(\xi_\alpha) = -\widehat{p'}_v\left(h^T\widehat{F'}P_\xi\widehat{F'}Te_\alpha + 0.5\hat{F}_\alpha h^T\widehat{F''}Vec(P_\xi)\right)$

$$+0.5\widehat{p''}_v\hat{F}_\alpha h^T\widehat{F'}P_\xi\widehat{F'}^T h + \hat{p}_v\left(\hat{F}_\alpha + 0.5e_\alpha^T\widehat{F''}Vec(P_\xi)\right)$$

and hence

$$\rho_{n+1}(\xi) = \left(\left(\rho_{n+1}(\xi_\alpha)\right)\right)$$

$$= \hat{p}_v\left(\hat{F} + 0.5\widehat{F''}Vec(P_\xi)\right) - \widehat{p'}_v\left(\widehat{F'}P_\xi\widehat{F'}^T h + 0.5\hat{F}h^T\widehat{F''}Vec(P_\xi)\right)$$

$$+0.5\widehat{p''}_v\hat{F}h^T\widehat{F'}P_\xi\widehat{F'}^T h$$

and $\tilde{\rho}_{n+1}(\xi) = -\widehat{p'}_v\sigma_w^2 h + \left(\sigma_w^2/2\right)\widehat{p''}_v\hat{F}$

Combining these, we get

$$\sigma_{n+1}(\xi) = \hat{p}_v\left(\hat{F} + 0.5\widehat{F''}\,\mathrm{Vec}(P_\xi)\right) - \widehat{p'}_v\left(\sigma_w^2 h + \widehat{F'}P_\xi\widehat{F'}^T h + 0.5\hat{F}h^T\widehat{F''}Vec(P_\xi)\right)$$

$$= +0.5\sigma_w^2\widehat{p''}_v\left(\hat{F} + \hat{F}h^T\widehat{F'}P_\xi\widehat{F'}^T h\right)$$

Now consider $\phi(\xi)=\xi_\alpha\xi_\beta$

We get $\rho_{n+1}(\xi_\alpha\xi_\beta)$

$$= \hat{p}_v\left(\hat{F}_\alpha\hat{F}_\beta + 0.5Tr\left(\left(e_\alpha e_\beta^T + e_\beta e_\alpha^T\right)\widehat{F'}P_\xi\widehat{F'}^T\right) + 0.5\left(\hat{F}_\alpha e_\beta^T + \hat{F}_\beta e_\alpha^T\right)\widehat{F''}Vec(P_\xi)\right)$$

$$-\widehat{p'}_v\left(h^T\widehat{F'}P_\xi\widehat{F'}^T\left(\widehat{F}_\alpha e_\beta+\widehat{F}_\beta e_\alpha\right)+0.5\widehat{F}_\alpha\widehat{F}_\beta h^T\widehat{F''}Vec\left(P_\xi\right)\right)$$

$$+0.5\widehat{p''}_v\widehat{F}_\alpha\widehat{F}_\beta h^T\widehat{F'}P_\xi\widehat{F'}^T h$$

and therefore,

$$\rho_{n+1}\left(\xi\xi^T\right) = \left(\left(\rho_{n+1}\left(\xi_\alpha\xi_\beta\right)\right)\right)$$

$$= \widehat{p}_v\left(\widehat{F}\widehat{F}^T+\widehat{F'}P_\xi\widehat{F'}^T+0.5\left(\widehat{F}Vec\left(P_\xi\right)^T\widehat{F''}^T+\widehat{F''}Vec\left(P_\xi\right)\widehat{F}^T\right)\right)$$

$$-\widehat{p'}_v\left(\left(\widehat{F}h^T\widehat{F'}P_\xi\widehat{F'}^T+\widehat{F'}P_\xi\widehat{F'}^T h\widehat{F}^T\right)+0.5\left(\widehat{F}\widehat{F}^T\right)h^T\widehat{F''}Vec\left(P_\xi\right)\right)$$

$$+0.5\widehat{p''}_v\widehat{F}\widehat{F}^T\left(h^T\widehat{F'}P_\xi\widehat{F'}^T h\right)$$

Further, with the approximation of neglecting higher powers in the noise covariances as well as in the state estimation error covariances than one, we have

$$\tilde{\rho}_{n+1}\left(\xi\xi^T\right) = -\sigma_w^2\widehat{p'}_v\left(\widehat{F}h^T+h\widehat{F}^T\right)+\widehat{p}_v R_w+\left(\sigma_w^2/2\right)\widehat{p''}_v\widehat{F}\widehat{F}^T$$

The expressions for $\rho_{n+1}(\xi\xi^T)$ and $\tilde{\rho}_{n+1}\left(\xi\xi^T\right)$ are added to get $\sigma_{n+1}(\xi\xi^T)$. Finally, we need to evaluate $\sigma_{n+1}(1)$.

2. Nonlinear filtering for LIP models in robotics: Consider the robot differential equations in noise

$$q'(t) = \omega(t) \qquad ...(2.1)$$

$$M\left(q(t),\theta\right)\omega'(t)+N\left(q(t),\omega(t),\theta\right) = \tau(t)+w(t) \qquad ...(2.2)$$

where $\tau(t)$ is the external torque, $w(t)$ is the noisy torque, $q(t)$ is the angular position vector and $\omega(t)$ is the angular velocity vector. M is the mass moment of inertia matrix and N is the combination of the centrifugal, coriolis, gravitational and frictional torque components. θ is the parameter vector comprising the masses of the links and their lengths. Assume that we have an initial estimate θ_0 of this parameter vector and that the parametric estimation error $\delta\theta = \theta - \theta_0$ is small. Then, we have approximately, writing $M_0(q) = M(q, \theta_0)$ and $N_0(q, \omega) = N(q, \omega, \theta_0)$ that

$$\left(M_0(q)+M_{01}(q)\left(\delta\theta\otimes I_2\right)\right)\omega'+N_0(q,\omega)+N_{01}(q,\omega)\delta\theta$$

$$= \tau(t)+w(t) \qquad ...(2.3)$$

where

$$M_{01}(q) = \frac{\partial M\left(q,\theta_0\right)}{\partial\theta} \qquad ...(2.4)$$

$$N_{01}(q) = \frac{\partial N\left(q,\omega,\theta_0\right)}{\partial\theta} \qquad ...(2.5)$$

(2.3) is an LIP model with respect to the parametric error $\delta\theta$. The measurement of the angular position vector can be described as

$$dz(t) = q(t)dt + dv(t) \quad ...(2.6)$$

and then we assume a dynamical model for $\delta\theta$ as

$$d\delta\theta(t) = d\varepsilon(t) \quad ...(2.7)$$

where $w(t)$, $v'(t)$, $\varepsilon'(t)$ are independent white noise processes and after discretization of this state cum measurement model, we can apply our theory of nonlinear filtering for estimating the state $q(t)$ as well as the parametric error $\delta\theta(t)$ from the noisy observations $z(s)$, $s \le t$. The discrete form of (2.3) is given by

$$\begin{aligned}&\Delta^{-2}M_0(q[n])(q[n+1]-2q[n]+q[n-1])\\&+\Delta^{-2}M_{01}(q[n])(I_p \otimes q[n]-2q[n-1]+q[n-2])\delta\theta\\&+N_0(q[n],\Delta^{-1}(q[n]-q[n-1]))\\&+N_{01}(q[n],\Delta^{-1}(q[n]-q[n-1]))\delta\theta\\=\;&\tau[n+1]+w[n+1]\end{aligned} \quad ...(2.7)$$

and this equation can be arranged in the form

$$\begin{aligned}&F_1(q[n+1],q[n],q[n-1],n+1)+F_2(q[n],q[n-1],q[n-2])\delta\theta\\=\;&w[n+1]\end{aligned} \quad ...(2.8)$$

which clearly shows the LIP structure. Here,

$$\begin{aligned}&F_1(q[n+1],q[n],q[n-1],n+1)\\=\;&\Delta^{-2}M_0(q[n])(q[n+1]-2q[n]+q[n-1])\\&+N_0(q[n],\Delta^{-1}(q[n]-q[n-1]))-\tau[n+1]\end{aligned} \quad ...(2.9)$$

and

$$\begin{aligned}&F_2(q[n],q[n-1],q[n-2])\\=\;&\Delta^{-2}M_{01}(q[n])(I_p \otimes (q[n]-2q[n-1]+q[n-2]))\\&+N_{01}(q[n],\Delta^{-1}(q[n]-q[n-1]))\end{aligned} \quad ...(2.10)$$

3. Nonlinear LIP filtering for estimating the linear and nonlinear susceptibilities in an electromagnetic channel.

From basic physics involving the classical motion of an electron in the electrostatic field of the nucleus and an external time varying electric field, it is easily shown that the displacement of the electron has the following form upto quadratic terms in the electric field:

$$\xi(t) = \int H_1(\tau)E(t-\tau)d\tau + \int H_2(\tau_1,\tau_2)\big(E(t-\tau_1)\otimes E(t-\tau_2)\big)d\tau_1 d\tau_2 \quad \text{...(3.1)}$$

where H_1, H_2 are matrix valued functions of the nuclear position. This equation leads us to the following constituent relation for the electric displacement vector with the electric field in the frequency domain:

$$D(\omega, r) = \varepsilon_0\big(E(\omega,r) + \delta\chi_1(\omega,r)E(\omega,r)$$

$$+\delta^2\int \chi_2(\omega_1,\omega-\omega_1,r)\big(E(\omega_1,r)\otimes E(\omega-\omega_1,r)\big)d\omega_1 \quad \text{...(3.2)}$$

Here, δ is a small perturbation parameter. The problem is to estimate the linear and nonlinear susceptibility functions χ_1 and χ_2. We can formally express (3.2) in the form

$$D = \varepsilon_0(1+\delta\chi_1).E + \delta^2\varepsilon_0\chi_2.(E\otimes E) \quad \text{...(3.3)}$$

It is clear that although the relationship between D and E is nonlinear, the parametric functions χ_1 and χ_2 appear linearly in the model and hence the LIP formalism of Sicuranza *et.al.* is applicable. To elaborate further on this, we consider the Maxwell equations

$$\text{curl}E = -j\omega\mu H,$$

$$\text{curl}H = j\omega\varepsilon_0\big(E + \delta\chi_1 E + \delta^2\chi_2.(E\otimes E)\big) \quad \text{...(3.4)}$$

We deduce from (3.4) by taking curl that

$$\nabla^2 E - \nabla(\nabla.E) + \omega^2\mu\varepsilon_0\big(E + \delta\chi_1 E + \delta^2\chi_2.(E\otimes E)\big) = 0 \quad \text{...(3.5)}$$

Further the Maxwell equation divD = 0 combined with (3.3) gives

$$\text{divE} = -\delta\text{div}(\chi_1 E) - \delta^2\text{div}\big(\chi_2.(E\otimes E)\big) \quad \text{...(3.6)}$$

and on substituting (3.6) into (3.5), we get

$$(\nabla^2 + k^2)E + \delta\big[\nabla\big(\text{div}(\chi_1 E)\big) + k^2\chi_1 E\big]$$

$$\delta^2\big[\nabla\big(\text{div}(\chi_2.(E\otimes E))\big) + K^2\chi_2.(E\otimes E)\big] = 0 \quad \text{...(3.7)}$$

where

$$k^2 = \omega^2\mu\varepsilon_0 \quad \text{...(3.8)}$$

(3.7) is a pde of the LIP form w.r.t. the parameter fields χ_1 and χ_2. To make this more explicit, we choose basis functions $\psi_n(r)$, $n = 1, 2 ..., N$ and expand

$$\chi_1(\omega, r) \approx \sum_{n=1}^{N}\chi_{1n}(\omega)\psi_n(r), \chi_2(\omega_1,\omega_2,r)$$

$$\approx \sum_{n=1}^{N}\chi_{2n}(\omega_1,\omega_2)\psi_n(r) \quad \text{...(3.9)}$$

The parameter matrices $\chi_{1n}(\omega)$ and $c^2(\omega_1, \omega_2)$ are respectively 3×3 and 3×9 in size. Substituting these expressions into (3.7). we get

$$(\nabla^2 + k^2)E_c + \delta\left(\sum_{a,b}[\chi_{1n}(\omega)]_{ab} E_{b,ac}(\omega, r) + k^2 \sum_a [\chi_{1n}(\omega)]_{ca} E_a(\omega, r)\right)$$

$$+ \delta^2 \sum_{a,b}\left[\int \chi_{2n}(\omega_1, \omega - \omega_1)\right]_{f,ab} (E_a(\omega_1, r)E_b(\omega - \omega_1, r))_{,fc} d\omega_1$$

To simplify matters further for MATLAB simulation of the channel i/o relation, we assume the nonlinear component to be zero then arrive at the LIP model

$$\left(\nabla^2 + k^2\right)E(\omega, r) + \delta\nabla(\nabla\chi(\omega, r), E(\omega, r)) + \delta k^2 \chi(\omega, r)E(\omega, r) = 0$$

We have

$$\text{div}E = -\delta\,\text{div}(\chi E) = -\delta((\nabla\chi, E) + \chi.\text{div}\,E)$$

so that with neglect of $O(\delta^2)$ terms, we get

$$\text{div}E = -\delta(\nabla\chi, E)$$

Writing

$$E = E_0 + \delta E_1 + O(\delta^2)$$

we get

$$(\nabla^2 + k^2)E_0 = 0,$$

$$(\nabla^2 + k^2)E_1 = -(\nabla(\nabla\chi, E_0) + k^2\chi E_0)$$

A particular solution for E_0 corresponding to propagation of a plane wave in the z direction is given by

$$E_0(\omega, r) = A.\exp(-jkz)\hat{x}$$

which in the time domain reads

$$E_0(t, z) = |A|\cos(\omega t - kz + \phi)$$

We take basis function $\psi_n(r)$ and expand

$$\chi(\omega, r) = \sum_{n=1}^{N} \chi_n(\omega)\psi_n(r)$$

Then,

$$(\nabla^2 + k^2)E_1 = -\sum_n \chi_n(\omega)\nabla(\nabla\psi_n, E_0) - k^2 \sum_n \chi_n(\omega)\psi_n E_0$$

We expand

$$E_1(w, r) = \sum_n \hat{y}_n(w)\psi_n(r)$$

and then get in time domain

$$-c^{-2}\sum_n y_n''(t)\psi_n(r) + \sum_n y_n(t)\nabla^2\psi_n(r) + \sum_n \chi_n(t) * (\nabla\psi_n(r), E_0(t, r))$$

$$+\sum_n \psi_n(r)\chi_n(t) * E_{0,tt}(t,r)$$

Making the approximation

$$\chi_n(t) = \sum_m \chi_n[m]\delta(t - m\Delta)$$

and assuming orthonormality of the basis function ψ_n we get from the above on taking inner products with ψ_n that

$$-c^{-2}y_n''(t) + \sum_m a(n,m)\,y_m(t) + \sum_{m,l} \chi_m[l]\,p_{nml}(t) + \sum_{ml} \chi_m[l]\,q_{nml}(t) = 0 \qquad ...(10)$$

where

$$p_{nml}(t) = \langle(\nabla\psi_m(r), E_0(t - l\Delta, r)), \psi_n(r)\rangle$$

and

$$q_{nml}(t) = \langle(\psi_m(r)\,E_{0,tt}(t - l\Delta, r)), \psi_n(r)\rangle$$

(10) shows explicitly the appearance of the unknown parameters $\{\chi_m[l]\}$ in LIP format and the estimation. The function $q_{nlm}(t)$ and $p_{nml}(t)$ can be computed using the input field $E_0(t, r)$ alone while $y_n(t)$ is obtained from the output/scattered field $E_1(t, r)$. (10) can be put in the form

$$\mathbf{y}''(t) + \mathbf{A}\mathbf{y}(t) + \theta^T\mathbf{x}(t) = \mathbf{w}(t)$$

where $\mathbf{w}(t)$ is white noise perturbing the system and θ is the parameter to be estimated. This can in the turn be case in extended state variable form

$$d\begin{pmatrix}\mathbf{y}(t)\\ \mathbf{y}'(t)\end{pmatrix} = \begin{pmatrix}\mathbf{0} & \mathbf{I}\\ -\mathbf{A} & \mathbf{0}\end{pmatrix}\begin{pmatrix}cc\mathbf{y}(t)\\ \mathbf{y}'(t)\end{pmatrix} + \begin{pmatrix}\mathbf{0}\\ \theta^T\mathbf{x}(t)\end{pmatrix} + \begin{pmatrix}\mathbf{0}\\ \mathbf{w}(t)\end{pmatrix}$$

and noisy output measurement and Kushner filtering theory, an EKF can be designed to estimate the channel parameters $\theta = \{\chi_n[r]\}$.

Appendix:

Here we outline an alternate method for discrete time nonlinear filtering based on discretizing the continuous time Kushner filter. In the Kushner filter, the state dynamics is given by

$$dx(t) = f(t, x(t))\,dt + g(t, x(t))\,dB(t),$$

and the measurement model is given by

$$dz(t) = h(t, x(t)) + dv(t)$$

where $B(.)$ is standard vector valued Brownian motion and $v(t)$ is $R^{1/2}$ times vector values Brownian motion independent of $B(.)$. The observations collected upto time t are given by

$$Z_t = \{z(s) : s \le t\}$$

and if $\phi_t = \phi(x(t))$ is an observable on the system state, the Kushner equation reads

$$d\pi_t(\phi_t) = \pi_t(L_t\phi)dt + \left(\pi_t(h_t\phi_t) - \pi_t(h_t)\pi_t(\phi_t)\right)^T R^{-1}\left(dz(t) - \pi_t(h_t)dt\right)$$

Here, L_t is the backward Kolmogorov generator of the diffusion $x(t)$ and is given by

$$L_t\phi(x) = -f(t,x)^T \nabla_x\phi(x) + \frac{1}{2}Tr\left(\nabla_x\nabla_x^T\left(gg^T(t,x)\phi(x)\right)\right)$$

For the discrete time state equation

$$x[n+1] - x[n] = f(n, x[n])\Delta + g(n, x[n])w[n+1]$$

with $w[n]$ iid $N(0, \sigma_w^2\Delta)$, with the measurement model

$$z[n+1] - z[n] = h[n, x[n]]\Delta + v[n+1]$$

where $v[n]$ is iid $N(0, \sigma_v^2\Delta)$. The observations upto time n are

$$Z_n = \{z[k] : k \le n\}$$

and if $\phi_n = \phi(x[n])$ is an observable on the state and

$$\mathbb{E}(\phi_n | Z_n) = \pi_n(\phi_n)$$

then the Kushner equation approximates to

$$\pi_{n+1}(\phi_{n+1}) - \pi_n(\phi_n) = \pi_n(L_n\phi)\Delta + \sigma_v^{-2}\left(\pi_n(h_n\phi_n) - \pi_n(h_n)\pi_n(\phi_n)\right)^T \left(z[n+1] - z[n] - \pi_n(h_n)\Delta\right)$$

where now L_n is the generator of the discrete time Markov process $x[n]$ and is to be taken as

$$\begin{aligned} L_n\phi(x) &= \Delta^{-1}\mathbb{E}\left[\phi(x[n+1]) - \phi(x[n]) | Z_n\right] \\ &= \Delta^{-1}\left[\left(2\pi\sigma_w^2\Delta\right)^{-p/2}\int\phi(x + f[n,x]\Delta + g[n,x]w)\right. \\ &\quad \left.\exp\left(-\|w\|^2/2\sigma_w^2\right)dw - \phi(x)\right] \end{aligned}$$

Quantum Belavkin filtering theory based on Hudson-Parthasarathy stochastic calculus: For filtering quantum signals, *i.e.* signals at the atomic level, a totally new calculus has to be used. It is called the quantum stochastic calculus and has been developed by Hudson and Parthasarathy [.]. The signals here are operator valued functions of time. Their physical interpretation occurs when we take their expected values in quantum states. The basic setup for describing quantum signals is (1) a system Hilbert space $\mathfrak{h}$, a Boson Fock space $\mathfrak{h}$ for noise

$$\Gamma_s(\mathcal{H}) = \mathbb{C} \oplus \bigoplus_{n \geq 1} H^{\oplus_s n}$$

where $\mathcal{H}$ is the noise Hilbert space and $\otimes_s$ denotes the symmetric tensor product. Finally, for describing time evolution, we use a spectral measure $\chi_{[0,\, t]}$, $t \geq 0$ with values in $P(\mathcal{H})$, the lattice of orthogonal projections in $\mathcal{H}$. For u $\in \mathcal{H}$, the exponential vector $e(u) \in \Gamma_s(\mathcal{H})$ is defined by

$$e(u) = 1 \oplus \bigoplus_{n \geq 1} \mathcal{H}^{\otimes n} / \sqrt{n!}$$

It is then easily seen by considering $\frac{d^n}{dt^n} e(tu)|_{t=0}$ that $U^{\otimes n}$ is recovered and hence $\{e(u) : u \in \mathcal{H}\}$ spans a dense subspace of $\Gamma_s(\mathcal{H})$. Further if $\mathcal{H}_t = \chi_{[0,\, t]}\mathcal{H}, t \geq 0$. Then $\mathcal{H}_t, t \geq 0$ forms an increasing family of Hilbert spaces that converges to $\mathcal{H}$. Thus, the spectral measure $\chi_{[0,\, t]}$ enables us to describe time evolution of quantum noise processes. Specifically, we take $\mathcal{H} = L^2(\mathbb{R}) \otimes \mathbb{C}^p$ and define operators $A_a(t), A_a(t)^*, \Lambda_b^a(t), 1 \leq a, b \leq p$ such that

$$\begin{aligned} A_a(t)e(u) &= u_a(t)e(u), A_a(t)^* e(u) \\ &= \frac{d}{d\alpha} e(u + \alpha.\chi_{[0,t]}e_a)|_{\alpha=0} \end{aligned}$$

where $\{e_a, 1 \leq a \leq p\}$ is an onb for $\mathbb{C}^p$. Also if X is an operator in $\mathcal{H}$, we define

$$\Lambda_X(t)e(u) = \frac{d}{d\alpha} e\left(e^{\alpha(\chi-[0,\, t]\otimes X)u}\right)\Big|_{\alpha=0}$$

we define

$$\Lambda_b^a(t) = \Lambda_{|e_a\rangle\langle e_b|}(t)$$

Finally, for notational convenience, we define

$$\begin{aligned} \Lambda_0^0(t) &= t, A_a(t) = \Lambda_0^a(t), A_a(t)^* \\ &= \Lambda_a^0(t), 1 \leq a \leq p \end{aligned}$$

Then, Hudson and Parthasarathy proved the quantum Ito formula

$$d\Lambda_b^a(t)d\Lambda_f^c(t) = \varepsilon_f^a d\Lambda_b^c(t), 0 \leq a, b, c, f \leq p$$

where ε_b^a is 1 if $a = b \geq 1$, zero if $a = b = 0$ or if $a, b \geq 1, a \neq b$. We now consider an Evans-Hudson diffusion

$$dj_t(X) = j_t\left(\theta_b^a(X)d\Lambda_a^b(t)\right)$$

where $X \in L(\hbar)$, *i.e.* X is a system operator and j_t is a homomorphism from $L(\hbar)$into $L(\hbar \otimes \Gamma_s(\mathcal{H}))$. For $\theta_b^a : \hbar \to \hbar$ are linear operators called the structure functions that satisfy certain relations for j_t to be a star unital homomorphism. $j_t(X)$ is the quantum signal at time t which contains noise since it is an operator in $L(\hbar \otimes \Gamma_s(\mathcal{H}))$. $j_0(X) = X$ is a pure signal observable. In physics, typically $j_t(X)$ is constructed using Schrodinger unitary evolutions with quantum noise, *i.e.*

$$dU(t) = \left[S_b^a(t)d\Lambda_a^b(t)\right]U(t)$$

where $S_b^a(t) \in L(h)$ and the Einstein summation is adopted, *i.e.*, summation over the repeated variables $a, b = 0, 1, \ldots, p$. The conditions on the coefficient operators $S_b^a(t)$ for $U(t)$ to be unitary are obtained from the quantum Ito formula

$$\begin{aligned} 0 &= d(U^*(t)U(t)) \\ &= dU^*(t).U(t) + U^*(t)dU(t) + dU^*(t)dU(t) \\ &= U^*(t)\left[S_b^a(t)^* + S_a^b(t) + S_b^d(t)^* S_a^c(t)\varepsilon_c^d\right]U(t)d\Lambda_b^a(t) \end{aligned}$$

Thus, the condition for unitary evolution is that

$$S_b^{a*} + S_a^b + S_b^{d*} S_a^c \varepsilon_c^d = 0$$

If X is a system observable, then it evolves after time t under the noisy unitary dynamics to

$$j_t(X) = U(t)^*(X \otimes I)U(t)$$

which is abbreviated to $U(t)^*XU(t)$ for short. The measurement process $Y(t), t \geq 0$ must be of the non-demolition type, *i.e*, measurement of $Y(s)$ should not affect the state $j_t(X)$ if $t > s$. This means that $[Y(s), j_t(X)] = 0$ for all $t > s$. For this we require that $Y(t)$ be of the form

$$Y(t) = U(t)^*(I \otimes Z(t))U(t)$$

where $Z(t)\varepsilon L(\Gamma_s(\mathcal{H}))$. It is then easy to show using the quantum Ito formula and the unitarity of $U(t)$ that

$$Y(s) = U(t)^*(I \otimes Z(s))U(t), t \geq s$$

and hence

$$[Y(s), j_t(X)] = U(t)^*[I \otimes Z(s), X \otimes I]U(t) = 0, t \geq s$$

We may assume that

$$Z(t) = \int_0^t c_b^a(s)d\Lambda_a^b(s)$$

where $c_b^a(t)$ are complex scalars. For Hermitianity of the measurements, we require $\bar{c}_b^a(t) = c_a^b(t)$. Equivalently, the measurement process is given by

$$dZ(t) = c_b^a(t)d\Lambda_a^b(t)$$

Application of the quantum Ito formula gives

$$\begin{aligned} dY(t) &= dU^*dZ.U + U^*dZdU + dZ \\ &U^*S_b^{a*}d\Lambda_b^c c_m^k d\Lambda_k^m U + U^*c_m^k d\Lambda_k^m S_b^a d\Lambda_a^b U + c_m^k d\Lambda_k^m \\ &U^*\left(c_m^k S_b^{a*}\varepsilon_k^c U d\Lambda_b^m + U^*\left(c_m^k S_b^a \varepsilon_a^m U d\Lambda_k^b + c_m^k d\Lambda_k^m\right)\right) \end{aligned}$$

$$U^*\left(c_m^k S_b^{a*}\varepsilon_k^c U d\Lambda_b^m + U^*\left(c_k^b S_m^a \varepsilon_a^k U d\Lambda_b^m + c_m^b d\Lambda_b^m\right)\right)$$

$$= \left[j_t\left(c_m^k S_b^{a*}\varepsilon_k^c + c_k^b S_m^a \varepsilon_a^k\right) + c_m^b\right] d\Lambda_b^m$$

$$= j_t\left(N_m^b(t)\right) d\Lambda_b^m(t)$$

say. Clearly, $N_b^m(t)$ are all system operators. We now consider a process $W(t)$ satisfying the qsde

$$dW(t) = \sum_{k\geq 1} f_k(t)W(t)(dY(t))^k$$

where $f_k(t)$ are complex valued functions. Then if $\pi_t(X)$ denotes the conditional expectations of $j_t(X)$ given $Y(s)$, $s \leq t$, *i.e*,

$$\pi_t(X) = \mathbb{E}(j_t(X) \mid \eta_t)$$

where

$$\eta_t = \text{sigma}\,(Y(s) : s \leq t)$$

(This conditional expectations is well defined since the operators $Y(s), s \geq 0)$ commute with each other and $Y(s), s \leq t$ commute with $j_t(X)$ (commuting observables are simultaneously measureable). We have by the orthogonality principle

$$\mathbb{E}\left[(j_t(X) - \pi_t(X))W(t)\right] = 0$$

We assume that

$$d\pi_t(X) = F_t(X)dt + \sum_{r\geq 1} G_{t,r}(X)(dW(t))^r$$

where $F_t(X), G_{t,r}(X)$ are all measurable w.r.t. the algebra $\sigma(Y(s) : s \geq t)$. Application of the quantum Ito formula gives

$$\mathbb{E}\left[(dj_t(X) - d\pi_t(X))W(t)\right] + \mathbb{E}\left[(j_t(X) - \pi_t(X))dW(t)\right]$$
$$+\mathbb{E}\left[(dj_t(X) - d\pi_t(X))dW(t)\right] = 0$$

We have

$$dj_t(X) = d(U(t)^* XU(t))$$
$$= dU^* XU + U^* XdU + dU^* XdU$$

Example of quantum Belavkin filtering. The quantum system is a two state system cosisting of a Hamiltonian matrix

$$H_0 = c_0 I_2 + c_1\sigma_x + c_2\sigma_y + c_3\sigma_z$$
$$= \begin{pmatrix} c_0 + c_3 & c_1 - ic_2 \\ c_1 + ic_2 & -c_3 \end{pmatrix}$$

where the $c_k's$ are real numbers. The Hamiltonian describe the motion of an electron as a two state spin system interacting with a constant magnetic field. The system Hilbert space is therefore $\mathfrak{h} = \mathbb{C}^2$. The noise Hilbert space is taken as $\Gamma_s(\mathbb{C}) = L^2(\mathbb{R}_+)$. The bath noise operators are taken as $a(t)$, $a^*(t)$ where $a(t)$, $a^*(t)$ are the annihilation and creation operators: $a(t) = a(\chi_{[0,t]})$. Let $u \in \mathbb{C}$ and consider

$$e(u) = 1 \oplus \underset{n \geq 1}{\otimes} u^n \sqrt{n!}$$

If we choose and onb $\phi_j(t)$, $j = 1, 2, ...$ for $L^2(\mathbb{R}_+)$, we can then write

$$e(u) = \sum_{n \geq 0} u^n \phi_n(t) / \sqrt{n!}$$

Then,

$$\langle e(u), e(v) \rangle = -\exp(\bar{u}v), u, v \in \mathbb{C}$$

Then,

$$a(t)\, e(u) = u\chi_{[0,t]} e(u)$$

Equivalently,

$$a(t) \sum_{n \geq 0} \left(u^n / \sqrt{n!}\right) \phi_n = \chi_{[0,t]} \sum_{n \geq 0} u^{n+1} \phi_n$$

so

$$a(t)\phi_n = \sqrt{n}\chi_{[0,t]}\phi_{n-1}$$

The quantum Ito formula reads

$$da(t)da(t)^* = dt, da(t)^* da(t) = 0, (da(t))^2 = (da(t)^*)^2 = 0$$

The noisy Schrodinger equation can now be expressed as

$$dU(t) = (-(iH_0 + 0.5LL^*)dt + Lda(t) - L^* da(t))U(t)$$

where $L \in \mathbb{C}^{2 \times 2}$. The noisy observations shall be taken as

$$Y(t) = U(t)^* Z(t)\, U(t)$$

where

$$Z(t) = a(t) + a^*(t)$$

In other words, we are taking measurement of a linear combination of the quantum noise processes and using these to construct a filtered estimate of a system observable $X \in \mathbb{C}^{2 \times 2}$ for which $X^* = X$. The system observable evolves as

$$j_t(X) = U(t)^* XU(t)$$

and the filtered estimate is

$$\pi_t(X) = \mathbb{E}(j_t(X) \mid \eta_t), \eta_t = \sigma(Y(s) : s \leq t)$$

Once we determine $\pi_t(X)$. We can choose a pure system state $|\psi\rangle \in \mathbb{C}^2$ and a pure noise state $|\phi\rangle \in L^2$ (∂ndR_+) and calculate the mean square error at time t

$$\langle \psi \otimes \phi | (j_t(X) - \pi_t(X))^2 | \psi \otimes \phi \rangle$$

Specifically, we choose $|\psi\rangle$ to be the exponential (coherent) state

$$|\psi\rangle = \exp(-|u|^2/2) \sum_{n=0}^{\infty} (u^n / \sqrt{n!}) |\phi_n\rangle$$

and perform these calculations.

❖❖❖❖❖

[74] Alfvenwavesinaplasma:Thedispersionrelationsformagnetohydrodynamics. The electric field is neglected. Then the relevant MHD equations are

$$v_{,t} + (v, \nabla)_v = -\nabla p / p + \nu \nabla^2 v + \alpha (v \times B) \times B,$$

$$\nabla . v = 0,$$

$$\nabla \times B = \sigma (v \times B)$$

We assume that v, p, B are small perturbations of constant vectors and scalars:

$$p = p_0 + \delta p(t, r), v = V_0 + \delta v(t, r), B = B_0 + \delta B(t, r)$$

where p_0 is a constant scalar and V_0, B_0 are constant vectors satisfying the zeroth order MHD equations:

$$(V_0 \times B_0) \times B_0 = 0$$

Then, the first order MHD equations are

$$\delta v_{,t} + (V_0, \nabla) \delta v$$

$$= -\nabla \delta p / p + \nu \nabla^2 \delta v + \alpha ((\delta v \times B_0) \times B_0 + (V_0 \times \delta B) \times B_0 + (V_0 \times B_0) \times \delta B),$$

$$\nabla . \delta v = 0,$$

$$\nabla \times \delta B = \sigma (\delta v \times B_0 + V_0 \times \delta B)$$

These equations describe MHD waves in the conducting fluid (plasma). To obtain the dispersion relations (relation between w and k), we assume that

$$\delta v(t, r) = \mathrm{Re}[\delta v_0 . \exp(i(wt - k.r))],$$

$$\delta B(t, r) = \mathrm{Re}[\delta B_0 . \exp(i(wt - k.r))],$$

$$\delta p(t, r) = \mathrm{Re}[\delta p_0 . \exp(i(wt - k.r))]$$

Then substituting these seven variables into the above seven linearized MHD relations gives

$$iw \delta v_0 - i(k, V_0) \delta v_0$$

$$= ik \delta p_0 / \rho - \nu k^2 \delta v_0 + \alpha ((\delta v_0 \times B_0) \times B_0 + (V_0 \times \delta B_0) \times B_0 + (V_0 \times B_0) \times \delta B_0)$$

$$(k, \delta v_0) = 0$$

$$ik \times \delta B_0 = \sigma(\delta v_0 \times B_0 + V_0 \times \delta B_0)$$

These are seven linear algebraic equations for the seven complex variables δv_0, δB_0, δp_0 and by setting the dterminant of the corresponding coefficient matrix to zero, we get the dispersion relation that relates w with k in terms of V_0, B_0, s, r, n $= \eta/\rho$. The details of the derivations are left to the reader.

❖❖❖❖❖

[75] Energy momentum tensor of the em field: Let p^α be the αth component of the momentum density for $\alpha = 1, 2, 3$ and p^0 the energy density. Let $\pi^{\alpha\beta}$ be the amount of the αth component of the field momentum flowing across a unit area perpendicular to the βth direction per unit time for $1 \le \alpha, \beta \le 3$ and let $\pi^{0\alpha}$ be the energy flowing per unit area per unit time along the αth direction. Clearly $\pi^{\alpha 0} = p^\alpha$, so $\pi^{00} = p^0$ is the energy density of the field. By momentum conservation,

$$\frac{d}{dt}\int_v p^k d^3r = -\int_S \pi^{km} n_m dS - \int_v (\rho E + J \times B)_k d^3r$$

Here, Greek symbols like α, β, μ *nu* assume values 0, 1, 2, 3 while roman symbols like k, m, j, l, n etc. assume values 1, 2, 3. Also by energy conservation

$$\frac{d}{dt}\int_v p^0 d^3r = -\int_S \pi^{0m} n_m dS - \int_v E.J d^3r$$

Here, v is a closed 3-D volume bounded by a closed surface S. Now

$$(\rho E + J \times B)_k = \varepsilon_0 E_k E_{m,m} + [(\text{curl}\, B / \mu_0 - \varepsilon_0 E_{,0}) \times B]_k$$

We have

$$E_{,0} \times B = (E \times B)_{,0} - E \times B_{,0}$$

$$= (E \times B)_{,0} + E \times (\text{curl}\, E)$$

$$= (E \times B)_0 + 0.5\nabla E^2 - (E, \nabla) E$$

Further,

$$(curl\, B) \times B = (B, \nabla) B - 0.5\nabla(B^2)$$

Combining these, and using $B_{m,m} = \text{div}\, B = 0$, we get for the spartial components of the four force density,

$$F^k = (\rho E + J \times B)_k$$

$$= \varepsilon_0 (E_k E_m)_{,m} - \varepsilon_0 0.5(E^2)_{,k} + \mu_0^{-1}(B_k B_m) - 0.5\mu_0^{-1}(B^2)_{,k} - \varepsilon_0 (E \times B)_{,0}$$

$$= -S^{k\alpha}_{,\alpha}$$

where

$$S^{k0} = p^k = \varepsilon_0 (E \times B)_0,$$

$$S^{km} = \pi^{km}$$
$$= ((\varepsilon_0 / 2)E^2 + B^2 / 2\mu_0)\delta_{km} - \varepsilon E_k E_m - B_k B_m / \mu_0$$

Likewise the time component of the four force density is

$$F^0 = J.E = (E, \nabla \times B) / \mu_0 - \varepsilon_0 (E, E_{,0})$$
$$= -\nabla(E \times B)/\mu_0 - (B, B_{,0})/\mu_0 - \xi_0(E, E_{,0}) = -S^{0\alpha}_{,\alpha}$$

where

$$S^{00} = p^0 = \varepsilon_0 E^2 / 2 + B^2 / 2\mu_0$$

is the energy density and

$$S^{0k} = \mu_0^{-1}(E \times B)_k$$

is the energy flux. Note that the energy flux is same as the momentum density upto a constant of proportionality. This is manisfested in the symmetry of the energy-momentum tensor of the electromagnetic field. We now assume $c = 1$ and express the energy-momentum tensor in terms of the em field tensor. The em field tensor is

$$F_{\mu\nu} = A_{\nu,\mu} - A_{\mu,\nu}$$

Consider

$$S^{\mu\nu} = \frac{1}{4} F^{\alpha\beta} F^{\alpha\beta} \eta^{\mu\nu} - F^{\mu}_{\alpha} F^{\nu\alpha}$$

We have

$$F_{0k} = A_{k,0} - A_{0,k} = -A^k_{,0} - A^0_{,k} = E_k,$$
$$F_{12} = A_{2,1} - A_{1,2} = -A^2_{,1} + A^1_{,2} = -B_3,$$
$$F_{23} = -B_1, F_{31} = -B_2$$

Thus,

$$F_{\mu\nu}F^{\mu\nu} = 2F_{12}^2 + 2F_{23}^2 + 2F_{31}^2 - \left(2F_{01}^2 + 2F_{02}^2 + 2F_{03}^2\right)$$
$$= 2B^2 - 2E^2$$
$$F^0_\alpha F^{0\alpha} = -F^{0k}F^{0k} = -F_{k0}F_{k0} = -E^2,$$
$$F^0_\alpha F^{k\alpha} = F^0_m F^{km} = F_{0m}F_{km} = E_m F_{km} = E_1 F_{k1} + E_2 F_{k2} + E_3 F_{k3}$$

For $k = 1$, this is

$$\text{s}\, E_2 F_{12} + E_3 F_{13} = -E_2 B_3 + E_3 B_2 = -(E \times B)_1$$

Likewise for the others. Thus,

$$F^0_\alpha F^{k\alpha} = -(E \times B)_k$$

This gives

$$S^{00} = (B^2 - E^2)/2 + E^2 = (E^2 + B^2)/2$$

which agrees with the energy density of the em field. Further,

$$S^{k0} = -F^k_m F^{m0} = -F_{km} F_{m0} = F_{0m} F_{mk} = -F_{01} F_{1k} - F_{02} F_{2k} - F_{03} F_{3k}$$

For $k = 1$, this is

$$S^{10} = -F_{02} F_{21} - F_{03} F_{31} = E_2 B_3 - E_3 B_2 = (E \times B)_1$$

and in general,

$$S^{k0} = (E \times B)_k$$

Further,

$$\begin{aligned} S^{11} &= 0.25 F_{\mu\nu} F^{\mu\nu} - F^1_\alpha F^{1\alpha} \\ &= -(B^2 - E^2)/2 - F_{10} F_{10} + F_{12} F_{12} + F_{13} F_{13} \\ &= -(B^2 - E^2)/2 - E_1^2 + B_3^2 + B_2^2 \\ &= \left(E_2^2 + E_3^2 - E_1^2\right)/2 + \left(B_2^2 + B_3^2 - B_1^2\right)/2 \end{aligned}$$

❖❖❖❖❖

[76] Relativistic MHD: The total energy momentum tensor of the matter fluid and radiation field is given by

$$\begin{aligned} K^{\mu\nu} &= T^{\mu\nu} - S^{\mu\nu} \\ &= (\rho + p) v^\mu v^\nu - p\eta^{\mu\nu} - 0.25 F_{\alpha\beta} F^{\alpha\beta} \eta^{\mu\nu} + F^{\mu\alpha} F^\nu_\alpha \end{aligned}$$

Conservation of energy and momentum of matter plus radiation requires that

$$K^{\mu\nu}_{,\nu} = 0$$

so that

$$T^{\mu\nu}_{,\nu} = F^\mu = -S^{\mu\nu}_{,\nu}$$

Now,

$$S^{\mu\nu}_{,\nu} = 0.5 \eta^{\mu\nu} F_{\alpha\beta,\nu} F^{\alpha\beta} - F^{\mu\alpha}_{,\nu} F^\nu_\alpha - F^{\mu\alpha} F^\nu_{\alpha,\nu}$$

Now,

$$\begin{aligned} &\eta^{\mu\nu} F_{\alpha\beta,\nu} F^{\alpha\beta} - 2 F^{\mu\alpha}_{,\nu} F^\nu_\alpha \\ &= F^{\alpha\beta}\left(2 F^\mu_{\alpha,\beta} + \eta^{\mu\nu} F_{\alpha\beta,\nu}\right) \\ &= F^{\alpha\beta} \eta^{\mu\nu} (2 F_{\nu\alpha,\beta} + F_{\alpha\beta,\nu}) \\ &= F^{\alpha\beta} \eta^{\mu\nu} (2 F_{\nu\alpha,\beta} - F_{\nu\beta,\alpha} + F_{\alpha\beta,\nu}) \\ &= F^{\alpha\beta} \eta^{\mu\nu} (F_{\nu\alpha,\beta} + F_{\beta\nu,\alpha} + F_{\alpha\beta,\nu}) = 0 \end{aligned}$$

by the two Maxwell equations not involving current or change densities. Thus,

$$S^{\mu\nu}_{,\nu} = F^{\mu\alpha}F^{\nu}_{\alpha,\nu}$$

Now,

$$F^{0\alpha}_{,\alpha} = F^{0k}_{,k} = -\sum_k F_{0k,k} = -\operatorname{div} E = -\rho/\varepsilon_0$$

Likewise,

$$F^{k\alpha}_{,\alpha} = -\mu_0 J^k$$

These can be combined into the single Maxwell equation

$$F^{\alpha\beta}_{,\beta} = -\mu_0 J^\alpha$$

(assuming $c = 1$) and hence choosing units so that $\mu_0 = 1$, we get

$$S^{\mu\nu}_{,\nu} = F^{\mu\alpha}J_\alpha$$

In particular,

$$S^{0\nu}_{,\nu} = F^{0k}J_k = F_{0k}J^k = (J, E)$$

$$S^{km}_{,m} = F^{km}J_m = -F_{km}J^m = F_{0k}J^0 + F_{1k}J^1 + F_{2k}J^2 + F_{3k}J^3$$

$$= \rho E_k + (J \times B)_k$$

The MHD equations

$$T^{\mu\nu}_{,\nu} = -S^{\mu\nu}_{,\nu}$$

give

$$\left[(\rho + p)v^\mu v^\nu - p\eta^{\mu\nu}\right]_{,\nu} = F^{\mu\nu}J_\nu \qquad ...(1)$$

In general relativity, this must be replaced by

$$\left[(\rho + p)v^\mu v^\nu - pg^{\mu\nu}\right]_{,\nu} = F^{\mu\nu}J_\nu$$

(1) must be supplemented by the Maxwell equations

$$F^{\mu\nu}_{,\nu} = -J^\mu$$

For a conducting fluid, we have

$$J^\mu = \sigma F^{\mu\nu}v_\nu$$

Taking for example $\mu = 1$ we get

$$J_1 = \sigma\left(F^{10}v_0 + F^{12}v_2 + F^{13}v_3\right)$$

$$= \sigma\left(v^0 E_1 + v^2 B_3 - v^3 B_2\right) = \gamma\sigma(E + \mu + B)_1$$

where $\mu^k = dx^k / dt, dt / d\tau = \gamma = (1 - \mu^2)^{-1/2}$

We note that for a conducting fluid with scalar homogeneous isotropic conductivity σ, Ohm's law is given by

$$J^\mu = -\sigma F^{\mu\nu} v_\nu$$

For example, taking $\mu = 1$ gives

$$J^1 = -\sigma\left(F^{10}v_0 + F^{12}v_2 + F^{13}v_3\right)$$

$$= \sigma\gamma\left(E_1 + u^2 B_3 - u^3 B_2\right) = \sigma\gamma(E + u \times B)_1$$

which agrees with the non-relativistic Ohm's law for small velocities *i.e.*, when $\gamma \approx 1$. The complete relativistic MHD equations are thus (for a non-viscous fluid)

$$\left[(\rho + p)v^\mu v^\nu - pg^{\mu\nu}\right]_{,\nu} = -\sigma F^{\mu\nu} F_{\nu\alpha} v^\alpha \qquad ...(2)$$

and
$$F^{\mu\nu}_{,\nu} = -\sigma F^{\mu\nu} v_\nu \qquad ...(3)$$

If tensor conductivity (inhomogeneous and anisotropic) is accounted for, then Ohm's law should be taken as

$$J^\mu = \sigma^\mu_{\alpha\beta\rho}(x) F^{\alpha\beta} v^\rho$$

where $\sigma^\mu_{\alpha\beta\rho}$ is a tensor field. Finally, if both curvature of space-time and viscous effects are accounted for, then the complete MHD equations become (assume that the metric of the background space-time $g_{\mu\nu}(x)$ is fixed)

$$\left[(\rho + p)v^\mu v^\nu - pg^{\mu\nu}\right]_{:\nu} = -\sigma F^{\mu\nu} F_{\nu\alpha} v^\alpha - \eta_0 g^{\alpha\beta} v^\mu_{:\alpha:\beta} \qquad ...(2)$$

and
$$F^{\mu\nu}_{:\nu} = z - \sigma F^{\mu\nu} v_\nu \qquad ...(3)$$

Small perturbations in p, v^μ, $g_{\mu\nu}$ and A^μ can be studied by linearizing the above equations about a nominal value.

❖❖❖❖❖

[77] Compound Poisson processes: Estimating parameters in dynamical systems driven by compound Poisson processes. Let Y_a, $a = 1, 2, \ldots$ be an idd sequence of random variables with values in $\mathbb{R}$, having distribution function F_Y and density $F_Y = f_Y$. Let $N(t)$, $t \geq 0$ be a Poisson process with intensity λ independent of $\{Y_a\}$. Consider the process

$$X(t) = \sum_{a=1}^{N(t)} Y_a$$

Then $X(t)$ is an independent increment process with moment generating function

$$\mathbb{E}[\exp(\alpha X(t))] = \exp\left(\lambda t \int (\exp(\alpha y) - 1)\, dFY(y)\right)$$

Let
$$\mathbb{E}\exp(\alpha Y_1) = \int \exp(\alpha y)\, dF_Y(y) = M_Y(\alpha)$$

the moment generating function of Y_1. Then, we get

$$M_X(\alpha, t) = \mathbb{E}\exp(\alpha X(t)) = \exp(\lambda t(M_Y(\alpha) - 1))$$

$$= \exp(-\lambda t)\sum_{n\geq 0}\left((\lambda t)^n / n!\right)M_Y(\alpha)^n$$

More generally,

$$\mathbb{E}\left[\exp\left(\int_0^T \alpha(t)\,dX(t)\right)\right] = \mathbb{E}\left[\exp\left(\int_0^T \alpha(t)Y_{N(t)+1}\,dN(t)\right)\right]$$

$$= \Pi_{t\in[0,T]}\mathbb{E}\left[\exp(\alpha(t)Y_1)\lambda dt + (1-\lambda dt)\right]$$

$$= \exp\left(\lambda\int_0^T (M_Y(\alpha(t))-1)\,dt\right)$$

Jerk noise in quantum systems: Let

$$X(t) = \sum_{k=1}^{N(t)} Y_k$$

Consider the noisy Schrodinger evolution equation

$$dU(t) = [(-iH_0(t)+P(t))\,dt - idX(t).V(t)]U(t)$$

We have $$dU(t) = Y_{N(t)+1}dN(t)$$

Thus, $$(dX(t))^2 = Y_{N(t)+1}^2 dN(t)$$

For $U(t)$ to be unitary, we require that

$$0 = d(U^*(t)U(t)) = dU^*.U + U^*.dU + dU^*.dU$$

$$= U^*((P(t)^* + P(t))\,dt + (dX)^2V^2)U$$

We therefore require that

$$P + P^* = -(dX)^2V^2 = -Y_{N(t)+1}^2 dN.V^2$$

We may for example choose

$$P(t) = -0.5Y_{N(t)+1}^2 V^2(t)\,dN(t)$$

❖❖❖❖❖

[78] Dirac Hamiltonian perturbed by quantum noise. The Dirac Hamiltonian for a free particle is given by

$$H_0 = (\alpha, p) + \beta m$$

where

$$(\alpha, p)\sum_{r=1}^{3}\alpha_r p_r, \alpha_r = \begin{pmatrix} 0 & \sigma_r \\ \sigma_r & 0 \end{pmatrix},$$

$$\beta = \begin{pmatrix} I_2 & 0 \\ 0 & -I_2 \end{pmatrix}$$

In other words, $$H_0 = \begin{pmatrix} mI_2 & (\sigma, p) \\ (\sigma, p) & -mI_2 \end{pmatrix}$$

When the Dirac Hamiltonian is perturbed by an electromagnetic field, described by a magnetic vector potential $A(t, r)$ and electric scalar potential $V(t, r)$, the perturbed Hamiltonian becomes

$$H(t) = (\alpha, p+eA)+\beta m - eV = H_0 + e((\alpha, A)-V)$$

In quantum field theory, A, V are expressed in terms of superpositions of creation and annihilation operators corresponding to the photon field. If we therefore assume that the bath em field that perturbs the atom (electron) is a quantum noise process, then we may express $A\,dt$, $V\,dt$ as superposition of creation and annihilation operators $da(t)$, $da^*(t)$ where $a(t,)$, $a^*(t)$ are operators in the Boson Fock space $\Gamma_s(L^2(\mathbb{R}_{++}))$ satisfying the Hudson-Parthasarathy quantum Ito formula

$$da.da^* = dt,\ da^*dt = 0,\ (da)^2 = (da^*)^2 = 0$$

We thus write

$$-iH(t)dt = -\left(iH_0 + \varepsilon^2 LL^*/2\right)dt + \varepsilon L.da(t) - \varepsilon.L^*da^*(t)$$

where the extra factor $\varepsilon LL^*/2$ is chosen so as to guarantee unitary evolution:

$$dU(t) = \left[-\left(iH_0 + \varepsilon^2 LL^*/2\right)dt + \varepsilon L.da(t) - \varepsilon.L^*da^*(t)\right]U(t)$$

i.e., $$0 = d\left(U^*(t)U(t)\right) = dU^*.U + U^*dU + dU^*dU$$

L acts on the atomic Hamiltonian Hilbert space which in the momentum domain is $\hbar = L^2(\mathbb{R}^3) \otimes \mathbb{C}^4$. We note that the unperturbed Dirac Hamiltonian H_0 is simply multiplication by 4×4 matrix valued function of p in $\hbar$. Since the vector and scalar potentials are functions of t, r and in the momentum domain, $r = q = i\nabla_p$, we can write

$$e\left((\alpha, A(t,r)) - V(t,r)\right) = \varepsilon\left(L(t, i\nabla p)da(t) - L(t, i\nabla p)^*da^*(t)\right)$$

Here, $L(t, r)$ is a 4×4 matrix valued function of t and $r = i\nabla_p$. Standard second order time dependent perturbation theory gives the unitary evolution of the radiation perturbed Dirac Hamiltonian as

$$U(t) = U_0(t) + \varepsilon U_1(t) + \varepsilon^2 U_2(t) + O\left(\varepsilon^3\right)$$

where $$U_0(t) = \exp\left(-itH_0\right) = \pounds^{-1}\left(sI_4 + iH_0\right)^{-1}$$

$$= \pounds^{-1}\frac{\left(sI_4 - iH_0\right)}{s^2 + E^2} = \cos(Et)I_4 - \left(iH_0/E\right)\sin(Et)$$

where $H_0 = H_0(p) = (\alpha, p) + \beta m,\ E = E(p) = \sqrt{m^2 + p^2}$

Equivalently,

$$U_0(t) = \begin{pmatrix} [\cos(Et)-(im/E)\sin(Et)]I_2 & -i(\sigma, p)\sin(Et)/E \\ -i(\sigma, p)\sin(Et)/E & [\cos(Et)+(im/E)\sin(Et)]I_2 \end{pmatrix}$$

$$U_1(t) = -\int_0^t U_0(t-\tau)(L_\tau da(\tau) - L_\tau^* da(\tau)^*)U_0(\tau)d\tau,$$

$$U_2(t) = \int_0^t U_0(t-\tau)\left(L_\tau L_\tau^*/2\right)U_0(\tau)d\tau$$

$$-\int_0^t U_0(t-\tau)\left(L_\tau da(\tau) - L_\tau^* da(\tau)^*\right)U_1(\tau)d\tau$$

$$= \int_0^t U_0(t-\tau)\left(L_\tau L_\tau^*/2\right)U_0(\tau)d\tau$$

$$+\int_{0<\tau'<\tau<t} U_0(t-\tau)\left(L_\tau da(\tau) - L_\tau^* da(\tau)^*\right)U_0(\tau-\tau')$$

$$\left(L_{\tau'} da(\tau') - L_{\tau'}^* da(\tau')^*\right)U_0(\tau')d\tau d\tau'$$

Assume that we wish to calculate the transition probability amplitude matrix element.

$$\langle g\phi(v)|U(t)|f\phi(u)\rangle$$

where $f(p) \in L^2(\mathbb{R}^3)$ is the initial momentum space wave function of the electron and $g(p) \in L^2(\mathbb{R}^3)$ is the final momentum space wave function of the electron. Here, $u, v \in L^2(\mathbb{R}_+)$ and

$$||\phi(u)\rangle = \exp\left(-\|u\|^2/2\right)\Big|e(u)>, e(u) \underset{n=0}{\overset{\infty}{\oplus}} u^{\otimes n}/\sqrt{n!}$$

$|\phi(u)\rangle$ is the initial state of the photon (radiation) field and $|\phi(u)\rangle$ is the final state of the photon field. When the photon state is $|\phi(u)\rangle$, it corresponds to the fact that with probability $\exp(-\|u\|^2)\|u\|^{2n}/n!$, there are n photons. We have

$$\langle g\phi(u)|U_1(t)|f\phi(u)\rangle/\langle\phi(u)|\phi(u)\rangle$$

$$= -\int_0^t \left[\langle g|U_0(t-\tau)L_\tau U_0(\tau)|f\rangle u(\tau) + \left\langle g\middle|U_0(t-\tau)L_\tau^* U_0(\tau)\middle|f\right\rangle \bar{u}(\tau)\right]d\tau$$

where we have used

$$da(t)|\phi(u)\rangle = u(t)dt\,|\phi(u),$$

and hence

$$\langle\phi(v)|da(t)|\phi(u)\rangle = u(t)dt\langle\phi(v)|\phi(u)\rangle$$

$$\left\langle\phi(v)\middle|da(t)^*\middle|\phi(u)\right\rangle = \bar{v}(t)dt\langle\phi(v)|\phi(u)\rangle$$

❖❖❖❖❖

[79] Brownian motion on a Riemannian manifold. Let the manifold $\mathcal{M}$ be imbedded in $\mathbb{R}^n$ as $x \to \left[x, F^T(x)\right]^T = H(x)$ where $x \in \mathbb{R}^p$ and $F(x) \in \mathbb{R}^{n-p}$. The p dimensional Riemannian manifold $\mathcal{M}$ has x as its coordinates. Its metric is given by given by

$$G(x) = H'(x)^T H(x) = I_p + F'(x)^T F(x) \in \mathbb{R}^{p\times p}$$

Let $P(x)$ be the orthogonal projection of $\mathbb{R}^n$ onto *Tmathcal* M_x. Thus,

$$P(x) = H'(x)G(x)^{-1}H'(x)^T \in \mathbb{R}^{n\times n}$$

has rank p. Note that $P(x) = P(x)^T = P(x)^2$. We write

$$P(x) = \begin{pmatrix} Q(x) \\ R(x) \end{pmatrix}$$

where

$Q(x) \in \mathbb{R}^{p\times n}$, $R(x) \in \mathbb{R}^{n-p\times n}$

Consider the sde

$$\partial X = Q(X)\partial B \qquad ...(1)$$

where $X(t) \in \mathbb{R}^P$ and $B(.)$ ia an n dimensional Brownian motion. Here ∂ denotes the Stratanovich differential. We claim that the process $X(t)$ is a Brownian motion on $\mathcal{M}$ in the sense that it is a diffusion process with generator given by one half times the Laplace-Beltrami operator on $\mathcal{M}$. Writing $((g_{\mu\nu}(x))) = G(x)$ and $((g^{\mu\nu}(x))) = G(x)^{-1}$, we also define $g(x) = det\,(G(x))$. Then the Laplace-Beltrami operator on $\mathcal{M}$ is given by

$$\Delta\psi = \left(g^{\mu\nu}\psi_{,\nu}\right)_{:\mu} = g^{-1/2}\left(g^{1/2}g^{\mu\nu}\psi_{,\nu}\right)_{,\mu}$$

$$= g^{\mu\nu}\psi_{,\mu\nu} + g^{-1/2}\left(g^{1/2}g^{\mu\nu}\right)_{,\mu}\psi_{,\nu}$$

(1) can be expressed as an Ito equation:

$$dX = Q(X)dB + \frac{1}{2}dQ(X)dB$$

$$= Q(X)dB + \frac{1}{2}Q_{,\alpha}(X)dX_\alpha dB$$

$$= Q(X)dB + \frac{1}{2}Q_{,\alpha}(X)Q_{\alpha\beta}(X)e_\beta dt$$

where e_β is then $n\times 1$ coluum vector with a one in the βth position. Let $(Q^T)_\beta$ denote the βth column of Q^T

$$Q_{,\alpha}Q_{\alpha\beta}e_\beta = Q_{,\alpha}\left(Q^T\right)_\alpha$$

Note that $$Q = G^{-1}H^T = G^{-1}\left[I_p, F'^T\right]$$

$$Q^T = H'G^{-1}$$

$$Q_{,\alpha}(Q^T)_\alpha = \left(G^{-1}H'^T\right)_{,\alpha} H'\left(G^{-1}\right)_\alpha \qquad ...(2)$$

where $(G^{-1})_\alpha$ is the αth column of G^{-1}. Let $y^r(x)$ denote the rth entry of the column vector $H(x)$. Note that $y^r(x) = x^r$, $r = 1, 2,..., p$ and $y^r(x) = F_r(x)$,. $r = p + 1...n$. Then the μth entry of

$Q_{,\alpha}(Q^T)_\alpha$

is from (2) given by

$$\left(g^{\mu\nu} y^n_{,\nu}\right), \alpha y^n_{,\beta} g^{\alpha\beta}$$

where summation over all the repeated indices is implied. This equals

$$g^{\mu\nu}_{,\alpha} g_\nu \beta g^{\alpha\beta} + g^{\mu\nu} g^{\alpha\beta} y^n_{,\nu\alpha} y^n_{,\beta}$$

where we have used

$$g_{\mu\nu} = y^n_{,\mu} y^n_{,\nu}$$

Moreover, from Dirac's book on general relativity, we know that the Connection coefficients for $\mathcal{M}$ in the coordinate system x are given by

$$\Gamma_{\mu\alpha\beta} = y^n_{,\mu} y^n_{,\alpha\beta}$$

Thus, we get for the μth entry of $Q_{,a}(Q^T)_\alpha$ the expression

$$g^{\mu\nu}_{,\alpha} + g^{\mu\nu} g^{\alpha\beta} \Gamma_{\beta\nu\alpha} = g^{\mu\nu}_{,\alpha} + g^{\mu\nu} \Gamma^\alpha_{\nu a}$$

The Ito equation

$$dX = Q(X)dB + \frac{1}{2} Q_{,\alpha}(X) Q_{\alpha\beta}(X) e_\beta dt$$

then implies that the generator of the process X is given by

$$L = \frac{1}{2} Tr\left(QQ^T \nabla\nabla^T\right) + \frac{1}{2}\left[g^{\mu\alpha}_{,\alpha} + g^{\mu\nu}\Gamma^\alpha_{\nu\alpha}\right]\partial_\mu$$

Now

$$QQ^T = G^{-1}H'^T H' G^{-1} = G^{-1}$$

Hence

$$L = \frac{1}{2} g^{\mu\nu} \partial_\mu \partial_\nu + \frac{1}{2}\left[g^{\mu\alpha}_{,\alpha} + g^{\mu\nu}\Gamma^\alpha_{\nu\alpha}\right]\partial_\mu$$

On the other hand, in terms of the covariant derivative, the Laplace-Beltrami operator on $\mathcal{M}$ can also expressed as

$$\Delta\psi = (g^{\mu\nu}\psi_{,\nu})_{:\mu}$$

$$= g^{\mu\nu}\psi_{,\mu:\nu} = g^{\mu\nu}\left(\psi_{,\mu\nu} - \Gamma^\alpha_{\mu\nu}\psi_{,\alpha}\right)$$

It therefore follows that $L = \frac{1}{2}\Delta$ provided that

$$= -g^{\alpha\nu}\Gamma^{\mu}_{\nu\alpha}$$

$$= g^{\mu\nu}_{,\alpha} + g^{\mu\nu}\Gamma^{\alpha}_{\nu\alpha}$$

The rhs is the same as

$$-g^{\mu\nu}g^{\alpha\beta}g_{\nu\beta,\alpha} + g^{\mu\nu}g^{\alpha\beta}\left(g_{\beta\nu,\alpha} + g_{\beta\alpha,\nu} - g_{\alpha\nu,\beta}\right)/2$$

$$= -\frac{1}{2}g^{\mu\nu}g^{\alpha\beta}\left(g_{\beta\nu,\alpha} + g_{\alpha\nu,\beta} - g_{\alpha\beta,\nu}\right)$$

$$-g^{\mu\nu}g^{\alpha\beta}\Gamma_{\nu\alpha\beta} = -g^{\alpha\beta}\Gamma^{\mu}_{\alpha\beta}$$

which is the same as the lhs. We have thus constructed a Brownian motion process on a Riemannian manifold with positive definite metric using the Stratanovich differential equation.

[80] Quantum stochastis processes in quantum field theory: The free Dirac wave function is

$$\psi(x) = \int\Big[\alpha_\sigma(P)u_\sigma(P)\exp(-ip.x)/\sqrt{E(P)}$$

$$+b_\sigma(P)^* v_\sigma(P)\exp(-ip.x)/\sqrt{E(P)}\Big]d^3P$$

where the repeated summation variable σ runs over 1, 2 corresponding two spin states of a spin 1/2 particle. $p = (p^\mu)$ is the four momentum,

$$p = (p^\mu) = (p^0, P), p^0 = E(P) = \sqrt{m^2 + P^2}$$

P is the 3-momentum. In order to satisfy the free Dirac equation

$$\left(i\gamma^\mu\partial_\mu - m\right)\psi(x) = 0$$

We require that u_σ and u_σ (these are 4 × 1 complex vectors) satisfy the algebraic eigen-equations

$$\left(\gamma^\mu p_\mu - m\right)u_\sigma(P) = 0$$

and

$$\left(\gamma^\mu p_\mu + m\right)v_\sigma(P) = 0$$

These equation are equivalent to

$$\left(E(P) - (\alpha, P) - \beta m\right)u_\sigma(P) = 0$$

and

$$\left(E(P) - (\alpha, P) - \beta m\right)v_\sigma(-P) = 0$$

for $\sigma = 1, 2$ and all $P \in \mathbb{R}^3$. In other words, $u_\sigma(P)$ is an eigenvector of the free Dirac Hamiltonian

$$H_{D0} = (\alpha, P) + \beta m$$

with positive energy eigenvalue $E(p) = \sqrt{m^2 + P^2}$ while $v_\sigma(-P)$ is an eigenvector of the free Dirac Hamiltonian with negative energy eigenvalue $-E(P) = -\sqrt{m^2 + P^2}$. Here,

$$\gamma^0 = \begin{pmatrix} I & 0 \\ 0 & -I \end{pmatrix} = \beta,$$

$$\gamma^r = \begin{pmatrix} 0 & \sigma_r \\ -\sigma_r & 0 \end{pmatrix}, r = 1, 2, 3$$

$$\alpha_r = \gamma^0\gamma^r = \begin{pmatrix} 0 & \sigma_r \\ \sigma_r & 0 \end{pmatrix}, r = 1, 2, 3$$

and $$(\alpha, P) = \sum_{r=1}^{3} \alpha_r \mathrm{P^r}$$

Note that $p^r = P^r$, r = 1, 2, 3. The second quantized Dirac Hamiltonian involves assuming $a_\sigma(P), b_\sigma(P)$ to be operators and satisfying the anticommutation relations

$$\left\{\alpha_\sigma(P), a_{\sigma'}(P')^*\right\} = \delta_{\sigma,\sigma'}\delta(P - P'),$$

$$\left\{b_\sigma(P), b_{\sigma'}(P')^*\right\} = \delta_{\sigma,\sigma'}\delta(P - P'),$$

and all other anicommutators are zero:

$$\{a_\sigma(P), a_{\sigma'}(P')\} = \{a_\sigma(P), b_{\sigma'}(P')\}$$
$$= \left\{a_\sigma(P), b_{\sigma'}(P')^*\right\} = 0$$

where

$$\{X, Y\} = XY + YX$$

The reason for introducing these anticommutation relations rather than commutations relations for bosons is a consequence of the Pauli exclusion principle. $a_\sigma(P)^* a_\sigma(P)$ is the number operator for electrons at momentum P and spin σ (No summation over σ). Likewise, $b_\sigma(\mathrm{P})^* b_\sigma(P)$ is the number operator for positrons at momentum P and spin σ. ($\sigma = 1$ corresponds to spin + 1/2 while $\sigma = 2$ corresponds to spin $-1/2$). $a_\sigma(P)$ annihilates an electron of momentum P and spin σ, $a_\sigma(P)^*$ creates an electron of momentum P and spin σ. $b_\sigma(P)$ annihilates a positron of momentum P and spin σ while $b_\sigma(P)^*$ creates a positron of momentum P and spin σ. The Pauli exclusion principle states that for any value of (P, σ), there cannot be more than one electron or one positron. Thus, $a_\sigma(P)^*\ a_\sigma(P)$ and $b_\sigma(P)^*$

$b_\sigma(P)$ can have eigenvalues only zero and one. Let $|0>$ be the vacuum. Then the Pauli exclusion principal implies that

$$a_\sigma(P)^{*2}|0> = 0, b_\sigma(P)^{*2}|0> = 0$$

A state in which there are electrons at momenta and spins (P_k, σ_k), $k = 1, 2, ..., d_1$ and positrons at momenta and spins $(P'_k \sigma'_k)$, $k = 1, 2, ... d_2$ is given by

$$P_k, \sigma_k, k = 1, 2, ..., d_1, P'_k, \sigma'_k, k = 1, 2, ..., d_2>$$
$$= \Pi_{k=1}^{d_1} a_{\sigma k}(P_k)^* \Pi_{k=1}^{d_2} b_{\sigma' k}(p'_k)^* |0\rangle$$

Suppose we had a commutation relation

$$a_\sigma(P) a_\sigma(P')^* - a_\sigma(P') a_\sigma(P) = \delta(P - P')$$

Discretization of this equation gives

$$a_\sigma(P) a_\sigma(P)^* - a_\sigma(P)^* a_\sigma(P) = 1$$

and hence if $|1>$ is a state with one electron at (P, σ), then

$$a_\sigma(P) a_\sigma(P)^* |1\rangle = 2|1>$$

in particular, $a_\sigma(p)^*|1>$ is non zero which means that we can have a state with two electrons at (p, σ) violating the Pauli exclusion principle.

The second quantized free Dirac Hamiltonian is

$$\tilde{H}_{D0} = \int \psi(x)^* ((\alpha, -i\nabla) + \beta m) \psi(x) d^3x$$

$$= \int (EE')^{-1/2} \alpha_\sigma(P)^* \alpha_{\sigma'}(P') u_\sigma(P)^* H_{D0}(P') u_{\sigma'}(P')$$
$$\exp(i(p - p').x) d^3x d^3p d^3p'$$
$$+ \int (EE')^{-1/2} \alpha_\sigma(P)^* b_{\sigma'}(P') u_\sigma(P)^* H_{D0}(-P') v_{\sigma'}(P')$$
$$\exp(i(p - p').x) d^3x d^3p d^3p'$$
$$+ \int (EE')^{-1/2} b_\sigma(P)^* b_{\sigma'}(P') v_\sigma(P)^* H_{D0}(-P') v_{\sigma'}(P')$$
$$\exp(-i(p - p').x) d^3x d^3p d^3p'$$
$$+ \int (EE')^{-1/2} b_\sigma(P)^* a_{\sigma'}(P') v_\sigma(P)^* H_{D0}(-P') u_{\sigma'}(P')$$
$$\exp(-i(p - p').x) d^3x d^3p d^3p'$$

$$= \int E^{-1} a_\sigma(P)^* a_{\sigma'}(P) u_\sigma(P)^* H_{D0}(P') u_{\sigma'}(P) d^3p$$
$$+ \int E^{-1} a_\sigma(P)^* b_{\sigma'}(-P) u_\sigma(P)^* H_{D0}(P) u_{\sigma'}(-P)$$
$$\exp(2iEt) d^3p$$

$$+\int E^{-1}b_{\sigma}(P)b_{\sigma'}(-P)^{*}v_{\sigma}(P)^{*}H_{D0}(-P)v_{\sigma'}(P)d^{3}p$$

$$+\int E^{-1}b_{\sigma}(P)^{*}a_{\sigma'}(-P)^{*}v_{\sigma}(P)^{*}H_{D0}(-P)u_{\sigma'}(P)d^{3}x$$

$$+\int E^{-1}b_{\sigma}(P)^{*}a_{\sigma'}(-P)^{*}v_{\sigma}(P)^{*}H_{D0}(-P)u_{\sigma'}(-P)$$

$$\exp\left[(-2iEt)d^{3}p\right.$$

apart from a multiplicative factor $(2\pi)^3$. We choose the following normaliztion for the eigenvectors of $H_{D0}(P)$

$$u_{\sigma}(P)^{*}u_{\sigma'}(P) = E(P)\delta_{\sigma\sigma'}$$
$$v_{\sigma}(P)^{*}v_{\sigma'}(p) = E(P)\delta_{\sigma\sigma'}$$

Note that since $u_{\sigma}(P)$ and $v_{\sigma'}(-P)$ belong to different eigenvalues $E(P)$ and $-E(P)$ of $H_{D0}(P)$, it follows that

$$u_{\sigma}(P)^{*}v_{\sigma'}(-P) = 0$$

Thus the above expression for the quantized Dirac Hamiltonian reduces to

$$\int E(P)a_{\sigma}(P)^{*}a_{\sigma}(P)d^{3}P$$

$$-\int E(P)b_{\sigma}(P)b_{\sigma}(P)^{*}d^{3}P$$

❖❖❖❖❖

Quantum Gravity, Lie Groups in Robotics, Field of Robots, Quantum Robots, Quantum Transmission Lines, Quantum Optimal Control

[81] Approximate quantization of the gravitational field. The Lagrangian density of the free gravitational field is given by

$$L = g^{\mu\nu}\sqrt{-g}R_{\mu\nu}$$

which can be shown to be equivalent to

$$L = g^{\mu\nu}\sqrt{-g}\left(\Gamma^{\alpha}_{\alpha\beta}\Gamma^{\beta}_{\alpha\beta} - \Gamma^{\alpha}_{\mu\beta}\Gamma^{\beta}_{\nu\alpha}\right)$$

We now assume that $g_{\mu\nu}$ is a small pertubation of the Minkowski metric:

$$g_{\mu\nu}(x) = \eta_{\mu\nu} + h_{\mu\nu}(x)$$

Then, we easily see that if we retain upto cubic terms in $h_{\mu\nu}$, $h_{\mu\nu,\rho}$ in L, we get

$$L = C_1(\mu\nu\alpha\beta\rho\sigma)h_{\mu\nu,\rho}h_{\alpha\beta,\sigma} + C_2(\mu\nu\alpha\beta\rho\sigma ab)h_{\mu\nu,\rho}h_{\alpha\beta,\sigma}\,h_{ab}$$

where C_1, C_2 are constants. If we retain upto quadratic orders in $h_{\mu\nu}$ and its partial derivatives, the Lagrangian aproximates to

$$L_0 = C_1(\mu\nu\alpha\beta\rho\sigma)h_{\mu\nu,\rho}h_{\alpha\beta,\sigma}$$

and the variational principle $\delta\int Ld^4x = 0$ results in the wave equation

$$C_1(\mu\nu\alpha\beta\rho\sigma)h_{\alpha\beta,\rho\sigma} = 0$$

Expanding $h_{\mu\nu}(x)$ as a superposition of plane waves

$$h_{\mu\nu}(x) = \int\left[u_{\mu\nu}(K)\exp(-ik.x) + \bar{u}_{\mu\nu}(K)\exp(ik.x)\right]d^3K$$

where

$$k.x = k_{\mu}x^{\mu} = k^0x^0 - \sum_{r=1}^{3}k^rx^r$$

we get the dispersion relation connecting k^0 to $K = \left(k^r\right)_{r=1}^{3}$ as

$$+\,C_1(\mu\nu\alpha\beta\rho\sigma)k_{\sigma}k_{\sigma}u_{\sigma\beta}(K) = 0$$

and for a nonzero solution $u_{\alpha\beta}(K)$ to exist, we require that

$$\det\left[\left(\left(C_1(\mu\nu\alpha\beta\rho\sigma)k_{\rho}k_{\sigma}\right)\right)_{(\mu\nu)(\alpha\beta)}\right] = 0$$

which is a 32th degree homogeneous polynomian equation in $k_{\mu.}$ This gives us in general 32 values of k_0 for a given K. We shall however assume two solutions for each K, namely $k^0 = \pm f(K)$ and proceed. The Lagrangian density after retaining only quadratic terms can be expressd as

$$L = C_1(\mu\nu\alpha\beta_{00})h_{\mu\nu,0}h_{\alpha\beta,0} + 2C_1(\mu\nu\alpha\beta_0 r)h_{\mu\nu,0}h_{\alpha\beta,r}$$
$$+ C_1(\mu\nu\alpha\beta rs)h_{\mu\nu,r}h_{\alpha\beta,s}$$

The position fields are the set of ten functions $h_{\mu\nu}(x)$ $0 \le m \le v \le 3$ and the momentum fields are the set of ten functions

$$\pi^{\mu\nu} = \frac{\partial L}{\partial h_{\mu\nu,0}}$$

$$= 2C_1(\mu\nu\alpha\beta_{00})h_{\alpha\beta\,0} + 2C_1(\mu\nu\alpha\beta_0 r)h_{\alpha\beta,r} \quad ...(1)$$

Let $D_1(\mu\nu\alpha\beta)$ denote the inverse matrix of $C_1(\mu\nu\alpha\beta_{00})$ in the sense that

$$C_1(\mu\nu\alpha\beta_{00})D_1(\alpha\beta\rho\sigma) = \delta_{\mu\rho}\delta_{\nu\sigma}$$

We write this equation as

$$D_1 = C_1^{-1}$$

Then writing (1) as

$$\pi = 2C_1 h_{,0} + 2\tilde{C}_1\nabla_h$$

we get

$$h_{,0} = C_1^{-1}(\pi/2 - \tilde{C}_1\nabla_h) = D_1(\pi/2 - \tilde{C}_1\nabla_h)$$

Note that L can also be expressed as

$L(h, h_{,0,} \nabla h)$

$$= h_{,0}^T C_1 h_{,0} + 2h_{,0}^T\tilde{C}_1\nabla h + \nabla h^T\hat{C}_1\nabla h$$

Then,

$$\pi = 2C_1 h_{,0} + 2\tilde{C}_1\nabla h$$

Finally, the Hamiltonian density is

$$H(h, \nabla h, h_{,0}) = \pi^T h_{,0}{}^{-L}$$
$$= h_{,0}^T C_1 h, 0 - \nabla h T\hat{C}\nabla h$$
$$= (\pi/2 - \hat{C}_1\nabla h)^T C_1^{-1}(\pi/2 - \hat{C}_1\nabla h) - \nabla h^T\hat{C}_1\nabla h$$
$$= H(h, \nabla h, \pi)$$

Now,

$$h_{\mu\nu}(x) = \int\left(u_{\mu\nu}(k)\exp(-ik.x)+\bar{u}_{\mu\nu}(k)\exp(ik.\,x)\right)d^3k$$

on in vector notation,

$$h(x) = \int\left(u(k)\exp(-ik.x)+\bar{u}_{\mu\nu}(k)\exp(ik.\,x)\right)d^3k$$

Here, $k^0 = f(K) = f(-K)$ is assumed.

$$\nabla h(x) = i\int\left(u(k)\otimes K\exp(-ik.x)-\bar{u}(k)\otimes K\exp(ik.\,x)\right)d^3k$$

$$h,0 = -i\int\left(u(k)k^0 e\text{xp}(-ik.x)-\bar{u}(k)k^0 e\text{xp}(ik.\,x)\right)d^3k$$

$$\int h^T_{,0}C_1h,0d^3x$$

$$= \int\left(-k^{02}u(K)^T C_1u(-K)\exp\left(-2ik^0t\right)+k^{02}u^T(K)C_1\bar{u}(K)\right.$$

$$\left.+k^{02}u(K)^* C_1u(K) - k^{02}u^*(K)C_1\bar{u}(-K)\exp\left(2ik^0t\right)\right)d^3k$$

❖❖❖❖❖

[82] Lie group-Lie algebra approach to modelling the motion of multiple 3-D robot links. The first link has its base point connected to the origin, the second link has its base point connected to the point p_1 on the top of the first link and in general, the kth link has its base point connected to the point p_{k-1} on the top the $(k-1)^{\text{th}}$ link. There are all d links. After time t all the d links experience a rotation $R_1(t)$ w.r.t the origin, followed by a rotation $R_2(t)$ experienced by the second to d^{th} link w.r.t. p_1 and so on till the dth link experiences the rotation $R_d(t)$ w.r.t. the $(d-1)^{\text{th}}$ link around p_{d-1}. Thus a point r in the kth link, after time t, moves to the point

$$r(t) = R_1(t)p_1 + R_2(t)R_0(t)(p_2-p_1) + \ldots R_k(t)R_{k-1}(t)\ldots R_1(t)(r-p_{k-1}) \qquad \ldots(1)$$

We define the rotation matrices

$$S_k(t) = R_k(t)R_{k-1}(t)\ldots R_1(t), 1\le k\le d$$

Then (1) can be exporessed as

$$r(t) = S_1(t)p_1 + S_2(t)(p_2-p_1)+\ldots+S_{k-1}(t)(p_{k-1}-p_{k-2})+S_k(t)(r-p_{k-1})$$

and hence the kinetic energy of the system is assuming that B_k is the region in $\mathbb{R}^3$ occupied by the kth link at time $t = 0$

$$K(t) = \sum_{k=1}^{d} \frac{\rho}{2} \int_{B_k} \left\| S_1'(t)q_1 + S_2'(t)q_2 + \ldots + S_{k-1}'(t)q_{k-1} + S_k'(t)(r - p_{k-1}) \right\|^2 d^3r$$

where

$$q_s = p_s - p_{s-1}, s \geq 1, p_0 = 0$$

The integration can be easily carried out leading to the following general expression for the total kinetic energy:

$$K(t) = \frac{1}{2} \sum_{k,m=1}^{d} Tr\left(S_k'(t) J_{km} S_m'(t)^T \right)$$

where J_{km} are 3×3 matrices with the block structured $3d \times 3d$ matrix $J = ((J_{km}))$ being positive definite. The potential energy of the system at time t has the form

$$V(t) = \sum_{k,=1}^{d} \rho g \int_{B_k} \left\langle e_3, S_1(t)q_1 + S_2(t)q_2 + \ldots + S_{k-1}(t)q_{k-1} + S_k(t)(r - p_{k-1}) \right\rangle d^3r$$

$$= \sum_{k,=1}^{d} m_k g \left\langle e_3, S_1(t)q_1 + S_2(t)q_2 + \ldots S_{k-1}(t)q_{k-1} + S_k(t)d_k \right\rangle$$

where

$$d_k = \int_{B_k} (r - p_{k-1}) d^3r / \mu(B_k)$$

is the centre of gravity of the kth link at time $t = 0$ relative to its base point p_{k-1} on the $(k - 1)^{th}$ link. In short, $V(t)$ is a linear function of the matrix elements of $S_1(t), \ldots, S_d(t)$ and we can write

$$V(t) = \sum_{k=1}^{d} \left\langle e_3, S_k(t)\xi_k \right\rangle$$

where the $\xi'_k s$ are constant 3-vectors. Now we come to the contribution to the Lagrangian by the machine torgue at the base link. The kth link is attached at the point p_{k-1} to the $(k-1)^{th}$ link where there ia a motor which provides a torgue descriobed by an antisymmetric 3×3 matrix $T_k(t)$. Since this motor is attached to a point on the $(k-1)^{th}$ link at the point p_{k-1}. This angular velocity tensor is easily see to be given by

$$\Omega_k(t) = R'_k(t)R_k(t)^{-1} = R'_k(t)R_k(t)^T$$

Now,

$$R_k = S_k S_{k-1}^{-1}$$
$$R'_k = S_k' S_{k-1}^T - S_k S_{k-1}^T S_{k-1}' S_{k-1}^T$$

so

$$\Omega_k = S_k' S_k^T - S_k S_{k-1}^T S_{k-1}' S_k^T$$

The total virtual work done on the robot by the machine torques during a small variation δS_k, $k = 1, 2, ..., d$ of the rotation matrices that define the robot configuration at any time is therefore given by

$$\delta W = -\frac{1}{2}\sum_{k=1}^{d} Tr\left(T_k\left(\delta S_k S_k^T - S_k S_{k-1}^T \delta S_{k-1} S_k^T\right)\right)$$

The equations of motion follow from the variational principle

$$\delta\int_0^T (K(t)-V(t))dt - \int_0^T \delta W(t)dt - \delta\int_0^T Tr\left(\Lambda_k(t)\left(S_k(t)^T S_k(t)-I\right)\right)dt = 0$$

where $\Lambda_k(t)$ is a 3×3 Lagrange multiplier matrix that the $S_k(t)'s$ are rotation matrices. The variations are to be calculated w.r.t. the $S_k(t)'s$ and the $\Lambda_k(t)'s$. The above variational principle yields after integrating by parts,

$$-2\sum_m J_{km}S_m''(t) - \sum_k \xi_k e_e^T - \frac{1}{2}S_k(t)^T T_k(t) + \frac{1}{2}S_{k+1}(t)^T T_{k+1}(t)S_{k+1}(t)S_k(t)^T$$
$$-\left(\Lambda_k(t)+\Lambda_k(t)^T\right)S_k(t)^T = 0$$

This equation is to be supplemented with the variation w.r.t. $\Lambda_k(t)$ set to zero which gives

$$S_k(t)^T S_k(t) = \mathrm{I}$$

Remark: Let $R(t)$ be a G valued function of t where G is a Lie group with Lie algebra g. Then writing $R(t) = \exp(X(t))$ where $X(t) \in g$. it follows that

$$R^{-1}R' = \frac{\left(I-\exp\left(-ad\left(X(t)\right)\right)\right)}{ad\left(X(t)\right)}\left(X'(t)\right) \in g$$

Again differentiating this equation and using the fact that g is a closed vector space, we get

$$-R^{-1}R'R^{-1}+R^{-1}R''\in g$$

and hence

$$R^{-1}R''\in g+g.g$$

where we assume that G is a matrix Lie group and hence g is a matrix Lie algebra. $g.g$ is the set of all products $A.\ B$ with $A, B \in g$.

❖❖❖❖❖

[83] Group algebras with multiplicative noise: Let G be a matrix Lie group and $\mathcal{A}$ be its group algebra. Let g be the Lie algebra of G. Consider a sequence g_n, $n = 0, 1, 2, \ldots$, of elements of $\mathcal{A}$ satisfying the stochastic difference equation

$$g_{n+1} = f_1(g_n, n)+f_2(g_n, n)w_{n+1}$$

where w_n is an iid g-valued random process and $f_m : \mathcal{A}\times\mathbb{Z}_+\rightarrow\mathcal{A}$, $m = 1, 2$. This is the state equation and the measurement equation is

$$z_n = h_1(g_n, n)+h_2(g_n, n)v_n$$

where K is a matrix Lie group with Lie algebra t and group algebra $\mathcal{B}$, h_m: $g\times\mathbb{Z}_+\rightarrow\mathcal{B}$, $m = 1, 2$ and v_n is a t valued iid stochastic sequence. The measurement sequence upto tiem n is given by $Z_n = \{z_k : k \le n\}$ and the filtering problem is to obtain $p(g_n|Z_n)$ recursively or equivalently, $\mathbb{E}(\phi(g_n|Z_n))$ for any measurable function $\phi : u\rightarrow\mathbb{R}$. Let $\{X_a : \text{a} = 1, 2, \ldots N\}$ be a basis for g and $\{Y_a{:}a = 1,2,\ldots\mathbb{M}\}$ be a basis for t. Then we can write

$$w_n = \sum_{a=1}^{N} w_n(a)X_a,\ v_n = \sum_{a=1}^{N} v_n(a)Y_a$$

where $w_n(a)v_n(a)$ are Gaussian random variables. From basic probability theory (conditional probability, Markov processes etc.) we have

$$p(g_{n+1}/Z_{n+1}) = p(g_{n+1}, Z_n, z_{n+1})/p(Z_n, z_{n+1})$$

$$= \int_u p(z_{n+1}|g_{n+1})p(g_{n+1}|g_n)p(g_n|Z_n)dg_n/D$$

where the denominator D equals the numerator integrated w.r.t. G_{n+1}. Here, we are assuming that $\mathcal{A}$ has a G invariant measure dg and $p(g_{n+1}|g_n)$ is the density of the conditional probability distribution of g_{n+1} given g_n with respect to this measure. We thus find that if

$$\pi_n(\phi(g)) = \mathbb{E}(\phi(g_n)|Z_n)$$

then

$$\pi_{n+1}(\phi(g)) = \frac{\sigma_{n+1}(\phi(g))}{\sigma_{n+1}(1)}$$

where

$$\sigma_{n+1}(\phi(g)) = \pi_n\left(\int\phi(g_{n+1})p(z_{n+1}|g_{n+1})p(g_{n+1}|g_n = g)dg_{n+1}\right)$$

$$= \pi_n\left(\int\phi(f_1(g,n)+f_2(g,n)w)\det\left(h_2\left(f_1(g,n)+f_2(g,n)w,n+1\right)\right)\right)^{-1}$$

$$\times pv\left(h_2\left(f_1(g,n)+f_2(g,n)w,n+1\right)^{-1}\left(z_{n+1}-h_1(g_{n+1},n+1)\right)\right)p_\omega(w)dw$$

❖❖❖❖❖

[84] Teleoperation robotic fields: $q(t, x) \in \mathbb{R}^d$ is the configuration of the robot at $x \in \mathbb{R}^3$ at time t. The equations of motion of this robot field are

$$M(q(t,x))q_{,tt}(t,x)+N(q(t,x),q_{,t}(t,x))$$

$$\tau(t,x)+\int K_p(x,y)(q(t,y)-q(t,x))d^3y$$

$$+\int K_d(x,y)(q_{,t}(t,y)-q_{,t}(t,x))d^3y \quad \text{...(1)}$$

Can this dynamical equation be derived from a Legrangian density ?. To do so, we consider a Lagrange multiplier function $\lambda(t, x) \in \mathbb{R}^3$ and define a Lagrangian

$$L(q,q_{,t},\gamma,\gamma_{,t}) = \int\Big[-\left(\gamma(t,x)^T M(q(t,x))\right)_{,tq,t}(t,x)$$

$$+\gamma(t,x)^T N(q(t,x),q,t(t,x))-\gamma(t,x)^T\tau(t,x)\Big]d^3x$$

$$+\int\Big[K_p(x,y)\lambda(t,x)^T(q(t,y)-q(t,x))$$

$$+K_d(x,y)\lambda(t,x)^T(q_{,t}(t,y)-q_{,t}(t,x))\Big]d^3xd^3y$$

The action functional is defined by

$$S(q, \lambda) = \int Ldt$$

It is easily seen that the variational equation

$$\delta_{\lambda(t, x)}S = 0$$

gives us after integration by parts, eqn. (1) Further, the variational $\delta_{q(t, x)}S = 0$ gives us after integration by parts, the "costate-equations"":

$$-\left(M(q(t,x))\lambda(t,x)\right)_{,tt}+\left(I\otimes q_{,tt}(t,x)^T\right)M'(q(t,x))\lambda(t,x)$$

$$+N_{,1}\left(q(t,x),q_{,t}(t,x)\right)^T\lambda(t,x)+\int K_p(y,x)\lambda(t,y)d^3y$$

$$-\left(\int K_p(x,y)d^3y\right)\lambda(t,x)$$

$$-\int K_d(y,x)\lambda_{,t}(t,y)d^3y+\left(\int K_d(x,y)d^3y\right)\lambda_{,t}(t,x$$

Now to quantize this Lagrangian, we must first perform the Legendre transformation and obtain the Hamiltonian.

❖❖❖❖❖

[85] Quantization of the dynamical system

$$q'(t) = f(q(t))$$

Define the Hamiltonian

$$H(q, p) = p^T f(q)$$

The Hamilton equations of motion give

$$q' = H_{,p} = f(q), p' = -H_{,q} = -[f'(q)]^T p$$

If $q(t)$ is the solution, then

$$p(t) = T\left\{\exp\left(-\int_0^t f'\left(q(s)^T ds\right)p(0)\right)\right.$$

where $T(.)$ is the time ordering operator.

❖❖❖❖❖

[86] Ito's formula of or quantum semi martingales: Let

$$dM = EdA + FdA^* + Gd\Lambda + Hdt$$

Then let

$$d(M^n) = E_n dA + F_n dA^* + G_n d\Lambda + H_n dt$$

We have

$$E_{n+1}dA + F_{n+1}dA^* + G_{n+1}d\Lambda + H_{n+1}dt = d(M.\ M^n)$$

$$= dM.M^n + M.d\left(M^n\right) + dM.d\left(M^n\right)$$

$$= \left(EdA + FdA^* + Gd\Lambda + Hdt\right)\left(M^n\right)$$

$$+M.\left(E_n dA + F_n dA^* + G_n d\Lambda + H_n dt\right)$$

$$+\left(EdA + FdA^* + Gd\Lambda + Hdt\right).\left(E_n dA + F_n dA^* + G_n d\Lambda + H_n dt\right)$$

$$\left(EM^n + ME_n + EG_n\right)dA + \left(FM^n + MF_n + GF_n\right)dA^*$$

$$+\left(GM^n + MG_n + GG_n\right)d\Lambda$$

$$\left(HM^n + HM_n + EF_n\right)dt$$

It follows that the processes E_m, F_n, G_n, H_n satisfy the recursion relations

$$E_{n+1} = EM^n + ME_n + E + EG_n, F_{n+1} = FM^n + MF_n + GF_{n'}$$

$$G_{n+1} = GM^n + MG_n + GG_{n'} H_{n+1} = HM^n + MH_n + EF_n$$

❖❖❖❖❖

[87] Quantization of the transmission line equations.

$$v_{,x}(t,x) + R(t,x)i(t,x) + (L(t,x)i(t,x))_{,t} = 0$$

$$i_{,x}(t,x) + G(t,x)v(t,x) + (C(t,x)v(t,x))_{,t} = 0$$

The Lagrangian density is obtained by introducing Lagrange multipliers $\lambda(t, x)$, $\mu(t, x)$:

$$£(t, x, v, i, \lambda, \mu) = \lambda\left(v_{,x} + Ri + (Li)_{,t}\right) + \mu\left(i_{,x} + Gv + (Cv)_{,t}\right)$$

The variational principle

$$\delta\int £dxdt = 0$$

gives the two line equations and the following costate-equations:

$$-\lambda_{,x} + G\mu - (C\mu)_{,t} = 0$$

$$R\lambda + L\lambda_{,t} - \mu_{,x} = 0$$

The canonical momentum densities coresponding to v, i, λ, are respectively

$$\pi_v = \frac{\partial £}{\partial v_{,t}} = \mu C,$$

$$\pi_i = \frac{\partial L}{\partial i_{,t}} = \lambda L,$$

$$\pi_\lambda = \frac{\partial £}{\partial \lambda_{,t}} = 0,$$

$$\pi_\mu = \frac{\partial £}{\partial \mu_{,t}} = 0$$

Thus, the Hamiltonian density is

$$\begin{aligned}\mathcal{H} &= \pi_v v_{,t} + \pi_i i_{,t} + \pi_\lambda \lambda_{,t} + \pi_\mu \mu_{,t} \\ &= \mu C v_{,t} + \lambda L i_{,t} \\ &\quad -\lambda\left(v_{,x} + Ri + (Li)_{,t}\right) - \mu\left(i_{,x} + Gv + (Cv)_{,t}\right) \\ &= -\lambda\left(v_{,x} + Ri\right) - \mu\left(i_{,x} + Gv\right) - \lambda L_{,t} i - \mu C_{,t} v \\ &= -(\pi_i/L)\left(v_{,x} + Ri\right) - (\pi_v/C)\left(i_{,x} + Gv\right) \\ &\quad -(\pi_i/L)L_{,t} i - \mu(\pi v/C)C_{,t} v\end{aligned}$$

$$= -\left[L^{-1}\pi_i v_{,x} + (R/L)\pi_i i + C^{-1}vi_{,x} + (G/C)\pi_v v + (\log L)_{,t}\pi_i i + (\log C)_{,t}\pi_v v \right]$$

$$= -\left[\left(G/C + (\log C)_{,t}\right)\pi_v v + \left(R/L + (\log L)_{,t}\right)\pi_i i + L^{-1}\pi_i v_{,x} + C^{-1}vi_{,x} \right]$$

$$= \mathcal{H}\left(v, i, v_{,x}, i_{,x}, \pi_v, \pi_i\right)$$

We leave it as an exercise to verify that the Hamilton equations

$$v_{,t} = \frac{\delta\mathcal{H}}{\delta\pi_v} = \frac{\partial\mathcal{H}}{\partial\pi_v}$$

$$i_{,t} = \frac{\delta\mathcal{H}}{\delta\pi_i} = \frac{\partial\mathcal{H}}{\partial\pi_i}$$

$$\pi_{v,t} = -\frac{\delta\mathcal{H}}{\delta v} = -\frac{\partial\mathcal{H}}{\partial v} + \frac{\partial}{\partial x}\left(\frac{\partial\mathcal{H}}{\partial v_{,x}}\right)$$

$$\pi_{i,t} = -\frac{\delta\mathcal{H}}{\delta i} = -\frac{\partial\mathcal{H}}{\partial i} + \frac{\partial}{\partial x}\left(\frac{\partial\mathcal{H}}{\partial i_{,x}}\right)$$

give the correct transmission line equations.

❖❖❖❖❖

[88] Quantum Lorentz transformations: $\psi_{p,\sigma}$ is the state corresponding to four momentum p and z component of spin σ. Let k be the four momentum $(1, 0, 0, 0)$. We assume that if Λ is a Lorentz transformation, then

$$D(\Lambda)\psi_{k,\sigma} = \psi_{p,\sigma}$$

where $p = \Lambda(k)$ i.e. $p^\mu = \Lambda_n^\mu k^v$. We wish to evaluate $D(\Lambda)\psi_{p,\sigma}$ for an arbitrary four momentum p and arbitrary Lorentz transformation Λ. Here, D is a representation of the Lorentz group in the Hilbeart space of wave functions $\psi_{p,\sigma}$. Let $L(p)$ be a Lorentz tranformation taking k to p, i.e, $L(p)k = p$, We have

$$D(\Lambda)\psi_{p,\sigma} = D(\Lambda)D\left(L(p)\right)\psi_{k,\sigma}$$

$$D\left(\Lambda L(p)\right)\psi_{k,\sigma} = D\left(L(\Lambda p)\right)DL\left(L(\Lambda p)^{-1}\Lambda L(p)\right)\psi_{k,\sigma}$$

Now consider the Lorentz transformation

$$W(\Lambda, p) = L(\Lambda p)^{-1}\Lambda L(p)$$

$$W(\Lambda, p)_k = L(\Lambda p)^{-1}\Lambda L = k$$

Hence W fixes k and hence acts on the spin component only, i.e, we can write

$$D(W(\Lambda, p))\psi_{k,\sigma} = \sum_{\sigma'} D_{\sigma',\sigma}\left(W(\Lambda, p)\right)\psi_{k,\sigma'}$$

it follows that

$$D(\Lambda)\psi_{k,\sigma} = D(L(\Lambda p))\sum_{\sigma'} D_{\sigma',\sigma}(W(\Lambda, p))\psi_{k,\sigma'}$$

$$= \sum_{\sigma'} D_{\sigma',\sigma}(W(\Lambda, p))\psi_{\Lambda(p),\sigma'}$$

(Reference: Steven Weinberg,"The quantum theory of fields", Volume 1)

❖❖❖❖❖

[89] Scalar Klein-Gordon equation in a perturbed metric: Background metric is $g^{(0)}_{\mu\nu}(x)$. Perturbed metric is

$$g_{\mu\nu}(x) = g^{(0)}_{\mu\nu}(x) + h_{\mu\nu}(x)$$

Upto linear orders in $g_{\mu\nu}$, we have

$$g^{\mu\nu} = g^{(0)\mu\nu} - h^{\mu\nu}$$

where

$$h^{\mu\nu} = g^{(0)\mu\alpha} g^{(0)nu\beta} h_{\alpha\beta}$$

$$g = \det((g_{\mu\nu})) = g^{(0)}(1 + h)$$

where

$$g^{(0)} = \det\left(\left(g^{(0)}_{\mu\nu}\right)\right)$$

and

$$h = g^{(0)\mu\nu} h_{\mu\nu}$$

Then

$$\sqrt{-g} = \sqrt{-g^{(0)}}(1 + h/2)$$

The scalar Klein-Gordon equation is

$$\left(g^{\mu\nu}\sqrt{-g}\psi_{,u}\right)_{,v} + m_0^2\sqrt{-g}\psi = 0$$

This equation is the same as

$$\left(g^{\mu\nu}\psi_{,u}\right)_{:v} + m_0^2\psi = 0$$

Noting that

$$g^{\mu\nu}\sqrt{-g} = \left(g^{(0)\mu\nu} - h^{\mu\nu}\right)\sqrt{-g^{(0)}}(1 + h/2)$$

$$= g^{(0)\mu\nu}\sqrt{-g^{(0)}} - \sqrt{-g^{(0)}}h^{\mu\nu} + g^{(0)\mu\nu}\sqrt{-g^{(0)}}h/2$$

and the above scalar wave equation upto linear orders in $h_{\mu\nu}$ becomes

$$\left[\left(g^{(0)\mu\nu}\sqrt{-g^{(0)}} - \sqrt{-g^{(0)}}h^{\mu\nu} + g^{(0)\mu\nu}\sqrt{-g^{(0)}}h/2\right)\psi_{,\mu}\right]_{,v}$$

$$m_0^2\sqrt{-g^{(0)}}\,(1+h/2)\psi = 0$$

Now writing

$$\psi = \psi^{(0)} + \psi^{(1)}$$

where $\psi^{(0)}$ is zeroth order in $h_{\mu\nu}$ and $\psi^{(1)}$ is first order in $h_{\mu\nu}$, we get equating zeroth and first order terms,

$$\left[g^{(0)\mu\nu}\sqrt{-g^{(0)}}\psi^{(0)}_{,\mu}\right]_{,\nu} + m_0^2\sqrt{-g^{(0)}}\psi^{(0)} = 0$$

$$\left[g^{(0)\mu\nu}\sqrt{-g^{(0)}}\psi^{(0)}_{,\mu}\right]_{,\nu} + m_0^2\sqrt{-g^{(0)}}\psi^{(1)}$$

$$+\left[\left(-\sqrt{-g^{(0)}}h^{\mu\nu} + g^{(0)\mu\nu}\sqrt{-g^{(0)}}\,h/2\right)\psi^{(0)}_{,\mu}\right]_{,\nu}$$

$$+ m_0^2\sqrt{-g^{(0)}}h\psi^{(0)}/2 = 0$$

Define the partial differential operators L_0 and L_1 by

$$L_0\psi = g^{(0)\mu\nu}\sqrt{-g^{(0)}}\psi_{,\mu}\Big]_{,\nu} + m_0^2\sqrt{-g^{(0)}}\psi$$

$$L_1\psi = \left[\left(-\sqrt{-g^{(0)}}h^{\mu\nu} + g^{(0)\mu\nu}\sqrt{-g^{(0)}}\,h/2\right)\psi_{,\mu}\right]_{,\nu}\; m_0^2\sqrt{-g^{(0)}}\,h\psi/2$$

Then our equation are

$$L_0\psi^{(0)} = 0,\; L_0\psi^{(1)} = -L_1\psi^{(0)}$$

For the Schwarzchild metric,

$$g_{00} = \alpha(r),\; g_{11} = -\alpha(r)^{-1},\; g_{22} = -r^2,\; g_{33} = -r^2\sin^2(\theta)$$

and hence

$$g^{00} = \alpha(r)^{-1},\; g^{11} = -\alpha(r)^{-1},\; g_{22} = -r^2,\; g^{33} = -r^{2(}\sin^2(\theta))^{-2}$$

$$\sqrt{-g} = r^2\sin^{(\theta)}$$

so

$$L_0\psi = \alpha(r)^{-1}r^2\sin(\theta)\psi_{,tt} - \sin(\theta)\left(\alpha(r)r^2\psi_{,r}\right)_{,r}$$

$$-\left(\sin(\theta)\psi_{,\theta}\right)_{,\theta} - \left(\sin(\theta)\right)^{-1}\psi_{,\phi\phi} + m_0^2r^2\sin(\theta)\psi = 0$$

The unperturbed radial Klein-Gordon equation in the Schwarzchild metric is then with $\psi = \psi(t, r)$

$$\alpha(r)^{-1}r^2\psi_{,tt} - \left(\alpha(r)r^2\psi_{,r}\right)_{,r} + m_0^2r^2\psi = 0$$

and replacing $\psi(t, r)$ by $\psi(r)\exp(j\omega t)$ gives the radial Klein-Gordon equation in the Schwarzchild metric:

$$\left(\alpha(r)r^2\psi'\right)' + \left(\omega^2\alpha(r)^{-1}r^2 + m_0^2 r^2\right)\psi = 0$$

This is the same as

$$\left(\alpha(r)/r^2\right)\left(\alpha(r)r^2\psi'\right)' + \left(\omega^2 + m_0^2\alpha(r)\right)\psi = 0$$

❖❖❖❖❖

[90] Optimal control of the wave function of an atomic system using eternally applied electromagnetic fields. The Hamitonian of the atom after interaction with the em field described by the potentials ($A(t, r)$, $\Phi(t, r)$) is given by

$$H(t) = \left(p + eA(t,r)\right)^2/2m + V(r) - e\Phi(t,r)$$

$$= H_0 + (e/2m)\left((A, p) + (p, A)\right) - e\Phi + e^2A^2/2m$$

where

$$H_0 = p^2/2m + V = -\nabla^2/2m + V$$

The wave function $\psi(t, r)$ satisfies Schrodinger's equation

$$\psi_{,t}(t, r) = -iH(t)\psi(t, r)$$

We have to design the potentials $A(t, r)$, $V(t, r)$ so that

$$\int_{\mathbb{R}^2\times[0,T]} \left|\psi_d(t,r) - \psi(t,r)\right|^2 d^3r dt$$

is a minimum. This is done using the Hamilton-Jacobi theory by incorporating the Schrodinger equation as a constraint using Lagrange multipiers. The functional to be minimized taking into account an additional energy constraint on the em field gives

$$S\left[\psi, \bar{\psi}, A, \Phi, \Lambda, \bar{\Lambda}, \lambda\right] = \int_{\mathbb{R}^2\times[0,T]} \left|\psi_d(t,r) - \psi(t,r)\right|^2 d^3r dt$$

$$- \int_{\mathbb{R}^3\times[0,T]} \mathrm{Re}\left(\Lambda(t,r)\left(\psi_{,t}(t,r) + iH(t)\psi(t,r)\right)\right) d^3r dt$$

$$-\lambda\left[\int_{\mathbb{R}^2\times[0,T]} \left(\left|\nabla\Phi + A_{,t}\right|^2 + \left|\nabla + A\right|^2\right) d^3r dt - E\right]$$

This action can be minimized using variational calculus resulting in (*a*) Schrodinger's equation (*b*) differential equations satisfied by A Φ and (c) costate equations for $\Lambda(t, r)$. We leave it as an exercise to the reader to carry out the ze computations. Note that the aim is to design the em field under the energy constraint so that the wave function tracks a given wave function over the region $\mathbb{R}^3 \times [0,T]$ of space-time.

❖❖❖❖❖

[91] Stochastic version of the previous problem: The wave function satisfies a stochastic differential equation

$$d\psi(t,r) = \left(\left(-H(t)+\sum_{k=1}^{p} L_k^2(t,r)\Big/2\right)dt - i\sum_{k=1}^{p} L_k(t,r)dB_k(t)\right)\psi(t,r)$$

where $B_k(t)$, $k = 1, 2,...p$ are independent Brownian motion processes and $L_k(t, r)$ $k = 1, 2, ..., p$ are real valued functions of t, r. Note that we can replace the multiplication operators $L_k(t, r)$ by linear self-adjoint differential operators $L_k(t)$ and still carry out our programme. The problem is then to minimize

$$\int_0^T \mathbb{E}\left|\psi_d(t,r)-\psi(t,r)\right|^2 dt$$

subject to the same energy constraint as above. This problem is solved by using the stochastic generalization of the Bellman-Hamilton-Jacobi dynamic programming algorithm by deriving an equation for

$$F(t,\psi(t,.)) = \min_{A(s,r),\,\Phi(s,r),\,t<s<T} \int_{R_2\times[t,T]} \mathbb{E}\left[\left|\psi_d(s,r)-\psi(s,r)\right|^2 \psi(t,.)\right]d^3rds$$

❖❖❖❖❖

[92] Quantum stochastic Bellman-Hamilton-Jacobi equation for optimal control of a quantum stochastic differential equation. The qsde for the state is in the Evans-Hudson form

$$d_{jt}(X) = j_t\left(\theta_b^{\alpha}(X,u(t))\right)d\Lambda_a^b(t)$$

where the summation is over a, $b = 0, 1, 2$ ... with the fundamental processes $\Lambda_b^a(t)$ satisfying the quantum Ito formula

$$A\Lambda_b^a(t), d\Lambda_d^c(t) = \varepsilon_d^a d\Lambda_b^c(t)$$

where ε_b^a is 1 if $a = b \geq 1$ and zero if either $a = b = 0$ or $a \neq b$, a, $b \geq 0$. $\Lambda_0^0(t) = t$, $\Lambda_0^a(t) = A_a(t), \Lambda_a^0(t) = A_a(t)^*$, $a \geq 1$ are respectively the annihilation and creation processes and $\Lambda_b^a(t)$, a, $b \geq 1$ are the conservation processes. $u(t)$ is a control input. We wish to select the control input $u(t)$, $0 \leq t \leq T$ so that the expected cost function

$$C_T(u) = Re\left(\langle fe(v), j_T(X)fe(v)\rangle\right)$$

is a minimum. We have

$$C_T(u) = \mathrm{Re}\int_0^T\left\langle fe(v), \int_0^T j_t\left(\theta_b^a(X,u(t))\right)d\Lambda_a^b(t)fe(v)\right\rangle$$

$$= \text{Re}\int_0^T \left\langle fe(v), j_t\left(\theta_b^a(X, u(t))\right) fe(v)\right\rangle v_b(t)\bar{v}_a(t)dt$$

Now assume that the structure functions $\theta_b^a(X, u)$ have the form

$$\sum_c \phi_c(u)\theta_{bc}^a(X)$$

With the notation that a repeated index means summation, we then have

$$C_T(u) = \int_0^T \text{Re}\left(\left\langle fe(v), j_t\left(\theta_{bc}^a(X)\right) fe(v)\right\rangle v_b(t)\bar{v}_a(t)\right)\phi_c(u(t))dt$$

Minimizing this w.r.t. the scalar input $u(t)$, $0 < t < T$ is a formidable task. We assume that the output measurable process $Y(t)$ is such that $\{Y(t), 0 < t < T\}$ form a commuting family of observables satisfying the non-demolition property, *i.e.*, $Y(t)$ commutes with $j_s(X)$ for all $s \geq t$ and system space observables X. Let $B_t = \sigma(Y(s), s \leq t)$. Then consider the problem of minimizing

$$\int_t^T \mathbb{E}\left(F\left(js(X), u(s)\right)\right)\Big| Y(t)ds$$

over $u(s)$, $t < s < T$ where $u(s)$ is assumed to be a function of $Y(s)$. F is a real number that is a function of the operators $j_s(X)$ and $u(s)$. Let the minimum of the above over $u(s)$, $t < s < T$ be denoted by $V(t, Y(t))$. Note that the conditional expectation $\mathbb{E}(F(j_s(X), u(s))|Y(t))$ is well defined since $Y(t)$ commutes with $j_s(X)$ for $s > t$ and also with $u(s)$, $s > t$ by the assumption the $u(s)$ is restricted to be a function of $Y(s)$ only. We have

$$V(t, Y(t)) = \min_{u(t)}\left[dt\mathbb{E}F\left(j_t(X), u(t)\right)\Big|Y(t) + \mathbb{E}\left(V\left(t + dt, Y(t) + dY(t)\right)\right)\Big|Y(t)\right]$$

from which, we deduce that

$$-V_{,t}(t, Y(t)) = \min_u\left[\mathbb{E}\left(F\left(j_t(X)_{,u}\right)\Big|Y(t)\right)\right]$$

$$= +\sum_{k=1}^{\infty} V_{,k}\left(t, Y(t)\right)\mathbb{E}\left(dY(t)^k\Big|Y(t)\right)\Big/dt\Big]$$

This is the quantum Bellman-Hamilton-Jacobe equation for quantum stochastic optimal control. Suppose for example that $Y(t)$ is given by

$$Y_0(t) = U(t)^* c_b^a\Lambda_a^b(t)U(t)$$

where c_b^a are complex scalars. The process $Y_i(t) = c_b^a\Lambda_a^b(t)$ takes values in the class of linear operators in the Boson Fock space $\Gamma_s(\mathcal{H})$ and $U(t)$ is unitary operator valued process satisfying the qsde

$$dU(t) = \left(L_b^a d\Lambda_a^b(t)\right)U(t)$$

Where L_b^a are system operators. Then if X is a system operator, we assume that

$$j_t(X) = U(t)^* X U(t)$$

It is then not hard to show that $[Y(s), j_t(X)] = 0$, $s \le t$ and $[Y(s), Y(t)] = 0 \forall s, t$. The proof is based on the condition that $U(t)$ is unitary and hence $d(U(t)^* U(t)) = 0$. The problem is that if we allow the system functions L_b^a to be functions of an input process $u(t)$ that is expressible as a function of $Y_i(t)$, the will the non-demolition property still be valid ? To this, we note that in this case, for $t > s > 0$,

$$d(U(t)^* U(t) = 0, d_t\left(U(t)^* Y_i(s) U(t)\right) = 0$$

since the operators $Y_i(t)$, $t \ge 0$ commute with the system operators and the L_p^a are expressible according to our hypothesis as functions of $Y_i(t)$ and system operators. Thus, it follows that

$$U(t)^* Y_i(s) U(t) = U(s)^* Y_i(s) U(s) = Y_0(s)$$

and hence $Y_0(s)$ commutes with $j_t(X) = U(t)^* X U(t)$ for all $t > s$. Thus we can formulate a non-trivial quantum stochastic optimal control problem.

❖❖❖❖❖

[93] A more rigorous formulation of quantum stochastic optimal control: $X \in \mathfrak{h}$, $L_b^a(t) \in \mathfrak{h} \otimes \Gamma_s(\mathcal{H}_{t]})$. We assume that $L_b^a(t)$ has the form

$$L_b^a(t) = F_b^a(S(t), u(t))$$

where F_b^a are ordinary analytic functinos, $S(t) = \left(S_b^a(t)\right)$ are system operators, *i.e.* in $\mathfrak{h}$, $u(t) = \psi(t, Y_i(t))$ where $Y_i(t) = c_b^a \Lambda_a^b(t)$ with $c_b^a \in \mathbb{C}$ is an ordinary analytic function. It is clear that $(Y_i(t), t \ge 0)$ commute with each other and with $S_b^a(t)$, $t \ge 0$ since Y_i are operators in $\Gamma_s(\mathcal{H})$ while $S_b^a(t)$ are operators in $\mathfrak{h}$. $U(t)$ satisfies

$$dU(t) = L_b^a(t) d\Lambda_a^b(t) U(t)$$

and for unitarity, we require that

$$L_b^a(t)^* d\Lambda_a^b(t) + L_b^a(t) d\Lambda_a^b(t) + L_b^a(t)^* L_d^c(t) d\Lambda_b^a(t) d\Lambda_c^d(t) = 0$$

or equivalently,

$$\left(L_b^a(t)^* + L_a^b(t)\right) d\Lambda_b^a(t) + L_b^a(t)^* L_d^c(t) \varepsilon_c^a d\Lambda_b^d(t) = 0$$

or equivalently,

$$L_b^a(t)^* + L_a^b(t) + L_b^d(t)^* + L_b^c(t) \varepsilon_c^d = 0$$

Now define

$$X(t) = j_t(X) = U(t)^* X U(t)$$

We get

$$dX(t) = U(t)^*\left(L_b^a(t)^* X + XL_a^b(t) + L_b^d(t)^* XL_a^c\right)U(t)d\Lambda_b^a(t)$$

The operator $L_b^a(t)^* X + XL_a^b(t) + L_b^d(t)^* XL_a^c$ can be expressed as $\theta_a^b\left(S(t), X, \psi(t, Y_i(t))\right)$ where θ_a^b is an ordinary function. We note that $u = \psi(t, Y_i(t))$ commutes with $S(t)$, X since $Y_i(t)$ satisfies this property. This we have

$$dX(t) = U(t)^*\theta_a^b\left(S(t), X, \psi(t, Y_i(t))\right)U(t)$$
$$= \theta_a^b\left(\tilde{S}(t), X(t), \psi(t, Y_0(t))\right)$$

where

$$\tilde{S}(t) = U(t)^* S(t)U(t) = j_t(S(t))$$
$$Y_0(t) = U(t)^* Y_i(t)U(t)$$

It is easy to see that for all $T \geq t$

$$Y_0(t) = U(T)^* Y_i(t)U(T)$$

by showing th at its differential w.r.t. T is zero making use of the unitarity property of $U(T)$ and the fact that $Y_i(t)$ commutes with $Y_i(s)$ for all s and with $S(s)$, X which imply that $Y_i(t)$ commuters with $L_b^a(s) = F_b^a\left(S(s), X, \psi_i(s, Y_i(s))\right)$ for all s. It follows that $w(t) = \psi(t, Y_0(t)) = U(s)^*\psi(t, Y_i(t))U(s)$ commutes with $\tilde{S}(s) = U(s)^* S(s)U(s)$, $X(s) = U(s)^* XU(s) \forall s \geq t$ (This property is called the non-demolition property in the sense of Belavkin). We now define the Abelian algebra

$$\mathcal{B}_{t]} = \sigma(Y_0(s), s \leq t)$$

and then consider for an analytic function G, the quantity

$$V(t, \mathcal{B}_{t]}) = \min_{\omega(s)=\psi(t, Y_0(s)), t\leq s\leq T} \mathbb{E}\left[\int_t^T G\left(X(s), w(s)\right)ds \middle| \mathcal{B}_{t]}\right]$$

We note that the conditional expectation is well defined since the observable $\int_t^T G\left(X(s), w(s)\right)ds$ commutes with $\mathcal{B}_{t]}$

We have

$$V(t, \mathcal{B}_{t]}) = \min \psi(t,.)\left[\mathbb{E}\left(G\left(X(t), \psi(t, Y_0(t))\right)\middle|\mathcal{B}_{t]}\right)dt\right.$$
$$\left. + \mathbb{E}\left[V\left(t + dt, \mathcal{B}_{t+dt]}\right)\middle|\mathcal{B}_{t]}\right]\right]$$

Now using Taylor series, let

$$\frac{\partial_k V\left(t, B_{t+dt]}\right)}{\partial dY_0(t)^k}\Big|_{dY0(t)=0} = V_k(t, \mathcal{B}_{t]}), k \geq 0$$

Then we get from the above

$$-\frac{\partial V(t,\mathcal{B}_t)}{\partial t}$$

$$= \min_{\psi(t,.)}\left[\mathbb{E}\left(G\left(X(t),\psi\left(t,Y_0(t)\right)\right)\middle|\mathcal{B}_{t]}\right)+\left|\sum_{k\geq 1}V_k\left(t,\mathcal{B}_t\right)\mathbb{E}\left(dY_0(t)^k\middle|B_{t]}\right)\right]$$

Now,

$$Y_0(t) = U(t)^* Y_i(t)U(t)$$

so using the unitarity of $U(t)$ and the commutativity of $Y_i(t)$ with system operators, we get

$$dY_0(t) = dY_i(t)+dU(t)^* dY_i(t)U(t)+U(t)^* dY_i(t)dU(t)$$

$$= dY_i(t)+U(t)^*\left(L_b^a(t)^* c_q^p d\Lambda_b^a(t)d\Lambda_p^q(t)\right.$$

$$\left.+c_q^p L_b^a(t)d\Lambda_p^q(t)d\Lambda_a^b(t)U(t)\right)$$

$$= dY_i(t)+j_t\left(c_q^p L_b^a(t)^* \varepsilon_p^a d\Lambda_p^q(t)+c_q^p L_b^a(t)(t)\varepsilon_a^q d\Lambda_p^b(t)\right)$$

$$= dY_i(t)+j_t\left(c_a^p\varepsilon_p^q L_b^q(t)^* +c_q^b L_a^p(t)\varepsilon_p^q\right)d\Lambda_b^a(t)$$

Define the process

$$R_a^b(t) = c_a^p\varepsilon_p^q L_b^q(t)^* +c_q^b L_a^p(t)\varepsilon_p^q$$

Then, we have

$$dY_0(t) = dY_i(t)+j_t\left(R_a^b(t)\right)d\Lambda_b^a(t) = \left(c_a^b + j_t\left(R_a^b(t)\right)\right)d\Lambda_b^a(t)$$

and we get

$$(dY_0(t))^k = \left[\Pi_{m=1}^k\left(c_{a_m}^{b_m} + j_t\left(R_{a_m}^{b_m}(t)\right)\right)\right]d\Lambda_{b_1}^{a_1}d\Lambda_{b_2}^{a_2}...d\Lambda_{b_k}^{a_k}$$

$$= \Pi\left(a_{a_m}^{b_m} + j_t\left(R_{a_m}^{b_m}(t)\right)\right)\left[\Pi_{m=0}^{k-1}\varepsilon_{b_{m+1}}^{a_m}\right]d\Lambda_{b_1}^{a_k}$$

and thus,

❖❖❖❖❖

[94] Problem on fractional delay nonlinear system design: Let $x(t)$ be the input signal and $y(t)$ the output. Design systems $h_k(r_1,..., r_k)$ and delays $\tau_1,..., \tau_M$ so that

$$z(t) = \sum_{k=1}^{p}\sum_{r_1,...,r_k=1}^{M} h_k(r_1,...,r_k)x(t-\tau_{r1})...x(t-\tau_{rk})$$

is a good approximation to the signal $y(t)$

Hint: The Fourier transform of $z(t)$ is given by

$$Z(\omega) = \sum_{k=1}^{p} (2\pi)^{-k+1} \sum H_k\left(\omega_1, ..., \omega_{k-1}, \omega - \omega_1 - ... - \omega_{k-1}; \tau_1, ..., \tau_M\right)$$

$$X(\omega_1)...X(\omega_{k-1})X(\omega - \omega_1 - ... - \omega_{k-1})d\omega_1...d\omega_{k-1}$$

where

$$H_k\left(\omega_1, ..., \omega_{k-1}, \omega_k; \tau_1, ..., \tau_M\right)$$

$$= \sum_{r_1,...,r_k} h_k(r_1, ..., r_k)\exp\left(-j\sum_{m=1}^{k} \omega_m \tau_m\right)$$

❖❖❖❖❖

[95] Commuting non-demolition measurement processes in quantum filtering. Let

$$Y_{in,j}(t) = c_b^a(j)\Lambda_a^b(t), c_b^a(j) \in \mathbb{C}$$

In order that the processes $Y_{inj}(t)$, $t \geq 0$, $j = 1, 2, ..., N$ form an Abelian family of operators, we require that

$$dY_{in,j}(t)dY_{in,k}(t) = dY_{in,k}(t)dY_{in,j}(t)$$

for all j, k, t. Equivalently,

$$c_b^a(j)c_q^p(k)d\Lambda_a^b(t)d\Lambda_p^q(t) = c_q^p(k)c_b^a(j)d\Lambda_p^q(t)d\Lambda_a^b(t)$$

or equivalently,

$$c_b^a(j)c_q^p(k)\varepsilon_p^b d\Lambda_a^q(t) = c_q^p(k)c_b^a(j)\varepsilon_a^q d\Lambda_p^b(t)$$

or equivalently,

$$c_q^p(j)c_b^a(k)\varepsilon_a^q = c_q^p(k)c_b^a(j)\varepsilon_a^q$$

or equivalently, in terms of the matrices $C(j) = \left(\left(c_q^p(j)\right)\right)$ and $\varepsilon = \left(\left(\varepsilon_b^q\right)\right)$, these conditions can be expressed as

$$C(j).\varepsilon C(k) = C(k).\varepsilon.C(j), 1 \leq k, j \leq N$$

Assuming this to be the case, we can define our input measurement algebra at time t as

$$\eta_{in,t} = \sigma\left(Y_{in,j}(s), 1 \leq j \leq N, s \leq t\right)$$

The quantum Kallianpur-Striebel formula gives

$$\mathbb{E}_t(X|\eta_{in,t}) = \frac{\mathbb{E}\left(F(t)^* XF(t)\middle|\eta_{in,t}\right)}{\mathbb{E}\left(F(t)^* F(t)\middle|\eta_{in,t}\right)}$$

where $F(t) \in \eta'_{in, t}$ is such that

$$\mathbb{E}_t(X) = \mathbb{E}(F(t)^* X F(t))$$

To see this, we let $Z \in \eta_{in, t}$ and note that since $[F(t), \eta_{in, t}] = 0$ we have $[F(t), Z] = [F^*(t), Z] = 0$,

$$\mathbb{E}_t\left[\frac{\mathbb{E}\left(F(t)^* XF(t)\middle|\eta_{in,t}\right)}{\mathbb{E}\left(F(t)^* F(t)\middle|\eta_{in,t}\right)}Z\right]$$

$$= \mathbb{E}\left[\frac{\mathbb{E}\left(F(t)^* XF(t)\middle|\eta_{in,t}\right)}{\mathbb{E}\left(F(t)^* F(t)\middle|\eta_{in,t}\right)}F(t)^* F(t) Z\right]$$

$$= \mathbb{E}\left[\frac{\mathbb{E}\left(F(t)^* XF(t)\middle|\eta_{in,t}\right)}{Z}\right]$$

$$= \mathbb{E}\left[F(t)^* XZF(t)\right] = \mathbb{E}_t(XZ)$$

This proves the quantum Kallianpur-Striebel formula. Now let $U(t)$ be unitary in $\hbar \otimes \Gamma_s(\mathcal{H})$ and satisfy the qsde

$$dU(t) = \left(L_b^a(t) d\Lambda_a^b(t)\right)U(t)$$

where $L_b^a(t) \in \pounds(\hbar)$. Then, we know that if $\eta_{out, t} = U(t)^* \eta_{in, t} U(t)$ then $\eta_{out, t}$ commutes with $j_s(X)$ for all $s \geq t$ where $j_t(X) = U(t)^* X U(t)$. Also, if we define the expectation $\mathbb{E}_t$ by the formula

$\mathbb{E}_t(X) = \mathbb{E}(U(t)^* X U(t))$

then

$$\mathbb{E}\left(U(t)^* XU(t)\middle|\eta_{out,t}\right) = U(t)^* \mathbb{E}_t\left(X\middle|\eta_{in,t}\right)U(t)$$

To see this let $Z \in \eta_{in, t}$. Then $U(t)^* Z U(t) \in \eta_{out, t}$ and we have

$$\mathbb{E}\left[U(t)^* \mathbb{E}_t\left(X\middle|\eta_{in,t}\right)U(t)\left(U(t)^* ZU(t)\right)\right]$$

$$= \mathbb{E}_t\left[\mathbb{E}_t\left(X\middle|\eta_{in,t}\right)Z\right] = \mathbb{E}_t(XZ)$$

$$= \mathbb{E}\left(U(t)^* XU(t)U(t)^* ZU(t)\right)$$

$$= \mathbb{E}\left(\mathbb{E}\left(U(t)^* XU(t) \middle| \eta_{out,t}\right) U(t)^* ZU(t)\right)$$

which proves the claim. Thus, we get the formula

$$\mathbb{E}(j_t(X)|\eta_{out,t}) = U(t)^* \left[\frac{\mathbb{E}\left(F(t)^* XF(t) \middle| \eta_{in,t}\right)}{\mathbb{E}\left(F(t)^* F(t) \middle| \eta_{in,t}\right)} \right] U(t)$$

which is actually, the final stge of the quantum Kallianpur Striebel formula. We now assume that

$$dF(t) = B_j(t) dY_{in,j}(t) F(t) = B_j(t) c_b^a(j) d\Lambda_a^b(t) F(t) \qquad ...(1)$$

where $B_j(t) \varepsilon \mathcal{L}(h)$ and calculate the values of these operators so that the relation

$$\mathbb{E}\left(U(t)^* XU(t)\right) = \left\langle f\phi(u) \middle| U(t)^* XU(t) \middle| f\phi(u) \right\rangle$$

$$= \mathbb{E}\left(F(t)^* XF(t)\right) = \left\langle f\phi(u) \middle| F(t)^* XF(t) \middle| f\phi(u) \right\rangle \qquad ...(2)$$

for all system operators X. Note that (1) implies that $F(t)$ is built out of the system operators $B_i(s)$, $s \le t$ and the noise operators Y_{in}, $j(s)$, $s \le t$ and all these operators commute with $\eta_{in,t}$. Thus $F(t)$ also commutes with $\eta_{in,t}$. Now we observe that

$$\frac{d}{dt}\left\langle f\phi(u) \middle| U(t)^* XU(t) f\phi(u) \right\rangle = \left\langle f\phi(u) \middle| U(t)^* \left(X L_b^a \bar{u}_a(t) u_b(t) + L_b^{a*} \bar{u}_b(t) u_a(t) \right.\right.$$

$$\left.\left. + L_b^{a*} X L_q^p \, \varepsilon_p^a u_q(t) \bar{u}_b(t) \right) U(t) \middle| f\phi(u) \right\rangle$$

On the other hand,

$$\frac{d}{dt}\left\langle f\phi(u) \middle| F(t)^* XF(t) \middle| f\phi(u) \right\rangle$$

$$= \left\langle f\phi(u) | F(t)^* \left(B_j(t)^* X \bar{c}_{cb}^{a}(j) u_a(t) \bar{u}_b(t) + c_b^a(j) X B_j(t) u_b(t) \bar{u}_a(t) \right.\right.$$

$$\left.\left. + \bar{c}_b^{\,a}(j) c_q^p(k) B_j(t)^* X B_k(t) \varepsilon_p^a u_q(t) \bar{u}_b(t) \right) | f\phi(u) \right\rangle$$

Thus, a unnecessary and sufficient condition on $F(t)$ for (2) to hold is that

$$B_j(t)^* X \bar{c}_b^{\,a}(j) u_a(t) \bar{u}_b(t) + c_b^a(j) X B_j(t) u_b(t) \bar{u}_a(t)$$

$$+ \bar{c}_b^{\,a}(j) c_q^p(k) B_j(t)^* X B_k(t) \in_p^a u_q(t) \bar{u}_b(t)$$

$$X L_b^a \bar{u}_a(t) u_b(t) + L_b^{a*} u_a(t) \bar{u}_b(t) + L_b^{a*} X L_q^p \in_p^a u_q(t) \bar{u}_b(t) \qquad ...(3)$$

be valid for all system operators X. A necessary and sufficient condition for (3) to be valid for all X is that

$$c_b^a(j)B_j(t)u_b(t)\bar{u}_a(t) = L_b^a(t)\bar{u}_a(t)u_b(t) \quad ...(4a)$$

and $$\left(c_b^m(j)u_b(t)B_j(t)\right)^* X\left(c_q^m(k)u_q(t)B_k(t)\right)$$

$$\left(u_b(t)L_b^m(t)\right)^* X\left(u_q(t)L_q^m(t)\right) \quad ...(4b)$$

for all system operators X. The summation range for m in (4b) is $m \geq 1$ as is for j, k while a, b, p, q have a summation range ≥ 0. For (4b) to hold for all system operators X we require that

$$c_q^m(k)u_q(t)B_k(t) = u_q(t)L_q^m(t), m \geq 0 \quad ...(4c)$$

Thus, (4a) and (4c) are the necessary and sufficient conditions for $F(t)$ to satisfy (2). Sufficient conditions for (4a) and (4c) to hold are that

$$c_q^m(k)u_q(t)B_k(t) = u_q(t)L_q^a(t), a \geq 0 \quad ...(5)$$

Note that $u_0(t) = 1$. The summation index k in (5) ranges over $k \geq 1$. Assuming that this has been done, we are now in a position to derive the Belavkin quantum filtering equations. We have

$$\pi_t(X) = \mathbb{E}(j_t(X)|\eta out,t)$$

$$= U(t)^* \frac{\mathbb{E}\left(F(t)^* XF(t)/\eta_{in,t}\right)}{\mathbb{E}\left(F(t)^* F(t)/\eta_{in,t}\right)} U(t)$$

$$= U(t)^* \frac{\sigma_t(X)}{\sigma_t(1)} U(t)$$

$$\sigma t(X) = \mathbb{E}\left(F(t)^* F(t)/\eta_{in,t}\right)$$

$$\sigma_{t+dt}(X) - \sigma_t(X) = \mathbb{E}\left(d\left(F(t)^* XF(t)\right)/\eta_{in,t+dt}\right) + \mathbb{E}\left(F(t)^* XF(t)/\eta_{in,t+dt}\right)$$

$$-\mathbb{E}\left(F(t)^* XF(t)/\eta_{in,t+dt}\right)$$

Assume that

$$E\left(F(t)^* XF(t)/\eta_{in,t+dt}\right) - \mathbb{E}\left(F(t)^* XF(t)/\eta_{in,t}\right)$$

$$= \mathcal{F}_t(X)dt + \mathcal{G}_{kt}(X)dY_{in,k}(t)$$

(summation over k) where $\mathcal{F}_t(X), \mathcal{G}_{k,t}(X) \in \eta_{in,t}$. Then, we get on taking conditional expectation on both sides given

$$0 = \mathcal{F}_t(X)dt + \mathcal{G}_{kt}(X)\mathbb{E}\left(dY_{in,k}(t)/\eta_{in,t}\right)$$

Also by multiplying by $dY_{in,j}(t)$ and taking conditional expectation given $\eta_{in,t}$, we get

$$\mathcal{G}_{k\,t}(X)\mathbb{E}\left(dY_{in,\,k(t)}dY_{in,\,t}(t)\middle|\eta_{in,\,t}\right)$$

$$= \mathbb{E}\left(F(t)^* XF(t)dY_{in,\,j}(t)\middle|\eta_{in,\,t}\right) - \sigma_t(X)\mathbb{E}\left(dY_{in,\,j}(t)\middle|\eta_{in,\,t}\right)$$

Now,

$$\mathbb{E}\left(dY_{in,\,k}(t)\middle|\eta_{in,\,k}\right)$$

$$= \mathbb{E}\left(c_b^a(j)B_j(t)d\Lambda_a^b(t)\middle|\eta_{in,\,t}\right)$$

$$= c_b^a(j)\left\langle f\middle|B_j(t)\middle|f\right\rangle u_b(t)\bar{u}_a(t)dt$$

since in the product state $|f\,\phi(u)>$, $B_j(t)$ is independent of $\eta_{in,\,t}$ and $d\Lambda_a^b(t)$ and $d\Lambda_a^b(t)$ is independent of $\eta_{in,\,t}$. Also

$$\mathbb{E}\left(dY_{in,\,k}(t)dY_{in,\,j}(t)\middle|\eta_{in,\,t}\right)$$

$$= \mathbb{E}\left(c_b^a(j)c_q^p(k)B_j(t)B_k(t)d\Lambda_a^b(t)d\Lambda_p^q(t)\middle|\eta_{in,\,t}\right)$$

$$= c_b^a(j)c_q^p(k)\left\langle f\middle|B_j(t)B_k(t)f\right|\rangle\varepsilon_p^b\mathbb{E}\left(d\Lambda_a^q(t)\middle|\eta_{in,\,t}\right)$$

$$= c_b^a(j)c_q^p(k)\left\langle f\middle|B_j(t)B_k(t)f\right|\rangle\varepsilon_p^b u_q(t)\bar{u}_a(t)dt$$

Thus, we get linear equations for the operators $\mathcal{F}_t(X)$ and $\mathcal{G}_{k,\,t}(X)$, which are easily solved. We further have

$$\mathbb{E}\left(d\left(F(t)^* XF(t)\middle|\eta_{in,\,t+dt}\right)\right.$$

$$= \mathbb{E}\left(dF(t)^* XF(t) + F(t)^* XdF(t)\right.$$

$$\left.+\ dF(t)^* XF(t)\middle|\eta_{in,\,t+dt}\right)$$

$$= \mathbb{E}\left(F(t)^*\left(c_b^{-a}(j)B_j(t)^* Xd\Lambda_b^a(t) + c_b^a(j)B_j(t)d\Lambda_a^b(t)\right.\right.$$

$$\left.c_b^{-a}(j)c_q^p(k)B_j(t)^* XB_k(t)d\Lambda_b^a(t)d\Lambda_p^q(t)\middle|\eta_{in,\,t+dt}\right)$$

❖❖❖❖❖

[96] Quantum particle in a spherically symmetric box. Solving the Schrodinger equation

$$\nabla^2\psi(r,\theta,\phi) = -2mE\psi(r,\theta,\phi)$$

for $0 \le r \le a$ with the boundary condition $\psi(a,\theta,\phi) = 0$ and determine the possible values of the energy levels E.

Hint:

$$\nabla^2 = r^{-1}\frac{\partial^2}{\partial r^2}r - L^2/r^2$$

Writing

$$\psi(r, \theta, \phi) = R(r)Y_{lm}(\theta, \phi)$$

and using

$$L^2Y_{lm} = l(l+1)Y_{lm}l = 0, 1, 2$$

show that $R(r)$ satisfies

$$r^{-1}\left(rR(r)\right)'' - l(l+1)R(r)/r^2 = -2mER(r)$$

which expands to

$$r^2R'' + 2rR' + \left(2mEr^2 - l(l+1)\right)R = 0$$

Now substitute

$$\mathrm{R(r)} = r^{-l/2}f(r)$$

and show that f satisfies the Bessel equation of order $l + 1/2$ in the variable $x = r\sqrt{2mE}$.

Now assume that the particle has a charge $-e$ and an external electric field $\delta.E(t)\hat{z}$ is applied. Calculate the evolution operator upto $O(\delta)$ in terms of the eigenfunctions of the unperturbed Hamiltonian and hence calculate upto $O(\delta^2)$, the probability of the particle making a transition from the state $|nlm>$ to $|n'l'm'>$ in time T.

❖❖❖❖❖

[97] The general relativistic equations for a conducting fluid are given by

$$R^{\mu\nu} - \frac{1}{2}R^{\mu\nu} = K\left((\rho + p)v^{\mu}v^{\nu} - pg^{\mu\nu} + S^{\mu\nu}\right)$$

$$S^{\mu\nu} = -\frac{1}{4}F_{\alpha\beta}F^{\alpha\beta}g^{\mu\nu} + F^{\mu\alpha}F^{\nu}_{\beta}$$

is the energy momentum tensor of the em field. Here,

$$F^{\mu\nu} = A_{\nu,\mu} - A_{\mu,\nu}$$

where A_{μ} is the covariant em four potential. From the above em field equations, we derive the mhd equations

$$\left((\rho + p)v^{\mu}v^{\nu} - pg^{\mu\nu} + S^{\mu\nu}\right)_{:\nu} = 0$$

which yield both the Navier-Stokes equation in an em field as well as the mass conservation equation. We note that

$$S^{\mu\nu}_{:\nu} = K_0F^{\mu\nu}J_{\nu}$$

where $$J_\mu = K_1 F^{\mu\nu}_{:\nu}$$

Using Ohm's law for the conducting fluid in the form

$$J_\mu = \sigma^{F\mu\nu} v_\nu$$

we can write

$$S^{\mu\nu}_{:\nu} = K_2 F^{\mu\nu} F_{\nu\alpha} v^\alpha$$

Now from the Navier-Stokes, mhd and Einstein field equations, by looking at first order perturbations in (v^μ, $g_{\mu\nu}$, A_μ, ρ, p), namely δv^μ,. $\delta g_{\mu\nu}$, δA_μ, $\delta\rho$, δp relative to the background values, we derive linear partial differential equations for these variables. Denoting the set of all these perturbed variables by $\xi_n(x)$, $n = 1, 2, ..., M$, we can write these equations in the form

$$C_{1nm\beta\rho}(x)\xi_{m,\beta\rho}(x) + C_{2nm\beta}(x)\xi_{m,\beta}(x) + C_{3nm}(x)\xi_m(x) = 0$$

where the functions C_1, C_2, C_3 are determined by the background values of $(v^\mu, g_{\mu\nu}, A_\mu, \rho, p)$. To determine the modes of oscillations of $\xi_n(x)$, we choose a set of basis functions $\psi_k(x)$, $k = 1, 2,...N$ and expand

$$\xi_n(x) = \sum_{k=1}^{N} a(n,k)\psi_k(x), n = 1, 2, ..., M$$

substitute these into the above linear pde, multiply by $\psi_m(x)$ and integrate w.r.t. d^4x to obtain a set of NM linear matrix equations for the coefficients $((a(n, k)))$. Setting the determinant of the associated matrix to zero gives the modes of oscillations of the gravitational field, fluid velocity field, fluid density and pressure field and em four potential field.

❖❖❖❖❖

[98] Evolution of inhomogeneities in a homogeneous and isotropic background universe as a model for galactic evolution: The Robertson-Walker metric is $g^{(0)}_{\mu\nu}(x)$. Specifically, the proper time corresponding to this metric is

$$d\tau_0^2 = dt^2 - \frac{S^2(t)dr^2}{1-kr^2} - S^2(t)r^2\left(d\theta^2 + \sin^2(\theta)d\phi^2\right)$$

Thus,

$$g^{(0)}_{00} = 1, g^{(1)}_{11} = -S^2(t)/(1-kr^2) = f(t, r)$$

say, $g^{(0)}_{22} = -S_2(t)r$, $g^{(0)}_{33} = -S^2(t)r^2\sin^2(\theta)$

We compute the elements of the Ricci tensor:

$$R_{00} = \Gamma^\alpha_{0\alpha,0} - \Gamma^\alpha_{00,\alpha} - \Gamma^\alpha_{00}\Gamma^\beta_{\alpha\beta} + \Gamma^\alpha_{0\beta}\Gamma^\beta_{0\alpha}$$

$$R_{11} = \Gamma^\alpha_{1\alpha,1} - \Gamma^\alpha_{11,\alpha} - \Gamma^\alpha_{11}\Gamma^\beta_{\alpha\beta} + \Gamma^\alpha_{1\beta}\Gamma^\beta_{1\alpha}$$

$$R_{22} = \Gamma^{\alpha}_{2\alpha,2} - \Gamma^{\alpha}_{22,\alpha} - \Gamma^{\alpha}_{22}\Gamma^{\beta}_{\alpha\beta} + \Gamma^{\alpha}_{2\beta}\Gamma^{\beta}_{2\alpha}$$

$$R_{33} = \Gamma^{\alpha}_{3\alpha,3} - \Gamma^{\alpha}_{33,\alpha} - \Gamma^{\alpha}_{33}\Gamma^{\beta}_{\alpha\beta} + \Gamma^{\alpha}_{3\beta}\Gamma^{\beta}_{3\alpha}$$

Now,

$$\Gamma^{\alpha}_{0\alpha,0} = \Gamma^{1}_{01,0} + \Gamma^{2}_{02,0} + \Gamma^{3}_{03,0}$$

$$\Gamma^{r}_{0r,0} = 0.5(g^{rr}g_{rr,0}),_0 = 0.5(\log(g_{rr}),_{00} = (S'/S)' = S''/S - S^2/S^2$$

So

$$\Gamma^{\alpha}_{0\alpha,0} = 3S''/S - 3S'^2/S^2$$

$$\Gamma^{\alpha}_{1\alpha,1} = \sum_r \Gamma^{r}_{1r,1} = 0.;5\sum_r \left(g^{rr}g_{rr,1}\right)_{,1}$$

$$= 0.5\sum_r \left(\log g_{rr}\right)_{,11} = k/\left(1-kr^2\right)^2 - 2/r^2$$

$$\Gamma^{\alpha}_{2\alpha,2} = \sum_r \Gamma^{r}_{2r,2} = 0.5\sum_r \log\left(g_{rr}\right)_{,22} = \cos t\,(\theta)$$

$$\Gamma^{\alpha}_{3\alpha,3} = 0.5\sum_r \left(\log g_{rr}\right)_{,33} = 0$$

$$\Gamma^{\alpha}_{00,\alpha} = 0$$

$$\Gamma^{\alpha}_{11,\alpha} = \Gamma^{\alpha}_{11,0} + \Gamma^{1}_{11,1} = 0.5\left(-g_{11,00} + \left(\log_{11}\right)_{,11}\right)$$

$$= \left(SS'\right)'/\left(1-kr^2\right) + k/\left(1-kr^2\right)^2$$

$$\Gamma^{\alpha}_{22,\alpha} = \Gamma^{\alpha}_{22,0} + \Gamma^{1}_{22,1} = 0.5\left(-g_{22,00} - \left(g^{11}g_{22,1}\right)_{,1}\right)$$

$$\Gamma^{\alpha}_{00} + \Gamma^{\beta}_{\alpha\beta} = 0.$$

❖❖❖❖❖

[99] Quantization of the two link robot:

$$H_0(q,p) = \frac{1}{2}p^T M(q)^{-1} p + V(q)$$

is the unperturbed Hamiltonian of the robot in the absence of external torques. Here, $M(q)$., the mass moment of inertia matrix of the robot has the form

$$M(q) = M_0 + \cos(q_2)M_1$$

where M_0, M_1 are constant real symmetric matrices and for $M(_q)$ to be positive definite for all q. we require that

$M_0 < M_1 < M_0$

or equivalently,

$$M_1^2 < M_0^2$$

V(q), the gravitational potential energy of the robot is given by

$$\begin{aligned} V(q) &= m_1 g l_1 \sin(q_1)/2 + m_2 g\left(l_1 \sin(q_1) + (l_2/2)\sin(q_1+q_2)\right) \\ &= a_1 \sin(q_1) + a_2 \sin(q_1+q_2) \end{aligned}$$

Since $|M_1| < M_0$ we can expand

$$\begin{aligned} M(g)^{-1} &= \left(M_0 + \cos(q_2) M_1\right)^{-1} \\ &= M_0^{-1/2}\left(I + \cos(q_2) M_0^{-1/2} M_1 M_0^{-1/2}\right)^{-1} M_0^{-1/2} \\ &= M_0^{-1/2} + \sum_{n=1}^{\infty} (-1)^n \cos(q_2)^n M_0^{-1/2}\left(M_0^{-1/2} M_1 M_0^{-1/2}\right)^n M_0^{-1/2} \\ &= M_0^{-1} + \sum_{n=1}^{\infty} (-1)^n \cos(q_2)^n \left(M_0^{-1} M_1\right)^n M_0^{-1} \\ &= M_0^{-1} + \sum_{n\geq 1} (-1)^n \cos(q_2)^n M_0^{-1}\left(M_1 M_0^{-1}\right)^n \end{aligned}$$

We can thus write

$$H_0 = H_{00} + H_{01}$$

where

$$H_{00} = \frac{1}{2} p^T M_0^{-1} p + \alpha^T q$$

where α is a constant 2×1 vector and

$$\begin{aligned} H_{01} = \frac{1}{2} p^T \sum_{n\geq 1} (-1)^n \cos(q_2)^n M_0^{-1}\left(M_1 M_0^{-1}\right)^n + a_1\left(\sin(q_1) - q_1\right) \\ + a_2\left(\sin(q_1+q_2) - q_1 - q_2\right) \end{aligned}$$

Here

$$\alpha_1 = a_1 + a_2,\ \alpha_2 = a_2$$

By applying a canonical orthogonal transformation to both q, p, we bring H_{00} to the decoupled form

$$H_{00} = P_1^2/2\mu_1 + P_2^2/2\mu_2 + \beta_1 q_1 + \beta_2 q_2$$

This is the sum of Hamiltonians of two independent quantum bouncing ball systems, *i.e.*, ball in a gravitational field and its bound states are obtained using

Airy's functions, namely solutions to the differential equation

$$f''(x) = xf(x)$$

This can be seen as follows. consider the one dimensional Schrodinger equation

$$-(2\mu_1)^{-1}\psi''(x)+\beta_1 x\psi(x) = E\psi(x)$$

It can be rearranged as

$$\psi''(x)+2\mu_1(E-\beta_1 x)\psi(x) \quad ...(1)$$

Writing

$$-2\mu_1(E-\beta_1 x) = \gamma y$$

we get

$$(2\mu_1\beta_1)^{-2}\frac{d^2}{dx^2} = \gamma^{-2}\frac{d^2}{dy^2}$$

and hence (1) becomes

$$(2\mu_1\beta/\gamma)^2\frac{d^2\psi}{dy^2} = y\psi$$

Choosing

$$\gamma = 2\mu_1\beta_1$$

gives

$$\frac{d^2\psi}{dy^2} = y\psi$$

which is Airy's equation. It is solved using power series. The stationary state eigenfunctions and eigenvalues of H_0 are then calculated approximately using the standard techniques of time independent perturbation theory. We denote these by $|n>E_n$, $n = 0, 1, 2, ...$ Now consider the effect of Levy noise on such a robot. The Schrodinger evolution equation is

$$dU(t) = -i\left(H0+\int_{x\in E} V(t,x,N(t)+1)N(dt,dx)\right)U(t)$$

where $V(.,.,.)$, $n = 1, 2...$ is an iid sequence of operator valued random fields independent of the spatial Poisson field $N(t, dx)$. For $U(t)$ to be a unitary operator, we require that

$$0 = d(U^*U) = dU^*.U + U^*.dU + dU^*dU$$

and this is satisfies provided

$$i\left(V(v,x,n)^* - V(t,x,n)\right) + V(t,x,n)^* V(t,x,n) = 0$$

for all n, t, x. Assuming that this is so, we define the random operator $W(t)$ by

$$U(t) = U_0(t)W(t),\ U_0(t) = \exp(-itH_0)$$

Then

$$W'(t) = -i\int_x U(t)^* V(t, x, N(t)+1)U(t)N(dt, dx)W(t)$$

and the approximate solution is

$$W(T) \approx I - i\int_{[0,T]\times E} U(t)^* V(t, x, N(t)+1)U(t)N(dt, dx)$$

We assume that

$$\mathbb{E}N(dt, dx) = \lambda.dt.dF(x)$$

Then, for $n \neq m$, we have that the approximate transition probability from $|\,m>$ to $|\,n>$ in time $[0, T]$ is given by

$$\mathbb{E}\left|\langle n|W.(T)|m\rangle\right|^2$$

$$= E\left[\left|\int_{[0,T]\times E} \exp(iE(m,n)t)\langle m|V(t, x, N(t)+1)|m\rangle N(dt, dx)\right|^2\right]$$

$$= \int_{[0,T]\times E} \mathbb{E}\left[\left|\langle m|V(t, x, N(t)+1)|n\rangle\right|^2\right]\lambda.dt.dF(x)$$

$$+ \int_{0\le t_1 \ne t_2 \le T,\, x\in E} \exp(iE(m,n)(t_1-t_2))\,\mathbb{E}\Big[\langle m|V(t, x, N(t_1)+1)|n\rangle$$

$$\langle n|V(t_2, x, N(t_2)+1)m|\rangle dN(t_1)dN(t_2)\Big] \quad ...(2)$$

Now for $t_1 < t_2$, it easy to see that

$$\mathbb{E}\big(f(N(t_1), N(t_2))dN(t_1)dN(t_2)\big) = \lambda dt_2\mathbb{E}\big(f(N(t_1), N(t_2))dN(t_1)\big)$$

$$= \lambda^2 dt_1 dt_2\mathbb{E}\big(f(N(t_1), N(t_2)+1)\big)$$

using which the second expectation in (2) is easily evaluated.

❖❖❖❖❖

[100] Computation of the perturbation in the Ricci tensor around the Robertson-Walker metric. This has applications in analyzing the propagation of non-uniformities in a homogeneous and isotropic cosmological background. The resulting linearized Einstein field equations are first order partial differential equations for the perturbations in the metric and they can be used to describe the evolution of isolated galaxies in the homogeneous and isotopic universe. The non-zero components of the Robertson-Walker metric are

$$g_{00} = 1, g_{11} = -S^2(t) / (1 - kr^2), g_{22} = -S^2(t)r^2, g_{33} = -S^2(t)r^2\sin^2(\theta)$$

Let $\delta g_{\mu\nu}$ be small pertubations to this metric. Then, the perturbation inducced in the Ricci tensor is

$$\delta R_{\mu\nu} = \delta\Gamma^{\alpha}_{\mu\alpha,\nu} - \delta\Gamma^{\alpha}_{\mu\nu,\alpha} - \delta\left(\Gamma^{\alpha}_{\mu\nu}\Gamma^{\beta}_{\alpha\beta}\right)\delta\left(\Gamma^{\alpha}_{\mu\beta}\Gamma^{\beta}_{\nu\alpha}\right)$$

❖❖❖❖❖

Random Feedback Control, Group Invariants and Pattern Classification, Quantum Mechanics in Levy Noise, State Observer and Trajectory Tracking

[101] Problem: Random perturbation in the feedback loop of an LTI system. Let $H(z)$, $A(z)$ be transfer functions, $H(z)$ is the open loop transfer function and $A(z)$ is the feedback transfer function. The closed loop transfer function is then

$$G(z) = \frac{H(z)}{1+H(z)A(z)}$$

Now suppose $A(z)$ gets perturbed to $A(z) + \varepsilon B(z)$ where $B(z) = \sum_{k\geq 0} b[k]z^{-k}$ with $\{b[k]\}$ a Gaussian random sequence with zero mean and know correlations:

$$\mathbb{E}(b\,[k]\,b\,[m]) = R_{bb}[k, m]$$

Then, $G(z)$ gets perturbed to

$$\begin{aligned} G(z) + \varepsilon\delta G(z) + O(\varepsilon^2) &= \frac{H(z)}{1+\left(A(z)+\varepsilon B(z)\right)H(z)} \\ &= \frac{H(z)}{1+A(z)H(z)+\varepsilon B(z)H(z)} \\ &= G(z)\left(1-\varepsilon\frac{B(z)H(z)}{1+A(z)H(z)}\right)+O\left(\varepsilon^2\right) \end{aligned}$$

Thus,

$$\delta G(z) = -\frac{B(z)H(z)}{1+A(z)H(z)}$$

Now using this expression, compute $\mathbb{E}\left(y_d(n)-y(n)\right)^2$ where $y_d(n)$ is a desired non-random output and $y(n)$ is the output of the randomly perturbed closed loop system to a given non-random input signal $x(n)$.

❖❖❖❖❖

[102] Let μ be a probability measure on a compact group G that is bi-invariant, *i.e.*, $\mu(gEg^{-1}) = \mu(E)$ for all $g \in G$ and all Borel sets E in G. Let $\widehat{G}$ be a complete set of inequivalent unitary irreducible representations of G. Consider the Fourier transform $\hat{\mu}(\pi)$ of μ where π ranges over $\widehat{G}$. Prover that

$$\hat{\mu}(\pi) = c(\pi) I_{\pi}, \pi \in \hat{G}$$

where $c(\pi) \in \mathbb{C}$ and I_{π} is the identity operator on the vector V_{π} space on which $\pi(g)$ acts.

❖❖❖❖❖

[103] Consider a random variable X assuming values in $\mathbb{R}^n$ that has an infinitely divisible probability disribution. The levy-Khintchine formula for the characteristic function of X is

$$\psi_X(t) = \mathbb{E}\left[\exp\left(i\langle t, X\rangle\right)\right] = \exp\left(-t^T Rt/2 + \int\left(\exp\left(i\langle t, x\rangle\right) - 1\right) dv(x)\right)$$

where v is a measure on $\mathbb{R}^n$. Equivalently, let $X(t)$ be a stationary independent increment stochastic process with values in $\mathbb{R}^n$. Its characteristic function then has the form

$$\psi_X(\omega, t) = \mathbb{E}\left[\exp\left(i\langle \omega, X(t)\rangle\right)\right] = \exp(t\phi_X(\omega))$$

where

$$\phi_X(w) = -\frac{1}{2}\omega^T R\omega + \int\left(\exp\left(i\langle \omega, x\rangle\right) - 1\right) dv(x)$$

Let $f: \mathbb{R}^n \to \mathbb{R}^n$ be a twice differentiable function. Then,

$$\lim_{h\to 0} h^{-1}\mathbb{E}\left(f(X(t+h)) - f(X(t)) \middle| X(t) = x\right)$$

$$= \frac{1}{2}\sum_{a, b} R(a, b) f_{,ab}(x) + \int\left(f(x+y) - f(x)\right) dv(y)$$

If we use the more general Levy-Khintchine formula,

$$\psi_X(\omega, t) = exp\ (t\phi_X(\omega))$$

where

$$\phi_X(\omega) = \exp\left(-\frac{1}{2}\omega^T R\omega + \int\left(\exp\left(i\langle \omega, x\rangle\right) - 1 - \frac{i\langle \omega, x\rangle}{1+|x|^2}\right) dv(x)\right)$$

then we ger for the generator of the independent increment process $X(t)$, the formula

$$\lim_{h\to 0} h^{-1}\mathbb{E}\left(f(X(t+h)) - f(X(t)) \middle| X(t) = x\right)$$

$$= \frac{1}{2}\sum_{a, b} R(a, b) f_{,ab}(x) + \int\left(f(x+y) - f(x) - \frac{i\langle y, \nabla f(y)\rangle}{1+|y|^2}\right) dv(y)$$

This formula was generalized by Hunt to independent increment processes with stationary increments (*i.e.*, Levy processes) on a Lip group. For a compact Lie group G, if $X(t)$, $t \geq 0$ is a Levy process with values in G, then for $t_0 < t_1 < t_2 < \ldots < t_n$, the group increments $X(t_{j-1})^{-1} X(t_j)$, $j = 1, 2, \ldots, n$ are independent random variables. By stationary increments, we mean that in addition, the distribution of $X(s)^{-1} X(t)$ for $s < t$ depends only on $t - s$. Hunt proved that the characteristic function of such a process has the from

$\mathbb{E}[\pi(X(t))] = \exp(t\phi_X(\pi))$

where for f: $G \to \mathbb{C}$, $\phi_X(\pi)\hat{f}(\pi)$ is the Fourier transform of the function

$$Tf(g) = \frac{1}{2}R(a,b)Z_aZ_bf(g) + \int_G (f(hg) - f(g))dv(h)$$

with $Z_1, \ldots, Z_n$ being some left invariant vector fields on G and v a measure on G.

❖❖❖❖❖

[104] GSA paper of Navneet: At the point r_0 in space is located a set of p infinitesimal current elements pointed along different directions. This current vector is described by

$$\mathbf{I}(t) = \sum_{k=1}^{p} \mathbf{I}_k p_k(t) d\mathbf{I}_k$$

We note that the corresponding current density field is given by

$$\mathbf{J}(t, \mathbf{r}) = \mathbf{I}(t)\delta^3(\mathbf{r})$$

The magnetic vector potential produced by this current at the origin is given by

$$\mathbf{A}(t, \mathbf{r}) = \frac{\mu}{4\pi r}\sum_{k=1}^{p} \mathbf{I}_k p_k(t) d\mathbf{I}_k$$

The electric and magnetic fields in the far field zone are computed using the formulae

$$\mathbf{E}(t, \mathbf{r}) = -\nabla\Phi(t, \mathbf{r}) - \frac{\partial \mathbf{A}(t, \mathbf{r})}{\partial t}$$

$$\mathbf{B}(t, \mathbf{r}) = \nabla \times \mathbf{A}(t, r)$$

and retaining only $O(1/r)$ terms. We note that the electric scalar potential $\Phi(t, \mathbf{r})$ is computed using the Lorentz gauge condition:

$$\Phi(t, r) = -c^2\int_0^t \text{div}\mathbf{A}(\tau, \mathbf{r})d\tau$$

When this elecrtromagnetic field falls on an atom described by an unperturbed Hamiltonian

$$H_0 = \mathbf{p}^2/2m + V(\mathbf{q})$$

where the atomic nucleus is assumed to be located at $\mathbf{r}_0$ and $\mathbf{r}_0 = \mathbf{r}_0 + \mathbf{q}$ is position of the electron relative to the origin, and $\mathbf{p} = -\nabla_{q,}$ the resulting perturbed Hamiltonian is given by

$$H(t, \mathbf{I}) = (\mathbf{p} + e\mathbf{A}(t, \mathbf{r}_0 + \mathbf{q}))^2 + V(\mathbf{q}) - e\Phi(t, \mathbf{r}_0 + \mathbf{q})$$

Since $\mathbf{A}$ and Φ are linear function of I_k, $k = 1, 2 \ldots,$ it follows that

$$H(t/I) = H_0 + e\sum_{k=1}^{p} I_k V_{1k}(t) + e^2 \sum_{k,m=1}^{p} I_k I_m V_{2km}(t)$$

where V_{1k} is first order time dependent linear partial differential operator in q (*i.e.*, the sum of a time dependent function of q and a time dependent vector field in q) while $V_{2km}(t)$ is a function of q, t, *i.e.*, a multiplication operator. Using time dependent perturbation theory, we calculate the Schrodinger evolution operator $U(T/I)$ generated by the Hamitonian $H(t/I)$ at tiome T, *i.e.*,

$$iU'(t \mid I) = H(t|I)U(t|I),\ t \geq 0,\ U(0 \mid I) = I$$

upto $O(e^3)$. This evolution operator has the form

$$U(T|I) = U_0 + e\sum_{k} I_k U_{1k} + e^2 \sum_{k,m} I_k I_m U_{2km}$$

$$+ e^3 \sum_{kml} I_k I_m I_l U_{3kml} + O\left(e^4\right)$$

where $_{0,}$ $U_{1k,}$ U_{2km}, U_{3kl} are operators that do not depend on the current amplitudes I_1, ...I_p. In other words, these operators are expressible entirely in terms of the operators H_0, $V_{1k}(t)$, $V_{2km}(t)$ with the operators $V_{1k}(t)$, $V_{2k}(t)$ completely independent of $I_1,\ldots, I_p$, *i.e.* expressible entirely in terms of the urrent pulses $p_k(t)$ and the length vectors $d\mathbf{I}_k$. We then take a given unitary gate U_g and express $E(I)$ $= \left\|U_g - U(T|I)\right\|_F$ upto $O(e^3)$ as a cubic polynomial in the $I'_k s$ and minimize this expression w.r.t. the $I'_k s$ using the GSA. If we try to minimize this expression by simply setting its partial derivative w.r.t. the I_k's to zero, we end up with a quadratic equation in the I_k's which is not solvable in closed form. The cubic polynomial approximation to the error energy can be expressed as

$$E(I) = \left\|U_g(T) - U(T|I)\right\|^2 = \left\|U_0(T)W(T) - U_g\right\|^2$$

$$= \left\|W(T) - U_0(-T)U_g\right\|^2$$

$$= \left\|W(T) - W_g\right\|^2 = \left\|W_g\right\|^2 + e\sum_{k} I_k S_1[k]$$

$$+e^2 \sum_{k,m} I_k I_m S_2[k,m] + e^3 \sum_{k,m,p} I_k I_m S_3[k,m,p] + O(e^4)$$

In our simulation studies, we have chosen H_0 as the 2 × 2 matrix, and have chosen p, the number of current elements as: By appropriately scaling down the physical quantities, we get the following table for the perturbing potential matrices V_{lk}, V_{2km}, k, m = 1, 2, ..., p. These matrices can be derived easily be choosing the following values for $d\{I\}_{\mathbf{k}}$ **k** = 1, 2, ..., **p** and the pulses $p_k(t)$, k = 1,..., p . Figures [] show plots of the error energy $E[I]$ as a function of the iteration number in the GSA experiments. The number of initial guess values for the current vector $I = [I_1,...I_p]^T$ has been chosen as q = :. The GSA algorithm implemented for the current vector $I = [I_1,..., I_p]^T$ has been chosen as q =: The GSA algorithm implemented for minimizing $E(I)$ as follows: Let at the iteration number n, the current vectors be $I[n, r]$, r = 1, 2,...,q. Here

$$I[n,r] = \left[I_1[n,r],...,I_p[n,r]\right]^T, r = 1, 2, ..., q$$

Then we compute

$$E_{max}[n] = max\{E[I[n, r]], r = 1, 2, ..., q\}$$

$$E_{min}[n] = min\{E[I[n, r]], r = 1, 2, ..., q\}$$

$$M[n, r] = \frac{E_{\max}[n] - E[I[n,r]]}{E_{\max}[n] - E_{\min}[n]}, r = 1, 2,...q$$

$M[n, r]$, r = 1, 2, ..., q are known as the masses at iteration n. Larger the value of the error energy $E[I[n, r]]$, smaller will be the value of $M[n, r]$. The GSA then guarantees that the "position vectors"$I[n, r]$, r = 1, 2, ..., q of the masses eventually converge to the position of the largest mass or equivalently, the smallest error energy. The GSA based on Newton's inverse square law of gravitation reads

$$I[n+1, r] = I[n, r] + \Delta V[n, r],$$

$$V[n+1, r] = V[n, r] + \Delta. \sum_{s=1, s\neq r}^{q} \frac{G(I[n,s] - I[n,r])}{\left(\|I[n,s] - I[n,r]\|^2 + \varepsilon^2\right)^{3/2}}$$

The factor ε has been chosen as () in order to guarantee that when two masses come very close, the gravitational force between them does not blow up to infinity. The gravitational constant G has been chosen as (). Our plots indicate very fast convergence of the error energy to the minimum. We further, display in Table (), the SNR corresponding to the approximated gate. if U_g is the desired gate and $U(T)$ is the approximated gate obtained after running the GSA, them the SNR is defined as

$$\text{SNR} = \frac{\left\|U_g\right\|^2}{\left\|U_g - U(T)\right\|^2}$$

If we use the quadratic approximation in the Dyson series, the minimization of the gate error energy leads to linear equations for the current vector but this erroor energy is significantly larger that obtained using the GSA applied to the cubic approximation in the Dyson series.

Significant contributions: (1) Accurate realization of non-separable quantum gates of large sizes. Non separable because, we can perturb the sum of independent Hamiltonians (direct sum) by a potential that acts on the tensor product all the component Hilbert spaces in a non-separable way.

(2) Smaller gate error energy because we can use high orders (2) in the Dyson series approximation for the Schrodinger unitary evolution an still minimize the gate error energy using the GSA. For Dyson series approximation > 2 closed form minimization by setting the partial derivatives w.r.t. the current vector to zero is not possible.

(3) Fast convergence of the error energy to zero than other iterative/search algorithms like gradient search because it relies on a natural physical principle that if n bodies interact with each other under Newton's inverse square law of gravitation, then eventually, very quickly, all the masses will towards the largest mass.

Why we use PAM signals in the design of the input current: PAM signals are time shifted versions of rectangular pulses and computing the integrals in the Dyson series for shifted PAM signals is very easy since they involve only integration from the beginning of the pulse to the end of the pulse whereas for non rectangular pulses, a weighted integration must be carried out which is computationally more expensive.

A problem for further research is in the design of the electric and magnetic fields perturbing a 3-D harmonic oscillator so that the gate evolved after time T is as close as possible to a given gate. In [] we have used the quadratic Dyson series approximation to design such a gate. It would be interesting to use a cubic approximation and apply the GSA method of this paper to design the optimal electromagnetic field. We would have to assume that the electric and magnetic fields are finite linear combinations of given basis / test functions of time and design the coeffcients to minimize the gate error energy defined by a cubic polynomial in the linear combination coeffcients.

Another open problem is to perturb an atomic system or harmonic oscillator with a quantum electromagnetic field which is modulated by complex scalar valued functions of time and then design the gate using the cubic approximation w.r.t. these complex function of time.

Conclusion: This paper addresses the problem of quantum gate design based on cubic approximation to the unitary evolution operator w.r.t. the current variables that define the electromagnetic field interacting with the atom. The gate error energy has been approximated by a cubic polynomial which cannot be minimized in closed from using the standard method of setting the partial derivatives to zero as these lead to quadratic equations in several variables. Therefore, this paper uses the GSA to minimize the cubic polynomial. Out simulation studies indicate that we get very fast converges of the gate error energy to the minimum and thus are able to easily design gates which approximate the given gates with greater accuracy then that possible using the standard quadratic approximation, In a future paper, we shall apply the GSA to design gates based on ion trap experiments, namely interaction of a quantum electromagnetic field with a spin system.

❖❖❖❖❖

[105] Generalized quantum filtering for non-demolition measurements in the sense of Belavkin

A. Summary of classical filtering: $x(t) \in \mathbb{R}^n$ is the state process, $z(t) \in \mathbb{R}^d$ is the measurement process:

$$dx(t) = f(t, x(t))dt + g(t, x(t))dB(t)$$

$B(t) \in \mathbb{R}^d$ is Brownian motion,

$dz(t) = h(h, x(t))dt + \sigma dv(t)$

$v(t) \in \mathbb{R}^d$ is Brownian motion independent of B (.).

$$Z_t = \{z(s):s \le t\}$$

is the observations collected upto time t.

$$p(x(t+dt)|Z_{t+dt}) = \frac{\int p(dz(t)|x(t))p(x(t+dt)|x(t))p(x(t)|Z_t)dx(t)}{\int \text{numerator } dx(t+dt)}$$

$$= \frac{\int \exp\left(\begin{matrix}-(dz(t)-h(t,x(t))dt)^T \\ (dz(t)-h(t,x(t))dt)/2\sigma_v^2\end{matrix}\right) p(x(t+dt)|x(t))p(x(t)|Z_t)dx(t)}{\int \text{numerator } dx(t+dt)}$$

$$= \frac{\int \exp\left(h(t,x(t))^T dz(t) \big/ \sigma_v^2 - \frac{1}{2\sigma_v^2} h(t,x(t))^T h(t,x(t))dt \right) p\left(x(t+dt) \middle| x(t)\right) p\left(x(t)Z_t|\right)dx(t)}{\int \text{numerator}dx(t+dt)}$$

Let

$$\pi_t(\phi) = \mathbb{E}\left(\Phi(x(t)) \middle| Z_t\right)$$

Then we get from the above,

$$\pi_{t+dt}(\phi) = \frac{\sigma_{t+dt}(\Phi)}{\sigma_{t+dt}(1)}$$

where

$$\sigma_{t+dt}(\phi) = \int \exp\left(h(t,x(t))^T dz(t) \big/ \sigma_v^2 - \frac{1}{2\sigma_v^2} h(t,x(t))^T h(t,x(t))dt \right)$$

$$\mathbb{E}\left(\Phi\left(x(t+dt) \middle| x(t)\right)\right) p\left(x(t) \middle| Zt\right)dx(t)$$

$$= \int \exp\left(h(t,x(t))^T dz(t) / \sigma_v^2 - \frac{1}{2\sigma_v^2} h(t,x(t))^T h(t,x(t)dt) \right)(\phi(x)(t))$$

$$+ dt.L_t(\phi)(x(t)))p(x)(t) \mid Z_t)dx(t)$$

so

$$\sigma_{t-dt}(\Phi(x)) = \pi_t \left[\exp\left((h(t,x))^T dz(t) \big/ \sigma_v^2 - \frac{1}{2\sigma_v^2} h(h,x)^T h(h,x)dt \right) \Phi(x) \right]$$

$$+ dt.\pi_t \left[\exp\left((h(t,x))^T dz(t) \big/ \sigma_v^2 - \frac{1}{2\sigma_v^2} h(h,x)^T h(h,x)dt \right) L_t\Phi(x) \right]$$

$$\pi_t \left[\phi(x) + h(t,x)^T dz(t)\phi(x) \big/ \sigma_v^2 + dtL_t\phi(x) \right]$$

using Ito's formula

$$dz(t)dz(t)^T = \sigma_v^2 I_d dt$$

$$\pi_{t+dt}(\phi) = \frac{\pi_t\left(\phi + h_t^T dz\phi/\sigma_v^2 + dtL_t\phi\right)}{1+\pi_t\left(h_t^T dz\right)/\sigma_v^2}$$

$$= \left[\pi_t(\phi)+\pi_t\left(\phi h_t^T\right)dz/\sigma_v^2 + \pi_t(L_t\phi)dt\right]$$

$$\left[1-\pi_t\left(h_t^T\right)dz/\sigma_v^2 + dt.\pi_t\left(h_t^T\right)\pi_t(h_t)/\sigma_v^2\right]$$

$$= \pi_t(\phi)+\sigma_v^{-2}\left[\pi_t\left(\phi.h_t^T\right)-\pi_t(\phi)\pi_t\left(h_t^T\right).dz - \pi_t(\phi)\pi_t\left(h_t^T\right)\pi_t(h_t)dt\right]$$

$$+\pi_t(L_t\phi)dt - \sigma_v^{-2}\pi_t\left(\phi.h_t^T\right)\pi_t(h_t)dt$$

$$d\pi_t(\phi) = \pi_t(L_t(\phi))dt + \sigma_v^{-2}\left(\pi_t(\phi.h_t)-\pi_t(\phi)\pi_t(h_t)\right)^T\left(dz - \pi_t(h_t)dt\right)$$

This is the Kushner filter. This filter can also be derived from the famous Kallianpur-Striebel formula. By considering second order approximations, we can derive the Extended Kalman filter from this obtained by taking $\phi(x) = x_a, \phi(x) = x_a x_a$ successively. The EKF reads

$$d\hat{x}(t) = f\left(t,\hat{x}(t)\right)dt + \sigma_v^{-2}P(t)H\left(t,\hat{x}(t)\right)^T\left(dz(t) - h\left(t,\hat{x}(t)\right)dt\right),$$

$$P'(t) = F\left(t,\hat{x}(t)\right)P(t)+P(t)F^T\left(t,\hat{x}(t)\right)+g\left(t,\hat{x}(t)\right)g\left(t,\hat{x}(t)\right)^T$$

$$-P(t)H\left(t,\hat{x}(t)\right)^T H\left(t,\hat{x}(t)\right)P(t)$$

where $H(t, x)$ is the $d \times n$ Jacobian matrix of the map $x \rightarrow h(t, x)$, *i.e.*,

$$H(t,x) = \frac{\partial h(t,x)}{\partial x}$$

where $F(t, x)$ is the $n \times n$ Jacobian matrix of the map $x \rightarrow f(t, x)$,

$$F(t,x) = \frac{\partial f(t,x)}{\partial x}$$

B. Summary of Quantum Stochastic Calculus and the Hudson-Parthasarathy Quantum I to formula

$\mathcal{H} = L^2(\mathbb{R}_+)$, $\Gamma_s(\mathcal{H})$ is the Boson Fock space over $\mathcal{H}$. $\Gamma_s(\mathcal{H})$ is the closure of the linear manifold generated by the exponential vectors

$$\mathrm{e}(u) = \bigoplus_{n=0}^{\infty} u^{\otimes n}\sqrt{n!}$$

We note that

$$u^{\otimes n} = \frac{d^n}{dt^2}e(tu)\big|_{t=0} = \sqrt{n!u}^{\otimes n}$$

❖❖❖❖❖

[106] Ph.D thesis summary of K. Gautam.

Abstract: In this work, we explore the idea of realizing quantum gates normally used in quantum computation using physical systems like atoms and oscillators perturbed by electric and magnetic fields. The basic idea around which the subject of this thesis revolves is that if a time independent Hamiltonian H_0 is perturbed by a time varying Hamiltonaian of the form $f(t)V$ where $f(t)$ is a scalar function of time and V is a Hermitian operator that does not commute with H_0. then a very large class of unitary operators can be realized *via* the Schrodinger evolution corresponding to the time varying Hamiltonian $H_0 + f(t)V$. H_0 by itself generates only a one dimensional class of unitary gates while $H_0+f(t)V$, $t \geq 0$ can generate an infinite dimensional manifold of unitary gates. This is a consequence of the Baker-Campbell-Hausdorff formula in Lie groups and Lie groups and Lie algebras []. Broadly speaking we treat two problems in this thesis based upon the above idea. First, we take a Harmonic oscillartor and perturb it with a time independent an harmonic term. The total Hamiltonian is then $H_1 = (q^2+p^2)/2 + \varepsilon q^3$. We then calculate $U_g = \exp(-iTH_1)$ and consider this to be the desired gate to be realized. We then perturb the harmonic Hamiltonian with a linear time dependent term so that the overall Hamiltonian becomes $H(t) = (q^2 + p^2)/2 + \varepsilon f(t)q$ and calculate the unitary evolution corresponding to $H(t)$ at time T. Using the time ordering operator T, this gate can be expressed as

$$U(T) = U(T, \varepsilon, f) = T\left\{\exp\left(-i\int_0^T H(t)dt\right)\right\}$$

$U(T)$ is calculated upto $O(\varepsilon^2)$ using time dependent perturbation theory and $f(t)$ is chosen so that $U(T, \varepsilon, f)$ is as close as possible in the Frobenius norm to U_g with a power constraint on $f(t)$. This optimization problem is solved by arriving at a linear integral equation for $f(t)$. This problem is equivalent to perturbing a charged Harmonic oscillator with a time varying electric field and using the electric field as our control process to generrate a gate as close as possible to the given gate. The an harmonic gate U_g is then replaced by a host of commonly used gates in quantum computation like controlled unitary gates, quantum Fourier transform gate etc. and the control electric field is then selected appropriately. We then apply the same formalism to Hamiltonians consisting of an atom described by a Pauli spin variable plus a quantum electromagnetic field Hamiltonian described by creation and annihilation operators and an interaction term between atom and field that is modulated by a scalar control function. This is particularly important since recently ion trap systems have been modelled in this way and quantum gates realized using this scheme [Nielsen and Chuang]. In the course of designing quantum gates using physical systems like atoms and oscillators perturbed by electric and magnetic fields, we have also addressed the controllability issue, *i.e.*, under what conditions does there exist a scalar real valued function of time $f(t)$, $0 \leq t \leq T$ such that if $|\psi_i\rangle$ is any initial wave function and $|\psi_f>$ is any final wave function, then $U(T,f)|\psi_i> = |\psi_f>$. We haves obtained a partial solution to this problem by replacing the unitary evolution kernel $U(T,f)$ by its Dyson series truncated version. In all our design procedures, the gates that actually appear are infinite dimensional, more precisely, they are of the form $\exp(-iH)$ where H is an unbounded Hermitian operator acting on an infinite dimensional Hilbert space. We have approximated the infinite dimensional problem by a finite dimensional one based on truncation. The primary novel feature of this thesis, is the design of quantum gates when the system consists of an atom / oscillator described by either position and momentum operators or creation and annihilation operators or spin matrices and a quantum electromagnetic field described by a sequence of creation and annihilation operators and there is an interaction between the atom and the em field that is modulated by a controllable function of time, like for example a spin interacting with a controllable quantum magnetic field.

❖❖❖❖❖

[107] Haar measure on some semisimple Lie groups:

(*a*) G = $SO(3)$. Any $R \in G$ can be expressed as

$$R = R_z(\phi)\, R_x(\theta)$$

$R_z(\psi) = \exp(\phi X_3)\exp(\theta X_1)\exp(\psi X_3)$

where X_1, X_2, X_3 are the standard generators of G satisfying the commutation relations

$$[X_1, X_2] = X_3\ [X_2, X_3] = X_1\ [X_3, X_1] = X_2$$

we have

$$R_{,\psi} = RX_3,$$

$$R_{,\theta} = \exp(\phi X_3)\exp(\theta X_1)\exp(\psi X_3)$$

$$R.\exp(-\psi ad\mathrm{X}_3)(\mathrm{X}_1)$$

$$= R.(X_1\cos(\psi) - X_2\sin(\psi))$$

so if $\tilde{X}_k$ is the left invariant vector field generated by X_k, i.e., for any smooth function $f: G \to \mathbb{C}$

$$\tilde{X}_k g(\phi,\theta,\psi) = \frac{d}{dt} f\left(R.\exp(tX_k)\right)\Big|_{t=0}$$

where $$g(\phi,\theta,\psi) = f(R)$$

then we have

$$\cos(\psi)\widehat{X}_1 - \sin(\psi)\widehat{X}_2 = \frac{\partial}{\partial\theta}$$

$$\widehat{X}_3 = \frac{\partial}{\partial\psi}$$

Finally, $$R_{,\phi} = X_3 R = \exp(\phi X_3)\, X_3 \exp(\theta X_1)\exp(\psi X_3)$$

$$= R.\exp(-\psi ad X_3).\ \exp(-\theta ad X_1)(X_3)$$

Now

$$\exp(-\theta ad X_1)(X_3) = X_3.\cos(\theta) + X_2.\sin(\theta)$$

$$\exp(-\psi ad X_3)(X_3.cos(\theta) + X_2.\sin(\theta))$$

$$= X_3.\cos(\theta) - \sin(\theta)(X_2\cos(\psi) + X_1.\sin(\psi))$$

Thus,

$$\frac{\partial}{\partial\phi} = -\sin(\theta)\sin(\psi)\tilde{X}_1 - \sin(\theta)\cos(\psi)\tilde{X}_2 + \cos(\theta)\tilde{X}_3$$

Using these formulae, we easily compute that

$$\frac{\partial}{\partial\psi}\wedge\frac{\partial}{\partial\theta}\wedge\frac{\partial}{\partial\phi}$$

$$= \sin(\theta)\tilde{X}_1\wedge\tilde{X}_2\wedge\tilde{X}_3$$

and hence that invariant Haar measure on $SO(3)$ parameterized by the Euler angles is

$$d\mu = \sin(\theta)d\psi d\theta d\phi$$

(*b*) Symplectic group $G=Sp(2n, \mathbb{R})$ with Lie algebra $g = sp(2n, \mathbb{R})$. Any $X \in g$ satisfies

$$X^T J + JX = 0$$

where

$$J=\begin{pmatrix} 0_n & I_n \\ -I_n & 0 \end{pmatrix}$$

$$X=\begin{pmatrix} X_1 & X_2 \\ X_3 & X_4 \end{pmatrix}$$

where the $X'_k s$ are all $n \times n$ matrices, we get

$$X^T J = \begin{pmatrix} -X_3^T & X_1^T \\ -X_4^T & X_2^T \end{pmatrix}$$

$$JX = \begin{pmatrix} X_3 & X_4 \\ -X_1 & -X_2 \end{pmatrix}$$

Thus the condition for $X \in g$ becomes

$$X_4 = -X_1^T, X_2^T = X_2, X_3^T = X_3$$

It follows that a basis for g is given by

$$F_{ij} = \begin{pmatrix} E_{ij} & 0 \\ 0 & -E_{ji} \end{pmatrix}$$

$$G_{ij} = \begin{pmatrix} 0 & E_{ij} + E_{ji} \\ 0 & 0 \end{pmatrix}$$

$$H_{ij} = G_{ij}^T = \begin{pmatrix} 0 & 0 \\ E_{ij} + E_{ji} & 0 \end{pmatrix}$$

where $1 \le i \le j \le n$. In particular, this implies that

$$dimsp(2n, \mathbb{R}) = 3n(n+1)/2$$

The Cartan algebra is spanned by the elements

$$h_i = \begin{pmatrix} E_{ii} & 0 \\ 0 & -E_{ii} \end{pmatrix}, i = 1, 2, ..., n$$

Exercise: Verify by calculating the commutators $[h_i, F_{kl}], [h_i, G_{kl}], [h_i, H_{kl}]$ that $\{G_{kl}, H_{kl}: 1 \le k \le l \le n\} \cup \{F_{kl}: 1 \le k \le l \le n\}$ are root vectors for g w.r.t. the Cartan subalgebra $\hbar$.

❖❖❖❖❖

[108] Invariants for functions defined on a semisimple Lie group. Let $f_k: G \to \mathbb{C}$, $k = 1, 2$ be two functions. Suppose X is the character of a representation of G. Then define

$$I_x(f_1, f_2) = \int_{G \times G} f_1(g) f_2(h^{-1}) \chi(gh^{-1}) dgdh$$

Let $x \in G$. Then

$$I_\chi(f_1 o x^{-1}, f_2 o x^{-1}) = \int f_1(x^{-1}g) f_2(h^{-1}x) \chi(gh^{-1}) dgdh$$

$$= \int f_1(g) f_2(h^{-1}) \chi(xgh^{-1}x^{-1}) dgdh$$

$$= I_\chi(f_1, f_2)$$

since

$$\chi\left(xgx^{-1}\right) = \chi(g), g, x \in G$$

This formula suggests a method for pattern classification when the objects/images are subject to transformations from the group G. The integration can be performed using Weyl's integration formula for Class functions.

❖❖❖❖❖

[109] Quantum Robotic teleoperation system analysis: Master robot dynamical equations:

$$dq_m(t) = \omega_m(t)dt,$$

$$d\omega_m(t) = F_m(q_m(t), \omega_m(t), t)dt + G(q_m(t))dB_m(t)$$
$$+\left[K_{mp}(q_s(t-T_s) - q_m(t)) + K_{md}(\omega_s(t-T_s) - \omega_m(t))\right]dt$$

$$d\omega_s(t) = F_s(q_s(t), \omega_s(t), t)dt + G(q_s(t))dB_s(t)$$
$$+\left[K_{sp}(q_m(t-T_m) - q_s(t)) + K_{sd}(\omega_m(t-T_m) - \omega_s(t))\right]dt$$

where $B_m(t)$ and $B_s(t)$ are independent two dimensional Brownian motion processes. Here,

$$F_m(q_m, \omega_m, t) = M_m(q_m)^{-1}(-N_m(q_m, \omega_m) + \tau_m(t)),$$

$$G_m(q_m) = M_m(q_m)^{-1}$$

and likewise for F_s and G_s. Here, T_s is the slave to master teleoperation delay and T_m is the master to slave teleoperation delay. Thus system can be put into the following abstract form:

$$d\xi(t)$$
$$= F(\xi(t), t)dt + G(\xi(t)dB(t)) + (K_0\xi(t) + K_2\xi(t-T_1) + K_3\xi(t-T_2))dt \quad ...(1)$$

where $\xi(t) \in \mathbb{R}^d$ and K_0, K_1, K_2 are $d \times d$ real matrices while $B(t)$ is standard R^p–valued Brownian motion. This is a stochastic differential equation with delay. We can approximately solve this using perturbation theory. For this, we introduce a perturbation parameter ε both into the noise and the feedback terms. We then write in place of (1),

$$d\xi(t)$$

$$= F(\xi(t),t)dt + \varepsilon G(\xi(t)dB(tt) + \varepsilon(K_0\xi(t) + K_2\xi(t-T_1) + K_3\xi(t-T_2))dt \quad ...(1)$$

Write the solution as

$$\xi(t) = \xi_0(t) + \varepsilon\xi_1(t) + \varepsilon^2\xi_2(t) + O(\varepsilon^3)$$

Then equating terms of equal powers of ε gives

$$d\xi_0(t) = F(\xi_0(t), t)dt,$$

$$d\xi_1(t) = F'(\xi_0(t),t)\xi_1(t)dt + G(\xi_0(t))dB(t)$$

$$+K_0\xi_0(t) + K_1\xi_0(t-T_1) + K_2\xi_0(t-T_2)$$

$$d\xi_2(t) = F'(\xi_0(t),t)\xi_2(t) + \frac{1}{2}F''(\xi_0(t),t)(\xi_1(t)\otimes\xi_1(t))$$

$$+G'(\xi_0(t))(\xi_1(t)\otimes I_p)dB(t) + K_0\xi_1(t)$$

$$+K_2\xi_1(t-T_1) + K_2\xi_2(t-T_2)$$

The equations satisfied by ξ_1 and ξ_2 are linear and easily expressible in terms of the state transition matrix corresponding to the forcing Jacobian matrix $F(\xi_0(t),t)$.

[110] Quantum filtering and control: Let $Y_{in}(t)$, $t \geq 0$ be the input measurement process. It forms an Abelian family of operators in the Boson Fock space $\Gamma_s(\mathcal{H})$ where $\mathcal{H} = L^2(\mathbb{R}_+)$. Let $\mathfrak{h}$ be the system space in contrast to $\Gamma_s(\mathcal{H})$ which is the bath/environment space. Write $\eta_{in}(t) = \sigma(Y_{in}(s) : s \leq t)$ be the Abelian Von-Neumann algebra generated by the measurement process upto time t. Let $U(t)$ be a unitary family of operators in the total space $\mathfrak{h} \otimes \Gamma_s(\mathcal{H})$ satisfying the qsde

$$dU(t) = \left(L^{\alpha}_{\beta}d\Lambda^{\alpha}_{\beta}(t) + k\left(X_d - \mathbb{E}_t\left(X|\eta_{in}(t)\right)\right)dt\right)$$

$$+\left(L^{\alpha}_{\beta}(t)\right)U(t) + kU(t)\left(j_t(X_d) - \pi_t(X)\right)dt$$

Note that for the unitarity of $U(t)$, we require that

$$0 = d\left(U(t)^* U(t)\right)$$

$$= dU(t)^* U(t) + U(t)^* dU(t) + dU(t)^* dU(t)$$

$$U(t)^*\left(L^{\beta*}_{\alpha} + L^{\alpha}_{\beta}\right)d\Lambda^{\beta}_{\alpha} + L^{\beta*}_{\alpha}L^{\mu}_{\nu}d\Lambda^{\beta}_{\alpha}d\Lambda^{\nu}_{\mu}$$

$$+ 2(\operatorname{Re} k)\left(X_d - \mathbb{E}_t\left(X|\eta_{in}(t)\right)dt\right)U(t)$$

we may ensure this by taking k as pure imaginary and choosing the system operators L_β^α so that

$$L_\alpha^{\beta *} + L_\beta^\alpha + L_\alpha^{\nu *} L_\nu^\mu \varepsilon_\mu^\nu = 0$$

Here, L_β^α , X, X_d are system observables. We define the state at time t as

$$j_t(X) = U(t)^* X U(t)$$

and its filtered estimate as

$$\pi_t(X) = \mathbb{E}\big(j_t(X) \mid \eta_{out}(t)\big) = U(t)^* \mathbb{E}_t(X \mid \eta_{in}(t)) U(t)$$

This provides the defining relation for the expectation $\mathbb{E}_t$, *i.e*,

$$U(t)^* \mathbb{E}_t(X) U(t) = \mathbb{E}(U(t)^* X U(t))$$

Here,

$$\begin{aligned} \eta_{out}(t) &= U(t)^* \eta_{in}(t) U(t) \\ &= \sigma(Y_{out}(s) : s \leq t) = U(t)^* \eta_{in}(t) U(t) \end{aligned}$$

where

$$Y_{out}(t) = U(t)^* Y_{in}(t) U(t)$$

is the output measurement process. It is easy to see that for all $T > t$,

$$Y_{out}(t) = U(T)^* Y_{in}(t) U(T)$$

This follows from the unitarity of $U(s)$, $s \geq 0$ and the fact that the system algebra $\mathcal{L}(h)$ and the input measurement algebra upto time T, $\eta_{in}(T)$, commutes with $Y_{in}(t)$. In other words, these facts imply that the differential of $U(T)^* Y_{in}(t) U(T)$ w.r.t. T is zero. $j_t(X_d)$ is the desired signal to be tracked by our state evolution. We see that

$$\begin{aligned} dj_t(X) &= dU(t)^* X U(t) + U(t)^* X dU(t) + dU(t)^* X dU(t) \\ &= j_t\Big(\Big(L_\alpha^{\beta *} X + X L_\beta^\alpha + L_\alpha^{\nu *} X L_\beta^\mu \varepsilon_\mu^\nu\Big) d\Lambda_\alpha^\beta \\ &\quad - k.j_t\Big(\big[X_d - \mathbb{E}\big(X \mid \eta_{in}(t), X\big)\big]\Big) dt \end{aligned}$$

assuming k to be purely imaginary.

❖❖❖❖❖

[111] Combination of state observer and state trajectory tracking in classical probability. $X(t)$ is the state at time t while $\widehat{X}(t)$ is the state observer output at time t. $X_d(t)$ is the desired non-random state trajectory. $K_t(\widehat{X}(t))$ is the state tracking feedback coefficient matrix and $L_t(\widehat{X}(t))$ is the output observer error feedback.

The sde's satisfied by these processes are

$$dX(t) = F(t, X(t))dt + G(t, X(t))dB(t) + K_t\left(\widehat{X}(t)\right)(X_d(t) - X(t))$$

$$d\widehat{X}(t) = F(t, \widehat{X}(t))dt + L_t(\widehat{X}(t))(dZ(t) - H(t, \widehat{X}(t))dt)$$

$$dX_d(t) = F(t, X_d(t))dt$$

$X_d(t)$ is the desired non-random trajectory. The output or measurement equation is

$$dZ(t) = H(t, X(t))dt + \sigma_v dV(t)$$

The tracking error at time t is

$$f(t) = X_d(t) - X(t)$$

and the state observer error at time t is

$$e(t) = X(t) - \widehat{X}(t)$$

We have approximately assuming that $e(t)$ and $f(t)$ have small amplitudes,

$$de(t) = F'(t, \widehat{X}(t))e(t)dt + K_t(\widehat{X}(t))f(t)$$

$$+ G(t, \widehat{X}(t))dB(t)$$

$$- L_t(\widehat{X}(t))(H'(t, \widehat{X}(t))e(t)dt + \sigma_v dV(t))$$

and

$$df(t) = F'(t, \widehat{X}(t))f(t)dt - G(t, \widehat{X}(t))dB(t) - K_t(\widehat{X}(t))f(t)dt$$

$$\left(F'(t, \widehat{X}(t)) - K_t(\widehat{X}(t))f(t)dt - G(t, \widehat{X}(t))dB(t)\right)$$

Problem: Derive algorithms for calculating $K_t(\widehat{X}(t))$ and $L_t(\widehat{X}(t))$ by minimizing a linear combination of $\mathbb{E}\left[d\|e(t)\|^2|\widehat{X}(t)\right]$, $\mathbb{E}\left[d\|f(t)\|^2|\widehat{X}(t)\right]$ and $\mathbb{E}\left[d\|e(t) - f(t)\|^2|\widehat{X}(t)\right]$ subject to quadratic energy constraints on K_t, L_t of the form $Tr(K_T(X(t))A_tK_t(X(t)^T) = E_t$ and $Tr\left(K_t(\widehat{X}(t))A_tK_t(\widehat{X}(t))^T\right) = N_t$. In doing so, you may assume that $e(t)$ is orthogonal to any Borel function of $\widehat{X}(t)$.

❖❖❖❖❖

[112] Let the metric be expressed upto second order of smallness as

$$g_{\mu\nu} = g_{0\mu\nu} + \varepsilon g_{1\mu\nu} + \varepsilon^2 g_{2\mu\nu}$$

In terms of matrices

$$G = G_0 + \varepsilon G_1 + \varepsilon^2 G_2$$

Then, upto $O(\varepsilon^2)$

$$G^{-1} = \left(I - \varepsilon G_0^{-1}G_1 + \varepsilon^2\left(G_1^2 - G_0^{-1}G_2\right)\right)G_0^{-1}$$

$$= G_0^{-1} - \text{epsilon } G_0^{-1}G_1G_0^{-1} + \varepsilon^2\left(G_1^2G_0^{-1} - G_0^{-1}G_2G_0^{-1}\right)$$

Thus in terms of components,

$$g^{\mu\nu} = g_0^{\mu\nu} - \varepsilon g_0^{\mu\alpha}\, g_0^{\nu\beta}\, g_{1\alpha\beta}$$

$$+ \varepsilon^2\left(g_{0\mu\alpha}\, g_{0\alpha\beta}\, g_0^{\alpha\beta} - g_0^{\mu\alpha}\, g_{2\alpha\beta} g_0^{\nu\beta}\right)$$

Using this metric tensor, we calculate the Einstien tensor

$$G^{\mu\nu} = R^{\mu\nu} - \frac{1}{2} R g^{\mu\nu}$$

upto $O(\varepsilon^2)$. We write this expansion as

$$G^{\mu\nu} = G_0^{\mu\nu} + \varepsilon G_1^{\mu\nu} + \varepsilon^2 G_2^{\mu\nu} + O(\varepsilon^3)$$

It is clear that $G_0^{\mu\nu}$ is a function of only $g_{0\mu\nu}$ and its first and second order partial derivatives, $G_1^{\mu\nu}$ is a function of only $g_{0\mu\nu}, g_{1\mu\nu}$ and their first two partial derivatives and moreover, it is linear in $g_{1\mu\nu}$ and its first and second partial derivatives. $G_2^{\mu\nu}$ is a function of $g_{0\mu\nu}, g_{1\mu\nu}, g_{2\mu\nu}$ and their first and second order partial derivatives and it is moreover linear in $g_{2\mu\nu}$ and its first two partial derivatives. The energy momentum tensor is

$$T^{\mu\nu} = T^{\mu\nu} + \varepsilon K^{\mu\nu}$$

where $\varepsilon K^{\mu\nu}$ is the contribution from external forces and

$$T^{\mu\nu} = T^{\mu\nu} + \varepsilon T_1^{\mu\nu} + \varepsilon T_1^{\mu\nu} + O(\varepsilon^3)$$

$$= (\rho + p)\, v^{\mu} v^{\nu} - p g^{\mu\nu}$$

We write

$$\rho = \rho_0 + \varepsilon\rho_1 + \varepsilon^2\rho_2 + O(\varepsilon^3),$$

$$p = p_0 + \varepsilon p_1 + \varepsilon^2 p_2 + O(\varepsilon^3),$$

$$v^{\mu} = v_0^{\mu} + \varepsilon v_1^{\mu} + \varepsilon^2 v_2^{\mu} + O(\varepsilon^3)$$

We have

$$T_0^{\mu\nu} = (\rho_0 + p_0)\, v_0^{\mu} v_0^{\nu} - p_0 g_0^{\mu\nu}$$

$$T_1^{\mu\nu} = (\rho_0 + p_0)\left(v_0^{\mu} v_1^{\nu} + v_1^{\mu} v_0^{\nu}\right) + (\rho_1 + p_1)\, v_0^{\mu} v_0^{\nu} - p_0 g_1^{\mu\nu} - p_1 g_0^{\mu\nu}$$

$$T_2^{\mu\nu} = (\rho_0 + p_0)\left(v_0^{\mu} v_2^{\nu} + v_2^{\mu} v_0^{\nu} + v_1^{\mu} v_1^{\nu}\right) + (\rho_1 + p_1)\left(v_0^{\mu} v_1^{\nu} + v_1^{\mu} v_0^{\nu}\right)$$

$$+ (\rho_2 + p_2) v_0^{\mu} v_0^{\nu} - p_0 g_2^{\mu\nu} - p_1 g_1^{\mu\nu} - p_2 g_0^{\mu\nu}$$

The Einstein field equations are

$$G^{\mu\nu} = k(T^{\mu\nu} + \varepsilon K^{\mu\nu})$$

The Bianchi identity

$$G^{\mu\nu}_{:\nu} = 0$$

gives the fluid dynamic equations in curved space-time:

$$T^{\mu\nu}_{:\nu} + \varepsilon K^{\mu\nu}_{:\nu} = 0$$

where the covarient derivative is w.r.t. the perturbed metric $g_{\mu\nu} = g_{0\mu\nu} + \varepsilon g_{1\mu\nu} + \varepsilon^2 g_{2\mu\nu}$. We can write

$$T^{\mu\nu}_{:\nu} = \left[T_0^{\mu\nu} + \varepsilon T_1^{\mu\nu} + \varepsilon^2 T_2^{\mu\nu}\right]_{:\nu}$$

$$T^{\mu\nu}_{0:\nu} = \left[T^{\mu\nu}_{0:\nu}\right]_0 + \varepsilon\left[T^{\mu\nu}_{0:\nu}\right]_1 + \varepsilon^2\left[T^{\mu\nu}0:\nu\right]_2$$

The various perturbation terms come due to the expansion of the Christoffel symbols appearing in the covarient derivatives in powers of ε. For example writing

$$\Gamma^{\alpha}_{\mu\nu} = \Gamma^{\alpha}_{0\mu\nu} + \varepsilon\Gamma^{\alpha}_{1\mu\nu} + \varepsilon^2\Gamma^{\alpha}_{2\mu\nu}$$

we have

$$\xi^{\mu\nu}_{:\alpha} = \xi^{\mu\nu}_{,\alpha} + \Gamma^{\mu}_{\alpha\rho}\xi^{\rho\nu} + \Gamma^{\nu}_{\alpha\rho}\xi^{\mu\rho}$$

Thus,

$$\left[\xi^{\mu\nu}_{:\alpha}\right]_0 = \xi^{\mu\nu}_{,\alpha} + \Gamma^{\mu}_{0\alpha\rho}\xi^{\rho\nu} + \Gamma^{\nu}_{0\alpha\rho}\xi^{\mu\rho},$$

$$\left[\xi^{\mu\nu}_{:\alpha}\right]_k = \Gamma^{\mu}_{k\alpha\rho}\xi^{\rho\nu} + \Gamma^{\nu}_{k\alpha\rho}\xi^{\mu\rho},\ k = 1, 2, 3, \ldots$$

Likewise,

$$\varepsilon T^{\mu\nu}_{1:\nu} = \varepsilon\left[T^{\mu\nu}_{1:\nu}\right]_0 + \varepsilon^2\left[T^{\mu\nu}_{1:\nu}\right]_1$$

with neglect of $O(\varepsilon^3)$ and finally,

$$\varepsilon^2 T^{\mu\nu}_{2:\nu} = \varepsilon^2\left[T^{\mu\nu}_{2:\nu}\right]_0$$

with neglect of $O(\varepsilon^3)$. So the equations for the zeroth order variables $g_{0\mu\nu}$, ρ_0, p_0, v_0^{μ} are

$$G_0^{\mu\nu} = kT_0^{\mu\nu}, \left[T^{\mu\nu}_{0:\nu}\right]_0 = 0$$

The equations for the second order variables $g_{1\mu\nu}$, ρ_0, p_1, v_1^μ are

$$G_1^{\mu\nu} = k\left(T_1^{\mu\nu} + K^{\mu\nu}\right),$$

$$\left[T_{0:\nu}^{\mu\nu}\right]_1 + \left[T_{1:\nu}^{\mu\nu}\right]_0 + \left[K_{:\nu}^{\mu\nu}\right]_0 = 0$$

The equations for the second order variables $g_{2\mu\nu}$, ρ_2, p_2, v_2^μ are

$$G_2^{\mu\nu} = kT_2^{\mu\nu},$$

$$\left[T_{0:\nu}^{\mu\nu}\right]_2 + \left[T_{1:\nu}^{\mu\nu}\right]_1 + \left[T_{2:\nu}^{\mu\nu}\right]_0 + \left[K_{:\nu}^{\mu\nu}\right]_1 = 0$$

❖❖❖❖❖

[113] Stability and robustness analysis for a state observer with state error feedback:

$$dX(t) = f(t, X(t), \theta_0)dt + g(t, X(t), \theta_0)dB(t) + K_t(X_d(t) - X(t))dt,$$

$X(t)$ is the state at time t, θ_0 is the true value of the parameter vector which is assumed to be unknown, $X_d(t)$ is the desired state at time t and $K_t = K_t(\widehat{X}(t))$ is the feedback coefficient which is a function of the observed state at time t. The desired state $X_d(t)$ satisfies

$$dX_d(t) = f(t, X_d(t), \theta_0)dt$$

and the state estimate/observed state at time t, namely $\widehat{X}(t)$ satisfies

$$d\widehat{X}(t) = f(t, \widehat{X}(t), \theta) + L_t(dZ(t) - H(t, \widehat{X}(t))dt)$$

where $\theta = \theta_0 + \delta\theta$ is the assumed value of the parameter θ_0. L_t is the output error feedback coefficient and is a function of the observed state $\widehat{X}(t)$ at time t. The output $Z(t)$ is given by

$$dZ(t) = H(t, X(t))dt + \sigma_v dV(t)$$

Here, $B(.)$ and $V(.)$ are standard independent vector valued Brownian motion processes. The state estimation error at time t is

$$e(t) = X(t) - \widehat{X}(t)$$

and the state tracking error at time t is

$$f(t) = X_d(t) - X(t)$$

Thus,

$$e(t) + f(t) = X_d(t) - \widehat{X}(t)$$

We have approximately,

$$de(t) = f_x(t, \hat{X}(t), \theta_0)e(t)dt - f_\theta(t, \hat{X}(t), \theta_0)\delta\theta\, dt + K_t f(t)dt$$

$$- L_t(H_x(t, \hat{X}(t))e(t)dt + \sigma_v dV(t))$$

$$+ g(t, \hat{X}(t), \theta_0)dB(t)$$

This equation can be expressed as

$$de(t) = (F_t - L_t H_t)e(t)dt + F_{\theta t}\delta\theta dt - \sigma_v L_t\, dV(t) + G_t dB(t)$$

where all the coefficient matrices are functions of t, θ_0 and $\hat{X}(t)$ only. Application of Ito's formula gives

$$E\left(d\left(e(t)e(t)^T\right)\middle|\hat{X}(t)\right) = \left(A_t P_t + P_t A_t^T + G_t G_t^T + \sigma_v^2 L_t L_t^T\right)dt$$

assuming that

$$\mathbb{E}\left(e(t) \mid \hat{X}(t)\right) = 0$$

or equivalently that

$$\mathbb{E}\left(X(t) \mid \hat{X}(t)\right) = \hat{X}(t)$$

Here,

$$A_t = F_t - L_t H_t$$

and

$$P_t = \mathbb{E}\left(e(t)e(t)^T \mid \hat{X}(t)\right)$$

If we assume that $e(t)$ is orthogonal to every function of $\hat{X}(t)$, then we get

$$P_t = \mathbb{E}\left(e(t)e(t)^T\right)$$

In this case, we then get

$$\frac{dP_t}{dt} = \mathbb{E}\left(A_t P_t + P_t A_t^T\right) + G_t G_t^T + \sigma_v^2 L_t L_t^T$$

and hence,

$$\frac{dTr(P_t)}{dt} = 2Tr(\mathbb{E}(A_t)P_t) + b(t)$$

where

$$b(t) = Tr\left(G_t G_t^T + \sigma_v^2 L_t L_t^T\right)$$

Assume that the maximum eigenvalue of $\mathbb{E}\left(A_t + A_t^T\right)$ over all times t is $-\lambda_0 < 0$. Then it follows that

$$\frac{dTr(P_t)}{dt} \leq -\lambda_0 Tr(P_t) + b(t)$$

and hence

$$Tr(P_t) \leq \exp(-\lambda_0 t)Tr(P_0)+\int_0^t \exp(-\lambda_0(t-s))b(s)ds$$

If we assume that $b(t) \leq b_0 < \infty \forall t$, then it follows from the above that

$$\lim_{t\to\infty} Tr\left(P_t\right) \leq b0/\lambda_0$$

and this proves that the state estimation error is bounded, *i.e*, the system is stable, Suppose we do not assume orthogonality of $e(t)$ with Borel functions of $\widehat{X}(t)$. Then we compute

$$\mathbb{E}\left(d\left((e(t)e(t)^T\right)\middle|\widehat{X}(t)\right) = \left(A_tP_t + P_tA_t^T + G_tG_t^T + \sigma_v^2 L_tL_t^T\right)dt$$

$$-E\left[\left(F_{\theta t}\delta\theta e(t)^T + e(t)\delta\theta^T F_{\theta t}^T\middle|\widehat{X}(t)\right)\right]dt$$

and hence taking trace gives us

$$\mathbb{E}\left[d\,\|e(t)\|^2|\,\widehat{X}(t)\right] = \left(Tr\left(\left(A_t + A_t^T\right)P_t\right)dt + c(t)dt\right) - 2\mathbb{E}\left[\left(e(t)^T F_{\theta t}\,|\,\widehat{X}(t)\right)\delta\theta\right]$$

where

$$\mu_e(t) = \mathbb{E}\left(e(t)\,|\,\widehat{X}(t)\right), P_t = \mathbb{E}\left[\left(e(t)\,e(t)^T\,|\,\widehat{X}(t)\right)\right]$$

$$e(t) = Tr\left(G_tG_t^T + \sigma_v^2 L_tL_t^T\right)$$

We shall now assume that the maximum eigenvalue of $A_t + A_t^T$ over all times is bounded by $-\lambda_0 < 0$. We shall also assume that the spectral norm of $F_{\theta t}$ is bounded by a finite constant f_0 over all times. Then we get from the above, on taking expectation and defining P_t as $\mathbb{E}\left(e(t)\,e(t)^T\right)$, and $b(t) = \mathbb{E}\left(e(t)\right) \leq b_0 < \infty$

$$dTr(P_t)\,|\,dt \leq -\lambda_0 Tr(P_t) + b(t) + f_0\sqrt{Tr(P_t)}\,\|\delta\theta\|$$

We write

$$E(t) = Tr(P_t) = E(\|e(t)\|^2)$$

and then get from the above,

$$E'(t) \leq -\lambda_0 E(t) + f_0\sqrt{E(t)}\,\|\delta\theta\| + b_0$$

From this equation, robustness is easily proved.

❖❖❖❖❖

[114] Quantum system perturbed by classical Levy noise: The Schrodinger evolution operator $U(t)$ satisfies the complex stochastic differential equation

$$dU(t) = \left[(-iH_0 + P)dt - i\sum_k L_k dB_k(t) - i\sum_k V_k X_{N_k(t)+1}(k)dN_k(t)\right]U(t)$$

where H_0, P, L_k, V_k are non-random operators in the Hilbert space and H_0 is Hermitian. $\{X_n(k)\}_{n,\,k=1,2,...}$ are independent random variables and $N_k(.)$, $k = 1, 2, ...$ are independent Poisson processes independent of $\{X_n(k)\}_{n,\,k=1,2,...}$ with respective rates λ_k, $k = 1, 2, ...$ $B_k(.)$, $k = 1, 2, ...$ are independent standard Brownian motion processes independent of the $X_n(k)$'s and $N_k(\cdot)$'s. Applying Ito's formula in the form

$$dN_k(t)dN_l(t) = \delta_{k,l}dN_k(t), dB_k(t)dB_l(t) = \delta_{k,l}dt$$

gives us

$$d(U(t)^*U(t)) = dU(t)^*U(t)+U(t)^*dU(t)+dU(t)^*dU(t)$$

$$= U(t)^*\left(\left(P^*+P+\sum_k L_k^*L_k\right)dt + i\sum_k\left(L_k^*-L_k\right)dB_k\right.$$

$$\left.+\sum_k\left(i\left(V_k^*-V_k\right)X_{N_k(t)+1}+V_k^*V_kX_{N_k(t)+1}\right)^2 dN_k(t)\right)U(t)$$

The condition for $U(t)$ to be unitary for all times given that it is unitary at time $t = 0$ is that $d(U(t)^*U(t)) = 0$ and this condition is equivalent to

$$P^*+P+\sum_k L_k^*L_k = 0,\ L_k^* = L_k, i\left(V_k^*-V_k\right)X_n(k)+V_k^*V_kX_n(k)^2 = 0$$

This implies that

$$L_k^* = L_k, P+P^*+\sum_k L_k^2 = 0, i\left(V_k^*-V_k\right)+V_k^*V_k = 0, X_n(k)^2 = X_n(k)$$

Thus, $X_n(k)$ are random variables that assume only the values zero and one and $P = Q + iR$ where Q, R are Hermitian and $Q = -\frac{1}{2}\sum_k L_k^2$. We may absorb R into H_0 and hence assume that $P = -\frac{1}{2}\sum_k L_k^2$. This gives us the equation

$$dU(t) = \left[-\left(iH_0+\frac{1}{2}\sum_k L_k^2\right)dt - i\sum_k L_kdB_k - i\sum_k V_kX_{N_k(t)+1}dN_k(t)\right]U(t)$$

where $S_k = I + iV_k$ is unitary (this is equivalent to $i\left(V_k^*-V_k\right)+V_k^*V_k = 0$) and the $L_k's$ are Hermitian. We take

$$P(X_n(k) = 1) = p(k), P(X_n(k) = 0) = 1 - p(k)$$

We now compute the evolution of the state of the system. The system starts in a non-random mixed state ρ_0 and after time t, its state evolves to

$$\rho_s(t) = \mathbb{E}(\rho(t)), \rho(t) = U(t)\rho_0U(t)^*$$

We have

$$d\rho(t) = dU(t)\rho_0(t)^* + U(t)\rho_0 dU(t)^* + dU(t)\rho_0 dU(t)^*$$

Taking expectations after applying Ito's formula, we get

$$\begin{aligned} d\rho_s(t) &= \mathbb{E}[d\rho(t)] \\ &= i[H_0, \rho_s(t)]dt - \frac{1}{2}\sum_k \left(L_k^2\rho_s(t) + \rho_s(t)L_k^2 - 2L_k\rho_s(t)L_k\right)dt \\ &\quad + \sum_k \lambda_k p(k)V_k^*\rho_s(t)V_k dt + i\sum_k \lambda_k p(k)\left(V_k^*\rho_s(t) - \rho_s(t)V_k\right)dt \end{aligned}$$

or equivalently, with $\mu(k) = \lambda_k p(k)$,

$$\begin{aligned} \rho_s'(t) &= i[H_0, \rho_s(t)] - \frac{1}{2}\sum_k \left(L_k^2\rho_s(t) + \rho_s(t)L_k^2 - 2L_k\rho_s(t)L_k\right) \\ &\quad + \sum_k \mu(k)V_k^*\rho_s(t)V_k + i\sum_k \mu(k)\left(V_k^*\rho_s(t) - \rho_s(t)V_k\right) \end{aligned}$$

This is a special case of the Sudarshan-Lindblad equation for quantum state evolution in the presence of bath noise. To get the most general from, we must describe the noise processes using the quantum stochastic calculus of Hudson and Parthasarathy.

❖❖❖❖❖

[115] (Pattern classification for the joint action of rotation and permutation groups): If the objects be located respectively at $\mathbf{r}_j$, $j = 1, 2, ..., N$. The signal received by such an object is $f(r_1, ..., r_N)$. If the objects are subject to rotation and permutation, then the received signal field is given by $f(R_1 r_{\sigma 1}, ..., R_N r_{\sigma N})$ where $R_1, ..., R_N \in SO(3)$ and $\sigma \in S_N$. We write

$$\psi_f(\sigma, R_1, ..., R_N) = f(R_1 r_{\sigma 1}, ..., R_N r_{\sigma N})$$

Then

$$\psi_f : S_n \times SO(3)^N \to \mathbb{R}$$

Now, suppose we permute the objects and rotate them by $r, S_1, ..., S_N$. Then the received signal field is

$$f'(r_1, ..., r_N) = f(S_1 r_{\tau 1}, ..., S_N r_{\tau N})$$

We wish to estimate $\tau, S_1, ..., S_N$ from measurement of the fields f, f'. We have

$$\psi_{f'}(\sigma, R_1, ..., R_N) = f'(R_1 r_{\sigma 1}, ..., R_N r_{\sigma N})$$

$$= f(S_1R_1r_{\tau\sigma 1},..., S_NR_Nr_{\tau\sigma N})$$

$$= \psi_f(\tau\sigma, S_1R_1,..., S_NR_N)$$

❖❖❖❖❖

[116] Design of quantum unitary gates by matching the generator: Let $H_0 + \varepsilon f(t)V$ be the Hamiltonian of a perturbed quantum system. The unitary evolution operator $U(t)$ satisfies the Schrondinger equation

$$U'(t) = -i(H_0 + \varepsilon f(t)V)U(t), t \geq 0$$

and we have the Dyson series

$$U(t) = U_0(t)W(t), U_0(t) = \exp(-itH_0),$$

$$W(T) = I - i\varepsilon \int_0^T f(t)\tilde{V}(t) - \varepsilon^2 \int_{0<t_2<t_1<T} f(t_1)f(t_2)\tilde{V}(t_1)\tilde{V}(t_2)dt_1\,dt_2 + O(\varepsilon^3)$$

where

$$\tilde{V}(t) = U_0(-t)VU_0(t)$$

is the interaction operator in the Heisenberg picture. We write

$$W(T) = \exp\left(-i\left(\varepsilon X_1 + \varepsilon^2 X_2\right)\right) + O(\varepsilon^3)$$

where X_1, X_2 are determined by equating the coefficient of ε and ε^2. Thus,

$$W(T) = I - i\varepsilon X_1 + \varepsilon^2(-X_2 - X_1^2/2)$$

and hence

$$X_1 = \int_0^T f(t)\tilde{V}(t)dt, iX_2 + X_1^2/2$$

$$= \int_{0<t_2<t_1<T} f(t_1)f(t_2)\tilde{V}(t_1), \tilde{V}(t_2)dt_1\,dt_2$$

$$X_2 = -i\int_{0<t_2<t_1<T} f(t_1)f(t_2)\left[\tilde{V}(t_1), \tilde{V}(t_2)\right]dt_1\,dt_2$$

We note that both X_1 and X_2 are Hermintian operators since $f(t)$ is a real valued function. Let $|n>$, $n = 1, 2,...$ be the eigenstates of H_0 with respective eigenvalue E_n, $n = 1, 2,...$ Let $U_g = \exp(-i\varepsilon H_g)$ be the gate to be designed. To design this, we match its generator εH_g with the evolved generator $\varepsilon X_1 + \varepsilon^2 X_2$. In other words, we choose the real function $f(t)$, $0 \leq t \leq T$ so that

$$E(f) = \| H_g - X_1 - \varepsilon X_2 \|^2$$

is a minimum upto $O(\varepsilon)$ terms. We have

$$E(f) = \| H_g \|^2 + \| X_1 \|^2 + 2\,\mathrm{Re}\, Tr(H_g X_1) - 2\varepsilon\,\mathrm{Re}\, Tr(H_g X_2)$$
$$+ 2\,\mathrm{Re}\, Tr(X_1 X_2) + O(\varepsilon^2)$$

❖❖❖❖❖

[117] Estimating the parameters of the Hamiltonian and the Sudarshan-Lindblad quantum noise from repeated measurement.

$$\rho'(t) = -i[H(\theta), \rho(t)] + K(\theta)((\rho))$$

where

$$K(\theta)(\rho) = -\frac{1}{2}\sum_k \left[L_k^*(\theta) L_k(\theta)(\rho) + \rho L_k(\theta)^* L_k(\theta) - 2L_k(\theta)\rho L_k(\theta)^* \right]$$

Denote the state evolution operator by $T_t(\theta)$. Thus,

$$\rho(t) = T_t(\theta)(\rho(0))$$

Approximate computation of $T_t(\theta)$:

$$\rho(t) \approx \exp(-it\,ad(H(\theta))(\rho(0)))$$
$$+ \int_0^t \exp(-i(t-s)ad(H(\theta)))K(\theta)(\exp(-is\,ad(H(\theta)))(\rho(0))\,ds$$

At time t_1, we make a POVM measurement $\{M_a : a = 1, 2,..., p\}$. Thus,

$$\sum_a M_a^* M_a = I$$

and the state at time $t_1 + 0$ is

$$\rho_1 = \frac{M_{a_1}\rho(t_1)M_{a_1}^*}{Tr(\rho(t_1)M_{a_1}^* M_{a_1})}$$

where a_1 is the outcome noted at time t_1. The system then evolves under the same Hamiltonian and noise operators to the state

$$\rho(t_2) = T_{t_2-t_1}(\theta)(\rho_1)$$

at time t_2 and again the POVM measurement $\{M_a\}$ is performed with the outcome a_2 noted. The system state after the measurement is

$$\rho_2 = \frac{M_{a_2}\rho(t_2)M_{a_2}^*}{Tr(\rho(t_2)M_{a_2}^* M_{a_2})}$$

Like this, measurement are taken at times $t_1 < t_2 < ... < t_N$ resulting in outcomes a_1, a_2,..., a_N respectively. The probability of this event is then

$$Pr(a_1,..., a_N, t_1,..., t_N) = Tr\left(M_{a_N} T_{t_N - t_{N-1}}(\theta)\left(M_{a_{N-1}} T_{t_{N-1}-t_{N-2}}\right)(\theta)...\right.$$

$$T_{t_2-t_1}(\theta)\left(M_{a_1}T_{t_1}(\theta)(\rho(0))M_{a_1}^*\right)...M_{a_N}^*\right)$$

and the state of the system just the Nth measurement is

$$\rho_N = \frac{M_{a_N}T_{t_N-t_{N-1}}(\theta)\left(M_{a_{N-1}}T_{t_{N-1}-t_{N-2}}(\theta)\right)...T_{t_2-t_1}(\theta)\left(M_{a_1}T_{t_1}(\theta)(\rho(0))M_{a_1}^*\right)...M_{a_N}^*}{Pr\left(a_1,...,a_N,t_1,...,t_N\right)}$$

❖❖❖❖❖

[118] Hilbert's Fourteen problem: Given a vector space V and the action of a group G on V (for example representation of G in V), an invariant of G is a function $f: V^m \to \mathbb{C}$ such that

$$f(g.v_1,...,g.v_m) = f(v_1,...,v_m)\forall v_1,...,v_m \in V \forall g \in G$$

Invariants of G obviously form a ring R_i. The problem is when is this ring finitely generated, *i.e*, when does there exist a finite subset E in R_i such that every element of R_i is a polynomial function of the elements in E?

❖❖❖❖❖

[119] Fractional delay system in discrete time. The aim of this work is to design a fractional delay second order Volterra filter that takes as input a discrete time sequence and produces an output that is as close as posible to the output of a given nonlinear system (which may have higher degree nonlinearities than two) in the least square sense. The basic reason for such a design is that rather than including higher than second degree nonlinearities in the designed system, we use the fractional delay degrees of freedom to approximate the given filter. The advantage is that we can get a better approximation of the given nonlinear system than is possible by using only integer delays (since we are giving more degrees of freedom via the fractional delays) and simultaneously we do not need to incorporate higher degree nonlinearities than two. This work hinges around the fact that if the input signal is a decimated version of another signal by a factor of M, then fractional delays can be regarded as delays by integer $< M$. Using the well known formula for the DTFT of a decimated signal, we then arrive at an expression for the DTFT of the output of the fractional delays final system in terms of the a yet unknown first and second order Volterra system coefficients and the fractional delays. The final energy function to be minimized is the norm square of the difference between the DTFT of the given output and the DTFT of the output of the fractional delay system. Minimization over the filter coefficients is a linear problems and thus the final problem is to minimize a highly nonlinear function of the fractional delays which is accomplished using search techniques like the GSA.

Given an input signal $x[n] = z[Mn]$, its fractional delay by r/M where r is an integer in 0, 1,..., $M-1$ is given by $x[n-r/M] = z[Mn-r]$. Let r_k be an integer of the form $Mk + s_k$ where $s_k \in 0, 1,..., M-1$, $k = 1, 2,..., p$. The output generated by passing the input signal $x[n]$ through a second order Volterra filter with fractional delays of $r_1,..., r_p$ is given by

$$y[n] = \sum_{k=1}^{p} h[k]x[n-r_k/M] + \sum_{k,m=1}^{p} g[k,m]x[n-r_k/M]x[n-r_m/M]$$

$$= \sum_{k} h[k]z[Mn-r_k] + \sum_{k,m} g[k,m]z[Mn-r_k]z[Mn-r_m] \quad ...(1)$$

Suppose in addition, there is noise. Then, (1) is an approximately ralation. We wish to determine the coefficients $h[k]$, $g[k, m]$ and the integers $r_1,..., r_p$ so that the difference between the lhs and rhs of (1) has minimum error energy. The Fourier transform (DTFT) of $z[Mn-r]$ is given by

$$M^{-1}\sum_{l=0}^{M-1} \exp(-jr(w-2\pi l)/M)Z((w-2\pi l)/M)$$

The Fourier transform of $y_1[n] = \sum_k h[k]z[Mn-r_k]$ is

$$Y_1(w) = M^{-1}\sum_{k,l} h[k]\exp(-j(w-2\pi l)r_k/M)Z((w-2\pi l)/M)$$

where in the sum, k ranges over 1, 2,..., p and l ranges over 0, 1,..., $M-1$. The Fourier transform of $y_2[n] = \sum_{k,m} g[k,m]z[Mn-r_k]z[Mn-r_m]$ is given by

$$Y_2(w) = M^{-1}\sum_{k,m,l} g[k,m]\int_{-\pi}^{\pi} \exp(-j(w_1 r_k + ((w-2\pi l)/M$$

$$-w_1)r_m))Z(w_1)Z((w-Mw_1-2l\pi)/M)dw_1$$

Let Ω be a discrete set of frequency in $[-\pi, \pi]$ which are equispaced and for each integer r define a column vector of size equal to the cardinality of Ω by:

$$e(r) = ((\exp(-jwr/M)))_{w\in\Omega}$$

Further define the diagonal matrix

$$D_Z[\alpha] = M^{-1}.diag[Z((w-\alpha)/M), w \in \Omega]$$

Assume that the inter-frequency spacing of Ω is Δ. Then we have

$$Y_1 = ((Y_1(w)))_{w\in\Omega}$$

$$= \sum_{k,l} h[k]\exp(j2\pi l r_k / M) D_Z[2\pi l] D(r_k) e(r_k)$$

$$D(r) = diag[\exp(-jwr/M) : w \in \Omega]$$

$$Y_2 = ((Y_2(w)))_{w\in\Omega}$$

$$= \Delta. \sum_{k,m,l,w_1} g[k,m]\exp(-jw_1(r_k - r_m)) Z(w_1)\exp(j2\pi l r_m / M)$$

$$D_Z(Mw_1 + 2l\pi) e(r_m)$$

Consider the vectors

$$Q[k, m|\mathbf{r}] = \Delta. \sum_{l,w_1} \exp(-jw_1(r_k - r_m)) Z(w_1)\exp(j2\pi l r_m / M) D_Z(Mw_1 + 2l\pi) e(r_m)$$

and

$$P[k|\mathbf{r}] = \sum_{l} \exp(j2\pi l r_k / M) D_Z(2l\pi] D(r_k] e(r_k]$$

where

$$\mathbf{r} = ((r_m))_{m=1}^{p}$$

Then,

$$Y_1 = \sum_{k} h[k] P[k|\mathbf{r}], Y_2 = \sum_{k,m} g[k,m] Q[k,m|\mathbf{r}]$$

In terms of the matrices

$$P(\mathbf{r}) = \text{Col}[P[k|\mathbf{r}]] : k = 1, 2,... p], Q[\mathbf{r}] = [Q[k, m|\mathbf{r}]] : k, m = 1, 2,...p]$$

Thus,

$$Y \approx Y_1 + Y_2 = P[\mathbf{r}]h + Q[\mathbf{r}]g$$

Here,,

$$h = ((h[k])) \in \mathbb{R}^p = Vec((g[k,m])) \in \mathbb{R}^{p^2}$$

h, g, $\mathbf{r}$ are estimated by minimizing

$$E[h, g, \mathbf{r}] = ||Y - P[\mathbf{r}]\, h - Q[\mathbf{r}]\, g||^2$$

Writing

$$\begin{pmatrix} h \\ g \end{pmatrix} = q \in \mathbb{R}^{p^2+p}$$

and

$$[P[\mathbf{r}] \| Q[\mathbf{r}]] = S[\mathbf{r}]$$

gives

$$E[q, \mathbf{r}] = \|Y - S[\mathbf{r}]\, q\|^2$$

This has to be minimized w.r.t. q, $\mathbf{r}$. Minimizing w.r.t. q gives

$$\hat{q}(\mathbf{r}) = (S[\mathbf{r}]^T S[\mathbf{r}])^{-1}\, S[\mathbf{r}]^T Y$$

Substituting into the expression for E gives

$$E[\mathbf{r}] = E[\hat{q}\mathbf{r}), \mathbf{r}] = \|Y\|^2 - \|P_{S[\mathbf{r}]}\, Y\|^2$$

Minimizing this w.r.t. $\mathbf{r}$ is equivalent to maximizing

$$F(\mathbf{r}) = \|P_{S[\mathbf{r}]}\, Y\|^2$$

w.r.t. $\mathbf{r}$. Here, $P_{S[\mathbf{r}]}$ is the orthogonal projection onto $\mathcal{R}(S[\mathbf{r}])$:

$$P_{S[\mathbf{r}]} = S[\mathbf{r}](S[\mathbf{r}]^T S[\mathbf{r}])^{-1}\, S[\mathbf{r}]^T$$

❖❖❖❖❖

[120] Proposed five year research plan:

[1] Develop CCD cameras that can be installed in the optical telescopes available in the institute astronomy clum to map the sky in different frequency renges-optical, Infared and radio. By forming the correlation between signals collected at two telescope sensors separated by variable distance, we can form the Fourier transform of this correlation as the sensor move on the earth's surface and using the well known Citt-Van-Zernike theorem, estimate the brightness distribution in the sky as a function of frequency and direction. Let $s\,(t, \mathbf{r})$ be the signal amplitude distribution per unit solid angle of radiation emitting sources in the sky. Assume that telescope with CCD cameras are located at $\mathbf{r}_1$ and $\mathbf{r}_2$. The signals received at these sensors are respectively given by

$$x(t{:}\ \mathbf{r}_k) = \int s(t - |\mathbf{r} - \mathbf{r}_k|/c, \hat{r})\, d\Omega(\hat{r}),\ k = 1, 2$$

The correlation between these two signals assuming uncorrelatedness of any two sources in the sky located at different directions is given by

$$\langle x(t, \mathbf{r}_1).x(t+\tau, \mathbf{r}_2)\rangle = \int R_{ss}(t + |\mathbf{r} - \mathbf{r}_1|/c - |\mathbf{r} - \mathbf{r}_2|/c, \hat{r})\, d\Omega(\hat{r})$$

Making the far field approximation

$$|\mathbf{r}-\mathbf{r}_k| \approx r-(r,\mathbf{r}_k), r = |\mathbf{r}|$$

we get

$$R_{xx}(\tau, \mathbf{r}_1, \mathbf{r}_2) = \int R_{ss}(\tau+(r, \mathbf{r}_2-\mathbf{r}_1)/c, \hat{r})d\Omega(\hat{r})$$

and its Fourier transform is

$$S_{xx}(w, \mathbf{r}_1, \mathbf{r}_2) = \int \exp(jw(r, \mathbf{r}_2-\mathbf{r}_1)/c)S_{ss}(w, \hat{r})d\Omega(\hat{r})$$

From this equation $\mathbf{r}_1$ and $\mathbf{r}_2$ vary, we can recover the sky brightness distribution $S_{ss}(w, \mathbf{r}_1, \mathbf{r}_2)\cdot$

[2] Develop CCD cameras for measuring the red-shift of galaxies shows elements like Hydrogen and Helium are known. From the redshift

$$w = w_0(1-v/c),$$

v, the redial velocity of galactic can be calculated using which, knowing the range r, Hubble's Constant $H(t) = v/r = S'(t)/S(t)$ is estimated where $S(t)$ is the redial scale factor of expansion of the universe. Hubble's constant enables us to get an accurate representation of the space-time metric using which galatic evolution can be studied.

[3] Use quantum gravity for developing quantum computers. The metric of space-time is expanded in terms of a background classical metric and a quantum fluctuation in this metric. By substituting this metric into the Lagrangian, we get upto second degree terms in the metric fluctuations and expressions for it. The metric quantum Fluctuations are represented as a finite superposition of basis functions of spatial position. The coefficient in the resulting expansion become quantum time dependent observables, The Lagrangian of the gravitational field thus becomes a linear-quadratic function of these observables and hence the corresponding Hamiltonian can be used to write down the Schrodinger equation for the gravitational field. Given an external electromagnetic field described by a four vectors potential $A^{\mu}(x)$, $\mu = 0, 1, 2, 3$, its interaction with the gravitational field is described by the Lagrangian density

$$L_1(g_{\mu\nu}, g_{\mu\nu,\alpha})$$

$$= F_{\mu\nu}F^{\mu\nu} = g^{\mu\alpha}g^{\nu\beta}(A_{\nu,\mu}-A_{\mu,\nu})(A_{\beta,\alpha}-A_{\alpha,\beta})$$

$$= 2g^{\mu\alpha}g^{\nu\beta}(A_{\nu,\mu}-A_{\mu,\nu})A_{\beta,\alpha}$$

$$= 2g^{\mu\alpha}g^{\nu\beta}\left(g_{\beta\rho}A^{\rho}\right)_{,\alpha}\left(\left(g_{\nu\sigma}A^{\sigma}\right)_{,\mu}-\left(g_{\mu\sigma}A^{\sigma}\right)_{,\nu}\right)$$

$$= 2g^{\mu\alpha}g^{\nu\beta}\left(g_{\beta\rho,\alpha}A^{\rho}+g_{\beta\rho}A^{\rho}_{,\alpha}\right)\left(\left(g_{\nu\sigma,\mu}-g_{\mu\sigma,\nu}\right)A^{\sigma}+g_{\nu\sigma}A^{\sigma}_{,\mu}-g_{\mu\sigma}A^{\sigma}_{,\nu}\right)$$

By controlling the em field, or equivalently the electromagnetic four potential A^{μ}, we can control the wave function evolution operator of the gravitational field so that a given wave function at time $t = 0$ evolves to given another wave function at time $t = T$. Thus, quantum gravity can be used to design quantum gates by making gravity interact with and electromagnetic field.

[4] Develop group representation theoretics algorithms for classification of moving image patterns. An object in motion can be describe by a sequence of Poincare transformations if realvistic effect are taken into account or as a sequence of Galilean transformations if non-realtivistic speeds are assumed. These transformations are specified by three translation parameters. three rotation parameters, one time delay parameter and three velocity parameters. The Poincare group or the Galilean group denoted by G thus acts on the original object described as a function of three space and one time variables. The set of all object can be viewed as a point in the Hilbert $\mathcal{H} = L^2(\mathbb{R}^4)$ on which the group G acts. The representations of the Euclidean motion group $E(3) = SO(3)\otimes_s \mathbb{R}^3$, namely the semidirect product of the rotation group and the translation group are easily obtained using the method of induced representations. The representations of the Galilean group involves in addition to $E(3)$ motion described by three velocity components and one time delay component. The action of the entire Galilean group on a point $(t, x) \in \mathbb{R}^4$ is given by

$$g(R, a, \tau, v)(t, x) = (t+\tau, Rx+a+vt),$$

where

$$R \in SO(3), a, v \in \mathbb{R}^3, \tau \in \mathbb{R}$$

In matrix form,

$$g\begin{pmatrix} t \\ x \\ 1 \end{pmatrix} = \begin{pmatrix} 1 & 0 & \tau \\ v & R & a \\ 0 & 0 & 1 \end{pmatrix}$$

The representation of the Galilean group can be obtained by successively inducing representations of $SO(3)$ on its semidirect product with Abelian groups and this can be used to obtain Galilean group invariants for classification. The general problem of pattern classification is to construct all invariants for the action of a group G on a vector space V. For image processing applications, V is a space of functions defined on a set X on which the group acts. Thus, if $f: X \to \mathbb{C}^n$ is an image field, then $g \in G$ acts of f to given the transformed image $g.f(x) = f(g^{-1}x)$. An invariant is a function $\psi: V^p \to \mathbb{C}$ that satisfies

$$\psi(g.f_1,...,g.f_p) = \psi(f_1,...,f_p), g \in G, f_1,...,f_p \in V$$

Hilbert's fourteenth problem out of the twenty three problems addressed by him in the international congress of mathematics in 1900 is that is the ring of invariants finitely generated? That is, do there exists a finite number N of invariants $\psi_k: V^{p_k} \to \mathbb{C}, k = 1, 2,..., N$ for G so that any invariant is a polynomial function of these N invariants. The answer depends on the nature of the group, the vector space and the group action on the vector space. If all the generating invariants can be constructed, then the pattern classification problem would be completely solved. One of the goals of the research in the image processing group at the NSIT is to work on the construction of these invariants.

[5] The next topic of research concerns developing algorithms for quantum filtering using the Hudson-Parthasarathy Belavkin approach. Noise. in quantam system is modelled using the Hudson-Parthasarathy quantum stochastic calculus based on the differentials of the creation annihilation and conservation processes which are simply families of operators in the infinite dimensional Boson Fock-space. Using this calculus the Hudson-Parthasarathy noisy Schrodinger equation can be set up. The Heisenberg evolution of observables based on this equation can then also be written down. This evolution appears as a continuous family of homomorphism on an initial Von-Neumann operator algebra. When measurements are carried out, the evolution of the state may get disturbed due to the Heisenberg uncertainty principle. In order to avoid this, Belavkin in a series of remarkable papers suggested how non-demolition measurement can be made on the system plus bath. These measurements consist of unitarily transforming input noise processes with the unitary transformation coming from the noisy Schrodinger equation. The Belavkin filter is essentially a real time implementation of the conditional expectation of the state of the system at time t (which is the Heisenberg evolved observable) given the measurement Abeliean, non-demolition Von-Neumann algebra upto time t. It is a generalized non-commutative version of the classical Kushner filter. The quantum information processing group at the NSIT shall be focussing on implementing the Hudon-Parthasarathy noisy Schrodinger equation as well as the Belavkin filter

using finite dimensional linear algebra and MATLAB. The Belavkin filter would enable us to get refined estimates of observables like the spin of the electron or its position, momentum, kinetic energy, potential energy and angular momentum form noisy meaurement of a subset of these variables.

❖❖❖❖❖

Radon Transform for Rotation Estimation, Interaction Picture in Quantum Mechanics, Curvature Tensor in GTR, Gauge Fields

[121] Estimating the rotation angle of an image using the Radon transform: $f(x, y)$ is the image field. Its Radon transform is given by

$$Rf(\theta, p) = \int f(x, y)\delta(p - x.\cos(\theta) - y.\sin(\theta))\, dx\, dy$$

The rotated and noise corrupted image field is given by

$$g(x, y) = f_\alpha(x, y) + w(x, y)$$

where $w(x, y)$ is a zero mean Gaussian noise field with autocorrelation

$$R_w(x, y \mid x', y') = \mathbb{E}(w(x, y)\, w(x', y')),$$

$$f_\alpha(x, y) = f(x.\cos(\alpha) + y.\sin(\alpha), -x.\sin(\alpha) + y.\cos(\alpha))$$

Thus, the Radon transform of g is

$$Rg(\theta, p) = Rf_\alpha(\theta, p) + Rw(\theta, p)$$

$$Rf(\theta + \alpha, p) + Rw(\theta, p)$$

The Fourier series coefficients of $Rf(\theta, p)$ w.r.t. θ are denoted by $c_f[n|p]$:

$$c_f[n|p] = \frac{1}{2\pi}\int_0^{2\pi} Rf(\theta, p)\exp(-jn\theta)\, d\theta$$

Then the Fourier series coefficient of $Rg(\theta, p)$ are

$$c_g[n|p] = c_f[n \mid p]\exp(jn\alpha) + c_w[n \mid p]$$

The least squares estimate of α is given by minimizing

$$E(\alpha) = \sum_{|n| \le N,\, p \in A} (Arg(c_g[n \mid p]) - Arg(c_f[n \mid p]) - n\alpha)^2$$

and it is calculated as

$$\hat{\alpha} = \left(M \sum_{|n| \le N} n^2\right)^{-1} \sum_{n, p} (n(Arg(c_g[n \mid p]) - Arg(c_f[n \mid p])))$$

where M is the number of elements in A. Let

$$K = M \sum_{|n| \le N} n^2$$

To evaluate the mean square error $\mathbb{E}(\hat{\alpha}-\alpha)^2$, we note that

$$\hat{\alpha} = K^{-1}\sum_{n,p}(n, Arg\, c_w[n\,|\,p])$$

Now

$$Arg(c_w[n\,|\,p]) = \tan^{-1}(c_{wi}[n\,|\,p]/c_{wr}[n\,|\,p])$$

where

$$c_{wr}[n\,|\,p] = \mathrm{Re}(c_w[n\,|\,p]), c_{wi}([n\,|\,p]) = \mathrm{Im}(c_w[n\,|\,p])$$

We also note that

$$c_w[n\,|\,p] = \int w(x,y)\delta(p - x.\cos(\theta) - y.\sin(\theta))\exp(-jn\theta)d\theta/2\pi$$

and hence

$$c_{wr}[n\,|\,p] = \int w(x,y)\delta(p - x.\cos(\theta) - y.\sin(\theta))\cos(n\theta)dx.dy.d\theta/2\pi,$$

$$= \int w(x,(p - x.\cos(\theta)/\sin(\theta))(\cos(n\theta)/\sin(\theta))dx.d\theta/2\pi)$$

$$c_{wi}[n\,|\,p] = -\int w(x,y)\delta(p - x.\cos(\theta) - y.\sin(\theta))\sin(n\theta)dx.dy.d\theta/2\pi,$$

$$= -\int w(x,(p - x.\cos(\theta))/\sin(n\theta))(\sin(n\theta)/\sin(\theta))dx.d\theta/2\pi$$

Thus, the random variables $\{c_{wr}[n\,|\,p], c_{wi}[n\,|\,p] : |n| \le N, p \in A\}$ are jointly zero mean Gaussian random variables. Further,

$$E(\Delta\alpha)^2 = K^{-2}\sum_{n,p,n',p'} nn'\mathbb{E}(Arg(c_w[n\,|\,p])\,Arg(c_w[n''\,|\,p']))$$

$$= K^{-2}\sum_{n,n',p',p'} nn\mathbb{E}[\tan^{-1}(c_{wi}[n\,|\,p]/c_{wr}[n\,|\,p]).(\tan^{-1}(c_{wi}[n'\,|\,p']/c_{wr}[n'\,|\,p']))]$$

Thus the problem of evaluating the mean square error of this estimate amounts to calculating

$$\mathbb{E}[(\tan^{-1}(Y_1/X_1).\tan^{-1}(Y_2/X_2))]$$

where $[X_1, Y_1, X_2, Y_2]^T$ is a zero mean Gaussian random vector with a known covariance matrix.

❖❖❖❖❖

[122] Gauge fields and field equations associated to a matrix Lie group. Let G be a matrix Lie group with t_a, $a = 1, 2, ...,...N$ a basis for its Lie algebra g. The commutation relations are

$$[t_a, t_b] = iC(abc)t_c$$

summation over the repeated index c being implied. $C(abc)$ are the structure constants of the Lie algebra associated with the basis $\{t_a\}$. Let $A_\mu^a(x)$ be the gauge fields and $\psi(x)$ the matter field. The gauge covariant derivative is given by

$$\nabla_\mu = \partial_\mu + ieA_\mu$$

where is a constant and

$$A_\mu(x) = A_\mu^a(x)t_a$$

The curvature field associated with the covariant derivative or equivalently with the gauge fields is given by

$$ieF_{\mu\nu} = [\nabla_\mu, \nabla_\nu] = ie\left[A_{\nu,\mu}^a - A_{\mu,\nu}^a\right]t_a - e^2 A_\mu^b A_\nu^c[t_b, t_c]$$

We shall be assuming the G is a subgrup of $U(N)$ so that the generators t_a can be chosen as $N \times N$ Hermitian matrices. Thus writing

$$F_{\mu\nu} = F_{\mu\nu}^a t_a$$

we get
$$F_{\mu\nu}^a = A_{\nu,\mu}^a - A_{\mu,\nu}^a - e.C(bca)A_\mu^b A_\nu^c$$

In case of the electromagnetic field, $N = 1$ and hence the structure constant vanishes so the above simplifies to the standard linear expression for the electromagnetic field tensor as the four dimensional curl of the four vector potential. The matter field $\psi : \mathbb{R}^4 \to \mathbb{C}_N$ has Lagrangian of the form

$$L_{mat}(\psi, \nabla_\mu\psi)$$

and we wish it be gauge invariant. The gauge transformation of the mater wave function $\psi(x)$ is given by $\psi(x) \to g(x)\psi'(x) = \psi(x)$ where $g: \mathbb{R}^4 \to G$. If we can ensure that under this gauge transformation of the matter field ψ, simultaneously, the gauge field A_μ also transforms in a way to A'_μ so that

$$g(x)\nabla_\mu\psi(x) = \nabla'_\mu\psi'(x) \qquad ...(1)$$

where

$$\nabla'_\mu = \partial_\mu + ieA'_\mu$$

then we select out matter Lagrangian $L_{mat}(\psi, \nabla_\mu\psi)$ so that it is G-invariant in the sense that

$$L_{mat}(g.psi, g\nabla_\mu\psi) = L_{mat}(\psi, \nabla_\mu\psi), g \in G$$

and that will ensure that

$$L_{mat}\left(\psi', \nabla'_\mu\psi'\right) = L_{mat}(\psi, \nabla_\mu\psi)$$

i.e, the matter Lagrangian is locally gauge invariant. Not, for (1) to hold, we require that

$$g(x)(\partial_\mu + ieA_\mu) = (\partial_\mu + ieA'_\mu)g(x)$$

or equivalently,

$$A'_\mu = g(x)A_\mu(x)g(x)^{-1} + ie^{-1}\left(\partial_\mu g(x)\right)g(x)^{-1} \quad \text{... (2}a\text{)}$$

or equivalently since $\partial_\mu(gg^{-1}) = 0$, we can also express (2) as

$$A'_\mu(x) = g(x)A_\mu(x)g(x)^{-1} - ie^{-1}g(x)\partial_\mu\left(g(x)^{-1}\right) \quad \text{...(2b)}$$

(2) is the required gauge transformation of the gauge fields. Let us compute an infinitesimal gauge transformation. Let

$$g(x) = I + i\,\varepsilon^a(x)t_a$$

where $\varepsilon^a(x)$ are of the first order of smallness. Then

$$g(x)t_b g(x)^{-1} = \left(I + i\varepsilon^a(x)t_a\right)t_b\left(I - e\varepsilon^c(x)t_c\right)$$

$$= t_b + i\varepsilon^a(x)[t_a, t_b] = t_b - \varepsilon^a(x)C(abc)t_c$$

where neglect of terms of the second order of smallness. Also,

$$(\partial_\mu g)g^{-1} = i\varepsilon^a_{,\mu}t_a$$

with neglect of second order of smallness terms. Thus, upto first order of smallness, the gauge field transforms as

$$A'_\mu = A_{mu} - \varepsilon^a C(abc)t_c A^b_\mu - e^{-1}\varepsilon^a_{,\mu}t_a$$

or equivalently, using

$$A'_\mu = A^{a'}_\mu t_a, A_\mu = A^a_\mu t_a$$

we get

$$A^{a'}_\mu = A^a_\mu - \varepsilon^c C(cba)A^b_\mu - e^{-1}\varepsilon^a_{,\mu}$$

The second term on the, rhs is zero if $G = U(I)$ (*i.e.* for the elecromagnetic field). We take as our total Lagrangian density for the matter and gauge fields

$$L_{tot}\left(\psi, \partial_\mu\psi, A^a_\mu, A^a_{\mu,\nu}\right)$$

$$= L_{mat}\left(\psi, \nabla_\mu \psi\right) + \frac{1}{2} Tr\left(F_{\mu\nu} F^{\mu\nu}\right) \quad ...(3)$$

Let us see how $F_{\mu\nu}$ transforms under a gauge transformation $g(x)$:

$$\nabla'_\mu = \partial_\mu + ieA'_\mu = \partial_\mu + ieg(x)A_\mu(x)g(x)^{-1} - \left(\partial_\mu g(x) g(x)^{-1}\right)$$

Equivalently, we have seen that

$$\nabla'_\mu = g(x)\nabla_\mu g(x)$$

and hence

$$F'_{\mu\nu}(x) = \left[\nabla'_\mu, \nabla'_\nu\right] = g(x)F_{\mu\nu}(x)g(x)^{-1} \quad ...(4)$$

It follows that the gauge field Lagrangian density

$$L_{gauge}\left(A^a_\mu, A^a_{\mu,\nu}\right) = \frac{1}{2} Tr\left(F_{\mu\nu}, F^{\mu\nu}\right)$$

is gauge invariant, *i.e.*,

$$L_{gauge}\left(A^{a'}_\mu, A^{a'}_{\mu,\nu}\right) = L_g\left(A^a_\mu, A^a_{\mu,\nu}\right)$$

Note that the gauge filed Lagrangian density is also Lorentz invariz. Suppose we assume that under Lorentz transformation the matter field $\psi(x)$ transforms according to a unitary representation S of the Lorentz group such that every element of this representation is in G. We then note that $\nabla_\mu\psi$ Transforms under a Lorentz transformation $\left(L^\mu_\nu\right)$ to

$$L^\nu_\mu \nabla_\nu S(L)\psi(x) = L^\nu_\mu S(L)\nabla_\nu\psi(x) = S(L)L^\nu_\mu \nabla_\nu\psi(x)$$

so that gauge invariance assures us that the matter Lagrangian will under the Lorentz transformation L to

$$L_{mat}\left(\psi, L^\nu_\mu \nabla_\nu \psi\right)$$

For this to equal $L_{mat}\left(\psi, \nabla_\mu \psi\right)$, we must further, assume that L_{mat} is Lorentz invariant. In this way, we are able to obtain a Lagrangian density for the matter and gauge fields that is both gauge invariant and Lorentz invariant.

❖❖❖❖❖

[123] Study topic as for Transmission lines and waveguides course:

1. The distributed parameters of a two wire transmission line, computation of resistance, conductance, capacitance and inductance per unit length (two lectures)

2. The transmission line differential equations-derivation from basic circuit theory (one lecture)

3. The solution to the transmission line equations: Notion of propagation constant and characteristic impedance (one lecture)

4. Reflection coefficient, input impedance, VSWR along a line, lossless lines (one lecture)

5. Constant r and x circles in the reflection coefficient plane, use of Smith chart to compute input impedance and reflection coefficient along a line (one lecture)

6. Using the Smith chart to find the input impedance along a stubbed line (Two one lectures)

7. The notion of matching a line to the load, quarter wave transformer, open and short circuited line (one lecture)

8. Solving the lossless transmission line in the time domain with a source voltage having a source resistance and an RLC circuit as the load impedance (two lectures)

9. The basic equation of a homogeneous waveguide of arbitrary cross section: Expressing the transverse components of the E and H fields in terms of their z components, boundary conditions on the fields. (two lectures)

10. Specialization to rectangular waveguides, TE and TM modes and power loss in the conducting walls of the guide (two lectures)

11. Specialization to cylindrical waveguides, TE and TM modes and power loss (two lectures).

12. Notion of cutoff frequency for a waveguide mode, plotting of lines of force using MATLAB (one lecture)

13. Solving waveguide problems for arbitrary cross section using analytic functions of a complex variable (one lecture)

14. The rectangular cavity resonator, characteristic frequencies of oscillation (one lecture)

15. The cylindrical cavity resonator, characteristic frequencies of oscillation (one lecture)

16. The cladded cylindrical cavity resonator-determination of the characteristic modes (one lecture)

17. Inhomogeneous waveguides–analysis using perturbation theory for partial differential equations (two lectures)

18. Miscellaneous problem solving related to transmission lines and waveguides:

(*a*) Smith chart usage, stub matching (two lectures)

(*b*) Drawing the diagram for time domain analysis of the line voltage and current (two lectures)

(*c*) Computing the waveguide fields given the source excitation via probes, surface current sheets and incident em waves (two lectures)

(*d*) Quantum mechanics of an atom / harmonic oscillator placed inside a waveguide-estimating the guide parameters from transition probability measurements an exercise in time dependent quantum mechanical perturbation theory (two lectures)

(*e*) Quantum mechanics of an atom/ harmonic oscillator due to the magnetic field produced by a transmission line (two lectures)

❖❖❖❖❖

[124] Schrodinger, Heisenbery and Dirac's interaction pictures of quantum dynamics:

Total Hamiltonan

$$H = H_0 + V$$

Schrodinger picture: Let X be an observable and $|\psi(t)\rangle$ the state. Then X remains constant while

$$|\psi(t)\rangle = \exp(-itH)|\psi(0)\rangle$$

Average value of X at time t is

$$\langle\psi(t)|X|\psi(t)\rangle = \langle\psi(0)|\exp(itH).X.\exp(-itH)|\psi(0)\rangle$$

We have

$$\frac{d}{dt}|\psi(t)\rangle = -iH|\psi(t)\rangle, \; dX/dt = 0$$

In the Heisenberg picture, the average value of X is the same as in the Schrodinger picture but we assume that the state $|\psi\rangle$ is a constant, *i.e.* = $|\psi(0)\rangle$ while the observable X changes with time to $X(t)$. To maintain the same average, we therefore require that

$$\langle\psi(0)|X|\psi(0)\rangle = \langle\psi(t)|X|\psi(t)\rangle = \langle\psi(0)|\exp(itH)X.\exp(-itH)|\psi(0)\rangle$$

Since this must be true for all states $|\psi(0)\rangle$ we require that

$$X(t) = \exp(itH)X.\exp(-itH)$$

or equivalently,

$$\frac{dX(t)}{dt} = i[H, X(t)]$$

We require that in both the Schrodinger and Heisenbery pictures, the averages of observables with must be the same since it is the observables that we physically measure.

The interaction picture: Observables evolve according to H_0, not according to H while states according to $V_0(t) = exp(itH_0)V.\ exp(-itH_0)$. The averages of observables then also remain the same as the following calculation shows: Let

$$\frac{d}{dt}|\psi_0(t)\rangle = iV_0(t)|\psi_0(t)\rangle,$$

$$\frac{dX_0(t)}{dt} = [H_0, X_0(t)]$$

Then,

$$\frac{d}{dt}\langle\psi_0(t)|X_0(t)|\psi_0(t)\rangle = \left(\frac{d}{dt}\langle\psi_0(t)|X_0(t)|\psi_0(t)\rangle + \left\langle\psi_0(t)|X_0(t)|\frac{d}{dt}|\psi_0(t)\right\rangle\right.$$

$$+\langle\psi_0(t)|X_0'(t)|\psi_0(t)\rangle$$

$$= i\langle\psi_0(t)|[V_0(t), X_0(t)] + [H_0, X_0(t)]|\psi_0(t)\rangle$$

$$= i\langle\psi_0(t)|\exp(itH_0)[H_0 + V, X]\exp(-itH_0)|\psi_0(t)\rangle$$

$$= i\langle\psi_0(t)|\exp(itH_0)[H, X]\exp(-itH_0)|\psi_0(t)\rangle$$

Now define

$$|\psi(t)\rangle = exp(-itH_0)|\psi_0(t)\rangle$$

Then, $|\psi(t)\rangle$ follows Schrodinger evolution since

$$\frac{d}{dt}|\psi(t)\rangle = -iH_0|\psi(t)\rangle - i\exp-(itH_0)V_0(t)|\psi_0(t)\rangle$$

$$= -i(H_0 + V)\exp(-itH_0)|\psi_0(t)\rangle = -iH|\psi(t)\rangle$$

Hence the rate of change of the average $\langle\psi_0(t)|X_0(t)|\psi_0(t)\rangle$ in the interaction picture coincides with $i\ \langle\psi(t)|[X,X]|\psi(t)\rangle$, *i.e.*, with that obtained in the Schrodinger or the Heisenberg pictures.

❖❖❖❖❖

[125] Generators of the homogeneous Lorentz group and its application to the construction of the curvature tensor of general relativity:

$$\varepsilon_{\mu\nu} = -\varepsilon_{\nu\mu}, 0 \le \mu\nu \le 3$$

An infinitesimal Lorentz transformation is described by

$$x^{\mu} \to x^{\mu} + \varepsilon^{\mu}_{\nu} x^{\nu}$$

where

$$\varepsilon^{\mu}_{\nu} = \eta^{\mu\rho}\varepsilon_{\nu\rho}$$

with

$$\left(\left(\eta^{\mu\nu}\right)\right) = \text{diag}\,[1, -1, -1, -1]$$

we have

$$\eta_{\mu\rho}\left(x^{\mu} + \varepsilon^{\mu}_{\nu}x^{\nu}\right)\left(y^{\rho} + \varepsilon^{\rho}_{\alpha}y^{\alpha}\right)$$

$$= x^{\mu}y_{\mu} + \varepsilon^{\mu}_{\nu}x^{\nu}y_{\mu} + \varepsilon_{\mu\alpha}x^{\mu}y^{\alpha} = x^{\mu}y_{\mu}$$

with neglect of $O(\varepsilon^2)$ terms.

❖❖❖❖❖

[126] Bloch wave functions in determining the band structure of semiconductors. The potential $V(\mathbf{r})$ is periodic with periods of d_x, d_y, d_z along the three directions *i.e.*,

$$V\left(\mathbf{r} + n_x d_x + n_y d_y + n_z d_z\right) = V(\mathbf{r}), n_x, n_y, n_z \in \mathbb{Z}$$

So V can developed into a three dimensional Fourier series:

$$V(\mathbf{r}) = \sum_{n_x, n_y, n_z \in \mathbb{Z}} V\left[n_x, n_y, n_z\right]\exp\left(j2\pi\left(n_x x/d_x + n_y y/d_y + n_z z/d_z\right)\right)$$

Schrodinger's wave equation for the wave function $\psi(\mathbf{r})$ is

$$\Delta\psi(\mathbf{r}) + 2m\left(E - V(\mathbf{r})\psi(\mathbf{r})\right) = 0$$

Replacing r by $\mathbf{r} + n_x d_x \hat{x} + n_y d_y \hat{y} + n_z d_z \hat{z}$ and using periodicity of V gives

$$\Delta\psi\left(\mathbf{r}+n_x d_x \hat{x}+n_y d_y \hat{y}+n_z d_z \hat{z}\right)$$
$$+2m\left(E-V(\mathbf{r})\psi\left(\mathbf{r}+n_x d_x \hat{x}+n_y d_y \hat{y}+n_z d_z \hat{z}\right)\right)=0$$

for all integers n_x, n_y, n_z. Thus we can write

$$\psi\left(\mathbf{r}+d_x\hat{x}\right) = C_x\psi(\mathbf{r}), \psi\left(\mathbf{r}+d_y\hat{y}\right) = C_y\psi(\mathbf{r})$$

$$\psi\left(\mathbf{r}+d_y\hat{z}\right) = C_z y(\mathbf{r}) \qquad ...(1)$$

for all $\mathbf{r}$ and some complex constants C_x, C_y, C_z of magnitude unity. Assuming that we have periodic boundary conditions at the ends of the cuboid crystal defined by $x = N_x d_x$, $y = N_y d_y$, $z = N_z d_y$ where N_x, N_y, N_z are positive integers, we get

$$\psi\left(N_x d_x \hat{x}\right) = \psi\left(N_y d_y \hat{y}\right)=\psi\left(N_z d_z \hat{z}\right) = \psi(\mathbf{0})$$

give $\quad C_x^{N_x} = C_y^{N_y} = C_z^{N_z} = 1$

and hence, we can write

$$C_x = \exp\left(j2\pi k_x/N_x\right) C_y = \exp\left(j2\pi k_y/N_z\right)$$

$$C_z = \exp\left(j2\pi k_z/N_z\right)$$

for some integers k_x, k_y, k_z which may, without loss of generality, be chosen to take values in {–0, 1,..., N_x–1} {0, 1,...N_y–1} and {0, 1, ..., N_z –1} respectively. Writing

$$\psi(\mathbf{r}) = u_k(\mathbf{r}).\exp\left(j2\pi\left(k_x x/N_x d_x + k_y y/N_y d_y + k_z z/N_z d_z\right)\right)$$

it follows that the conditions (1) are satisfied provided that

$$u_k\left(\mathbf{r}+d_z\hat{x}\right) = u_k\left(\mathbf{r}+d_y\hat{y}\right)=u_k\left(\mathbf{r}+d_z\hat{z}\right) = u_k(\mathbf{r})$$

for all $\mathbf{r}$, or in other words, $u_k(\mathbf{r})$ is periodic along the three coordinate axes with periods of d_x, d_y, d_z respectively. Here, $k = (k_x, k_y, k_z)$ indexes the wave function. Thus, u_k can be developed into a Fourier series

$$u_k(x, y, z) = \sum_{n_x, n_y, n_z \in \mathbb{Z}} u_k\left[n_x, n_y, n_z\right]\exp\left(j2\pi\left(n_x x/d_x + n_y y/d_y + n_z z/d_z\right)\right)$$

Since V is also periodic with the same periods, it can also be developed into a Fourier series

$$V(x, y, z) = \sum_{n_x, n_y, n_z \in \mathbb{Z}} V\left[n_x, n_y, n_z\right]\exp\left(j2\pi\left(n_x x/d_x + n_y y/d_y + n_z z/d_z\right)\right)$$

We write these equations as

$$u_k(\mathbf{r}) = \sum_{\mathbf{n}\in\mathbb{Z}^3} u_k[\mathbf{n}]\exp(j2\pi(\mathbf{n}, \mathbf{Ar}))$$

$$V(\mathbf{r}) = \sum_{\mathbf{n}\in\mathbb{Z}^3} V[\mathbf{n}]\exp(j2\pi(\mathbf{n}, \mathbf{Ar}))$$

where

$$\mathbf{A} = \mathbf{D}^{-1}\mathbf{D} = diag[d_x, d_y, d_z]$$

❖❖❖❖❖

[127] Robot hitting a stretched elastic sheet, dynamics, End point of the robot at time t has coordinates

$$\eta_x(q) = l_1\cos(q_1)+l_2\cos(q_1+q_2)$$

$$\eta_y(q) = l_1\sin(q_1)+l_2\sin(q_1+q_2), \eta_z = 0$$

The band is parallel to the yz plane. Let u (t, y, z) denote the displacement of the elastic sheet (along the x direction) The robot dynamical equations are

$$M(q)q'' = N(q, q') = \tau_f(t) + \tau(t)$$

where $\tau(t)$ is the external machine torque. $\tau_f(t)$ is the torque on the robot produced by the elastic sheet force $-f(t)$ along the x direction. The equation of motion of the sheet is

$$u_{,tt}(t, y, z)+\gamma u_{,t}(t, y, z)-c^2u_{,yy}(t, y, z)-c^2u_{,zz}(t, y, z)$$
$$= f(t)p(y-n_y(q), z)$$

where p (y, z) is a constant σ^{-1} for $|y|, |z| \le \delta$ and zero otherwise, The region $(y, z) \in [\eta_y-\delta, \eta_y+\delta] \times [-\delta, \delta]$ is the square region which is hit by the robot end effector which is assumed to have a cross sectional area $4\delta^2$. The torque on the robot due to impact with the elastic sheet is given by

$$\tau_f(t) = -f(t)\eta'_x(q)$$

Eliminating $f(t)$ gives us

$$-\left[u_{,tt}(t, y, z)+\gamma u_{,t}(t, y.z)-c^2u_{,yy}(t, y, z)-c^2u_{,zz}(t, y, z)\right]\eta_x'(q)$$

$$= \left[M(q)q''+N(q,q')-\tau(t)\right]p(y-\eta_y(q), z)$$

Estimating parameters of quantum system.

❖❖❖❖❖

[128] In this paper, we explore the possibility of estimating parameters of a quantum system when the Hamiltonian depends on the unknown parameters. Specifically, if θ is the parameter vector and $H(\theta)$ the Hamiltonian, then an observable X evolves after time t to

$$X(t) = \exp(itH(\theta))X.\exp(-itH(\theta))$$

Let $\lambda_1, ..., \lambda_n$ denote the distinct eigenvalues of $H(q)$. Then, we can write

$$exp(itH(\theta)) = \sum_{k=1}^{n} \exp(-i\lambda_k t)P_k$$

where P_k is the orthogonal projection on the eigensubspace $\mathcal{N}(H(\theta)-\lambda_k I)$ of $H(\theta)$ corresponding to the eigenvalue λ_k. We note that both λ_k and P_k are functions of θ. It follows then that

$$X(t) = \sum_{k,m=1}^{n} \exp(it(\lambda_k - \lambda_m))P_k X P_m$$

and hence the time average of $X(t)$ equals

$$X_{av} = \lim_{T\to\infty} T^{-1}\int_0^T X(t)dt = \sum_{k=1}^{n} P_k X P_k$$

where we make use of the fact that if λ is nonzero, then

$$\lim_{T\to\infty} T^{-1}\int_0^T \exp(it\lambda)dt = 0$$

It follows that if $|e_1\rangle$, $|e_2\rangle$ are two vectors in the Hilbert space of the quantum system, then

$$\begin{aligned} 2\,\mathrm{Re}\langle e_1|X_{av}|e_2\rangle &= \langle e_1|X_{av}|e_2\rangle + \langle e_2|X_{av}|e_1\rangle \\ &= \langle e_1+e_2|X_{av}|e_1+e_2\rangle - \langle e_1|X_{av}|e_1\rangle - \langle e_2|X_{av}|e_2\rangle \end{aligned}$$

and replacing $\langle e_1|X_{av}|e_2\rangle$, we can obtain a similar expression for $\mathrm{Im}\langle e_1|X_{av}|e_2\rangle$. Thus, the matrix element $\langle e_1|X_{av}|e_2\rangle$ can obtained by computing the time average of the quantum average of $X(t)$ in four different states. Since we are assuming our system to be finite dimensional, we can thus obtain the matrix X_{av} relative to any orthonormal basis. In view of the preceding discussion, we can for any observable X, obtain a good estimate of the matrix

$$F(X) = \sum_{k=1}^{N} P_k X P_k$$

or equivalently, of

$$Vec(F(X)) = \sum_{k=1}^{N} \left(P_k \otimes P_k^T\right) Vec(X)$$

Doing so for different obsevables X, we obtain using the least squares method, a good estimate of $\sum_{k=1}^{N} P_k \otimes P_k^T$ form which, by taking the transpose of the second component in the Hilbert space $\mathcal{H} \otimes \mathcal{H}$, we can calculate $\sum_{k=1}^{N} P_k \otimes P_k$. Alternately, we have

$$F(X) + iF(Y) = \sum_k P_k (X + iY) P_k$$

where X, Y are arbitrary Hermitian matrices. Denoting $X + iY$ by Z, we thus obtain for an arbitrary complex matrix Z using quantum and time averages, the matrix

$$F(Z) = \sum_k P_k Z P_k$$

Taking $Z = |e_1\rangle\langle e_m|$ where $\{|e_1\rangle : k = 1, 2, ...,N\}$is an onb for the Hilbert space $\mathcal{H}$, we thus obtain

$$\left\langle e_r \middle| F\left(|e_l\rangle\langle e_m|\right) \middle| e_s \right\rangle = \sum_k \langle e_r | P_k | e_l \rangle \langle e_m | P_k | e_s \rangle$$

or equivalently, we obtain an estimate of the matrix $Q = \Sigma_k P_k \otimes P_k$. The problem is then to estimate θ from the matrix Q. In our simulation studies, we have taken a Hamiltonian that is expressible as a linear function of the parameter vector θ with the coefficients being Hermitian matrices and have applied the GSA algorithm to estimate θ from Q. The paper also presents an alternate scheme assuming the Hamiltonain to be of the form

$$H(\delta\theta) = H_0 + \sum_{k=1}^{p} \delta\theta_k H_k$$

where the $\delta\theta'_k$ are small. The projections $P_k(\delta\theta)$ of $H(\delta\theta)$ have been computed in terms of those of H_0 and the matrix elements of the $H'_k s$ upto $O(\delta\theta)$ using first order perturbation theory and then Q has been expressed approximately as an affine linear function of , following which an elementary least squares algorithm has been used to extract $\delta\theta$ from Q. Our paper presents a situation in which the quantum averages measured are corrupted with white Gaussian noise and we then evaluate the performance of our algorithm in the presence of such noise, *i.e.* evaluate the mean square error of the estimate of $\delta\theta$ *i.e.* $\mathbb{E}\left(\left\|\hat{\delta\theta} - \delta\theta\right\|^2\right)$

❖❖❖❖❖

[129] Feynman diagrams for interaction between matter (electrons and positrons), photons and gravitons. The Lagrangian density of the gravitational field is

$$L_1 = K_1 g^{\mu\nu} \sqrt{-g} \left(\Gamma^{\alpha}_{\mu\nu} \Gamma^{\beta}_{\alpha\beta} - \Gamma^{\alpha}_{\mu\beta} \Gamma^{\beta}_{\nu\alpha} \right)$$

Writing

$$g_{\mu\nu} = \eta_{\mu\nu} + \varepsilon h_{\mu\nu}(x)$$

We can expand L_1 upto $O(\varepsilon^3)$ to get

$$L_1 = \varepsilon^2 C_1(\mu\nu\alpha\rho\sigma\beta) h_{\mu\nu,\alpha} h_{\rho\sigma,\beta}$$

$$+ \varepsilon^3 C_2(\mu\nu\alpha\rho\sigma\beta\gamma\delta) h_{\mu\nu,\alpha} h_{\rho\sigma,\beta} h_\gamma \delta$$

upto $O(\varepsilon^2)$, the Euler-Lagrange field equations for the gravitational field are

$$C_1(\mu\nu\alpha\rho\sigma\beta) h_{\rho\sigma,\beta\alpha} = 0$$

For $K \in \mathbb{R}^3$, define $k_0 \in \mathbb{R}$ so that if $(k^\mu) = k = (k_0, K) \in \mathbb{R}^4$, then

$$C_1(\mu\nu\alpha\rho\sigma\beta) k_\beta k_\alpha H_{\rho\sigma}(K) = 0$$

has a solution for $H_{\rho\sigma}(K)$. This is equivalent to requiring that the determinant of the 16×16 matrix

$$A = (a(\mu\nu|\rho\sigma))$$

where

$$a(\mu\sigma/\rho\sigma) = C_1(\mu\nu\alpha\rho\sigma\beta) k_\beta k_\alpha$$

vanishes. This is a 32th degree equation for k_0 and we assume that it has two solutions of the form

$$k_0 = \pm E(K)$$

Then the gravitational wave solution is given by

$$h_{\mu\nu}(x) = \int H_{\mu\nu}(K)$$

❖❖❖❖❖

[130] EKF for obtained for the filtered estimate of the pure state of a noisy quantum system based on noisy measurements of the average of an observable:

$$\psi'(t) = \left[-\left(iH + V^2\right)dt/2 - iVdB(t)\right]\psi(t)$$

write

$$\psi(t) = \psi_R(t) + i\psi_I(t), H = H_R + iH_I, V = V_R + iV_I,$$

$$V^2 = \left(V^2\right)_R + i\left(V^2\right)_I, \left(V^2\right)_R = V_R^2 - V_I^2, \left(V^2\right)_I = -2V_R V_1$$

❖❖❖❖❖

[131] Dirac's equation in curved space-time. Let J^{ab} be the generators of the Lorentz group in the Dirac representation D. Thus, if is Λ is a Lorentz transformation, then $D(L)$ can be expressed as $exp(\omega_{ab}J^{ab})$. If $\Lambda(x)$ is a Lorentz transformation for each $x \in \mathbb{R}^4$, *i.e.* $\Lambda(x)$ is a local Lorentz transformations, then we can write

$$D(L(x)) = exp\,(\omega_{ab}(x)J^{ab})$$

An infinitesimal local Lorentz transformation in the Dirac representation has the form

$$D(I + \omega(x)) = I + \omega_{ab}(\mathrm{x})J^{ab}$$

where $\omega_{ab}(x)$ are quantities of the first order of smallness. Note that $\omega_{ab} = -\omega_{ba}$. Consider the antisymmetric matrices K^{ab} with matrix elements

$$K^{ab}\,(c, d) = \delta^a_c\delta^b_d - \delta^a_d\delta^b_c$$

We have (with summation over the repeated indices n, q being implied

$$\left(K^{ab}\eta K^{cd}\right)(m, p)$$

$$= K^{ab}(m,n)\eta(n,q)K^{cd}(q,p)$$

$$= \eta(n,q)\left(\delta^a_m\delta^b_n - \delta^a_n\delta^b_m\right)\left(\delta^c_q\delta^d_p - \delta^c_p\delta^d_q\right)$$

$$= \delta^a_m\delta^d_p\eta(b,c) - \delta^a_m\delta^c_p\eta(b,d) - \delta^b_m\delta^d_p eta(a,c) + \delta^b_m\delta^c_p\eta(a,d)$$

So,

$$\left(k^{ab}\eta K^{cd} - K^{cd}\eta K^{ab}\right)(m,p) = \left[k^{ab}, K^{cd}\right]_\eta (m,p)$$

$$= K^{ab}(m,n)\eta(n,q)K^{cd}(q,p) - K^{cd}(m,n)\eta(n,q)K^{ab}(q,p)$$

$$\delta^a_m\delta^d_p\eta(b,c) - \delta^a_m\delta^c_p\eta(b,d) - \delta^b_m\delta^d_p eta(a,c) + \delta^b_m\delta^c_p\eta(a,b)$$

$$-\left[\delta^c_m\delta^b_p\eta(d,a) - \delta^c_m\delta^a_p\eta(b,d) - \delta^d_m\delta^b_p eta(a,c) + \delta^d_m\delta^a_p\eta(c,b)\right]$$

$$= \left[\eta(b,c)K^{ad} + \eta(b,d)K^{ca} + \eta(a,d)K^{bc} + \eta(a,c)K^{db}\right](m,p)$$

Thus,

$$[K^{ab}, K^{cd}]_\eta = K^{ad}\eta(b,c) + K^{bc}\eta(a,d) + K^{db}\eta(a,c) + K^{ca}\eta(b,d)$$

Define

$$\tilde{K}^{ab} = \eta(a,c)K^{cb}$$

We note that

$$(\eta K^{ad})(m,p) = \eta(m,n)K^{ab}(n,p) = \eta(m,n)\left(\delta^a_n\delta^b_p - \delta^a_p\delta^b_n\right)$$

$$= \eta(a,m)\delta_p^b - \eta(b,m)\delta_p^q$$

on the one hand while on the other

$$\eta(a,c)K^{cb}(m,p) = \eta(a,c)\left(\delta_m^c\delta_p^b - \delta_p^c\delta_m^b\right)$$

$$= \eta(a,m)\delta_p^b - \eta(a,p)\delta_m^b$$

In particular, ηK^{ad} is not skew-symmetric while $\eta(a, c)K^{cd}$ is skew-symmetric. Now writing

$$I + \omega_{ab}J^{ab} = D\left(I + \omega_{ab}K^{ab}\right)$$

we have

$$\tilde{J}^{ab} = \eta(a,c)J^{cb} = D\left(\tilde{K}^{ab}\right)$$

We have proved that if

$$\left(K^{ab}\right)_d^c = \eta^{ac}\delta_d^b - \eta^{ba}\delta_d^a$$

or equivalently,

$$(K^{ab})^{cd} = \eta^{ac}\eta^{bd} - \eta^{bc}\eta^{ad}$$

then

$$\left[K^{ab}, K^{cd}\right]_q^p = \left[K^{ab}\right]_r^p\left(K^{cd}\right)_q^r - \left(K^{cd}\right)_r^p\left(K^{ab}\right)_q^r$$

$$= \left(K^{ab}\right)_q^p\eta^{bc} + \left(K^{bc}\right)_q^p\eta^{ad} + \left(K^{db}\right)_q^p\eta^{ac} + \left(K^{ca}\right)_q^p\eta^{bd}$$

which we formally write as

$$\left[K^{ab}, K^{cd}\right] = k^{ad}\eta^{bc} + K^{bc}\eta^{ad} + K^{db}\eta^{ac} + K^{ca}\eta^{bd}$$

Now let $V_a^\mu(x)$ be a tetrad for the metric tensor $g_{\mu\nu}(x)$. Thus

$$\eta^{ab}V_\alpha^\mu(x)V_b^\nu(x) = g^{\mu\nu}(x)$$

Then consider a covariant derivative

$$D_a = V_a^\mu(x)\left(\partial_\mu + \Gamma_\mu(x)\right)$$

where $\Gamma_\mu(x)$ acts in the representation space of a representation π of the Lorentz group. Let $\delta\pi(K^{ab}) = J^{ab}$. Then J^{ab} satisfy the same Lie algebra commutation relations as the K^{ab}. We wish that $\Gamma_\mu(x)$ transforms into $\tilde{\Gamma}_\mu(x)$ under local Lorentz transformations in the sense that

$$\tilde{D}_a\pi\left(\Lambda(x)\right) = \pi\left(\Lambda(x)\right)D_a$$

where

$$\tilde{D}_a = V_a^\mu\left(\partial_\mu + \tilde{\Gamma}_\mu\right)$$

Thus, we require that

$$\tilde{\Gamma}_\mu(x) = -\left(\partial_\mu \pi(\Lambda(x))\right)\pi(\Lambda(x))^{-1} + \pi(\Lambda(x))\Gamma_\mu(x)\pi(\Lambda(x))^{-1}$$

Taking $\Lambda(x)$ to be an infinitesimal Lorentz tranformation

$$\Lambda(x) = I + \omega(x)$$

we get

$$\tilde{\Gamma}_\mu(x) = \Gamma_\mu(x) + \omega_{ab}(x)\left[J^{ab}, \Gamma_\mu(x)\right] - \omega_{ab,\mu}(x)J^{ab}$$

We need to search for a matrix function $\Gamma_\mu(x)$ that satisfies this transformation law under an infinitesimal local Lorentz transformation. Under such a transformation $w(x)$, it is clear that

$$V_a^\mu(x) \to V_a^\mu(x) - \omega_a^b(x)V_b^\mu(x)$$

Thus, we get on neglecting quadratic orders in ω,

$$V_a^\mu V_{\mu b,\nu} \to \left(V_a^\mu - \omega_a^c V_c^\mu\right)\left(V_{\mu b} - \omega_b^d V_{\mu d}\right)_{,\nu}$$

$$= V_a^\mu V_{\mu b,\nu} - \omega_a^c V_c^\mu V_{\mu b,\nu} - V_c^\mu\left(\omega_b^d V_{\mu d}\right)_{,\nu}$$

Thus defining

$$\theta_\nu = \frac{1}{2}J^{ab}V_a^\mu V_{\mu b,\nu}$$

we have

$$\theta_\nu \to \theta_\nu - J^{ab}\omega_a^c V_c^\mu V_{\mu b,\nu}/2 - J^{ab}V_a^\mu\left(\omega_b^d V_{\mu d}\right)_{,\nu}/2$$

On the other hand, we have

$$\omega_{ab}\left[j^{ab}, \theta_\nu\right] = \omega_{ab}\left[j^{ab}, \frac{1}{2}J^{cd}V_c^\mu V_{\mu d,\nu}\right]$$

$$= (\omega_{ab}/2)V_c^\mu V_{\mu d,\nu}\left[J^{ab}, J^{cd}\right]$$

$$= (1/2)\omega_{ab}V_c^\mu V_{\mu d,\nu}\left(J^{bc}\eta^{ad} + J^{ad}\eta^{bc} + J^{ca}\eta^{bd} + J^{db}\eta^{ac}\right)$$

$$= \frac{1}{2}V_c^\mu V_{\mu d,\nu}\left(J^{bc}\omega_b^d - J^{ad}\omega_a^c - J^{ca}\omega_a^d + J^{ab}\omega_b^c\right)$$

$$= V_c^\mu V_{\mu d,\nu}J^{bd}\omega_b^c - V_c^\mu V_{\mu d,\nu}J^{ca}\omega_a^d$$

where we have used the fact that $V_c^{\mu} V_{\mu d,\nu}$ is antisymmetric in the indices c, d since $\eta_{cd} = V_c^{\mu} V_{\mu d}$ are scalar constants.

Thus,

$$\theta_\nu(x) + \omega_{ab}(x)\left[J^{ab}, \theta_\nu(x)\right] - \omega_{ab,\nu}(x) J^{ab}$$

$$= \theta_\nu + V_c^{\mu} V_{\mu d,\nu} J^{db} \omega_b^c - V_c^{\mu} V_{\mu d,\nu} J^{ca} \omega_a^d - \omega_{ab,\mu} J^{ab}$$

Now,

$$\theta_\nu \to \theta_\nu - J^{ab} \omega_a^c V_c^{\mu} V_{\mu d,\nu}/2 - J^{ab} V_a^{\mu}\left(\omega_b^d V_{\mu d}\right)_{,\nu}/2$$

$$= \theta_\nu + V_{\mu d,nu} V_c^{\mu} J^{db} \omega_b^c/2 - J^{ca} V_c^{\mu} V_{\mu d,\nu} \omega_a^d/2 - J^{ab} \omega^{ab,\nu}/2$$

❖❖❖❖❖

[132] Velocity and rotation angle estimation in two dimensional images. The original image field is $f(x, y)$. The rotated and moving image field is

$$g(t, x, y) = f\left(x.\cos(\theta) + y.\sin(\theta) - v_x t, -x.\sin(\theta) + y.\cos(\theta) - v_y t\right) + w(t, x, y)$$

The aim is to identify θ, v_x, v_y from the measured field $g(t, x, y)$, $t, x, y, \in \mathbb{R}$. In the absence of noise, the spatial Fourier transform of g is given by

$$\hat{g}(t, \xi, \eta) = \int g(t, x, y)\exp(-i(\xi x + \eta y))\,dxdy$$

$$= \int f(x', y')\exp(-i(\xi((x' + v_x t)\cos(\theta)(y' + v_y t)\sin(\theta))$$

$$+ \eta((x' + v_x t)\sin(\theta) + (y' + v_y t)\cos(\theta)))\,dx'dy'$$

$$= \hat{f}(\xi.\cos(\theta) + \eta, \sin(\theta), -\xi.\sin(\theta) + \eta, \cos(\theta)).\exp$$

$$(-it(v_x(\xi.\cos(\theta) + \eta.\sin(\theta))))$$

$$+ v_y(-\xi.\sin(\theta) + \eta.\cos(\theta))$$

It follows that

$$\left|\hat{g}(t, \xi, \eta)\right| = \left|\hat{g}(\xi.\cos(\theta) + \eta.\sin(\theta), -\xi.\sin(\theta) + \eta.\cos(\theta))\right|$$

and thus the rotation angle θ can be determined from measurements of $|\hat{g}|$ using the Radon transform. Having thus determined θ, we can determine v_x, v_y using measurements of g or equivalently $\hat{g}$ and the equation

$$\exp\left(-it\left(v_x\left(\xi.\cos(\theta) + \eta.\sin(\theta) + v_y\left(-\xi.\sin(\theta) + \eta.\cos(\theta)\right)\right)\right)\right)$$

$$= \hat{g}(t, \xi, \eta)/\hat{f}(\xi.\cos(\theta)+\eta.\sin(\theta), -\xi.\sin(\theta)+\eta.\cos(\theta))$$

Taking the Fourier transform of the rhs of the above equation (whose estimate is known from measurements of *f*, *g* and the estimated angle of rotation θ), we get

$$\delta\left(X - v_x t.\cos(\theta) + v_y t.\sin(\theta), Y - v_x t.\sin(\theta) - v_y t.\cos(\theta)\right)$$

A plot of this function of *X*, *Y* shows a peak at $\left(v_x t.\cos(\theta) - v_y t.\sin(\theta),\right.$ $\left.v_x t.\sin(\theta) - v_y t.\cos(\theta)\right)$ from which the velocity (v_x, v_y) can be estimated.

[133] Newtonian cosmology can reproduce all the results of general relativistic cosmology:

Consider a particle of mass *m* on the surface of the universe of radius *R*(*t*). Let ρ(*t*) be the matter within the universe and *p*(*t*) the pressure. The pressure does not act on the particle because it is on the surface. Thus, conservation of the energy of the particle gives

$$mR'(t)^2/2 - 4\pi Gm\rho(t)R(t)^3/3R(t) = E$$

Equivalently letting $k = -2E/m$, we can write the above eqution as

$$R'(t)^2 + k = 8\pi G\rho(t)R^2(t)/3 \qquad ...(1)$$

which is precisely produced by the general theory of relativity. The other equation is the energy conservation equation for the sphere. The rate at which the energy of the sphere containing matter increases with time equals the rate at which pressure force does work on the sphere:

$$\frac{d}{dt}\left(4\pi\rho(t)R^3(t)/3\right) = -p(t).4\pi R^2(t)$$

or equivalently,

$$\frac{d}{dt}\left(\rho(t)R^3(t)\right) = -3p(t)R^2(t) \qquad ...(2)$$

(1) and (2) are the fundamental equations of cosmology. In the primordial era, radiation dominated matter and hence regarding the gas that fills the sphere as adiababic, we had an equation of state $p(t) = C.\,rho(t)^\gamma$. Then the above differential equations for the evolution of the universe become

$$R'^2 + k = 8\pi G\rho R^2/3$$

$$(\rho R^3)' = -3R^2 C\rho^\gamma$$

❖❖❖❖❖

[134] Chandrasekhar limit on the size of a white dwarf:

Equation of a polytrope: $p = K\rho^\gamma$ relates the pressure and density. This is the equation of state of an adiabatic gas. $\rho = \rho(r)$ where r is the distance from the origin. The equilibrium between the pressure force and the gravitational force gives the following equations: Let $V(r)$ denote the gravitational potential at a distance r from the origin. Then, $\nabla^2 V(r) = 4\pi G\rho(r)$ and hence by Gauss' theorem, if $F(r) = -V'(r)$, then

$$F(r)4\pi r^2 = -4\pi G\int_0^r \rho(x)4\pi x^2 dx = -4\pi GM(r)$$

so that

$$F(r) = -GM(r)/r^2$$

$F(r)$ is the gravitational force (radial) experienced by a unit mass placed at a distance r from the origin. For equilibrium with the pressure forces, we require that

$$-p'(r)+\rho(r)V'(r) = 0$$

on equivalently,

$$p'(r) = \rho(r)F(r)$$

which is the same as

$$p'(r) = -GM(r)\rho(r)/r^2$$

Noting that

$$M'(r) = 4\pi r^2\rho(r)$$

it follows that

$$p'(r) = -GM(r)M'(r)/4\pi r^4$$

and using

$$p'(r) = K\gamma\rho(r)^{\gamma-1}\rho'(r)$$

the condition for equilibrium can be expressed as

$$K\gamma\rho(r)^{\gamma-2}\rho'(r)r^2 + 4\pi G\int_0^r \rho(x)x^2 dx = 0$$

which gives on differentiation,

$$(4\pi G/K)\rho r^2 + \gamma\left(\rho^{\gamma-2}\rho' r^2\right)' = 0$$

If we assume the gas to be a degenerate Fermi gas with each particle having mass m, then with the Fermi level E_F, we get for the total number of particles per unit volume the formula

$$n(r) = 8\pi \int_0^{p_F} p^2 dp/h^3 = \left(8\pi p_F^3/3h^3\right)$$

where by special relativity,

$$E_F = c\sqrt{p_F^2 + m^2c^2}$$

p_F and E_F are functions of r. The total internal energy of the gas per unit volume (also a function of r) is give by

$$U(r) = \int_0^{p_F} \left(8\pi p^2/h^3\right)\left[c\sqrt{p^2 + m^2c^2} - mc^2\right]dp$$

The pressure as a function of r can also be calculated for this degenerate adiabatic Fermi gas. It is obtained using the thermodynamic relation

$$d(U/\rho) + pd(1/\rho) = 0$$

(assuming the gas to be adiabatic, *i.e.* $dS = 0$ where S is the entropy per unit mass). Thus,

$$p = \rho dU/d\rho - U$$

to obtain U as a function of ρ, we must use $\rho(r) = n(r)m$ and eliminate p_F between the above equation for n and U. The result is an expression for p as a function of ρ and this can be used in the equilibrium equation

$$p'(r) = p'(\rho(r))\rho'(r) - GM(r)M'(r)/4\pi r^4$$

where

$$M'(r) = 4\pi r^2 \rho(r)$$

to obtain ρ as a function of r. The point $r = R$ where $\rho(r)$ vanishes determines the radius of the star that has collapsed into a white dwarf and the mass of the white dwarf can then be evaluated as $M(R) = \int_0^R 4\pi r^2 \rho(r)dr$

❖❖❖❖❖

[135] Dirac's equation in curved space-time is given by

$$\left[i\gamma_a V_a^\mu\left(\partial_\mu + \Gamma_\mu\right) - m_0\right]\psi = 0$$

where

$$\Gamma_\mu = \frac{1}{2} V_{va} V_{vb:\mu} J^{ab}$$

with the metric of space-time being given by

$$g_{\mu\nu} = \eta_{ab}V^a_\mu V^b_\nu$$

i.e. V^a_μ, $a = 0, 1, 2, 3$ from a tetrad basis of covariant vectors. $J^{ab} = \frac{1}{4}\left[\gamma^a, \gamma^b\right]$ are representing Lie algebra elements of the Dirac representation . We can verify their commutation relations:

$$[\gamma^a, \gamma^b] = 2\gamma^a\gamma^b - 2eta^{ab} = F^{ab}$$

say, So

$$[F^{ab}, F^{cd}] = 4\left[\gamma^a\gamma^b, \gamma^c\gamma^d\right]$$

Now,

$$\begin{aligned}
\gamma^a\gamma^b\gamma^c\gamma^d &= \gamma^a\left(2\eta^{bc} - \gamma^c\gamma^b\right)\gamma^d \\
&= 2\eta^{bc}\gamma^a\gamma^b - \gamma^a\gamma^c\gamma^b\gamma^d \\
&= 2\eta^{bc}\gamma^a\gamma^d - \left(2\eta^{ac} - \gamma^c\gamma^a\right)\gamma^b\gamma^d \\
&= 2\eta^{bc}\gamma^a\gamma^d - 2\eta^{ac}\gamma^b\gamma^d + \gamma^c\gamma^a\gamma^b\gamma^d \\
&= 2\eta^{bc}\gamma^a\gamma^d - 2\eta^{ac}\gamma^b\gamma^d + \gamma^c\gamma^a\left(2\eta^{bd} - \gamma^d\gamma^b\right) \\
&= 2\eta^{bc}\gamma^a\gamma^d - 2\eta^{ac}\gamma^b\gamma^d + 2\eta^{bd}\gamma^c\gamma^a - \gamma^c\gamma^a\gamma^d\gamma^b \\
&= 2\eta^{bc}\gamma^a\gamma^d - 2\eta^{ac}\gamma^b\gamma^d + 2\eta^{bd}\gamma^c\gamma^a - \gamma^c\left(2\eta^{ad} - \gamma^d\gamma^a\right)\gamma^b \\
&= 2\eta^{bc}\gamma^a\gamma^d - 2\eta^{ac}\gamma^b\gamma^d + 2\eta^{bd}\gamma^c\gamma^a - 2\eta^{ad}\gamma^c\gamma^b + \gamma^c\gamma^d\gamma^a\gamma^b
\end{aligned}$$

Thus,

$$\begin{aligned}
\frac{1}{4}\left[F^{ab}, F^{cd}\right] &= 4\left[\gamma^a\gamma^b, \gamma^c\gamma^d\right] \\
&= 8\left(\eta^{bc}\gamma^a\gamma^d - \eta^{ac}\gamma^b\gamma^d + \eta^{bd}\gamma^c\gamma^a - \eta^{ad}\gamma^c\gamma^b\right) \\
&= 4\left(\eta^{bc}F^{ad} - \eta^{ac}F^{bd} + \eta^{bd}F^{ca} - \eta^{ad}F^{cb}\right) \\
&\quad + 8\left(\eta^{bc}\eta^{ad} - \eta^{ac}\eta^{bd} + \eta^{bd}\eta^{ca} - \eta^{ad}\eta^{cb}\right) \\
&= 4\left(\eta^{bc}F^{ad} - \eta^{ac}F^{bd} + \eta^{bd}F^{ca} - \eta^{ad}F^{cb}\right)
\end{aligned}$$

Now, $J^{ab} = F^{ab}/4$ and so, we get

$$[J^{ab}, J^{cd}] = \eta^{bc}J^{ad} + \eta^{ac}J^{db} + \eta^{bd}J^{ca} + \eta^{ad}J^{bc}$$

This proves that $\{J^{ab} : 0 \le a < b \le 3\}$ form a standard basis for the Lie algebra of a representation of the Lorentz group.

❖❖❖❖❖

[136] Post-Newtonian approximation in general relativity: The velocity of matter is regarded as being of the first order of smallness. The velocity v is of the order $\sqrt{GM/r}$ and the gravitational potential GM/r is thus of the second order of smallness. We thus seek expansions of the metric tensor of the from

$$g_{00} = 1 + g_{00}(2) + g_{00}(4) + \ldots,$$

$$g_{ij} = -\delta_{ij} + g_{ij}(2) + g_{ij}(4) + \ldots,$$

$$g_{0i} = -g_{0i}(3) + g_{0i}(5) + \ldots$$

and likewise for $g^{\mu\nu}$. The equations $g^{\mu\nu} g_{va} = \delta^{\mu}_{\alpha}$ give

$$g^{00} g_{0i} + g^{0j} g_{ij} = 0, \quad g^{ij} g_{jk} + g^{i0} g_{0j} = \delta^i_k$$

$$g^{00} g_{00} + g^{oi} g_{i0} = 1$$

These give

$$\left(1 + g^{00}(2) + g^{00}(4) + \ldots\right)\left(g_{0i}(3) + g_{0i}(5) + \ldots\right)$$

$$+\left(g^{0j}(3) + g^{0j}(5) + \ldots\right)\left(-\delta_{ij} + g_{ij}(2) + g_{ij}(4) + \ldots\right) = 0,$$

$$\left(-\delta_{ij} + g^{ij}(2) + g^{ij}(4) + \ldots\right)\left(-\delta_{ij} + g_{jk}(2) g_{jk}(4) \ldots\right)$$

$$\left(g^{i0}(3) + g^{i0}(5) + \ldots\right)\left(g_{0j}(3) + g_{0j}(5) + \ldots\right) = \delta_{ik},$$

$$\left(1 + g^{00}(2) + g^{00}(4) + \ldots\right)\left(1 + g_{00}(2) + g_{00}(4) + \ldots\right)$$

$$+\left(g^{0i}(3) + g^{0i}(5) + \ldots\right)\left(g_{0i}(3) + g_{0i}(5) \ldots\right) = 1$$

Equating coefficients of same orders gives us

$$g_{0i}(3) - g^{0i}(3) = 0, \quad g^{00}(2) g^{0i}(3) + g_{0i}(5) + g^{0i}(5) + g_{ij}(2) g^{0j}(3) = 0$$

$$g_{ik}(2) + g^{ik}(2) = 0, \quad - g_{ik}(4) + g^{ij}(2) g_{jk}(2) - g^{ik}(4) = 0$$

$$g_{00}(2) g^{00}(2) = 0, \quad g_{00}(2) g^{00}(2) + g_{00}(4) + g^{00}(4) = 0$$

Thus,

$$g^{00}(2) = -g_{00}(2), g^{00}(4) = -g_{00}(4) + (g_{00}(2))^2$$

$$g^{ik}(2) = -g_{ik}(2), g^{ik}(4) = g_{ik}(4) + g_{ij}(2) g_{jk}(2)$$

$$g^{0i}(3) = g_{0i}(3)$$

etc. We now derive from the Einstein field equations, using perturbation theory, solutions for some of the approximates of $g_{\mu\nu}. \Gamma^{\alpha}_{\mu\nu}, \Gamma_{\alpha\mu\nu}$ have perturbation series beginning with at least the second order terms in the velocity. We may see this as follows:

$$\Gamma_{ijk} = \frac{1}{2}(gi_{j,k} + g_{ikj} - g_{jk,i})$$

which starts at second order since $g_{ij} = -\delta_{ij} + g_{ij}(2) + g_{ij}(4) +$ ... Likewise,

$$\Gamma_{0jk} = \frac{1}{2}(g_{0j,k} + g_{0kj} - g_{jk,0})$$

which starts at third order since g_{0j} starts at third order and any time derivative increases the order by one. Thus, g_{jk0} starts at $g_{ik,0}(2)$ which is of third order. Finally,

$$\Gamma_{00k} = \frac{1}{2}(g_{00,k})$$

which starts with second order. Thus, upto second order, the spatial components R_{ij} of the Ricci tensor are given by

$$R_{ij}(2) + R_{ij}(4) + ... = \Gamma^{\alpha}_{i\alpha,j} - \Gamma^{\alpha}_{ij,\alpha}$$

Now, upto second order we have

$$\Gamma^{k}_{ij} = g^{km}\Gamma_{mij} + g^{k0}\Gamma_{0ij} = -\Gamma_{kij}(2)$$

$$= -\frac{1}{2}(g_{ki,j}(2) + g_{kj,i}(2) - g_{ij,k}(2))$$

$$\Gamma^{0}_{ij} = 0$$

upto second order. Thus,

$$R_{ij}(2) = \Gamma^{0}_{i0,j}(2) + \Gamma^{k}_{ik,j}(2) - \Gamma^{k}_{ij,k}(2)$$

where

$$\Gamma^{0}_{i0,j}(2) = g_{00,ij}(2)$$

$$\Gamma^k_{ij,k}(2) = -\frac{1}{2}\left(g_{ki,kj}(2)+g_{kj,ki}(2)-g_{ij,kk}(2)\right)$$

$$\Gamma^k_{ik,j}(2) = -\frac{1}{2}\left(g_{ki,kj}(2)+g_{kj,ij}(2)-g_{ik,jk}(2)\right)$$

Thus,

$$R_{ij}(2) = g_{00,ij}(2)/2-\frac{1}{2}\left(g_{ki,kj}(2)+g_{kk,ij}(2)-g_{ik,jk}(2)\right)$$

$$+\frac{1}{2}\left(g_{ki,kj}(2)+g_{kj,ki}(2)-g_{ij,kk}(2)\right)$$

$$= \frac{1}{2}\left(g_{00,ij}(2)+g_{ki,kj}(2)+g_{kj,ki}(2)-g_{kk.i}j(2)-g_{ij,kk}(2)\right)$$

We now calculate $R_{00}(2)$ and $R_{00}(4)$. Clearly,

$$R_{00}(2) = \Gamma^\alpha_{0\alpha,0}(2)-\Gamma^\alpha_{00,\alpha}(2)$$

since $\Gamma^\alpha_{0\alpha}$ does not have smaller than third order terms, it follows that $\Gamma^\alpha_{0\alpha,0}$ does not have smaller than fourth order term and hence $\Gamma^\alpha_{0\alpha,0}(2) = 0$ Further, Γ^0_{00} begins with third order terms and hence $\Gamma^0_{00,0}$ begins with fourth order terms. Hence

$$\Gamma^\alpha_{00,\alpha}(2) = \Gamma^k_{00,k}(2)$$

But

$$\Gamma^k_{00}(2) = (-\delta_{kr})\Gamma_{r00}(2) = -\Gamma_{k00} = g_{00,k}(2)/2$$

and hence,

$$\Gamma^\alpha_{00,\alpha}(2) = g_{00,kk}(2)/2 = \nabla^2 g_{00}(2)/2$$

Thus,

$$R_{00}(2) = -\nabla^2 g_{00}(2)/2$$

Further,

$$R_{00}(4) = \Gamma^\alpha_{0\alpha,0}(4)-\Gamma^\alpha_{00,\alpha}(4)-\Gamma^\alpha_{00}(2)\Gamma^\beta_{\alpha\beta}(2)+\Gamma^\alpha_{0\beta}(2)\Gamma^\beta_{0\alpha}(2)$$

We have

$$\Gamma^\alpha_{0\alpha}(3) = \Gamma^\alpha_{00}(3)+\Gamma^k_{0k}(3)$$

$$= \frac{1}{2}g_{00}(2)_{,0}-\Gamma_{k0k}(3) = \frac{1}{2}g_{00}(2)_{,0}-\frac{1}{2}g_{kk}(2)_{,0}$$

so that

$$\Gamma^{\alpha}_{0\alpha,0}(4) = \Gamma^{\alpha}_{0\alpha,0}(3)_{,0}$$

$$= \frac{1}{2}g_{00}(2)_{,00} - \frac{1}{2}g_{kk}(2)_{,00}$$

$$\Gamma^{\alpha}_{00,\alpha}(4) = \Gamma^{0}_{00}(3)_{,0} + \Gamma^{k}_{00}(4)_{,k}$$

$$= \frac{1}{2}g_{00}(2)_{,00} - \frac{1}{2}\left(2g_{k0}(3)_{,0k} - g_{00}(4)_{,kk}\right)$$

$$\Gamma^{\alpha}_{00}(2)\Gamma^{\beta}_{\alpha\beta}(2)$$

$$= \Gamma^{0}_{00}(2)\Gamma^{\beta}_{0\beta}(2) + \Gamma^{k}_{00}(2)\Gamma^{\beta}_{k\beta}(2)$$

$$= \Gamma^{0}_{00}(2)\Gamma^{0}_{00}(2) + \Gamma^{0}_{00}(2)\Gamma^{k}_{0k}(2)$$

$$+ \Gamma^{k}_{00}(2)\Gamma^{0}_{k0}(2) + \Gamma^{k}_{00}(2)\Gamma^{r}_{kr}(2)$$

$$= \Gamma^{k}_{00}(2)\Gamma^{0}_{k0}(2) + \Gamma^{k}_{00}(2)\Gamma^{r}_{kr}(2)$$

$$= \left(g_{00}(2)_{,k}/2\right)\left(g_{00}(2)_{,k}/2\right) - \left(g_{00}(2)_{,k}/4\right)g_{rr,k}(2)$$

Further,

$$\Gamma^{\alpha}_{0\beta}(2)\Gamma^{\beta}_{0\beta}(2) = 2\Gamma^{k}_{00}(2)\Gamma^{0}_{0k}(2) = \frac{1}{2}\left(g_{00,k}(2)\right)^2$$

Combining all these expressions, we get

$$R_{00}(4) = \frac{1}{2}g_{00}(2)_{,00} - \frac{1}{2}g_{kk}(2)_{,00}$$

$$-\frac{1}{2}g_{00}(2)_{,00} + \frac{1}{2}\left(2_{gk0}(3)_{,0k} - g_{00}(4)_{,kk}\right)$$

$$= -\left(g_{00}(2)_{,k}/2\right)\left(g_{00}(2)_{,k}/2\right) + \left(g_{00}(2)_{,k}/4\right)g_{rr,k} + \frac{1}{2}\left(g_{00,k}(2)\right)^2$$

$$= -g_{kk}(2)_{,00}/2 + g_{k0}(3)_{,0k} - g_{00}(4)_{,kk}/2$$

$$+\frac{1}{4}\left(g_{00}(2)_{,k}\right)^2 + g_{00}(2)_{,kgrr,k}/4$$

The components of the energy momentum tensor are as follows:

$$T_{\mu\nu} = (\rho + p)v_{\mu}v_{\nu} - pg_{\mu\nu}$$

The Einstein field equations are

$$R_{\mu\nu} = K\left(T_{\mu\nu} - \frac{1}{2}Tg_{\mu\nu}\right)$$

where

$$T = g^{\mu\nu}T_{\mu\nu} = \rho - 3p$$

$$T_{00} = T_{00}(2) + T_{00}(4) + \ldots$$

$$T_{0k} = T_{0k}(3) + T_{0k}(5) + \ldots$$

$$T_{rs} = T_{rs}(4) + T_{rs}(6) + \ldots$$

where

$$T_{00}(2) = \rho v_0^2, T_{00}(4) = pv_0^2 - pg_{00}(2)$$

$$T_{00}(3) = \rho v_0 v_k, T_{rs}(4) = pv_r v_s + p\delta_{rs}$$

We define

$$S_{\mu\nu} = T_{\mu\nu} - \frac{1}{2}Tg_{\mu\nu}$$

Thus,

$$S_{00}(2) = T_{00}(2) - \frac{1}{2}\rho = \rho v_0^2 - \rho/2$$

upto second order terms, this is the same as ρ/2.

$$S_{ok}(3) = T_{0k}(3) = \rho v_0 v_k,$$

$$S_{rs}(2) = \frac{1}{2}\rho\delta_{rs}$$

$$S_{00}(4) = T_{00}(4) - \frac{1}{2}\rho g_{00}(2)$$

$$S_{rs}(4) = T_{rs}(4) - \frac{1}{2}\rho g_{rs}(2) - (3p/2)\delta_{rs}$$

$$= p/2 - \rho g_{rs}(2)/2 - (3p/2)\delta_{rs}$$

Thus, the field equations give

$$R_{00}(2) = KS_{00}(2), R_{00}(4) = KS_{00}(4)$$

$$R_{0k}(3) = KS_{0k}(3), R_{rs}(2) = KS_{rs}(2)$$

$$R_{rs}(4) = KS_{rs}(4)$$

These equations give

$$\nabla^2 g_{00}(2) = -K\rho,$$

$$\frac{1}{2}\left(g_{00,ij}(2)+g_{ki,kj}(2)+g_{kj,ki}(2)-g_{kk,ij}(2)-g_{ij,kk}(2)\right)$$

$$(K/2)\rho\delta_{ij}$$

We now choose harmonic coordinates, *i.e.*, coordinates such that

$$g^{\mu\nu}\Gamma^{\alpha}_{\mu\nu} = 0$$

For $\alpha = 0$, we get from this

$$g^{00}\Gamma^{0}_{00} + 2g^{ok}\Gamma_{k00} + g^{km}\Gamma^{0}_{km} = 0$$

The third order contribution of this equation is

$$\left.g_{00}(2)_{,0} - 2g_{0k}(3)_{,k} + g_{kk}(2)_{,0}\right) = 0$$

For $\alpha = r = 1, 2, 3$, we get from this

$$g^{00}\Gamma^{r}_{00} + 2g^{0k}\Gamma^{r}_{ok} + g^{km}\Gamma^{r}_{km} = 0$$

and the second order contribution of this equation is

$$g_{00}(2)_{,r} + 2g_{rk}(2)_{,k} - g_{kk}(2)_{,r} = 0$$

The $R_{ij}(2)$ therefore becomes

$$-\frac{1}{2}\nabla^2 g_{ij}(2)+\frac{1}{2}g_{00}(2)_{,ij}-\frac{1}{2}g_{kk}(2)_{,ij}$$

$$-\frac{1}{2}\left(g_{00}(2)_{,ij}-g_{kk}(2)_{,ij}\right)=KS_{ij}(2)$$

or after making the cancellations,

$$\nabla^2 g_{ij}(2) = -2KS_{ij}(2)$$

In the absence of pressure, we thus get

$$\nabla^2 g_{ij}(2) = -K\rho\delta_{ij}$$

Thus,

$$g_{ij}(2) = (2 + g_{00}(2))\delta_{ij}$$

So

$$g_{00}(2) = 1+2\phi,\ g_{ij}(2)=(-1+2\phi)\delta_{ij}$$

where ϕ is the Newtonian gravitational potential and

$$K = -8\pi G$$

This ϕ has the interpretation of being the Newtonian gravitational potential follows from the geodesic equation in the limit of small velocities and weak gravitational fields:

$$\frac{dv^r}{d\tau} \approx \Gamma^r_{00} \approx \Gamma_{r00} = -g_{00,r}/2$$

i.e. $$dv/dt = -\nabla g_{00}/2$$

which means that $g_{00}/2$ must approximately equal the Newtonian gravitational potential plus a constant that is unity to get agreement with the Minkowskian metric at large distances from matter. Now, we've seen that

$$R_{00}(4) = -g_{kk}(2)_{,00}/2 + g_{k0}(3)_{,0k} - g_{00}(4)_{,kk}/2$$

$$+\frac{1}{4}\left(g_{00}(2),_{k}\right)^2 + g_{00}(2),_{kgrr,k}(2)/4$$

$$= 3\phi_{,00} + g_{k0}(3)_{,0k} - \nabla^2 g_{00}(4)/2 + 4|\nabla\phi|^2$$

and using the harmonic coordinate constraint, this reduces to

$$R_{00}(4) = -\phi_{,00} + \frac{1}{2}\left(g_{00}(2)_{,00} + g_{kk}(2)_{,00}\right) - \nabla^2 g_{00}(4)/2 + 4|\nabla\phi|^2$$

$$= 3\phi_{,00} - \nabla^2 g_{00}(4)/2 + 4|\nabla\phi|^2 = KS_{00}(4)$$

❖❖❖❖❖

[137] Spherically symmetric metrics: Let

$$d\tau^2 = A(t,r)dt^2 - B(t,r)dr^2 - C(t,r)\left(d\theta^2 + \sin^2(\theta)d\phi^2\right)$$

Thus,

$$g_{00} = A, g_{11} = -B, g_{22} = -C, g_{33} = -C\sin^2(\theta)$$

Then, the nonzero Christoffel symbols are:

$$\Gamma^0_{00} = A_{,0}/2A, \Gamma^0_{01} = \Gamma^0_{01} = A_{,1}/2A$$

$$\Gamma^0_{11} = B_{,1}/2A, \Gamma^1_{00} = A_{,1}/2B, \Gamma^1_{01} = \Gamma^1_{10} = B_{,0}/2B$$

❖❖❖❖❖

[138] Action functional for a robot moving in a gravitational field in the presence of an electromagnetic field and the electron-positron field. All the field equations, namely, the robot dynamics, the gravitational field dynamics as seen in the general theory of relativity, the electromagnetic field as seen in the gtr and the electron-positron Dirac field are to be set up. The total action in, assuming the robot to be a rigid body occupying the region B of space at time $t = 0$ is given by

$$S\left[g_{\mu\nu}, \phi, \theta, \chi, \phi', \theta', \psi', A_\mu, \psi\right]$$

$$= K_1 \int g^{\mu\nu} \sqrt{-g}\left(\Gamma^\alpha_{\mu\nu}\Gamma^\beta_{\alpha\beta} - \Gamma^\alpha_{\mu\beta}\Gamma^\beta_{\nu\alpha}\right) d^4x$$

$$+ K_2 \int F_{\mu\nu} F^{\mu\nu} \sqrt{-g} d^4x$$

$$-\rho_0 \int_{B\times\mathbb{R}} \sqrt{-g\left(t, R(t)\xi\right)}\left(g_{00}\left(t, R(t)\xi\right)\right.$$

$$+2g_{0k}\left(t, R(t)\xi\right) R^{k'}_m(t)\xi^m$$

$$\left.+ g_{km}\left(t, R(t)\xi\right) R^{k'}_r(t) R^{m'}_s(t)\xi^r\xi^s\right)^{1/2} d^3\xi dt$$

$$+ K_3 \operatorname{Re}\left(\psi^*(x)\gamma^0\gamma^c\left(iV^\mu_c\partial_\mu + \frac{i}{2}J^{ab}V^\nu_a V_{\nu b:\mu} + eA_\mu - m_0\right)\right.$$

$$\left.\psi(x)\sqrt{-g(x)} d^4x\right)$$

Here, $R(t) = \left(R^k_m(t)\right)_{1\le k,\, m\le 3}$ is the 3×3 rotation matrix satisfying the condition $R(t)^T = R(t)^{-1}$. We can express it in terms of the Euler angles $\phi(t)$, $\phi(t)$, $\psi(t)$ as follows:

$$R(t) = R_z(\phi) R_x(\theta) R_z(\psi)$$

Further,

$$J^{ab} = \frac{1}{4}\left[\gamma^a, \gamma^b\right]$$

are the matrices of the Dirac representation of the Lie algebra of the Lorentz group. The dynamical and field equations are after introducing the constraint Lagrangian $\int_{\mathbb{R}} Tr\left(\Lambda(t)\left(R(t)^T R(t)\right)\right) dt$ and replacing the Euler angles by the 3×3 matrix $R(t)$

$$\delta_{R(t)} S = 0,\ \delta_\psi S = 0,\ \delta_{g_{\mu\nu}} S = 0,\ \delta_{A_\mu} S = 0$$

We propose to set up these equations.

❖❖❖❖❖

[139] Radial time independent metric with spherical distribution of matter and pressure.

$$d\tau^2 = A(r)dt^2 - B(r)dr^2 - r^2\left(d\theta^2 + \sin^2(\theta)d\theta^2\right)$$

the nonzero components of the Ricci tensor are given by

$$R_{00} = -A''/2B + A'B'/4B^2 + A'^2/4AB - A'/rB$$

$$R_{11} = A''/2A - A'^2/4A^2 - A'B'/4AB - B'/rB$$

$$R_{22} = 1/B - 1 - rB'/2B^2 + r\,A'/2AB$$

$$R_{33} = R_{22}\sin^2(\theta)$$

We assume that the three velocity of the fluid is zero, *i.e.* $v^k = dx^k/d\tau = 0$, $k = 1, 2, 3$. This implies that

$$v^0 = g_{00}^{-1/2} = A^{-1/2}$$

since

$$g_{\mu\nu}v^{\mu}v^{\nu} = 1$$

Hence

$$v_0 = g_{00}v^0 = A^{1/2}$$

Thus the nonzero components of the energy-momentum tensor of the fluid are

$$T_{00} = (\rho + p)v_0^2 - pg_{00} = \rho A$$

$$T_{11} = pB, T_{22} = pr^2, T_{33} = pr^2\sin^2(\theta)$$

We define

$$T = g^{\mu\nu}T_{\mu\nu}, S_{\mu\nu} = T_{\mu\nu} - \frac{1}{2}Tg_{\mu\nu}$$

Then,

$$T = \rho - 3p,$$

$$S_{00} = \rho A - (\rho - 3p)A/2 = (\rho + 3p)A/2$$

$$S_{11} = \rho B - (\rho - 3p)B/2 = (\rho - 3p)B/2$$

$$S_{22} = \rho r^2 + (\rho - 3p)r^2/2 = (\rho - p)r^2/2$$

$$S_{33} = S_{22}r^2\sin^2(\theta)$$

The non-trivial Einstein field equations are then

$$R_{00} = KS_{00}, R_{11} = KS_{11}, R_{22} = KS_{22}$$

It is assumed that ρ, p are functions of only r. We shall derive the fundamental general relativistic equation of stellar structure from these field equations. The non-trivial field equations are

$$-A''/2B + A'B'/4B^2 + A'^2/4AB - A'/rB$$

$$= K(\rho + 3p)A/2 \quad \text{...(1)}$$

$$A''/2A - A'^2/4A^2 - A'B'/4AB - B'/rB$$

$$= K(\rho - p)B/2 \quad \text{... (2)}$$

$$1/B - 1 - rB'/2B^2 + rA'/2AB$$

$$= K(\rho - p)r^2/2 \quad \text{...(3)}$$

Eliminating A'' between the first two equations gives

$$-Kr.B^2 A(\rho + 3p)/2 - Kr.\, AB^2(\rho - p)/2$$

$$= (AB' + A'B) \quad \text{...(4)}$$

On the other hand, (3) can be expressed as

$$2AB/r - 2AB^2/r - K.\, AB^2(\rho - p)r$$

$$= (AB' + A'B) \quad \text{...(5)}$$

Adding (4) and (5), we get

$$2B' = (-3Kr/2)B^2(\rho - p) + (2B/r)(1 + B) - KrB^2(\rho + 3p)/2$$

$$= 2B(1 - B)/r - 2KrB^2\rho$$

or equivalently,

$$B' = B(1 - B)/r - 2KrB^2\rho$$

We substitute

$$B(r) = (1 - 2GM(r)/r)^{-1} = r/(r - 2GM)$$

into this equation and get on noting that $K = -8\pi G$,

$$(r - 2GM) - r(1 - 2GM') = -2GM + 8\pi G r^3 \rho$$

so

$$M' = 4\pi r^2\rho$$

which gives

$$M(r) = \int_0^r 4\pi r^2 \rho(x)dx$$

i.e. $M(r)$ equals the total mass of the star contained within a sphere of radius r. Substituting, this solution into (3) *i.e.*

$$2/r - 2B/r - B'/B + A'/A = KB(\rho - \rho)r \qquad ...(3)'$$

we can integrate and obtain an integral expression for A in terms of $\rho(r)$ and $p(r)$. Speciffcally, (3)' gives

$$log(A) = K\int_0^r B(\rho)(\rho(r) - p(r))rdr + \log B(B(r))$$

$$+2\int_0^r B(r)dr/r - 2\log(r)$$

Having determined A, B we can substitute these expressions into either (1) or (2) to derive a relation between the functions $p(r)$ and $\rho(r)$ and this is the general relativistic equation of stellar structure.

❖❖❖❖❖

[140 *a*] Optimal control of electromagnetic fields within a box by control currents in another box. Let V_1, V_2 be two disjoint volumes in $\mathbb{R}^3$. We apply a current density field $J(\omega, r)$ in the first box V_1 and the magnetic vector potential generated in the box V_2 is given by

$$A(r, \omega) = \int_{V_1} G(r - r', \omega)J(r', \omega)d^3r'$$

where the green's function is given by

$$G(r, \omega) = \frac{\mu}{4\pi|r|}\exp(-j\omega|r|/c)$$

The magnetic field in V_2 is then

$$H(r, \omega) = \mu^{-1}\nabla \times A(r, \omega) = \mu^{-1}\int_{V_1} \nabla G(r - r', \omega) \times J(r', \omega)d^3r'$$

and the electric field

$$j\omega E(r, \omega) = \nabla \times H(r, \omega)$$

$$\mu^{-1}\int \nabla\times\left(\nabla G\left(r-r',\omega\right)\times J\left(r',\omega\right)\right)d^3r'$$

where the gradient and curl operations are taken w.r.t. the position vector r.

❖❖❖❖❖

[140 *b*] Channel capacity for a C_q channel. A finite alphabet A is encoded into density matrices $\rho(x)$, $x \in A$. For n a positive integer and a sequence $u \in A^n$, we define

$$E(u, n, \delta) = \underset{x \in A}{\otimes} E\left(\rho(x)^{\otimes N(x/u)}, \delta\right)$$

where $N(x|u)$ is the number of time x appears in the sequence u and where if ρ is any density matrix with spectral representation

$$\rho = \sum_j |j\rangle p_\rho(j)\langle j|$$

then

$$E(\rho^{\otimes m}, \delta) = \sum_{\left|m^{-1}\sum_{m=1}^{m}\log\left(p_\rho(j_k)\right)+H(\rho)\right|\langle\delta} |j_1 \ldots j_m\rangle\langle j_1 \ldots j_m|$$

We note that

$$H(\rho) = -Tr\left(\rho.\log(\rho)\right) = -\sum_j p_\rho(j)\log\left(p_\rho(j)\right)$$

is the Von-Neumann entropy of ρ. We define for any probability distribution p on A, the set $T(n, \delta, p)$ on δ typical sequences of length as

$$T(n, \delta, p) = \left\{\left(j_1, \ldots, J_n\right) \in A^n : \left|n^{-1}\sum_{k=1}^{n}\log\left(p\left(j_k\right)\right) + H(p)\right|\langle\delta\right\}$$

Then, by the law of large numbers,

$$p^{\otimes n}\left(T(n, \delta, p)\right) \to 1, n \to \infty$$

Now, for $(j_1, \ldots, j_n) \in T(n, \delta, p)$, we have

$$p^{\otimes n}\left(j_1, \ldots j_n\right) = p\left(j_1\right)\ldots p\left(j_n\right) = 2^{-n\left(H(p)+\delta 0(j1, \ldots, jn)\right)}, \left|\delta_0\left(j_1, \ldots, j_n\right)\right|\langle\delta, k$$

$$= 1, 2, \ldots, n$$

and hence

$$2^{-n\left(H(p)+\delta\right)} \leq p^{\otimes n}\left(j_1, \ldots, j_n\right) \leq 2^{-n\left(H(p)^{-\delta}\right)}$$

summing this equation over $T(n, \delta, p)$, we get

$$2^{-n(H(p)+\delta)}\mu(T(n, \delta, p)) \le p(T(n, \delta, p)) \le 2^{-n(H(p)-\delta)}\mu(T(n, \delta, p))$$

where $\mu(F)$ denotes cardinality of the set F. It follows that for any $\varepsilon > 0$ there exists an $N(\varepsilon)$ such that for all $n > N(\varepsilon)$

we have

$$(1-\varepsilon)2^{n(H(p)-\delta)} \ge \mu(T(n, \delta, p)) \le 2^{n(H(p)+\delta)}$$

in particular,

$$H(p)-\delta \le \lim in \int_{n\to\infty} n^{-1}.\log(\mu(T(n, \delta, p))) \le$$

$$\lim\sup n^{-1}.\log(\mu(T(n, \delta, p))) \le\le H(p)+\delta$$

and hence,

$$\lim_{\delta\to\infty} \lim in \int_{n\to\infty} n^{-1}.\log(\mu(T(n, \delta, p)))$$

$$= \lim_{\delta\to 0} \lim\sup_{n\to\infty} n^{-1}.\log(\mu(T(n, \delta, p)))$$

we write

$$\rho(u) = \otimes_{x\varepsilon u}\rho(x) = \otimes_{x\varepsilon A}\rho(x)^{\otimes N(x|u)}$$

Then,

$$\rho(u)E(u, n, \delta) = \otimes_{x\in A}\left[\rho(x)^{\otimes N(x|u)} E\left(\rho(x)^{\otimes N(x|u)}, \delta\right)\right]$$

Now,

$$\rho(x)^{\otimes N(x|u)} E\left(\rho(x)^{\otimes N(x|u)}, \delta\right)$$

$$= \sum_{(j1, \ldots, jN_{(x|u)})\in T(N(x|u), \delta, p\rho(x))} p_{\rho(x)}^{\otimes N(x|u)}\left(J_1, \ldots j_{N(x|u)}\right)$$

$$\left|J_1, \ldots j_{N(x|u)}\right\rangle\left\langle J_1, \ldots, j_{N(x|u)}\right|$$

and hence,

$$2^{-N(x|u)(H(\rho(x))+\delta)} E\left(\rho(x)^{\otimes N(x|u)}, \delta\right) \le \rho(x)^{\otimes N(x|u)} E\left(\rho(x)^{\otimes N(x|u)}, \delta\right)$$

$$\le 2^{-N(x|u)(H(\rho(x))-\delta)} E\left(\rho(x)^{\otimes N(x|u)}, \delta\right)$$

Taking the tensor product over all $x \in A$ and noting that $\sum_{x\in A} N(x|u) = n$, we get

$$2^{-n\left(\sum_{x\in A} Pu(x)H(\rho(x))+\delta\right)}E(u,n,\delta)$$

$$\leq \rho(u)E(u,n,\delta) \leq 2^{-n\left(\sum_{x\in A} pu(x)H(\rho(x))-\delta\right)}E(u,n,\delta)$$

where

$$p_u(x) = N(x|u)/n$$

is the empirical distribution of the alphabets in the sequence u. We now define Bernoulli typical sequences $\tilde{T}(n,\delta,p)$ of length n for a probability distribution p on A as the set sequences $u \in A^n$ for which $|N(x|u) - np(x)|$ $< \delta\sqrt{np(x)(1-p(x))}\forall x \in A$. Then, if u is such a sequence, we have with $u = (u_1,...u_n)$,

$$p(u) = \Pi_{x\in} p(x)^{np(x)+\delta_0(x)} \sqrt{np(x)(1-p(x))}$$

where $|\delta_0(x)| < \delta \forall x \in A$. Equivalently,

$$P(u) = 2^{-nH(p)+\sqrt{n}\sum_{x\in A}\delta_0(x)} \sqrt{p(x)(1-p(x))}\log(p(x))$$

Let $\mu(A) = a$ and let $max_{x\in A}\left|\delta_0(x)\sqrt{p(x)(1-p(x))}\log(p(x))\right| = b$, Then, we get

$$2^{-nH(p)-b\sqrt{n}} \leq p(u) \leq 2^{-nH(p)+b\sqrt{n}}$$

Also be the central limit theorem, if the elements of $u = u[n]$ are independently chosen with probability $p(.)$, then $(N(x|u[n]) - np(x))/\sqrt{np(x)(1-p(x))}$ converges in distribution to an $N(0, 1)$ distribution. Thus,

$$\lim_n P\left(N(x|u[n]) - np(x)|\sqrt{n}\ \delta\sqrt{p(x)(1-p(x))}\right) = \phi(\delta) - \phi(-\delta), x \in A$$

The multivariate central limit theorem states that the random a variables $(N(x|u[n]) - np(x))/\sqrt{np(x)(1-p(x))}, x \in A$ converge to a joint normal distribution and hence, we can more generally assert that

$$\lim_n P\left(|N|(x|u[n]) - np(x)|/\sqrt{n} \leq \delta\sqrt{p(x)(1-p(x))}, \forall_x \in A\right) = \alpha$$

for some $0 < \alpha < 1$. Thus,

$$P\left(\tilde{T}(n,\delta,p)\right) > \alpha/2$$

for sufficiently large n and hence

$$2^{-nH(p)+b\sqrt{n}}\mu\left(\tilde{T}(n,\delta,p)\right) > \alpha/2$$

and hence

$$\mu\left(\left(\tilde{T}(n, \delta, p)\right)\right) > (\alpha/2).2^{nH(p)-b\sqrt{n}}$$

for all sufficiently large n. Also

$$2^{-nH(p)-b\sqrt{n}}\mu\left(\tilde{T}(n, \delta, p)\right) \leq 1$$

and so

$$\mu\left(\tilde{T}(n, \delta, p)\right) \leq 2^{nH(p)+b\sqrt{n}}$$

Thus we get

$$\lim in\int_n n^{-1}.\log\left(\mu\left(\tilde{T}(n, \delta, p)\right)\right) \leq \lim\sup_n n^{-1}.\log\left(\mu\left(\tilde{T}(n, \delta, p)\right)\right) = H(p)$$

or equivalently,

$$\lim_n n^{-1}.\log\left(\mu\left(\tilde{T}(n, \delta, p)\right)\right) = H(p)$$

We define $\tilde{E}\left(\rho^{\otimes m}, \delta\right)$ and likewise $\tilde{E}(u, n, \delta)$ by using $\tilde{T}(n, \delta, p)$ in place of $T(n, \delta, p)$, *i.e.*,

$$\tilde{E}\left(\rho^{\otimes m}, \delta\right) = \sum_{(j_1,\ldots j_m)\in\tilde{T}(n, \delta, p)} |j_1\ldots j_m\rangle\langle j_1\ldots j_m|$$

and for $u \in \tilde{T}(n, \delta, p)$

$$\tilde{E}(u,n,\delta) = \otimes_{x\in A}\tilde{E}\left(\rho(x)^{\otimes N(x|u)}, \delta\right)$$

The above inequality then gets modified to

$$2^{-n\sum_x pu(x)H(\rho(x))-c\sqrt{n}}\tilde{E}(u, n, \delta)$$

$$\leq \rho(u)\tilde{E}(u, n, \delta) \leq 2^{-n\sum xpu(x)H(\rho(x))+c\sqrt{n}}\tilde{E}(u, n, \delta)$$

But now, we have an additional feature. Since $u \in \tilde{T}(n, \delta, p)$ it follows that

$$p_u(x) = N(x|u)/n \in [p(x)-\delta, p(x)+\delta]$$

and hence,

$$\sum_x\left[p(x)H(\rho(x)) - \delta\sum_x p(x)H(\rho(x))\right]$$

$$\leq \sum_x p_u(x)H(\rho(x)) \leq \sum_x \left[p(x)H(\rho(x)) + \delta\sum_x H(\rho(x))\right]$$

and hence if we define

$$\delta' = \delta\sum_x p(x)H(\rho(x))$$

we have

$$2^{-n\sum_x p(x)H(\rho(x)) - n\delta' - c\sqrt{n}}\tilde{E}(u,n,\delta)$$

$$\leq \rho(u)\tilde{E}(u,n,\delta) \leq 2^{-n\sum_x p(x)H(\rho(x)) + n\delta' + c\sqrt{n}}$$

$$\forall u \in \tilde{T}(n,\delta,p)$$

Quantum Greedy algorithm: A is a finite alphabet and $\rho(x), x \in A$ is a family of density matrices, all defined on the Hilbert space. Let p be a probability distribution on A. Let us assume that messages $u_1, \ldots u_M A^n$ have been chosen and positive operators $D_1,\ldots D_M$ have also been chosen so that (1) $D_k \geq 0, \tilde{D} = \sum_{k=1}^{M} D_k \leq I$ (2) $TrD_k \leq TrE(u_k, n, \delta)$, (3) $Tr(\rho(u_k)D_k) > 1 - r = 1, 2, \ldots, M$ and finally that M is maximal subject to these constraints. Then we claim that $Tr(\rho(u)\tilde{D}) > \gamma\forall u \in T$ (n, δ, p) where $\gamma > 0$ depends only on δ, $\in$ and γ is sufficiently small. For suppose that $Tr(\rho(u)\tilde{D}) \leq \gamma$ for some $u \in T(n, \delta, p)$. Then let $D_{M+1} = \sqrt{I-\tilde{D}}E(u,n,\delta)\sqrt{I-\tilde{D}}$. Then, the first two conditions are satisfied by D_{M+1} and further,

$$\begin{aligned} Tr(\rho(u)D_{M+1}) &= Tr(\rho(u)E(u,n,\delta)) \\ &\quad -Tr\left(\rho(u)\left(E(u,n,\delta) - \sqrt{I-\tilde{D}}E(u,n,\delta)\sqrt{I-\tilde{D}}\right)\right) \\ &\geq 1 - \phi_n(\delta) - \left\|\rho(u) - \sqrt{I-\tilde{D}}\rho(u)\sqrt{I-\tilde{D}}\right\|_1 \\ &\geq 1 - \phi_n(\delta) - \psi_n(\gamma) \end{aligned}$$

where $\phi_n(\delta) \to 0$ as $n \to \infty$ and $\psi_n(\gamma)$ decreases to zero as γ decreases to zero . The gentle operator lemma has been used which roughly states that if $Tr\left(\left(\rho(u)(I-\tilde{D})\right)\right)$ is large, then $\left\|\rho(u) - \sqrt{I-\tilde{D}}\rho(u)\sqrt{I-\tilde{D}}\right\|$ is small. In particular it follows that $Tr(\rho(u)D_{m+1}) > 1 - \varepsilon$ for n sufficiently large which contradicts the maximality of M. Note;

$$Tr(\rho(u)E(u,n,\delta)) = \Pi_{x\in A}Tr\left(\rho(x)^{\otimes N(x|u)}E\left(\rho(x)^{\otimes N(x|u)},\delta\right)\right)$$

$$= \Pi_{x\in A}p_{\rho(x)}^{\otimes N(x|u)}\left(T\left(N(x\,|\,u),p_{\rho(x)}\right)\right)$$

Now by Chebyshev's inequality,

$$p_{\rho(x)}^{\otimes N(x|u)}\left(T\left(N(x\,|\,u),\delta\,p_{\rho(x)}\right)\right) \geq 1-Var\left(\log\left(p_{\rho(x)}(j)\right)\right)/n\delta^2$$

We note that

$$Var\left(\log\left(p_{\rho(x)}\right)(j)\right) = Tr\left(\rho(x)\left(\log(\rho(x))\right)^2\right)-H(\rho(x))^2 = v(x)$$

say, Then, we get

$$Tr(\rho(u)E(u,b,\delta)) \geq 1-\left(\sum_{x\in A}v(x)\right)/n\delta^2 = 1-v/n\delta^2$$

where v $=\sum_{x\in A}v(x)$ and this has been used in the above proof. Now we shall derive a lower bound on M. We use the following inequality: Let $0<\theta<1$ and Z, T be operators such that $0\leq Z, T\leq I$. If r is density matrix, $[\rho, T]=0$ and $\rho T\leq\theta T$, then

$$Tr(Z)\geq\theta^{-1}\left(Tr(\rho Z)-Tr(\rho(I-T))\right)$$

Example: If T is the projection onto $\mathcal{R}(r)$, then $\rho(I-T)=0$ and $\rho T\leq\theta T$ implies $\rho\leq\theta I$. Thus we trivially have $Tr(\rho Z)\leq\theta Tr(Z)$. On the other hand, if T is the projection onto $\mathcal{N}(\rho)$, then $\rho(I-T)=\rho$ and hence

$$Tr(\rho Z)\leq 1 = Tr(\rho(I-T))$$

and so the above inequality trivially holds. Now, define

$$\bar{\rho} = \sum_{x\in A}p(x)\rho(x)$$

Then, by the above inequality, we have

$$Tr(\bar{D})\geq\theta^{-1}\left(Tr\left(\bar{\rho}^{\otimes n}\tilde{D}\right)-Tr\left(\bar{\rho}\left(I-E\left(\bar{\rho}^{\otimes n},\delta\right)\right)\right)\right)$$

where

$$\theta = 2^{-n(H(\bar{\rho})-\delta)}$$

Hence,

$$Tr(\tilde{D}) \geq 2^{n(H(\bar{p})-\delta)}\left(Tr\left(\bar{\rho}^{\otimes n}\tilde{D}\right)-\beta_n\right)$$

where $\beta_n \to 0$ as $n. \to \infty$ Now,

$$Tr\left(\bar{\rho}^{\otimes n}\tilde{D}\right) = \sum_{u\in A^n} p^{\otimes n}(u)Tr\left(\rho(u)\tilde{D}\right)$$

$$\geq \sum_{u\in T(n,\,\delta,\,p)} p^{\otimes n}(u)Tr\left(\rho(u)\tilde{D}\right)$$

$$\geq \gamma p\left(T(n,\delta,p)\right) = \gamma(1-\alpha_n)$$

where $\alpha_n \to 0$ as $n \to \infty$. Thus,

$$Tr\left(\tilde{D}\right) \geq \gamma(1-\alpha_n-\beta_n)2^{n\left(H(\bar{\rho})-\delta\right)}$$

on the other hand

$$Tr\left(\tilde{D}\right) \leq \sum_{k=1}^{M} Tr\left(E\left(u_k, n, \delta\right)\right)$$

However, for any $u \in T(n, d, p)$, we have

$$Tr(E(u,n,d)) = \Pi_{x\in A}Tr\left(E\left(\rho(X)^{\otimes N(x|u)}, \delta\right)\right)$$

$$= \Pi_{x\in A}\mu\left(T\left(N(x\,|\,u), \delta, p\right)\right) \leq \Pi_{x\in A}2^{N(x|u)\left(H\left(\rho(x)\right)+\delta\right)}$$

$$= 2^{n\left(\sum_x pu(x)H\left(\rho(x)\right)+\delta\right)}$$

Hence,

$$\mathrm{M} \geq \gamma(1-\alpha_n-\beta_n)2^{n\left(H(\bar{\rho})-\sum_x pu(x)H\left(\rho(x)\right)-2\delta\right)}$$

This is the required lower bound. We remark that if $u \in T\,(n, \delta, p)$, then $p_u(x)$ is likely to be close to $p(x)$ for all x. Suppose $u \in \tilde{T}\,(n, \delta, p)$. Then for any $x \in A$ we have

$$\left|N(x\,|\,u)/n - p(x)\right|\left\langle\delta/\sqrt{n}\right.$$

and hence

$$N(x|u) = np(x) + \delta_x/\sqrt{n}$$

for $|\delta_x| < \delta$. It follows that

$$p(u) = \Pi_{j=}^{n} p\left(u_j\right) = \Pi_{x\in A}p(x)^{N(x|u)} = \Pi_{x\in A}p(x)^{np(x)+\delta x\sqrt{n}}$$

and hence

$$n-1\sum_{j=1}^{n}\log p\left(u_j\right) = \sum_{x\in A}\left(p(x)+\delta_x/\sqrt{n}\right)\log\left(p(x)\right)$$

$$= -H(p) + n^{-1/2} \sum_{x \in A} \delta_x . \log(g(x))$$

Hence for all sufficiently large *n*, we have

$$\left| n^{-1} \sum_{j=1}^{n} \log\left(p\left(u_j\right)\right) + H(p) \right| \langle \delta / \sqrt{n}$$

which implies that $u \in T\left(n, \delta/\sqrt{n}, p\right) \subset T(n, \delta, p)$. In addition, we have that $p_u(x) = N(x|u)/n = p(x) + \delta_x / \, sqrtn$
and hence

$$|p_u(x) - p(x)| \le \delta/\sqrt{n}, x \in A$$

So in the greedy algorithm, if we replace *T* by $\tilde{T}$, then we get for all sufficiently large *n* the result that

$$M \ge K_n . 2^n \left(H(\bar{\rho}) - \sum_{x \in A} P(x) H(\rho(x))\right) - 2\delta\sqrt{n}$$

for all sufficiently large *n* where $K_n \to 1$ as $n \to \infty$. We define the mutual information of the C_q channel as

$$I(\rho, p) = H(\bar{\rho}) - \sum_{x \in A} p(x) H(\rho(x))$$

$$= H\left(\sum_{x \in A} p(x)\rho(x)\right) - \sum_{x \in A} p(x) H(\rho(x))$$

Then we can write

$$M \ge K_n . 2^{n - I(\rho, p) - 2\delta\sqrt{n}}, K_n \to 1$$

Writing M_n for M, it follows that

$$\lim in \int_n n^{-1} \log(M_n) \ge I(\rho, p)$$

Upper bound on M: Consider the spectral decomposition

$$\bar{\rho} = \sum_j |j\rangle p_{\bar{\rho}}(i) \langle j|$$

where now

$$\bar{\rho} = \sum_j p_u(x) \rho(x)$$

with $u \in \tilde{T}(n, \delta, p)$ being fixed. We also define

$$p(x|j) = \langle j | \rho(x) | j \rangle$$

Also define

$$\tilde{\rho}(x) = \sum_j p(x \mid j)|j\rangle\langle j| = \sum_j |j\rangle\langle j|\rho(x)|j\rangle\langle j|$$

Then we have

$$\sum_x p_u(x)\tilde{\rho}(x) = \bar{\rho}$$

We note that the matrices $\tilde{\rho}(x), x \in A$ all commute with each other and with $\bar{\rho}$. Then consider

$$F(u, n, \delta) = \otimes_{x\in A}\tilde{E}\left(\tilde{\rho}(x)^{\otimes N(x|u)}, \delta\right)$$

where

$$\tilde{E}\left(\tilde{\rho}(x)^{\otimes m}, \delta\right) = \sum_{(j_1,\ldots,j_m)\in\tilde{T}(m,\delta,p_u)} |j_1,\ldots,j_m\rangle\langle j_1,\ldots,j_m|$$

Note that $|j_1,\ldots,j_m\rangle$ is an eigenvector of $\bar{\rho}^{\otimes m}$ with eigenvalue $p_{\bar{\rho}}(j_1)\ldots p_{\bar{\rho}}(j_m)$ an also an eigenvector of $\bar{\rho}(x)^{\otimes m}$ with eigenvalue $p(x|j_1)\ldots p(x|j_m)$. We have

$$\sum_{x\in A} N\left(j \mid j_1,\ldots,j_{N(x|u)}\right) = N(j|v)$$

where $v = (k_1,\ldots k_n)$ is obtained by arranging $(j_1, \ldots, j_N(x|u))$, in a sequential order. Now suppose

$$\left|N\left(j|j_1,\ldots,J_{N(x|u)}\right) - N(x \mid u)p_{\tilde{\rho}(x)}(j)\right| < \delta\sqrt{N(x \mid u)p_u(j)(1-p_u(j))}$$

This is the same as

$$\left|N\left(j \mid j_1,\ldots,j_{N(x|u)}\right) - N(x \mid u)p(x \mid j)\right| < \delta\sqrt{N(x \mid u)p_u(j)(1-p_u(j))}$$

In other words, we are assuming that

$$\left(j_1,\ldots,j_{N(x|u)}\right) \in \tilde{T}\left(N(x \mid u), \delta, p_u\right), x \in A$$

Then,

$$\left|N(j \mid v) - np_{\bar{\rho}}(j)\right| = \left|\sum_x N\left(j \mid j_1,\ldots,j_{N(x|u)}\right) - np_{\bar{\rho}}(j)\right|$$

$$= \left|\sum_x N(j \mid j_1,\ldots,j_{N(x|u)}) - n\sum_x p_u(x)p(x \mid j)\right|$$

$$= \left|\sum_x N(j \mid j_1,\ldots,j_{N(x|u)}) - \sum_x N(x\mid u)\,p(x\mid j)\right|$$

$$\leq \sum_x |N(j \mid j_1,\ldots,j_{N(x|u)}) - N(x\mid u)\,p(x\mid j)|$$

$$\leq \delta \sum_x \sqrt{N(x\mid u)\,p_u(j)(1-p_u(j))}$$

$$\leq \delta \left(\sum_x N(x\mid u)\right)^{1/2} \cdot \sqrt{a p_u(j)(1-p_u(j))}$$

$$= \delta\sqrt{n a p_u(j)(1-p_u(j))}$$

by the Cauchy-Schwarz inequality. This shows that

$$v \in \tilde{T}(n, \delta\sqrt{a}, p_u)$$

It follows that

$$F(u,n,\delta) \leq \tilde{E}(\rho^{-\otimes n}, \delta\sqrt{a})$$

Then,

$$\tilde{\rho}(u) = \sum |j_1,\ldots,j_n\rangle\langle j_1,\ldots,j_n|\rho(u)|j_1,\ldots,j_n\rangle\langle j_1,\ldots,j_n|$$

which we write as

$$\tilde{\rho}(u) = \sum_\pi \pi\rho(u)\pi$$

Then, it follows that

$$Tr(\tilde{\rho}(u)F(u,n,\delta)) = \sum_\pi Tr(\pi\rho(u)\pi F(u,n,\delta))$$

$$= \sum_\pi Tr(\rho(u)\pi F(u,n,\delta)\pi)$$

$$= Tr(\rho(u)F(u,n,\delta))$$

since $\pi = |j_1,\ldots,j_n\rangle\langle j_1,\ldots,j_n|$ commutes with $F(u, n, \delta)$ and $\sum \pi = I$. Since $u \in \tilde{T}(n, \delta, p_u)$, we have

$$Tr(\tilde{\rho}(u)F(u,n,\delta)) \geq 1 - \eta_n(\delta)$$

where $\eta_n(\delta) \to 1$. Thus

$$Tr(\rho(u)F(u,n,\delta)) \geq 1 - \eta_n(\delta)$$

and hence

$$Tr(\rho(u)\tilde{E}(\bar{\rho}^{\otimes n}, \delta\sqrt{a})) \geq 1 - \eta_n(\delta)$$

Let

$$D'_k = \tilde{E}(\bar{\rho}^{\otimes n}, \delta\sqrt{a})D_k\tilde{E}(\bar{\rho}^{\otimes n}, \delta\sqrt{a})$$

Then, it follows from the gentle operator lemma that

$$Tr(\rho(u_k)D'_k) = Tr(\rho(u_k)D_k) - Tr(\rho(u_k)(D_k - \tilde{E}(\bar{\rho}^{\otimes n}, \delta\sqrt{a})D_k\tilde{E}(\bar{\rho}^{\otimes n}, \delta\sqrt{a}))$$

$$\geq 1-\varepsilon-\left\|\rho(u_k)-\tilde{E}(\bar{\rho}^{\otimes n},\delta\sqrt{a})\rho(u_k)E(\bar{\rho}^{\otimes n},\delta\sqrt{a})\right\|_1$$

$$\geq 1-\varepsilon-\lambda_n$$

where $\lambda_n \to 0$. Now,

$$\rho(u_k)\tilde{E}(u_k,n,\delta)\leq 2^{-n\left(\sum_x p(x)H(\rho(x))-\delta\right)+c\sqrt{n}}\,\tilde{E}(u_k,n,\delta)$$

since $u_k \in \tilde{T}(n,\delta,p)$. It follows that

$$Tr(D'_k)\geq 2^{n\left(\sum_x p(x)H(\rho(x))-\delta\right)-c\sqrt{n}}\,(Tr(\rho(u_k)D'_k)$$

$$-Tr(\rho(u_k)(I-E(u_k,n,\delta)))$$

$$\geq 2^{n\left(\sum_x p_u(x)H(\rho(x))-\delta\right)-c\sqrt{n}}\,(1-\varepsilon-\gamma_n-\rho_n)$$

where $\rho_n \to 0$. Summing over k gives

$$Tr(\tilde{E}(\bar{\rho}^{\otimes n},\delta\sqrt{a}))\geq\sum_k Tr(D'_k)\geq M(1-\varepsilon-\gamma_n-\rho_n)2^{n\left(\sum_x p(x)H(\rho(x))-\delta\right)-c\sqrt{n}}$$

and hence

$$2^{n(H(\bar{\rho})+\delta\sqrt{an}+c\sqrt{n})}\geq M(1-\varepsilon-\gamma_n-\rho_n)2^{n\left(\sum_x p(x)H(\rho(x))-\delta\right)-c\sqrt{n}}$$

from which the desired upper bound on M is obtained.

❖❖❖❖❖

Spin-field Interaction, Viscous and Thermal Effects in GTR, Channel Capacity, Filtering and Control for Robots, Discrete Time Nonlinear Stochastic Filter

[141] Viscous and thermal effects in a special and general relativistic fluid: First consider the special relativistic case. The general relativistic case can be handled by replacing all partial derivatives by covariant derivatives the Minkowski metric $\eta^{\mu\nu}$ by the metric tensor $g^{\mu\nu}(x)$ of space-time. The energy momentum tensor of the fluid is

$$T^{\alpha\beta} + \Delta T^{\alpha\beta}$$

where

$$T^{\alpha\beta} = (\rho + p)U^{\alpha}U^{\beta} - pg^{\alpha\beta}$$

The conservation equation

$$T^{\alpha\beta}_{,\beta} + \Delta T^{\alpha\beta}_{,\beta} = 0$$

gives after contracting with U_{α},

$$((\rho + p)U^{\beta})_{,\beta} - U_{\alpha}p^{,\alpha} + U_{\alpha}\Delta T^{\alpha\beta}_{,\beta} = 0$$

Letting n denote the number density of the particles, its conservation equation is

$$(nU^{\alpha})_{,\alpha} = 0$$

Finally, letting σ denote the entropy per particle, the first law of thermodynamics for energy conservation is

$$Td\sigma = d(\rho / n) + pd(1/n)$$

or equivalently,

$$\begin{aligned} T\sigma_{,\alpha}U^{\alpha} &= ((\rho + p)/n)_{,\alpha}U^{\alpha} - p_{,\alpha}U^{\alpha}/n \\ &= [(\rho + p)U^{\alpha}/n]_{,\alpha} - (\rho + p)U^{\alpha}_{,\alpha}/n - p_{,\alpha}U^{\alpha}/n \\ &= [(\rho + p)U^{\alpha}]_{,\alpha}/n + (\rho + p)U^{\alpha}n_{,\alpha}/n^2 - (\rho + p)U^{\alpha}_{,\alpha}/n - p_{,\alpha}U^{\alpha}/n \\ &= [(\rho + p)U^{\alpha}]_{,\alpha}/n - p_{,\alpha}U^{\alpha}/n \\ &= -U_{\alpha}\Delta T^{\alpha\beta}_{,\beta}/n \end{aligned}$$

This can be rearranged as

$$n\sigma_{,\alpha}U^{\alpha} = -U_{\alpha}\Delta T^{\alpha\beta}_{,\beta}/T$$

or since $(nU^{\alpha})_{,\alpha} = 0$, we get

$$(n\sigma U^{\alpha})_{,\alpha} = -[U_{\alpha}\Delta T^{\alpha\beta}/T]_{\beta} + \Delta T^{\alpha\beta}(U_{\alpha,\beta}/T - U_{\alpha}T_{,\beta}/T^2)$$

or

$$[n\sigma U^{\alpha} + U_{\beta}\Delta T^{\alpha\beta}/T]_{,\alpha} = \Delta T^{\alpha\beta}(U_{\alpha,\beta}/T - U_{\alpha}T_{,\beta}/T^2) \quad ...(1)$$

Left side is a perfect four divergence. By considering a co-moving frame in which $U^i = 0$, we can interpret

$$S^{\alpha} = n\sigma U^{\alpha} + U_{\beta}\Delta T^{\alpha\beta}/T$$

as the four entropy flux (S^0 is the entropy densities and (S^i) is the entropy flux). Thus, the lhs of (1) is the net rate of entropy generation per unit volume which must be non-negative by the second law of thermodynamics. In a co-moving frame, $\Delta T^{00} = 0$ and hence the rhs of (1) in such a frame is equals

$$\Delta T^{ij}(U_{i,j}/T) + \Delta T^{i0}[U_{i,0}/T - T_{,i}/T^2]$$

$$= \Delta T^{ij}U_{i,j}/T - \Delta T^{i0}[U^i_{,0}/T + T_{,i}/T^2]$$

For this to be non-negative, we must have the following general forms ΔT^{ij} and ΔT^{0i}:

$$\Delta T^{ij} = \chi_1(U_{i,j} + U_{j,i} + (2/3)U^k_{,k}\delta_{ij}) - \chi_2 U^k_{,k}\delta_{ij}$$

$$\Delta T^{i0} = \chi_3[U^i_{,0}T + T_{,i}]$$

where χ_1, χ_2, χ_3 are positive functions of T. Note that the temperature T is a scalar field. Note that $\Delta T^{\alpha\beta} = \Delta T^{\beta\alpha}$. These equations are valid in a co-moving frame. To get a tensor equation for $\Delta T^{\alpha\beta}$ that reduces to the above expressions in a co-moving frame, we must have

$$\Delta T^{\alpha\beta} = \chi_1 H^{\alpha\rho}H^{\beta\sigma}[(U_{\rho,\sigma} + U_{\sigma,\rho} - (2/3)U^{\mu}_{,\mu}\eta_{\rho\sigma} + \chi_2 U^{\mu}_{,\mu}\eta_{\rho\sigma})]$$

$$-[H^{\alpha\rho}U^{\beta} + H^{\beta\rho}U^{\alpha}]Q_{\rho}$$

where

$$Q_{\rho} = \chi_3[\mathrm{T}_{,\rho} - TU_{\rho,\mu}U^{\mu}]$$

and

$$H^{\mu\nu} = \eta^{\mu\mu} - U^{\mu}U^{\nu}$$

Note that in a co-moving frame,

$$H_{0i} = H^{i0} = H^{00} = 0, H^{ij} = -\delta_{ij}$$

so that in such a frame, we get agreement with the previous formulas. Combining all these facts, we finally find that the energy momentum tensor of matter in general relativity is given by

$$\begin{aligned}\tilde{T}^{\alpha\beta} &= (p+\rho)U^{\alpha}U^{\beta} - pg^{\alpha\beta} \\ &+ \chi_1 H^{\alpha\rho}H^{\beta\sigma}[(U_{\rho:\sigma} + U_{\sigma:\rho} - (2/3)U^{\mu}_{:\mu}\eta_{\rho\sigma}) + \chi_2 U^{\mu}_{:\mu}\eta_{\rho\sigma}] \\ &-[H^{\sigma\rho}U^{\beta} + H^{\beta\rho}U^{\alpha}]Q_{\rho}\end{aligned}$$

where

$$Q_{\rho} = \chi_3[T_{,\rho} - TU_{\rho:\mu}U^{\mu}]$$

❖❖❖❖❖

[142] Interaction between spin of an atom and a quantum magnetic field. The Coulomb gauge is assumed, *i.e.*, the magnetic vector potential $A(t, r)$ is given by

$$\begin{aligned}A(t,r) &= \sum_{k,s}[a(k,s)u(k,s)\exp(i(Kct - k.r)) \\ &+a(k,s)^*\bar{u}(k,s)\exp(-i(Kct-k.r))]\end{aligned}$$

where $K = |k|$. $a(k, s)$, $s = 1, 2$ are annihilation operators and $u(k, s), s = 1, 2$ are $\mathbb{R}^3$ valued vectors such that the Coulomb gauge condition $\text{div}A = 0$ is satisfied, *i.e.*, $(k, u(k, s)) = 0$, $s = 1, 2$. The commutation relations satisfied by the creation and annihilation operators are

$$[a(k,s), a(k',s')^*] = \delta_{s,s'}\,\delta_{k,k'}$$

The quantum magnetic field is given by

$$\begin{aligned}B(t,r) &= \nabla\times A(t,r) \\ &= -i\sum_{k,s} a(k,s)(k\times u(k,s))a(k,s).\exp(i(Kct-k.r)) \\ &-(k\times\bar{u}(k,s))a(k,s)^*\exp(-i(Kct-k.r)))\end{aligned}$$

The interaction energy between the spin of the atom located at r_0 and the quantum magnetic field is

$$V(t) = K_0(\sigma, B(t, r_0))$$

Writing

$$-ik\,Xu(k,s)\exp(-ik, r_0) = v(k,s)$$

and assuming that the magnetic field has been filtered by an LTI filter having transfer function $H(\omega) = \int_{\mathbb{R}} h(t)\exp(-i\omega t)\,dt$, it follows that the output magnetic field is given by

$$B_{out}(t) = \int_{\mathbb{R}} h(t')B(t-t', r_0)\,dt'$$
$$= \sum_{k,s}[H(Kc)a(k,s)v(k,s)\exp(iKct)$$
$$+\tilde{H}(Kc)a(k,s)^*\exp(-iKct)]$$

and the intraction of this output magnetic field with the atomic spin is given by

$$V_{out}(t) = K_0(\sigma, B_{out}(t))$$
$$= K_0\sum_{k,s}[H(Kc)a(k,s)(\sigma, v(k,s))\exp(iKct)$$
$$+\bar{H}(Kc)a(k,s)^*(\sigma, \bar{v}(k,s))\exp(-iKct)]$$

We may approximate this interaction Hamiltonian by a sum of N terms, *i.e.*,

$$V_{out}(t) \approx K_0 \sum_{1\le r\le N} [H[r]a[r]\otimes P[r]\exp(i\omega[r]t)$$
$$+H[r]a[r]^*\otimes P[r]^*\exp(-i\omega[r]t)]$$

where $P[r]$ are 2×2 complex matrices and $a[r]$ are operators in an infinite dimensional Hilbert space satisfying

$$[a[r], a[m]^*] = \delta[r-m]$$

$H[r]$ are complex numbers that are in our control. We can control them so that the designed gate generator $\int_0^T \tilde{V}_{out}(t)\,dt$ is as close as possible to a given Hermitian generator H_d.

❖❖❖❖❖

[143] Some concepts in cosmology:
Robertson-Walker metric: Spatial region $\mathbf{r} = (x, y, z)$ is defined by $r^2 + z^2 = S^2$. The spatial line element is then

$$dl^2 = (d\mathbf{r})^2 + dz^2 = dr^2 + r^2 d\Omega^2 + (rdr/sqrtS^2 - r^2)^2 = S^2 dr^2/(S^2 - r^2) + r^2 d\Omega^2$$

where

$$d\Omega^2 = d\theta^2 + \sin^2(\theta)d\phi^2$$

The metric dl^2 as well as the surface equation $r^2 + z^2$ are invariant under $SO(4)$, *i.e.*,

$$\mathbf{r} \to \mathbf{r}' = \mathbf{A}\mathbf{r} + \mathbf{b}z,$$
$$z \to z' = \mathbf{a}^T\mathbf{r} + \lambda\mathbf{z}$$

where

$$T = \begin{pmatrix} \mathbf{A} & \mathbf{b} \\ \mathbf{a}^T & \lambda \end{pmatrix}$$

is a 4 × 4 real orthogonal matrix, *i.e.*,

$$AA^T + bb^T = I_3, Aa + \lambda b = 0, |a|^2 + \lambda^2 = 1$$

In particular, the Robertson-Walker metric is invariant under spatial rotations as well as spatial quasi-translations. By spatial rotations, we mean that $b = 0$, $a = 0$, $\lambda = 1$, $A \in SO(3)$. By quasi-translations, we mean that

$$A = \sqrt{I_3 - bb^T}, |b| \leq 1$$

Since $z = \sqrt{S^2 - r^2}$, this quasi-translation amounts to

$$\mathbf{r} \to \mathbf{r}' = \sqrt{I_3 - \mathbf{b}\mathbf{b}^T\mathbf{r}} + \mathbf{b}\sqrt{S^2 - r^2}$$

Now, let

$$I - bb^T = (I - cc^T)^2 = I + (|c|^2 - 2)cc^T$$

Equivalently,

$$b = \sqrt{2 - |c|^2 c}$$

so that

$$|b|^2 = (2 - |c|^2)|c|^2$$

$$|c|^4 - 2|c|^2 + |b|^2 = 0$$

$$(|c|^2 - 1)^2 + |b|^2 - 1 = 0$$

$$1 - |c|^2 = \sqrt{1 - |b|^2}$$

$$c = b/(1 + \sqrt{1 - |b|^2})$$

$$\sqrt{I - bb^T} = I - cc^T = I - bb^T/(2 - |b|^2 + 2\sqrt{1 - |b|^2})$$

$$= \frac{2I(1 + \sqrt{1 - |b|^2}) - |b|^2 I - bb^T}{2(1 + \sqrt{1 - |b|^2}) - |b|^2}$$

$$= \frac{2I(1 + \sqrt{1 - |b|^2}) - |b|^2 I - bb^T}{(1 + \sqrt{1 - |b|^2})^2}$$

Another way to derive a formula for $\sqrt{I - bb^T}$ is by using the spectral theorem. $I - bb^T$ is a real Hermitian matrix. Let x be an eigenvector of this matrix with eigenvalue λ. Then $x - (b^T x)b = \lambda x$, so $x = (b^T x)b/(1 - \lambda)$. Thus,

$$b^T x = |b|^2 (b^T x)/(1 - \lambda)$$

If $b^Tx \neq 0$, then $\lambda = 1 - |b|^2$. If $b^Tx = 0$, then $\lambda = 1$. Thus, $I - bb^T$ has three eigenvalues, namely, 1 repeated twice and $1 - |b|^2$. The eigen-subspace corresponding to eigenvalue 1 is $b^\perp$ which is of dimension two and the normalized eigenvalue corresponding to eigenvalue $1 - |b|^2$ is $b/|b|$. Thus,

$$\sqrt{I - bb^T} = (I - bb^T / |b|^2) + \sqrt{1 - |b|^2}\, bb^T / |b|^2$$

Writing

$$n = b/|b|$$

we have

$$\sqrt{I - bb^T} = I - nn^T + \sqrt{1 - a^2 nn^T} = I - (1 - \sqrt{1 - a^2})nn^T$$

where

$$a = |b| \leq 1$$

Thus, quasi translations are defined by the unit vector n and a real number a smaller than unity in magnitude and are given by

$$\mathbf{r}' = \mathbf{r} + \mathbf{n}(a\sqrt{S^2 - r^2} - (1 - \sqrt{1 - a^2})\mathbf{n}^T\mathbf{r})$$

Diagonalization of the $t - r$ part of a spherically symmetric metric: Let

$$d\tau^2 = A(t, r)dt^2 - B(t, r)dr^2 + 2C(t, r)dt\,dr - r^2 d\Omega^2$$

Consider the transformation

$$t = t' + f(t', r'), r = r', \theta = \theta', \phi = \phi'$$

Then,

$$dt = dt' + f_{,1}dt' + f_{,2}dr', dr = dr', d\theta = d\theta', d\phi = d\phi'$$

and we get

$$Adt^2 - Bdr^2 + 2Cdtdr$$

$$= A(1 + f_{,1})^2 dt'^2 + (Af_{,2}^2 - B + 2Cf_{,2})dr'^2 + 2(C + f_{,2})dt'dr'$$

We choose f so that

$$C + f_{,2} = 0$$

Equivalently,

$$C(t' + f(t', r'), r') + f_{,2}(t', r') = 0$$

Then the metric comes to the standard spherically symmetric form (with t' being renamed as t):

$$d\tau^2 = A(t, r)dt^2 B(t, r)dr^2 - r^2 d\Omega^2$$

❖❖❖❖❖

[144] Consider

$$g(\phi, t, \theta) = u_\phi a_t u_\theta \in SL(2, \mathbb{R})$$

where

$$u_\theta = \begin{pmatrix} \cos(\theta) & -\sin(\theta) \\ \sin(\theta) & \cos(\theta) \end{pmatrix}$$

$$a_t = \begin{pmatrix} \exp(t) & 0 \\ 0 & \exp(-t) \end{pmatrix}$$

We define

$$X = \begin{pmatrix} 0 & 1 \\ 0 & 0 \end{pmatrix}, Y = \begin{pmatrix} 0 & 0 \\ 1 & 0 \end{pmatrix},$$

$$H = \begin{pmatrix} 1 & 0 \\ 0 & -1 \end{pmatrix}$$

Then $\{H, X, Y\}$ is a basis for the Lie Algebra $sl(2,\mathbb{R})$ of $SL(2,\mathbb{R})$. They satisfied the commutation relations

$$[H, X] = 2X, [H, Y] = -2Y, [X, Y] = H$$

We have

$$\begin{aligned} g_{,t} &= u_\phi a_t H u_\theta = u_\phi a_t u_{-\theta} H = g u_{-2\theta} H \\ &= g\left(\cos(2\theta) H + \sin(2\theta)(X+Y)\right) \end{aligned}$$

$$g_{,\theta} = g(X+Y),$$

$$\begin{aligned} g_{,\phi} &= u_\phi (X+Y) a_t u_\theta = g u_{-\theta} a_{-t} (X+Y) a_t u_\theta \\ &= g.(\sin(2\theta)\cosh(2t) H + (\exp(-2t)\cos^2(\theta) \\ &\quad -\exp(2t)\sin^2(\theta))(X+Y)) \\ &= g.(\sin(2\theta)\cosh(2t) H + (\sinh(2t) + \cos(2\theta)\cosh(2t))(X+Y)) \end{aligned}$$

Formally, these relations are expressed as

$$\frac{\partial}{\partial t} \to \cos(2\theta) H + \sin(2\theta)(X+Y),$$

$$\frac{\partial}{\partial \theta} \to X+Y,$$

$$\frac{\partial}{\partial \phi} \to \sin(2\theta)\cosh(2t) H + (\sinh(2t) + \cos(2\theta)\cosh(2t))(X+Y)$$

Remark: The above decomposition of $SL(2,\mathbb{R})$ is the well known KA_+K decomposition of a semisimple Lie group with K a maximal compact subgroup and A_+ the positive Weyl chamber corresponding to a maximal Abelian subgroup A (see Helgason's book). Equivalently, in matrix theory, this is called the singular value decomposition.

❖❖❖❖❖

[145] A survey of the work of some great Indian Scientists:

We shall briefly discuss here the contributions of a few of the famous Indian Scientists to world science. The scientists discussed here are (1) Shrinivasa Ramanujan, (2) Harish Chandra, (3) Satyendranath Bose, (4) Subramaniam Chandrasekher, (5) S.R.Srinivasa Varadhan, (6) V.S. Varadarajan and (7) K.R. Parthasarathy.

(1) Ramanujan was an untrained mathematical genius. He did not qualify for university education and so after finishing his school and college, worked as a clerk in the Madras Port Trust, However, while in school, he came across a book called Carr's Synopsis in which several identities regarding infinite series, elliptic integrals,continued fractions etc. were stated serially without proof. This triggered Ramanujan and he, independently of any teacher, discovered a variety of new identities. He used to work on a slate, record the final result in a notebook and then erase the contents on the slate with his elbow. In this way he filled up several pages of his notebooks with unproved formulae for π and its functions and several other formulae for trigonometrical functions, hyperbolic functions, infinite series, continued fractions etc. all in the style of Carr's synopsis. Today, these famous notebooks have been carefully edited by well known mathematicians all over the world like Bruce Brendt and Artin with proofs supplied for most of the theorems. However, Ramanujan remains an enigma because nobody knows how he proved all these results. While working as a clerk on a salary of ₹ 20 per month, Ramanujan sent his formulae to many well established mathematicians in England like Hobson and Baker who were fellows of the Royal society, but they all rejected his work saying that without proofs we cannot be sure about these results. Finally, Ramanujan under the advice of P.V.Seshu Aiyar, his mentor in Madras sent an envelope of formulas to G.H. Hardy at the Trinity College, Cambridge. Hardy, one day, on receiving this thick envelope with a lot of Indian stamps put it aside and went for his usual game of tennis and cricket watching. However, in the evening, he took out the envelope from his table and out of curiosity went through its contents. Ramanujan had written in a covering letter that he required Hardy's recommendation for an increase in his salary so that he could continue with his mathematical passion. Hardy tried to work through the formulae given by Ramanujan, some of them he could see were well known results, some he could himself prove and some went just tangential to him. He summoned his colleague Littlewood and all through the night, they worked on the results of Ramanujan. Most of the identities were correct, some needed minor corrections and some other identities were so fascinating but impossible for them to prove. By early next morning, Hardy and Littlewood arrived at the conclusion that Ramanujan was a genius and must be called to England so that they could collaborate with him. So Hardy dispatched another fellow mathematician Neville Scott to Madras to

persuade Ramanujan's family to allow Ramanujan to come to England. Neville lectured on elliptic functions and related topics at the Madras University and Ramanujan sitting on the last bench asked a few questions and stated several more related identities. Neville knew that this was Ramanujan and within a short time, despite having no degree from Madras University, Ramanujan was on his way to England. Ramanujan's mother had infact agreed to Nevill's suggestions since she claimed to have had a dream in which her family godess had told her that her son would become famous all over the world. Ramanujan on reaching Trinity was given a room and he cooked vegetarian meals for himself, not taking part in the public dinners where Hardy and Littlewood were present. Hardy used to collaborate with Ramanujan in the Institute Library and it was there that the two of them wrote their famous paper on asymptotic formulas for the partition function $p(n)$– the number of ways of writing a positive integer n as a sum of smaller positive integers. Hardy has written that Ramanujan would make conjecture after conjecture and Hardy would then struggle to prove these using the "western methods of proof" of which Ramanujan was totally unfamiliar. Another important conjecture of Ramanujan was regarding the τ function which are defined as the coefficients of different powers of z^n in the Taylor series expansion of $\left[\prod_{j\geq 1}(1-z^j)\right]^{24}$. Ramanujan made several conjectures abnout the behaviour of $\tau(n)$ for large values of n and these were only recently proved by great Bulgarian mathematician Pierre Deligne using sophisticated tools of algebric geometry. Deligne's proof of Ramanujan's conjectures runs into around 3000 printed pages and for this work, he was awarded the Field's medal, the highest honour given to a mathematician below the age of forty years. Hardy has written about Ramanujan, "I have made many significant discoveries in mathematics but my greatest discovery has been Ramanujan. Towards the end of his short life, Ramanujan discovered the Mock-Theta functions which behave like the theta functions but differ in some respects. Ramanujan's work finds applications today in statistical mechanics and string theory.

(2) The next mathematical star after Ramanujan was Harish Chandra who single handedly created the entire theory of representations of semisimple Lie groups. His most famous work is the Plancherel formula for real semisimple Lie group which was developed by him in a series of two remarkable papers on the discrete series. Earlier, following the work of Cartan who classified all the semisimple Lie algebras and H.Weyl who supplied the beautiful character formula for semisimple Lie groups using Cartan's theory of roots, Gelfand and Naimark had constructed the representation theory of complex semisimple Lie groups using a great deal of analytical machinery. Gelfand introduced the principal and supplementary series using which he was also able to obtain the Plancherel formula for complex semisimple Lie groups which is basically Fourier analysis on such a group. By the Plencherel formula, it is meant that the Diarc δ function on the Lie group be expressible as a linear superposition of its irreducible characters. Gelfand's method failed to work for real semisimple Lie groups like $SL(n, \mathbb{R})$ and so he was unable to obtain the complete Plancherel formula for these groups for the simple reason that

real semisimple Lie group have more than one non-conjugate Cartan subgroups and hence the representations appearing in the decomposition of the left regular representations into irreducibles in general will not contain all the representations. It was only later that Harish Chandra solved the problem completely for $SL(n, \mathbb{R})$ by introducing the Lie algebraic method. Bargmann had obtained the discrete series for $SL(2, \mathbb{R})$ which has two non-conjugate Cartan subgroups, namely

$\begin{pmatrix} a & 0 \\ 0 & a^{-1} \end{pmatrix}, a > 0$ and $\begin{pmatrix} \cos(\theta) & -\sin(\theta) \\ \sin(\theta) & \cos(\theta) \end{pmatrix}, \theta \in [0, 2\pi)$. These Cartan subgroup are

respectively known as the hyperbolic and elliptic Cartan groups. Harish Chandra using the infinitesimal, *i.e.*, the Lie algebraic machinery solved this problem and in the process created several fundamental concepts in representation theory like spherical functions, Harish Chandra transform, distribution character etc. Harish Chandra's work stands out as one of the most towering achievements of twentieth century mathematics.

(3) In physics, the greatest star who spent his entire life in India, apart from Sir C.V. Raman was Satyendranath Bose. When Bose was young, the quantum theory had just been created by Max Planck who suggested that the energy of the electromagnetic field at a given frequency ν appeared in integral multiples of $h\nu$ where h is a universe constant called Planck's constant. Planck was forced to introduce this hypothesis in order to get agreement with the experimental results of Rubens and Kurlbaum on the spectrum of black-body radiation. However, Planck was unable to give a sarisfactory derivation of his law of black-body radiation using the basic principles of statistical mechanics even after the introduction of his hyposthesis. Bose assumed that according to Planck's quantum hypothesis, radiation came in packets of particles called photons which were indistinguishable particles and moreover that a given state can have any number of these particles. He assumed that a given energy level E_k had a degeneracy of g_k and that n_k indistinguishable photons are to be distributed over these g_k for each k = 1, 2, ... with each state having any number of photons. From basic combinatorics, he calculated the total number of ways in which these particles can be distributed is $\prod_k \binom{g_k + n_k - 1}{n_k}$ and then applying the Stirling approximation, he maximized these number of ways of distributing n_k indistinguishable photons over g_k states for k = 1, 2... with the total energy constraint $\sum_k g_k E_k = E$. Bose noted that maximizing these number of ways is equivalent to maximizing the probability distribution of particles which according to Boltzmann, is equivalent to maximizing the entropy of the distribution. The result of this maximization was $n_k = \frac{g_k}{\exp(-\beta E_k) - 1}$ which today

is called the Bose-Einstein statistics for reasons which we will soon explain. Bose's next step was to use Planck's quantum hypothesis replacing Ek by $h\nu$ and g_k by the classical formula for the number of states between energy $E = h\nu$ and $E + dE = h(\nu + d\nu)$ per unit volume. The latter is obtained as $d^3p/h^3 = 4\pi p^2dp/h^3$ where $E = cp$ is Einstein's energy momentum relation for the photon (zero mass particle). Thus, Bose arrived at Planck's celebrated law of Black body radiation which states that the total energy of radiation per unit volume per unit frequency is $\frac{(8\pi h\nu^3/c^3)}{\exp(\beta h\nu)-1}$ with the Lagrange multiplier β being identified as $1/kT$ where T is the temperature and k is Boltzmann's constant. Bose described this new derivation of Planck's black-body radiation law in a letter to Einstein written at a time when the quantum theory was just in its infancy and a rigorous derivation of Planck's law was yet to be made. Einstein, on receiving this letter from Bose, immediately recognized the significance of Bose's discovery and with a footnote added to it saying that Bose's way of looking at photons was to consider them as being indistinguishable with no restriction on the number particles in a state, had it translated into German and published in the Annalen-Der-Physik, the most prestigious journal in physics at that time. The future generators described this statistics discovered by Bose as the Bose-Einstein and Diarc later on named particles obeying these statistics as Bosons. Today we know that elementary particles either have integral spin and they obey the Bose statistics and therefore are bosons or they have half integral spin and they obey the Fermi-Diarc statistics and are known as Fermions. n_k bosons can be distributed amongst g_k states in $\binom{g_k+n_k-1}{g_k}$ ways since no restriction is made on the number of particles that can occupy each state. On the other hand, n_k Fermions can be distributed amongst g_k states in $\binom{g_k}{n_k}$ ways since they are also indistinguishable but by the Pauli exclusion principle, no state can be occupied by more than one particle. Maximizing the probability $\prod_k \binom{g_k}{n_k}$ for Fermions with energy and total particle number constraint leads to the Fermi-Diarc statistics: $n_k = \frac{g_k}{\exp(\alpha+\beta E_k)+1}$ where the Lagrange multipliers α, β are identified as $\beta = 1/kT$, $\alpha = -E_F/kT$ with E_F as the Fermi energy.

❖❖❖❖❖

[146] A brief survey of estimation, filtering and control theory applied to the d-link robot system.

1. **Introduction:** The dynamical equations of a *d*-link robot. A *d*-link robot has

$d + 1$ degrees of freedom, namely rotations of each link in the same plane about its joint with the previous link and a rotation of the entire plane containing the d links about the z axis. The position $P(\xi)$ of a point on the kth link at a distance ξ from its joint with the $(k-1)^{th}$ link is given by

$$R_k(t,\xi) = \hat{x}\left[\sum_{j=1}^{k-1} \cos(q_j) + \xi\cos(q_k)\right]\cos(q_{d+1})$$

$$+\hat{y}\left[\sum_{j=1}^{k-1} l_j\cos(q_j) + \xi.\cos(q_k)\right]\sin(q_{d+1})$$

$$+\hat{z}\left[\sum_{j=1}^{k-1} l_j\sin(q_j) + \xi.\sin(q_k)\right]$$

where q_j is the angle made by the jth link with the xy plane, $j = 1, 2,\ldots d$ and q_{d+1} is the angle made by the plane containing all the links with the xz plane. We note that $q_j - q_{j-1}$ is the angle made by the jth link with the $(j-1)^{th}$ link $j = 1, 2,\ldots d$. The total kinetic energy of the system is

$$K(t) = \sum_{k=1}^{d} \frac{\sigma}{2}\int_0^{l_k}\left|\frac{\partial R_k(t,\xi)}{\partial t}\right|^2 d\xi$$

where σ is the linear mass density of the links. An easy calculation shows that

$$\left|\frac{\partial R_k(t,\xi)}{\partial t}\right|^2 = \left[\sum_{k=1}^{k-1} l_j\sin(q_j)q_j' + \xi q_k'\cos(q_k)\right]^2$$

$$+\left[\sum_{j=1}^{k-1} l_j\cos(q_j) + \xi\cos(q_k)\right]^2 q_{d+1}'^2$$

and integration gives the kinetic energy of the kth link as

$$K_k(t) = \frac{\sigma}{2}\int_0^{l_k}\left|\frac{\partial R_k(t,\xi)}{\partial t}\right|^2 d\xi$$

$$= \frac{\sigma}{6} q_{d+1}'^2 \sec(q_k)\left[\left(\sum_{j=1}^{k} l_j\cos(q_j)\right)^3 - \left(\sum_{j=1}^{k-1} l_j\cos(q_j)\right)^3\right]$$

$$+\frac{\sigma}{6} q_k'^{-1}\sec(q_k)\left[\left(\sum_{j=1}^{k} l_j\cos(q_j)q_j'\right)^3 - \left(\sum_{j=1}^{k-1} l_j\cos(q_j)q_j'\right)^3\right]$$

$$+\frac{\sigma}{6}q_k'^{-1}\text{cosec}(q_k)\left[\left(\sum_{j=1}^{k} l_j\sin(q_j)q_j'\right)^3-\left(\sum_{j=1}^{k-1} l_j\sin(q_j)q_j'\right)^3\right]$$

It easily follows from this expressions using $a^3 - b^3 = (a^2 + ab + b^2)(a - b)$ that the total kinetic energy $K = \sum_{k=1}^{d}$ of the robot has the form

$$K = \frac{1}{2}q'^T M(q)q'$$

where

$$q = [q_1,\ldots, q_{d+1}]^T \in \mathbb{R}^{d+1}$$

and

$$M : \mathbb{R}^{d+1} \to \mathbb{R}^{d+1\times d+1}$$

is positive definite. The gravitational potential energy of the system is easily found to be

$$V(q) = \sum_{k=1}^{d} m_k g\left(\sum_{j=1}^{k-1} l_j\sin(q_j)+l_k\sin(q_k)/2\right) = \sum_{k=1}^{d}\alpha_k\sin(q_k)$$

and hence the total Lagrangian of the system in the presence of motor torque at the joints is given by

$$\begin{aligned}L(t,q,q') &= K - V + \sum_{k=1}^{d}\tilde{\tau}_k(t)(q_k - q_{k-1}) + \tilde{\tau}_{d+1}(t)q_{d+1}\\ &= \frac{1}{2}q'^T M(q)q' - \sum_{k=1}^{d}\alpha_k\sin(q_k)\\ &\quad+\sum_{k=1}^{d}\tilde{\tau}_k(t)(q_k - q_{k-1}) + \tilde{\tau}_{d+1}(t)q_{d+1}\end{aligned}$$

where $\tilde{\tau}_k(t)$ is the torque supplied by the motor at the kth joint which rotates the kth link relative to the $k - 1^{\text{th}}$ link by an angle $q_k - q_{k-1}$. The Eular-Lagrange equations of the system are

$$\frac{d}{dt}\frac{\partial L}{\partial q_k'} = \frac{\partial L}{\partial q_k}, 1 \le k \le d+1$$

and these give

$$\frac{d}{dt}\sum_{j=1}^{d+1} M_{kj}(q)q = \frac{1}{2}\sum_{j,m=1}^{d+1} q_j'q_m'\frac{\partial M_{jm}(q)}{\partial q_k} - \alpha_k\cos(q_k) + \tau_k(t)$$

or equivalently,

$$\sum_{j=1}^{d+1} M_{kj}(q)q_j'' + \sum_{j,m=1}^{d+1}\Gamma_{kjm}(q)q_j'q_m' - \alpha_k\cos(q_k) = \tau_k(t)$$

where

$$\tau_k(t) = \tilde{\tau}_k(t) - \tilde{\tau}_{k+1}(t), \Gamma_{kjm}(q)$$

$$= \frac{1}{2}(M_{kj,m}(q) + M_{km,j}(q) - M_{jm,k}(q))$$

The term $\Gamma_{kjm}(j)$ in differential geometry is called the Christoffel symbol and plays a vital role in the general theory of relativity. The standard notation in robotics is to denote

$$N_k(q, q') = \Gamma_{j,m=1}^{d+1}\Gamma_{kjm}(q)q'_j q'_m - \alpha_k \cos(q_k)$$

and then the above dynamical equations can be expressed as

$$M(q)q'' + N(q, q') = \tau(t)$$

2. Parameter Estimation based on block data: When the torque vector process $\tau(t)$ contains a tremor component modelled as WGN, we write $\tau(t) + w(t)$ in place of $\tau(t)$. Further, we write $M(q, \theta), N(q, q', \theta)$ in place of $M(q), N(q, q')$ respectively, to emphasize the dependence of the matrices M, N on the parameter vector

$$\theta = [m_1, ..., m_d, l_1, ..., l_d]^T \in \mathbb{R}^{2d}$$

The dynamical equation assume the form

$$M(q, \theta)q'' + N(q, q', \theta) = \tau(t) + w(t)$$

If we assume that w is white, *i.e.*

$$\mathbb{E}(w(t)w(t')^T) = R_w\delta(t - t')$$

then the negative of the log-likelihood function of the random process $\{q(t), 0 \le t \le T\}$ given θ is, apart from additive constants,

$$L(q(t) : 0 \le t \le T \mid \theta) = \frac{1}{2}\int_0^T \|M(q(t), \theta)q''(t) - N(q(t) - q'(t), \theta) - \tau(t)\|_{R_w}^2 \, dt$$

where for any vector $\xi \in \mathbb{R}^{d+1}$, we write

$$\|\xi\|_{R_w}^2 = \xi^T R_w^{-1} \xi$$

θ may now be estimated by minimizing this function using a gradient search scheme:

$$\hat{\theta}[k+1] = \hat{\theta}[k] - \mu \frac{\partial}{\partial \theta} L(q(t) : 0 \le t \le T \mid \hat{\theta}[k])$$

An approximate way to overcome this nonlinear optimization is to assume an initial guess parameter estimate θ_0 and set $\theta = \theta_0 + \delta\theta$ to be the corrected parameter estimate based on the measured position data. Then, linearizing around θ_0 assuming that $\delta\theta$ and the noise $w(.)$ are of the same orders of smallness, we have with $q_0(t)$ denoting the solution for $\delta\theta = 0$, $w = 0$ and $q_0(t) + \delta\theta(t)$ the solution for $\theta_0 + \delta\theta$ and $w \neq 0$,

$$M(q_0(t), \theta_0) + N(q_0(t), q_0'(t), \theta_0) = \tau(t),$$

$$\frac{\partial M(q_0(t), \theta_0)}{\partial q}(\delta q(t) \otimes q_0''(t)) + \frac{\partial M(q_0(t), \theta_0)}{\partial \theta}(\delta\theta \otimes q_0''(t))$$

$$+ M(q_0(t), \theta_0)\delta q''(t) + \frac{\partial N}{\partial q}(q_0(t), q_0'(t), \theta_0)\delta q(t)$$

$$+ \frac{\partial N}{\partial q'}(q_0(t), q_0'(t), \theta_0)\delta q'(t) + \frac{\partial N}{\partial \theta}(q_0(t), q_0'(t), \theta_0)\delta\theta = w(t)$$

Such a perturbed model is based on the assumption that the actual perturbed trajectory is caused by noise and imprecise knowledge of the parameter vector. We note that $q_0(t)$ is calculated by actually solving its differential equation and $\delta q(t)$ is measured (which is equivalent to measuring $q(t) = q_0(t) + \delta q(t)$. Thus, the mle of $\delta\theta$ is given by

$$\hat{\delta\theta} = \arg\min_{\delta\theta} \int_0^T (f(t) + F(t)\delta\theta)^T R_w^{-1} (f(t) + F(t)\delta\theta) dt$$

$$= -\left[\int_0^T F(t)^T R_w^{-1} F(t) dt)\right]^{-1} \left[\int_0^T F(t)^T R_w^{-1} f(t) dt\right]$$

where

$$f(t) = \left[\frac{\partial M}{\partial q}(q_0(t), \theta_0)(I_{d+1} \otimes q_0''(t)) + \frac{\partial N}{\partial q}(q_0(t), q_0'(t), \theta_0)\right]\delta q(t)$$

$$+ \frac{\partial N}{\partial q_0}(q_0(t), q_0'(t), \theta_0)\delta q'(t)$$

and

$$F(t) = \frac{\partial M}{\partial \theta}(q_0(t), \theta_0)(I_p \otimes q_0''(t)) + \frac{\partial N}{\partial \theta}(q_0(t), \theta_0)$$

where

$$p = 2d$$

We note that $q_0(t)$ is a non-random process while $\delta q(t)$ is a random process and therefore, $f(t)$ is a random vector valued process while $F(t)$ is a non-random matrix valued process. We have

$$f(t) = w(t) - F(t)\delta\theta$$

and so,

$$\mathbb{E}[f(t)] = -F(t)\delta\theta, Cov(f(t), f(t')) = R_w \delta(t - t')$$

Hence,

$$\mathbb{E}(\hat{\delta\theta}) = \delta\theta,$$

$$Cov(\hat{\delta\theta}) = \left[\int_0^T F(t)^T R_w^{-1} F(t)\,dt\right]^{-1}$$

and the mean square parameter estimation error is

$$\mathbb{E}[\|\hat{\theta}\|^2 = Tr\left(\left[\int_0^T F(t)^T R_w^{-1} F(t)\,dt\right]^{-1}\right)$$

Remark: This estimation procedure is a book processing algorithm without any measurement noise. In a later section, we shall be introducing real time processing in the presence of measurement noise. That algorithm is based on the EKF (Exteded Kalman Filter) which is an approximation to the optimal Kushner filter. In the EKF, as and when more and more noisy measurements are collected, we keep improving our parametric estimates and also keep recursively constructing the estimate of the evolving state vector given the entire past measurement data. The EKF also keeps recursively computing the state and parameter estimation error covariance matrix and uses this estimate to update the estimates of the parameters and the evolved state.

3. Lyapunov Energy based methods for Parameter estimation and trajectory tracking:

If $q_g(t)$ is the desired/given trajectory to be tracked, then we give a feedback torque proportional to the trajectory error $e(t) = q_g(t) - q(t)$ and $e'(t)$ to make the trajectory of the actual robot get closer to $q_g(t)$. The resulting dynamical system is

$$M(q)q'' + N(q,q') = \widehat{M}(q)(q'' + e'' + K_p e'' + K_d e') + \widehat{N}(q,q')$$

$$= \widehat{M}(q)(q_d'' + K_p e + K_d e') + \widehat{N}(q,q') \quad ...(\alpha)$$

Here, θ_0 is the true value of the unknown parameter vector and $\hat{\theta}(t)$ is its estimate at time t. The notation used is

$$M(q) = M(q,\theta_0), \widehat{M}(q) = M(q,\hat{\theta}(t)),$$

$$N(q,q') = N(q,q',\theta_0), \widehat{N}(q,q') = N(q,q',\hat{\theta}(t))$$

We note that the computed torque on the right side of (α) uses only $\hat{\theta}$. We cannot use θ_0 to supply the computed torque since it is unknown. Linearization of (α) w.r.t. θ around θ_0 around then gives

$$e'' + K_p e + K_d e' = W(t)\delta\theta, \delta\theta(t) = \hat{\theta}(t) - \theta_0 \quad ...(\beta)$$

where $W(t)$ is a matrix involving q, q', q_g''. We can ignore like $e \otimes \delta\theta$ and $e' \otimes \delta\theta$, these being of the second order of smallness and then $W(t)$ involves only $q_g(t), q_g'(t), q_g''(t)$. In short, $W(t)$ can be treated as a known function of time. We choose our parametric update equation as

$$\delta\theta = -F\delta\theta + H\tilde{e}(t)$$

where

$$\tilde{e}(t) = \begin{pmatrix} e(t) \\ e'(t) \end{pmatrix}$$

The matrices F, H are chosen in accord with forcing a Lyapunov energy function

$$V(\tilde{e}, \delta\theta) = \frac{1}{2}\tilde{e}^T Q_1 q\tilde{e} + \frac{1}{2}\delta\theta^T Q_2 \delta\theta \qquad ...(\gamma)$$

to satisfy

$$\frac{dV}{dt} < 0$$

Here, Q_1, Q_2 are positive definite matrices. We note that the trajectory tracking error equation (β) can be put in the following convenient form:

$$\frac{d\tilde{e}(t)}{dt} = A\tilde{e}(t) + \tilde{W}(t)\delta\theta(t)$$

where

$$A = \begin{pmatrix} 0 & I \\ -K_p I & -K_d I \end{pmatrix}$$

where all the four blocks are of size $d + 1 \times d + 1$ and

$$\tilde{W}(t) = \begin{pmatrix} 0 \\ W(t) \end{pmatrix}$$

Straightforward differentiation leads to

$$\frac{dV}{dt} = \frac{d\tilde{e}^T}{dt} Q_q \tilde{e} + \frac{d\delta\theta^T}{dt} Q_2 \delta\theta$$

$$= \frac{1}{2}\tilde{e}^T (Q^T Q_1 + Q_1 A)\tilde{e} + \delta\theta^T \tilde{W}(t) Q_1 \tilde{e} + (H\tilde{e} - F\delta\theta)^T Q\delta\theta$$

So choosing H, F, K_p, K_d so that

$$A^T Q_1 + Q_1 A < 0, Q_2 H^T = -\tilde{W} Q_1, F^T Q_2 + Q_2 F > 0$$

we ensure that

$$\frac{dV}{dt} = \frac{1}{2}\tilde{e}^T (A^T Q_1 + Q_1 A)\tilde{e} - \frac{1}{2}\delta\theta^T (F^T Q_2 + Q_2 F)\delta\theta \le 0$$

We note that

$$H = -Q_1 \tilde{W}^T Q_2^{-1}$$

depends on time.

4. Lyapunov energy method for trajectory tracking and parameter estimation in the presence of noise: When noise is present both in the dynamics and in the parameter estimation algorithm, then the respective equation of the previous section get replaced by stochastic differential equations. Singla *et.al*[] have proposed the use of expected value of the Lyapunov energy function to assess the performance to trajectory tracking and parameter updates when WGN corrupts both the robot dynamics as well as the parameter estimation algorithm. In such a case the error dynamcis is given by

$$d\tilde{e}(t) = A\tilde{e}(t)dt + \tilde{W}(t)\delta\theta(t)dt + C_1 dB_1(t),$$

$$d\delta\theta(t) = -F\delta\theta(t)dt + H\tilde{e}(t)dt + C_2 dB_2(t)$$

where B_1 and B_2 are independent vector valued Brownian motion processes. The Lyapunov energy function is defined as usual, *i.e.* eqn(γ). Then, we find by the standard Ito rule

$$dV = d\tilde{e}^T Q_1 \tilde{e} + d\tilde{e}^T Q_1 d\tilde{e} + d\delta\theta^T Q_2 \delta\theta + d\tilde{\theta} Q_2 d\tilde{\theta}$$

and

$$d\tilde{e}.d\tilde{e}^T = C_1 C_1^T dt, \; d\delta\theta d\delta\theta^T = C_2 C_2^T dt$$

so

$$\frac{d}{dt}\mathbb{E}(V(t)) = \mathbb{E}(\tilde{e}^T (Q_1 A + A^T Q_1)\tilde{e}) - \mathbb{E}(\delta\theta^T Q_2 F + F^T Q_2)\delta\theta)$$

$$+\frac{1}{2}Tr(C_1^T Q_1 C_1 + C_2^T Q_2 C_2)$$

provided that as before, we select

$$H = -Q_2^{-1}\tilde{W}^T Q_1$$

It follows that if $-\lambda_1 < 0$ is the maximum eigenvalue of $Q_1 A + A^T Q_1$ and $\lambda_2 > 0$ is the minimum eigenvalue of $Q_2 F + F^T Q_2$, then

$$\frac{d}{dt}\mathbb{E}(V(t)) \leq -\lambda_1 \mathbb{E}(\| e(t)\|^2 - \lambda_2 \mathbb{E}(\| \delta\theta(t)\|^2) + \alpha$$

where

$$\alpha = \frac{1}{2}Tr(C_1^T Q_1 C_1 + C_2^T Q_2 C_2)$$

Now if $\lambda_0 > 0$ is the minimum of all the eigenvalue of Q_1, Q_2 put together and λ_0' $= \min(\lambda_1, \lambda_2)$, then

$$V(t) \geq \frac{1}{2}\tilde{e}^T Q_1 \tilde{e} + \frac{1}{2}\delta\theta^T Q_2 \delta\theta$$

$$\geq \lambda_0(\|\tilde{e}\|^2 + \|\delta\theta\|^2)$$

and so we deduce that if we define

$$E(t) = \mathbb{E}(\|\tilde{e}(t)\|^2 + \|\delta\theta(t)\|^2)$$

then,

$$E(t) \leq -(2\lambda_0' / \lambda_0)\int_0^t E(s)ds + 2\alpha t / \lambda_0$$

Let

$$\xi(t) = \int_0^t E(s)ds$$

Then, we have

$$\xi'(t) \leq -(2\lambda_0' / \lambda_0)\xi(t) + 2\alpha t / \lambda_0$$

which gives on using the integrating factor method,

$$\xi(t) \le (2\alpha/\lambda_0)\int_0^t (s.\exp(-2\lambda_0'(t-s)/\lambda_0))ds$$

Now,

$$\int_0^t s\exp(as)ds = t\exp(at)/a - (\exp(at)-1)/a^2$$

so that with $a = 2\lambda_0'/\lambda_0$, we have

$$\xi(t) \le (2\alpha/\lambda_0)(t/a - 1/a^2) + \exp(-at)/a^2$$

which for large t is bounded by $2\alpha t/a\lambda_0 = at/\lambda_0'$. Thus $E(t)$, for large t is bounded by a/λ_0'. This proves boundeness of the mean square trajectory error cum parametric estimation error energy.

Section 5: Lyapunov energy approach to disturbance observer design and its performance evaluation

Consider the d-link robot with disturbance torque $d(t)$:

$$M(q)q'' + N(q,q') = \tau(t) + d(t)$$

The following model has been proposed for estimating the disturbance $d(t)$ based on real time measurements of $q(t)$ and $q'(t)$:

$$\hat{d}(t) = z(t) + p(q'(t))$$

$$a'(t) = L(q(t), q'(t))(-\tau(t) + N(q(t), q'(t)) - \hat{d}(t))$$

The function $p(q')$ and $L(q, q')$ are selected as follows: First note that the above equation for $\hat{d}(t)$ gives on differentiation,

$$\hat{d}'(t) = L(-\tau + N - \hat{d}) + p'(q')q''$$

$$= L(L^{-1}p'(q')q'' + N - \tau - \hat{d})$$

So if we choose the function $L = L(q, q')$ such that

$$L^{-1}(q,q')p'(q') = M(q) \quad ...(\delta)$$

then we would get on account of the dynamical equations,

$$\hat{d}' = L(d - \hat{d}) \quad ...(\rho)$$

We may for example take

$$p(q') = Cq'$$

where C is a constant matrix and then eqn. (δ) becomes

$$L = CM(q)^{-1}$$

once we note that the Jacobian matrix of p is $p'(q') = C$. From eqn. (ρ), it follows that if $d(t) \to d(\infty)$ a constant vector and $d'(t) \to 0$, then for large t,

$$d' - \hat{d}' \approx -L(d - \hat{d})$$

so if as $t \to \infty$, $L(q(t), q'(t)) \approx L(\infty)$ where $L(\infty)$ has all eigenvalue with positive real part, then

$$d(t)-\hat{d}(t)\to 0$$

which means that the disturbance will be well tracked. In the general case, we have with

$$f(t) = d(t)-\hat{d}(t)$$

that

$$f' = d' - L_t f$$

so if $L_t \approx L(\infty)$ for all $t > T$ sufficiently large (L_t is an abbreviation for $L(q(t), q'(t))$), then

$$f(t) \approx \exp(-(t-T)L(\infty))f(T)+\int_T^t \exp(-(t-\tau)L(\infty)); (\tau)d\tau \quad ...(\sigma)$$

So if $L(\infty)$ has all eigenvalue with positive real part then with λ_0 being the minimum real part of all its eigenvalue, it follows that provides $d'(t)$ remains bounded as $t \to \infty$, we get from (σ) that

$$\lim\sup_{t\to\infty} \| f(t) \| \leq B / \lambda_0$$

where

$$B = \lim\sup_{t\to\infty} \| d'(t) \|$$

Thus, the distribution estimation error remains bounded as $t \to \infty$. The distribution error energy can also be analyzed through the Lyapunov method. Let

$$V(t) = \frac{1}{2} f(t)^T J(q(t)) f(t)$$

where $J(q)$ is a positive definite matrix valued function on $\mathbb{R}^{d+1}$. We wish to choose the function $J(q)$ so that when the angular velocity q' falls in a certain bounded region, $V'(t) < 0$. This would guarantee that the disturbance observer error $f(t)$ decays down to zero. A straightforward differentiation gives

$$V'(t) = f'(t)^T J(q(t)) f(t) + \frac{1}{2} f(t)^T \sum_{\alpha=1}^{d+1} J_{,\alpha}(q) q'_\alpha f(t)$$

$$= (d' - Lf)^T Jf + \frac{1}{2} f^T \left(\sum_\alpha J_{,\alpha} q'_\alpha \right) f$$

$$= -\frac{1}{2} f^T (L^T J + JL - \sum_\alpha q'_\alpha J_{,\alpha}) f + d'^T Jf$$

Now, $L = CM(q)^{-1}$ so

$$V' = -\frac{1}{2} f^T \left(M^{-1} C^T J + JCM^{-1} - \sum_\alpha q'_\alpha J_{,\alpha} \right) f + d'^T Jf$$

For convenience of convergence analysis, we choose $J(q)$ so that

$$M(q)^{-1} C^T J(q) = A$$

or

$$J(q) = C^{-T}M(q)A$$

where A is a constant matrix that is positive definite and such that $J(q)$ is also positive definite. (For example, we can take $A = C^{-1}$). Then we find that

$$V' = -\frac{1}{2}f^T(A + A^T - \sum_{\alpha} q'_{\alpha} J_{,\alpha})f + d'^T C^{-T} M(q)Af$$

Let $q'_{\max}$ be the maximum allowable link angular velocity magnitude and let

$$\max_{q \in \mathbb{R}^{d+1}} q'_{\max} \sum_{\alpha=1}^{d+1} \| J_{,\alpha}(q) \| = \beta$$

where

$$\beta < \lambda_{\min}(A + A^T)$$

Then we get

$$V' = (-\alpha/2)\| f \|^2 + b\| f \|$$

where

$$\alpha = \lambda_{\min}(A + A^T) - \beta > 0$$

and

$$b = \max_{q \in \mathbb{R}^n} \| d' \| . \| C^{-T}M(q)A \|$$

b is finite provided that we assume that $\|d'\|$ is bounded. Now,

$$V(t) = \frac{1}{2}f(t)^T J(q(t))f(t) \geq (\lambda_{\min}(J)/2)\| f(t) \|^2$$

where $\lambda_{\min}(J)$ is the smallest eigenvalue of $J(q)$ minimized over all $q \in \mathbb{R}^{d+1}$. Note that everywhere $q \in \mathbb{R}^{d+1}$ can be replaced by $q \in [0, 2\pi)^{d+1}$ since all the function that appear are periodic in q_j with period 2π. Indeed, these functions of q appears as trigonometric functions. No we get

$$(\lambda_{\min}(J)/2)\| f(t) \|^2 \leq -\frac{\alpha}{2}\int_0^t \| f(s) \|^2 ds + b\int_0^t \| f(s) \| ds$$

Setting

$$E(t) = \| f(t) \|^2$$

gives

$$E(t) \leq (-\alpha/\lambda_{\min}(J))\int_0^t E(s)ds + b\int_0^t \sqrt{E(s)ds}$$

A more useful inequality is obtained by noting that

$$V' \leq -\frac{\alpha}{2}\| f \|^2 + b\| f \|$$

$$\leq -\frac{\alpha}{\lambda_{\max}(J)}V + \left(b\sqrt{2}/\sqrt{\lambda_{\min}(J)}\right)\sqrt{V}$$

in view of the facts

$$V = \frac{1}{2} f^T Jf \leq (\lambda_{\max}(J)/2 \| f \|^2$$

and

$$V \geq (\lambda_{\min}(J)/2) \| f \|^2$$

Writing $V = \xi^2$ gives

$$2\xi\xi' = -\alpha\xi^2 / \lambda_{\min}(J) + b\sqrt{2/\lambda_{\min}(J)\xi}$$

so that

$$\xi'(t) \leq -\alpha\xi(t)/2\lambda_{\max}(J) + b/\sqrt{2\lambda_{\min}(J)}$$

This gives

$$\xi(t) \leq \exp(-\alpha t/2\lambda_{\max}(J))\xi(0) + \left(b\sqrt{2}\lambda_{\max}(J)\right.$$

$$\left. a\sqrt{\lambda_{\min}(J)}(1-\exp(-\alpha t/2\lambda_{\max}(J)))\right)$$

so that

$$\limsup_{t\to\infty} V(t) \leq 2b^2\lambda_{maqx}(J)^2 / \alpha^2\lambda_{\min}(J)$$

proving boundedness of the Lyapunov energy.

Remark: A certain kind of inequality called the Gronwall inequality is being used repeatedly here. For the sake of completeness, we shall state this inequality here with a proof: Suppose

$$V(t) \leq a(t)V(t) + b(t), t \geq 0$$

where a, b, V are real valued functions of t. Then, we get on multiplying both sides of this inequality by the positive function $\exp\left(\int_0^t a(s)ds\right)$,

$$\frac{d}{dt}\left[\exp\left(\int_0^t a(s)ds\right)V(t)\right] \leq b(t)\exp\left(\int_0^t a(s)ds\right)$$

from which we deduce that

$$V(t) \leq V(0) + \int_0^t \Phi(t,s)b(s)ds, t \geq 0$$

Section 6: Joint parameter estimator and disurbance observer

The construction of the disturbance observer discussed in the previous section depended on knowledge of the functions M, N. If however, we do not have accurate knowledge of θ_0, the true parameter vector, then we would use $\widehat{M}, \widehat{N}$ based on substituting $\hat{\theta}$ in place θ_0. Once we substitute $\hat{\theta}(t)$ in place of θ_0, we can construct the disturbance estimate $\hat{d}(t+dt)$ at time $t + dt$ and then subtract off this estimate from the dynamics and then update and then $\hat{\theta}(t)$ to $\hat{\theta}(t + dt)$. Thus, disturbance estimation and parameter estimation can be carried out hand in hand on a real time bases. The recursion in brief is as follows:

$$L(q(t), q'(t)) = CM(q(t), \hat{\theta}(t))^{-1}$$

$$z(t+dt) = z(t) + dt.L(q(t), q'(t))(-\tau(t) + N(q(t), q'(t), \hat{\theta}(t)) - \hat{d}(t))$$

$$\hat{d}(t+dt) = z(t+dt) + p(q'(t+dt))$$

Then,

$$\hat{\theta}(t+dt) = \hat{\theta}(t) + \delta\hat{\theta}(t)$$

where

$$\delta\hat{\delta}(t) = \arg\min_{\delta\theta} \sum_{k=1}^{t/dt} \| M(q((k+1)dt, \hat{\theta}(t) + \delta\theta)(q'((k+1)dt) - q'(kdt)) / dt$$

$$-N(q((k+1)dt), q'((k+1)dt), \hat{\theta}(t) + \delta\theta) - \tau((k+1)dt) - \hat{d}((k+1)dt) \|^2$$

Here, it is being assumed that t takes values in integer multiples of dt. The last step of minimizing w.r.t. $\delta\theta$ can be carried out using a linear least squares method by linearizing the functions M, N around $\hat{\theta}(t)$, *i.e.* be setting

$$M(q, \hat{\theta} + \delta\theta) \approx M(q, \hat{\theta}) + \frac{\partial M(q, \hat{\theta})}{\partial\theta}\delta\theta$$

$$N(q, q', \hat{\theta} + \delta\theta) \approx N(q, q', \hat{\theta}) + \frac{\partial M(q, q', \hat{\theta})}{\partial\theta}\delta\theta$$

Section 7: Design of Extended state and disturbance observers with simultaneous control for trajectory tracking

It is known that given a stochastic differential equation for the state $x(t)$:

$$dx(t) = f(t, x(t))\, dt + g(t, x(t))\, dB(t)$$

with a measurement model

$$dz(t) = h_t(x(t))\, dt + \sigma_v dv(t)$$

where $B(.)$, $v(.)$ are independent vector valued Brownian motions, the optimum MMSE of $\phi(x(t))$ based on the observations $Z_t = \{z(s) : s \le t\}$ collected upto time t, namely,

$$\hat{\phi}_t = \mathbb{E}[\phi(x(t)) \,|\, Z_t] = \pi_t(\phi)$$

satisfies the Kushner nonlinear filtering equation

$$d\pi_t(\phi) = \pi_t(L_t\phi)dt + \sigma_v^{-2}(\pi_t(\phi h) - \pi_t(\phi)\pi_t(h))^T(dz(t) - \pi_t(h)dt)$$

This is proved in [Jazwinsky, Sage and Melsa] using Bayes rule, Markovian property of the state and the Ito formula for Brownian motion. L_t is the backward Kolmogorov operator:

$$L_t = f(t, x)^T \frac{\partial}{\partial x} + \frac{1}{2} Tr\left(g(t, x).g(t, x)^T \frac{\partial^2}{\partial x \partial x^T} \right)$$

Further, if we write

$$x(t) = \pi_t(x) + \delta x(t),$$

$$P(t) = \text{Cov}(\delta x(t)) = \mathbb{E}[(x(t) - \pi_t(x))(x(t) - \pi_t(x)^T)]$$

$$= \mathbb{E}[(x(t) - \pi_t(x))(x(t) - \pi_t(x)) \mid Z_t]$$

(the last equality following from the orthogonality property of the MMSE, namely $x(t) - \pi_t(x)$ is independent of Z_t, then $\pi_t(x)$ and $P(t)$ satisfy approximation the EKF equations which are derived from the Kushner filtering equation using approximation such as

$$\phi(x(t)) \approx \phi(\pi_t(x)) + \phi_{,\alpha}(\pi_t(x))\,\delta x_\alpha(t) + \frac{1}{2}\phi_{,\alpha\beta}(\pi_t(x))\,\delta x_\alpha(t)\,\delta x_\beta(t)$$

(namely, Taylor expansion about the conditional mean with the Einstein summation convention over repeated indices being used). The EKF equation are

$$d\pi_t(x) = f(t, \pi_t(x))dt + \sigma_v^{-2} P(t) H(t)^T (dz(t) - h_t(\pi_t(x))dt) \quad ...(\delta 1)$$

where

$$H(t) = h_t'(\pi_t(x))^T$$

with $h_t'(x)$ being the Jacobian matrix of the map $x \to h_t(x)$, and

$$\frac{dP(t)}{dt} = f'(t, \pi_t(x))P(t) + P(t) f'^T(t, \pi_t(x))$$

$$-\sigma_v^{-2} P(t) H(t) H(t)^T P(t)$$

$$+ g(t, \pi_t(x)) g(t, \pi_t(x))^T \quad ...(\delta 2)$$

The actual EKF is used to estimate state, namely both the state and the unknown parameters of the state model. The extended state equation are

$$dx(t) = f(t, x(t), \theta(t))dt + g(t, x(t), \theta(t))dB(t)$$

$$d\theta(t) = \sum_\theta dB_\theta(t)$$

where $B(.)$ and $B_\theta(.)$ are independent Brownian motion processes. The extended state vector is

$$\xi(t) = \begin{pmatrix} x(t) \\ \theta(t) \end{pmatrix}$$

and the extended state equations can be expressed as

$$d\begin{pmatrix} x(t) \\ \theta(t) \end{pmatrix} = \begin{pmatrix} f(t, x(t), \theta(t)) \\ \theta(t) \end{pmatrix} dt$$

$$+ \begin{pmatrix} g(t, x(t), \theta(t)) & 0 \\ 0 & \sum_\theta \end{pmatrix} d\begin{pmatrix} B(t) \\ B_\theta(t) \end{pmatrix}$$

The output measurement model is

$$dz(t) = h_t(x(t))dt + \sigma_v dv(t)$$

where $v(.)$ is another vector Brownian motion independent of $B(.)$, $B_\theta(.)$. The extended Kalman filtering equations are then

$$d\begin{pmatrix} \pi_t(x) \\ \pi_t(\theta) \end{pmatrix} = \begin{pmatrix} f(t,\pi_t(x),\pi_t(\theta)) \\ 0 \end{pmatrix} dt$$

$$+\sigma_v^{-2}\begin{pmatrix} P_{xx}(t) & P_{x\theta}(t) \\ P_{\theta x}(t) & P_{\theta\theta}(t) \end{pmatrix}\begin{pmatrix} h_t'(\pi_t(x))^T \\ 0 \end{pmatrix}(dz(t) - h_t'(\pi_t(x))dt),$$

$$\frac{d}{dt}\begin{pmatrix} P_{xx}(t) & P_{x\theta}(t) \\ P_{\theta x}(t) & P_{\theta\theta}(t) \end{pmatrix}$$

$$= \begin{pmatrix} f_{,x}(t,\pi_t(x),\pi_t(\theta)) & f_{,\theta}(t,\pi_t(x),\pi_t(\theta)) \\ 0 & 0 \end{pmatrix}\begin{pmatrix} P_{xx}(t) & P_{x\theta}(t) \\ P_{\theta x}(t) & P_{\theta\theta}(t) \end{pmatrix}$$

$$+\begin{pmatrix} P_{xx}(t) & P_{x\theta}(t) \\ P_{\theta x}(t) & P_{\theta\theta}(t) \end{pmatrix}\begin{pmatrix} f_{,x}(t,\pi_t(x),\pi_t(\theta))^T & 0 \\ f_{,\theta}(t,\pi_t(x),\pi_t(\theta))^T & 0 \end{pmatrix}$$

$$+\sigma_v^{-2}\begin{pmatrix} P_{xx}(t) & P_{x\theta}(t) \\ P_{\theta x}(t) & P_{\theta\theta}(t) \end{pmatrix}\begin{pmatrix} h_t'(\pi_t(x))^T \\ 0 \end{pmatrix}\begin{pmatrix} h_t'(\pi_t(x)) & 0 \end{pmatrix}\begin{pmatrix} P_{xx}(t) & P_{x\theta}(t) \\ P_{\theta x}(t) & P_{\theta\theta}(t) \end{pmatrix}$$

$$+\begin{pmatrix} g(t,\pi_t(x),\pi_t(\theta)) & 0 \\ 0 & \sum_\theta \end{pmatrix}\begin{pmatrix} g(t,\pi_t(x),\pi_t(\theta))^T & 0 \\ 0 & \sum_\theta \end{pmatrix}$$

Based on the idea of using an output error feedback in the EKF, we can propese a general state observer and controller that uses the output error feedback for the state observer and the state observer error feedback that generalizes pd control to the state equations. Let $x_d(t)$ be the desired state and $\hat{x}(t)$ the state observer. Let $x(t)$ be the actual state. Then, to cause the state $x(t)$ to track $x_d(t)$, we modify the state equations to

$$dx(t) = f(t,x(t))dt + g(t,x(t))dB(t) + K(t)(x_d(t) - \hat{x}(t))$$

and the state observer to

$$d\hat{x}(t) = f(t,\hat{x}(t)) + L(t)(dz(t) - h_t(\hat{x}(t))dt)$$

where the output/measurement process $z(t)$ is given by

$$dz(t) = h_t(x(t)) + \sigma_v dv(t)$$

For analysis of the joint process

$$\begin{pmatrix} x(t) \\ \hat{x}(t) \end{pmatrix}$$

we can write the above equations as

$$d\begin{pmatrix} x(t) \\ \hat{x}(t) \end{pmatrix} = \begin{pmatrix} f(t, x(t)) + K(t)(x_d(t) - \hat{x}(t)) \\ f(t, \hat{x}(t)) + L(t)(h_t(x(t)) - h_t(\hat{x}(t))) \end{pmatrix} dt$$

$$+\begin{pmatrix} g(t, x(t)) & 0 \\ 0 & \sigma_v L(t) \end{pmatrix} d\begin{pmatrix} B(t) \\ v(t) \end{pmatrix}$$

If the observer used is the EKF observer, then we choose our observer gain

$$L(t) = \sigma_v^{-2} P(t) h'(\hat{x}(t))^T$$

where $P(t)$ satisfies the Riccati equation (δ2).
We now consider the following method for constructing a state observer and state tracker. The state equations are

$$dx(t) = f(t, x(t))dt + g(t, x(t))dB(t) + K(t)(x_d(t) - \hat{x}(t))$$

where the output equation is

$$dz(t) = h(t, x(t))dt + \sigma_v dv(t)$$

and the state observer is

$$d\hat{x}(t) = f(t, \hat{x}(t))dt + L(t)(dz(t) - h(t, \hat{x}(t))dt)$$

The state observer error is

$$e(t) = x(t) - \hat{x}(t)$$

and the state tracking error is

$$f(t) = x_d(t) - x(t)$$

We assume that the desired state $x_d(t)$ follows the non-random component of the state dynamics, *i.e.*,

$$dx_d(t)/dt = f(t, x_d(t))$$

We have approximately,

$$\begin{aligned} de(t) &= dx(t) - d\hat{x}(t)(f(t, x(t)) - f(t, \hat{x}(t)))dt + g(t, x(t))dB(t) \\ &\quad +K(t)(x_d(t) - \hat{x}(t))dt - L(t)(h(t, x(t)) - h(t, \hat{x}(t)))dt + \sigma_v dv(t)) \\ &\approx f'(t, \hat{x}(t))e(t)dt + g(t, \hat{x}(t))dB(t) + K(t)(f(t) + e(t))dt \\ &\quad - L(t)h'(t, \hat{x}(t))e(t)dt - \sigma_v L(t)dv(t) \\ &= (F(t) - L(t)H(t) + K(t))e(t)dt + K(t)f(t)dt \\ &\quad G(t)dB(t) - \sigma_v L(t)dv(t) \end{aligned}$$

where

$$F(t) = f'(t, \hat{x}(t)) = f_{,x}(t, \hat{x}(t)), G(t) = g(t, \hat{x}(t))$$

Also

$$df(t) = dx_d(t) - dx(t)$$
$$= (f(t, x_d(t)) - f(t, x(t))dt - K(t)(x_d(t) - \hat{x}(t))dt - g(t, x(t))dB(t)$$
$$= F(t)f(t) - K(t)(e(t) + f(t))dt + G(t)dB(t)$$
$$= (F(t) - K(t))f(t)dt - K(t)e(t)dt - G(t)dB(t)$$

So

$$d\begin{pmatrix} e(t) \\ f(t) \end{pmatrix} = \begin{pmatrix} F(t) - L(t)H(t) + K(t) & K(t) \\ -K(t) & F(t) - K(t) \end{pmatrix} \begin{pmatrix} e(t) \\ f(t) \end{pmatrix}$$
$$+ \begin{pmatrix} G(t) & -\sigma_v L(t) \\ -G(t) & 0 \end{pmatrix} d\begin{pmatrix} B(t) \\ v(t) \end{pmatrix}$$

We note that $F(t) = f'(t, \hat{x}(t))$ and $G(t) = g(t, \hat{x}(t))$ are known functions of $(t, \hat{x}(t))$ while $L(t)$ and $K(t)$ are unknown functions of $(t, \hat{x}(t))$. These unknown functions must be designed so that

$$\int_0^T \mathbb{E}(\| e(t) \|^2 + \| f(t) \|^2) dt$$
$$= \mathbb{E} \int_{0<\tau<t<T} Tr[\Phi(T, \tau)Z(\tau)Z(\tau)^T \Phi(T, \tau)^T] dt\, d\tau$$

is a minimum where $\Phi(t, \tau)$ satisfies the matrix differential equation

$$\frac{\partial \Phi(T, \tau)}{\partial t} = \begin{pmatrix} F(t) - L(t)H(t) + K(t) & K(t) \\ -K(t) & F(t) - K(t) \end{pmatrix} \Phi(T, \tau), t \ge \tau$$

$$\Phi(\tau, \tau) = I,$$

$$Z(t) = \begin{pmatrix} G(t) & -\sigma_v L(t) \\ -G(t) & 0 \end{pmatrix}$$

More generally, if Q is a positive definite matrix, we may select the control gain $K(t)$ and the observer gain $L(t)$ so that

$$\int_0^T \mathbb{E}\left[(e(t)^T, f(t)^T) Q \begin{pmatrix} e(t) \\ f(t) \end{pmatrix}\right] dt$$

$$= Tr\left[Q\mathbb{E} \int_{0<\tau<t<T} \Phi(t, r)Z(\tau)Z(\tau)^T \Phi(T, \tau)^T dt d\tau\right]$$

is a minimum. This problem is difficult to solve as it involves a highly nonlinear optimization.

Robustness of the observer: Suppose the actual state equation with desired observer error feedback are

$$dx(t) = f(t, x(t), \theta_0)dt + g(t, x(t), \theta_0)dB(t)$$

$$+K(t)(x_d(t)-\hat{x}(t))dt$$

and

$$\frac{dx_d(t)}{dt} = f(t, x_d(t), \theta_0)$$

θ_0 is unknown and so our observer uses a given value $\theta_0 + \delta\theta = \theta$ for its construction. The observer dynamics is then

$$d\hat{x}(t) = f(t, \hat{x}(t), \theta_0)dt + L(t)(dz(t) - h(t, \hat{x}(t))dt)$$

where the output is

$$dz(t) = h(t, x(t))dt + \sigma_v dv(t)$$

we have approximately with

$$\begin{aligned} e(t) &= x(t)-\hat{x}(t), f(t)=x_d(t)-x(t), e(t)+f(t)=x_d(t)-\hat{x}(t) \\ de(t) &= [f(t, x(t), \theta_0) - f(t, \hat{x}(t), \theta)]dt + g(t, x(t), \theta_0)dB(t) \\ &\quad +K(t)(e(t)+f(t))dt \\ &\quad -L(t)((h(t, x(t)) - h(t, \hat{x}(t)))dt + \sigma_v dv(t)) \\ &\approx [f_{,x}(t, \hat{x}(t), \theta_0)e(t) + f_{,\theta}(t, \hat{x}(t), \theta_0)\delta\theta]dt \\ &\quad + g(t, \hat{x}(t), \theta_0)dB(t) + K(t)(e(t)+f(t))dt \\ &\quad -L(t)(h_{,x}(t, \hat{x}(t))e(t)dt + \sigma_v dv(t)) \end{aligned}$$

where $\delta\theta \otimes d(t), \delta\theta \otimes f(t)$ etc. terms are neglected. Further,

$$\begin{aligned} df(t) &\approx f_{,x}(t, \hat{x}(t), \theta_0)f(t)dt - g(t, \hat{x}(t), \theta_0)dB(t) \\ &\quad -K(t)(e(t)+f(t)) \end{aligned}$$

so combining these two error evolution equations we get

$$\begin{aligned} d\begin{pmatrix} e(t) \\ f(t) \end{pmatrix} &= \begin{pmatrix} F_x(t)+K(t)-L(t)H(t) & K(t) \\ -K(t) & F_x(t)-K(t) \end{pmatrix}\begin{pmatrix} e(t) \\ f(t) \end{pmatrix}dt \\ &\quad +\begin{pmatrix} G(t) & -\sigma_v L(t) \\ -G(t) & 0 \end{pmatrix}d\begin{pmatrix} B(t) \\ v(t) \end{pmatrix} \\ &\quad +\begin{pmatrix} F_\theta(t) \\ 0 \end{pmatrix}\delta\theta \end{aligned}$$

where

$$F_x(t) = f_{,x}(t, \hat{x}(t), \theta_0), F_\theta(t) = f_{,\theta}(t, \hat{x}(t), \theta_0)$$

In this formation, we're not using the EKF to estimate the parameter θ_0. Instead, we are assuming a guess value θ and are calculating the effect of the guess parameter error $\delta\theta$ upon the tracking error energy $\mathbb{E}(\| f(t) \|^2)$ and the observer error $\mathbb{E}(\| e(t) \|^2)$. Solving the above equation gives

$$\begin{pmatrix} e(t) \\ f(t) \end{pmatrix} = \int_0^t \Phi(t, \tau) Z(\tau) d \begin{pmatrix} B(\tau) \\ v(\tau) \end{pmatrix}$$

$$+ \left[\int_0^t \Phi(t, \tau) \begin{pmatrix} F_\theta(\tau) \\ 0 \end{pmatrix} d\tau \right] d\theta$$

In the computation of $\Phi(t, \tau), Z(\tau), F_\theta(\tau)$, we may replace $\hat{x}(t)$ by $x_d(t)$ since second order terms line $(x_d(t) - \hat{x}(t)) \otimes \begin{pmatrix} B(\tau) \\ v(\tau) \end{pmatrix}$ and $(x_d(t) - \hat{x}(t)) \otimes \delta\theta$ are negligible. Then, with this approximation, $\Phi(t, \tau)$ and $Z(t)$ become non-random functions of t, τ. Then, we get

$$\mathbb{E}\left[(e(t)^T, f(t)^T) Q \begin{pmatrix} e(t) \\ f(t) \end{pmatrix} \right]$$

$$= \int_0^t Tr(Q\Phi(t, \tau) Z(\tau) Z(\tau)^T \Phi(t, \tau)^T) d\tau$$

$$+ \delta\theta^T \left(\int_0^t \Phi(t, \tau) \begin{pmatrix} F_\theta(\tau) \\ 0 \end{pmatrix} d\tau \right)^T Q \left(\int_0^t \Phi(t, \tau) \begin{pmatrix} F_\theta(\tau) \\ 0 \end{pmatrix} d\tau \right) \delta\theta$$

Suppose

$$\lim_{t \to \infty} \int_0^t \Phi(t, \tau) Z(\tau) Z(\tau)^T \Phi(t, \tau)^T d\tau = P_1$$

and

$$\lim_{t \to \infty} \int_0^t \Phi(t, \tau) \begin{pmatrix} F_\theta(\tau) \\ 0 \end{pmatrix} d\tau = P_2$$

are finite matrices. Then, we get

$$\lim_{t \to \infty} \mathbb{E}\left[(e(t)^T, f(t)^T) Q \begin{pmatrix} e(t) \\ f(t) \end{pmatrix} \right]$$

$$\approx Tr(P_1) + \delta\theta^T P_2^T Q P_2 \delta\theta \leq Tr(P_1) + \lambda_{\max}(P_2 Q P_2) \| \delta\theta \|^2$$

proving robustness of the errors with respect to noise and parametric uncertainties.

Section 8. Robot motion in s strong gravitational field

To describe the motion of a robot in a strong gravitational field, we need to use the formalism of general relativity in which the gravitational field is described by a metric tensor of space-time:

$$d\tau^2 = g_{\mu\nu}(x) dx^\mu dx^\nu$$

where

$$x = (x^{\mu}) = (t, x, y, z)$$

We use the notation $x = (t. \mathbf{r})$ and then if $R(t)$ is the rotation matrix applied after time t to a single 3-D link robot, the Lagrangian of the robot is assuming that B is the region of space occupied by the robot at time $t = 0$,

$$L(R(t), R'(t), t) = \int_B (g_{00}(t, R(t)\xi) + 2g_{0k}(t, R(t)\xi)(R'(t)\xi)^k + g_{km}(t, R(t)\xi)(R(t)\xi)^k (R(t)\xi)^m)^{1/2} d^3\xi$$

We express $R(t)$ in terms of the Eular angles:

$$R(t) = R_z(\phi(t))R_x(\theta(t))R_z(\psi(t))$$

and then the Lagrangian can be expressed as $L(t, \phi, \theta, \psi, \phi', \theta', \psi')$ and the Eular-Lagrange equations can be set up alternately, we can use the variational calculus to extremize $\int L(R(t), R'(t), t)\, dt$ after incorporating the constraint $R(t)^T R(t) = I$ using Lagrange multiplier function. Specifically, we extremize

$$S(L(R, R', \Lambda)) = \int L(R(t), R'(t), t)\, dt - \int Tr(\Lambda(t)(R(t)^T R(t) - I))\, dt$$

We leave it to the reader to work out the details using integration by parts wherever $\delta R'(t)$ occurs in the variational principle

$$\delta_R S = 0$$

Section 9: Lie group theotric analysis of a d link robot with 3-D links

Assume that we have d 3 dimensional links with the first link attached at a point $p_0 = 0$ to origin. The kth link is attached at the point p_{k-1} to the $k - 1^{\text{th}}$ link at time $t = 0$. After time t, a rotation $R_1(t)$ has been applied to the entire system of links about p_0, w.r.t. the rest frame and in general, a rotation $R_k(t)$ has been applied to the system comprising the kth, $k + d^{\text{th}}$ links about the point p^{k-1}. Take a point ξ on the kth link at time $t = 0$. After time t, it has moved to the point

$$P_k(\xi, t) = R_1(t)p_1 + R_2(t)R_1(t)(p_2 - p_1) + \ldots + R_{k-1}(t)R_{k-2}(t) \ldots R_1(t)(p_{k-1} - p_{k-2}) + R_k(t)\ldots R_1(t)(\xi - p_{k-1})$$

where $k = 1, 2, \ldots, d$. Writing

$$S_k(t) = R_k(t)R_{k-1}(t)\ldots R_1(t)$$

we have

$$P_k(\xi, t) = \sum_{j=1}^{k-1} S_j(t)(p_j - p_{j-1}) + S_k(t)(\xi - p_{k-1})$$

and the kinetic energy of the entire system is

$$K(t) = \sum_{k=1}^{d} \frac{\rho}{2} \int_{B_k} \left| \frac{\partial P_k(\xi, t)}{\partial t} \right|^2 d^3\xi$$

where B_k is region of $\mathbb{R}^3$ occupied by the kth link at time $t = 0$. It is easy to see that $K(t)$ can be expressed as a quadratic form in $S_k'(t), k = 1, 2, \ldots, d$:

$$K(t) = \frac{1}{2}\sum_{j,k=1}^{d} Tr(S'_j(t)J_{jk}S'_k(t)^T)$$

where the moment of inertia matrices J_{jk} depend only on the geometry, *i.e.*, shape of the different links. Further, the potential energy of the system can be expressed as

$$V(t) = m_1 g\langle e_3, R_1(t)d_1\rangle + m_2 g\langle e_3, R_1(t)p_1 + R_2(t)R_1(t)d_2\rangle + ...$$

$$+\langle e_3, R_1(t)p_1 + R_2(t)(p_2 - p_1) + ... + R_{d-1}(t)...$$

$$R_1(t)(p_{d-1} - p_{d-2}) + R_d(t)...R_1(t)d_d\rangle$$

$$= \sum_{k=1}^{d}\langle e_3, S_k(t)f_k\rangle$$

where $e_3 = (0, 0, 1)^T$ and d_k is the position vector of the centre of mass of the kth link relative to p_{k-1}, the point of contact of the kth link with the k – 1th link at time $t = 0$. The $f_k's$ are constant vectors expressible in terms of the $m_k's$, g, the $p_k's$ and the $d_k's$. Taking into account the constraint $S_k(t)TS_k(t) = I$ using Lagrangian multiplier matrices $\Lambda_k(t)$, the Lagrangian of the overall system can be expressed as

$$L(S_k, S'_k, \Lambda_k, k = 1, 2, ..., d) = K(t) - V(t) - \sum_{k=1}^{d} Tr(\Lambda_k(t)(S_k(t)^T S_k(t) - I))$$

$$= \frac{1}{2}\sum_{j,k=1}^{d} Tr(S'_j(t)J_{jk}S'_k(t)^T) - \sum_{k=1}^{d} e_3^T S_k(t)f_k$$

$$-\sum_{k=1}^{d} Tr(\Lambda_k(t)(S_k(t)^T S_k(t) - I))$$

and the variational principle

$$\delta s_j \int L dt = 0$$

taking into account the symmetry $J_{jk}^T = J_{kj}$ gives after integration by parts, the differential equations

$$-2\sum_k J_{jk}S''_k(t)^T - f_j e_3^T - (\Lambda_j(t) + \Lambda_j(t)^T)S_j(t)^T = 0,\ 1 \le j \le d$$

We can also take into account motor torque applied at the joints $p_1,..., p_d$. Denoting these torque tensor by $T_k(t)$ (these are 3 × 3 antisymmetric matrices), the torque Lagrangian to be added is given by

$$L_{\text{torque}} = \sum_{k=1}^{d}\sum_{j=1}^{3} T_{kj}(t)\theta_{kj}(t)$$

where T_{k1}, T_{k2}, T_{k3} are the nonzero components of $T_k(t)$ around the 3 Eular angles of the kth link relative to the k – 1th link attached at p_{k-1}. θ_{k1}, θ_{k2}, θ_{k3} are the Eular angles that specify the rotation $R_k(t)$ of the kth link relative to the $k - 1^{th}$ link attached at p_{k-1}. The torque T_k to the kth link is provided by motors attached at p_{k-1}.

Section 10: The dynamics of Quantum Robots based on the Schrodinger and Diarc equations

The unperturbed Hamiltonian of the robot in a gravitational field has the form

$$H_0(q, p) = \left(-iH_0 dt + \frac{1}{2} LL\ dt + L da(t) - L\ da\ (t)\right) U(t)$$

$$-\frac{1}{2} p^T M(g)^{-1} p + V(q)$$

Assume that its eigenfunctions and eigenvalues are $\phi_n(q), E_n, n = 1, 2, \ldots$ respectively, *i.e.*,

$$H_0(q, -i\partial_q)\phi_n(q) = E_n \phi_n(q), n \geq 1$$

$$\langle \phi_n, \phi_m \rangle = \int_0^{2\pi} \bar{\phi}_n(q) \phi_m(q) d^d q = \langle \phi_n, \phi_m \rangle = \delta_{n,m}$$

These can be determined approximately using time independent perturbation theory by noting that $M(q)$ has the form

$$M(q) = M_0 + \sum_{1 \leq k \leq r \leq d} M_{kr}^{(1)}(q) + \sum_{1 \leq k \leq r \leq d} M_{kr}^{2}(q)$$

where M_0 is a constraint $d \times d$ matrix, $M_{kr}^{(1)}(q)$ is a linear function of $\cos(q_k)\cos(q_r)$ and $M_{kr}^{(2)}(q)$ is a linear function of $\cos(q_k - q_r)$. We write the above as

$$M(q) = M_0 + \varepsilon M_1(q)$$

Then we have the expansion

$$M(q)^{-1} = M_0^{-1/2}(I + \varepsilon M_0^{-1/2} M_1(q) M_0^{-1/2}) M_0^{-1/2}$$

$$= M_0^{-1/2}\left(I + \sum_{k=1}^{\infty} \varepsilon^k (M_0^{-1/2} M_1(q) M_0^{-1/2})^k\right) M_0^{-1/2}$$

$$= M_0^{-1} + \left(\sum_{k=1}^{\infty} \varepsilon^k M_0^{-1/2} (M_0^{-1/2} M_1(q) M_0^{-1/2})^k M_0^{-1/2}\right)$$

and hence, the kinetic energy of the robot is given by

$$K = K_0 + \sum_{k=1}^{\infty} \varepsilon^k K_k$$

Remark: Using the formulae in section 1, we note that the kinetic energy of the kth link can be expressed as

❖❖❖❖❖

For Manisha khulba

[147] Wave propagation in a plane using Boltzmann eqn. maxwell's Morally equation.

$$\nabla \times E = \frac{\rho}{\varepsilon_0}, \rho\,(t, \underline{r}) = \int f_1(t, \underline{r}, \underline{v}) d^3V$$

$$\nabla \times E = -B,_t\, \underline{J}(t, \underline{r}) = \int q\underline{V}\, f_1(t, \underline{r}_1, \underline{v}) d^3V$$

$$\nabla \times B = \mu_0 \underline{J} + \frac{1}{c^2}\underline{E},_t$$

$$\nabla \times B = 0$$

$$f_0(\underline{V}) = \left(\frac{m}{2\pi kT}\right)^{3/2} \exp\left(\frac{-mv^2}{\alpha kT}\right)$$

$$\nabla(\nabla \cdot E - \nabla^2 E) = -\underline{\nabla}\times\underline{B},_t = -\mu\underline{J}, t - \frac{1}{c^2}\underline{E},_{tt}$$

or

$$\nabla^2\underline{E} - \frac{1}{c^2}\underline{E},_{tt} = \underline{\nabla}(\,\underline{\nabla}\cdot\underline{E}) + \mu_0\underline{J},_t = \frac{\nabla\rho}{\epsilon_0} + \mu_0\underline{J},_t$$

$$f_{1,t} + (\underline{v}, \underline{\nabla}_r) f_1 + \frac{q}{m}(\underline{E} + \underline{v}\times\underline{B}, \underline{\nabla}_v) f_0 = -\frac{f_1}{\tau}$$

$$= f_{1,t} + (\underline{v}, \underline{\nabla}_r) f_1 - \frac{qm}{KT}(\underline{v}\,\underline{E}) f_0(\underline{v}) = -\frac{f_1}{\tau}$$

Let

$$\int f_1(t, \underline{r}, \underline{v}) \exp(-i\,\underline{k}\cdot\underline{r}) d^3r = \hat{f}_1(t, \underline{k}, \underline{v})$$

Then

$$\hat{f}_{1,t} + i(\underline{k}, \underline{v})\hat{f}_1 = -\frac{qm}{kT}\left(\underline{V}\hat{\underline{E}}(t, \underline{h})\right) f_0(v) = -\frac{f_1}{\tau}$$

So, $$\hat{f}_1(t, \underline{k}, \underline{v}) = \left[\frac{qm}{kT}\int_0^t \exp\left\{\underline{i}(t-s)\left(i(\underline{k}, \underline{V}) + \frac{1}{\tau}\right)\left(\underline{V}, \hat{\underline{E}}(s, \underline{k}) ds\right)\right\} f_0(\underline{V})\right]$$

Then,

$$\hat{f}_1(t, \underline{k}, \underline{v}) = \frac{qm}{kT} f_0(\underline{V}) \int_0^t \exp\left(-\frac{(t-s)}{\tau}\right)\left(\underline{V}, \underline{E}(s, \underline{r} - (t-s)\underline{V})\right) ds$$

Assume $\tau = \tau(\underline{V})$. Then,

$$\hat{\rho}(t, \underline{k}) = \int \hat{f}_1(t, \underline{h}, \underline{v}) d^3V$$

$$\frac{qm}{kT}\int_0^t ds\left(\hat{\underline{E}}(s, \underline{k}), \int_{\mathbb{R}^3} \underline{V} \exp\{-i(\underline{k}, \underline{V})(t-s)\} \exp\left\{-\frac{t-s}{\tau(\underline{V})}\right\} f_0(\underline{V}) d^3V\right)$$

Equivalently,

$$\rho(t, r) = \frac{qm}{kT}\int_0^t ds \int_{\mathbb{R}^3} f_0(\underline{V})\cdot \exp\left(-\frac{(t-s)}{\tau(\underline{V})}\right)\cdot\left(\underline{V}, \underline{E}\left(s, \underline{r}-(t-s)\underline{V}\right)\right)d^3V$$

Problems in group theory.

❖❖❖❖❖

[148] Let G be a group and μ, v probability meaning on G. Then if

$$\rho(E) = \int_G \mu\left(E\, g^{-1}\right)dv(g), E \in \mathbb{B}(G)$$

ρ is also probability measure on G.

Let π be representation of G, then

$$\hat{\rho}(\pi) = \int_G \pi(g)d\rho(g)$$

$$= \int_{G\times G} \pi(g)d\mu\left(g\, h^{-1}\right)dv(k)$$

$$= \int_{G\times G} \pi(gh)d\mu(g)dv(h)$$

$$= \left(\int_G \pi(g)d\mu(g)\right)\left(\int_G \pi(h)dv(h)\right)$$

$$= \hat{\mu}(\pi)\hat{v}(\pi)$$

We write $r = \mu \times v$

and so

$$\widehat{\mu\times v}(\pi) = \hat{\mu}(\pi)\hat{v}(\pi)$$

i.e.

$$\widehat{\mu\times v} = \hat{\mu}\,\hat{v}$$

❖❖❖❖❖

[149] Let $g \in G$ have probability dissipation μ.

Let $F : G \to \mathbb{R}$. Let W: $G \to \mathbb{R}$ be a real valued random filled on G.

Define

$X(x) = f\left(g^{-1}x\right) + W(x), x \in G$

Assuming $W(.)$ to be a zero mean Gaussian random field with covariance

$K\omega\ (x, y) = E\ (W(x), W(y))$

Calculate the MAP of g (Maximum apostenory probability estimate) based on the date $\{X(x), x \in G\}$. Assume μ has density $\frac{d\mu}{dL}$ with respect to the left invariant

❖❖❖❖❖

[150] Haar measure on G.

Boltzmann equation in a strong constant magnetic field

$$\frac{\partial f(t,\underline{r},\underline{V})}{\partial t} + (\underline{V}, \underline{\nabla}_r) f(t,\underline{r},\underline{V}) + \frac{q}{m}(\underline{V} \times \underline{B}, \underline{\nabla}_V) f(t,\underline{r},\underline{V})$$

$$= \left(f_0(\underline{V}) - f(t,\underline{r},\underline{V})\right)/\tau$$

$$f(t, \underline{r}, \underline{V}) = f_0(\underline{V}) + f_1(t,\underline{r},\underline{V})$$

$$f_1(t, \underline{r}, \underline{V}) = f_1(\underline{V}) \exp\{i(\omega t - \underline{k} \cdot \underline{r})\}$$

$$i(w - (\underline{k}, \underline{V}))f_1(\underline{V}) + \Omega_0 \left(\underline{V} \times \hat{z}, \underline{\nabla}_V\right) f1(\underline{V}) = -f_1(\underline{V})/\tau$$

$$\underline{V} = V_z \hat{z} + V_\rho \hat{\rho} + V_\phi \hat{\phi}$$

$$V_\rho = V_X \cos\phi + V_Y \sin\phi$$

$$V_\phi = -V_X \sin\phi + V_Y \cos\phi$$

$$V_\rho^2 + V_\phi^2 = V_\perp^2 = V_X^2 + V_Y^2$$

$$\underline{V} \times \hat{z} = -V_X \hat{Y} + V_Y \hat{X}$$

$$\underline{V} \times \hat{z} = -V_\rho \hat{\phi} + V_\theta \hat{\rho}$$

$$\left(\underline{V} \times \hat{z}, \underline{\nabla}_V\right) f_1(\underline{\nabla}) = -V_X \frac{\partial f_1}{\partial V_Y} + V_Y \frac{\partial f_1}{\partial V_X}$$

$$V_\rho^2 + V_\phi^2 = V_X^2 + V_Y^2, = V_\perp^2$$

$$V_X = V_\perp \cos\psi$$

$$V_Y = V_\perp \sin\psi \, f_1(V_z, V_\perp, \psi)$$

$$\frac{\partial V_\perp}{\partial V_Y}\frac{\partial f_1}{\partial V_\perp} + \frac{\partial f_1}{\partial \psi}\frac{\partial \psi}{\partial V_Y} = \frac{\partial f_1}{\partial V_Y}$$

$$\frac{\partial V_\perp}{\partial V_X}\frac{\partial f_1}{\partial V_\perp} + \frac{\partial \psi}{\partial V_X}\frac{\partial f_1}{\partial \psi} = \frac{\partial f_1}{\partial V_X}$$

$$\frac{\partial V_\perp}{\partial V_X} = \frac{\partial_X}{\partial_\perp} = \cos\psi, \quad \frac{\partial V_\perp}{\partial V_Y} = \frac{V_Y}{V_\perp} = \sin\psi$$

$$\frac{\partial\psi}{\partial V_X} = \frac{\partial}{\partial V_X}\tan^{-1}\left(\frac{V_Y}{V_X}\right) = -\frac{V_Y}{V_\perp^2} = -\frac{\sin\psi}{V_\perp}$$

$$\frac{\partial\psi}{\partial V_Y} = \frac{\partial}{\partial V_Y}\tan^{-1}\left(\frac{V_Y}{V_X}\right) = \frac{V_X}{V_\perp^2} = \frac{\cos\psi}{V_\perp}$$

$$-V_X\frac{\partial f_1}{\partial V_Y} + V_Y\frac{\partial f_1}{\partial V_X} = -V_\perp\cos\psi\left\{\sin\psi\frac{\partial f_1}{\partial V_\perp} + \frac{\cos\psi}{V_\perp}\frac{\partial f_1}{\partial\psi}\right\}$$

$$+V_\perp\sin\psi\left\{\cos\psi\frac{\partial f_1}{\partial V_\perp} - \frac{\sin\psi}{V_\perp}\frac{\partial f_1}{\partial\psi}\right\} = -\frac{\partial f_1}{\partial\psi}$$

$$i\left(w - (\underline{k}, \underline{V})\right)f_1(\underline{V}) - \Omega_0\frac{\partial f_1}{\partial\psi} = -\frac{f_1}{\tau}$$

$$(\underline{k}, \underline{V}) = k_zV_z + \left(k_X\cos\psi + k_Y\sin\psi\right)V_\perp$$

$$\frac{\partial f_1}{\partial\psi}(V_z, V_\perp, \psi) + \frac{i}{\Omega_0}\left(k_zV_z + k_\perp\cos(\psi - \theta_k)V_\perp - w\right)f_1 = \frac{f_1}{\Omega_0\tau}$$

$$\frac{\partial f_1}{\partial\psi} = \left[\frac{1}{\Omega_0\tau} + \frac{i}{\Omega_0}\left(w - k_zV_z - k_\perp V_\perp\cos(\psi - \theta_k)\right)\right]f_1$$

❖❖❖❖❖

[151] $g(\underline{r}) = f\left(R^{-1}(\underline{r} - \underline{a})\right) + w(\underline{r})$

$R \in So(3)$, $\underline{a} \in \mathbb{R}^3$

So, (3), $\quad \mathbb{R}^3 = \left\{(R, a) \middle| R \in So(3), a \in \mathbb{R}^3\right\}$

$(R_2, G_2)\cdot(R_1, a_1) = (R_2R_1, R_2a_1 + a_2)$

Let dR = Haar measure on So (3)

Find an explio formal for dR in terms of the parameters α, β, γ

$$R = \exp\left(\alpha L_1 + \beta L_2 + \gamma L_3\right)$$

$$\left(w_1L_1 + w_2L_2 + w_3L_3\right) = \underline{r} = \underline{w}\times\underline{r}$$

$$= \left(w_2X_3 - w_3X_2, w_3X_1 - w_1X_3, w_1X_2 - w_2X_1\right)$$

$$= \begin{pmatrix} 0 & -w_3 & w_2 \\ w_3 & 0 & -w_1 \\ -w_2 & w_1 & 0 \end{pmatrix}$$

So

$$L_1 = \begin{pmatrix} 0 & 0 & 0 \\ 0 & 0 & -1 \\ 0 & 1 & 0 \end{pmatrix}, L_2 = \begin{pmatrix} 0 & 0 & 1 \\ 0 & 0 & 0 \\ -1 & 0 & 0 \end{pmatrix}, L_3 = \begin{pmatrix} 0 & -1 & 0 \\ 1 & 0 & 0 \\ 0 & 0 & 0 \end{pmatrix}$$

Left invariant vector field enacted to

$$w.\, L = w_1 L_1 + w_2 L_2 + w_3 L_3$$

$$X_g^{(w)}(f) = \frac{d}{du} f\left(g \cdot \exp\left(t\, w \cdot L\right)\right)\Big|_{t=0}, g \in So\,(3), f : So\,(3) \to \mathbb{C}$$

$$\frac{d}{dt} g \cdot \exp\left(t\, w \cdot L\right) \cdot \underline{r}\Big|_{t=0}$$

$$g(w.\, L).\, r = g(w \times r). \text{ Let } \underline{S} \in \mathbb{R}^3$$

$$\left\langle \underline{S}, g \cdot \left(\underline{w} \times \underline{r}\right)\right\rangle = \left\langle \underline{g}^T \underline{S}, \underline{w} \times \underline{r}\right\rangle = -\left\langle \underline{w} \times \underline{g}^T \underline{S}, \underline{r}\right\rangle$$

$$= -\left\langle \underline{g}^T \left(\underline{g}\, \underline{w} \times \underline{S}\right), \underline{r}\right\rangle$$

$$g(w \times r) = g.w \times g.r$$

$$= (g\, w).L.(gr) = (g\, w, L)g.r$$

$$\frac{d}{dt} g \cdot \exp\left(t\, w \cdot L\right)\Big|_{t=0} = (g\, w, L)g$$

$$= \sum_{\alpha=1}^{3} (g\, w)_\alpha L_\alpha\, g = \sum_{\alpha, \beta=1}^{3} g_{\alpha\beta} w_\beta \left(L_\alpha g\right)$$

$$= \sum_\beta w_\beta \sum_\alpha g_{\alpha\beta} L_{\alpha g}$$

Then,

$$\frac{d}{dt} g \cdot \exp\left(t\, w\, L\right)\Big|_{t=0} = \sum_{\alpha\beta} g_{\alpha\beta} w_\beta L_\alpha g$$

Equivalently,

$$g.\exp(t\, w.\, L) = \exp\left(t \sum_\alpha w_\alpha g L_\alpha g^{-1}\right) \cdot \underline{g}$$

So,

$$\frac{d}{dt} g \cdot \exp\left(t\, w\, L\right)\Big|_{t=0} = \sum_\alpha w_\alpha g L_\alpha g^{-1} \cdot g$$

$$= \sum_\alpha w_\alpha g \cdot L_\alpha$$

So,

$$\frac{d}{dt} f\left(g \cdot \exp\left(t\, w \cdot L\right)\right)\Big|_{t=0} = \sum_{1 \le \alpha \mu \nu \le 3} w_\alpha \left(g \cdot L_\alpha\right)_{\mu\nu} \frac{\partial f}{\partial g_{\mu\nu}}(g)$$

Taking w = (1, 0, 0), (0, 1, 0), (0, 0, 1), we get 3 left invariant vector fields on $So(3)$ namely

$$X_k(g) = \sum_{\mu\nu=1}^{3} \left(gL_k\right)_{\mu\nu} \frac{\partial}{\partial g_{\mu\nu}},\ k = 1, 2, 3$$

Now,

$$d \exp\,(w.\,L) = \exp(w \cdot L) \frac{\left(I - \exp\left(-ad\left(w \cdot L\right)\right)\right)}{ad\left(w \cdot L\right)} (dw \cdot L)$$

$$= \sum_k D_k\left(w\right) dw_k$$

Let

$$g = \exp\,(w.L) = g(w).$$

Then

$$df(g) = \sum_{\mu\nu} \frac{\partial f\left(q\right)}{\partial g_{\mu\nu}} dg_{\mu\nu}\left(w\right)$$

$$= \sum_{\mu\nu k} \frac{\partial f}{\partial g_{\mu\nu}} \frac{\partial g_{\mu\nu}}{\partial w_k} dw_k$$

$$\frac{\partial}{\partial w_k} f\left(g\left(w\right)\right) = \sum_{\mu,\nu} \frac{\partial g_{\mu\nu}}{\partial w_k} \frac{\partial f\left(g\right)}{\partial g_{\mu\nu}}$$

$$= \sum_{\mu,\,\nu} \left(D_k\left(w\right)\right)_{\mu\nu} \frac{\partial f}{\partial g_{\mu\nu}}(g)$$

$$= \frac{d}{dt} f\left(e^{w \cdot L} e^{t\alpha \cdot L}\right)\Big|_{t=0}$$

$$= \frac{d}{dt} f\left(e^{w \cdot L}\left(1 + t\,\alpha \cdot L\right)\right)\Big|_{t=0} e^{\left(\underline{w} + \delta \underline{w}\right) \cdot L}$$

$$= e^{w \cdot L} + e^{w \cdot L}\left(\frac{\underline{T} - e^{-ad\left(\underline{w} \cdot L\right)}}{ad\left(\underline{w} \cdot L\right)}\right)(\delta w \cdot L)$$

$$= e^{w \cdot L} e^{\delta\alpha \cdot L}$$

$$\delta\alpha.L = \sum_k \delta\alpha_k L_k$$

$$= \sum_m \delta w_m \sum_{r=0}^{\infty} \frac{(ad(\underline{w} \cdot \underline{L}))^r}{\lfloor r+1} (-1)^r (L_m)$$

$$ad(w \cdot L)(\delta_v . L_m) = [w \cdot L, \delta w \cdot L]$$

$$\sum_{r=0}^{\infty} \frac{(ad(w \cdot L))^r}{\lfloor r+1} (-1)^r (\alpha \cdot L) = \beta . L$$

Find α

$(-1)^r$ an$(ad(W. L))^r(\alpha. L)$

$$= ((W. L)^r. \alpha)_m \mathrm{L}_m$$

$$\sum_{r=0}^{\infty} \frac{((w \cdot L)^r \cdot \alpha)_m}{\lfloor r+1} = \beta_m$$

$ad(w . L)(\alpha . L)$

$$= [w . L, \alpha, L]$$

$$= w_k \alpha_j \in (kjm) L_m$$

$$= \alpha_j (w . L)_{jm} L_m$$

$$= -((w . L)_{mi}) L_m$$

$$= -((w. L)\underline{\alpha})_m L_m$$

$$\left(\frac{\exp\{(w \cdot L)\} - 1}{(w \cdot L)} \right) \cdot \underline{\alpha} = \beta$$

$$\underline{\alpha} = \frac{(w. L)}{\exp(w. L) - 1} (\underline{\beta})$$

$$dt \left[\frac{w \cdot L}{\exp\{(w \cdot L)\} - 1} \right]^{-1} d^3 w$$

Haar measure on $So(3)$,

$$X_1^{(g)} \Lambda \cdot\cdot \Lambda X_n(g)$$

$$\int f(g) w_1(g) \Lambda \cdot\cdot \Lambda w_n(g)$$

$$w_1(x^{-1} g) \Lambda \cdot\cdot \Lambda w_n(x^{-1} g)$$

$$(L_X w_1)(g) X \cdot\cdot X (L_X w_n)(g)(L^{-1} x)$$

Haar measure:

$$dt\left|\frac{d\underline{\alpha}}{d\underline{\beta}}\right|^{-1} dw = dt\left(\frac{e^{w\cdot L}-1}{w\cdot L}\right)d^3w$$

❖❖❖❖❖

[152] Approximate nonlinear filtering eqn. in discrete time

$$\underline{\xi}[n+1] = \underline{F}_n\left(\underline{\xi}[n]\right)+\underline{W}[n+1] \text{ State model}$$

$$z[n] = \underline{H}_n\left(\underline{\xi}[n]\right)+\underline{V}[n] \text{ Measurement model}$$

$$\{\underline{W}[n]\} = iid\text{-}p(W),\ \{\underline{V}[n] iid\{p_v(\underline{V})\}\}$$

$$\{\underline{W}[n]\}\ \{\underline{V}[n]\} \text{ are mutually independent}$$

$$Z_n = \{z[k]: k\le n\}$$

$$\pi_n\left(\bar{\Phi},(X)\right) = \mathbb{E}\left\{\bar{\phi}\left(\underline{\xi}(n)\right)\middle|Z_n\right\}$$

$$p\left(\xi(n+1)\middle|Z_{n+1}\right) = p\left(\xi(n+1),Z_n,z(n+1)/p\left(Z_n,z(n+1)\right)\right)$$

$$\frac{\int p\left(z[n+1]\middle\|\xi[n+1]\right)p\left(\xi[n+1]\middle\|\xi(n)\right)\ p\left(\xi[n]\middle|Z_n\right)d\xi[n]}{\left(\int \text{numerator}\, d\xi[n+1]\right)}$$

$$\therefore\ \pi_{n+1}\left(\bar{\Phi}(X)\right) = \frac{\sigma_{n+1}\left(\bar{\underline{\Phi}}(X)\right)}{\sigma_{n+1}(1)}$$

where

$$\sigma_{n+1}\left(\bar{\underline{\Phi}}(X)\right) = \int\bar{\underline{\Phi}}(\xi[n+1])\,p\left(z[n+1]\middle\|\xi[n+1]\right)p\left(\xi[n+1]\middle\|\xi[n]\right)$$

$$p\left(\xi[n]\middle\|Z_n\right)d\xi[n]d\xi[n+1]$$

$$= \int\bar{\underline{\Phi}}(\xi[n+1])\,p_v\left(z[n+1]-H_{n+1}\xi[n+1]\right)$$

$$p_w\left(\underline{\xi}[n+1]-\underline{F}_n\left(\xi[n]\right)\right)\ p\left(\xi[n]\middle\|Z_n\right)d\xi[n]d\xi[n+1]$$

$$\int\bar{\underline{\Phi}}\left(F_n\left(\xi[n]\right)+\underline{W}[n+1]\right)$$

$$p_V\left(z[n+1]-H_{n+1}\left(F_n\left(\xi[n]\right)+W[n+1]\right)\right)$$

$$p_w\left(W[n+1]\right)p\left(\xi[n]\middle\|Z_n\right)dW[n+1]d\xi[n]$$

$$= \pi_n\left\{\int\bar{\underline{\Phi}}\left(F_n(X)+W\right)p_V\left(z[n+1]-H_{n+1}\left(F_n(X)+W\right)\right)\right.$$

$$\left.p_w(W)dW\right\}$$

$$\sigma_{n+1}\{1\} = \pi_n\left\{\int p_V\left(z[n+1]-H_{n+1}\left(F_n(X)+W\right)\right)p_W(W)dW\right\}$$

So the nonlinear filtering eqn. become

$$\pi_{n+1}\left(\underline{\Phi}(X)\right) = \frac{\pi_n\left\{\int \underline{\Phi}\left(F_n(X)+W\right)\times p_V\begin{pmatrix} z[n+1]- \\ H_{n+1}\left(F_n(X)+W\right)\end{pmatrix} p_W(W)dW\right\}}{\pi_n\left\{\int p_V\left(z[n+1]-H_{n+1}\left(F_n(X)+W\right)\right)p_W(W)dW\right\}}$$

More generally suppose $\xi(n)$ is Markov with generator K_n *i.e.*

$$\mathbb{E}\left[\varphi(\xi[n+1])\middle|\xi(n)=X\right] = (K_n\,\varphi)(X)$$

Then, $$\pi_{n+1}\{\underline{\Phi}(X)\} = \frac{\pi_n\left\{K_n\left(\underline{\Phi}(X)\,p_V\left(z[n+1]-H_{n+1}(X)\right)\right)\right\}}{\pi_n\left\{K_n\left(p_V\left(z[n+1]-H_{n+1}(X)\right)\right)\right\}}$$

These an excat expansion without any approximation is:

$$\phi\left(F_n(X)+W\right)p_V\left(z[n+1]-H_{n+1}\left(F_n(X)+W\right)\right)$$

$$= \sum_{r=0}^{\infty}\frac{\left(\underline{W}^{\otimes r}\right)^T}{\underline{|r}} D_\xi^r\left(\underline{\Phi}(\xi)\,p_V\left(z[n+1]-H_{n+1}(\xi)\right)\right)\Big|_{\xi=F_n(X)}$$

So putting

$$\int \underline{W}^{\otimes r} p_W(\underline{W})d\underline{W} = \underline{M}_r^{(W)}$$

We get

$$\pi_{n+1}\left(\underline{\Phi}(X)\right) = \frac{\sum_{r=0}^{\infty}\frac{1}{\underline{|r}}\underline{Mr}_{\pi_n}^{(W)^T}\left\{D_\xi^r\left(\underline{\Phi}(\xi)\,p_V\begin{pmatrix} z[n+1] \\ -H_{n+1}(\xi)\end{pmatrix}\right)\Big|_{\xi=F_n(X)}\right\}}{\sum_{r=0}^{\infty}\frac{1}{\underline{|r}}\underline{Mr}_{\pi_n}^{(W)^T}\left\{D_\xi^r p_V\left(z[n+1]-H_{n+1}(\xi)\right)\Big|_{\xi=F_n(X)}\right\}}$$

For Naman

$$y[n] = \underline{\psi}^T\left(\underline{y}_n, n\right)\underline{a} + W(n)$$

$$\underline{y}_n = \underline{y}_n(n-1:n-p), \underline{a}\in\mathbb{R}^p$$

$$\underline{\hat{a}}[n+1] = \underline{\hat{a}}[n] + z\mu\left(y[n]-\underline{\psi}^T\left(\underline{y}_n, n\right)\underline{\hat{a}}[n]\right)\underline{\psi}\left(\underline{y}_n, n\right)$$

$$= \underline{\hat{a}}[n] + 2\mu\left(\underline{\psi}^T\left(\underline{y}_n, n\right)\left(\underline{a}-\underline{\hat{a}}[n]\right) + W[n]\right)\underline{\psi}\left(\underline{y}_n, n\right)$$

$$\underline{\mu a}[n+1] \approx \underline{\mu a}[n] - 2\mu\mathbb{E}\left\{\underline{\psi}\left(\underline{y}_n, n\right)\underline{\psi}^T\left(\underline{y}_n, n\right)\right\}\left(\underline{\mu a}[n]-\underline{a}\right)$$

Let

$$\underline{\theta}[n] = \underline{\mu a}[n] - a$$

Then,

$$\underline{\theta}[n+1] = \underline{\theta}[n] - 2\mu\mathbb{E}\left(\underline{\psi}_n\underline{\psi}_n^T\right)\underline{\theta}[n]$$

So $\underline{\theta}[n] \to 0$ if for all sufficiently larger n, the largest eigenvalue of $\mathbb{E}\left(\underline{\psi}_n\underline{\psi}_n^T\right)$ $\lambda_n(max)$ is such that $\lambda_n(max) < \dfrac{1}{2\mu}$

Covariance evolution

Let

$$\hat{\underline{a}}[n] - \underline{\mu}[n] = \delta\underline{a}[n]. \text{ Then, } \delta\underline{a}[n+1]\,\delta\underline{a}[n] + 2\mu\left(\underline{\psi}_n\underline{\psi}_n^T - \mathbb{E}\left(\underline{\psi}_n\underline{\psi}_n\right)^T\right)\underline{a}$$

$$-2\mu\,\underline{\psi}_n\underline{\psi}_n^T\,(\underline{\mu}a[n] + \delta\underline{a}[n]) - \mathbb{E}\left(\underline{\psi}_n\underline{\psi}_n^T\right)\underline{\mu}a(n) + 2\mu\underline{\psi}_n W[n]$$

Note $\quad \underline{\psi}_n = \underline{\psi}\left(\underline{y}_n, n\right).$

Then if $\rho_a[n] = \mathrm{cov}\left(\hat{\underline{a}}[n]\right) = \mathrm{cov}\left(\hat{\underline{a}}[n] - \underline{\mu}a[n]\right)$

then $\quad \delta\underline{a}[n+1] = \left(I - 2\mu\underline{\psi}_n\underline{\psi}_n^T\right)\delta\underline{a}[n]$

$$+2\mu\underline{\underline{X}}_n\underline{a} - 2\mu\underline{\underline{X}}_n\underline{\mu}a[n] - 2\mu\underline{\psi}_n\underline{\psi}_n^T\,\delta\,\underline{a}[n] + 2\mu\underline{\psi}_n W[n]$$

When $\quad \underline{\underline{X}}_n = \underline{\psi}_n\underline{\psi}_n^T - \mathbb{E}\left(\underline{\psi}_n\underline{\psi}_n^T\right)$

We are assuming $\hat{\underline{a}}[n]$ (or equivalently $\delta\underline{a}[n]$) to be independent of $\underline{y}_n = \underline{y}[n-1:n-p]$

Thus,

$$\delta\,\underline{a}[n+1]\left(\underline{I} - 2\mu\underline{\psi}_n\underline{\psi}_n^T\right)\delta\underline{a}[n]$$

$$+\,2\mu\underline{\underline{X}}_n\left(\underline{a} - \underline{\mu}a[n]\right) - 2\mu\underline{\psi}_n\underline{\psi}_n^T\delta\underline{a}[n] + 2\mu\underline{\psi}W[n]$$

$$\mathrm{Cov}(\underline{a}[n+1]) = \mathrm{Cov}(\delta\underline{a}[n+1]) = \underline{\underline{P}}_a[n+1]$$

$$= \mathbb{E}\left\{\left(I - 2\mu\underline{\psi}_n\underline{\psi}_n^T\right)P_a[n]\left(I - 2\mu\underline{\psi}_n\underline{\psi}_n^T\right)\right\}$$

$$+\,4\mu^2\mathbb{E}\left\{\underline{\underline{X}}_n\left(\underline{a} - \underline{\mu}a[n]\right)\left(\underline{a} - \underline{\mu}a[n]\right)^T\underline{\underline{X}}_n\right\}$$

$$+\,4\mu^2\mathbb{E}\left(\underline{\psi}_n\,\underline{\psi}_m^T\,P_a[n]\underline{\psi}_n\,\underline{\psi}_n^T\right) + 4\mu^2\mathbb{E}\left(\underline{\psi}_n\underline{\psi}_n^T\right)\sigma_{w^2}$$

$$-\,2\mu\left\{\mathbb{E}\left(\left(I - 2\mu\underline{\psi}_n\,\underline{\psi}_n^T\right)P_a[n]\underline{\psi}_n\,\underline{\psi}_n^T\right)\right.$$

$$\left.+\,\mathbb{E}\left(\underline{\psi}_n\underline{\psi}_n^T p_a[n]\left(I - 2\mu\underline{\psi}_n\underline{\psi}_n^T\right)\right)\right.$$

Approximate calculation of $\mathbb{E}\left\{\underline{\psi}_n\underline{\psi}_n^T\underline{\underline{X}}\underline{\psi}_n\underline{\psi}_n^T\right\}$

$$\mathbb{E}\left\{\underline{\psi}_n\underline{\psi}_n^T\underline{\underline{X}}\underline{\psi}_n\underline{\psi}_n^T\right\} = \mathbb{E}\left\{\underline{\psi}_n\underline{\psi}_n^T\left(\psi_n^T X\psi_n\right)\right\}$$

$$\mathbb{E}\left\{\underline{\psi}_n\underline{\psi}_n^T\underline{\underline{X}}\underline{\psi}_n\underline{\psi}_n^T\right\} = \sum_{\alpha\beta\rho}\mathbb{E}\left\{\left(\underline{\psi}_n\underline{\psi}_n^T\right)_{\alpha\beta}\left(\underline{\psi}_n\underline{\psi}_n^T\right)_{\rho\alpha}\right\}X_{\beta\rho}$$

$$= \mathbb{E}\left\{\left(\underline{\psi}_n\underline{\psi}_n^T\right)_{\alpha\beta}\left(\underline{\psi}_n\underline{\psi}_n^T\right)_{\rho\alpha}\right\}$$

$$= \mathbb{E}\left\{\underline{\psi}_{n\alpha}\underline{\psi}_{n\beta}\underline{\psi}_{n\rho}\underline{\psi}_{n\alpha}\right\}$$

$$\psi_{n\alpha} = \underline{\psi}_\alpha\left(\underline{y}[n-1:n-q],n\right)$$

$$\approx \psi_\alpha\left(\underline{\mu}[n-1:n-q],n\right)+\psi_{\alpha,\beta}\left(\underline{\mu}[n-1:n-q],n\right)\delta y_\beta[n]$$

$$+\frac{1}{2}\psi_{\alpha,\beta,\rho}\left(\underline{\mu}[n-1:n-q],n\right)\delta y_\beta[n]\delta y_\rho[n]$$

where

$$\delta y_\beta[n] = y[n-\beta]-\mu[n-\beta]+\frac{1}{2}\psi_{\alpha,\beta\rho\sigma}(n)\delta y_\beta[n]\delta y_\rho[n]\delta y_\sigma[n]$$

$$+\frac{1}{24}\psi_{\alpha\beta\rho\sigma\mu}(n)\delta y_\beta(n)\delta y_\rho(n)\delta y_\sigma(n)\delta y_\mu(n)$$

$$\mathbb{E}\left[\psi_{n\alpha}\psi_{n\beta}\psi_{n\rho}\psi_{ny}\right]$$

$$\approx \psi_{n,\beta_1}(n)\psi_{\beta,\beta_2}(n)\psi_{\rho,\beta_3}(n)\psi_{\gamma,\beta_4}(n)$$

$$\mathbb{E}\left\{\delta y[n-\beta_1]\delta y[n-\beta_2]\delta y[n-\beta_3]\delta y[n-\beta_4]\right\}$$

$$+\frac{1}{24}\psi_{\alpha,\beta_1\beta_2\beta_3\beta_4}(n)\psi_\beta(n)\psi_\rho(n)\psi_\gamma(n)$$

$$\mathbb{E}\left\{\delta y[n-\beta_1]\delta y[n-\beta_2]\delta y[n-\beta_3]\delta y[n-\beta_4]\right\}$$

$$+\text{ permutation }+\frac{1}{6}\psi_{\alpha,\beta_1\beta_2\beta_3}(n)\psi_{\beta,\beta_4}(n)\psi_\rho(n)\psi_\gamma(n)$$

$$+\mathbb{E}\left\{\delta y[n-\beta_1]\delta y[n-\beta_2]\delta y[n-\beta_3]\delta y[n-\beta_4]\right\}$$

$$\begin{cases}1111\\4400\\3100\ \frac{1}{4}\psi_{\alpha,\beta_1,\beta_2}(n)\psi_{\beta,\beta_1,\beta_2}(n)\psi_\rho(n)\psi_\gamma(n)\\220\\211\end{cases}$$

$$\mathbb{E}\{\delta y[n-\beta_1]\delta y[n-\beta_2]\delta y[n-\beta_3]\delta y[n-\beta_4]\} + \text{permutation}$$

$$+\frac{1}{2}\psi_{\alpha,\beta_1,\beta_2}(n)\psi_{\beta,\beta_3}(n)\psi_{\rho,\beta_4}(n)\psi_{\gamma}(n)$$

$$\mathbb{E}\{\delta y[n-\beta_1]\delta y[n-\beta_2]\delta y[n-\beta_3]\delta y[n-\beta_4]\} + \text{permutation}$$

$$+\frac{1}{2}\psi_{\alpha,\beta_1\beta_2}(n)\psi_{\beta}(n)\psi_{\rho}(n)\psi_{\gamma}(n)$$

$$\mathbb{E}\{\delta y[n-\beta_1]\cdot\delta y[n-\beta_2]\} + \text{permutation}$$

$$+\frac{1}{2}\psi_{\alpha,\beta_1}(n)\psi_{\beta,\beta_2}(n)\psi_{\rho}(n)\psi_{\gamma}(n)$$

$$\mathbb{E}\{\delta y[n-\beta_1]\delta y[n-\beta_2]\} + \text{permutation}$$

$$+\frac{1}{6}\psi_{\alpha\beta_1\beta_2\beta_3}(n)$$

For Rohit Sinha

❖❖❖❖❖

[153] Estimating the time varying delay. Teleoperation with time varying delay:

$$\underline{\xi}'(t) = \underline{F}(\underline{t},\underline{\xi}(t)) + \underline{\underline{G}}(\underline{\xi}(t))\underline{\xi}(t-T(t))$$

Aim: To estimate $T(t)$ ≃ time varying delay. assume that $T(t)$ is constant are $[n\Delta, (n+1)\Delta]$

i.e. T is price with constant. Then

$$\underline{\xi}'(t) = \underline{F}(\underline{t},\underline{\xi}(t)) + \underline{\underline{G}}(\underline{\xi}(t))\underline{\xi}(t-T_n),\ n\Delta \le t < (n+1)\Delta, n = 0, 1. 2,...$$

$$\approx \underline{F}(t,\underline{\xi}(t)) + \underline{\underline{G}}(\underline{\xi}(t))\sum_{k=0}^{N}\underline{\xi}^{(k)}(t)T_n^k(-1)^k t \in [n\Delta, (n+1)\Delta]$$

$$\sum_{t\in[n\Delta,(n+1)\Delta]}\left\|\underline{\xi}'(t) - \underline{F}(t,\underline{\xi}(t)) - \sum_{k=0}^{N}(-1)^k\underline{\underline{G}}(\underline{\xi}(t))\underline{\xi}^{(h)}(t)T_n^h\right\|^2$$

$$= \xi(T_n)$$

Minimize $\xi(T_n)$ using the gradient to descent scheme. Exact optimizing :

$$\sum_{t\in[n\Delta,(n+1)\Delta]}\left\langle\underline{\xi}'(t) - \underline{F}(t,\underline{\xi}(t))\right. \quad \sum_{t\in[n\Delta,(n+1)\Delta]}\left\langle\underline{\xi}'(t) - \underline{F}(t,\underline{\xi}(t))\right.$$

$$\left.\sum_{k=0}^{N} k\,\underline{\underline{G}}(\underline{\xi}(t))\underline{\xi}^{(k)}(-1)^k(t)T_n^{k-1}\right\rangle = 0$$

$$\sum_t \left\{ \left\langle \underline{\xi}'(t) - \underline{F}(t), \underline{\xi}(t), \sum_{k=0}^{N} k\, \underline{\underline{G}}\left(\underline{\xi}(t)\right)(-1)^k \underline{\xi}^{(k)}(t) \right\rangle T_n^{k-1} \right\}$$

$$= \sum_{t,k,m} \left\langle \underline{\underline{G}}\left(\underline{\xi}(t)\right)\underline{\xi}^{(k)}(t), (-1)^k\, m \underline{\underline{G}}\left(\underline{\xi}(t)\right)(-1)^m \underline{\xi}^{(m)}(t) \right\rangle T_n^{k+m-1}$$

or

$$= \sum_{k=1}^{N} T_n^{k-1} \sum_{t \in [n\Delta, (n+1\Delta)]} \left\langle \underline{\xi}'(t) - \underline{F}\left(t, \underline{\xi}(t)\right), \right.$$

$$\left. \underline{\underline{G}}\left(\underline{\xi}(t)\right)\underline{\xi}^{(k)}(t) \right\rangle \left((-1)^k k\right)$$

$$= \sum_{k,m} T_n^{k+m-1} \left\{ \sum_{t \in [n\Delta, (n+1)\Delta]} \left\langle \underline{\underline{G}}\left(\underline{\xi}(t)\right)\underline{\xi}^{(k)}(t), \right. \right.$$

$$\left. \left. \underline{\underline{G}}\left(\underline{\xi}(t)\right)\underline{\xi}^{(m)}(t) \right\rangle (-1)^{k+m} m \right\}$$

Polynomial equation for T_n of degree $2N - 1$.

Fourier analysis based time delay estimate:

$$\underline{\xi}(t - T(t)) = \psi(\underline{\xi}(t)\, \underline{\xi}'(t))$$

$$\underline{\underline{\Delta}}\, \underline{\underline{G}}\left(\underline{\xi}(t)\right)^{-1}\left(\underline{\xi}'(t) - \underline{F}\left(t, \underline{\xi}(t)\right)\right) \equiv \eta(t)$$

STFT based analysis

$$h(t) = \text{window}$$

$$T(t) \simeq \text{Constant over } t \in [n\, T_0, (n+1), T_0]$$

$$h(t) = 0 \text{ for } t < 0 \text{ and } T > T_0$$

$$\underline{\xi}(t - T(t))h(t - n\, T_0)$$

$$= \underline{\xi}(t - T[n])\, h(t - n\, T_0)$$

Assume $T(t) = T[n]$ for $t \in [n\mathrm{T}_0, n + 1]\, T_0)$

$$\int_{\mathbb{R}} \underline{\xi}\left(t - T(t)\right)h\left(t - n\, T_0\right)e^{-j\omega t}dt$$

$$= \int_{\mathbb{R}} \underline{\xi}\left(t - T[n]\right)h\left(t - n\, T_0\right)e^{-j\omega t}dt$$

$$= \frac{1}{2\pi} \int_{\mathbb{R}\times\mathbb{R}} \underline{\xi}\left(t - T[n]\right)H\left(\omega'\right)e^{j(\omega' - \omega)t}e^{-j\omega T_0 n}d\omega' dt$$

$$= \frac{1}{2\pi} \int_{\mathbb{R}} \hat{\underline{\xi}}\left(\omega - \omega'\right)e^{j}\left(\omega' - \omega\right)T[n]H\left(\omega'\right)e^{-j\omega T_0 n}d\omega'$$

$$= \left\{ \frac{1}{2\pi} \int_{\mathbb{R}} \hat{\underline{\xi}}(\omega') e^{j\omega' T[n]} H(\omega' - \omega) d\omega' \right\} e^{-j\omega T_0 n}$$

Assume $\underline{\xi}$ has dominant frequency components at frequencies ω_k, $k = 1, 2, p$. Then assume $\omega_{k+1} - \omega_k = \Delta$,

$$\int \underline{\xi}(t - T(t)) h(t - nT_0) e^{-j\omega t} dt$$

$$\approx \frac{1}{2\pi} e^{-jwT_0 n} \Delta \cdot \sum_k \hat{\underline{\xi}}(\omega_k) H(\omega - \omega_k) e^{-j\omega k T[n]}$$

$$\equiv \sum_{k=1}^{p} F[\omega, k, n] e^{-j\omega_k T[n]}$$

$$= \int \eta(t) h(t - nT_0) e^{-j\omega t} dt$$

$$\equiv (S_h \eta)(n, \omega)$$

Consisting this equation for different ω, we get a system of linear algebraic equation for the vector

$$\left[e^{-j\omega_1, T[n]}, e^{-j\omega_2 T[n]}, \ldots, e^{-j\omega_p T[n]} \right]$$

which can be solved and $T[n]$ determined there form.

Delay differential eqn. with noise

$$\frac{d\underline{\xi}(t)}{dt} = \underline{F}_1(t, \underline{\xi}(t)) + \underline{\underline{F}}_{21}(t, \underline{\xi}(t)) \underline{\xi}(t - T_1)$$

$$+ \underline{\underline{F}}_{22}(t, \underline{\xi}(t)) \underline{\xi}(t) \underline{\xi}(t - T_2) + \underline{\underline{F}}_3(t, \underline{\xi}(t)) \underline{W}(t)$$

$$T_1 = M.\Delta,\ T_2 = N.\Delta,\ M \le N.$$

Approximate

$$\underline{\xi}[n+1] - \underline{\xi}[n] = \Delta.\ \underline{F}_1[n, \underline{\xi}[n]]$$

$$+ \Delta \cdot \underline{\underline{F}}_{21}[n, \underline{\xi}[n]] \cdot \underline{\xi}[n - M]$$

$$+ \Delta \cdot \underline{\underline{F}}_{22}[n, \underline{\xi}[n]] \cdot \underline{\xi}[n - N]$$

$$+ \underline{\underline{F}}_3[n, \underline{\xi}[n]] \sqrt{\Delta}\ \underline{W}[n+1]$$

State vactor:

$$\begin{bmatrix} \underline{\xi}[n] \\ \underline{\xi}[n-1] \\ \vdots \\ \underline{\xi}[n-N] \end{bmatrix} \equiv \underline{\eta}[n]$$

$$\underline{\eta}[n+1] = \underline{\underline{\psi}}\left(\underline{\eta}[n], n\right) + \underline{\underline{\Phi}}\left(\underline{\eta}[n], n\right)\underline{W}[n+1]$$

Estimating the state from noisy measurements

$$\underline{Z}[n] = \underline{h}\left(\underline{\eta}[n], n\right) + \underline{V}[n]$$

$$dZ_t = h(x_t)dt + dv_{t,}\ Y_t = \{Z_s : s \leq t\}$$

$$p(X_{t+dt} \mid Y_{t+dt}) = \frac{p\left(x_{t+dt}, Y_t, dz_t\right)}{p\left(Y_t, dz_t\right)}$$

$$= \int p\left(X_{t+dt} \middle| X_t\right) p\left(X_t \middle| Y_t, dz_t\right) dX_t$$

$$= \frac{\int p\left(X_{t+dt} \middle| X_t\right) p\left(Xz_t \middle| X_t\right) p\left(X_t \middle| Y_t\right)}{p\left(dz_t \middle| Y_t\right)} dX_t$$

$$\hat{\varphi}_{t+dt} = \pi_{t+dt}\varphi$$

$$= \frac{\int p_{dV_t}\left(dz_t - h\left(X_t\right)dt\right) p\left(X_{t+dt} \middle| X_t\right) p\left(X_t \middle| Y_t\right) dX_t dX_{t+dt} \varphi\left(X_{t+dt}\right)}{\int \text{numerator } \varphi = 1}$$

$$= \frac{\sigma_{t+dt}(\varphi)}{\sigma_{t+dt}(1)}.$$

$$\sigma_{t+dt}(\varphi) = \int\left(\varphi\left(X_t\right) + \left(\mathcal{L}_t\varphi\right)\left(X_t\right)dt\right)\left(1 + dt\hat{\psi}(\xi)\right)$$

$$i\left\langle \xi, dz_t - h\left(X_t\right)dt\right\rangle p\left(X_t \middle| Y_t\right) dX_t d\xi$$

$$= \int \varphi(X) p\left(t, X \middle| Y_t\right) \delta\left(dz_t - h(x)dt\right) dX$$

$$+ dt \int \mathcal{L}_t\varphi(X) \psi\left(dz_t - h(X)dt\right) p\left(t, X \middle| Y_t\right) dX$$

$$\int \delta\left(dz_t - h\left(X_t\right)dt\right) + dt\, \psi\left(dz_t - h\left(X_t\right)dt\right)$$

$$p\left(X_t \middle| z_t\right)\left[\varphi\left(X_t\right) + dt \cdot \mathcal{L}_t\varphi\left(X_t\right)\right] dX_t$$

$$\hat{\psi}(\xi) = \int \varphi\left(X_t\right) \delta\left(dz_t - h\left(X_t\right)dt\right) p\left(X_t \middle| z_t\right) dX_t$$

$$+ dt\left\{\varphi\left(X_t\right) \psi\left(dz_t - h\left(X_t\right)dt\right) p\left(X_t \middle| z_t\right) dX_t\right.$$

$$\frac{\left. + \int\left(\mathcal{L}_t\varphi\left(X_t\right)\right) p\left(X_t \middle| z_t\right) \delta\left(dzt - h\left(X_t\right)\right) dX_t\right\}}{(\text{numerator with } \phi = 1)}$$

$$= \hat{\varphi}_{t+dt} \middle| t + dt$$

$$\int (\pi_{t+dt}\varphi)(z_{t+dt}) X(dz_t) ddz_t$$

$$\int \varphi(X_t) X(h(X_t)dt) p(X_t|z_t) dx_t + dt$$

$$\mathbb{E}\, e^{i\xi^T v_t} = e^{\psi(\xi)}$$

$$C.\int e^{t\psi(\xi)} e^{-i\xi^T x} d\xi = e^{t\psi(\xi)}$$

$$p_{v_t}(X) = C.\int e^{t\psi(\xi)} e^{-i\xi^T x} d\xi$$

$$= C\, e^{t\psi(iD_x)} \int e^{-i\xi^T x} d\xi$$

$$= e^{i\psi(iD_x)} \delta(x)$$

$$p(X_{t+dt}|Y_{t+dt})$$

$$\pi_{t+dt}(\varphi) = \frac{\int p_{dv_t}(dz_t - h(x_t)dt)\varphi(X_{t+dt}) p(X_{t+dt}|X_t) p(X_t|Y_t) dX_t dX_{t+dt}}{(11\,\varphi = 1)}$$

$$\int (\varphi(X_t) + dt\mathcal{L}_t\varphi(X_t))(\varphi(dz - h(X_t)dt)$$

$$dt(\psi(D)\delta(dz - h(X_t)dt)) p(X_t|Y_t) dX_t$$

$$\pi_t\{\varphi(X)\delta(dz_t - h(X)dt)\}$$

$$+ dt\{\delta(dZ_t - h(X)dt)\mathcal{L}_t\varphi(X) + \varphi(X)(\psi(D)\delta)(dZ_t - h(X)dt)\}$$

$$\frac{\delta''(\xi)}{\delta(\xi)} = \frac{\int e^{i\xi t} t^2 dt}{\int e^{i\xi t} dt}$$

$$\frac{\delta''(dz)}{\delta(dz)} = \frac{\int x^2 e^{ixdz} dx}{\int e^{ixdz} dx} = \frac{\int x^2 dx \left(1 + ixdz - x^2 \frac{dt}{2}\right)}{\int \left(1 + ixdz - x^2 \frac{dt}{2}\right) dx}$$

$$\delta'(dz) = \int x e^{ixdz} dx = \int x \left(1 + i\,x\,dz - x^2 \frac{dt}{2}\right) dx$$

$$\pi_{t+dt}(\varphi) = i\frac{N^3 dz}{3}$$

$$\frac{\pi_t(\varphi) - dt\left[\frac{\delta'(dz)}{\delta(dz)}\pi_t(\varphi h) + \frac{\psi(dz)}{\delta(dz)}\pi_t(\mathcal{L}_t\varphi)\right]}{1 - dt\left[\frac{\delta'(dz)}{\delta(dz)}\pi_t(h)\right]}$$

$$= \pi_t(\varphi) + dt\left\{\frac{\delta'(dz)}{\delta(dz)}\left(\pi_t(\varphi)\pi_t(h) - \pi_t(\varphi h)\right) - \frac{\psi(dz)}{\delta(dz)}\pi_t(\mathcal{L}_t\varphi)\right\}$$

$$d\pi_t(\varphi) = -\frac{\psi(dz)}{\delta(dz)}\pi_t(\mathcal{L}_t\varphi)dt$$

Approximate condition mean and covariance propagation

$Y_n = \{z_k : k \le n\}$ in non-linier filtering theory in discrete time.

$$Y_{n+1} = \{Y_n, z_{n+1}\}$$

$$\underline{z}_n = \underline{h}(X_n) + \underline{V}_n$$

$$p\left(X_{n+1} \middle| Y_{n+1}\right) = \frac{p\left(X_{n+1}, Y_n, z_{n+1}\right)}{p\left(Y_n, z_{n+1}\right)}$$

$$= \frac{p\left(z_{n+1} \middle| X_{n+1}\right)p\left(X_{n+1} \middle| Y_n\right)}{\int p\left(z_{n+1} \middle| X_{n+1}\right)p\left(X_{n+1} \middle| Y_n\right)dX_{n+1}}$$

$$= \frac{p_v\left(z_{n+1} - h\left(X_{n+1}\right)\right)p\left(X_{n+1} \middle| Y_n\right)}{\int (11)\, dX_{n+1}}$$

$$\widehat{X}_{n+1|n+1} = \frac{\int X_{n+1}p_v\left(z_{n+1} - h\left(X_{n+1}\right)\right)p\left(X_{n+1} \middle| Y_n\right)dX_{n+1}}{\int p_v\left(z_{n+1} - h\left(X_{n+1}\right)\right)p\left(X_{n+1} \middle| Y_n\right)dX_{n+1}}$$

$$\text{Numerator} \approx \int X_{n+1}p_v\left(z_{n+1} - h\left(\widehat{X}_{n+1|n}\right) - h'\left(\widehat{X}_{n+1|n}\right)e_{n+t|n}\right)$$

$$p\left(X_{n+1} \middle| Y_n\right)dX_{n+1}$$

$$\approx \int \underline{X}_{n+1}\left\{p_V\left(z_{n+1} - h\left(\widehat{X}_{n+1|n}\right)\right)\right.$$

$$\left. + p_V'\left(z_{n+1} - h\left(\widehat{X}_{n+1|n}\right)\right)h'\left(\widehat{X}_{n+1|n}\right)e_{n+1|n}\right\}$$

$$p\left(X_{n+1} \middle| Y_n\right)dX_{n+1}$$

$$= p_V\left(z_{n+1} - h\left(\widehat{X}_{n+1|n}\right)\right)\widehat{X}_{n+1|n}$$

$$+ p_V'\left(z_{n+1} - h\left(\widehat{X}_{n+1|n}\right)\right)\int \underline{e}_{n+1|n}X_{n+1}p\left(X_{n+1} \middle| Y_n\right)dX_{n+1}$$

$$= p\left(z_{n+1} - h\left(\widehat{X}_{n+1|n}\right)\right)\widehat{X}_{n+1|n}$$

$$+\int X_{n+1} e_{n+1|n} T h'\left(\widehat{X}_{n+1|n}\right)^T p_{V}{}'\left(z_{n+1} - h\left(\widehat{X}_{n+1|n}\right)\right)$$

$$p\left(X_{n+1} \middle| Y_n\right) dX_{n+1}$$

$$= p\left(z_{n+1} - h\left(\widehat{X}_{n+1|n}\right)\right)\widehat{X}_{n+1|n}$$

$$+P_{n+1|n} h'\left(\widehat{X}_{n+1|n}\right)^T p_V'\left(z_{n+1} - h\left(\widehat{X}_{n+1|n}\right)\right)$$

Denominator

$$\int p_V\left(z_{n+1} - h\left(X_{n+1}\right)\right) p\left(X_{n+1} \middle| Y_n\right) dX_{n+1}$$

$$\approx \int\Big[p_V\left(z_{n+1} - h\left(\widehat{X}_{n+1|n}\right)\right) - p_V'\left(z_{n+1} - h\left(\widehat{X}_{n+1|n}\right)\right)^T$$

$$\left(h'\left(\widehat{X}_{n+1|n}\right) e_{n+1|n} + \frac{1}{2} e_{n+1|n}^T h''\left(\widehat{X}_{n+1|n}\right)^T e_{n+1|n}\right)$$

$$+\frac{1}{2} e_{n+1|n}^T h''\left(\widehat{X}_{n+1|n}\right)^T p_V''\left(z_{n+1} - h\left(\widehat{X}_{n+1|n}\right)\right) h'\left(\widehat{X}_{n+1|n}\right) e_{n+1|n}\Big]$$

$$p\left(X_{n+1} \middle| Y_n\right) dX_{n+1}$$

$$= p_V\left(z_{n+1} - h\left(\widehat{X}_{n+1|n}\right)\right)$$

$$-\frac{1}{2} p_V'\left(z_{n+1} - h\left(\widehat{X}_{n+1|n}\right)\right)^T h''\left(\widehat{X}_{n+1|n}\right) V_{ec}\left(P_{n+1|n}\right) \quad \widehat{X}_{n+1|n+1}$$

$$\approx \frac{\left\{\widehat{X}_{n+1|n} + P_{n+1|n} h'\left(\widehat{X}_{n+1|n}\right)^T \nabla \log p_V\left(z_{n+1} - h\left(\widehat{X}_{n+1|n}\right)\right)\right\}}{\left\{1 - \frac{1}{2} \underline{\nabla} \log p_V\left(z_{n+1} - h\left(\widehat{X}_{n+1|n}\right)^T\right) h''\left(\widehat{X}_{n+1|n}\right) Vec\left(p_{n+1|n}\right) + \frac{1}{2} Tr\left(p_V''\left(z_{n+1} - h\left(\widehat{X}_{n+1|n}\right)\right) h'\left(\widehat{X}_{n+1|n}\right) P_{n+1|n} h'\left(\widehat{X}_{n+1|n}\right)^T\right)\right\}}$$

$$\mathbb{E}\left[X_{n+1|} Y_n\right] = \widehat{X}_{n+1|n} = \int X_{n+1} p\left(X_{n+1|} Y_n\right) dX_n dX_{n+1}$$

$$= \int X_{n+1} p\left(X_{n+1|} X_n\right) p\left(X_n \mid Y_n\right) dX_n\, dX_{n+1}$$

$$= \int f\left(X_n\right) p\left(X_n \mid Y_n\right) dX_n$$

$$\left(\text{where } f\left(X_n\right) = \int X_{n+1} p\left(X_{n+1} \mid X_n\right) dX_{n+1}\right)$$

$$= \int f\left(\widehat{X}_{n|n} + e_{n|n}\right) p\left(X_n \mid Y_n\right) dX_n$$

$$\approx \int\left(f\left(\widehat{X}_{n|n}\right) + f'\left(\widehat{X}_{n|n}\right)e_{n|n} + \frac{1}{2}f''\left(\widehat{X}_{n|n}\right)\left(e_{n|n} \otimes e_{n|n}\right)\right)p\left(X_n \mid Y_n\right)dX_n$$

$$= \int\left(\widehat{X}_{n|n}\right) + \frac{1}{2}f''\left(\widehat{X}_{n|n}\right)Vec\left(p_{n|n}\right)$$

In general put $\psi(X)$ be any function on $\mathbb{R}^d$ when $X_n \in \mathbb{R}^d\ \forall n$. Then

$$\begin{aligned}
\mathbb{E}\left\{\psi\left(X_{n+1}\right)\middle|Y_n\right\} &= \widehat{\psi}_{n+1|n} \\
&= \mathbb{E}\left\{\psi\left(\widehat{X}_{n+1|n} + C_{n+1|n}\right)\middle|Y_n\right\} \\
&\approx \psi\left(\widehat{X}_{n+1|n}\right) + \frac{1}{2}Tr\left(\psi\left(\widehat{X}_{n+1|n}\right)P_{n+1|n}\right) \\
\widehat{\psi}_{n+1|n+1} &= \mathbb{E}\left\{\psi\left(X_{n+1}\right)\middle|Y_{n+1}\right\} \\
&= \frac{\int \psi\left(X_{n+1}\right)p_V\left(z_{n+1} - h\left(X_{n+1}\right)\right)p\left(X_{n+1}\middle|Y_n\right)dX_{n+1}}{\int p_V\left(z_{n+1} - h\left(X_{n+1}\right)\right)p\left(X_{n+1}\middle|Y_n\right)dX_{n+1}}
\end{aligned}$$

$$\begin{aligned}
F_m(s) &= C_m(s)\,(X_s(s) - X_m(s)) \\
F_s(s) &= C_s(s)\,(X_m(s) - X_s(s)) \\
F_m + F_h &= s((ms + b) + Z_h(s))X_m = \frac{X_m(s)}{G_m(s)} \\
C_m(X_s - X_m) + F_h &= \frac{X_m}{G_m} \\
C_s(X_m - X_s) &= \frac{X_s}{G_s}
\end{aligned}$$

$$G_m(s) = \frac{1}{s\left(m_m s + b_m + Z_h\right)} \quad G_s(s) = \frac{1}{s\left(m_s s + b_s + Z_s\right)}$$

$$\begin{aligned}
\left(m_m s^2 + b_m s + sZ_h\right)X_m &= F_h + C_m\left(X_s - X_m\right) \\
\left(m_s s^2 + b_s s + sZ_e\right)X_s &= F_e + C_s\left(X_m - X_s\right)
\end{aligned}$$

$$\begin{aligned}
X_s &= \frac{\left(C_s X_m + F_e\right)}{\left(m_s s^2 + b_s s + sZ_e + C_s\right)} = \frac{\left(C_s X_m + F_e\right)}{\frac{1}{G_s} + C_s} \\
&= \frac{G_s\left(C_s X_m + F_e\right)}{1 + G_s C_s}
\end{aligned}$$

$$X_m\left(\frac{1}{G_m} + C_m\right) = F_h + \frac{C_m G_s\left(C_s X_m + F_e\right)}{1 + G_s C_s}$$

$$X_m\left[(1+G_mC_m)(1+G_sC_s)-C_mG_mC_sG_s\right]$$

$$= G_mF_h(1+G_sC_s)+C_mG_sF_eG_m$$

$$X_m = \frac{G_mF_h(1+G_sC_s)+C_mG_sF_eG_m}{1+G_mC_m+G_sC_s}$$

Random fluctuation in the feedback transfer functions

$$C_m(s)\to C_m^{(0)}(s)+\delta C_m(s)$$

$$C_s(s)\to C_s^{(0)}(s)+\delta C_s(s)$$

$$\left[m_ms^2+b_ms+sZ_h+C_m^{(0)}(s)+\delta C_m(s)\right]X_m(s)$$

$$= \left(F_h(s)+C_m^{(0)}(s)+\delta C_m(s)\right)X_s(s)$$

$$\frac{\delta X_m}{G_m} = C_m(X_s-X_m)+C_m(\delta X_s-\delta X_m)$$

$$\frac{\delta X_s}{G_s} = \delta C_s(X_m-X_s)+C_s(\delta X_m-\delta X_s)$$

$$(1+G_mC_m)\delta X_m - G_mC_m\delta X_s$$

$$= G_m(X_s-X_m)\delta C_m$$

$$(1+G_sC_s)\delta X_s - G_sC_s\delta X_m = G_s(X_m-X_s)\delta C_s$$

$$\begin{bmatrix}1+G_mC_m & -G_mC_m\\ -G_sC_s & 1+G_sC_s\end{bmatrix}\begin{bmatrix}\delta X_m\\ \delta X_s\end{bmatrix}$$

$$= \begin{bmatrix}G_m(X_s-X_m) & 0\\ 0 & G_m(X_m-X_s)\end{bmatrix}\begin{bmatrix}\delta C_m\\ \delta C_s\end{bmatrix}$$

$$\begin{bmatrix}\delta X_m\\ \delta X_s\end{bmatrix} = (X_m-X_s)\frac{\begin{bmatrix}1+G_sC_s & G_mC_m\\ G_sC_s & 1+G_mC_m\end{bmatrix}\begin{bmatrix}-G_m & 0\\ 0 & G_s\end{bmatrix}\begin{bmatrix}\delta_m\\ \delta C\end{bmatrix}}{1+G_mC_m+G_sC_s}$$

$$\delta X_m = \frac{(X_m-X_s)\{-(1+G_sX_s)G_m\delta C_m+G_mC_mG_s\delta(s)\}}{(1+G_mC_m+G_sC_s)}$$

$$C_m(s) = \sum_n\left(C_m[n]e^{-snT}\right)$$

$$Cs(s) = \sum_n \left(C_s[n]e^{-snT}\right)$$

$$\delta C_m(s) = \sum_n \delta C_m[n]e^{-snT}$$

$$\delta C_s(s) = \sum_n \delta C_s[n]e^{-snT}$$

$$-\frac{(1+G_sX_s)G_m(X_m-X_s)}{1+G_mC_m+G_sC_s} = A_{11}(s)$$

$$\frac{G_mC_mG_s(X_m-X_s)}{1+G_mC_m+G_sC_s} = A_{12}(s)$$

$$\delta X_m(s) = A_{11}(s)\delta C_m(s) + A_{12}(s)\delta C_s(s)$$

$$\delta x_m(\mathrm{s}) = \sum_n \left\{\delta C_m[n]a_{11}(t-nT) + \delta C_s[n]a_{12}(t-nT)\right\}$$

$$\delta x_s(t) = \sum_n \left\{\delta C_m[n]a_{21}(t-nT) + \delta C_s[n]a_{22}(t-nT)\right\}$$

Stochastic differential eqn. with delay term linear case

$$d\underline{X}(t) = \left(\underline{\underline{A}}_1(t)\underline{X}(t) + \underline{\underline{A}}_2(t)\underline{X}(t-T)\right)dt + \underline{\underline{G}}(t)d\underline{B}(t)$$

$\underline{\underline{G}}(t)$, $\underline{\underline{A}}_1(t)$, $\underline{\underline{A}}_2(t)$ are n in t with period T

$$\int \underline{X}(t)e^{-st}dt = \widehat{\underline{X}}(s)$$

$$s\widehat{\underline{X}}(s) = \sum_n \underline{\underline{A}}_{1n}\widehat{\underline{X}}(s - jn\,w_0) + \sum_n \underline{\underline{A}}_{2n}\widehat{\underline{X}}(s - jn\,w_0)e^{-sT} + \sum_n \underline{\underline{G}}_n\widehat{\underline{W}}(s - jnw_0)$$

Solve for $\widehat{\underline{X}}(s)$.

Approximate solution using perturbation theory.

$$\widehat{W}(s) = \int_{\mathbb{R}} e^{-st}d\underline{B}(t)$$

Put $s = \sigma + jmw_0$. Then

$$(\sigma + jmw_0)\widehat{\underline{X}}(\sigma + jmw_0) = \sum_n \left(\underline{\underline{A}}_{1n} + e^{-\sigma T}\underline{\underline{A}}_{2n}\right)\widehat{\underline{X}}(\sigma + j(m-n)w_0) + \sum_n \underline{\underline{G}}_n\widehat{\underline{W}}(\sigma + j(m-n)w_0)$$

or equivalently

$$(\sigma + jmw_0)\underline{\widehat{X}}(\sigma + jmw_0)$$

$$= \sum_n \left(\underline{\underline{A}}_{1m-n} + e^{-\sigma T}\underline{\underline{A}}_{2m-n}\right)\underline{\widehat{X}}(\sigma + jnw_0)$$

$$+\sum_n \underline{\underline{G}}_{m-n}\underline{\widehat{W}}(\sigma + jnw_0)$$

Define

$$\underline{\xi}_m(\sigma) = \underline{\widehat{X}}(\sigma + jmw_0),$$

$$\underline{\xi}_m(\sigma) = (\underline{\xi}_m(\sigma))\ m \in \mathbb{Z}$$

Then with

$$\underline{\underline{D}}(\sigma) = diag[\sigma + jmw_0 : m \in \mathbb{Z}],$$

we have

$$\underline{\underline{D}}(\sigma)\underline{\xi}(\sigma) = \underline{\underline{P}}(\sigma)\underline{\xi}(\sigma) + \underline{\underline{G}}\underline{\tilde{W}}(\sigma)$$

Where

$$\underline{\tilde{W}}(\sigma) = \left(\left(\underline{\widehat{W}}(\sigma + jnw_0)\right)\right)_{n \in \mathbb{Z}}$$

Then,

$$\underline{\xi}(\sigma) = \left(\underline{\underline{D}}(\sigma) - \underline{\underline{P}}(\sigma)\right)^{-1}\underline{\underline{G}}\underline{\tilde{W}}(\sigma)$$

where

$$\underline{\underline{P}}(\sigma) = \left(\left(\underline{\underline{A}}_{1,\, m-n} + e^{-\sigma T}\underline{\underline{A}}_{2,\, m-n}\right)\right)_{(m,\, n) \in \mathbb{Z}^2}$$

$$= \underline{\underline{\tilde{A}}}_1 + e^{-\sigma T}\underline{\underline{\tilde{A}}}_2$$

where

$$\underline{\underline{\tilde{A}}}_1 = \left(\left(\underline{\underline{\tilde{A}}}_{1,\, m-n}\right)\right)_{(m,\, n) \in \mathbb{Z}^2},$$

$$\underline{\underline{\tilde{A}}}_2 = \left(\left(\underline{\underline{\tilde{A}}}_{2,\, m-n}\right)\right)_{(m,\, n) \in \mathbb{Z}^2}$$

Note that

$$\mathbb{E}\left[\underline{\widehat{W}}(s_1)\overline{\underline{\widehat{W}}(s_2)}\right] = \mathbb{E}\left\{\left(\int_0^\infty e^{-s_1 t} d\underline{B}(t)\right)\left(\int_0^\infty e^{-\bar{s}_2 t} d\underline{B}^T(t) dt\right)\right\}$$

$$= \int_0^\infty e^{-(s_1 + \bar{s}_2)t} dt$$

$$\underline{\underline{I}} = (s_1 + \bar{s}_2)^{-1}\underline{\underline{I}}.$$

❖❖❖❖❖

[154] Propagation of em waves in wave guide having non-uniform and anisotropic permittivity and permeability.

$$\underline{\nabla} \times \underline{E} = -jw\underline{\underline{\mu}}\,\underline{H}$$

$$\underline{\nabla} \times \underline{H} = jw\underline{\underline{\varepsilon}}\,\underline{E}$$

$$\underline{\underline{\mu}} = \underline{\underline{\mu}}(X, y, w) \quad \underline{\underline{\varepsilon}} = \underline{\underline{\varepsilon}}(X, Y, w)$$

$$\underline{E}(X, Y, Z, w) = \underline{E}(X, Y, w)\exp(-\gamma Z)$$

$$\underline{H}(X, Y, Z, w) = \underline{H}(X, Y, w)\exp(-\gamma Z)$$

$$\nabla_\perp E_Z \times \hat{Z} - \gamma\hat{Z} \times \underline{E}_\perp = -jw\left(\underline{\underline{\mu}} \cdot \underline{H}\right)_\perp$$

$$\nabla_\perp H_Z \times \hat{Z} - \gamma\hat{Z} \times \underline{H}_\perp = -jw\left(\underline{\underline{\varepsilon}} \cdot \underline{E}\right)_\perp$$

$$\underline{\nabla}_\perp E_Z + \gamma\underline{E}_\perp = -jw\hat{Z}X(\mu H)_\perp$$

$$jw\left(\varepsilon_{1:2} E_Z + \underline{\underline{\varepsilon}}_\perp \cdot E_\perp\right) = -\gamma\hat{Z} \times \underline{H}_\perp + \underline{\nabla}_\perp H_Z \times \hat{Z}$$

$$\underline{\underline{\varepsilon}} = \begin{bmatrix} \underline{\varepsilon}_\perp & \underline{\varepsilon}_{1:2} \\ \underline{\varepsilon}_{1:2}^T & \varepsilon_3 \end{bmatrix}$$

$$\underline{E}_\perp = \frac{-\gamma}{j\omega}\underline{\underline{\varepsilon}}_\perp^{-1} \cdot \left(\hat{Z} \times \underline{H}_\perp\right) + \frac{1}{j\omega}\underline{\underline{\varepsilon}}_\perp^{-1}\underline{\nabla}_\perp H_Z \times \hat{Z} - \underline{\underline{\varepsilon}}_\perp^{-1}\,\underline{\varepsilon}_{1:2}\,\underline{E}_Z$$

So,

$$\underline{\nabla}_\perp E_Z + j\omega\left(\hat{Z}X(\mu \cdot H)_\perp\right) = -\gamma\underline{E}_\perp$$

$$= \frac{\gamma^2}{j\omega}\underline{\underline{\varepsilon}}_\perp^{-1}\left(\hat{Z} \times \underline{H}_\perp\right) - \frac{\gamma}{j\omega}\underline{\underline{\varepsilon}}_\perp^{-1}\left(\nabla_\perp H_Z \times \hat{Z}\right)$$

$$+\gamma\,\underline{\underline{\varepsilon}}_\perp^{-1}\underline{\varepsilon}_{1:2}E_Z$$

$$(\mu \,.\, H)_\perp = \underline{\mu}_{1:2}H_Z + \underline{\underline{\mu}}_\perp \cdot \underline{H}_\perp$$

$$\hat{Z} \times \underline{H}_\perp = \begin{bmatrix} H_Y \\ -H_X \end{bmatrix} = \begin{bmatrix} 0 & 1 \\ -1 & 0 \end{bmatrix}\underline{H}_\perp = \underline{\underline{J}} \cdot \underline{H}_\perp$$

So $$\underline{\nabla}_\perp E_Z + j\omega\underline{\underline{J}}_\mu H_\perp = \frac{Y^2}{j\omega}\underline{\underline{\varepsilon}}_\perp^{-1}\underline{\underline{J}} \cdot \underline{H}_\perp - \frac{\gamma}{j\omega}\underline{\underline{\varepsilon}}_\perp^{-1}\left(\underline{\nabla}_\perp H_Z \times \hat{Z}\right)$$

$$+\gamma\,\underline{\underline{\varepsilon}}_\perp^{-1}\,\underline{\varepsilon}_{1:2}E_Z$$

or

$$\underline{\nabla}_\perp E_Z + j\omega\left\{\left(\underline{\underline{J}}\underline{\mu}_{1:2}\right)H_Z + \underline{\underline{J}}\,\underline{\underline{\mu}}_\perp\underline{H}_\perp\right\}$$

$$= \frac{\gamma^2}{j\omega}\underline{\underline{\varepsilon}}_{\perp}^{-1} J H_{\perp} + \frac{Y}{j\omega}\underline{\underline{\varepsilon}}_{\perp}^{-1}\underline{\underline{J}}\underline{\nabla}_{\perp} H_Z + \gamma\underline{\underline{\varepsilon}}_{\perp}^{-1}\underline{\underline{\varepsilon}}_{1:2} E_Z$$

or

$$\frac{\left(Y^2\underline{\underline{\varepsilon}}_{\perp}^{-1}\underline{\underline{J}} + w^2\underline{\underline{J}}\underline{\underline{\mu}}_{\perp}\right)}{j\omega}\underline{H}_{\perp}$$

$$= -\gamma\underline{\underline{\varepsilon}}_{\perp}^{-1}\underline{\underline{\varepsilon}}_{1:2} E_Z - \frac{\gamma}{j\omega}\underline{\underline{\varepsilon}}_{\perp}^{-1}\underline{\underline{J}}\underline{\nabla}_{\perp} H_Z$$

$$= \underline{\nabla}_{\perp} E_Z + j\omega\underline{\underline{J}}\underline{\mu}_{1:2} H_Z$$

So,

$$\underline{H}_{\perp} = -jw\gamma\left(\gamma^2\underline{\underline{\varepsilon}}_{\perp}^{-1}\underline{\underline{J}} + \omega^2\underline{\underline{J}}\cdot\underline{\underline{\mu}}_{\perp}\right)^{-1}\underline{\underline{\varepsilon}}_{\perp}^{-1}\underline{\underline{\varepsilon}}_{1:2} E_Z$$

$$-\gamma\left(\gamma^2\underline{\underline{\varepsilon}}_{\perp}^{-1}\underline{\underline{J}} + \omega^2\underline{\underline{J}}\cdot\underline{\underline{\mu}}_{\perp}\right)^{-1}\underline{\underline{\varepsilon}}_{\perp}^{-1}\underline{\underline{J}}\underline{\nabla}_{\perp} H_Z$$

$$+ j\omega\left(\gamma^2\underline{\underline{\varepsilon}}_{\perp}^{-1}\underline{\underline{J}} + w^2\underline{\underline{J}}\cdot\underline{\underline{\mu}}_{\perp}\right)^{-1}\underline{\nabla}_{\perp} E_Z$$

$$-\omega^2\left(\gamma^2\underline{\underline{\varepsilon}}_{\perp}^{-1}\underline{\underline{J}} + \omega^2\underline{\underline{J}}\cdot\underline{\underline{\mu}}_{\perp}\right)^{-1}\underline{\underline{J}}\underline{\mu}_{1:2} H_Z$$

❖❖❖❖❖

[155] Entropy of zero mean Gauss-Markov processes

$$\underline{X}_{n+1} = \underline{\underline{A}}\,\underline{X}_n + \underline{W}_{n+1}$$

$$\{\underline{W}_n\}\, iid\ N\left(0, \underline{\underline{\Sigma}}_W\right)$$

$$X_n = \sum_{k=1}^{N} A^{n-k}\underline{W}_k$$

$$\text{Cov}(X_n) = \underline{\underline{\Sigma}}_X[n] = \sum_{k=1}^{n}\underline{\underline{A}}^{n-k}\underline{\underline{\Sigma}}_W\underline{\underline{A}}^{Tn-k}$$

$$= \sum_{k=1}^{N} A^{n-k}\underline{W}_k$$

Let

$$Y_n = \underline{C}^T\underline{X}_n$$

$$\sigma y_n^2 = \text{Var}(y_n) = \underline{C}^T\underline{\underline{\Sigma}}_X[n]\underline{C}$$

$$= \sum_{k=0}^{n+1}\underline{C}^T\underline{\underline{A}}^k\underline{\underline{\Sigma}}_W\underline{\underline{A}}^{T^k}\underline{C}$$

Entropy of $\{Y_n\}$

$$S_y[n] = -\mathbb{E}\log p(y_n)$$

$$= \mathbb{E}\left[\frac{y_n^2}{2\sigma y_n^2} + \log\left(\sigma y_n^2\right)\right] + \log\sqrt{2\pi}$$

$$= \overbrace{\frac{1}{2}\log(2\pi e)}^{\alpha} + \log\sigma y_n^2$$

$$= \alpha + \log\left\{\sum_{k=0}^{n-1} \underline{C}^T \underline{\underline{A}}^k \underline{\underline{\Sigma}}_W \underline{\underline{A}}^{T\,k} \underline{C}\right\}$$

Clearly, $S_y[n] \uparrow$ as $n \uparrow$

Let $\{\underline{X}_n\}$ be any zero mean Gauss. Markov process in $\mathbb{R}^p$. For $n_3 \geq n_2 \geq n_1 \geq 0$

$$\mathbb{E}\left\{\mathbb{E}\left(\underline{X}_{n3} \mid \underline{X}_{n2}\right) \mid \underline{X}_{n1}\right\} = \mathbb{E}\left\{\underline{X}_{n3} \mid \underline{X}_{n1}\right\}$$

$$\mathbb{E}\left\{\underline{X}_{n3} \mid \underline{X}_{n2}\right\} = FX_{n_2}$$

$$F\sum{}_X[n_2] = \Sigma_X[n_3, n_2]$$

$$F = \sum{}_X[n_3, n_2]\cdot \sum{}_X[n_2]^{-1}$$

So,

$$\sum{}_X[n_3, n_2]\sum{}_X[n_2]^{-1}\sum{}_X[n_2, n_1]\sum{}_X[n_1]^{-1}$$
$$= \sum{}_X[n_3, n_1]\sum{}_X[n_1]^{-1}$$

or

$$\sum{}_X[n_3, n_2]\sum{}_X[n_2]^{-1}\sum{}_X[n_2, n_1]$$
$$= \sum{}_X[n_3, n_1]$$

Suppose $\{\underline{X}_n\}$ is zero mean stationary. Then

$\underline{\underline{\Sigma}}_X[n_2] = \underline{\underline{\Sigma}}_X[0]$ is a constant matrix,

and

$$\sum{}_X[n_2, n_1] \equiv \tilde{\Sigma}_X[n_2 - n_1].\ \text{So,}$$

$$\tilde{\Sigma}_X[n]\Sigma_X[0]^{-1}\tilde{\Sigma}_X[m] = \tilde{\Sigma}_X[n+m].$$

Let $$R_X[X] = \Sigma_X[0]^{-1/2}\tilde{\Sigma}_X[m]\Sigma_X[0]^{-1/2}$$

Then

$$R_X[n+m] = R_X[n].\ R_X[m]$$

So, $$R_X[n] = R_X[1]^n,\ n \ge 0$$

$$\tilde{\underline{\underline{\Sigma}}}_X[n] = \underline{\underline{\Sigma}}_X[0]^{1/2} \cdot \underline{\underline{R}}_X[1]^n \cdot \underline{\underline{\Sigma}}_X[0]^{1/2},\ n \ge 0$$

The general non-stationary case:

$$= \Sigma_X[n_3, n_2] \cdot \Sigma_X[n_2]^{-1} \cdot \Sigma_X[n_2, n_1]$$

$$= \Sigma_X[n_3, n_1]$$

Let $$\Sigma_X[n_2]^{-1/2} \Sigma_X[n_2, n_1] \Sigma_X[n_1]^{-1/2} \quad n_3 \ge n_2 \ge n_1$$

$$= R_X[n_2, n_1]$$

Then, $$R_X[n_3, n_1] = R_X[n_3, n_2] R_X[n_2, n_1]$$

Let $$R_X[n+1, n] = \tilde{R}_X[n]$$

Then for $n \ge m$,

$$R_X[n, m] = \tilde{R}_X[n-1] \cdots \tilde{R}_X[m+1] \tilde{R}_X[m] R_X[m, m]$$

$$x_{n+1} = ax_n + \xi_n,\ n \ge 0,\ |\alpha| < 1,$$

x_0 independent of $\{\xi_n\}$ $\{\xi_n\} \triangle$ iid $N(0, 1)$

$$\frac{x_{n+1}}{\alpha^{n+1}} - \frac{x_n}{\alpha^n} = \frac{\xi_n}{\alpha^{n+1}}$$

$$\frac{x_{n+l}}{\alpha^{n+l}} - \frac{x_n}{\alpha^n} = \sum_{k=n}^{n+l-1} \frac{\xi_k}{\alpha^{k+1}}$$

$$x_{n+l} = \alpha^l x_n + \sum_{k=n}^{n+l-1} \alpha^{n+l-k-1} \xi_k$$

$$x_{n+1} | x_n \simeq N\left(\alpha^l x_n, \sum_{k=n}^{n+l-1} \alpha^{2(n+l-k-1)} \right)$$

$$\sum_{k=n}^{n+l-1} \alpha^{2(n+l-k-1)} = \sum_{r=0}^{l-1} \alpha^{2n} = \frac{1-\alpha^{2l}}{1-\alpha^2} \underline{\underline{\triangle}}\, \sigma_l^2$$

$$\mu_l \underline{\underline{\triangle}}\, \alpha^l \cdot \pi^l(x, dy) = \left(2\pi\sigma_l^2\right)^{-1/2} \cdot \exp\left(-\frac{1}{2\sigma_l^2} \left(y - \alpha^l x\right)^2 \right) dy$$

$$\equiv P\left\{x_{n+l} \in dy \mid x_n = x\right\}.$$

❖❖❖❖❖

[156] Ordinary of stochastic differential equation with delay

$$\frac{dX(t)}{dt} = f(t, X(t-\text{t}))$$

$$= f\big(t, X(t) - (X(t) - X(t-\tau))\big)$$

$$= \sum_{r=1}^{\infty} f^{(r)} f(t, X(t))(X(t) - X(t-\tau))^r (-1)^r / \underline{|r} + f(t, X(t))$$

$$\equiv f(t,X(t)) - \varepsilon\big(f(t,X(t)) - f(t, X(t-\tau))\big) \; (\varepsilon = 1)$$

Let $X_0\,(t)$ satisfy

$$\frac{dX_0(t)}{dt} = f(t, X_0(t))$$

and let $X(t) = X_0(t) + \varepsilon X_1(t) + O(\varepsilon^2)$

Then,

$$\frac{dX_1(t)}{dt} = -\big(f(t, X_0(t)) - f(t, X_0(t-\tau))\big) + f_X\big(t, X_0(t) X_1(t)\big)$$

$$X_1(t) = -\int_0^t \overline{\underline{\Phi}}(t,s)\big(f(s, X_0(s))\big) - f(s, X_0(s-\tau))\,ds$$

$$\Phi(t, s) = \exp\left(\int_s^t f_X(t', X_0(t'))\,dt'\right)$$

Side with delay

$$\frac{dX(t)}{dt} = f(t, X(t), X(t,\tau)) + g(X(t))W(t)$$

$$\simeq f_0(t, X(t)) + \varepsilon\{g(X(t))W(t) - f_0(t, X(t))$$

$$+ f(t, X(t), X(t-\tau))\} \; (\varepsilon = 1)$$

$f_0(t, X) = f(t, X, X)$. $X(t) = X_0(t) + \varepsilon X, (t) + O(\varepsilon^2)$

$$X_1(t) = \int_0^t g(X_0(t'))W(t') + \int_0^t \big(f(t', X_0(t'), X_0(t'-\tau))\big)$$

$$-f_0(t', X_0(t'))\,dt'$$

$$+\int_0^t f_{oX}(t', X_0(t'))X_1(t')\,dt'$$

$$X_1(t) = \int_0^t \bar{\underline{\Phi}}(t,s)g(X_0(s))W(s)ds$$

$$+\int_0^t \bar{\underline{\Phi}}(t,s)\left(f(s,X_0(s),X_0(s-\tau))\right)-f_0(s,X_0(s))ds$$

$$\mathbb{E}[X_1(t)] = \int_0^t \bar{\underline{\Phi}}(t,s)\left(f(s,X_0(s),X_0(s-\tau))\right)-f_0(s,X_0(s))ds$$

$$\mathbb{E}[X_1(t_1)X_1(t_2)] = \int_0^{t_1}\int_0^{t_2} \bar{\underline{\Phi}}(t_1,s_1)\bar{\underline{\Phi}}(t_2,s_2)g(X_0(s_1))g(X_0(s_2))$$

$$\mathbb{E}\{W(s_1)W(s_2)\}ds_1\, ds_2$$

$$\frac{dX_2}{dt} = f_{0X}(t,X_0(t))X_2(t)$$

$$+\frac{1}{2}f_{oXX}(t,X_0(t))X_1(t)^2$$

$$+g'(X_0(t))X_1(t)W(t)$$

$$+f_X(f,X_0(t),X_0(t-\tau))X_1(t)$$

$$+f_y(f,X_0(t),X_0(t-\tau))X_1(t-\tau)$$

$$-f_{0X}(t,X_0(t))X_1(t)$$

So,

$$X_2(t) = \int_0^t \bar{\underline{\Phi}}(t,s)\left(\frac{1}{2}f_{0XX}(s,X_0(s))\right)X_1(s)^2$$

$$+g'(X_0(s))X_1(s)W(s)+f_X(s,X_0(s),X_0(s-\tau))X_1(s)$$

$$+f_y(s,X_0(s),X_0(s-\tau_1)X_1(s-\tau)-f_{0X}(s,X(s))X_,(s))$$

Passivity of a stochastic differential system.

$$dX(t) = f(X(t),u(t))dt+g(X(t),u(t))dB(t)$$

For passivity, we require

$$\int_0^T \mathbb{E}\{u(t)dX(t)\}\geq 0 \;\forall\; T\geq 0$$

or assume the in put for a $\{u(t)\}$ to be non-randam

$$\int_0^T u(t)\mathbb{E}\{f(X(t),u(t))\}dt \geq 0$$

❖❖❖❖❖

[157] Born -Oppenheimer approximation

$\{\underline{R}_i\}_{i=1}^N \simeq$ nucleic position

$\{\underline{r}_{ik}\}_{k=1}^Z \simeq$ electron position of ith nucleus

$$H = T_N + T_e + V_N + V_e + V_{Nc}$$

T_N = Nuclei kinetic energy

$$= -\frac{\hbar^2}{2M}\sum_{i=1}^N \nabla_{R_i}^2$$

T_e = Electron kinetic energy

$$= -\frac{\hbar^2}{2m}\sum_{1\leq k\leq Z,\, 1\leq i\leq N} \nabla_{rik}^2$$

$V_N \simeq$ Nucleic interaction potential energy

$$= \sum_{i\neq j}\frac{Z^2e^2}{z\left|\underline{R}_i - \underline{R}_j\right|}$$

$V_e \simeq$ electron interaction potential energy

$$= \sum_{(i,j)\neq(k,l)}\frac{e^2}{2\,|\,\underline{r}_{ij} - \underline{r}_{kl}\,|}$$

$$V_{Ne} = -\sum_{i,k,l}\frac{Ze^2}{|\,\underline{R}_i - \underline{r}_{kl}\,|}$$

Since nuclear kinetic energy is small compound to electron kinetic energy

(By momentum conservation)

$$\sum_{i=1}^N \underline{P}_i = -\sum_{\substack{i\leq i\leq N,\\ 1\leq k\leq Z}} P_{ik}$$

where

P_i = momentum of ith nucleus,

p_{ik} = momentum of kth electron of ith nucleic,

$|\underline{p}_i| \approx Z\,|\underline{p}_{ik}|$

So $$\frac{|P_i|^2}{2Mi} \approx \frac{Z^2}{M}\frac{|p_{ik}|^2}{2M_i} = \frac{Z^2 m}{M}\frac{|P_{ik}|^2}{2M_i}$$

or $T_{\text{nucleus}} \approx \dfrac{Z^2 m}{M}\, T_{\text{electron}}$

or $\dfrac{T_{\text{nucleus}}}{T_{\text{electron}}} \approx \dfrac{Z^2 m}{M} << 1$

since $m << M$

❖❖❖❖❖

[158] Robots with periodic inputs $\simeq$ Fourier series analysis

$$\left(M\left(\underline{q}\right)\underline{q}'\right) - V'(q) = \underline{\tau}(\text{t})$$

$$\Rightarrow\ M\left(\underline{q}\right)\underline{q}'' + \underline{N}\left(\underline{q}, \underline{q}'\right) = \underline{\tau}(\text{t})$$

$$\underline{N}\left(\underline{q}, \underline{q}'\right) = M'\left(\underline{q}\right)\left(q' \otimes \underline{q}'\right) - V'(q)$$

$\simeq$ centrifugal, gravitational Coriolis forces.

$$\underline{\tau}(t) = \sum_{n\in\mathbb{Z}} \underline{a}_n e^{jnw_0 t} \quad \underline{a}_n \in \mathbb{C}^2 \in \mathbb{R}^2 . \underline{a} - n = \bar{\underline{a}}_n$$

Torqne is periodic with period w_0.

$$\underline{\underline{M}}\left(\underline{q}\right) = \begin{pmatrix} a + b\cos q_2 & c + d\cos q_2 \\ c + d\cos q_2 & f \end{pmatrix}$$

$$M'(q) = \left[\frac{\partial M}{\partial q_1} \,\Big|\, \frac{\partial M}{\partial q_2}\right]$$

$$= -\left[\underline{\underline{O}}_2 \,\Big|\, \begin{pmatrix} b & d \\ d & 0 \end{pmatrix}\right]\sin(q_2)$$

$$M'(q)\left(\underline{\dot{q}} \otimes \underline{\dot{q}}\right) = -\sin(q_2)\left[\begin{pmatrix} b \\ d \end{pmatrix}\dot{q}_1\,\dot{q}_2 + \begin{pmatrix} d \\ 0 \end{pmatrix}\dot{q}_2^2\right]$$

$$V(q) = \frac{m_1 g L_1}{2}\sin q_1 + m_2 q\left(L_1 \sin\dot{q}_1 + \frac{L^2}{2}\sin\dot{q}_2\right)$$

$$V'(q) = \begin{pmatrix} \dfrac{\partial_V}{\partial q_1} \\ \dfrac{\partial V}{\partial q_2} \end{pmatrix} = \begin{pmatrix} \left(\dfrac{m_1}{2} + m_2\right) g L_1 \cos q_1 \\ \dfrac{m_2 g L_2}{2}\cos q_2 \end{pmatrix}$$

equation of motion

$$\begin{pmatrix} a+b\cos q_2 & c+d\cos q_2 \\ c+d\cos q_2 & f \end{pmatrix}\begin{pmatrix} q_1'' \\ q_2'' \end{pmatrix}$$

$$= -\sin(q_2)\begin{pmatrix} b\dot{q}_1\dot{q}_2 + d\dot{q}_2^2 \\ d\dot{q}_1\dot{q}_2 \end{pmatrix} + \begin{pmatrix} \left(\frac{m_1}{2}+m_2\right) & g L_1 \cos q_1 \\ \frac{m_2 g L_2}{2} & \cos q_2 \end{pmatrix} + \begin{pmatrix} \tau_1(t) \\ \tau_2(t) \end{pmatrix}$$

or

$$\begin{pmatrix} a & c \\ c & f \end{pmatrix}\begin{pmatrix} q_1'' \\ q_2'' \end{pmatrix} + \cos(q_2)\begin{pmatrix} b & d \\ d & 0 \end{pmatrix}\begin{pmatrix} q_1'' \\ q_2'' \end{pmatrix}$$

$$= -\sin(q_2)\begin{pmatrix} b\dot{q}_1\dot{q}_2 + d\dot{q}_2^2 \\ d\dot{q}_1\dot{q}_2 \end{pmatrix} - \begin{pmatrix} \alpha_1 & \cos q_1 \\ \alpha_2 & \cos q_2 \end{pmatrix}\begin{pmatrix} \tau_1(t) \\ T_2(t) \end{pmatrix}$$

More generally, consider

$$\underline{\xi}'(t) = \underline{\underline{A}}\,\underline{\xi}(t) + \varepsilon \underline{F}(\underline{\xi}(t)) + \underline{u}(t)$$

where $\underline{\xi}(t) \in \mathbb{R}^n$, $\underline{F}: \mathbb{R}^n \to \mathbb{R}^n$,

$$\underline{u}(t) \in \mathbb{R}^n, \underline{u}(t+T) = \underline{u}(t), T = \frac{2\pi}{W_0}$$

$$\underline{\xi}(t) = \underline{\xi_0}(t) + \varepsilon \underline{\xi_1}(t) + \varepsilon^2 \underline{\xi_2}(t) + O(\varepsilon^3)$$

$$\underline{\xi}'_0(t) = \underline{\underline{A}}\,\underline{\xi_0}(t) + \underline{u}(t),$$

$$\underline{\xi}'_1(t) = \underline{\underline{A}}\,\underline{\xi_1}(t) + \underline{F}(\underline{\xi_0}(t)),$$

$$\underline{\xi}'_2 = \underline{\underline{A}}\,\underline{\xi_2}(t) + \underline{\underline{F}}'(\underline{\xi_0}(t))\underline{\xi_1}(t)$$

Let $\xi(t)$, $t \geq 0$ be a diffusion process in R^d

$$d\xi(t) = \underline{\mu}(\underline{\xi}(t))dt + \sqrt{\varepsilon}\,\underline{\underline{\sigma}}(\underline{\xi}(t))d\underline{B}(t)$$

$\underline{B}(t) \in \mathrm{R}^d$ B.M.

Rate function for $\underline{\xi}(\cdot)$ assume $\underline{\underline{\sigma}}(\underline{\xi})$ is non-singular is

$$\frac{T}{T}(\xi) = \frac{1}{2}\int_0^T \left\| \underline{\underline{\sigma}}^{-1}(\underline{\xi})\left(\overset{0}{\underline{\xi}} - \underline{\mu}(\underline{\xi})\right)\right\|^2 dt$$

Let $G \subset \mathbb{R}^d$ be a connected open let,

∂G its boundary and $x \in G$, $y \in \partial G$

Define $V_T(x, y) = \underline{\xi} \in \mathbb{C}^1\left(\mathbb{R}^x, \mathbb{R}^d\right) \frac{1}{2}\int_0^T \left\| \underline{\underline{\sigma}}\left(\underline{\xi}\right)^{-1}\left(\overset{0}{\underline{\xi}} - \underline{\mu}\left(\underline{\xi}\right)\right)\right\|^2 dt$

$$\underline{\xi}(0) = \underline{x},\ \underline{\xi}(T) = \underline{y}$$

Let $\tau(x) = \inf\{t \geq 0 \mid \underline{\xi}(t) = y,\ \underline{\xi}(0) = x\}$

$$P\left\{\underline{\xi}(t) = \underline{f}(t), 0 \leq t \leq T\right\} = P_T\left\{\underline{\xi}(\cdot) = \underline{f}(\cdot)\right\}$$

$$\approx \exp\left(-\frac{1}{\varepsilon} I_T(f)\right)$$

Martingale

$$\exp\left(\left\langle \underline{\lambda}, \underline{\xi}(t)\right\rangle - \int_0^t \left\langle \underline{\lambda}, \mu\left(\underline{\xi}(t)\right)\right\rangle dt - \frac{\varepsilon}{2}\int_0^t \left\langle \underline{\lambda}, \underline{\underline{g}}\left(\underline{\xi}(\tau)\right)\underline{\lambda}\right\rangle d\tau\right) = M_\lambda(t)$$

where $\underline{\underline{g}}\left(\underline{\xi}\right) = \underline{\underline{\sigma}}\left(\underline{\xi}\right)\underline{\underline{\sigma}}^T\left(\underline{\xi}\right)$

$$dM_\lambda \left\langle \underline{\lambda}, \underline{\underline{\sigma}}\left(\underline{\xi}\right) d\underline{B}\right\rangle$$

Hence, M_λ is a Martingale (Exponential Martingale)

In particular.

$$d\left\|\underline{\xi}(t)\right\|^2 = 2\left\langle \underline{\xi}, d\underline{\xi}\right\rangle + d\underline{\xi}^T d\underline{\xi}$$

$$= 2\left\langle \underline{\xi}, \underline{\mu}(\underline{\xi})\right\rangle dt + Tr\left(\sigma\left(\underline{\xi}\right)\sigma\left(\underline{\xi}\right)^T\right) dt + d(\text{Martingale})$$

So by Doob's optional stopping theorem,

$$\mathbb{E}\left\{\left\|\underline{\xi}(\tau_\varepsilon)\right\|^2 - 2\int_0^{\tau_\varepsilon} \left\langle \underline{\xi}(t), \underline{\mu}\left(\underline{\xi}(t)\right)\right\rangle dt - \int_0^{\tau_\varepsilon} a\left(\underline{\xi}(t)\right) dt\right\} = \left\|\underline{x}\right\|^2$$

where $a(\underline{\xi}) = Tr\left(\underline{\underline{\sigma}}\left(\underline{\xi}\right)\underline{\underline{\sigma}}\left(\underline{\xi}\right)^T\right) = Tr\left(\underline{\underline{g}}\left(\underline{\xi}\right)\right)$

Let $\psi(\underline{\xi}) = 2\left\langle \underline{\xi}, \underline{\mu}\left(\underline{\xi}\right)\right\rangle + a\left(\underline{\xi}\right)$

Then

$$\mathbb{E}\left\{\left\|\underline{\xi}(\tau_\varepsilon)\right\|^2 - \int_0^{\tau_\varepsilon} \psi\left(\underline{\xi}(t)\right) dt\right\} = \left\|\underline{x}\right\|^2$$

So,

❖❖❖❖❖

[159] Large deviation Theory for image point of an object taken by a camera executing random motion,

2. Problem statement

Objects (point) are located at $\underline{r}_j \,.., \underline{r}_N$.

The screen is the $X - Z$ plane. The camera is located at the tip of a d-link Robot whose analyses undergo random motion. We assume that the links move in a plane parallel to the $X- Z$ plane but with the base point at $\underline{r}_0$. Let $\underline{r}_k = (X_k, Y_k, Z_k)$, $0 \leq k \leq N$. Let q_1, q_2, q_d denote the link analysis. Writing $q = (q_1, ..., q_d) \in \mathbb{R}^d$, they are governed by the equation of motion.

$$\underline{\underline{M}}(\underline{q})\ddot{\underline{q}} + N(\underline{q}, \dot{\underline{q}}) = \underline{\tau}_0(t) + \varepsilon\, \underline{w}(t)$$

where $\tau_0(t)$ is a non-random torque signal ($\varepsilon\mathbb{R}^d$) and $\underline{w}(t)$ is an $\mathbb{R}^d$-valued zero-mean correlated Gaussian noise process.

$$= (Z_c Y_k - Z_k Y_c) / {}_c Y_k - Y_c$$

Now we wish to calculate the fluctuating component ($O(\varepsilon)$ component) of $(\xi_k(t), \eta_k(t)) \forall k$ and device a rate function for this component, where $(\xi_k(t), \eta_k(t))$ are the coordinates of the image of $\underline{r}$k projected on the sources by the corenera.

Let
$$\underline{q}(t) = \underline{q}^{(0)}(t) + \varepsilon\delta\underline{q}(t) + O(\varepsilon^2)$$

$$\underline{\underline{M}}(\underline{q}^{(0)})\ddot{\underline{q}}^{(0)} + \underline{N}(\underline{q}^{(0)}, \dot{\underline{q}}^{(0)}) = \tau^0(t),$$

$$\underline{\underline{M}}'(\underline{q}^{(0)})(\delta\, \underline{q} \otimes \ddot{\underline{q}}^{(0)}) + \underline{\underline{M}}(\underline{q}^{(0)})\delta\ddot{\underline{q}}$$

$$+ \underline{\underline{N}}_{,1}(\underline{q}^{(0)}, \dot{\underline{q}}^{(0)})\delta\underline{q} + \underline{\underline{N}}_{,2}(\underline{q}^{(0)}, \dot{\underline{q}}^{(0)})\delta\dot{\underline{q}} = \underline{w}(t)$$

Equivalently,

$$\underline{\underline{A}}_0(t)\delta\ddot{\underline{q}}(t) + \underline{\underline{A}}_1(t)\delta\dot{\underline{q}}(t) + \underline{\underline{A}}_2(t)\delta\underline{q}(t) = \underline{w}(t)$$

where,

$$\underline{\underline{A}}_0(t) = \underline{\underline{M}}(\underline{q}^{(0)}(t))$$

Representation of the Lorentz group for the constructor of invariant for moving images.

$$f(t, X) \to f\left(\frac{t + vX}{\sqrt{1 - v^2}}, \frac{X + vt}{\sqrt{1 - v^2}}\right)$$

$$SL(2,\mathbb{C})\,.\; T = \begin{pmatrix} a & b \\ c & d \end{pmatrix} \; ad - bc = 1$$

$$X \to aX + bY, Y \to CX + dY$$

$\left\{X^i Y^{m-i}.0 \le i \le m\right\} \simeq$ Substratum of the representation.

Let $\varphi_i^{(X,Y)} = X^i Y^{m-i}$

$$\varphi_i\left(T^{-1}\begin{pmatrix} X \\ Y \end{pmatrix}\right) = \varphi_i\left(\begin{pmatrix} d & -b \\ -c & a \end{pmatrix}\begin{pmatrix} X \\ Y \end{pmatrix}\right)$$

$$= \varphi_i\left(dX - bY, aY - cX\right)$$

$$= \left(dX - bY\right)^i \left(aY - cX\right)^{m-i}$$

$$= \sum_{r,s}\binom{i}{r}\binom{m-i}{s} d^r\left(-b\right)^{i-r} a^s \left(-c\right)^{m-i-s} X^{r+m-i-s} Y^{i-r+s}$$

$$= \sum_{r,s}\binom{i}{r}\binom{m-i}{s} a^s b^{i-r} c^{m-i-s} d^r \left(-1\right)^{m-s-r} \varphi_{r+m-i-s}\left(X,Y\right)$$

$$= \sum_{r,s}\binom{i}{i+j+s-m}\binom{s+j-r}{S}$$

$$a^s b^{m-s-j} c^{j-r} d^{i+j+s-m}$$

$$(-1)^{2m-i-j}\varphi_j(X,Y)$$

$$= \sum_j \left[\pi_m(T)\right]_{ji} \varphi_j\left(X,Y\right)$$

where

$$\left[\pi_m(T)\right]_{ji} = \sum_s \binom{i}{i+j+s-m}\binom{s+j-r}{s} a^s b^{m-s-j} c^{j-r} d^{i+j+s-m}$$

Invariant measure on $SL(2,\mathbb{C})$:

$$T = \begin{pmatrix} a & b \\ c & d \end{pmatrix}, S = \begin{pmatrix} p & q \\ r & s \end{pmatrix}$$

$$\int_{SL(2,\mathbb{C})} f^{-1}\left(TS\right)\varphi\left(s\right) dp\, dq\, dr = \int_{SL(2,\mathbb{C})} f\left(S\right)\varphi\left(s\right) dp\, dq\, dr$$

$$= \int f(S)\varphi(TS)\,dp\,dq\,dr\,|T'(pqr)|$$

$$= \int f(S)\varphi(S)\,dp\,dq\,dr$$

Hance,

$$|T'(pqr)| = \frac{\varphi(S)}{\varphi(TS)}$$

So, $$\varphi(T) = \frac{1}{|T'(pqr)|}$$

$$TS = \begin{pmatrix} a & b \\ c & d \end{pmatrix}\begin{pmatrix} p & q \\ r & s \end{pmatrix}$$

$$= \begin{pmatrix} ap+br & aq+bs \\ cp+dr & cq+ds \end{pmatrix}$$

So, $$T(p\,q\,r) = (ap+br, aq+bs, cp+dr)$$

$$a = a_1 + i\dot{a}_2, b = b_1 + ib_2, c = c_1 + ic_2, d = d_1 + id_2$$

$$p = p_1 + ip_2, q = q_1 + iq_2, r = r_1 + ir_2, s = s_1 + is$$

$$(p\,q\,r) \equiv (p_1, p_2, q_1, q_2, r_1, r_2)$$

$$T(pq.r) \equiv (a_1p_1 - a_2p_2 + b_1r_1 - b_2r_2, a_1p_2 + a_2p_1 + b_1r_2 + b_2r_1,$$

$$a_1q_1 - a_2q_2 + b_1s_1 - b_2s_2, a_1q_2 + a_2q_1 + b_1s_2 + b_2s_1$$

$$c_1p_1 - c_2p_2 + d_1r_1 - d_2r_2, a_1p_2 + c_2p_1 + d_1r_2 + d_2r_1)$$

$$\equiv (t_1, t_2, t_3, t_4, t_5, t_6)$$

$$T^1(p\,q\,x) = \frac{\partial(t_1, t_2, \ldots, t_6)}{\partial(p_1p_2\,q_1q_2\,r_1r_2)}$$

❖❖❖❖❖

[160] Feynman diagram for scattering absorption and emission. Lagrangian density for the free Dirac field. $\bar{\psi} = \psi + \psi^{+}\gamma^{0}$

$$\mathcal{L}(\psi, \bar{\psi}, \partial_\mu\psi) = i\psi^{+}(X)\gamma^{0}\gamma^{\mu}\partial_\mu\psi(X) - m\psi^{+}(X)\gamma^{0}\psi(X)$$

$$\gamma^{0+} = \gamma^{0}, \gamma^{r+} = -\gamma^{r}, 1 \le r \le 3,$$

$$\gamma^{\mu}\gamma^{\nu} + \gamma^{\nu}\gamma^{\mu} = 2\eta^{\mu\nu}$$

$$((\eta^{\mu\nu})) = ((\eta_{\mu\nu})) = \text{diag}\,[1, -1, -1, -1]$$

$$\mathcal{L} = i\psi^{+}\partial_0\psi + i\psi^{+}\gamma^0\gamma^r\partial_r\psi$$

$$-m\psi^{+}\gamma^0\psi$$

$$\int \mathcal{L}d^4X \text{ is real since}\left(\int \psi^{+}\partial_0\psi d^4X\right)^{+}$$

$$= \int\left(\partial_0\psi^{+}\right)\psi d^4X$$

$$= -\int \psi^{+}\partial_0\psi d^4X$$

(integration by parts w.r.t. t),

$$\left(\int \psi^{+}\gamma^0\gamma^r\partial_r\psi d^4X\right)^{+} = \int \partial_r\psi^{+}\gamma^{r^{+}}\gamma^{0^{+}}\psi d^4X$$

$$= -\int \partial_r\psi^{+}\gamma^r\gamma^0\psi d^4X$$

$$\left(\gamma^{r^{+}} = -\gamma^r, \gamma^{0^{+}} = \gamma^0, \gamma^r\gamma^0 = -\gamma^0\gamma^r\right)$$

$$= \int \psi^{+}\gamma^r\gamma^0\partial_r\psi d^4X$$

$$= -\int \psi^{+}\gamma^0\gamma^r\partial_r\psi d^4X = \left(\psi^{+}\gamma^{0^{+}}\psi\right)^{+} = \psi^{+}\gamma^0\psi = \psi + \gamma^0\psi$$

Electron propagator:

$$S[\psi, \bar{\psi}] = \int \mathcal{L}d^4X$$

$$= \int\left(i\bar{\psi}(X)\gamma^{\mu}\partial_{\mu}\psi(X) - m\bar{\psi}(X)\psi(X)d^4X\right.$$

$$\mathcal{L}\int \bar{\hat{\psi}}(p)\gamma^{\mu}p_{\mu}\hat{\psi}(p) - m\bar{\hat{\psi}}(p)\hat{\psi}(p)\Big)d^4X$$

$$= \int \bar{\hat{\psi}}(p)\left(\gamma^{\mu}p_{\mu} - m\right)\hat{\psi}(p)d^4X$$

$$\int \psi_{\alpha}(X)\bar{\psi}_{\beta}(Y)\exp\left(\frac{i}{\hbar}S[\psi, \bar{\psi}]\right)(\mathbb{P})\psi(\mathbb{P})\bar{\psi}$$

$$\left\langle Vac\middle|T\left\{\psi_{\alpha}(X)\bar{\psi}_{\beta}(Y)\right\}\middle|Vac\right\rangle \triangleq D_{\alpha\beta}(X-Y)$$

or more previously

$$D_{\alpha\beta}(X-Y) = \left\langle Vac\middle|T\left\{\psi_{\alpha}(X).\overline{\psi_{\beta}(Y)}\right\}\middle|Vac\right\rangle$$

$$= \frac{\int \psi_{\alpha}(X)\overline{\psi_{\beta}(Y)}\exp(iS[\psi, \bar{\psi}])(\mathbb{R})\psi(\mathbb{R})\bar{\psi}}{\int \exp(iS[\psi, \bar{\psi}])(\mathbb{R})\psi(\mathbb{R})\bar{\psi}}$$

$$= \int \left(\left(\gamma^{\mu} p_{\mu} - m \right)^{-1} \right)_{\alpha\beta} \exp\left(i p (X - Y)\right) d^4 p$$

Electron propagator in momentum domain.

$$\int \langle Vac | T\{\psi(X)\bar{\psi}(Y)\} | Vac \rangle \exp(-ip.(X-Y)) d^4 (X-Y)$$

$$= \frac{1}{\gamma^{\mu} p_{\mu_{+i\varepsilon}} - m} \equiv \frac{1}{\gamma \cdot p - m + i\varepsilon}$$

Direct operator through deviation

$$\psi_l(X) = \int \left[a(p,\sigma) u_l(p,\sigma) \exp(-i p \cdot X) + b^{+}(p,\sigma) v_l(ip,\sigma) \exp(ip \cdot X) \right] d^3 p$$

This satisfy the Dirac eqn.

$$\left(i \gamma^{\mu} \partial_{\mu} - m\right) \psi = 0$$

provided that

$$\left(\gamma^{\mu} p_{\mu} - m\right) \underline{v}(p,\sigma) = 0,$$

$$\left(\gamma^{\mu} p_{\mu} + m\right) \underline{v}(p,\sigma) = 0$$

From the Lagrangian density

$$\pi_l(X) = \frac{\partial \mathcal{L}}{\partial \partial_t \psi_l} = i \psi_l^{+}(X)$$

$$\{\psi_l(X), \pi_m(Y)\} = i \delta_{lm} \delta(X - Y)$$

$$\Rightarrow \quad \{\psi_l(X), \pi_m^{+}(Y)\} = \delta_{lm} \delta(X - Y)$$

($X^0 = Y^0$ anti commutation rules at some times)

So,
$$\delta_{lm} \delta^3 (X - Y) = \{\psi_l(X), \psi_m^{+}(Y)\}$$

$$= \int \{a(p,\sigma,) a^{+}(p',\sigma')\} u_l(p,\sigma,) \bar{u}_m(p',\sigma')$$

$$\exp(-i(p \cdot X - p' \cdot Y)) d^3 p d^3 p'$$

$$+ \int \{b^{+}(p,\sigma), b(p',\sigma')\} v_l(p,\sigma) \bar{v}_m(p',\sigma')$$

$$\exp(-i(p \cdot X - p' \cdot Y)) d^3 p d^3 p'$$

Let
$$\{a(p,\sigma), a^{+}(p',\sigma')\}$$

$$= F(p) \delta^3 (p - p') \delta_{\sigma,\sigma'}$$

$$\{b(p,\sigma), b^+(p',\sigma')\} = G(p)\delta^3(p-p')\delta_{\sigma,\sigma'}$$

Then, $\int F(p)u_l(p,\sigma)\bar{u}_m(p,\sigma)\exp(-i\,p\cdot(X-Y))d^3p$

$$-\int G(p)v_l(p,\sigma)\bar{v}_m(p,\sigma)\exp(-i\,p\cdot(X-Y))d^3p$$

$$= \delta_{lm}\delta(X-Y)$$

Hence

$$F(p)\sum_{\sigma=\pm} u_l(p,\sigma)\bar{u}_m(p,\sigma) - G(p)\sum_{\sigma=\pm} v_l(p,\sigma)\bar{v}_m(p,\sigma)$$

$$= \frac{\delta_{lm}}{(2\pi)^3}$$

Take $F(p) = G(p) = 1$

Then

$$\{a(p,\sigma), a^+(p',\sigma')\} = \delta_{\sigma\sigma'}\delta^3(p-p')$$

$$\{b(p,\sigma), a^+(p',\sigma')\} = \delta_{\sigma\sigma'}\delta^3(p-p')$$

$$\sum_{\sigma=\pm}\left(u_l(p,\sigma)\bar{u}_m(p,\sigma) - v_l(p,\sigma)\bar{v}_m(p,\sigma)\right) = \frac{\delta_{lm}}{(2\pi)^3}$$

Interaction between electrons, positions and photon

$$\mathcal{L}_{int} = J^\mu A_\mu$$

$$S_{int} = \int \mathcal{L}_{int} d^4X = -e\int \psi^+(X)\gamma^0\gamma^\mu\psi(X)A_\mu(X)d^4X$$

$$= -e\int \bar{\psi}(X)\gamma^\mu\psi(X)A_\mu(X)d^4X$$

$$A_\mu(X) = \int\big(a(p,\sigma)e_\mu(p,\sigma)\exp(-i\,p\cdot X)$$

$$+ a^+(p,\sigma)e_u^*(p,\sigma)\exp(i\,p\cdot X)\big)d^3p$$

$$\partial^\mu A_\mu = 0$$

$$\Rightarrow \quad p^\mu e_\mu(p,\sigma) = 0, \sigma = 1, 0, -1$$

$$\mathcal{L} = \frac{1}{2}F_{\mu\nu}F^{\mu\nu} = \frac{1}{2}\left(A_{\nu,\mu} - A_{\mu,\nu}\right)\left(A^{\nu,\mu} - A^{\mu,\nu}\right)$$

$$= 2F_{\mu\nu}F^{\nu,\mu}$$

$$\frac{\partial \mathcal{L}}{\partial A^{\mu}_{,0}} = \pi_{\mu}(X)$$

$$\pi_0(X) = 0, \pi_r(X) = \left(A_{r,0} - A_{0,r}\right) - A^r_{,0} - A^0_{,r} = E_r$$

$$\left[A^r(X), \pi_s(Y)\right] = i\delta^r_s \delta^3(X-Y)$$

Let

$$\left[a(p,\sigma), a^+(p',\sigma')\right] = \delta_{\sigma\sigma'}\delta^3(p-p')$$

Then,

$$[A_{\mu}(X), A_r(Y)]\,(X^0{=}Y^0) = \int \delta_{\sigma,\sigma'}\delta^3(p-p')e_{\mu}(p,\sigma)e^*_{\nu}(p',\sigma')$$

$$\exp(-i(p\cdot X - p'\cdot Y))d^3p d^3p'$$

$$-\int \delta_{\sigma,\sigma'}\delta^3(p-p')e^*_{\mu}(p,\sigma)e_{\nu}(p',\sigma')$$

$$\exp(i(p\cdot X - p'\cdot Y))d^3p d^3p$$

$$= \int\left\{\left(\sum_{\sigma} e_{\mu}(p,\sigma)e^*_{\nu}(p,\sigma)\right)\exp(-ip\cdot(X-Y))\right.$$

$$\left. - \int\left(\sum_{\sigma} e^*_{\mu}(p,\sigma)e_{\nu}(p,\sigma)\right)\exp(i\,p\cdot(X-Y))\right\}d^3p$$

$$(X^0 = Y^0) \equiv \int G_{\mu\nu}(p)\exp(-i\,p\cdot(X-Y))d^3p$$

$$\left[A^r(X), \left(-A^s_{,0} - A^0_{,s}\right)(Y)\right] = i\delta_{rs}\delta^3(X-Y)$$

$$-\frac{\partial}{\partial Y_0}\left[A^r(X), A^s(Y)\right]$$

So,

$$\int G_{rs}(p) i\,|p|\exp(-i\,p\cdot(X-Y))d^3p$$

$$(p^0 = |p|) = i\delta_{rs}\delta^3(X-Y)$$

Thus,

$$|p|\,G_{rs}(p) = (2\pi)^{-3}\,\delta_{rs}$$

Note that $G_{rs}(p)$

$$= \sum_{\sigma=1,0,-1}\left\{e_r(p,\sigma)e^*_s(p,\sigma) - e^*_r(p,\sigma)e_s(p,\sigma)\right\}$$

$$= \frac{(2\pi)^{-3}\delta_{rs}}{|p|}$$

Let $|0\rangle$ be the vacumn state

$a^+(p,\sigma)|0\rangle = |p,\sigma\rangle$ is a state having one electron of moment p and spin σ.

$$\alpha_0 = \left\langle p',\sigma' \middle| \exp\left(i\int J^\mu A_\mu d^4X\right) \middle| p,\sigma \right\rangle$$

$$\approx \left\langle p',\sigma' \middle| i\int J^\mu A_\mu d^4X \middle| p,\sigma \right\rangle$$

is the transition probability amplitude for an electon to go from the state $|p,\sigma\rangle$ to $|p',\sigma'\rangle$ under interaction with an external photon field A_m = (c-number field)

Here

$$J^\mu = -e\bar{\psi}\gamma^\mu\psi$$

$$-i_{os} = \left\langle 0 \middle| a(p',\sigma')\left(\int \bar{\psi}(X)\gamma^\mu\psi(X)A_\mu(X)d^4X\right)a^+(p,\sigma) \middle| 0 \right\rangle$$

$$= \int d^4X \left\langle 0 \middle| a(p',\sigma')\bar{\psi}(X)\gamma^\mu\psi(X)A_\mu(X)a^+(p,\sigma) \middle| 0 \right\rangle$$

$$\psi(X) = \int\left(a(p,\sigma)\underline{u}(p,\sigma)\exp(-ip\cdot X) + b^+(p,\sigma)\underline{v}(p,\sigma)\exp(ip\cdot X)\right)d^3p$$

$$\left\langle 0 | a(p',\sigma')\bar{\psi}(X)\gamma^\mu_\psi(X)a^+(p,\sigma) \neq 0 \right\rangle$$

$$\langle 0|\int a(p',\sigma')\left(a^+(p_1,\sigma_1)\bar{u}(p_1,\sigma_1)\exp(i\,p_1\cdot X)\right.$$

$$\left. + b(p_1,\sigma_1)\bar{v}(p_1,\sigma_1)\exp(-i\,p_1\cdot X)\right)\gamma^\mu$$

$$Xa(p_2,\sigma_2)\mu(p_2,\sigma_2)\exp(-ip_2 . X)$$

$$+ b_+(p_2,\sigma_2)v(p_2,\sigma_2)\exp(ip_2\cdot X))$$

$$a^+(p,\sigma)d^3p_1, d^3p_2|0\rangle$$

$$= \int\left\{-\delta(p'-p_1)\delta_{\sigma'\sigma_1}\bar{u}(p_1,\sigma_1)\exp(i\,p_1\cdot X)\gamma^\mu\right.$$

$$\left.\times\delta(p_2-p)\delta_{\sigma\sigma_2}u(p_2,\sigma_2)\exp(-i\,p_1\cdot X)\right\}d^3p_1d^3p_2$$

$$= \bar{u}(p',\sigma')\gamma^\mu u(p,\sigma)\exp(i(p'-p)\cdot X)$$

The amplitude of scattering in this proportional

$$\int \bar{u}(p',\sigma')\gamma^\mu u(p,\sigma)e^{i(p-p')\cdot X}A_\mu(X)d^4X$$

$$= \bar{u}(p',\sigma')\gamma^\mu u(p,\sigma)\hat{A}_\mu(p'-p)$$

❖❖❖❖❖

Quantum Filtering in Robotics, Information, Feynman Path Integrals, Levy Noise, Haar Measure on Groups, Gravity and Robots, Canonical Quantum Gravity, Langevin Equation, Antenna Current in a Field Dependent Medium

[161] For Dr. Rajveer

$$H_Z = 0, E_Z \neq 0$$

$$\nabla \times E = -j\omega\mu H, \nabla \times H = j\omega \in E$$

$$\nabla_\perp E_Z \times \hat{Z} - \gamma \hat{Z} \times E_\perp = -j\omega\mu H_\perp$$

$$-\gamma \hat{Z} \times H_\perp = j\omega\mu\varepsilon E_\perp$$

So,

$$\nabla_\perp E_Z + \gamma E_\perp = -j\omega\mu \hat{Z} \times H_\perp = -\frac{\omega^2 \mu \varepsilon}{\gamma} E_\perp$$

So

$$\left(\frac{\gamma^2 + \omega^2 \mu \varepsilon}{\gamma}\right) E_\perp = \nabla_\perp E_Z$$

$$E_\perp = -\frac{\gamma}{h^2} \nabla_\perp E_Z \equiv \frac{1}{h^2} \frac{\partial}{\partial Z} \nabla_\perp E_Z$$

$$\gamma H_\perp = j\omega\varepsilon \hat{Z} \times E_\perp$$

$$H_\perp = \frac{j\omega\varepsilon}{\gamma} \hat{Z} \times E_\perp = \frac{-j\omega\varepsilon}{h^2} \hat{Z}$$

$$E_X = \frac{1}{h^2} \frac{\partial^2 E_Z}{\partial X \partial_Z}, \ E_Y = \frac{1}{h^2} \frac{\partial^2 E_Z}{\partial Y \partial_Z}$$

$$H_X = \frac{j\omega\varepsilon}{h^2} \frac{\partial E_Z}{\partial Y}, H_Y = \frac{-j\omega\varepsilon}{h^2} \frac{\partial E_Z}{\partial X}$$

εE_X is continuous at $X = 0, a.$
εE_Y is continuous at $Y = 0, b.$
H_Y is continuous at $X = 0, a.$
H_X is continuous at $Y = 0, b.$
$\approx E_X$ is continuous at $X = 0, a.$
εE_Y is continuous at $X = 0, b.$

$$u_1(X, Y \mid k_X, k_Y) = \cos(k_X X)\cos(k_Y Y)$$

$$u_2(X, Y \mid k_X, k_Y) = \cos(k_X X)\sin(k_Y Y)$$

$$u_3(X, Y \mid k_X, k_Y) = \sin(k_X X)\cos(k_Y Y)$$

$$u_4(X, Y \mid k_X, k_Y) = \sin(k_X X)\sin(k_Y Y)$$

$$E_Z(X, Y, Z) = \cos\left(\frac{p\pi Z}{d}\right)\sum_{m=1}^{4} C_m u_m(X, Y \mid k_Y, R_Y), 0 < X < a, 0 < Y < b.$$

$$\cos\left(\frac{p\pi Z}{d}\right)\sum_{m=1}^{4} C'_m u_m(X, Y \mid k'_Y, k'_Y) X > 0 \text{ or } X > a, \text{ or } Y < 0 \text{ or } Y > b.$$

$$k_X^2 + k_Y^2 = h^2 = \omega^2\mu\varepsilon_1 - (p\pi/d)^2$$

$$k_X'^2 + k_Y'^2 = h'^2 = \omega^2\mu\varepsilon_2 - (p\pi/d)^2$$

Boundary continuous

$\frac{\varepsilon}{h^2}\frac{\partial E_Z}{\partial X}$ continuous at $X = 0, a$

$\frac{\varepsilon}{h^2}\frac{\partial E_Z}{\partial Y}$ continuous at $X = 0, b$

$$\frac{\varepsilon_1}{h^2}\sum_m C_m \frac{\partial u_m(0, Y \mid k_X, k_Y)}{\partial X} = \frac{\varepsilon_2}{h'^2}\sum_m C'_m \frac{\partial u_m(0, Y \mid k'_X, k'_Y)}{\partial X} \quad (1)$$

$$\frac{\varepsilon_1}{h^2}\sum_m C_m \frac{\partial u_m}{\partial X}(a, Y \mid k_X, k_Y) = \frac{\varepsilon_2}{h'^2}\sum_m C'_m \frac{\partial um}{\partial X}(a, Y \mid k'_X, k'_Y) \quad (2)$$

$$= \frac{\varepsilon_1}{h'^2}\sum_m C_m \frac{\partial u_m}{\partial Y}(X \mid k_X, k_Y) = \frac{\varepsilon_2}{h'^2}\sum_m C'_m \frac{\partial u_m}{\partial Y}(X, 0 \mid k'_X, k'_Y) \quad (3)$$

$$= \frac{\varepsilon_1}{h^2}\sum_m C'_m \frac{\partial u_m}{\partial Y}(X, b \mid k_X, k_Y) = \frac{\varepsilon_2}{h'^2}\sum_m C'_m \frac{\partial u_m}{\partial Y}(X, b \mid k'_X, k'_Y) \quad (4)$$

$(1) \Rightarrow$ $$\frac{\varepsilon_1}{h^2}\{C_3 k_X \cos(k_Y Y) + C_4 k_X \sin(k_Y Y)\}$$

$$\frac{\varepsilon_2}{h'^2}\left\{C_3'k_X'\cos\left(k_Y'Y\right)+C_4'k_X'\sin\left(k_Y'Y\right)\right\}$$

$$\approx k'_y = k_y$$

$$\frac{\varepsilon_1 C_4 k_X}{h^2} = \frac{\varepsilon_2 C_3' k_X'}{h'^2} \qquad (\alpha_1)$$

$$\frac{\varepsilon_1 C_4 k_X}{h^2} = \frac{\varepsilon_2 C_4' k_X'}{h'^2} \qquad (\alpha_2)$$

$$\Rightarrow \qquad \frac{C_4}{C_3} = \frac{C_4'}{C_3'}$$

$$(2) \Rightarrow \frac{\varepsilon_1}{h^2}\left\{-C_1 k_X \sin\left(k_X a\right)\cos\left(k_Y Y\right) - C_2 k_X \sin\left(k_X a\right)\sin\left(k_Y Y\right)\right.$$

$$\left. + C_3 k_X \cos\left(k_X a\right)\cos\left(k_Y Y\right) + C_4 k_X \cos\left(k_X a\right)\sin\left(k_Y Y\right)\right\}$$

$$\underline{k} \leftrightarrow \underline{k}', \varepsilon_1 \leftrightarrow \varepsilon_2 h \to h'$$

$$= 11 \;\; \underline{k} \leftrightarrow \underline{k}', \varepsilon_1 \leftrightarrow \varepsilon_2 h \to h'$$

$$\Rightarrow \qquad \frac{\varepsilon_1 k_X}{h^2}\left\{C_1 \sin\left(k_X a\right) - C_3 \cos\left(k_X a\right)\right\}$$

$$= \frac{\varepsilon_1 k_X'}{h'^2}\left\{C_1' \sin\left(k_X' a\right) - C_3' \cos\left(k_X' a\right)\right\} \qquad (\alpha_3)$$

and

$$\frac{\varepsilon_1 k_X}{h^2}\left\{C_2' \sin\left(k_X' a\right) - C_4 \cos\left(k_X a\right)\right\}$$

$$\frac{\varepsilon_1 k_X'}{h'^2}\left\{C_2' \sin\left(k_X' a\right) - C_4' \cos\left(k_X' a\right)\right\} \qquad (\alpha_4)$$

$$k'_Y = K_Y,\; k_X'^2 + k_Y'^2 = h'^2 \Rightarrow K'_X = \sqrt{h'^2 - k_Y^2}$$

$$= \sqrt{\omega^2\mu\varepsilon_2 - \left(\frac{p\pi}{d}\right)^2 - k_Y^2}$$

$$k_X = \sqrt{h^2 - k_Y^2} = \sqrt{\omega^2\mu\varepsilon_1 - \left(\frac{p\pi}{d}\right)^2}$$

(α_2) and $(\alpha_4) \Rightarrow$

$$\frac{\varepsilon_1 C_2}{h^2}\sin\left(k_X^* a\right) = \frac{\varepsilon_2 C_2'}{h'^2}\sin\left(k_X' a\right) \qquad (\alpha_5)$$

(α_1) and $(\alpha_3) \Rightarrow$

$$\frac{\varepsilon_1 C_1}{h^2}\sin\left(k_X a\right) = \frac{\varepsilon_2 C_1'}{h'^2}\sin\left(k_X' a\right) \qquad (\alpha_6)$$

$$(3) \Rightarrow \frac{\varepsilon_1}{h^2}\{C_2 k_Y \cos(k_X X) + C_4 k_Y \sin(k_X X)\}$$

$$= \frac{\varepsilon_2}{h'^2}\{C_2' k_Y' \cos(k_X' X) + C_4' k_Y' \sin(k_X' X)\}$$

Since $k'_X \neq k_X k'_Y = k_Y$, we get

$$C_2 = C'_2 = 0$$
$$C_4 = C'_4 = 0$$

$$(4) \Rightarrow \frac{\varepsilon_1}{h^2}\{-C_1^* k_Y \cos(k_X X)\sin(k_Y b) + C_2 k_Y \cos(k_X X)\mathrm{Cos}(k_Y b)$$
$$-C_3 k_Y \sin(k_X X)\sin(k_Y b)$$
$$+C_4 k_Y \sin(k_X X)\cos(k_Y b)\}$$

= 11 with $\varepsilon_1 \to \varepsilon_2, h \to h', \underline{k} \to \underline{k}'$

Since $C_2 = C'_2 = 0$ and $C_4 = C'_4 = 0$ and $k_x \neq k'_x, k'_Y = k_Y$

it follows that

$$-\frac{\varepsilon_1 C_1}{h^2}\sin(k_Y b) = 0 \; \frac{\varepsilon_1 C_3}{h^2}\sin(k_Y b)$$

$$\Rightarrow \quad \sin(k_Y b) = 0, k_Y = \frac{n_Y \pi}{b}, n_Y \varepsilon \mathbb{Z}$$

So,

$$C_2 = C'_2 = 0 = C_4 = C'_4$$

$$k_Y = \frac{n_Y \pi}{b} = k'_Y$$

$$k_X = \left(h^2 - k_Y^2\right)^{1/2}, k_X' = \left(h'^2 - k_Y^2\right)^{1/2}$$

$$\therefore \quad C'_1 = \frac{\varepsilon_1}{\varepsilon_2}\frac{h'^2}{h^2}\frac{\sin(k_X a)}{\sin(k_X' a)}C_1$$

❖❖❖❖❖

[162] Quantum filtering with robotics application

$$d\pi_t(X) = F_{kt}(X)dt + \sum_{k=0}^{\infty} G_{kt}(X)(dY_t)^k$$

$$F_t(X), G_{kt}(X) \in \sigma\{Ys : s \le t\} = \eta_{t]}$$

$$dC_t = f(t)C_t(dY_t)$$

$$d'j_t(X) = j_t\left(\theta^{\alpha}_{\beta}(X)\right)d\Lambda^{\beta}_{\alpha}(t)$$

$$\mathbb{E}\{j_t(X)\mid\eta_t\} = \pi_t(X)$$

$$\mathbb{E}\left\{\left(j_t(X)-\pi_t(X)\right)C_t\right\} = 0$$

$$\mathbb{E}\left\{\left(dj_t(X)-d\pi_t(X)\right)C_t+\left(d_t(X)-\pi_t(X)\right)dC_t\right.$$

$$\left.+\left(dj_t(X)-d\pi_t(X)\right)dC_t\right\} = 0$$

$$\pi_t\left(\theta^{\alpha}_{\beta}(X)\right)\mathbb{E}\left\{d\Lambda^{\beta}_{\alpha}(t)\right\}$$

$$-F_t(X)dt-\sum_{k=1}^{\infty}G_{kt}(X)\mathbb{E}\left\{\left(dY_t\right)^k\right\} = 0 \qquad ...(1)$$

$$\pi_t\left(\theta^{\alpha}_{\beta}(X)\right)\mathbb{E}\left\{d\Lambda^{\beta}_{\alpha}(t)dY_t\right\}-\sum_{K}G_{kt}(X)\mathbb{E}\left(dY_t\right)^{k+1} = 0 \qquad ...(2)$$

$$dG_{kt} = f_k(t)C_{kt}\left(dY_t\right)^k$$

$$\mathbb{E}\left\{j_t(X)-\pi_t(X)C_{kt}\right\} = 0$$

$$\mathbb{E}\left\{\left(dj_t(X)-d\pi_t(X)\right)C_{kt}\right\}$$

$$+\mathbb{E}\left\{\left(j_t(X)-\pi_t(X)\right)dC_{kt}\right\}$$

$$+\mathbb{E}\left\{\left(dj_t(X)-d\pi_t(X)\right)dC_{kt}\right\} = 0$$

$$\begin{cases}\pi_t\left(\theta^{\alpha}_{\beta}(X)\right)\mathbb{E}\left\{d\Lambda^{\beta}_{\alpha}(t)\right\}-F_t(X)dt-\sum\limits_{k\geq 1}G_{kt}(X)\mathbb{E}\left\{\left(dY_t\right)^k\right\}=0\\ \pi_t\left(\theta^{\alpha}_{\beta}(X)\right)E\left\{d\Lambda^{\beta}_{\alpha}(t)\left(dY_t\right)^k\right\}-\sum\limits_{m\geq 1}G_{kt}(X)\mathbb{E}\left\{\left(dY_t\right)^{m+k}\right\}=0\end{cases}$$

$$Y_t = U_t^*\left(c^{\alpha}_{\beta}\Lambda^{\beta}_{\alpha}(t)\right)U_t$$

$$= j_t\left(c^{\alpha}_{\beta}\Lambda^{\beta}_{\alpha}(t)\right)$$

$$dY_t = c^{\alpha}_{\beta}d\Lambda^{\beta}_{\alpha}(t)+c^{\alpha}_{\beta}dU_t^*d\Lambda^{\beta}_{\alpha}(t)U_t+c^{\alpha}_{\beta}U_t^*d\Lambda^{\beta}_{\alpha}dU_t$$

$$dU_t = \left(L^{\alpha}_{\beta}d\Lambda^{\beta}_{\alpha}(t)\right)U(t)$$

$$dU^*(t)d\Lambda^{\beta}_{\alpha}(t) = U^*(t)L^*_{\sigma}d\Lambda^{\sigma}_{\rho}d\Lambda^{\beta}_{\alpha}$$

$$= U^*(t)L^{\rho}_{\sigma}\varepsilon^{\sigma}_{\rho}d\Lambda^{\beta}_{\rho}$$

$$dY_t = c^\alpha_\beta d\Lambda^\beta_\alpha + c^\alpha_\beta U^* L^{\rho^*}_\sigma U\, \varepsilon^\sigma_\alpha d\Lambda^\beta_\rho + c^\alpha_\beta U^* L^\rho_\sigma U\, \varepsilon^\beta_\rho d\Lambda^\sigma_\alpha$$

$$d\Lambda^\beta_\alpha dU = d\Lambda^\beta_\alpha L^\rho_\sigma d\Lambda^\sigma_\rho U = L^\rho_\sigma U \varepsilon^\beta_\rho d\Lambda^\sigma_\rho$$

$$dY_t = c^\alpha_\beta d\Lambda^\beta_\alpha + c^\alpha_\beta \varepsilon^\sigma_\alpha j_t\left(L^{\rho^*}_\sigma\right) d\Lambda^\beta_\rho + c^\alpha_\beta \varepsilon^\beta_\rho j_t\left(L^\rho_\sigma\right) d\Lambda^\sigma_\alpha$$

$$dj_t(X) = d\left(U_t^* X U_t\right) = dU_t^* X U_t + U_t^* X dU_t + dU_t^* X\, dU_t$$

$$= U_t^*\left\{L^{\alpha^*}_\beta X d\Lambda^\alpha_\beta + X L^\alpha_\beta d\Lambda^\beta_\alpha + L^\alpha_\beta X L^\rho_\sigma d\Lambda^\alpha_\beta d\Lambda^\sigma_\rho\right\} U_t$$

$$= -j_t\left(\theta^\alpha_\beta(X)\right) d\Lambda^\beta_\alpha$$

$$\theta^\alpha_\beta(X) = L^{\beta^*}_\alpha X + X L^\alpha_\beta + L^{\sigma^*}_\beta X L^{\sigma^*}_\alpha \varepsilon^{\alpha\sigma}_\rho$$

Approximate solution to the Even's Hudson flow:

$$dj_t(X) = j_t\left(\theta^\alpha_\beta(X)\right) d\Lambda^\beta_\alpha(t)$$

$$= j_t\left(\theta^0_0(X)\right) dt + j_t\left(\theta^0_k(X)\right) dA_k(t) + j_t\left(\theta^k_0(X)\right) dA_k^* + j_t\left(\theta^k_m(X)\right) d\Lambda^m_k(t)$$

$$T_{r_2}(\rho j_t(X)) = \rho_s(t) \equiv Tr_2\left[j_t^*(\sigma) X\right] = Tr_2\left(j_t^*(\rho)\right) X$$

$$d\rho_s(t) = T_{r_2}(\rho_d j_t(X))$$

$$= Tr_2\left\{\rho\, j_t\left(\theta^\alpha_\beta(X)\right) d\Lambda^\beta_\alpha(t)\right\}$$

$$\rho = \rho_s(0) \otimes \sum_k \lambda_k |e(u_k)\rangle\langle e(u_k)|$$

$$\sum_k \lambda_k \exp\left(\|u_k\|^2/2\right) = 1$$

$$d\rho_s(t) = Tr_2\left\{\left(\rho_s(0) \otimes \sum_k \lambda_k |e(u_k)\rangle\langle e(u_k)|\right) j_t\left(\theta^\alpha_\beta(X)\right) d\Lambda^\beta_\alpha(t)\right\}$$

$$= Tr_2\left\{\left(\rho_s(0) \otimes \rho_e(0)\right) j_t\left(\theta^\alpha_\beta(X)\right) d\Lambda^\beta_\alpha(t)\right\}$$

$$= Tr_2\left\{U_t\left(\rho_s(0) \otimes \rho_e(0)\right) U_t^* \theta^\alpha_\beta(X) dX^\beta_\alpha(t)\right\}$$

$$= Tr_2\left\{\rho_s(0)\sum_k \lambda_k \left\langle e(u_k)\middle| j_t\left(\theta^\alpha_\beta(X)\right)d\Lambda^\beta_\alpha(t)\middle|e(u_k)\right\rangle\right\}$$

$$= dt.Tr_2\left\{\rho_s(0)\sum_k \lambda_k u_{k\beta}(t)\overline{u_{k\alpha}(t)}\left\langle e(u_k)\middle| j_t\left(\theta^\alpha_\beta(X)\right)\middle|e(u_k)\right\rangle\right\}$$

$$Tr\left\{\left(\rho_s(0)\otimes\rho_e(0)\right)j_t(X)\right\} \triangleq Tr\left[\rho_s(t)X\right]$$

$$dt.Tr\left[\rho'_s(t)X\right] = Tr\left\{\left(\rho_s(0)\otimes\rho_e(0)\right)j_t\left(\theta^\alpha_\beta(X)\right)d\Lambda^\beta_\alpha(t)\right\}$$

$$= dt\cdot\sum_{k,\alpha,\beta}\lambda_k u_{k\beta}(t)\overline{u_{k\alpha}(t)}Tr\left\{\rho_s(0)\otimes|e(u_k)\rangle\langle e(u_k)|\,j_t\left(\theta^\alpha_\beta(X)\right)\right\}$$

$$= dt\cdot\sum_{k,\alpha,\beta}\lambda_k u_{k\beta}(t)\overline{u_{k\alpha}(t)}Tr\left\{\rho_{sk}(t)\theta^\alpha_\beta(X)\right\}$$

where $\rho_{sk}(t) = Tr\left\{\left(\rho_s(0)\otimes|e(u_k)\rangle\langle e(u_k)|\right)U\right\}U_t^*$

So $\rho'_s(t) = \sum_{k\alpha\beta}\lambda_k u_{k\beta}(t)\overline{u_{k\alpha}(t)}\theta^{\alpha*}_\beta\left(\rho_{sk}(t)\right)$

$$dU_t = \left(L^\alpha_\beta(t)d\Lambda^\beta_\alpha(t)\right)U_t$$

$$U_t^*U_t = I$$

$$d\Lambda^\beta_\alpha\cdot d\Lambda^\rho_\sigma = \varepsilon^\beta_\sigma\, d\Lambda^\rho_\alpha$$

$$j_t(X) = U_t^*XU_t$$

$$dj_t(X) = j_t\left(\theta^\alpha_\beta(X)\right)d\Lambda^\beta_\alpha(t)$$

$$Tr(\rho_s(t)X) = Tr\left(\left(\rho_s(0)\otimes|\varphi(u)\rangle\langle\varphi(u)|\right)j_t(X)\right)$$

$$|\varphi(u)| = \exp\left(-\frac{\|u\|^2}{2}\right)|e(u)\rangle$$

$$dt.Tr(\rho'_s(t)X) = Tr\left(\rho_s(0)\otimes|\varphi(u)\rangle\langle\varphi(u)|\right)j_t\left(\theta^\alpha_\beta(X)\right)d\Lambda^\beta_\alpha(t)$$

$$= Tr\left(\rho_s(0)\otimes|\varphi(u)\rangle\langle\varphi(u)|\right)j_t\left(\theta^\alpha_\beta(X)\right)u_\beta(t)\overline{u_\alpha(t)}dt$$

$$= u_\beta(t)\overline{u_\alpha(t)}dt.Tr\left(\rho_s(t)\theta^\alpha_\beta(X)\right)$$

$$= u_\beta(t)\overline{u_\alpha(t)}dt.Tr\left(\theta^{\alpha^*}_\beta\left(\rho_s(t)\right)X\right)$$

Hence, our generalized Sudarshan Lindblad equation is

$$\rho'_s(t) = \sum_{\alpha,\beta} u_\beta(t)\overline{u_\alpha(t)}\,\theta_\beta^{\alpha^*}(\rho_s(t))$$ (Noisy Schrödmger equation)

❖❖❖❖❖

[163] Problem from Thomann Quantum General Relativity

$$\left[\vec{J}(\vec{N}_1), \vec{J}(\vec{N}_2)\right] = \left[\vec{J}(\vec{N}_1), \vec{J}(\vec{N}_2)\right]$$

$$= \int d^3X d^3X' \left[p^{ab}(X)\mathcal{L}_{N_1}q_{ab}(X), P^{cd}(X')\mathcal{L}_{N_2}q_{cd}(X')\right]$$

$$= \int d^3X\, d^3X' \left\{P^{cd}(X')\left[P^{ab}(X), \mathcal{L}_{N_2}q_{cd}(X')\right]\mathcal{L}_{N_1}q_{ab}(X)\right.$$

$$\left. + P^{ab}(X)\left[\mathcal{L}_{N_1}q_{ab}(X), p^{cd}(X')\right]\mathcal{L}_{N_2}q_{cd}(X')\right\}$$

$$= \int d^3X\, d^3X' \left\{P^{cd}(X')\mathcal{L}'_{N_2}\left(\delta(X-X')\delta_c^a\,\delta_d^b\right)\mathcal{L}_{N_1}q_{ab}(X)\right.$$

$$\left. - P^{ab}(X)\mathcal{L}_{N_1}\left(\delta(X-X')\delta_a^c\,\delta_b^d\right)\mathcal{L}_{N_2}q_{cd}(X')\right\}$$

$$= \int d^3X \left\{-\mathcal{L}_{N_2}P^{cd}(X)\mathcal{L}_{N_1}q_{cd}(X)\right.$$

$$\left. + \mathcal{L}_{N_1}P^{ab}(X)\mathcal{L}_{N_2}q_{ab}(X)\right\}$$

$$= \int d^3X\, P^{ab}\left\{\mathcal{L}_{N_2}\mathcal{L}_{N_1} - \mathcal{L}_{N_1}\mathcal{L}_{N_2}\right\}q_{ab}$$

$$= \int d^3X\, P^{ab}\mathcal{L}_{[N_2,\,N_1]}q_{ab} = J\left(\mathcal{L}_{N_2}N_1\right)$$

$$N_2(X)N_1^a(X')\left(\frac{S}{\sqrt{q}}\left[q_{ac}q_{bd} - \frac{1}{D-1}q_{ab}q_{cd}\right]P^{ab}P^{cd} + \sqrt{q}\,R\right)(X)_3$$

$$q_{\rho\sigma}D_\mu P^{\mu\sigma}(X')dX\,dX'$$

$$q_{\rho\sigma}(X')\left[q_{ab}(X)q_{cd}(X), D_\mu P^{\mu\sigma}(X')\right]P_{(X)}^{ab}P^{cd}(X)$$

$$q_{ab}(X)q_{cd}(X)\left[q_{(X)}^{ab}P^{cd}(X), q_{\rho\sigma}(X')\right]D_\mu P^{\mu\sigma}(X')$$

❖❖❖❖❖

[164] Parameter estimation in non-linear models?

$$\underline{X}_n = \left[\underline{\underline{H}}_{on} + \varepsilon\underline{\underline{H}}_{1n}\left(X_{n-1:n-p}\right)\right]\underline{\theta}$$

$$+\left[G_{on} + \varepsilon\underline{\underline{G}}_{1n}\left(X_{n-1:n-p}\right)\right]\underline{V}_n,\ n = p, p+1,.,N$$

Hence $\underline{X}_{n-1:\,n-p} = \begin{bmatrix} \underline{X}_{n-1} \\ \underline{X}_{n-p} \end{bmatrix} \in \mathbb{R}^{dp}$

$\underline{\underline{H}}_{on}$ and $\underline{\underline{G}}_{on}$, $n = p, p+1,.., N$ are constant matrices.

$\underline{\underline{H}}_{1n}$ and $\underline{\underline{G}}_{1n}$ are function of $\underline{X}_{n-1:\,n-p}$.

$\{\underline{V}_n\}_{n \ge p}$ is an iid $N\left(\underline{0}, \underline{\underline{\Sigma}}_V\right)$ sequence

Writing $\underline{\underline{H}}_n\left(X_{n-1:n-p}\right) = \underline{\underline{H}}_{on} + \varepsilon \underline{\underline{H}}_{1n}$

$$\underline{\underline{G}}_n\left(X_{n-1:n-p}\right) = \underline{\underline{G}}_{on} + \varepsilon \underline{\underline{G}}_{1n}$$

Then assuming $\underline{\underline{G}}_n$ to be non-singular and square,

$$p\left(\left\{\underline{X}_n\right\}_{n=p}^{N} \middle| \left\{\underline{X}_n\right\}_{n=0}^{p-1}, \underline{\theta}\right)$$

$$= K \cdot \exp\left\{-\frac{1}{2}\sum_{n=p}^{N}\left(\underline{X}_n - \underline{\underline{H}}_n\left(\underline{X}_{n-1:n-p}\right)\underline{\theta}\right)^T\right.$$

$$\left(\underline{\underline{G}}_n\left(X_{n-1:n-p}\right)\Sigma_V G_n^T\left(X_{n-1:n-p}\right)\right)^{-1}$$

$$\left.\left(\underline{X}_n - \underline{\underline{H}}_n\left(\underline{X}_{n-1:n-p}\right)\underline{\theta}\right)\right\}$$

so $$\underline{\hat{\theta}}_{ML}(N) = \underset{\underline{\theta}}{\arg\min}\left(-\log p\right)$$

$$= \underset{\underline{\theta}}{\arg\min}\sum_{n=p}^{N}\left(\underline{X}_n - \underline{\underline{H}}_n\underline{\theta}\right)^T\left(G_n\Sigma_V G_n^T\right)^{-1}\left(\underline{X}_n - \underline{\underline{H}}_n\underline{\theta}\right)$$

$$= \left[\sum_{n=p}^{N}\underline{\underline{H}}_n^T\left(G_n\Sigma_V G_n^T\right)^{-1}\underline{\underline{H}}_n\right]^{-1}\left[\sum_{n=p}^{N}H_n^T\left(G_n\Sigma_V G_n^T\right)^{-1}\underline{X}_n\right]$$

$$= \underline{\theta} + \underline{\underline{F}}_N^{-1}\sum_{n=p}^{N}\underline{\underline{L}}_n\underline{V}_n, \text{ where}$$

$$\underline{\underline{L}}_n = \underline{\underline{H}}_n^T\left(\underline{\underline{G}}_n\underline{\underline{\Sigma}}_V\underline{\underline{G}}_n^T\right)^{-1}\underline{\underline{G}}_n$$

$$\underline{\underline{F}}_N = \sum_{n=p}^{N}\underline{\underline{H}}_n^T\left(\underline{\underline{G}}_n\underline{\underline{\Sigma}}_V\underline{\underline{G}}_n^T\right)^{-1}\underline{\underline{H}}_n$$

$$= \sum_{n=p}^{N}\underline{\underline{H}}_n^T\underline{\underline{G}}_n^{-T}\underline{\underline{\Sigma}}_V^{-1}\underline{\underline{G}}_n^{-1}H_n$$

Now,

Maximam likelihood with non-Gaussian noise

$$p_W(W) = \frac{\exp\left(-\frac{w^2}{2}\right)}{\sqrt{2\pi}}\left(1+\sum_{k=1}^{p} C_k H_k(w)\right)$$

$$H_k(W) = (-1)^n e^{w^2/2} \frac{d^n}{dw^n} e^{-w^2/2}, n \geq 0 \simeq \text{Hermite polynomials}$$

$$\underline{\xi}[n+1] = \underline{f}\left(n, \underline{\xi}[n], \underline{\theta}\right) + \varepsilon \underline{W}[n+1]$$

where $\underline{W}[n] = \begin{pmatrix} W_1[n] \\ W_2[n] \end{pmatrix}$ and $W_1[n]$, $W_2[n]$, $n \geq 1$

are *iid* with pdf $p_W(W)$. Then,

$$p\left(\underline{\xi}[1], \cdot, \underline{\xi}[N] \middle| \underline{\xi}[0], \underline{\theta}\right)$$

$$= \prod_{n=0}^{N-1} p_{\underline{W}}\left(\underline{\xi}[n+1] - \underline{f}\left(n, \underline{\xi}[n], \underline{\theta}\right)\right)$$

where $p_{\underline{W}}\left(\underline{W}\right) = p_W\left(W_1\right) p_W\left(w_2\right)$

$$\underline{W} = \begin{pmatrix} W_1 \\ W_2 \end{pmatrix}$$

$$\hat{\theta}(N) \approx \underset{\underline{\theta}}{\arg\min}\left\{-\frac{1}{2}\sum\left\|\underline{\xi}[n+1] - \underline{f}\left(n_1\, \underline{\xi}[n]\right)\underline{\theta}\right\|^2 + \sum_{k=1}^{p} C_k H_k\left(\underline{\xi}[n+1] - \underline{f}\left(n, \underline{\xi}[n], \underline{\theta}\right)\right)\right\}$$

❖❖❖❖❖

[165] Mark Wilde Quantum Information Theory

$$I\left(A; B \middle| XU\right) = -H\left(A \mid B\,X\,U\right) + H\left(A \mid XU\right)$$

$$I(X; B|U) = H\left(X \mid U\right) - H\left(X \mid BU\right)$$

$$H(A|B\,X\,U) = -(1-\lambda)\log(1-\lambda) + (1-\lambda)H(A \mid B)\mathcal{N}(\theta)$$

$$\left(-\lambda p_X(x)\log\left(\lambda p_X(x)\right) + \lambda p_X(x) H(A \mid B) \mid 0 >< 0 \otimes \mathcal{N}\left(\psi_x\right)\right)$$

$$H(A|\,X\,U) = -(1-\lambda)\log(1-\lambda) + (1-\lambda)H(A)\mathcal{N}(\phi)$$

$$-\sum \lambda p_X(x)\log\left(\lambda p_X(x)\right) + \sum_x \lambda p_X(x) H(A)_{\mathcal{N}(\psi_x)}$$

$$H(A|X\,U) - H(A|B\,X\,U) = (1-\lambda)\left(H(A)_{\mathcal{N}(\phi)} - H(A\,|\,B)_{\mathcal{N}(\phi)}\right)$$

$$+\lambda\sum p_X(x)\left(H(A)_{\mathcal{N}(\psi_x)} - H(A\,|\,B)_{\mathcal{N}(\psi_x)}\right)$$

$$= (1-\lambda)I(A;B)_{\mathcal{N}(\phi)} + \lambda\sum p_X(x)I(A;B)\{p_X(x)N(\psi_X)\}$$

$$H(X\,|\,U) = -(1-\lambda)\log(1-\lambda) - \sum \lambda p_X(x)\log(\lambda p_X(x))$$

$$= -(1-\lambda)\log(1-\lambda) - \lambda\log\lambda - \lambda\sum p_X(x)\log p_X(x)$$

$$= -(1-\lambda)\log(1-\lambda) - \lambda\log\lambda + \lambda H(X)_{\{p_X(x),\,\psi_x\}}\,H(X|BU)$$

where $$W = W(\underline{p},\underline{U}) = \left|\sum_{k=1}^{M}\right| p_k U_k^T \,|\otimes U_k^*$$

$$\rho'_1(t) = -i[H_1,\rho_1] + \varepsilon\theta_{L_1}(\rho_1)$$

$$\rho'_2(t) = -i[H_2,\rho_2] + \varepsilon\theta_{L_2}(\rho_2)$$

$$(\rho_1\otimes\rho_2)' = \rho_1'\otimes\rho_2 + \rho_1\otimes\rho_2'$$

$$= \left(-i[H_1,\rho_1] + \varepsilon\theta_{L_1}(\rho,1)\right)\otimes\rho_2 + \rho_1\otimes\left(-i[H_2,\rho_2] + \varepsilon\theta_{L_2}(\rho_2)\right)$$

$$= -i[H_1\oplus H_2,\rho_1\otimes\rho_2]$$

$$+\varepsilon\left\{-\frac{1}{2}\left(L_1^*L_1\rho_1 + \rho_1L_1^*L_1 - 2L_1\rho_1L_1^*\right)\otimes\rho_2\right.$$

$$\left.-\frac{1}{2}\rho_1\otimes\left(L_2^*L_2\rho_2 + \rho_2L_2^*L_2 - 2L_2\rho_2L_2^*\right)\right\}$$

$$= -i[H_1\oplus H_2,\rho_1\otimes\rho_2] - \frac{\varepsilon}{2}\left\{\left(L_1^*L_1\otimes I\right)(\rho_1\otimes\rho_2)\right.$$

$$+(\rho_1\otimes\rho_2)\left(L_1^*L_1\otimes I\right) - 2(L_1\otimes I)(\rho_1\otimes\rho_2)\left(L_1^*\otimes I\right)$$

$$+\left(I\otimes L_2^*L_2\right)(\rho_1\otimes\rho_2) + (\rho_1\otimes\rho_2)\left(I\otimes L_2^*L_2\right)$$

$$\left.-2(I\otimes L_2)(\rho_1\otimes\rho_2)\left(I\otimes L_2^*\right)\right\}$$

$$= -i[H_1\oplus H_2,\rho_1\otimes\rho_2] + \varepsilon\left(\theta_{L_1}\oplus\theta_{L_2}\right)(\rho_1\otimes\rho_2)$$

Here the two system undergo decoupled dynamics. Coupling takes place due to our interaction between the two system spaces as well as bath spaces.

The coupled dynamics in

$$\rho'_{12} = -i[H_1 \oplus H_2 + \delta V_{12}, \rho_{12}]$$
$$-\frac{1}{2}\sum\{M_k^* M_k \rho_{12} + \rho_{12} M_k^* M_k - 2M_k \rho_{12} M_k^*\}$$

where

$$M_1 = L_1 \otimes I, M_2 = I \otimes L_2,$$
$$M_3 = \delta \cdot \sum_{k=1}^{p} M_{1k} \otimes M_{2k}$$

$\delta \simeq$ perturbation parameter.

Let $\rho_{12} = \rho_1 \otimes \rho_2 + \delta \cdot g_{12} + O(\delta^2)$

Then $g'_{12} = -i[H_1 \oplus H_2, g_{12}] - i[V_{12}, \rho_1 \otimes \rho_2]$
$$+\{\theta_{M_1}(g_{12}) + \theta_{M_2}(g_{12}) + \theta_{M_3}(\rho_1 \otimes \rho_2)\}$$

Write $\theta = -i\,ad(H_1 \oplus H_2) + \theta_{M_1} + \theta_{M_2}$

Then, $g'_{12} = \theta(g_{12}) - i[V_{12}, \rho_1 \otimes \rho_2] + \theta_{M_3}(\rho_1 \otimes \rho_2)$

so that $g_{12}(t) = \int_0^t \exp\{(t-\tau)\theta\}\{-i[V_{12}(\tau), \rho_1(\tau) \otimes \rho_2(\tau)]$
$$+ \theta_{M_3}(\rho_1(\tau) \otimes \rho_2(\tau))\}d\tau$$

❖❖❖❖❖

[166] Stochastic differential equation driven by jump processes

$N(t, d\xi) \simeq$ Poisson random measure

$\int_{[0,t]xX} f(s,\xi)N(ds, d\xi) \simeq$ Compound process

Let $\mathbb{E}\{N(t,E)\} = F(t, E) = E \in \mathcal{F}_X(X, \mathcal{F}_X)$ a measurable space.

$$\mathbb{E}\left\{\exp\left(\int_{[0,T]\times X} \varphi(t,\xi)N(dt, d\xi)\right)\right\}$$

$$\exp\left\{\int_{[0,T]\times X} (\exp(\varphi(t,\xi)) - 1)dF(t,\xi)\right\}$$

sde $\quad d\underline{x}(t) = \underline{f}(t, x(t))dt + \int_{\xi \in X} \underline{g}(t, \underline{x}(t), \xi)dN(t, \xi)$

Fokker-Planck equation.

$$d\psi(\underline{x}(t)) = \psi'(\underline{x}(t))^T \underline{f}(t, \underline{x}(t))dt + \int_{\xi \in X} \left(\psi(\underline{x}(t) + \underline{g}(t, \underline{x}(t), \xi)) - \psi(\underline{x}(t))\right)dN(t, \xi)$$

This in Ito,s formula for the jump driven Markov process $\underline{x}(t)$

$$\frac{d}{dt}\mathbb{E}\{\psi(\underline{x}(t))\} = \mathbb{E}\left\{\psi'(\underline{x}(t))^T \underline{f}(t, \underline{x}(t))\right\} + \int_{\xi \in X} \mathbb{E}\left\{\psi(\underline{x}(t) + \underline{g}(t, \underline{x}(t), \xi)) - \psi(\underline{x}(t))\right\} f(t, d\xi)$$

where $\quad f(t, E) = \dfrac{\partial F(t, E)}{\partial t}$

Let $p(t, x)$ denote the pdf of $\underline{x}(t)$. Then the above gives

$$\frac{d}{dt}\int \psi(\underline{x})p(t, \underline{x})d\underline{x} = \int \psi'(\underline{k})^T \underline{f}(t, \underline{x})p(t, x)d\underline{x} + \int_{X \, X \, \mathbb{R}^n} \left(\psi(\underline{x} + \underline{g}(t, \underline{x}, \underline{\xi})) - \psi(\underline{x})\right) f(t, d\underline{\xi})p(t, \underline{x})d\underline{x}$$

Let $\underline{\eta}(t, y, \xi)$ denote the inverse of the map $\underline{x} \to \underline{x} + \underline{g}(t, \underline{x}, \underline{\xi})$ *i.e.*

$$\underline{\eta}(t, \underline{y}, \xi) + \underline{g}(t, \underline{\eta}(t, \underline{y}, \xi), \xi) = y$$

Then we get the Fokker – Planck eqn.:

$$\frac{\partial p(t, \underline{x})}{\partial t} = -\underline{\nabla}_x^T(\underline{f}(t, \underline{x})p(t, \underline{x})) + \int_X f(t, d\xi)p(t, \underline{\eta}(t, \underline{x}, \xi)) - f(t, X)p(t, x)$$

❖❖❖❖❖

[166] Nano robots noise described using the quantum Langevin equation.

[1] $dj_t(X) = j_t(\theta_\beta^\alpha(X))d\Lambda_\alpha^\beta(t)$

$$\Lambda_0^0(t) = t,\ \Lambda_0^\alpha(t) = \Lambda_\alpha(t), \Lambda_\alpha^0(t) = \Lambda_\alpha^+(t),\ \alpha \geq 1$$

$$d\Lambda_\beta^\alpha(t)d\Lambda_\sigma^\rho(t) = \varepsilon_\sigma^\alpha d\Lambda_\beta^\rho(t)$$

$$\Lambda_\beta^\alpha(t) = \Lambda_{|e_\beta\rangle\langle e_\alpha|}(t), \alpha, \beta \geq 1$$

$$\varepsilon_0^0 = 0, \varepsilon_0^\alpha = \varepsilon_\alpha^0 = 0, \alpha \geq 1, \varepsilon_0^\alpha = \varepsilon_\alpha^0 = 0, \alpha \geq 1$$

$$\langle e_\alpha | e_\beta \rangle = \delta\alpha\beta.$$

Example:

$$dU(t) = \left((-iH_0 + \rho)dt + \sum_\alpha L_\alpha dA_\alpha + M_\alpha dA_\alpha^+ + \sum_{\alpha, \beta \geq 1} S_\beta^\alpha d\Lambda_\alpha^\beta \right) U(t)$$

or

$$dU(t) = \left(-iH_0 dt + \sum_{\alpha, \beta \geq 0} S_\beta^\alpha d\Lambda_\alpha^\beta \right) U(t)$$

For $U(t)$ to be unitary

$$0 = d(U^* U) = dU^* \cdot U + U^* \cdot dU + dU^* dU$$

$$= U^* \left[S_\beta^{\alpha^*} d\Lambda_\beta^\alpha + S_\beta^\alpha d\Lambda_\alpha^\beta + S_\beta^{\alpha^*} S_\sigma^\rho d\Lambda_\alpha^\beta d\Lambda_\rho^\sigma \right] U$$

$$= U^* \left[\left(S_\beta^{\alpha^*} + S_\alpha^\beta \right) d\Lambda_\beta^\alpha + S_\beta^{\alpha^*} S_\sigma^\rho \varepsilon_\rho^\beta d\Lambda_\alpha^\sigma \right] U$$

So we require $\quad S_\beta^{\alpha^*} + S_\alpha^\beta + S_\alpha^{\beta^*} S_\alpha^\rho \varepsilon_\rho^\alpha$

❖❖❖❖❖

[167] Teleoperation of system in the presence of shot noise. Let $N_1(t)$, $N_2(t)$ be two independent Poisson presence with rates λ_1 and λ_2 repeatedly. Define

$$\xi_k(t) = \int_0^t h_k(t-s) dN_k(s), k = 1, 2$$

Then $\mathbb{E}[\xi_k(t)] = \lambda_k \int_0^t h_k(s) ds$, $k = 1, 2,$

$$\mathbb{E}[\xi_k(t)\xi_k(t')] = \lambda_k \int_0^{t \Lambda t'} h_k(t-s) h_k(t'-s) ds + \lambda_k^2 \left(\int_0^t h_k(s) ds \right) \left(\int_0^{t'} h_k(s) ds \right)$$

Then,

$$\lim_{t \to \infty} \mathbb{E}[\xi_k(t)] = \lambda_k \int_0^\infty h_k(s) ds = \lambda_k H_k(0),$$

$$\lim_{t\to\infty} \mathbb{E}\left[\xi_k(t+\tau)\xi_k(t)\right] = \lambda_k \int_0^\infty h_k(t)h_k(t+\tau)dt + \lambda_k^2\left(\int_0^\infty h_k(t)dt\right)^2$$

$$S_k(w) \triangleq \int_{\mathbb{R}} e^{-j\omega\tau}d\tau \lim_{t\to\infty} \mathbb{E}\left[\xi_k(t+\tau)\cdot\xi_k(t)\right]$$

$$= 2\pi\lambda_k^2 H_k(0)^2 \delta(\omega)\lambda_k \left|H_k(\omega)\right|^2,, k = 1, 2$$

Teleopration with shot noise

$$\frac{X_1(s)}{G_1(s)} = F_1(s) + C_1(s)\left(X_2(s) - X_1(s)\right) + \hat{\xi}_1(s)$$

$$\frac{X_2(s)}{G_2(s)} = F_2(s) + C_2(s)\left(X_1(s) - X_2(s)\right) + \hat{\xi}_2(s)$$

In the absence of external forcing $F_1 = F_2 = 0$, we get

$$(1+G_1C_1)X_1 - G_1C_1X_2 = G_1\hat{\xi}_1$$

$$(1+G_2C_2)X_2 - G_2C_2X_1 = G_2\hat{\xi}_2$$

$$\Rightarrow \quad X_1(s) = A_{11}(s)\hat{\xi}_1(s) + A_{12}(s)\hat{\xi}_2(s)$$

$$X_2(s) = A_{21}(s)\hat{\xi}_1(s) + A_{22}(s)\hat{\xi}_2(s)$$

where

$$\begin{bmatrix} A_{11} & A_{21} \\ A_{21} & A_{22} \end{bmatrix} = \underline{\underline{A}}(s) = \begin{bmatrix} 1+G_1C_1 & -G_1C_1 \\ -G_2C_2 & 1+G_2C_2 \end{bmatrix}^{-1} \times \begin{bmatrix} G_1 & 0 \\ 0 & G_2 \end{bmatrix}$$

Let

$$K_{11}(s) = A_{11}(s).H_1(s)$$

$$K_{12}(s) = A_{12}(s).H_2(s)$$

$$K_{21}(s) = A_{21}(s)\, H_1(s)$$

$$K_{22}(s) = A_{22}(s)H_2(s)$$

Then,

$$x_1(t) = \int_0^t k_{11}(t-\tau)dN_1(\tau) + \int_0^t k_{12}(t-\tau)dN_2(\tau)$$

$$x_2(t) = \int_0^t k_{21}(t-\tau)dN_1(\tau) + \int_0^t k_{22}(t-\tau)dN_2(\tau)$$

Joint Law of x_1 (.) and x_2(.)

$$M_x(f_1, f_2) = \mathbb{E}\exp\left(\int \left(f_1(t)x_1(t) + f_2(t)x_2(t)\right)dt\right)$$

$$\mathbb{E}\left\{\exp\left(\int dN_1(\tau)\left(\int k_{11}(t-\tau)f_1(t)dt + \int k_{21}(t+\tau)f_2(t)dt\right)\right)\right.$$

$$\left. + \int dN_2(\tau)\left(\int k_{12}(t-\tau)f_1(t)dt + \int k_{22}(t-\tau)f_2(t)dt\right)\right\}$$

Let

$$\varphi_1(\tau) = \int k_{11}(t-\tau)f_1(t)dt + \int k_{21}(t-\tau)f_2(t)dt$$

$$\varphi_2(\tau) = \int k_{12}(t-\tau)f_1(t)dt + \int k_{22}(t-\tau)f_2(t)dt$$

Then

$$M_x(f_1, f_2) = \mathbb{E}\left\{\exp\left(\int \varphi_1(\tau)dN_1(\tau) + \varphi_2(\tau)dN_2(\tau)\right)\right\}$$

$$= \exp\left\{\lambda_1 \int \left(e^{\varphi_1(\tau)} - 1\right)d\tau + \lambda_2 \int \left(e^{\varphi_2(\tau)} - 1\right)d\tau\right\}$$

$$\hat{\varphi}_1(\omega) = \overline{\hat{k}_{11}(\omega)}\hat{f}_1(\omega) + \overline{\hat{k}_{21}(\omega)}\hat{f}_2(\omega)$$

$$\hat{\varphi}_2(\omega) = \overline{\hat{k}_{12}(\omega)}\hat{f}_1(\omega) + \overline{\hat{k}_{22}(\omega)}\hat{f}_2(\omega)$$

$$\underline{\varphi}(\omega) = \begin{pmatrix} \hat{\varphi}_1(\omega) \\ \hat{\varphi}_2(\omega) \end{pmatrix} = \underline{\underline{\hat{k}}}(\omega)^T \underline{\hat{f}}(\omega)$$

Two link robots in the presence of shot noise:

$$M(q_1)q_1'' + N(q_1, q_1') = \xi_1(t) + C_1(t) + \left(q_2(t) - q_1(t)\right)$$

$$M(q_2)q_2'' + N(q_2, q_2') = \xi_2(t) + C_2(t) + \left(q_1(t) - q_2(t)\right)$$

❖❖❖❖❖

[168] 1. Nano-robotics in a strong gravitational field.

2. Controlling the quantum gravitational filed for properly a nono-robot.

$$\int R\sqrt{-g}d^4x \approx \int g^{\mu\nu}\sqrt{-g}\left(\Gamma^{\alpha}_{\mu\nu}\Gamma^{\beta}_{\alpha\beta} - \Gamma^{\alpha}_{\mu\beta}\Gamma^{\beta}_{\nu\alpha}\right)d^4x$$

$$g_{\mu\nu} = \eta_{\mu\nu} + \varepsilon h_{\mu\nu}(x)$$

$$g^{\mu\nu} = \eta_{\mu\nu} + \varepsilon h^{\mu\nu}(x) + O\left(\varepsilon^2\right)$$

$$\Gamma^{\alpha}_{\mu\nu} = g^{\alpha\beta}\Gamma_{\beta\mu\nu} = \frac{\varepsilon}{2}\eta_{\alpha\beta}\left(h_{\beta\mu,\nu} + h_{\beta\nu,\mu} - h_{\mu\nu,\beta}\right) + O\left(\varepsilon^2\right)$$

$$\Gamma^{\beta}_{\alpha\beta} = \frac{\varepsilon}{2}\eta_{\mu\beta}\left(h_{\mu\alpha,\beta} + h_{\mu\beta,\alpha} - h_{\alpha\beta,\mu}\right) + O\left(\varepsilon^2\right)$$

$$\Gamma^{\alpha}_{\mu\nu}\Gamma^{\beta}_{\alpha\beta} = \frac{\varepsilon^2}{4}\eta_{\alpha\rho}\eta_{\beta\sigma}\left(h_{\rho\mu,\nu} + h_{\rho\nu,\mu} - h_{\mu\nu,\rho}\right)\left(h_{\alpha\sigma,\beta} + h_{\beta\sigma,\alpha} - h_{\alpha\beta,\sigma}\right) + O\left(\varepsilon^3\right)$$

$$= \frac{\varepsilon^2}{4}\left(h_{\rho\mu,\nu} + h_{\rho\nu,\mu} - h_{\mu\nu,\rho}\right)\left(h^{\rho\beta}_{,\beta} + h^{,\rho} - h^{\rho\beta}_{,\beta}\right) + O\left(\varepsilon^3\right)$$

$$= \frac{\varepsilon^2}{4} h^{,\rho}\left(h_{\rho\mu,\nu} + h_{\rho\nu,\mu} - h_{\mu\nu,\rho}\right) + O\left(\varepsilon^3\right)$$

$$g^{\mu\nu}\Gamma^{\alpha}_{\mu\nu}\Gamma^{\beta}_{\alpha\beta} = \frac{\varepsilon^2}{4} h^{,\rho}\left(2h^{\nu}_{\rho,\nu} - h_{,\rho}\right) + O\left(\varepsilon^3\right)$$

$$\Gamma^{\alpha}_{\mu\beta}\Gamma^{\beta}_{\nu\alpha} = \frac{\varepsilon^2}{4}\eta_{\alpha\rho}\eta_{\beta\sigma}\left(h_{\rho\mu,\beta} + h_{\rho\beta,\mu} - h_{\mu\beta,\rho}\right)$$

$$\times\left(h_{\sigma\nu,\alpha} + h_{\sigma\alpha,\nu} - h_{\alpha\nu,\sigma}\right) + O\left(\varepsilon^3\right)$$

$$= \frac{\varepsilon^2}{4}\left(h^{\alpha,\sigma}_{\mu} + h^{\alpha\sigma}_{,\mu} - h^{\sigma,\alpha}_{\mu}\right)\left(h_{\sigma\nu,\alpha} + h_{\alpha\sigma,\nu} - h_{\alpha\nu,\sigma}\right) + O\left(\varepsilon^3\right)$$

$$= \frac{\varepsilon^2}{4}\left\{h^{\alpha,\sigma}_{,\mu} h_{\alpha\sigma,\nu} + \left(h^{\alpha,\sigma}_{\mu} - h^{\sigma,\alpha}_{\mu}\right)\left(h_{\sigma\nu,\alpha} - h_{\alpha\nu,\sigma}\right)\right\} + O\left(\varepsilon^3\right)$$

$$g^{\mu\nu}\Gamma^{\alpha}_{\mu\beta}\Gamma^{\beta}_{\nu\alpha} = \frac{\varepsilon^2}{4}\left\{h^{\alpha\sigma}_{,\mu} h^{,\mu}_{\alpha\sigma} + 2h^{\mu}_{\sigma,\alpha}\left(h^{\alpha,\sigma}_{\mu} - h^{\sigma,\alpha}_{\mu}\right)\right\} + O\left(\varepsilon^3\right)$$

$$= \frac{\varepsilon^2}{4}\left\{-h^{\mu}_{\sigma,\alpha} h^{\sigma,\alpha}_{\mu} + 2h^{\mu}_{\sigma,\alpha} h^{\alpha,\sigma}_{\mu}\right\} + O\left(\varepsilon^3\right)$$

$$= \frac{\varepsilon^2}{4} h^{\mu}_{\sigma,\alpha}\left\{2h^{\alpha,\sigma}_{\mu} - h^{\sigma,\alpha}_{\mu}\right\} + O\left(\varepsilon^3\right)$$

$$£ = \frac{\varepsilon^2}{4}\sqrt{-g}\left\{h^{,\rho}\left(2h^{\nu}_{e,\nu} - h_{,\rho}\right) - h^{\mu}_{\sigma,\alpha}\left(2h^{\alpha,\sigma}_{\mu} - h^{\sigma,\alpha}_{\mu}\right)\right\} + O\left(\varepsilon^3\right)$$

$$= \frac{\varepsilon^2}{4}\left\{h^{,\rho}\left(2h^{\nu}_{\rho,\nu} - h_{,\rho}\right) - h^{\mu}_{\sigma,\alpha}\left(2h^{\alpha,\sigma}_{\mu} - h^{\sigma,\alpha}_{\mu}\right)\right\} + O\left(\varepsilon^3\right)$$

Exercises in group theory applied to physical problems

❖❖❖❖❖

[169] Calculate the Haar measures a $SL(2, \mathbb{R})$ and $SL(2, \mathbb{C})$

Hint: Let $g = \begin{bmatrix} g_{11} & g_{12} \\ g_{21} & g_{22} \end{bmatrix} \in SL(2, \mathbb{R})$

Parametric g by (g_{11}, g_{12}, g_{22}), so

$$g_{21} = (g_{11}\, g_{22} - 1)/g_{12}$$

$$X = \begin{bmatrix} 0 & 1 \\ 0 & 0 \end{bmatrix}, Y = \begin{bmatrix} 0 & 0 \\ 1 & 0 \end{bmatrix}, H = \begin{bmatrix} 1 & 0 \\ 0 & -1 \end{bmatrix}$$

Calculate = $g\,(1 + t\,X)$, $g(1 + t\,Y)$, $g(1 + t\,H)$

and hence derive expansion for $\tilde{X}, \tilde{Y}, \tilde{H}$
as linear partial differential operator in (g_{11}, g_{12}, g_{22})
For the complex case, choose the parameter
$R(g_{\alpha\beta})$, $I(\delta_{\alpha\beta})$, $1 \le \alpha \le \beta \le 2$
and calculate $g(1+tX), g(1+itX), g(1+tY), g(1+itY)$,

$g(1+tH), g(1+itH)$ and here derive

expressions for $\tilde{X}, \widetilde{iX}, \tilde{Y}, \widetilde{iY}$,

$\tilde{Z}, \widetilde{iZ}$ as linear partial differential is

$R(g_{\alpha\beta}), I(g_{\alpha\beta}), 1 \le \alpha \le \beta \le \alpha$

Write for the real case,

$$\begin{bmatrix} \widetilde{X} \\ \tilde{Y} \\ \widetilde{H} \end{bmatrix} = A(g) \begin{bmatrix} \frac{\partial}{\partial g_{11}} \\ \frac{\partial}{\partial g_{12}} \\ \frac{\partial}{\partial g_{22}} \end{bmatrix}$$

and for the complex case,

$$\begin{bmatrix} \widetilde{X} \\ \widetilde{iX} \\ \tilde{Y} \\ \widetilde{iY} \\ \widetilde{H} \\ \widetilde{iH} \end{bmatrix} = B(g) \begin{bmatrix} \frac{\partial}{\partial R(g_{11})} \\ \frac{\partial}{\partial I(g_{11})} \\ \frac{\partial}{\partial R(g_{12})} \\ \frac{\partial}{\partial I(g_{12})} \\ \frac{\partial}{\partial R(g_{22})} \\ \frac{\partial}{\partial I(g_{22})} \end{bmatrix}$$

Then the Haar measure is the real case is

$$|A(g)|^{-1} dg_{11} dg_{12} dg_{22}$$

and is the complex case is

$$|B(g)|^{-1} \prod_{1 \le \alpha \le \beta \le 2} dR(g_{\alpha\beta})\, dI(g_{\alpha\beta})$$

(2) Consider the trans minor line eqn.

$$\frac{\partial V}{\partial Z} + RI + L\frac{\partial I}{\partial t} = 0$$

$$\frac{\partial I}{\partial Z} + GV + C\frac{\partial V}{\partial t} = 0$$

or equivalently

$$\frac{\partial I}{\partial Z}\begin{pmatrix} V \\ I \end{pmatrix} + \begin{pmatrix} 0 & L \\ C & 0 \end{pmatrix}\frac{\partial}{\partial t}\begin{pmatrix} V \\ I \end{pmatrix} + \begin{pmatrix} 0 & R \\ G & 0 \end{pmatrix}\begin{pmatrix} V \\ I \end{pmatrix} = 0$$

Write this as $$\frac{\partial \underline{\xi}}{\partial Z} + \underline{\underline{A}}_1 \frac{\partial \underline{\xi}}{\partial t} + \underline{\underline{A}}_2 \underline{\xi} < 0$$

were $$\underline{\xi} = \begin{pmatrix} V \\ I \end{pmatrix} \in \mathbb{R}^2$$

Consider a group of transformation $g \in G$ acting linearly on $\begin{pmatrix} t \\ Z \end{pmatrix}$

The calculate the induced informations on the matrices $\underline{\underline{A}}_1$ and $\underline{\underline{A}}_2$

$$\begin{pmatrix} t \\ Z \end{pmatrix} \to \begin{pmatrix} t' \\ Z' \end{pmatrix} \begin{pmatrix} g_{11} & g_{12} \\ g_{21} & g_{22} \end{pmatrix}\begin{pmatrix} t \\ Z \end{pmatrix} = \underline{\underline{g}}\begin{pmatrix} t \\ Z \end{pmatrix}$$

Then $$\frac{\partial}{\partial t} = \frac{\partial t'}{\partial t}\frac{\partial}{\partial t'} + \frac{\partial Z'}{\partial t}\frac{\partial}{\partial Z'}$$

$$= g_{11}\frac{\partial}{\partial t'} + g_{21}\frac{\partial}{\partial Z'}$$

$$\frac{\partial}{\partial Z} = \frac{\partial t'}{\partial Z}\frac{\partial}{\partial t'} + \frac{\partial Z'}{\partial Z}\frac{\partial}{\partial Z'}$$

$$= g_{12}\frac{\partial}{\partial t'} + g_{22}\frac{\partial}{\partial Z'}$$

i.e. $$\begin{pmatrix} \dfrac{\partial}{\partial t} \\ \dfrac{\partial}{\partial Z} \end{pmatrix} = \underline{\underline{g}}^T \begin{pmatrix} \dfrac{\partial}{\partial t'} \\ \dfrac{\partial}{\partial Z'} \end{pmatrix}$$

The line equation becomes

$$\left(g_{12}\frac{\partial}{\partial t'}+g_{22}\frac{\partial}{\partial Z'}\right)\underline{\xi}+\underline{\underline{A}}_1\left(g_{11}\frac{\partial}{\partial t'}+g_{21}\frac{\partial}{\partial Z'}\right)\underline{\xi}+\underline{\underline{A}}_2\underline{\xi}=0$$

$$\text{or } \left(g_{22}\underline{\underline{I}}_2+g_{21}\underline{\underline{A}}_1\right)\frac{\partial\underline{\xi}}{\partial Z'}+\left(g_{12}\underline{\underline{I}}_2+g_{11}\underline{\underline{A}}_1\right)\frac{\partial\underline{\xi}}{\partial t'}+\underline{\underline{A}}_2\underline{\xi}=0$$

So under G, $\left|\begin{matrix}\underline{\underline{H}}_1 \\ \underline{\underline{A}}_2\end{matrix}\right|\begin{matrix}g_{22}\underline{\underline{I}}_2 \\ g_{12}\underline{\underline{I}}_2\end{matrix}\left|\begin{matrix}+g_{21}\underline{\underline{A}}_1 \\ +g_{11}\underline{\underline{A}}_1\end{matrix}\right.$

or $\underline{\underline{A}}_1\to\left(g_{22}I_2+g_{21}A_1\right)^{-1}\cdot\left(g_{12}I_2+g_{11}A_1\right)$,

$\underline{\underline{A}}_2\to\left(g_{22}I_2+g_{21}A_1\right)^{-1}\underline{\underline{A}}_2$

Invariance of the Tx line equation under G requires

$$\begin{aligned}\left(g_{22}I_2+g_{21}A_1\right)A_1 &= g_{12}I_2\div g_{11}A_1\ \forall g\in G\\ \left(g_{22}I_2+g_{21}A_1\right)A_2 &= \left(g_{22}I_2+g_{21}A_1\right)A_2\end{aligned}$$

first is sameos

$$\text{or}\quad\left[\begin{pmatrix}g_{22}&0\\0&g_{22}\end{pmatrix}+g_{21}\begin{pmatrix}0&L\\C&0\end{pmatrix}\right]\begin{pmatrix}0&L\\C&0\end{pmatrix}=\begin{pmatrix}g_{12}&0\\0&g_{22}\end{pmatrix}+g_{11}\begin{pmatrix}0&L\\C&0\end{pmatrix}$$

$$\text{or}\quad\begin{pmatrix}g_{22}&g_{21}L\\g_{21}C&g_{22}\end{pmatrix}\begin{pmatrix}0&L\\C&0\end{pmatrix}=\begin{pmatrix}g_{12}&g_{11}L\\g_{11}C&g_{22}\end{pmatrix}$$

$$\text{or}\quad\begin{pmatrix}g_{21}LC&g_{22}L\\g_{22}C&g_{21}LC\end{pmatrix}=\begin{pmatrix}g_{12}&g_{11}L\\g_{11}C&g_{22}\end{pmatrix}$$

$$\text{or}\quad\frac{g_{12}}{g_{21}}=LC_1\frac{g_{22}}{g_{21}}=LC,\ g_{11}=g_{22}$$

$$\left(\left(x_{\alpha\beta}\right)\right)_{1\le\alpha\le n,\,1\le\beta\le r}\in\mathbb{M}_{n,r}(K)$$

$$\underline{x}_\beta=\left(\left(x_{\alpha\beta}\right)\right)_{\alpha=1}^{n}\in K^n.,\beta=1,2,.\,r$$

$$\begin{aligned}\underline{x}_1\Lambda\underline{x}_2\Lambda\cdot\cdot\Lambda x_r &= x_{\alpha_1}1e_{\alpha_1}\Lambda x_{\alpha_2}2\,e_{\alpha_2}\Lambda\cdots\Lambda x_{\alpha_r}e_\alpha\alpha_r\\ &= x_{\alpha_1}x_{\alpha_{22}}\cdots x_{\alpha_r}r\,e_{\alpha_1}\Lambda e_{\alpha_2}\Lambda\cdot\cdot\Lambda e_{\alpha_r}\end{aligned}$$

$$\sum_{1\le\alpha_1<\alpha_2<\cdot\cdot<\alpha_r\le n}\left(\sum_{\sigma\in S_r}\begin{matrix}S_{qn}(\sigma)x_{\alpha_{\sigma_1}}1x_{\alpha\sigma2}2\cdots x_{\alpha\sigma r}n_r\\ e_{\alpha_1}\Lambda e_{\alpha_2}\Lambda\cdot\cdot\Lambda e_{\alpha_r}\end{matrix}\right)$$

Plucker co-ordinates

$$= \sum_{1\le\alpha_1<\alpha_2<\cdot<\alpha_r\le n}\left(\sum_{\sigma\in S_r} \frac{S_{qn}(\sigma)x_{\alpha_1}\sigma^{-1}1x_2\sigma^{-1}2\cdots x_{\alpha_r}\sigma^{-1}r}{e_{\alpha_1}\Lambda e_{\alpha_2}\Lambda\cdot\cdot\Lambda e_{\alpha_r}}\right)$$

$$= \sum_{1\le\alpha_1<\alpha_2<\cdot\cdot<\alpha_r\le n} \det\left(\left(x_{\alpha_i}\cdot j\right)\right) e_{\alpha_1}\Lambda e_{\alpha_2}\Lambda\cdot\cdot\Lambda e_{\alpha_r}$$

Exercise Grassmanian expressed in term of plucker coordinators

❖❖❖❖❖

[170] Four vector potential of a current source:

$$A^{\mu}(t,\underline{r}) = \int \frac{J^{\mu}\left(t-\frac{|\underline{r}-\underline{r}'|}{1}\underline{r}'\right)}{|\underline{r}-\underline{r}'|}d^3r'$$

If the current source suffers a lorentz transformation *i.e.* a rotation and a boost, then the transformed vector potential es is

$$\tilde{A}(\underline{x}) = \underline{\underline{L}}\,\underline{A}\,\underline{\underline{L}}^{-1}(\underline{x})$$

$$L^{-1}(X) = \tilde{x} = \begin{bmatrix}\tilde{t}\\ \tilde{\underline{r}}\end{bmatrix}$$

$\underline{X} = (t, \underline{r})$, $\underline{A} = (\underline{A}_{\mu})^3$, $\mu = 0$

$$\Rightarrow \qquad \tilde{t} = \left(L^{-1}\right)_{00}t + \sum_{k=1}^{3}\left(L^{-1}\right)_{ok}X^k$$

$$\tilde{X}^k = \left(L^{-1}\right)_{k0}t + \sum_{m=1}^{3}\left(L^{-1}\right)_{km}X^m, \; 1\le k\le 3$$

$$|\tilde{r}-r'|^2 = \sum_{k=1}^{3}\left(\tilde{X}^k - X'^k\right)^2$$

$$\sum_k\left(\left(L^{-1}\right)_{k0}t + \sum_{ms\,1}^{3}\left(L^{-1}\right)_{km}X^m - X'^k\right)^2$$

$$\tilde{t} - |\tilde{r}-r'|/C = \left(L^{-1}\right)_{00}t + \sum_k\left(L^{-1}\right)_{ok}X^k$$

$$-\frac{1}{C}\left(\sum_k\left(\left(L^{-1}\right)_{k0}t + \sum\left(L^{-1}\right)_{km}X^m - X'^k\right)^2\right)^{1/2}$$

The problem is to identify L from measurements of the fields $\underline{\underline{A}}(\underline{X})$ and $L\,A\,L^{-1}$ $(\underline{X})$ Let $\{\varphi_l\}_{l=1}^{\infty}$ be an ONB for $L^2(\mathbb{R}^4)$

$$\text{Let} \quad \varphi_l(L^{-1}X) = \sum_{m=1}^{\infty}[\pi(L)]_{ml}\,\varphi_m(x)$$

So, $L \to \pi(L)\varepsilon \cdot GL(\mathbb{R})$ is a representation of the lorentz group G. Let $\chi_\pi(L) = Tr(\pi(L))$ be finite. Consider for a fixed $\xi \in \mathbb{R}^4$,

$$I_\pi(A) = \int_{G\times G} A^\mu(L_1^{-1}\xi)A_\mu(L_2^{-1}\xi)\chi_\pi(L_1^{-1}L_2)d\mu(L_1)d\mu(L_2)$$

We claim that this is an invariant for the transformation

$A \to L\,A\,L^{-1}, L \in G$

Indeed, let $B = L\,A\,L^{-1}$, $L \in G$ fixed

Then

$$= B^\mu(L_1^{-1}\xi)B_\mu(L_2^{-1}\xi)$$

$$= \left\langle B(L_1^{-1}\xi), B(L_2^{-1}\xi)\right\rangle$$

$$\equiv \left\langle L\,A\,L^{-1}L_1^{-1}(\xi), L\,A\,L^{-1}L_2^{-1}(\xi)\right\rangle$$

$$= \left\langle A(L^{-1}L_1^{-1}\xi), A(L^{-1}L_2^{-1}\xi)\right\rangle$$

$$= \left\langle A((L_1L)^{-1}\xi), A((L_2L)^{-1}\xi)\right\rangle$$

So,

$$= \int\left\langle B(L_1^{-1}\xi), B(L_2^{-1}\xi)\right\rangle\chi_\pi(L_1^{-1}L_2)d\mu(L_1)d\mu(L_2)$$

$$= \int\left\langle A((L_1L)^{-1}\xi), A((L_2L)^{-1}\xi)\right\rangle\chi_\pi(L_1^{-1}L_2)d\mu(L_1)d\mu(L_2)$$

$$= \int\left\langle A(L_1^{-1}\xi), A(L_2^{-1}\xi)\right\rangle\chi_\pi(LL_1^{-1}L_2L^{-1})d\mu(L_1)d\mu(L_2)$$

$$= \int\left\langle A(L_1^{-1}\xi), A(L_2^{-1}\xi)\right\rangle\chi_\pi(L_1^{-1}L_2)d\mu(L_1)d\mu(L_2)$$

This equation implies that

$$I_\pi(L\,A\,L^{-1}) = I_\pi(A),\ L \in G.$$

❖❖❖❖❖

[171] $dU(t) = \left(L_\beta^\alpha\,d\Lambda_\alpha^\beta\right)U(t)\ L_\beta^\alpha \in \mathcal{L}(\hbar)$.

$$
\begin{aligned}
d\Lambda^{\alpha}_{\beta}\, d\Lambda^{\mu}_{\nu} &= \varepsilon^{\alpha}_{\nu} d\Lambda^{\mu}_{\beta} \\
\rho(0) &= \rho_s(0) \otimes |\varphi(u)\rangle\langle\varphi(u)| \\
\varphi(u) &= \exp(-\|u\|^{2/2})|e(u)\rangle \\
Tr_2\left[U(t)\rho(0)U^*(t)\right] &= \rho_s(t) \\
d\rho_s(t) &= Tr_2\left[dU\rho(0)U^*\right] + Tr_2\left[U\rho(0)dU^*\right] \\
&\quad + Tr_2\left[dU\rho(0)dU^*\right] \\
&= Tr_2\left[L^{\alpha}_{\beta}\, d\Lambda^{\beta}_{\alpha}\, U\rho(0)U^* d\Lambda^{\mu}_{\nu}\, L^{\mu}_{\nu}\right] \\
&= L^{\alpha}_{\beta}\, Tr_2\left[U d\Lambda^{\beta}_{\alpha}\rho(0) d\Lambda^{\mu}_{\nu} U^*\right] L^{\mu}\, \nu^*
\end{aligned}
$$

Since $\left[d\Lambda^{\alpha}_{\beta}(t), U(t) = 0\right]$

$$
\begin{aligned}
Tr_2\left(d\Lambda^{\beta}_{\alpha}\rho(0)d\Lambda^{\mu}_{\nu}\right) &= \rho_s(0)Tr\left[d\Lambda^{\beta}_{\alpha}|\varphi(u)\rangle\langle\varphi(u)|d\Lambda^{\mu}_{\nu}\right] \\
&= \rho_s(0)\left\langle\varphi(u)\middle|d\Lambda^{\mu}_{\nu} d\Lambda^{\beta}_{\alpha}\middle|\varphi(u)\right\rangle \\
&= \rho_s(0)\varepsilon^{\mu}_{\alpha}\left\langle\varphi(u)\middle|d\Lambda^{\beta}_{\alpha}\middle|\varphi(u)\right\rangle \\
&= \rho_s(0)\varepsilon^{\mu}_{\alpha} u_{\beta}(t)\overline{U_{\nu}(t)}\, dt
\end{aligned}
$$

So,

$$
\begin{aligned}
\frac{d\rho_s(t)}{dt} &= \varepsilon^{\mu}_{\alpha} u_{\beta}(t)\overline{U_{\nu}(t)} L^{\alpha}_{\beta}\, \rho_s(t) L^{\mu*}_{\nu} \\
&= \sum_{\beta,\, \nu\ge 0} u_{\beta}(t)\overline{u_{\nu}(t)} \sum_{j\ge i} L^{j}_{\beta}(t)\rho_s(t)L^{*}_{\nu}(t)
\end{aligned}
$$

u_0(t) = 1: This is the generalized Sudarshan Linblad equation

Simulation: Let $u_1, u_2, .., u_N \in L_2(\mathbb{R}_t) = \mathcal{H}$

and let
$$\xi_r = \sum_{\alpha=1}^{r} c(r,\alpha)e(u_{\alpha}), 1 \le r \le N$$

be an ONB for span $\{u_1, u_2 ... u_N\}$ (Gram-Schmidt on process). Let $\{f_1, f_2, .., f_M\}$ be an ONB for the system Hilbert space $\mathfrak{h}$

$$
\begin{aligned}
& d\left\langle f_a \otimes \xi_r \middle| U(t) f_b \otimes \xi_s \middle|\right\rangle \\
&= \left\langle f_a \otimes \xi_r \middle| U(t) \middle| f_b \otimes \xi_s \right\rangle
\end{aligned}
$$

$$= \left\langle f_a \xi_r \middle| L^{\alpha}_{\beta}\, d\Lambda^{\beta}_{\alpha} U(t) \middle| f_b \xi_s \right\rangle$$

$$= \sum_{\alpha,\beta} \overline{c(r,\alpha)} c(s,\beta) \left\langle f_a c(u_\alpha) \middle| L^{\alpha}_{\beta}\, d\Lambda^{\beta}_{\alpha} U \middle| f_b e(u_\beta) \right\rangle$$

$$= \sum_{\alpha,\beta} \overline{c(r,\alpha)} c(s,\beta) \left\langle f_a e(u_\alpha) \middle| L^{\alpha}_{\beta} U \middle| f_b e(u_\beta) \right\rangle u_\beta(t) \overline{u_\alpha(t)}$$

We have the approximate eqn.

$$\left\langle f_a \xi_r \middle| L^{\alpha}_{\beta}\, d\Lambda^{\beta}_{\alpha} U \middle| f_b \xi_s \right\rangle = \sum_{c,k} \left\langle f_a \xi_r \middle| L^{\alpha}_{\beta}\, d\Lambda^{\beta}_{\alpha} \middle| f_c \xi_k \right\rangle \left\langle f_c \xi_k \middle| U(t) \middle| f_b \xi_s \right\rangle$$

$$= \sum_{c,k} \left\langle f_a \middle| L^{\alpha}_{\beta} \middle| f_c \right\rangle \left\langle \xi_r \middle| d\Lambda^{\beta}_{\alpha} \middle| \xi_k \right\rangle \left\langle f_c \xi_r \middle| U(t) \middle| f_p \xi_s \right\rangle$$

$$\left\langle \xi_r \mid d\Lambda^{\beta}_{\alpha} \mid \chi_k \right\rangle$$

$$= \sum_{\mu,v} \overline{c(r,\mu)} C(k,v) \left\langle e(u_\mu) \middle| d\Lambda^{\beta}_{\alpha}(t) \middle| e(u_v) \right\rangle$$

$$= \sum_{\mu,v} \overline{c(r,\mu)} C(k,v) u_{v\beta}(t) \overline{u_{\mu\alpha}(t)} \exp\left(\left\langle u_\mu \mid u_v \right\rangle\right) dt$$

Thus, we have approximately,

$$\frac{d}{dt} \left\langle f_a \xi_\alpha \middle| U(t) \middle| f_b \xi_s \right\rangle$$

$$= \sum_{c,k,\mu,\alpha,\beta} \left\langle f_a \middle| L^{\alpha}_{\beta} \middle| f_c \right\rangle \overline{c(r,\mu)}\, c(k,v) u_{v\beta}(t) \overline{u_{\mu\alpha}(t)}$$

$$\exp\left(\left\langle u_\mu \middle| u_v \right\rangle\right) \left\langle f_c \xi_k \middle| U(t) \middle| f_b \xi_s \right\rangle$$

Note that $u_{\mu 0}(t) = 1, \forall \mu$

❖❖❖❖❖

[172] Quantum filtering and control

$$dU(t) = \left(L^{\alpha}_{\beta}(t) d\Lambda^{\beta}_{\alpha}(t)\right) U(t) + K\left(X_d(t) - \mathbb{E}_t\left(X \middle| \eta^{in}_{t]}\right) U(t)\right) dt \quad 1$$

$$\eta^{in}_{t]} = \sigma\left\{Y^{in}_s, s \le t\right\}$$

$$Y^{in}_t \in \mathcal{L}\left(\Gamma s\left(\mathcal{H}_{t]}\right)\right)$$

$$U^*(t) \mathbb{E}_t\left(X \middle| \eta^{in}_{t]}\right) U(t) = \mathbb{E}\left(j_t(X) \middle| \eta^{out}_{t]}\right)$$

$$\eta^{out}_{t]} = U^*(t)\, \eta^{in}_{t]}\, U(t) \quad j_t(X) = U^*(t)\, X\, U(t)$$

$$Y^{out}_t = U^*(t)\, Y^{in}_t\, U(t) = U^*(T) Y^{in}_t U(T)$$

$\therefore\ \eta_{t]}^{in}$ is Abelian and $\left[\eta_{t]}^{in}, \hbar\right] = 0$, where $\hbar$ is the system Hilbert space.

(1) can also be expressed as

$$dU(t) = \left(L_{\beta}^{\alpha}(t)\, d\Lambda_{\alpha}^{\beta}(t)\right)U(t) + K\, U(t)\left(\tilde{X}_d(t) - \pi_t(X)\right)dt$$

where $\quad \pi_t(X) = \mathbb{E}\left(j_t(X)\middle|\eta_{t]}^{out}\right)$

$$\tilde{X}_d(t) = U^*(t)X_d(t)U(t)$$

$$\tilde{X}_d(t) = \cancel{X_d}(t) \in \eta_t^{in}\, so\ \cancel{\tilde{X}_d}(t) \in \cancel{\eta_{t]}^{out}}$$

$$X_d(d)\varepsilon\mathcal{L}\left(\hbar \otimes \eta_{t]}^{in}\right) \text{ is assumed}$$

Let $\quad X_d(t) = V^*(t)\chi_d V(t)\chi_d \varepsilon\mathcal{L}(\hbar)$

where $\quad dV(t) = \left[\left(-iH_0 + \frac{1}{2}L_0^2\right)dt - i\,L_0 d\,B_{out}(t)\right]V(t)$

where $L_0^* = L_0$, $B(t) = \underline{A}_t + A_t$ so $(dB)^2\ dt$.

Let

$$\mathbb{E}\left(j_t(X)\middle|\eta_t^{out}\right) = \pi_t(x)$$

$$d\pi_t(X) = F_t(X)dt + G_t(X)d\,B_{out}(t)$$

We're assuming that $Y_t^{in} = B(t)$

$$Y_t^{out} = B_{out}(t)$$

$$= U^*(t)B_{in}(t)U(t)$$

so, $\quad Y_t^{out} = U^*(t)\, B(t)\, U(t)$

$$\mathbb{E}\left[j_t(X) - \pi_t(N)C_t\right] = 0$$

$$B_{out}(t) = U^*(t)\, B(t)U(t)$$

where $dC_t = f(t)C\chi dB_{out}(t)$

Then $\ C_t \in \eta_t^{out} = \sigma\{B_{out}(s), s \le t\}$

$$\mathbb{E}(\xi) \triangleq \langle f\varphi(u)|\xi|f\varphi(u)\rangle$$

$$\varphi(u) = \exp\left(-\|u\|^2/2\right)e(u)$$

$$dj_t(X) = d(U^*(t)XU(t)) = dU^*(t)\times U(t) + U^*(t)\times dU(t)dU^*(t)\times dU(t)$$

$$= U^*(t)\Big\{L_\beta^{\alpha *} X d\Lambda_\beta^\alpha + X L_\beta^\alpha d\Lambda_\alpha^\beta + \left(\underline{\Phi}_t^* X + X\underline{\Phi}_t\right)dt$$

$$+ L_\beta^{\alpha *} X L_\nu^\mu d\Lambda_\beta^\alpha d\Lambda_\mu^\nu\Big\}U(t)$$

$$= U^*(t)\Big\{\left(L_\beta^{\alpha *} X + X L_\alpha^\beta\right)d\Lambda_\beta^\alpha + L_\beta^{\alpha *} X L_\nu^\mu \varepsilon_\mu^\alpha d\Lambda_\beta^\nu$$

$$+ dt\left(\Phi_t^* X + X\Phi_t\right)\Big\}U(t)$$

$$= U^*(t)\left(L_\beta^{\alpha *} X + X L_\alpha^\beta + L_\beta^{\nu *} X L_\alpha^\mu \varepsilon_\mu^\nu\right)d\Lambda_\beta^\alpha + \left(\Phi_t^* X + X\Phi_t\right)dt\Big\}U(t)$$

Here,

$$\Phi_t = K\left(X_d(t) - E\left(X \mid \eta_{t]}^{in}\right)\right)$$

We write this as

$$dj_t(x) = j_t\left(\theta_\beta^\alpha(X)\right)d\Lambda_\alpha^\beta(t) + j_t\left(\Phi_t^* X + X\Phi_t\right)dt$$

We write

$$\tilde{\theta}_t(X) = \Phi_t^* X + X\Phi_t$$

Thus, $\tilde{\theta}_t(X) \in \mathcal{L}\left(h \otimes \eta_{t]}^{in}\right)$

Note: $\eta_{t]}^{in} \subset \Gamma s(\mathcal{H}_{t]})$

So, $dj_t(X) = j_t\left(\theta_\beta^\alpha(X)\right)d\Lambda_\alpha^\beta + j_t\left(\tilde{\theta}_t(X)\right)dt$

$$= j_t\left(\left(\theta_0^0 + \tilde{\theta}_t\right)(X)\right)dt + \sum_{(\alpha,\beta)\neq(0,0)} j_t\left(\theta_\beta^\alpha(X)\right)d\Lambda_\alpha^\beta$$

Note that $\tilde{\theta}_t$ depend on both signal and noise spaces while θ_β^α depend only on signal space. Let $X_{\beta_t}^\alpha = \theta_\beta^\alpha\,(\alpha,\beta) \neq 0,\ X_{0_t}^0 = \theta_0^0 + \tilde{\theta}_t$

$$\left(dj_t(X) - d\pi_t(X)\right)C_t = j_t\left(X_{\beta_t}^\alpha(X)\right) \subset_t^- d\Lambda_\alpha^\beta$$

$$-F_t(X)C_t dt - G_t(X)C_t dB_{out}(t)$$

$$\mathbb{E}\left\{\left(dj_t(X) - d\pi_t(X)C_t\right)\right\}$$

$$u_\beta(t)\overline{u_\alpha(t)}\,dt\ \mathbb{E}\left[j_t\left(X_\beta^\alpha(X)C_t\right)\right]$$

$$-\mathbb{E}\left[F_t(X)C_t\right]dt - \mathbb{E}\left[G_t(X) \subset_t dB(t)_{out}\right]$$

$$= \mathbb{E}\left[j_t(X) - \pi_t(X) d\,C_t\right] = f(t)\mathbb{E}\left[j_t(X) - \pi_t(X)C_t\,dB(t)_{out}\right]$$

$$= f(t)\left(u_1(t)+\overline{u_1(t)}\right)dt\,\mathbb{E}\left[j_t(X)-\pi_t(X)C_t\right]$$

$$B_{out}(t) = U^*(t)\,B(t)\,U(t)$$

Note: $F_t(X), G_t(X) \in \eta_{t]}^{out}$ is assumed

$$dB_{out}(t) = dB(t) + dU^* dB\,U + U^* dBdU$$

$$= dB\,U^* L_\beta^{\alpha *} d\Lambda_\beta^\alpha dBU + U^* L_\beta^\alpha dBd\Lambda_\alpha^\beta\,dU$$

$$d\Lambda_\beta^\alpha\,dB = d\Lambda_\beta^\alpha\left(dA_1 + dA_1^*\right)$$

$$= d\Lambda_\beta^\alpha\left(d\Lambda_0^1 + d\Lambda_1^0\right) = \delta_1^\alpha d\Lambda_\beta^0 = \begin{cases}\delta_1^\alpha dA_\beta^*, \beta \geq 1\\ \delta_1^\alpha dt, \beta \leq 0\end{cases}$$

$$dB\,d\Lambda_\alpha^\beta = \left(dA_1 + dA_\alpha^*\right)d\Lambda_\alpha^\beta = \left(d\Lambda_0^1 + d\Lambda_1^0\right)d\Lambda_\alpha^\beta$$

$$= \varepsilon_\alpha^1 d\Lambda_0^\beta = \delta_\alpha^1\,d\Lambda_0^\beta$$

So,
$$dB_{out} = dB + j_t\left(L_\beta^{1*} d\Lambda_\beta^0 + L_\beta^1 d\Lambda_0^\beta\right)$$

$$= dB + j_t\left(L_\alpha^1\right)d\Lambda_0^\alpha + j_t\left(L_\alpha^{1*}\right)d\Lambda_\alpha^0$$

So,

$$\mathbb{E}\left[\left(j_t(X)-\pi_t(X)\right)dC_t\right] = f(t)\,\mathbb{E}\left[\left(j_t(X)-\pi_t(X)\right)C_t\right]\left(u_1(t)+\overline{U_1(t)}\right)dt$$

$$+\,\mathbb{E}\left[\left(j_t\left(X\,L_\alpha^1\right)-\pi_t(X)\,j_t\left(L_\alpha^1\right)\right)C_t\right]U_\alpha(t)$$

$$+\,\mathbb{E}\left[\left(j_t\left(X\,L_\alpha^1\right)-\pi_t(X)\,j_t\left(L_\alpha^{1*}\right)\right)C_t\right]\overline{U_\alpha(t)}dt$$

$$= \mathbb{E}\left[\left(j_t\left(X\,L_\alpha^1\right)-\pi_t(X)\,j_t\left(L_\alpha^1\right)\right)C_t\right]U_\alpha(t)dt$$

$$+\,\mathbb{E}\left[\left(j_t\left(X\,L_\alpha^{1*}\right)-\pi_t(X)\,j_t\left(L_\alpha^{1*}\right)\right)C_t\right]\overline{U_\alpha(t)}$$

Finally,

$(dj_t(\chi) - d\pi_t(X))dc_t$

$$= \left(j_t\left(X_\beta^\alpha(X)\right)d\Lambda_\alpha^\beta - F_t(X)dt - G_t(X)dB_{out}(t)\right)f(t)C_t dB_{out}(t)$$

$$f(t)\left[\,j_t\left(X_\beta^\alpha(X)\right)d\Lambda_\alpha^\beta - F_t(X)dt - G_t(X)\left(dB + j_t\left(L_\alpha^1\right)d\Lambda_0^\alpha\right)\right.$$

$$+ j_t\left(L_\alpha^{1*}\right)d\Lambda_\alpha^0\Big]\left(dB + j_t\left(L_\rho^1\right)d\Lambda_0^\rho + j_t\left(L_p^{1*}\right)d\Lambda_\rho^0\right)C_t$$

$$= f(t)\Big[j_t\left(\chi_\beta^\alpha(X)\right)d\Lambda_\alpha^\beta dB + j_t\left(\chi_\beta^\alpha(X)L_\rho^{1*}\right)\varepsilon_\rho^\beta d\Lambda_\alpha^0$$

$$-Gt(X)\left(dt + j_t\left(L_\rho^1\right)\left(dBd\Lambda_0^\rho + d\Lambda_0^\rho dB\right) + jt\left(L_\rho^{1*}\right)\right)$$

$$\left(dB\,d\Lambda_\rho^0 + d\Lambda_\rho^0\,dB\right) + j_t\left(L_\alpha^1 L_\rho^{1*}\right)\varepsilon_\rho^\alpha dt\Big]C_t$$

$$= F(t)\Big[j_t\left(\chi_\beta^\alpha(X)\right)\delta_1^\beta d\Lambda_\alpha^0 + j_t\left(\chi_\beta^\alpha(X)L_\rho^{1*}\right)\varepsilon_\rho^\beta d\Lambda_\alpha^0$$

$$-G_t(X)\left(dt + j_t\left(L_\rho^{1^*}\right)\right)\delta_\rho^1\,dt + j_t\left(L_\rho^1\right)\delta_1^\rho dt + j_t\left(L_j^1 L_j^{1*}\right)dt\Big]C_t$$

So,

$$\mathbb{E}\left[\left(dj_t(X) - d\pi_t(X)d\,C_t\right)\right] = f(t)\mathbb{E}\Big[j_t\left(X_1^\alpha(X)\right)C_t\,\overline{u_\alpha(t)}$$

$$+j_t\left(\chi_j^\alpha(X)L_j^{1*}\right)C_t\overline{u\alpha(t)}$$

$$-G_t(X) - G_t(X)j_t\left(L_1^1 + L_1^{1*} + L_j^1 L_j^{1*}\right)C_t\Big]dt$$

Going back to page (5), we get

$$\mathbb{E}\left[G_t(X)C_t dB_{out}^{(t)}\right] = \mathbb{E}\left[G_t(X)C_t\left(dB + j_t\left(L_\alpha^1\right)d\Lambda_0^\alpha + j_t\left(L_\alpha^{1*}\right)d\Lambda_\alpha^0\right)\right]$$

$$= \left(u_1(t) + \overline{u_1(t)}\right)dt\,\mathbb{E}\left(G_t C_t\right) + \mathbb{E}\left(G_t j_t\left(L_\alpha^1\right)C_t\right)u_\alpha(t)dt$$

$$+\mathbb{E}\left(G_t j_t\left(L_\alpha^*\right)C_t\right)\overline{u_\alpha(t)}\,dt$$

Thus, we get using $d\ \mathbb{E}\left[\left(j_t(X) - \pi_t(X)C_t\right)\right] = 0$. Using arbitrariness of $f(t)$ that

$$u_\beta(t) + \overline{u_\alpha(t)}\,\pi_t\left(\chi_\beta^\alpha(X)\right) - F_t(X) - \left(u_1(t) + \overline{u_1(t)}\right)G_t(X)$$

$$-G_t(X)\pi_t\left(u_\alpha(t)L_\alpha^1 + \overline{u_\alpha(t)}\,L_\alpha^{1*}\right) = 0, \tag{1}$$

$$\pi_t\left(X\left(L_\alpha^{1*}\overline{u_\alpha(t)} + L_\alpha^1 u_\alpha(t)\right)\right) - \pi_t(X)\pi_t\left(L_\alpha^{1*}\overline{u_\alpha(t)} + L_\alpha^1 u_\alpha(t)\right)$$

$$+\pi_t\left(\chi_1^\alpha(X) + \chi_j^\alpha(X)L_j^{1*}\right)\overline{u_\alpha(t)}$$

$$-G_t(X) - G_t(X)\pi_t\left(L_1^1 + L_1^{1*} + L_j^1 L_j^{1*}\right) = 0 \tag{2}$$

❖❖❖❖❖

[173] $g^{\mu\nu} = q^{\mu\nu} + n^{\mu}n^{\nu}$ (quantum gravity)

$$g_{\mu\nu} = q_{\mu\nu} + n_{\mu}n_{\nu}$$

$$K_{\mu\nu} = q_{\mu}^{\alpha} q_{\nu}^{\beta} \nabla_{\alpha} n_{\beta}$$

$$K_{\mu\nu} - K_{\mu\nu} = q_{\mu}^{\alpha} q_{\nu}^{\beta} \left(\nabla_{\alpha} n_{\beta} - \nabla_{\beta} n_{\alpha}\right)$$

$n_{\alpha} = F\, \psi,_{\alpha}$(unit normal to a surface ψ_3 constant)

$$\nabla_{\alpha} n_{\beta} = F,_{\alpha}\psi,_{\beta} + F\nabla_{\alpha}\psi,_{\beta}$$

$$\nabla_{\alpha} n_{\beta} - \nabla_{\beta} n_{\alpha} = F,_{\alpha}\psi,_{\beta} - F,_{\beta}\psi,_{\alpha} + F(\psi,_{\alpha\beta} - \psi,_{\beta\alpha})$$

(symmetry of the Riemannian Connector)

$$= \frac{F,_{\alpha}}{F} n_{\beta} - \frac{F,_{\beta}}{F} n_{\alpha}$$

$$= (\log F),_{\alpha} n_{\beta} - (\log F),_{\beta} n_{\alpha}$$

Hence

$$q_{\nu}^{\beta} n_{\beta} = 0,$$

$$q_{\mu}^{\alpha} n_{\alpha} = 0$$

$$\Rightarrow \quad K_{\mu\nu} - K_{\nu\mu} = 0, \textit{i.e.}\, K_{\mu\nu} = K_{\nu\mu}$$

Note:

$$n_{\alpha} = K_0 \frac{\partial X^0}{\partial X^{\alpha}} \quad n_{\alpha} X^{\alpha}_{,a} = K_0 \frac{\partial X^0}{\partial X^{\alpha}} = 0$$

$$g^{\alpha\beta} n_{\alpha} n_{\beta} = 1$$

$$\Rightarrow \quad \tilde{g}^{00} K_0^2 = 1, K_0 = \frac{1}{\sqrt{\tilde{g}^{00}}}$$

So $$F = K_0 = \frac{1}{\sqrt{\tilde{g}^{00}}}, \psi = x_0$$

$\therefore$ (n_{α}) is the unit normal to the surface x_0 = constant Spatial curvature tensor of spotio-tamporal curvature tensor

Let u_{μ} be a spatial vetor, Thus, $u_{\mu}n^{\mu} = 0$ or $u^{\mu}n_{\mu} = 0$ where $u^{\mu} = g^{\mu\nu}u_{\nu} = g^{\mu\nu}u_{\nu}$

$D_{\mu}u_{\nu} \triangleq q_{\mu}^{\alpha} q_{\nu}^{\beta} \nabla_{\alpha} u_{\beta}$ D_{μ} = Spatial covariant derivative.

Let u_{μ}, u_{μ} be spatial vector

$$D_{\rho}(u_{\mu}v_{\nu}) \triangleq (D_{\rho}u_{\mu})\, v_{\nu} + u_{\mu}\, D_{\rho}u_{\nu}$$

$$= q_{\rho}^{\alpha} q_{\mu}^{\beta} \left(\nabla_{\alpha} u_{\beta}\right) u_{\nu} + q_{\rho}^{\alpha} q_{\nu}^{\beta} u_{\mu} \nabla_{\alpha} u_{\beta}$$

$$= q_\rho^\alpha q_\mu^\beta q_v^\gamma \left(\nabla_\alpha u_\beta\right) u_v + q_\rho^\alpha q_v^\beta q_\mu^\gamma \left(\nabla_\alpha u_\beta\right) u_\gamma$$

$$= q_\rho^\alpha q_\mu^\beta q_v^\gamma \nabla_\alpha u_\beta \left(\left(u_\beta u_\lambda\right) + \left(\nabla_\alpha v_\gamma\right) u_\beta\right)$$

$$= g_\rho^\alpha q_v^\beta q_v^\gamma \nabla_\alpha \left(u_\beta v_\gamma\right)$$

Thus, since $D_\mu u_v$ is a spatial tensor,

$$D_\rho D_\mu u_v = D_\rho q_\mu^\alpha q_v^\beta \nabla_\alpha u_\beta$$

$$= q_\mu^{\mu'} q_v^{v'} q_\rho^{\rho'} \nabla \rho' \left(q_{\mu'}^\alpha q_{v'}^\beta \nabla_\alpha u_\beta\right)$$

$$= q_\mu^{\mu'} q_v^{v'} q_\rho^{\rho'} \nabla \rho' \left(\left(\delta_{\mu'}^\alpha - n^\alpha n_{\mu'}\right)\left(\delta_{v'}^\beta - n^\beta v'\right) \nabla_\alpha u_\beta\right)$$

$$= q_\mu^{\mu'} q_v^{v'} q_\rho^{\rho'} \nabla_{\rho'} \left(\nabla_{\mu'} u_{v'} - n^\alpha n_{\mu'} \nabla_\alpha u_{v'}\right.$$

$$\left. - n^\beta n_{v'} \nabla_{\mu'} u_\beta + n^\alpha n^\beta n_{\mu'} n_{v'} \nabla_\alpha u_\beta\right)$$

$$= q_\mu^{\mu'} q_v^{v'} q_\rho^{\rho'} \left[\nabla_{\rho'} \nabla_{\mu'} \mu_{v'}\right.$$

$$\left. - n^\alpha \left(\nabla_{\rho'} n_{\mu'}\right)\left(\nabla_\alpha u_{v'}\right) - n^\alpha \left(\nabla_{\rho'} n_{v'}\right)\left(\nabla_{\mu'} u_\beta\right)\right]$$

Since $\nabla \rho'$ is a derivation and $q_\mu^{\mu'} n_{\mu'} = 0$.

Now, $X \triangleq q_\mu^{\mu'} q_v^{v'} q_\rho^{\rho'} n^\alpha \left(\nabla_{\rho'} n_{\mu'}\right)\left(\nabla_\alpha u_{v'}\right)$

$$= K_{\rho\mu} q_v^{v'} n^\alpha \nabla_\alpha u_{v'}$$

$$q_v^{v'} \nabla_\alpha u_{v'} = \nabla_\alpha u_v - n^{v'} n_v \nabla_\alpha u v'$$

$$= \nabla_\alpha u_v + n_v \left(\nabla_\alpha n v'\right) u_{v'} \qquad \left(\because \eta^{v'} u_{v'} = 0\right)$$

So,

$$X = K_{\mu\rho} n^\alpha \left(\nabla_\alpha u_v + n_v u_{v'} \nabla_\alpha n^{v'}\right)$$

❖❖❖❖❖

[174] Nonlinear Pocklington integral equation

Medium is described by $\underline{D}(r, \underline{r}) = \underline{F}\left(\underline{E}(t-\tau, \underline{r}'), \tau \geq 0, \underline{r}' \in \mathbb{R}^3, t, \underline{r}\right)$

For example,

$$\underline{D}(r, \underline{r}) = \sum_{k=1}^{p} \int \underline{\underline{H}}_k \left(t, \tau,' \underline{r}, \underline{r}_1 \ldots, \underline{r}_k\right)$$

$$\left(\underline{E}\left(t - \tau_1^{r_1}\right) \otimes \underline{E}\left(t \quad \tau_2, \underline{r}_2\right) \otimes \ldots \otimes \underline{E}\left(t - \tau_k, \underline{r}_2\right)\right)$$

$$d\tau_1 ... d\tau_k \, d^3\underline{r}_1 ... d^3\underline{r}_k$$

$$\underline{B}(t, \underline{r}) = \mu_0 \, \underline{H}(t, \underline{r})$$

Maxwells equation:

$$\nabla \times \underline{D} = 0$$

$$\nabla \times \underline{E} = -\mu \frac{\partial \underline{H}}{\partial t}$$

$$\underline{\nabla} \times \underline{H} = \underline{J} + \frac{\partial \underline{D}}{\partial t}$$

$$\underline{\nabla} \times \underline{H} = 0.$$

Thus,

$$\nabla (\nabla \cdot E) - \nabla^2 E = -\mu \frac{\partial}{\partial t} \underline{\nabla} \times \underline{H}$$

$$= -\mu \frac{\partial \underline{J}}{\partial t} - \mu \frac{\partial^2 \underline{D}}{\partial t^2}$$

or

$$\nabla^2 \underline{E} - \mu \frac{\partial^2}{\partial t^2} F(\underline{E}, t, \underline{r})$$

$$-\nabla (\nabla \cdot E) = \mu \frac{\partial \underline{J}}{\partial t} \qquad (1)$$

$$\underline{\nabla} \times F(\underline{E}, t, \underline{r}) = 0 \qquad (2)$$

Taking this diagram of (1) gives

$$-\mu \frac{\partial^2}{\partial t^2} (\nabla \cdot F) = \mu \frac{\partial}{\partial t} (\nabla \cdot \underline{J}) = 0$$

Since $\rho = 0$ is assumed

Thus (1) $\nabla \cdot F = 0 \Rightarrow (2)$

Let the solution to (1) be experiment as

$$\underline{\mathrm{E}}(t, \underline{r}) = \psi\left(\underline{J}_s (t - \tau, \underline{r}'), \tau \geq 0, \underline{r}' \in \mathbb{R}^3, t, \underline{r}\right)$$

$$\equiv \psi\left(\underline{J}_s, t, \underline{r}\right)$$

Then if at any point $\underline{r}$ on the antenna surface S two linearly independence tangent vector are $\underline{G}(\underline{r})$ and $\underline{e}_2(\underline{r})$, we have

$$\left(\underline{E}(t, \underline{r}), \underline{e}_k(\underline{r})\right) = 0, \underline{r} \in S^1\{r_0\}$$

$$J_s(t, \underline{r}) = J_{0s}(t)$$

$\underline{r}_0$ being the source in S.

$\underline{J}_2$ is the surface current density on the antenna surface.

$$\underline{J}(t, \underline{r}) = \int \underline{J}_s\left(t, \underline{R}(v,u)\right)\delta\left(\underline{r} - \underline{R}(u,v)\right)dS(u,v)$$

Where the antenna surface parametrized by (u, v) *i.e.* $\underline{R} = \underline{R} = (u, v)$

So,

$$\int \underline{J}_s(t, \underline{r})d^3r = \int_{(u,v):\, \underline{R}(u,v)\in v} \underline{J}_s\left(t, \underline{R}(u,v)\right)dS(u,v)$$

For any $u \subset \mathbb{R}^3$ bonel measurable. The nonlinear pocklington equation for J_s is

$$\left(\underline{\psi}\left(\underline{J}_s t, \underline{r}\right)\underline{e}_k(\underline{r})\right) = 0,\ \underline{R}(v,u), \underline{R}(u,v) \neq \underline{r}_0$$

or

$$\left(\underline{\psi}\left(\underline{J}_s\left(t', \underline{R}(u,v), t' \le t, \underline{R}(u,v) \in S, t, \underline{R}(u',v')\right)\underline{e}_k\left(\underline{R}(u',v')\right)\right)\right) = 0,\ k = 1,2,$$

$$\underline{R}\,(u',v')r^0 \neq \underline{r}_0$$

Vollerne formulation and approximation using perturbation theory.

❖❖❖❖❖

[175] Compulation of the Haar measure for some groups.

So (3). Let X_1, X_2, X_3 be the standard generation.

$R = R_Z(\phi)R_X(\theta)R_Z(\psi)$ Euler only representation

$$= e^{\phi X_3}e^{\theta X_1}e^{\psi X_3}$$

$$\frac{d}{dt} f\left(\mathrm{Re}^{tX_k}\right)\Big|_{t=0} = \left(\tilde{X}_k f\right)(R)$$

$\tilde{X}_k \simeq$ Left invariant vector field corresponding to X_k.

Let
$$f(R) = X(\phi, \theta, \psi)$$

$$Re^{\delta t X_1} \approx R(I + \delta t \mathrm{X}_1)$$

$$RX_1 = e^{\phi X_3}e^{\theta X_1}e^{\psi X_3}X_1$$

$$= e^{\phi X_3}e^{\theta X_1}e^{\psi ad X_3}X_1 e^{\psi X_3}$$

$$= e^{\theta X_3}e^{\theta X_1}\left(X_1 \mathrm{Cos}\psi + X_2 \mathrm{Sin}\psi\right)e^{\psi X_3}$$

$$= \cos\psi\frac{\partial R}{\partial \theta} + \sin\psi e^{\theta X_2}e^{\theta X_1}X_2 e^{\psi X_3}$$

$$e^{\theta X_3}e^{\theta X_1}X_2e^{\psi X_3} = e^{\theta X_3}e^{\theta adX_1}(X_2)e^{\theta X_1}e\psi X_3$$

$$= e^{\theta X_3}(X_2\text{Cos}\theta + X_3\text{Sin}\theta)e^{\theta X_1}e^{\psi X_3}$$

$$= \sin\theta\frac{\partial R}{\partial\theta} + \cos\theta \times e^{\theta X_3}X_2e^{\theta X_1}e^{\psi X_3}$$

$$= \sin\theta\frac{\partial R}{\partial\theta} + \cos\theta(X_2\cos\theta - X_1\sin\theta)R$$

equivalently, $e^{\theta X_3}e^{\theta X_1}X_2e^{\psi X_3} = e^{\theta X_3}e^{\theta X_1}e\psi X_3 e - \psi^{adX_3}(X_2)$

$$= R(X_2\cos\psi + X_1\sin\psi)\ \tilde{X}_1$$

So, $\tilde{X}_1 = \cos\psi\frac{\partial}{\partial 0} + \sin\psi(\tilde{X}_1\sin\psi + \tilde{X}_2\cos\psi)$

$$\tilde{X}_3 = \frac{\partial}{\partial\psi}$$

$$\tilde{X}_2 f(R) = \frac{d}{dt}f\left(\text{Re}^{tX_2}\right)\Big|_{t=0}$$

$$RX_2 = e^{\phi X_3}e^{\theta X_1}e^{\psi X_3}X_2$$

$$= e^{\phi X_3}e^{\theta X_1}e^{\psi adX_3}(X_2)e^{\psi X_3}$$

$$= e^{\phi X_3}e^{\theta X_1}(X_2\text{Cos}\ \psi - X_1\text{Sin}\ \psi)e^{\psi X_3}$$

$$= -\sin\psi\frac{\partial R}{\partial\theta} + \cos\psi e^{\phi X_3}\ e^{\theta X_1}X_2e^{\psi X_3}$$

$$= -\sin\psi\frac{\partial R}{\partial\theta} + \cos\psi R(X_2\cos\psi + X_1\sin\psi)$$

So, $\tilde{X}_2 = -\sin\psi\frac{\partial}{\partial\theta} + \cos^2\psi\tilde{X}^2 + \sin\psi\cos\psi\tilde{X}_1$

Then, $-\frac{\partial}{\partial\theta} - \sin\psi\tilde{X}_2 + \cos\psi\tilde{X}_1 = 0$

which is the same as before.

Consider $\frac{\partial R}{\partial\phi} = X_3R = X_3e^{\phi X_3}e^{\theta X_1}e^{\psi X_3}$

$$= e^{\phi X_3}X_3e^{\theta X_1}e^{\psi X_3}$$

$$= e^{\phi X_3}e^{\theta X_1}(X_3\cos\theta + X_2\sin\theta)e\psi X_3$$

$$= RX_3\cos\theta + \sin\theta R(X_2\cos\psi + X_1\sin\psi)$$

Thus, $$\frac{\partial}{\partial\phi} = \tilde{X}_3\cos\theta + \sin\theta\cos\psi\tilde{X}_2 + \sin\theta\sin\psi\tilde{X}_1$$

Then, $$\begin{bmatrix} \frac{\partial}{\partial\psi} \\ \frac{\partial}{\partial\theta} \\ \frac{\partial}{\partial\phi} \end{bmatrix} = \begin{bmatrix} 0 & 0 & 1 \\ \cos\psi & -\sin\psi & 0 \\ \sin\theta\sin\psi & \sin\theta\cos\psi & \cos\theta \end{bmatrix} \begin{bmatrix} \tilde{X}_1 \\ \tilde{X}_2 \\ \tilde{X}_3 \end{bmatrix}$$

So,

$$\frac{\partial}{\partial\psi}\Lambda\frac{\partial}{\partial\theta}\Lambda\frac{\partial}{\partial\phi} = det\begin{bmatrix} 0 & 0 & 1 \\ \cos\psi & -\sin\psi & 0 \\ \sin\theta\sin\psi & \sin\theta\cos\psi & \cos\theta \end{bmatrix}\tilde{X}_1\Lambda\tilde{X}_2\Lambda\tilde{X}_3$$

$= \sin\theta\, \tilde{X}_1\Lambda\tilde{X}_2\Lambda\tilde{X}_3$. Hena Haar measure on $S_0(3)$ in $\sin\theta\, d\theta\, d\phi d\psi$

Symplectic group $sp(2, 1)$.

$$g \in Sp(2, \mathbb{R}) \Leftrightarrow g^T\begin{bmatrix} 0 & 1 \\ -1 & 0 \end{bmatrix}, g = \begin{bmatrix} 0 & 1 \\ -1 & 0 \end{bmatrix}$$

Non-compact, Non, Abelian group.

Lie algebra element : $sp(z, \mathbb{R}) = g(spC, \mathbb{R})$

$$X = \begin{bmatrix} X_{11} & X_{12} \\ X_{21} & X_{22} \end{bmatrix} \in g(sp(2, \mathbb{R}))$$

$$\Leftrightarrow \quad \begin{bmatrix} 0 & -1 \\ -1 & 0 \end{bmatrix}\begin{bmatrix} X_{11} & X_{12} \\ X_{21} & X_{22} \end{bmatrix} + \begin{bmatrix} X_{11} & X_{12} \\ X_{12} & X_{22} \end{bmatrix}\begin{bmatrix} 0 & -1 \\ -1 & 0 \end{bmatrix} = 0$$

$$\Leftrightarrow \quad \begin{bmatrix} X_{21} & X_{22} \\ -X_{11} & -X_{12} \end{bmatrix} + \begin{bmatrix} -X_{21} & X_{11} \\ -X_{22} & X_{12} \end{bmatrix} = 0$$

$$\Leftrightarrow \quad X_{11} + X_{22} = 0$$

So, $$sp(2, R) = \left\{\begin{bmatrix} a & b \\ c & -a \end{bmatrix} : a, b, c \in \mathbb{R}\right\}$$

dim $sp(z, \mathbb{R}) = 3$. Generators

$$X_1 = \begin{bmatrix} 1 & 0 \\ 0 & -1 \end{bmatrix}, X_2 = \begin{bmatrix} 0 & 1 \\ 0 & 0 \end{bmatrix}, X_3 = \begin{bmatrix} 0 & 0 \\ 1 & 0 \end{bmatrix}$$

Note: $sp(2, \mathbb{R}) = sl(2, \mathbb{R})$

$$\text{Let } g(x_1, x_2, x_3) = e^{x_1}X_1 e^{x_2}X_2 e^{x_3}X_3 \equiv g$$

$$e^{tX_1} = \begin{bmatrix} e^t & 0 \\ 0 & e^{-t} \end{bmatrix}$$

$$ge^{\delta t X_1} \approx g(I + \delta t X_1)$$

$$gX_1 = e^{x_1X_1}e^{x_2X_2}e^{x_3X_3}X_3$$

$$= \frac{\partial g}{\partial x_3}$$

$$\frac{\partial g}{\partial x_2} = e^{x_1X_1}e^{x_2X_2}X_2e^{x_3X_3}$$

$$= ge^{-x_3adX_3}\left(X_2\right)$$

$$adX_3(X_2) = [X_3, X_2] = -X_1$$

$$(adX_3)^2(X_2) = -[X_2, X_1] = [X_1, X_2] = 2X_2$$

$$(adX_3)^3(X_2) = 2[X_3, X_2] = -2X_1 \text{ etc}$$

$$(adX_3)^{2m}(X_2) = 2^m X_2, m = 0, 1, 2, ...$$

$$(adX_3)^{2m+1}(X_2) = -(2)^m X_1, m = 0, 1, 2, ...$$

$$\therefore \quad e^{-x_3adX_3}(X_2) = \sum_{m=0}^{\infty} \frac{x_3^{2m}2^m X_2}{\underline{|2m}} - \sum_{m=0}^{\infty} \frac{x_3^{2m+1}(2)^m X_1}{\underline{|2m+1}}$$

$$= \cosh(3x_3)X_2 - \sinh(2x_3)X_1$$

$$\therefore \quad \frac{\partial g}{\partial x_2} = g(\cosh(2x_3)X_2 - \text{Sin}h(2x_3)X_1)$$

$$\text{So,} \quad \frac{\partial g}{\partial x_2} = \cos\left(2x_3\right)\tilde{X}_2 - \sin\left(2x_3\right)\tilde{X}_1$$

$$\frac{\partial G}{\partial x_1} = e^{x_1X_1}e^{x_2X_2}e^{x_3X_3}$$

$$= e^{x_1X_1}e^{x_2X_2}e^{-x_2adX_2}\left(X_1\right)e^{x_3X_3}$$

$$adX_2(X_1) = [X_2, X_1] = -X_2$$

$$\therefore \quad (adX_2)^m(X_1) = 0, \ m \geq 2$$

$$\therefore \quad e^{-x_2adX_2}(X_1) = X_2 - X_2X_2$$

So,

$$\frac{\partial g}{\partial x_1} = e^{x_1 X_1} e^{x_2 X_2} (X_1 - x_2 X_2) e^{x_3 X_3}$$

$$= g e^{x_3 ad X_3} (X_1 - x_2 X_2)$$

orbit XX^*, $X \in O(n, \mathbb{C})$

$$X^T X = I_n$$

$$\Rightarrow \quad X^* X^{*T} = I_n$$

$$\Rightarrow \quad XX^* X^{*T} X^T = XX^T = I_n$$

$$\Rightarrow \quad XX^* (XX^*)^T = I_n$$

Let $O = \{XX^* | X \in O(n, \mathbb{C})\}$

Then we have $A \in O \Rightarrow AA^T \varepsilon O$

$A \in O \Rightarrow A^T \in O \Rightarrow A^T A \varepsilon O$

XX^* is a Hermitian + ve definite matrix.

Let $X = U\Sigma V^*$ be the svd of X.

Then $X^T X = I \Rightarrow V^{T*} \Sigma U^T U \Sigma V^* = I$

$$\Rightarrow \quad \Sigma U^T U \Sigma = V^T V$$

$$XX^* = U\Sigma^2 U^*$$

$\left\|\Sigma U^T U \Sigma\right\|_S \leq \sigma_{\max}^2$ where $\sigma_{max} = \{\Sigma ii_{\,1 \leq \iota \leq n}\}$

$\sigma_{max} \geq 1$

Let $X \in O(n, \mathbb{C})$

$$X = Y + i Z,\ Y, Z \in \mathbb{M}_n(\mathbb{R}),$$

$$X^T X = I \Rightarrow (Y^T + i Z^T)(Y + iZ) = I$$

$$\Rightarrow \quad Y^T Y - Z^T Z = I,$$

$$Y^T Z + Z^T Y = 0$$

$$XX^* = (Y + i Z)(Y^T - iZ^T)$$

$$YY^T + ZZ^T + i(ZY^T - YZ^T)$$

$$X^* X = (Y^T - iZ^T)(Y + iZ)$$

$$Y^T Y + Z^T Z + i(Y^T Z - Z^T Y)$$

$$I + 2Z^T Z + 2iY^T Z$$

Lie algebra generation of $sp\,(2n, \mathbb{R})$

$$X^T J_{2n} + J_{2n} X = 0$$

$$X = \begin{pmatrix} X_1 & X_2 \\ X_3 & X_4 \end{pmatrix} \cdot \begin{pmatrix} X_1^T & X_3^T \\ X_2^T & X_4^T \end{pmatrix} \begin{pmatrix} 0 & I_n \\ -I_n & 0 \end{pmatrix} + \begin{pmatrix} 0 & I_n \\ -I_n & 0 \end{pmatrix} \begin{pmatrix} X_1 & X_2 \\ X_3 & X_4 \end{pmatrix} = 0$$

$$\Leftrightarrow \quad X_1^T + X_4 = 0, \; -X_3^T + X_3 = 0, X_2^T \quad X_2 = 0 \cdot$$

$$\text{So,} \quad sp\,(2n, \mathbb{R}) = \left\{ \begin{pmatrix} X_1 & X_2 \\ X_3 & -X_1^T \end{pmatrix} : X_1, X_2, X_{3 \in} \mathbb{M}_n(\mathbb{R})\; X_2, X_3 \text{ Symmetrix} \right\}$$

A basis for *sp* (2*n*, ℝ) is

$$= \left\{ \begin{pmatrix} Eij & E_{kl} + E_{lk} \\ E_{pq} + E_{qp} & -E_{ji} \end{pmatrix} : 1 \le i, j, k, l, p, q \le n, \iota \le jk \le l, p \le q \right\}$$

$$\text{Write} \quad F_{ij} = \begin{pmatrix} Eij & 0 \\ 0 & -E_{ji} \end{pmatrix}$$

$$G_{kl} = \begin{pmatrix} 0 & E_{kl+Elk} \\ 0 & 0 \end{pmatrix}$$

$$H_{pq} = \begin{pmatrix} 0 & 0 \\ E_{pq} + E_{qp} & 0 \end{pmatrix}$$

Then, $\{F_{ij}, G_{kl}, H_{pq} | 1 \le i \le j \le n, 1 \le k \le l \le n, 1 \le p \le q \le n\}$

is a basis for *sp*(2*n*, ℝ)

So dim *sp*(2*n*, ℝ)

$$= 3 \times \left(\frac{n(n-1)}{2} + n \right)$$

$$= \frac{3}{2} n(n+1)$$

Cartan sub algebra:

$$\hbar = h_n \begin{pmatrix} E_{ii} & 0 \\ 0 & -E_{ii} \end{pmatrix} : 1 \le i \le n$$

$$\text{Let} \quad \hbar_\iota = \begin{pmatrix} E_{ii} & 0 \\ 0 & -E_{ii} \end{pmatrix}$$

$$[\hbar_i, F_{kl}] = \left[\begin{pmatrix} E_{ii} & 0 \\ 0 & -E_{ii} \end{pmatrix} \begin{pmatrix} E_{kl} & 0 \\ 0 & -E_{lk} \end{pmatrix} \right]$$

$$= \begin{pmatrix} E_{ii}\, E_{kl} - E_{kl}\, E_{ii} & 0 \\ 0 & E_{ii}\, E_{lk} - E_{lk}\, E_{ii} \end{pmatrix}$$

$$= \begin{pmatrix} E_{iki}\ Ei - \delta_{il}\ E_{ki} & 0 \\ 0 & \delta_{il}\ E_{lk} - \delta_{ki}\ E_{li} \end{pmatrix}$$

$$= \begin{pmatrix} \delta_{ik}\ E_{kl} - \delta_{il}\ E_{ki} & 0 \\ 0 & \delta_{il}\ E_{lk} - \delta_{ik}\ E_{lk} \end{pmatrix}$$

$$= (\delta_{ik} - \delta_{il}) \begin{pmatrix} E_{kl} & 0 \\ 0 & -E_{lk} \end{pmatrix}$$

$$(\delta_{ik} - \delta_{il}) F_{kl}$$

So, F_{kl} is a root vector.

$$[G_i, G_{kl}] = \begin{pmatrix} 0 & E_{ii}(E_{kl} + E_{lk}) + (E_{kl} + E_{lk})E_{ii} \\ 0 & 0 \end{pmatrix}$$

$$= \begin{pmatrix} 0 & E_{il} + \delta_{il} E_{ik} + \delta_{il} E_{ki} + \delta_{ik} E_{li} \\ 0 & 0 \end{pmatrix}$$

$$= \delta_{ik}\, G_{\iota l} + \delta_{il} G_{\iota k}$$

$$= \delta_{\iota k}\, G_{kl} + \delta_{\iota l} G_{lk}$$

$$= (\delta_{\iota k} + \delta_{\iota l})\, G_{kl}$$

So, G_{kl} is also a root vector of $SP\ (2n, \mathbb{R})$

G $\simeq$ a semiruple lie group with H a eartan subgroup.

$$G^{veg} = \bigcup_{g \in G} g H g^{-1}$$

$\pi \simeq$ a*representatnon of G with character X.

$$\chi(ghg^{-1}) = \chi(h),\ h \in H,\ g \in G.$$

$\int_G f(g)\pi(g) \equiv G_\pi(f)$. f has conpect support

$$G_\pi\,(f_0 x^{-1}) = \int_G f\left(x^{-1}g\right)\pi(g)\,dg$$

Let $$x \in G(fox^{-1}) = \int_a f(x^{-1}g) \int_G f(x^{-1}g)\gamma(g)2g$$

$$\int_G f(g)\pi(g)\,dg = \pi(x) G_\pi(f)$$

$$= \int_G f\left(x^{-1}g\right)\pi\left(g^{-1}\right)dg$$

$$= \int_G f(g)\pi\left(g^{-1}x^{-1}\right)dg$$

$$= \left(\int_G f(g)\pi\left(g^{-1}\right)dg\right)\pi\left(x^{-1}\right) = H_x(f)\pi(x^{-1})$$

where $$H_\pi(t) = \int_G f(g)\pi\left(g^{-1}\right)dg$$

Let

$$t_r(H_\pi(f)\,G_\pi(f)) = I_\pi(f) = [_\pi$$

Then

$$I_\pi(f) = \int_{G\times G} f(g_1)\,t_r\left(\pi\left(g_1^{-1}g_2\right)\right)f(g_2)\,dg_1 dg_2$$

$$= \int_{G\times G} f(g_1)+(g_2)X_\pi\left(g_1^{-1}g_2\right)dg_1 dq_2$$

We have

$$H_\pi(fox^{-1}) \times G_\pi(fox^{-1}) = H_\pi(f) \times G_\pi(f)$$

In particular, $$I_\pi(fox^{-1}) = I_\pi(f)\ \delta$$

$\forall\, x \in G$

$f: G \to \mathbb{C}$

Practical computation of $I_\pi(f)$:

$$I_\pi(f) = \int_{G\times G} f(g_1)+(g_1\,g)\times_\pi(g)\,dg_1 dg$$

$$= \int_{G\times G_{/H}\times H} f(g_1)f\left(g_1\,aha^{-1}\right) - X_\pi(h)\,dg_1 da\,dh$$

Consider $J_\pi(f) = \int_{G\times G\times H} f(g_1)f\left(g_1\,aha^{-1}\right) - X_\pi(h)\,dg_1 da\,dh$

Then, for $x \in G$, $J_\pi(fox^{-1}) = \int f\left(x^{-1}g_1\right)f\left(g^{-1}g_1 aha^{-1}\right)X_\pi(h)\,dg_1 da\,dh$

$sp\,(2n, \mathbb{R})$ Haar measure computable.

Certain subalgebra basis.

$$\hbar i = \begin{bmatrix} E_{ii} & 0 \\ 0 & -E_{ii} \end{bmatrix} 1 \le i \le n$$

Root vectors

$$F_{ij} = \begin{bmatrix} 0 & E_{ij} + E_{ji} \\ 0 & 0 \end{bmatrix}, 1 \le i \le j \le n,$$

$$G_{ij} = \begin{bmatrix} 0 & 0 \\ E_{ij} + E_{j\iota} & 0 \end{bmatrix}, 1 \le i \le j \le n,$$

$$H_{ij} = \begin{bmatrix} E_{ij} & 0 \\ 0 & -E_{j\iota} \end{bmatrix}, 1 \le i \le j \le n$$

In general for any semi-simple Lie algebra g with $G = \exp(g)$, we have the root space decomposition

$$g = \hbar \oplus \underset{\alpha \in \Delta}{\oplus} g_\alpha = \hbar \oplus \underset{\alpha \in \Delta_+}{\oplus} (g_\alpha \oplus g_\alpha)$$

$$[H, X_\alpha] = \alpha(H) X_\alpha, X_\alpha \in g_\alpha, \dim g\alpha = 0$$

Let $\{\alpha_1, \alpha_2 ., \alpha_p\}_r$ be the positive roots

Let $\{\alpha_1, \alpha_2 ., \alpha_r\}_r$ be the simple roots.

Thus, $$\alpha_k = \sum_{j=1}^{r} m(k, j) \alpha_{ji} \quad r + 1 \le k \le p$$

Where $m(k, j) \in \{0, 1, 2, ...\}$.

Let $\{H_1, H_2 ., H_r\}$ be a basis for $\hbar$ such that

$$\alpha_i(H_j) = \delta_{ij}, 1 \le i, j \le r.$$

Let $$g = \left(\prod_{j=1}^{r} e^{t_j H_j} \right) \left(\prod_{j=1}^{r} e^{s_j X_j} \right) \left(\prod_{j=1}^{r} e^{r_j Y_j} \right)$$

$t_j, S, r_s \in \mathbb{R}$, where $Y_j \in g_{-\alpha_j}$, $X_j \in g_{\alpha j}$,

$$[X_j, Y_m] = \delta_{jm} H_j, \; [H_j, X_k] = \alpha_k(H_j) X_k = a_{kj} X_k$$

$$[H_j, Y_k] = -\alpha_k(H_j) Y_k = -a_{kj} Y_k$$

Note: By $\prod_{j=1}^{m} \Lambda_j$ we mean the ordered product $\Lambda_1 \Lambda_2 ... \Lambda_m$

$$\frac{\partial g}{\partial r_r} = g Y_r . \text{ So } \frac{\partial}{\partial r_r} = \tilde{Y}_r,$$

$$\frac{\partial g}{\partial r_{r-1}} = g\exp(-r_r\, ad\, Y_r)(Y_{r-1})$$

$= g\xi_{r-1}$ where

$$\xi_{r-1} = \exp(-r_r\, ad\, Y_r)(Y_{r-1}) = Y_{r-1} + \sum_{m=1}^{\infty}\frac{(-1)^m r_r^m}{\lfloor m}(ad\, Y_r)^m(Y_{r-1})$$

Now either

$$(ad\, Y_r)^m(Y_{r-1}) = 0 \text{ or } (ad\, Y_r)^m(Y_{r-1})$$

is in $g_{-\alpha}$ with $\alpha = \alpha_{r-1} + m\alpha_r$

❖❖❖❖❖

[176] Some important equations in quantum gravity.

$$K_{\mu v} = q_\mu^\alpha q_v^\beta \nabla_\alpha n_\beta$$

$$K_{\mu v} = K_{ru}$$

$$q_\mu^\alpha n_\alpha = 0, q_\mu^\alpha n^\mu = 0$$

$$\text{Let } u_\rho n^\rho = 0$$

$$D_\mu D_v U_r = q_\mu^{\mu'} q_v^{v'} q_\rho^{\rho'} \nabla_{\mu'} D_{v'} \mu_{\rho'}$$

$$= q_\mu^{\mu'} q_v^{v'} q_\rho^{\rho'} \nabla_{\mu'}\left(q_{v'}^\alpha q_{\rho'}^\beta \nabla_\alpha \mu_\beta\right)$$

$$= q_\mu^{\mu'} q_v^{v'} q_\rho^{\rho'} q_{v'}^\alpha q_{\rho'}^\beta \nabla_{\mu'} \nabla_\alpha u_\beta$$

$$+ q_\mu^{\mu'} q_v^{v'} q_\rho^{\rho'}\left(\nabla_\alpha u_\beta\right)\nabla_{\mu'}\left(q_{v'}^\alpha q_{\rho'}^\beta\right)$$

$$q_v^{v'} q_{v'}^\alpha = \left(\delta_v^{v'} - n^{v'} n_v\right) q_{v'}^\alpha = q_v^\alpha$$

So, $$D_\mu D_v u_\rho = q_\mu^{\mu'} q_\rho^\beta q_v^\alpha \nabla_{\mu'} \nabla_\alpha u_\rho$$

$$+ q_\mu^{\mu'} q_v^{v'} q_\rho^{\rho'}\left(\nabla_\alpha u_\beta\right)\nabla_{\mu'}\left\{\left(\delta_{v'}^\alpha - n^\alpha n_{v'}\right)\left(\delta_{\rho'}^\beta - n^\beta n_{\rho'}\right)\right\}$$

$$\equiv T_1 + T_2 \text{ say}$$

$$T_2 = -\left(\nabla_\alpha u_\beta\right) q_\mu^{\mu'} q_v^\alpha q_\rho^{\rho'} \nabla_{\mu'}\left(n^\beta n_{\rho'}\right) - \left(\nabla_\alpha u_\beta\right) q_\mu^{\mu'} q_v^{v'} q_\rho^\beta \nabla_{\mu'}\left(n^\alpha n_{v'}\right)$$

$$= -T_{21} - T_{22} \text{ say}$$

$$T_{21} = n^\beta\left(\nabla_\alpha u_\beta\right) q_\mu^{\mu'} q_\rho^\alpha q_\rho^{\rho'} \nabla_\mu n\rho'$$

$$= n^\beta\left(\nabla_\alpha u_\beta\right) q_v^\alpha k_{\mu\rho}$$

$$= -u_\beta q_v^\alpha K_{\mu\rho}\left(\nabla_\alpha n^\beta\right) \qquad \left(\because u_\beta = q_\beta^\rho u_\sigma\right)$$

$$= -u_\sigma q_\beta^\sigma q_v^\alpha K_{\mu\rho}\left(\nabla_\alpha n^\beta\right) \qquad (n^\sigma n_\beta u^\sigma = 0)$$

$$= -u_\sigma q^{\beta\sigma} q_v^\alpha \left(\nabla_\alpha n_\beta\right) K_{\mu\sigma} = -\,u^\sigma K_{v\sigma} K_{\mu\rho}$$

$$T_{22} = n^\alpha \left(\nabla_\alpha u_\beta\right) q_\mu^{\mu'} q_v^{v'} q_\rho^\beta \nabla_{\mu'} n_{v'}$$

$$= n^\alpha \left(\nabla_\alpha u_\beta\right) q_\rho^\beta K_{\mu v}$$

$$= -u_\beta q_\rho^\beta \left(\nabla_\alpha n^\alpha\right) K_{\mu v} = -u_\rho K_{\mu v} \nabla_\alpha n^\alpha$$

Then, $[D_\mu, D_v]u_\rho = q_\mu^{\mu'} q_\rho^\beta q_v^\alpha \left[\nabla_{\mu'}, \nabla_\alpha\right]\mu_\beta + \mu^\sigma \left(K_{v\sigma}K_{\mu\rho} - K_{\mu\sigma}K_{v\rho}\right)$

$$= q_\mu^{\mu'} q_\rho^\beta q_v^\alpha R_{\rho\,\mu'}^\sigma u_\sigma + \left(K_v^\sigma K_{\mu\rho} - K_\mu^\sigma K_{v\sigma}\right) u_\sigma$$

$$= \left[R_{\beta\alpha\mu'}^{\sigma'} q_\mu^{\mu'} q_v^\alpha q_\rho^\beta q_\sigma^\sigma + K_v^\sigma K_{\mu\sigma} - K_\mu^\sigma K_{v\rho}\right] u_\sigma$$

Let $\quad \tilde{R}_{\rho v\mu}^\beta \mu_\beta = [D_\mu, D_v]u_\rho$ for all spacial vector u_ρ *i.e.* $u_\rho n_\rho = 0$
Then we deduce

$$\tilde{R}_{\rho v\mu}^\rho = R_{\rho' v'\mu'}^{\sigma'} q_{\sigma'}^\sigma q_\sigma^{\sigma'} q_v^{v'} q_\mu^{\mu'} + K_v^\sigma K_{\mu\sigma} - K_\mu^\sigma K_{v\sigma}$$

Contracting σ and μ gives

$$R_{\rho' v'} q_\sigma^{\sigma'} q_v^{v'} = \tilde{R}_{\rho v} + K K_{v\rho} - K_v^\mu K_{\mu\rho}$$

Note: $\quad \rho_{\sigma'}^\sigma q_\sigma^{\mu'} = \left(\delta_{\sigma'}^\sigma - n^\sigma n_{\sigma'}\right)\left(\delta_\sigma^{\mu'} - n^{\mu'} n_\sigma\right)$

$$= \delta_{\sigma'}^{\mu'} - n^{\mu'} n_{\sigma'} - n_{\mu'} n_{\sigma'} + n^{\mu'}{}_v n_{\sigma'} \qquad (\because n^\rho n^\sigma = 1)$$

$$= \delta_{\sigma'}^{\mu'} - n^{\mu'} n_{\sigma'} = \delta_{\sigma'}^{\mu'}$$

Now,

$$q^{\mu v} = \tilde{q}^{ab} X_{,a}^\mu X_{,b}^v$$

$$g^{\mu v} = q^{\mu v} + n^\mu n^v$$

$$= \tilde{q}^{\alpha b} X_{,\alpha}^\mu X_{,\beta}^v$$

$$= \tilde{g}^{\alpha b} X_{,\alpha}^\mu X_{,\beta}^v + \tilde{g}^{0a}\left(X_{,0}^\mu X_{,a}^v + X_{,a}^\mu X_{,0}^v\right) + \tilde{g}^{00} X_{,0}^\mu X_{,0}^v$$

$$= \tilde{g}^{\alpha b} X_{,a}^\mu X_{,b}^v + \tilde{g}^{0a}\left(Nn^\mu + N^\mu\right) X_{,a}^v$$

$$+\ \tilde{g}^{0a}\left(Nn^v + N^v \mid X_{,a}^\mu + \tilde{g}^{00}\left(Nn^\mu + N^\mu\right)\left(Nn^\mu + N^\mu\right)\right)$$

Spatial part + temporal part + mixed part mixed part

$$= \tilde{g}^{0a} N\left(n^{\nu} X^{\nu}_{,a} + n^{\nu} X^{\mu}_{,a}\right) + \tilde{g}^{0a} N\left(n^{\mu} N^{\nu} + N^{\mu} n^{\nu}\right)$$

$$= N\left\{\tilde{g}^{0a}\left(n^{\mu} X^{\nu}_{,0} + \tilde{g}^{00}\left(n^{\mu} N^{\nu} + N^{\mu} n^{\nu} + n^{\nu} X^{\mu}_{,a}\right)\right)\right\} = \psi^{\mu\nu} \text{ say}$$

Then $\frac{1}{N}\psi^{\mu\nu} n_{\nu} = \tilde{g}^{00} N^{\mu} + \tilde{g}^{0a} X^{\mu}_{,a}$

$$B_{20}\left(X^{\mu}_{,0} - N^{\mu}\right) X^{\nu}_{,b} g_{\mu\nu} = 0$$

$\Rightarrow$ $\tilde{g}_{ob} - N^{\mu} X^{\nu}_{,b}\, g_{\mu\nu} = 0$

$\Rightarrow$ $\tilde{g}_{ob} - C^{a} X^{\mu}_{,a} X^{\nu}_{,b} g_{\mu\nu} = 0$

$\Rightarrow$ $\tilde{g}_{ob} - C^{a} \tilde{g}_{ab}$

On the other hand,

$$\tilde{g}^{00} N^{\mu} + \tilde{g}^{oa} X^{\nu}_{,a}$$

$$= \tilde{g}^{00} C^{b} X^{\mu}_{,b} + \tilde{g}^{0a} X^{\mu}_{,a}$$

$$= X^{\mu}_{,b}\left[c^{b} \tilde{g}^{00} + \tilde{g}^{0a} \delta^{b}_{a}\right]$$

$$\left(c^{b} \tilde{g}^{00} + \tilde{g}^{0a} \delta^{b}_{a}\right) \tilde{g}_{bc}$$

$$= c^{b} \tilde{g}^{00} \tilde{g}_{bc} + \tilde{g}^{oa} \tilde{g}_{ac}$$

$\Rightarrow$ $c^{b} \tilde{g}_{bc} \tilde{g}^{00} + \tilde{g}^{o\mu} \tilde{g}_{\mu c} - \tilde{g}^{00} \tilde{g}_{oc}$

$\Rightarrow$ $\tilde{g}^{00}\left(c^{b} \tilde{g}_{bc} - \tilde{g}_{0c}\right) - \delta^{0}_{0} = 0$

Thus since $\left(\left(\tilde{g}_{bc}\right)\right) \in \mathbb{R}^{3\times 3}$ is non singulars it follows that $C^{b} \tilde{g}^{00} + \tilde{g}^{0a} \delta^{b}_{a} = 0$ and hence $\tilde{g}^{00} N^{\mu} + \tilde{g}^{0a} X^{\mu}_{,a} = 0$

i.e. $\psi^{\mu\nu} n_{\nu} = 0$

Again,

$$\frac{1}{N}\psi^{\mu\nu} N_{\nu} = \tilde{g}^{00} n^{\mu} N^{\nu} N_{\nu} + \tilde{g}^{oa} n^{\mu} X^{\nu}_{,a} N_{\nu}$$

$$= n^{\mu} N_{\nu}\left[\tilde{g}^{00} N^{\mu} + \tilde{g}^{oa} X^{\nu}_{,a}\right] = 0 \text{ as proved above}$$

Hence $\psi^{\mu\nu} n_{\nu} = \psi^{\mu\nu} N_{\nu} = 0$

So, $\underline{\psi^{\mu\nu} = 0}$, *i.e.* (mixed porb of $g^{\mu\nu}$) = 0

Hence $\quad g^{\mu\nu} = g^{\mu\nu} + n^{\mu}n^{\nu},\ q^{\mu\nu}n_{\nu} = 0$

$$R_{\rho'\nu'}q_{\rho}^{\rho'}q_{\nu}^{\nu'} = \tilde{R}_{\rho\nu} + KK_{\nu\rho} - K_{\nu}^{\mu}K_{\mu\rho}$$

So, $\quad q^{\rho\nu}R_{\rho\nu} = \tilde{R} + Kq^{\nu\rho}K_{\nu\rho} - q^{\nu\rho}K_{\nu}^{\mu}K_{\mu\rho}$

$$K_{\mu\nu} = q_{\mu}^{\mu'}q_{\nu}^{\nu'}\nabla_{\mu'}n_{\nu'}$$

$\Rightarrow \quad K_{\mu\nu}n^{\nu} = 0,\ K_{\mu\nu} = K_{\nu\mu}$

So, $\quad q^{\nu\rho}K_{\nu\rho} = g^{\nu\rho}K_{\nu\rho} = K$ and $q^{\nu\rho}K_{\nu}^{\mu}K_{\mu\rho} = K^{m\rho}K_{\mu\rho}$

So, $\quad q^{\rho\nu}\mathrm{R}_{\rho\nu} = \tilde{R} + K^2 - K^{\mu\rho}K_{\mu\rho}$

Thus,

$$R - n^{\rho}n^{\nu}R_{\rho\nu} = \tilde{R} + K^2 - K^{\mu\rho}K_{\mu\rho}$$

$$n^{\rho}n^{\nu}R_{\rho\nu} = n^{\rho}n^{\nu}R_{\rho\nu\beta}^{\beta}$$

Let $\quad u_{\beta}n^{\beta} = 0$, then $n^{\rho}n^{\nu}R_{\nu\rho\alpha}^{\beta}u_{\beta}$

$$= n^{\rho}n^{\nu}\left[\nabla_{\rho},\nabla_{\nu}\right]u_{\beta}$$

$$n^{\rho}n^{\nu}R_{\rho\nu\beta}^{\beta} = n^{\rho}n^{\nu}g^{\beta\alpha}R_{\alpha\rho\nu\beta}$$

$$= g^{\beta\alpha}n^{\nu}R_{\alpha\beta\nu}^{\rho}\,n_{\rho}$$

$$= g^{\beta\alpha}n^{\nu}\left(n_{\alpha:\beta:\nu} - n_{\alpha:\nu:\beta}\right)$$

$$= g^{\beta\alpha}n^{\nu}\left[\nabla_{\nu},\nabla_{\beta}\right]n_{\alpha}$$

$$g^{\beta\alpha}n^{\nu}\nabla_{\nu}\nabla_{\beta}n\alpha$$

$$= n^{\nu}\nabla_{\nu}\nabla_{\beta}n^{\beta}$$

$$= \nabla_{\nu}\left(n^{\nu}\nabla_{\beta}n^{\beta}\right) - \left(\nabla_{\nu}n^{\nu}\right)\left(\nabla_{\beta}n^{\beta}\right) \qquad \because \nabla_{\nu}g^{\beta\alpha} = 0$$

$$g^{\beta\alpha}n^{\nu}\nabla_{\beta}\nabla_{\nu}\overset{*}{n_{\alpha}}$$

$$= n^{\nu}\nabla_{\beta}\nabla_{\nu}n^{\beta} = \nabla_{\beta}\left(n^{\nu}\nabla_{\nu}n^{\beta}\right) - \left(\nabla_{\beta}n^{\nu}\right)\left(\nabla_{\nu}n^{\beta}\right)$$

$$= \nabla_{\beta}\nabla_{\nu}\left(n^{\nu}n^{\beta}\right) - \nabla_{\beta}\left(\left(\nabla_{\nu}n^{\nu}\right)n^{\beta}\right) - \left(\nabla_{\beta}n^{\nu}\right)\left(\nabla_{\nu}n^{\beta}\right)$$

$$g^{\beta\alpha}n^{\nu}\nabla_{\nu}\nabla_{\beta}v_{\alpha} = \nabla_{\nu}\nabla_{\beta}\left(n^{\nu}v_{\alpha}\right) - \nabla_{\nu}\left(\left(\nabla_{\beta}n^{\nu}\right)n^{\beta}\right) - \left(\nabla_{\nu}n^{\nu}\right)\left(\nabla_{\beta}n^{\beta}\right)$$

So,

$$g^{\beta\alpha}n^{\nu}\left[\nabla_{\nu},\nabla_{\beta}\right]n_{\alpha}$$

$$= \left(\nabla_\nu n^\nu\right)\left(\nabla_\beta n^\beta\right)\left(\nabla_\beta n^\nu\right)\left(\nabla_\nu n^\beta\right)$$

$$+\nabla_\beta\left(\left(\nabla_\nu n^\nu\right)n^\beta\right)-\nabla_\nu\left(\left(\nabla_\beta n^\nu\right)n^\beta\right)$$

$$\approx \left(\nabla_\nu n^\nu\right)\left(\nabla_\beta n^\beta\right)-\left(\nabla_\beta n^\nu\right)\left(\nabla_\nu n^\beta\right)$$

where $\approx$ means that perfect divergences are neglected since they do not contribute to the action integral.

Thus,

$$\int R\sqrt{-g}d^4X = \int \tilde{R}\sqrt{-g}d^4X$$

$$+\int\left[\left(K^2-K^{\mu\rho}K_{\mu\rho}\right)+\left(\nabla_\nu n^\nu\right)\left(\nabla_\beta n^\beta\right)-\left(\nabla_\beta n^\nu\right)\left(\left(\nabla_\nu n^\beta\right)\right)\right]\sqrt{-g}d^4X$$

$$\nabla_\nu n^\nu = g^{\nu\alpha}\nabla_\nu n_\alpha$$

$$= \left(q^{\nu\alpha}+n^\nu n^\alpha\right)\nabla_\nu n_\alpha$$

$$= q^{\nu\alpha}\nabla_\nu n_\alpha$$

$$= q^\nu_\rho q^{\rho\alpha}\nabla_\nu n_\alpha = K^\rho_\rho = K$$

$$\left(\nabla_\beta n^\nu\right)\left(\nabla_\nu n^\beta\right) = g^\rho_\nu g^\beta_\sigma\left(\nabla_\beta n^\nu\right)\left(\nabla_\rho n^\sigma\right)$$

$$\left(q^\beta_\sigma+n^\beta n_\sigma\right)X\left(q^\beta_\sigma+n^\beta n_\sigma\right)\left(\nabla_\beta n^\sigma\right)\left(\nabla_\rho n^\sigma\right)$$

$$= q^\beta_\nu q^\beta_\sigma\left(\nabla_\beta n^\nu\right)\left(\nabla_\rho n^\sigma\right)$$

$$\left(\because n_\sigma\nabla_\rho n^\sigma = 0 \text{ and } n_\nu\nabla_\beta n^\nu = 0\right)$$

$$= \left(\nabla_\beta n^\nu\right)K^\beta_\nu = K^{\beta\nu}\nabla_\beta n_\nu = K^{\beta\nu}\left(\delta^{\beta'}_\beta\ \delta^{\nu'}_\nu\right)\nabla_{\beta'}n_{\nu'}$$

$$= K^{\beta\nu}\left(\delta^{\beta'}_\beta+n^{\beta'}n_\beta\right)\left(q^{\nu'}_\nu+n^{\nu'}n_\nu\right)\nabla_{\beta'}n_{\nu'}$$

$$= K^{\beta\nu}q^{\beta'}_\beta q^{\nu'}_\nu\nabla_{\beta'}n_{\nu'}$$

$$\left(\because n_\beta K^{\beta\nu} = 0, n_\nu K^{\beta\nu} = 0\right)$$

$$= K^{\beta\nu}K_{\beta\nu}$$

Thus,

$$\int R\sqrt{-g}d^4X = \int\left(\tilde{R}+2\left(K^2-K^{\mu\rho}K_{\mu\rho}\right)\right)\sqrt{-g}d^4X$$

❖❖❖❖❖

[177] Symplectic invariants

$$\begin{pmatrix} \underline{q} \\ \underline{p} \end{pmatrix} \in \mathbb{R}^{2n}$$

$$\begin{pmatrix} \underline{q} \\ \underline{p} \end{pmatrix} = \underline{\xi} \underset{\to}{T} \underline{\underline{T}} \underline{\xi} \quad T \text{ Symplectic } i.e.$$

$$T^t J_{2n} T = J_{2n}, \text{ where } J_{2n} = \begin{pmatrix} \underline{\underline{0}} & \underline{\underline{I}}_n \\ -\underline{\underline{I}}_n & \underline{\underline{0}} \end{pmatrix}$$

Find $f: \mathbb{R}^{2n} \to \mathbb{C}$ s.t. $f(T\xi) = f(\xi) \forall \xi \in \mathbb{R}^{2n} \forall T \in Sp(2n, \mathbb{R})$

$$Sp(2n, \mathbb{R}) = \left\{T \in \mathbb{R}^{2n \times 2n} \mid T^t J_{2n} T = J_{2n}\right\}$$

= Symplectic group of order $2n$.

Character formula for Sp $(2n)$:

$$S_\lambda(\underline{x}) = A_{l_1 l_2, \dots l_n}(\underline{x}) / \Delta(\underline{x})$$

$$A_{l_1 l_2, \dots l_n}(\underline{x}) = \sum_{\sigma \in S_n} \in(\sigma) \prod_{i=1}^{n} \left(x_{\sigma(i)}^{li} - x_{\sigma(i)}^{-li}\right) \quad l_1 > l_2 .. > l_n > 0 + \text{ve integers}$$

$$\lambda = (l_1, l_2 \dots l_n)$$

and $$\Delta(\underline{x}) = A_{nn-1 \cdots 21}(\underline{x})$$

$x = (x_1, .., x_n)$, are the eigenvalues of $T. x_1^{-1}, ..., x_n^{-1})$
Let $X \in g(Sp(n))$

Then $X^t J + JX = 0 \quad \underline{\underline{J}} = \begin{pmatrix} \underline{\underline{0}} & \underline{\underline{I}}_n \\ -\underline{\underline{I}}_n & \underline{\underline{0}} \end{pmatrix}$

Let λ be an eigenvalne of $\underline{\underline{X}}$ with eigenvector $\underline{u}$. Thus,

$$\underline{\underline{X}} \underline{u} = \lambda \underline{u}$$

Then,

$$u^t X^t J u + u^t J X u = 0$$

$$\Rightarrow \quad \lambda^{u^t} J u + \lambda^{u^t} J u = 0 \text{ or } \lambda^{u^t} J u = 0$$

Either $$\lambda = 0 \text{ or } u^t J u = 0$$

Let m *i.e.* another eigavalue of X with eigevector v.

Then $$\underline{\underline{X}} \underline{v} = \mu \underline{v}$$

Then,

$$v^t X^t J u + v^t J X u = 0$$

$$\Rightarrow \qquad (\mu + \lambda) v^t J u = 0$$

if $\qquad v^t J u \neq 0, \mu = -\lambda$

Writing $\qquad X = \begin{bmatrix} X_{11} & X_{12} \\ X_{21} & X_{22} \end{bmatrix}$ gives

$$\begin{bmatrix} X_{11}^t & X_{21}^t \\ X_{12}^t & X_{22}^t \end{bmatrix}\begin{bmatrix} 0 & I \\ -I & 0 \end{bmatrix} + \begin{bmatrix} 0 & I \\ -I & 0 \end{bmatrix}\begin{bmatrix} X_{11} & X_{12} \\ X_{21} & X_{22} \end{bmatrix} = 0$$

or $\qquad X_{21}^t = X_{21},\ X_{12}^t = X_{12},$

$$X_{11}^t + X_{22} = 0$$

Thus any $X \in g(\mathrm{Sp}(n))$ has the form

$$X = \begin{bmatrix} L & M_1 \\ M_2 & -L^t \end{bmatrix} \text{ where } M_1^t = M_1, M_2^t = M_2$$

Eigenvalues of X: $det(\lambda I_{2n} - X) = dt \begin{bmatrix} \lambda I_n - L & -M_1 \\ -M_2 & \lambda I_n + L^t \end{bmatrix}$

$$\Rightarrow \qquad |\lambda I_n - L| \left| \lambda I_n + L^t - M_2 (\lambda I_n - L)^{-1} M_1 \right| = 0$$

Provided λ is not a eigenvalue of L.

Let $\qquad Xu = \lambda u\ X \in g(sp(2n, \mathbb{R}))$

$$X\bar{u} = \bar{\lambda}\bar{u}$$

$$X^t J + JX = 0$$

$$\Rightarrow \qquad \bar{u}^t X^t J u + \bar{u}^t J X u = 0$$

$$\Rightarrow \qquad (\bar{\lambda} + \lambda)\bar{u}^t J u = 0$$

$$\Rightarrow \qquad \bar{\lambda} = -\lambda \text{ if } \bar{u}^t J u \neq 0$$

So the eigenvalue of $e^X \in Sp(2n, \mathbb{R})$ one of the form Z where $|Z|^2 = Z\bar{Z} = 1$, *i.e.* the eigenvalue of $g \in sp(2n, \mathbb{R})$ are on the unit terms

Let $f(T) = \psi(\underline{X})$ where $\underline{X} \in \Pi^n$ are eigenvalue of T and so are $X_i^{-1} \in \Pi^n, i = 1,2,.,n$ where $\underline{X} = (X_i)_{i=1}^n$ ψ in an arbtrary function on $\mathbb{T}^n$.

That let $S_\lambda(\underline{x})$ be the character of a finite dimensional representation π_λ of $sp(2n\ \mathbb{R})$ corresponding to dominant integrate weight $\lambda = (l_1, l_2, ..l_n)$

i.e. $\qquad S_\lambda(x) = Tr(\Pi_\lambda(T)) = \dfrac{A_{\lambda+\rho}(\underline{u})}{\Delta(\underline{x})}$

Let f: $R^{2n} \to \mathbb{C}$ have bounded subport $\xi_0 \mathbb{R}^{2n}$ fixed

Then

$$\int_{Sp(2n,\mathbb{R})} f(T\underline{\xi}_0)\Pi_\lambda(T)dT = F\lambda(f)$$

(dT = Left invariant Haar measure on the semi simply lie group $Sp(2n, \mathbb{R})$)

We have for $S \in Sp(2n, \mathbb{R})$, $F_\lambda(-foS^{-1})$

$$= \int_{Sp(2n,\mathbb{R})} \left(foS^{-1}\right)(T\xi_0)\Pi_\lambda(T)dT$$

$$= \int f\left(S^{-1}T\xi_0\right)\Pi_\lambda(T)dT$$

$$= \int f(T\xi_0)\Pi_\lambda(ST)dT$$

$$= \Pi_\lambda(S)F_\lambda(f)$$

$$Tr(F_\lambda(f)) = \int f(T\xi_0)X_\lambda(\underline{x}(T))dt$$

$$Tr(F_\lambda(f)) = Tr(\Pi_\lambda(S)F_\lambda(t))$$

$$F_\lambda\left(g_0 s^{-1}\right) = \Pi_\lambda(S)F_\lambda(g)$$

$$F_\lambda\left(g_0 s^{-1}\right)F_\lambda\left(fos^{-1}\right) = F_\lambda(g)^{-1}F_\lambda(f)$$

$$\Rightarrow \qquad Tr\left(F_\lambda\left(g0s^{-1}\right)^{-1}F_\lambda\left(FoS^{-1}\right)\right) = Tr\left(F_\lambda(g)^{-1}F_\lambda(f)\right)$$

∴ For 2 function f, g on $\mathbb{R}^{2n}$,

$$I_\lambda(f, g) = F_\lambda(g)^{-1}F_\lambda(f)$$

is an $Sp(2n, \mathbb{R})$ invariant

Let $\quad G_\lambda(f) = \int f(T\xi_0)\Pi_\lambda\left(T^{-1}\right)dT$

Then,

$$G_\lambda(fos^{-1}) = \int f\left(S^{-1}T\xi_0\right)\Pi_\lambda\left(T^{-1}\right)dT$$

$$= \int f(T\xi_0)\Pi_\lambda\left(T^{-1}S^{-1}\right)dT$$

$$= G_\lambda(f)\,\Pi_\lambda(S)^{-1}$$

So,

$$Tr\left(G_{\lambda}\left(GoS^{-1}\right)F_{\lambda}\left(foS^{-1}\right)\right)$$

$$= Tr\left(G_{\lambda}(g)F_{\lambda}(f)\right)$$

$$= Tr \int_{S\,p(n)X\,S\,p(n)} g(T\,\xi_0)\Pi_{\lambda}\left(T^{-1}\right)f(S\,\xi_0)\Pi_{\lambda}(S)dT\,dS$$

$$= \int g(T\,\xi_0)f(S\,\xi_0)\chi_{\lambda}\left(T^{-1}S\right)dTdS$$

$$= \int g(T\,\xi_0)f(TS\,\xi_0)\chi_{\lambda}(S)dTdS$$

$$= \int g(T\,\xi_0)f(TS\,\xi_0)\chi_{\lambda}(x(S))dTdS$$

Let $\psi(S) = \int g(T\,\xi_0)f(TS\,\xi_0)dT$

Than for $K \in Sp(2n)$

$$\psi_{\xi 0}(KSK^{-1}) = \int g(T\,\xi_0)f\left(T\,K\,S\,K^{-1}\xi_0\right)dT$$

$$= \int g\left(T\,K^{-1}\xi_0\right)f\left(T\,S\,K^{-1}\xi_0\right)dT$$

$$= \psi_{K^{-1}\xi_0}(S)$$

Consider

$$\underline{\Phi}_{\xi_0}(S) = \int_{SP(n)} \psi_{\xi_0}\left(K\,S\,K^{-1}\right)dK$$

Then $\underline{\Phi}_{\xi_0}$ is a T function *i.e.*

$$\underline{\Phi}_{\xi_0}\left(K\,S\,K^{-1}\right) = \underline{\Phi}_{\xi_0}(S)\forall K \in Sp(2n)$$

Then, $\forall K \in Sp(2n)$

$$Tr\left(G_{\lambda}(g)F_{\lambda}(f)\right)$$

$$= \int \psi_{\xi 0}(S)\chi_{\lambda}(S)dS$$

$$= \int \psi_{\xi 0}(S)\chi_{\lambda}(S)dS$$

$$= \int \psi_{\xi 0}\left(K^{-1}SK\right)\chi_{\lambda}(S)dS$$

It follows that since g, f have compact support, say S_g and S_z respectively, we have

$$\underline{\Phi}_{\xi_0}(S) = \int \psi_{\xi 0}\left(K\,S\,K^{-1}\right)dS$$

$$= \int g\left(T\,K^{-1}\xi_0\right)f\left(T\,S\,K^{-1}\xi_0\right)dT\,dK$$

$$T K^{-1} \in S_g$$

$$TS K^{-1} \in S_f$$

$$SdT \int g\left(T K^{-1}\xi_0\right) f\left(T S K^{-1}\xi_0\right) dK$$

$$K^{-1} \varepsilon T^{-1} S_g S^{-1} T^{-1} S_f$$

❖❖❖❖❖

[178] Canonical quantum gtr.

$X^\mu(t,x), x = \left(x^a\right)_{a=1}^3$ $a, b, c, d, e, f \in \{1, 2, 3\}$ $\mu, \nu, \rho, \sigma, \alpha, \beta, \lambda \in \{0, 1, 2, 3\}$

Mention $g_{\mu\nu} dX^\mu dX^\nu$

$$X^\mu_{,0} = Nn^\mu + N^\mu$$

Where $\left(N^\mu\right)_\mu \in$ span $\left\{\left(X^\mu_{,a}\right)_\mu : a = 1, 2, 3\right\}$

$$g_{\mu\nu} n^\mu X^\nu{,a} = 0 \; i.e. \; n\mu X^\mu_{,a} = 0, a = 1, 2, 3$$

$$g_{\mu\nu} n^\mu n^\nu = 1, \; i.e. \; n^\mu n^\mu = 1 \text{ (normalization)}$$

$$n^\mu = \left(X^\mu_{,0} - N^\mu\right) / N$$

Let $$N^\mu = C^a X^\mu{,a}$$

Then (C) are determine from

$$g_{\mu\nu}\left(X^\mu_{,0} - C^a X^\mu_{,a}\right) X^\nu_{,0} = 0$$

or

$$\tilde{g}_{ab} C^a = g_{\mu\nu} X^\nu_{,b} X^\mu_{,0}$$

$$a, b = 1, 2, 3,$$

$$\tilde{g}_{ab} = g_{\mu\nu} X^\mu_{,a} X^\nu_{,b}$$

Then N^μ is determined. Then,

$$n^\mu = \frac{X^\mu_{,0} - N^\mu}{N}$$

$\Rightarrow g_{\mu\nu}\left(X^\mu_{,0} - N^\mu\right)\left(X^\nu_{,0} - N^\nu\right) = N^2$ determined N. up to a sign

Let $\left(\left(\tilde{g}^{ab}\right)\right) = \left(\left(\tilde{g}^{ab}\right)\right)^{-1}$ (3 × 3 matrices)

Then, let

$$q^{\mu\nu} = \tilde{g}^{ab} X^{\mu}_{,a} X^{\nu}_{,b}, \left(\left(\tilde{g}^{ab}\right)\right) = \left(\left(\tilde{g}^{ab}\right)\right)^{-1}$$

We get $q^{\mu\nu} n_{\nu} = q^{\mu\nu} n_{\mu} = 0 \; q^{\mu\nu} = q^{\nu\mu}$

Note that

$$\tilde{g}_{ab} = \tilde{g}_{\mu\nu} X^{\mu}_{,a} X^{\nu}_{,b}$$

$$\begin{aligned}
\tilde{g}_{\mu\nu} &= \tilde{g}_{\alpha\beta} X^{\alpha}_{,\mu} X^{\beta}_{,\nu} \\
&= g_{\mu\nu}\left(g^{\nu\alpha} + n^{\nu} n^{\alpha}\right) \\
&= g_{\mu\nu}\left(\tilde{g}^{ab} X^{\nu}_{,a} X^{\alpha}_{,b} + n^{\nu} n^{\alpha}\right) \\
&= g_{\mu\nu}\left(\tilde{g}^{ab} X^{\nu}_{,a} X^{\alpha}_{,b} + \frac{\left(X^{\nu}_{,0} - N^{\nu}\right)\left(X^{\alpha}_{,0} - N^{\alpha}\right)}{N^2}\right)
\end{aligned}$$

$= T^{\alpha}_{\mu}$ say. Then

$$\begin{aligned}
T^{\alpha}_{\mu} X^{\mu}_{,\beta} &= g_{\mu\nu} X^{\mu}_{,\beta} X^{\nu}_{,a} \tilde{q}^{ab} X^{\alpha}_{,b} + n^{\nu} n^{\alpha} X^{\mu}_{,\beta}\, g_{\mu\nu} \\
&= \tilde{q}_{\beta\nu} \tilde{q}^{ab} X^{\alpha}_{,b} + n_{\mu} n^{\alpha} X^{\mu}_{,\beta}
\end{aligned}$$

Thus,

for $\beta = 0$,

$$\begin{aligned}
T^{\alpha}_{\mu} X^{\mu}_{,0} &= \tilde{g}_{0a} \tilde{q}^{ab} X^{\alpha}_{,b} + n^{\alpha} n_{\mu}\left(N n^{\mu} + N^{\mu}\right) \\
&= \tilde{g}_{0a} \tilde{q}^{ab} X^{\alpha}_{,b} + N n^{\alpha} \qquad \left(\because n_{\mu} N^{\mu} = 0\right)
\end{aligned}$$

Note:

$$\tilde{g}_{0a} \tilde{q}^{ab} + \tilde{g}_{00}\, \tilde{g}^{0b} = 0$$

So,

$$\tilde{g}_{0a} \tilde{q}^{ab} = -\tilde{g}_{00} \tilde{g}^{0b}$$

Assuming $\tilde{g}^{0a} = 0$, we get $\tilde{g}_{0a} = 0$ and so,

$$T^{\alpha}_{\mu} X^{\mu}_{,0} = N^{n^{\alpha}}\; X^{\alpha}_{,0} - N^{\alpha}$$

Further,

$$\begin{aligned}
T^{\alpha}_{\mu} X^{\mu}_{,a} &= X^{\mu}_{,a}\left(g_{\mu\nu}\right)\left(\tilde{q}^{bc} X^{\nu}_{,b} X^{\alpha}_{,c} + n^{\nu} n^{\alpha}\right) \\
&= n^{\alpha} n_{\mu} X^{\mu}_{,a} + \tilde{q}^{bc} \tilde{q}_{ab} X^{\alpha}_{,c} \\
&= n^{\alpha} n_{\mu} X^{\mu}_{,a} + \delta^{c}_{a} X^{\alpha}_{,c}
\end{aligned}$$

$$= X^{\alpha}_{,c} \quad \left(n_{\mu} X^{\mu}_{,a} = 0\right)$$

Let $(g_{\mu\nu})$ be the metric relative to X^{μ}

and $\left(\tilde{g}_{\mu\nu}\right)$ relative to X^{μ}.

Consider
$$f_{\mu} = \frac{\partial X'^{0}}{\partial X^{\mu}}$$

We have,

$$f_{\mu} X^{\mu}_{,a} = \frac{\partial X^{0}}{\partial X^{\mu}} \frac{\partial X^{\mu}}{\partial X^{a}} = \frac{\partial X^{0}}{\partial X^{a}} = \delta^{0}_{a} = 0$$

Hence, $(f_{\mu}) \propto (n_{\mu})$

$$g_{\mu\nu} = \tilde{g}_{\alpha\beta} \frac{\partial X^{\alpha}}{\partial X^{\mu}} \frac{\partial X^{\beta}}{\partial X^{\nu}}$$

$$= \tilde{g}_{00} f_{\mu} f_{\nu} + \tilde{g}_{ab} \frac{\partial X^{\alpha}}{\partial X^{\mu}} \frac{\partial X^{b}}{\partial X^{\nu}} + 2\tilde{g}_{0a} \frac{\partial X^{0}}{\partial X^{\mu}} \frac{\partial X^{a}}{\partial X^{\nu}}$$

$$\frac{\partial X^{\alpha}}{\partial X^{\nu}} \frac{\partial X^{\nu}}{\partial X^{0}} = \frac{\partial X^{\alpha}}{\partial X^{0}} = 0$$

$$\Rightarrow \quad \frac{\partial X^{\alpha}}{\partial X^{\nu}} \left(N n^{\nu} + N^{\nu}\right) = 0$$

$$N^{\nu} = C^{b} \frac{\partial X^{\nu}}{\partial X^{b}}$$

$$\Rightarrow \quad C^{b} \frac{\partial X^{a}}{\partial X^{\nu}} \frac{\partial X^{\nu}}{\partial X^{b}} + N n^{\nu} \frac{\partial X^{a}}{\partial X^{\nu}} = 0$$

$$\Rightarrow \quad C^{a} = -N n^{\nu} \frac{\partial X^{a}}{\partial X^{\nu}}$$

So
$$\frac{\partial X^{a}}{\partial X^{\nu}} n^{\nu} \neq 0 \text{ } i.e \text{ } \frac{\partial X^{a}}{\partial X^{\nu}} f^{\nu} \neq 0$$

We assume $\tilde{g}_{oa} = 0$, $a = 1, 2, 3$.

Thus,
$$\underline{\underline{\tilde{g}}} = \begin{bmatrix} \tilde{g}_{00} & 0 & 0 & 0 \\ 0 & \tilde{g}_{11} & \tilde{g}_{12} & \tilde{g}_{13} \\ 0 & \tilde{g}_{21} & \tilde{g}_{22} & \tilde{g}_{23} \\ 0 & \tilde{g}_{31} & \tilde{g}_{32} & \tilde{g}_{33} \end{bmatrix}$$

Thus,
$$g_{\mu\nu} = \tilde{g}_{00} f_{\mu} f_{\nu} + \tilde{g}_{ab} \frac{\partial X^{a}}{\partial X^{\mu}} \frac{\partial X^{b}}{\partial X^{\nu}}$$

$$g^{\mu\nu}f_\mu f_\nu = \tilde{g}_{00} = \frac{1}{\tilde{g}_{00}}. \text{ So } n_\mu = \sqrt{\tilde{g}_{00}}\, f_\mu$$

and writing

$$q_{\mu\nu} = \tilde{g}_{ab}\frac{\partial X^a}{\partial X^\mu}\frac{\partial X^b}{\partial X^\nu}, \text{ we get}$$

$$g_{\mu\nu} = n_\mu n_\nu + q_{\mu\nu}$$

Note that

$$\frac{\partial X^a}{\partial X^\mu}n^\mu = -\frac{C^a}{N}$$

So,

$$\tilde{g}_{ab}\frac{\partial X^a}{\partial X^\mu}\frac{\partial X^b}{\partial X^\nu}n^\mu = -\frac{C^a}{N}\tilde{g}_{ab}\frac{\partial x^b}{\partial X_\nu}$$

$$= \tilde{g}_{ab}\frac{\partial X^a}{\partial X^\mu}\frac{\partial X^b}{\partial X^\nu}n^\mu$$

$$= \left(\tilde{g}_{\alpha\beta}\frac{\partial X^\alpha}{\partial X^\mu}\frac{\partial X^\beta}{\partial X^\nu} - \tilde{g}_{00}\frac{\partial X^0}{\partial X^\mu}\frac{\partial X^0}{\partial X^\nu}\right)n^\mu$$

$$= \left(g_{\mu\nu} - \tilde{g}_{00}f_\mu f_\nu\right)n^\mu$$

$$\left(g_{\mu\nu} - n_\mu n_\nu\right)n^\mu = n_\nu - n_\nu = 0$$

Thus, $q_{\mu\nu}n^\nu = 0$.

Hence we have the decomposition

$$g_{\mu\nu} = q_{\mu\nu} + n_\mu n_\nu,$$

$$q_{\mu\nu}n^\nu = 0,\ q_{\mu\nu}n^\mu = 0$$

Let $q_{\mu\nu} = q_{\nu\mu}$.

$$K_{\mu\nu} = q^\alpha_\mu q^\beta_\nu \nabla_\mu n_\nu$$

$$g^{\mu\nu} = \tilde{g}^{\alpha\beta}X^\mu_{,\alpha}X^\nu_{,\beta}$$

$$= \tilde{g}^{ab}X^\mu_{,\alpha}X^\nu_{,b} + \tilde{g}^{00}X^\mu_{,0}X^\nu_{,0} \qquad \left(\because \tilde{g}^{a0} = 0\right)$$

$$X^\mu_{,0} = Nn^\mu + N^\mu$$

$$= Nn^\mu + C^a X^\mu_{,a}$$

So, $$g_{\mu\nu} = \tilde{g}^{ab}X^\mu_{,a}X^\nu_{,b} + \tilde{g}^{00}\left(Nn^\mu + N^\mu\right)\left(Nn^\nu + N^\nu\right)$$

$$= \tilde{g}^{ab} X^{\mu}_{,a} X^{v}_{,b} + \tilde{g}^{00}\left(N^2 n^{\mu} n^{v} + N^{\mu} N^{v} + N\left(n^{\mu} N^{v} + N^{\mu} N^{v}\right)\right)$$

or $$g^{\mu v} = \left(\tilde{g}^{ab} X^{\mu}_{,a} X^{v}_{,b} + \tilde{g}^{00} N^{\mu} N^{v}\right) + N\tilde{g}^{00}\left(N^{\mu} n^{v} + N^{v} n^{\mu}\right) + \tilde{g}^{00} N^2 n^{\mu} n^{v}$$

$$= q^{\mu v} + s\, n^{\mu} n^{v}, \text{ say}$$

$$X^{\mu}_{,0} = Nn^{\mu} + N^{\mu}$$

$$\Rightarrow X^{\mu}_{,0}\, n_{\mu} = N$$

$$= X^{\mu}_{0} \sqrt{\tilde{g}_{00}} f_{\mu} = \sqrt{\tilde{g}_{00}} X^{\mu}_{,0} \frac{\partial X^0}{\partial X^{\mu}} = \sqrt{\tilde{g}_{00}} = \frac{1}{\sqrt{\tilde{g}_{00}}}$$

So,

$$q^{\mu v} = \tilde{g}^{ab} X^{\mu}_{,a} X^{v}_{,b} + \tilde{g}^{00} N^{\mu} N^{v} + \frac{1}{N}\left(N^{\mu} N^{v} + N^{v} N^{\mu}\right) + (1-s) n^{\mu} n^{v}$$

We require $q^{\mu v}\eta_n = 0$. Thus, since $X^{v}_{,b} n_v = 0, N^v n_v = 0$

We get $(1-s)n^{\mu} + \frac{1}{N} N^{\mu} = 0$ which impossible.

$$\tilde{g}_{a0} = 0 \Leftrightarrow g_{\mu v} X^{\mu}_{,a} X^{v}_{,0} = 0$$

$$\Leftrightarrow \quad q_{\mu v} X^{\mu}_{,a} X^{v}_{,0} = 0 \quad \left(g_{\mu v} = q_{\mu v} + n_{\mu} n_{v}\right)$$

$$\Leftrightarrow \quad q_{\mu v} X^{\mu}_{,a} N^{v} = 0$$

Suppose we do not make the annuxcture $\tilde{g}_{a0} = 0$.

Then,

$$X^{\mu}_{,0} = Nn^{\mu} + N^{\mu}$$

$$N^{\mu} = C^a X^{\mu}_{,a},\ g_{\mu v} n^{\mu} n^{v} = 1,$$

$$X^{\mu}_{,a} n_{\mu} = 0 \text{ hence } N^{\mu} n_{\mu} = 0.$$

$$f_{\mu} = \frac{\partial X^0}{\partial X^{\mu}}$$

$$f_{\mu} X^{\mu}_{,a} = 0 \Rightarrow f_{\mu} = K.\, n_{\mu}$$

$$g^{\mu v} n_{\mu} n_{v} \Rightarrow K^2 = g^{\mu v} f_{\mu} f_{v} = \tilde{g}^{00}$$

$$K = \sqrt{\tilde{g}^{00}}$$

$$X^{\mu}_{,0} n_{\mu} = N$$

$$\Rightarrow \quad X^{\mu}_{,0} n_{\mu} = KN$$

$$\Rightarrow \quad X^{\mu}_{,0} \frac{\partial X^0}{\partial X^{\mu}} = KN$$

$$\Rightarrow \quad 1 = KN \Rightarrow K = \frac{1}{N}$$

$$g_{\mu\nu} = \tilde{g}_{\alpha\beta} \frac{\partial x^{\alpha}}{\partial X^{\mu}} \frac{\partial x^{\beta}}{\partial X^{\nu}}$$

$$= \tilde{g}_{00} f_{\mu} f_{\nu} + \tilde{g}_{0\alpha} \left(f_{\mu} \frac{\partial x^{\alpha}}{\partial X^{\nu}} + f_{\nu} \frac{\partial x^{\alpha}}{\partial X^{\mu}} \right) + \tilde{g}_{ab} \frac{\partial x^a}{\partial X^{\mu}} \frac{\partial x^b}{\partial X^{\nu}}$$

$$= \frac{\partial x^a}{\partial X^{\mu}} \frac{\partial x^{\mu}}{\partial X^0} = 0 \Rightarrow \frac{\partial x^a}{\partial X^{\mu}} \left(N^{\mu} + N n^{\mu} \right) = 0 \Rightarrow \frac{\partial x^a}{\partial X^{\mu}} N^{\mu}$$

$$= -N \frac{\partial x^a}{\partial X^{\mu}} n^{\mu} = -\frac{N}{K} \frac{\partial x^a}{\partial X^{\mu}} g^{\mu\alpha} \frac{\partial X^0}{\partial X^{\alpha}}$$

$$= -N^2 \tilde{g}^{a0} \neq 0$$

We choose our coordinate system to that

$$g_{\mu\nu} = q_{\mu\nu} + s n_{\mu} n_{\nu}$$

where $\quad q_{\mu\nu} n^{\mu} = q_{\mu\nu} n^{\nu} = 0,\ q_{\mu\nu} = q_{\mu\nu}$

Let $\quad \tilde{q}_{ab} = q_{\mu\nu} X^{\mu}_{,a} X^{\nu}_{,b}$

$$= g_{\mu\nu} X^{\mu}_{,a} X^{\nu}_{,b} = \tilde{g}_{ab}$$

Let $\quad \left(\left(\tilde{g}^{ab} \right) \right) = \left(\left(\tilde{g}^{ab} \right) \right)^{-1}$

$$\tilde{g}_{a0} = g_{\mu\nu} X^{\mu}_{,a} X^{\nu}_{,0}$$

$$= q_{\mu\nu} X^{\mu}_{,a} X^{\nu}_{,0}$$

$$= q_{\mu\nu} X^{\mu}_{,a} N^{\nu}$$

Let $\quad \underline{\underline{X}} = \left(\left(X^{\mu}_{,a} \right) \right) \in \mathbb{R}^{4\times 3}$

$$\underline{\underline{Q}} = \left(\left(q_{\mu\nu} \right) \right) \in \mathbb{R}^{4\times 4}$$

$$\underline{\underline{\tilde{Q}}} = \left(\left(\tilde{q}_{ab} \right) \right) \in \mathbb{R}^{3\times 3}$$

Then,
$$\underset{3\times3}{\tilde{Q}} = \underset{3\times4}{X^T}\ \underset{4\times4}{Q}\ \underset{4\times3}{X}$$

$$\underset{3\times3}{\tilde{Q}^{-1}} = \underset{3\times4}{X^-}\ \underset{4\times4}{Q^-}\ \underset{4\times3}{X^{-T}} \quad \text{(where } A^- \text{ is the generalized inverse of A)}$$

$$X\tilde{Q}^{-1}X^T = XX^-Q^-X^{-T}X^T$$

$$= XX^-Q^-\left(XX^-\right)^T$$

In terms of matrix element

$$\tilde{q}^{ab}X^{\mu}_{,a}X^{v}_{,b} = q^{\mu v}$$

XX^- = orthogonal projection and $R(X)$.

$$\underline{\underline{\tilde{g}}} = \begin{bmatrix} \tilde{g}_{00} & \tilde{g}_{0,1:3} \\ \tilde{g}_{1:3,0} & \tilde{q} \end{bmatrix} = \left(\left(\tilde{g}_{\mu v}\right)\right)$$

$$= \left(\left(g_{\alpha\beta}X^{\alpha}_{,\mu}X^{\beta}_{,v}\right)\right)$$

$$\underline{\underline{\tilde{g}}}^{-1} = \left(\left(\tilde{g}^{\mu v}\right)\right) = \begin{bmatrix} \alpha & \underline{\beta}^T \\ \underline{\beta} & \underline{\underline{\Gamma}} \end{bmatrix} \text{ say}$$

Then

$$\begin{bmatrix} \alpha & \underline{\beta}^T \\ \underline{\rho} & \underline{\underline{\Gamma}} \end{bmatrix} \times \begin{bmatrix} \tilde{g}_{00} & \tilde{g}_{0,1:3} \\ \tilde{g}1:3,0 & \underline{\underline{\tilde{g}}} \end{bmatrix} = \begin{bmatrix} 1 & \underline{0}^T \\ \underline{0} & \underline{\underline{I}}_3 \end{bmatrix}$$

$$\Rightarrow \qquad \alpha\tilde{g}_{00} + \underline{\beta}^T\tilde{g}_{1:3,0}\ \underline{\beta}\tilde{g}_{00} + \underline{\Gamma}\,\tilde{g}_{1:3,0} = \underline{0}$$

$$\underline{\beta}\tilde{g}^T_{01:3,0} + \underline{\underline{\Gamma}}\underline{\underline{\tilde{q}}} = I_3$$

$$\Rightarrow \qquad \underline{\beta}^T = -\alpha\, \underline{g}^T_{1:3,0}\underline{\underline{\tilde{g}}}^{-1}$$

$$\underline{\beta} = -\alpha\,\underline{\underline{\tilde{g}}}^{-1}\underline{g}_{1:3,0}$$

$$\alpha\left(\tilde{g}_{00} - \underline{\tilde{g}}^T_{1:3,0}\underline{\underline{\tilde{q}}}^{-1}\underline{\tilde{g}}_{1:3,0}\right) = 1$$

$$\alpha = \frac{1}{\tilde{g}_{00} - \underline{\tilde{g}}^T_{1:3,0}\underline{\underline{\tilde{q}}}^{-1}\underline{\tilde{g}}_{1:3,0}}$$

$$\underline{\underline{\Gamma}} = \left(I_3 - \underline{\beta}\,\underline{\tilde{g}}^T_{1:3,0}\right)\underline{\underline{\tilde{q}}}^{-1}$$

$$\underline{\beta}\left(\tilde{g}_{00} - \underline{\tilde{g}}^T_{1:3,0}\,\underline{\underline{\tilde{q}}}^{-1}\underline{\tilde{g}}_{1:3,0}\right) + \underline{\underline{\tilde{q}}}^{-1}\underline{\tilde{g}}_{1:3,0} = 0$$

$$\beta = \frac{-\tilde{\underline{\underline{q}}}^{-1}\, \tilde{\underline{g}}_{1:3,0}}{\left(\tilde{g}_{00} - \tilde{\underline{g}}_{1:3,0}^{T}\tilde{\underline{\underline{q}}}^{-1}\tilde{\underline{g}}_{1:3,0}\right)}$$

Since $Q = \left(\left(q_{\mu\nu}\right)\right)$ is the spacial component $\left(\left(g_{\mu\nu}\right)\right)$ and XX^{-} is the orthogonal projection on to the spatial $\left\{\left(X^{\mu}_{,a}\right)_{\mu=0}^{3} : 0 \le \mu \le 3\right\}$, it follows that

$$XX^{-}Q^{-}\left(XX^{-}\right)^{T} = Q^{-} \text{ and hence}$$

$$Q^{-} = X\tilde{Q}^{-1}X^{T}\text{. Thus, } \left(\left(q^{\mu\nu}\right)\right) = Q^{-} = \left(\left(q_{\mu\nu}\right)\right)$$

Note:

$$\begin{aligned} g^{\mu\nu} &= \tilde{g}^{\alpha\beta}X^{\mu}_{,\alpha}X^{\nu}_{,\beta} \\ &= \tilde{g}^{ab}X^{\mu}_{,a}X^{\nu}_{,b} + \tilde{g}^{0a}\left(X^{\mu}_{,0}X^{\nu}_{,a} + X^{\mu}_{,c}X^{\nu}_{,0}\right) + \tilde{g}^{00}X^{\mu}_{,0}X^{\nu}_{,0} \\ &= \tilde{g}^{ab}X^{\mu}_{,a}X^{\nu}_{,b} + \tilde{g}^{0a}\left(\left(Nn^{\mu} + N^{\mu}\right)X^{\nu}_{,a} + \left(Nn^{\nu} + N^{\nu}\right)X^{\mu}_{,a}\right) \\ &\quad + \tilde{g}^{00}\left(Nn^{\mu} + N^{\mu}\right)\left(Nn^{\nu} + N^{\nu}\right) \\ &= \left[\tilde{g}^{ab}X^{\mu}_{,a}X^{\nu}_{,b} + \tilde{g}^{oa}\left(N^{\mu}X^{\mu}_{,a} + N^{\nu}X^{\mu}_{,a} + \tilde{g}^{00}N^{\mu}N^{\nu}\right)\right] \\ &\quad + \tilde{g}^{00}N^{2}n^{\mu}n^{\nu} \\ &\quad + \tilde{g}^{0a}N\left(n^{\mu}X^{\nu}_{,a} + n^{\nu}X^{\mu}_{,a}\right) + \tilde{g}^{00}N\left(N^{\mu}N^{\nu} + N^{\mu}n^{\nu}\right) \end{aligned}$$

Now $$N = \frac{1}{K} = \frac{1}{\sqrt{\tilde{g}^{00}}}$$

So, $$g^{\mu\nu} = q^{\mu\nu} + n^{\mu}n^{\nu}$$

where $$g^{\mu\nu} = \tilde{g}^{ab}X^{\mu}_{,a}X^{\nu}_{,b} + \tilde{g}^{0a}\left(N^{\mu}X^{\nu}_{,a} + N^{\nu}X^{\mu}_{,a}\right) + \tilde{g}^{00}N^{\mu}N^{\nu}$$

provided that $\tilde{g}^{0a}N\left(n^{\mu}X^{\nu}_{,a} + n^{\nu}X^{\mu}_{,a}\right) + \tilde{g}^{00}N\left(n^{\mu}N^{\nu} + N^{\mu}n^{\nu}\right) = 0$ (α)

This also condition (α) gives on multiply by ν_{μ},

$$\tilde{g}^{0a}NX^{\nu}_{,a} + \tilde{g}^{00}NN^{\nu} = 0 \qquad (\beta_1)$$

or $$\tilde{g}^{0a}X^{\nu}_{,a} + \tilde{g}^{00}N^{\nu} = 0$$

or $$\tilde{g}^{0a}X^{\nu}_{,a} + \tilde{g}^{00}C^{a}X^{\nu}_{,a} = 0$$

So (β_1) requires $\tilde{g}^{0a} + \tilde{g}^{00} C^a = 0$

$$C^a = \frac{-\tilde{g}^{0a}}{\tilde{g}^{00}}$$

$$X^{\mu}_{,0} = N^{\mu} + Nn^{\mu} = C^a X^{\mu}_{,a} + Nn^{\mu}$$

$\Rightarrow$

$$C^a X^{\mu}_{,a} g_{\mu\nu} X^{\nu}_{,b} = X^{\mu}_{,0} g_{\mu\nu} X^{\nu}_{,b}$$

or

$$\tilde{g}_{ab} C^a = \tilde{g}_{0b}$$

So, for (β_1) we require

$$\tilde{g}_{ab}\tilde{g}^{0a} + \tilde{g}^{0b}\tilde{g}^{00} = 0$$

or

$$\tilde{g}_{ab}\,\tilde{g}^{\mu 0} = 0$$

which is always valid. Also for (α) we get on multiply by $g_{\mu\rho} X^{\rho}$ from

$$\tilde{g}^{0a}\left(n^{\mu} X^{\nu}_{,a} + n^{\nu} X^{\mu}_{,a}\right) g_{\mu\rho} X^{\rho}_{,b} + \tilde{g}^{00}\left(n^{\mu} N^{\nu} + n^{\nu} N^{\mu}\right) g_{\mu\rho} X^{\rho}_{,b} = 0 \quad (\beta_2)$$

Since

$$n_{\rho} X^{\rho}_{,b} = n^{\mu} g_{\mu\rho} X^{\rho}_{,b}$$

(β_2) is the same as

$$\tilde{g}_{ab}\tilde{g}^{0a} n^{\nu} + \tilde{g}^{00} g_{\mu\rho} n^{\nu} N^{\mu} X^{\rho}_{,b} = 0$$

or

$$\tilde{g}_{ab}\tilde{g}^{0a} + \tilde{g}^{00} g_{\mu\rho} N^{\mu} X^{\rho}_{,b} = 0$$

or

$$\tilde{g}_{ab}\tilde{g}^{0a} + \tilde{g}^{00} g_{\mu\rho} c^a X^{\mu}_{,a} X^{\rho}_{,b} = 0$$

or

$$\tilde{g}_{ab}\tilde{g}^{0a} + \tilde{g}^{00} g_{ab} c^a = 0$$

or

$$\tilde{g}_{ab}\,\tilde{g}^{0a} + \tilde{g}^{00}\tilde{g}_{0b} = 0$$

or $\tilde{g}_{b\mu}\,\tilde{g}^{\mu 0} = 0$, which is always ture.

Hence, we get the decomposition

$$g^{\mu\nu} = q^{\mu\nu} + n^{\mu} n^{\nu}$$

where

$$q^{\mu\nu} n_{\nu} = 0, \; q^{\mu\nu} n_{\mu} = 0$$

Lowering the indirect using $g_{\mu\nu}$ gives

$$g_{\mu\nu} = q_{\mu\nu} + n_{\mu} n_{\nu}$$

where

$$q_{\mu\nu} n^{\nu} = 0, \; q_{\mu\nu}\, n^{\mu} = 0$$

Spatials covariant derivation.

❖❖❖❖❖

[179] Path integral approach to q.m.

Hamiltonian $H(q, p, t)$

$$
\begin{aligned}
(q''|q') &= \left\langle q'' \middle| \exp\left(-i\Delta t H(q,p,t)\right) \middle| q' \right\rangle \\
&= \int \langle q''|p'\rangle \, dp' \left\langle p' \middle| \exp\left(-i\Delta t H(q,p,t)\right) \middle| q' \right\rangle \\
&= \int \langle q''|p'\rangle \, dp' \cdot \exp\left(-i\,\Delta t H(q',p',t)\right) \langle p'|q'\rangle
\end{aligned}
$$

provided that H is arranged so that all component of q appear on its right and all p component appear on tis left.

Thus,

$$
\begin{aligned}
K_{\Delta t}(q''\,|\,q') &= C\left(\int \exp\left(-i\,p' \times (q'-q'')\right) \exp\left(-i\,\Delta t H(q',p')t\right)\right) dp' \\
&= C\left(\int \exp\left(-i\Delta t\left(-q' \cdot \frac{(q'-q'')}{\Delta t} - H(q',p',t)\right)\right)\right) dp' \\
&= C\left(\int \exp\left(-i\Delta t\left(p' \times \frac{(q''-q')}{\Delta t} - H(q',p',t)\right)\right)\right) dp'
\end{aligned}
$$

By Considering

$$
K_{N\Delta t}\left(q_N' \middle| q_1'\right) = \int \prod_{K=1}^{N-1} K\Delta t\left(q_{k+1}' \middle| q_k'\right) \prod_{K=2}^{N-1} dq_k'
$$

and its limit as $\Delta \leftrightarrow 0 N \to \infty\ N\Delta t \to t$ the required path integral formula is obtained.

$$
K_T\left(q_\alpha' \middle| q_1'\right) = C \times \int \exp\left(i\int_0^T \left(\underline{p}(t), \underline{\dot{q}}(t)\right) - H\left(\underline{q}(t), \underline{p}(t), t\right) dt \right) \prod_{t\in[0,T]} \left(d\underline{q}(t) \times d\,\underline{p}(t)\right)
$$

❖❖❖❖❖

[180] Levy process driven classical and quantum models of a two link robot arm. In the classical cost, We simply write down the sde for the robot dynamics with external torque plug Levy noise and make an approximate statistical analysis of the same using perturbation theory considering the noise to be weak. We also set up expressions for the averge rate of change of a Lyapunov function and indicate schemes for simulating Levy noise of various types and robot dynamic driven by Levy noise. In the quantum contract we outline a mental for calculating the stationary states of the robot is a gravitational field by applying time independent perturbation theory to the bouncing ball problem and the by applying time dependent perturbation theory calculate stationary states of the robot including the transition probability between stationary for a levy noise perturbed robot Finally, we make an analysis of classical and quantum of robots dravin by with Gaussian noise plus a compound process which is a more general model taking into account continuous and jerk tremor.

$$dq(t) = d\dot{q}(t)\,dt,$$

$$d\dot{q}(t) = \underline{F}(t, q(t), \dot{q}(t))dt + \underline{\underline{G}}(q(t))\left(\sigma_1 d\underline{B}(t) + Y_{N(t)+1} dN(t)\right)$$

$$= \underline{F}(t, q(t), \dot{q}(t)) - \underline{\underline{M}}(\underline{q}(t))^{-1} \underline{N}(\underline{q}(t), \underline{\dot{q}}(t)) + \underline{\underline{M}}(q(t))^{-1} \underline{\tau}(t)$$

This models the robot with both kinds of tremon continuous gaussian plus discrete non Gaussian poisson sliky

Linearization:

$$\underline{q}(t) = \underline{q}_0(t) + \delta\underline{q}(t)$$

Where $\underline{q}_0(t)$ is solely the motion due to non random torque $\underline{\tau}(t)$ and $\delta\underline{q}(t)$ is true perturbation produced by the Levy noise $w(t)$ where

$$W(t) = \sigma_1 \underline{B}(t) + \sum_{h=1}^{N(t)} \underline{Y}_k$$

$\underline{B}(t) \simeq 2-D$ Brownian motion $\{Y_k\}$ iid *r.v's* $N(t) \simeq$ Poisson process with rate λ.

$$\ddot{\underline{q}}_0(t) = \underline{F}(t, \underline{q}_0(t), \dot{\underline{q}}_0(t))$$

$$\delta\ddot{\underline{q}}(t) = \underline{\underline{F}}_{11}(t)\delta\underline{q}(t) + \underline{\underline{F}}_{12}(t)\delta\dot{\underline{q}}(t) + \underline{\underline{G}}(t)\left(\sigma_1 \dot{\underline{B}}(t) + \underline{Y}_{N(t)+1}\dot{N}(t)\right)$$

$$\underline{\underline{F}}_{11}(t) = \frac{\partial \underline{F}}{\partial \underline{q}}(t, \underline{q}_0(t), \dot{\underline{q}}_0(t))$$

$$\underline{\underline{F}}_{12}(t) = \frac{\partial \underline{F}}{\partial \dot{\underline{q}}}(t, \underline{q}_0(t), \dot{\underline{q}}_0(t))$$

and
$$\underline{\underline{G}}(t) = \underline{\underline{G}}(\underline{q}_0(t))$$

In state variable notation

$$d\begin{bmatrix} \delta\underline{q}(t) \\ \delta\dot{\underline{q}}(t) \end{bmatrix} = \begin{bmatrix} \underline{\underline{0}} & \underline{\underline{T}} \\ \underline{\underline{F}}_1(t) & \underline{\underline{F}}_{12}(t) \end{bmatrix} \begin{bmatrix} \delta\underline{q}(t) \\ \delta\dot{\underline{q}}(t) \end{bmatrix} dt + \begin{bmatrix} \underline{\underline{O}} \\ \underline{\underline{G}}(t) \end{bmatrix} \left(\sigma_1 d\underline{B}(t) + \underline{Y}_{N(t)+1} dN(t)\right)$$

Solution: Let $\underline{\underline{\Phi}}(t,\tau)$ be the state transition matrix of $\begin{bmatrix} 0 & \underline{\underline{I}} \\ \underline{\underline{F}}_{,1}(t) & \underline{\underline{F}}_{,2}(t) \end{bmatrix}$. Then, assuming $\mathbb{E}\left(\underline{Y}, \underline{Y}_1^T\right) = \alpha^2 \underline{\underline{I}}_2$,

$$\delta \underline{q}(t) = \int_0^t \underline{\underline{\Phi}}_{12}(t,\tau)\underline{\underline{G}}(\tau)\left(\sigma_1 d\underline{B}(\tau) + \underline{Y}_{N(\tau)+1} dN(\tau)\right)$$

Thus,

$$\mathbb{E}\left[\delta \underline{q}(t)\delta \underline{q}(t)T\right] = \left(\int_0^t \underline{\underline{\Phi}}_{12}(t,\tau)\underline{\underline{G}}(\tau)\underline{\underline{G}}^T(\tau)\underline{\underline{\Phi}}_{12}^T(T,\tau)d\tau\right)\left(\sigma_1^2 + \alpha^2\right)$$

$$+ \int_{0<\tau_1 \neq \tau_2 < t} \underline{\underline{\Phi}}_{12}(t,\tau_1)\underline{\underline{G}}(\tau_1)(\mathbb{E})\underline{Y}_{N(\tau_1)+1}\underline{Y}_{N(\tau_2)+1} dN(\tau_1)dN(\tau_2)$$

$$G(\tau_2)^T \underline{\underline{\Phi}}_{12}(t,\tau_2)^T$$

For

$$\tau_1 < \tau_2,\ \mathbb{E}\left[\underline{Y}_{N(\tau_1)+1}.\ \underline{Y}_{N(\tau 2)+1}^T dN(\tau_1)dN(\tau_2)\right]$$

$$= \lambda d\tau_2 \mathbb{E}\left[\underline{Y}_{N(\tau_1)+1}\underline{Y}_{N(\tau 2)+1}^T dN(\tau_1)\right]$$

$$= \lambda^2 d\tau_2 d\tau_1 \mathbb{E}\left(\underline{Y}_1\right)\mathbb{E}\left(\underline{Y}_1^T\right) = \lambda^2 d\tau_1 d\tau_2\, \underline{\mu}_Y\, \underline{\mu}_Y^T$$

likewise for $\tau_2 < \tau_1$. Thus,

$$\mathbb{E}\left[\delta \underline{q}(t)\delta \underline{q}(t)^T\right] = \left(\sigma_1^2 + \alpha^2\right)\int_0^t \underline{\underline{\Phi}}_{12}(t,\tau)\underline{\underline{G}}(\tau)\underline{\underline{G}}(\tau)^{\mathrm{T}}\underline{\underline{\Phi}}_{12}(t,\tau)T\tau$$

$$+\lambda^2\left(\int_0^t \underline{\underline{\Phi}}_{12}(t,\tau)G(\tau)d\tau\right)\underline{\mu}_Y\, \underline{\mu}_Y^T$$

$$\left(\int_0^t \underline{\underline{\Phi}}_{12}(t,\tau)G(\tau)d\tau\right)^T$$

The quantum case:

$$dU(t) = -i\left(H_0 + P(t)dt + V_1(t)dB(t) + \int_{x\in E} V_2(t,x,N(t)+1)N(dt,dx)\right)$$

where $V_2(t, x, n)$ is a random field independent of $N(.,.)$ and $V_2(.,.,n)$ $n = 1,2,...$ iid random fields.

❖❖❖❖❖

Perturbation Theory in Cosmology, Hamiltonian of Charged Particles in Curved Space-time, Brownian Local Time, Statistics of Quantum Observables,Quantization of Stochastic Dynamical Systems, Quantum Relative Entropy Evolution for Open Systems, Quantum Gates in Levy Noise

[181] Evolution of inhomogeneties lives in a homogeneous and isotropic universe. Analysis is based on small perturbation of the Robertson Walker matrix.

$$d\tau^2 = dt^2 - \frac{S^2(t)}{1-r^2}dr^2 - S^2(t)r^2\left(d\theta^2 + \sin^2\theta d\varphi^2\right)$$

$$X^2 + Y^2 + Z^2 + u^2 = S^2$$

$$u = \sqrt{S^2 - r^2} \quad du = \frac{-rdr}{\sqrt{S^2 - r^2}}$$

$$dl^2 = du^2 + dr^2 + r^2\left(d\theta^2 + \sin\theta d\varphi^2\right)$$

$$= \frac{S^2 dr^2}{S^2 - r^2} + r^2\left(d\theta^2 + \sin\theta d\varphi^2\right)$$

$$r = Sr_1 \quad dl^2 = \frac{S^2 dr_1^2}{S^2 - r_1^2} + S^2 r_1^2\left(d\theta^2 + \sin^2 d\varphi^2\right)$$

$\simeq$ Spheroid universe replacing the equation for space by

$$K\left(X^2Y^2 + Z^2\right) + u^2 = S^2$$

with $K = 1, 0, -1$ we get respectively spheroid, flat and hyperbolic universe

Let $$\frac{S^2(t)}{1-kr^2} = f(t, r),\ S^2(t) = \psi(t)$$

Then

$d\tau^2 = dt^2 - f(t,r)dr^2 - \psi(t)r^2\left(d\theta^2 + \sin^2\theta\, d\varphi^2\right)$. This is the Robertson walker matrix. $\delta(t)$, $\rho_0(t)$, $p_0(t)$ satisfy the Einstein field eqn.

$$R^{\mu\nu} - \frac{1}{2}Rg^{\mu\nu} = K\left(\rho(t) + p_0(t)V^{\mu}V^{\nu} - p_0(t)g^{\mu\nu}\right)$$

with $(V^{\mu}) = (1, 0, 0, 0)$

Let

$$\begin{aligned}
v^{\mu}(t, \underline{r}) &= V^{\mu} + \delta v^{\mu}(t, \underline{r}) \\
p(t, \underline{r}) &= p_0(t) + \delta p(t, \underline{r}) \\
\rho(t, \underline{r}) &= \rho_0(t) + \delta\rho(t, \underline{r}) \\
g_{\mu\nu}(t, \underline{r}) &= g_{\mu\nu}^{(0)}(t, \underline{r}) + \delta g_{\mu\nu}(t, \underline{r})
\end{aligned}$$

where $g_{\mu\nu}^{(0)} = 0\ \mu \neq \nu$

$$\begin{aligned}
g_{00}^{(0)} &= g_{11}^{(0)} = -f_1\, g_{22}^{(0)} = -\psi(t)r^2, \\
g_{33}^{(0)} &= -\psi(t)r^2\sin^2\theta
\end{aligned}$$

$$R^{\mu\nu} - \frac{1}{2}R\,g^{\mu\nu} = K\left((\rho + p)u^{\mu}u^{\nu} - pg^{\mu\nu}\right)$$

$$\Rightarrow \qquad -R = K(r + p - p) = K(\rho - 3p)$$

$$\begin{aligned}
R^{\mu\nu} &= -\frac{1}{2}K(\rho - 3p)g^{\mu\nu} + K\left((\rho + p)v^{\mu}v^{\nu} - pg^{\mu\nu}\right) \\
&= K\left((\rho + p)v^{\mu}v^{\nu} + \frac{1}{2}(p - \rho)g^{\mu\nu}\right)
\end{aligned}$$

$$\begin{aligned}
\therefore \qquad \delta R^{\mu\nu} &= K\Big((\delta\rho + \delta p)V^{\mu}V^{\nu} + (\rho_0 + p_0)\left(V^{\mu}\delta u^{\nu} + V^{\nu}\delta u^{\mu}\right) \\
&\quad + \frac{1}{2}(\delta p - \delta\rho)g^{\mu\nu}(0) + \frac{1}{2}(p_0 - \rho_0)\delta g^{\mu\nu}\Big)
\end{aligned}$$

or equivalently

$$\begin{aligned}
\delta R_{\mu\nu} &= K\Big((\delta\rho + \delta p)V_{\mu}V_{\nu} + (\rho_0 + p_0)\left(V_{\mu}\delta u_{\nu} + V_{\nu}\delta v_{\mu}\right) \\
&\quad + \frac{1}{2}(\delta p - \delta\rho)g_{\mu\nu}^{(0)} + \frac{1}{2}(p_0 - \rho_0)\delta g_{\mu\nu}\Big)
\end{aligned}$$

$$\delta R_{\mu\nu} = \delta\Gamma^{\alpha}_{\mu\alpha,\nu} - \delta\Gamma^{\alpha}_{\mu\nu,\alpha} - \delta\left(\Gamma^{\alpha}_{\mu\nu}\Gamma^{\beta}_{\mu\beta}\right) + \delta\left(\Gamma^{\alpha}_{\mu\beta}\Gamma^{\beta}_{\nu\alpha}\right)$$

Let $S_{\mu\nu}$ be rhs of (1). then

$$\begin{aligned} S_{00} &= K\left(\delta\rho + \delta p + 2(\rho_0 + p_0)V_0\delta v_0 \right. \\ &\quad \left. + \frac{1}{2}g^{(0)}_{00}(\delta p - \delta\rho) + \frac{1}{2}(p_0 - \rho_0)\delta g_{00}\right) \\ &= K\left(\delta\rho + \delta p + 2(\rho_0 + p_0)\delta v_0 \right. \\ &\quad \left. + \frac{1}{2}(\delta p - \delta\rho) + \frac{1}{2}(p_0 - \rho_0)\delta g_{00}\right) \\ &= K\left(\frac{1}{2}(3\delta p + \delta\rho) + 2(p_0 + \rho_0)\delta v_0 + \frac{1}{2}(p_0 - \rho_0)\delta g_0\right) \end{aligned}$$

$$S_{0k} = K\left((p_0 + \rho_0)\delta v_k + \frac{1}{2}(p_0 - \rho_0)\delta g_{0k}\right)$$

$$S_{km} = \frac{1}{2}K\left(g^{(0)}_{km}(\delta p + \delta\rho) + (p_0 - \rho_0)\delta g_{k\mu}\right)$$

$$g^{\mu\nu}v_\mu v_\nu = 1$$

$$\Rightarrow \qquad (\delta g^{\mu\nu})V_\mu V_\nu + 2g^{\mu\nu(0)}V_\mu \delta v_\nu = 0$$

Thus $\qquad \delta g^{00} + 2g^{0\nu(0)}\delta v_\nu = 0$

or $\qquad \delta g^{00} + 2\delta v_0 = 0$

i.e. $\qquad \delta v_0 = -\frac{1}{2}\delta g^{00}$

$(V_\mu = g_{\mu\nu}V^\nu = g_{\mu 0} \Rightarrow (V_\mu) = (1, 0, 0, 0) = (V^\mu))$

$$S_{km} = \frac{1}{2}K(p_0 - \rho_0)\delta g_{km}, \; k \neq m,$$

$$S_{11} = \frac{1}{2}K(-f(t, \underline{r})(\delta p - \delta\rho) + (p_0 - \rho_0)\delta g_{11})$$

$$S_{22} = \frac{1}{2}K(-\psi(t)r^2(\delta p - \delta\rho) + (p_0 - \rho_0)\delta g_{22})$$

$$S_{33} = \frac{1}{2}K(-\psi(t)r^2\sin^2\theta(\delta p - \delta\rho) + (p_0 - \rho_0)\delta g_{33})$$

❖❖❖❖❖

[182] Hamiltonian for a single charged particle moving in curved space time in the presence of an external em field.

$$\mathcal{L} = -m_0\left(g_{\mu\nu}(X)u^{\mu}v^{\nu}\right)^{\frac{1}{2}} + ev^{\mu}A_{\mu}(X)$$

$$v^{\mu} = \frac{dX^{\mu}}{dt}, v^0 = 0, v^r = \frac{dX^r}{dt}, 1 \le r \le 3$$

$$\mathcal{L}\left(t, x^r, v^r\right) = -m_0\left(g_{00}(x) + 2g_{or}(x)v^r + g_{rs}(x)v^r u^s\right)^{1/2} + eA_0 + ev^r A_r$$

$$p_r = \frac{\partial\mathcal{L}}{\partial v^r} = \frac{-m_0\left(g_{0r} + g_{rs}u^s\right)}{\tau'} + eA_r$$

$$\tau' = \frac{d\tau}{dt} = \left(g_{\mu\nu}v^{\mu}v^{\nu}\right)^{1/2}$$

Let $$v^{\mu} = \frac{dx^{\mu}}{d\tau} = \frac{dt}{d\tau}\frac{dx^{\mu}}{dt}, \gamma v^{\mu} = \gamma = \frac{d\tau}{dt} = \frac{1}{\tau'} = u^0$$

Let $$p_0 = \frac{-m_0\left(g_{00} + g_{0s}u^s\right)}{\tau'} + eA_0$$

Then, $$p_{\mu} = -m_0 g_{\mu\nu}u^{\nu} + eA_{\mu} = -m_0 u_{\mu} + eA_{\mu}$$

$$u_{\mu} = -\frac{1}{m_0}\left(p_{\mu} - eA_{\mu}\right)$$

$$H = p_r v^r - \mathcal{L} = -m_0 g_{r\mu}u^{\mu}v^r + ev^r A_r + m_0\left(g_{\mu\nu}u^{\mu}v^{\nu}\right)^{1/2} - ev^{\mu}A_{\mu}$$

$$= -m_0\, g_{r\mu}u^{\mu}\, u^r/\gamma + m_0\left(g_{\mu\nu}v^{\mu}v^{\nu}\right)^{1/2} - eA_0$$

$$= \frac{-m_0}{\gamma}\left(1 - g_{0\mu}u^0u^{\mu}\right) + \frac{m_0}{u^0} - eA_0$$

$$= m_0 g_{0\mu}u^{\mu} - eA_0 = -p_0$$

$$u_{\mu} = -\frac{1}{m_0}\left(p_{\mu} - eA_{\mu}\right)$$

$$g^{\mu\nu}u_{\mu}u_{\nu} = 1$$

$$\Rightarrow \quad g^{\mu\nu}\left(p_{\mu} - eA_{\mu}\right)\left(p_{\nu} - eA_{\nu}\right) = m_0^2$$

Thus,

$$g_{00}\left(\rho^0 + eA^0\right)^2 + 2g_{0r}\left(H + eA^r\right)\left(P^r + eA^r\right)$$

$$+ g_{rs}\left(P^r + eA^r\right)\left(P^s + eA^s\right) = m_0^2$$

where $\rho^r = -g^{rv}p_v$ *i.e.* $p^\mu = -g^{\mu v}p_v$ $\rho^0 = -g^{ov}p_v$

The correct Hamiltonian $H = -p_0$ satisfies the quantum equation

$$g^{00}(H + eA_0)^2 - 2g^{or}(H + eA_0)(p_r - eA_r)$$

$$+ g^{rs}(p_r - eA_r)(p_s + eA_s) - m_0^2 = 0$$

or

$$H + eA_0 = \Big\{g^{0r}(p_r - eA_r) + \Big[g^{0r}g^{0s}(p_r - eA_r)(p_s - eA_s)$$

$$-g^{00}g^{rs}(p_r - eA_r)(p_s - eA_s) + g^{00}m_0^2\Big]^{1/2}\Big\}g^{00}$$

❖❖❖❖❖

[183] Moment generating function for quantum observables in boson Fock space.

Let $[A, B] = C$ where $[A, C] = [B, C] = 0$

Then

$$\exp(A + B) = \exp(A)\cdot\exp(B)\cdot\exp\left(\frac{-C}{2}\right)$$

$$= \exp\left(\frac{-C}{2}\right)\cdot\exp(A)\cdot\exp(B)$$

Proof:

Let $F(t) = \exp t\,(A + B)$

Then $F'(t) = (A + B)\,F(t)$

Let $F(t) = \exp(tA)\,.\,G(t)$

Then

$$G'(t) = \exp(-tA)\,.\,B\,.\,\exp(tA)\,G(t)$$

$$= \exp(-t\,ad\,A)(B)\,G(t)$$

$$= (B - tC)\,G(t)$$

Since

$$(\text{ad } A)^n(B) = \begin{cases} B, & n = 0 \\ C, & n = 1 \\ 0, & n \geq 2 \end{cases}$$

So, $$G(t) = \exp(tB)\cdot\exp\left(-t^2\frac{C}{2}\right) \quad ([B, C] = 0)$$

Hence, $$F(t) = \exp(tA)\cdot\exp + tB\cdot\exp\left(-t^2\frac{C}{2}\right)$$

$$\exp\left(a(u)+a^{+}(v)\right) = \exp\left(a^{+}(v)\right)\exp\left(a(u)\right)\cdot\exp\left(\frac{1}{2}\left[a(u),a^{+}(u)\right]\right)$$

$$= \exp\left(\frac{1}{2}\langle u,v\rangle\right)\cdot\exp\left(a^{+}(v)\right)\cdot\exp\left(a(u)\right)$$

Let $$\varphi(w) = \exp\left(-\frac{\|w\|^2}{2}\right)\times e(w)$$

Then, $$\left\langle\varphi(w),\exp\left(a(u)+a^{+}(v)\right)\varphi(w)\right\rangle$$

$$= \exp\left(\frac{1}{2}\langle u,u\rangle\right)\exp\left(-\|w\|^2\right)\left\langle e(w),\exp\left(a^{+}(v)\right)\cdot\exp\left(a(u)\right)\cdot e(w)\right\rangle$$

$$= \exp\left\{\langle u,w\rangle+\langle w,v\rangle+\frac{1}{2}\langle u,v\rangle\right\}$$

This if $f(t)$ and $g(t)$ and complex functions of time, then left

$$\xi_T = \exp\left(\int_0^T f(t)da\left(u_{t]}\right)+g(t)da^{+}\left(v_{t]}\right)\right)$$

$$= \exp\left(\int_0^T f(t)da_t(u)+g(t)da_t^{+}(v)\right)$$

$$= \exp\left(a\left(\int_0^T \bar{f}(t)da_{t]}\right)+a^{+}\left(\int_0^T g(t)dv_{t]}\right)\right)$$

$$= \exp\left(a\left(\bar{f}\,u\,X_{[0,T]}\right)+a^{+}\left(\bar{f}\,v\,X_{[0,T]}\right)\right)$$

(Note the Boson Fock space have is $\Gamma_s\left(L^2\left(\mathbb{R}_+\right)\right)$)

This, $\left\langle\varphi(w),\xi_T\varphi(w)\right\rangle$

$$= \exp\left\{\left\langle\bar{f}uX_{[0,T]},w\right\rangle+\left\langle w,gvX_{[0,T]}\right\rangle+\frac{1}{2}\left\langle\bar{f}u,gv\right\rangle\right\}$$

$$\frac{1}{T}\log\left\langle\varphi(w),\xi_T\varphi(w)\right\rangle = \frac{1}{T}\int_0^T\left(f(t)\bar{u}(t)w(t)+g(t)v(t)\bar{w}(t)\right.$$

$$+\frac{1}{2}f(t)g(t)\bar{u}(t)v(t)\Big)dt$$

i.e.
$$\lim_{x\to\infty}\frac{1}{T}\log\mathbb{E}_{\varphi(w)}\left[\exp\left(\int_0^T f(t)dA_t(u)+g(t)dA_t^+(u)\right)\right]$$

$$=\left\langle f\bar{u}\,w+g\,v\,\bar{w}+\frac{1}{2}f\,g\,\bar{u}\,v\right\rangle$$

where < · > denote time average.

This is a quantum version of the Gartner - ellis limiting moment generally function in deviation theory.

Let

$$\eta_T=\sum_{k=1}^{p}\int_0^T f_k(t)d\Lambda_t(H)_k a\equiv\sum_k\int_0^T f_k(t)d\lambda\left(H_k\chi_{[0,t]}\right)$$

$$=\lambda\left(\sum_{k=1}^{p}f_kH_k\chi_{[0,T]}\right)$$

f_k is a multiplication operator on $L^2(\mathbb{R}_+)$ and H_k is a linear operator in $L^2(\mathbb{R}_+)$.

$$\langle\varphi(w),\eta_T\varphi(w)\rangle=\exp\left(-\|w\|^2\right)\left\langle e(w),\lambda\left(\sum_{k=1}^{p}t_kH_kX_{[0,t]}\,\chi_{[0,T]}e(w)\right)\right\rangle$$

$$=\left\langle w_1\sum_{k=1}^{p}f_kH_k\chi_{[0,T]}w\right\rangle$$

It is assumed that H_k commutes with $\chi_{[0,T]}.\forall\ T$. In kernal notation,

$$\langle\varphi(w),\eta_T\,\varphi(w)\rangle=\int_0^T f_k(t_1)\overline{w(t_1)}(H_kw)(t_1)dt_1$$

$$=\int_0^T\int_0^T H_k(t_1,t_2)f_k(t_1)w(t_2)\overline{w(t_1)}\,dt_1dt_2$$

Let
$$P_T=\exp(\eta_T)=\exp\left(\lambda\left(\sum_{k=1}^{p}f_kH_k\chi_{[0,T]}\right)\right)$$

Then,
$$=\langle\varphi(w),P_T\varphi(w)\rangle$$

$$=\exp\left(-\|w\|^2\right)\left\langle e(w),e\left(\exp\left(\sum_{k=1}^{p}f_k\chi_{[0,T]}H_k\right)w\right)\right\rangle$$

$$= \exp\left(\left\langle w, \exp\left(\sum_{k=1}^{p} f_k \chi_{[0,T]} H_k\right) w\right\rangle\right) \cdot \exp\left(-\|w\|^2\right)$$

Suppose the $H_k's$ commute with each other. Then they have a common spatial measure $E(u)$ say. Then, in kernal notation,

for $\quad t_1 \le t_2 T, = \left(\sum_{k=1}^{p} f_k \chi_{[0,T]} H_k\right)(t_1, t_2)$

$$= \sum_{k=1}^{p} f_k(t_1) H_k(t_1, t_2)$$

$$= \sum_{k=1}^{p} f_k(t_1) \int_{x \in \mathbb{R}} \alpha_k(x) d\mathbb{E}(x, t_1, t_2)$$

α_k: $\mathbb{R} \to \mathbb{R}$.

The multiplication operator f_k commutes with H_j too for all $k \neq j$, So,

$$\exp\left(\sum_{k=1}^{p} f_k \chi_{[0,T]} H_k\right) = \prod_{k=1}^{p} \exp\left(f_k \chi_{[0,T]} H_k\right)$$

$$= \int_{\mathbb{R}} \left[\prod_{k=1}^{p} \exp\left(f_k \alpha_k(x) \chi_{[0,T]}\right)\right] dE(x)$$

$$\log\langle \varphi(w), P_T(w)\rangle = \int_0^T \overline{w}(t)\left(\exp\left(\sum_{k=1}^{p} f_k(t)\alpha_k(x) - 1\right) d_x(E(x)w)(t)\right) dt$$

$$= \int_0^T \int_0^T \overline{w}(t_1) w(t_2) d_x E(x, t_1, t_2)$$

$$\exp\left(\sum_{k=1}^{p} f_k(t_1)\alpha_k(x) - 1\right) dt_1 dt_2$$

Note: For a single observable H in L^2 ($\mathbb{R}$),

$$= \left\langle \varphi(w), \exp\left(\int_0^T f(t), d\Lambda_t(H)\right) \varphi(w)\right\rangle$$

$$= \left\langle \varphi(w), \exp\left(\lambda\left(\int_0^T f(t), H d\chi_{[0,t]}\right)\right) \varphi(w)\right\rangle$$

$$= \langle \varphi(w), \exp(\lambda(f_T H)) \varphi(w)\rangle \qquad ([H, f] = 0 \text{ assume})$$

(where $f_T = f\chi_{[0,\,T]}$)

Let $$H = \int_{\mathbb{R}} x\,dE(x)$$

Then, $$= \left\langle \varphi(w), \exp\left(\int_0^T f(t)\,d\Lambda_t(H)\right)\varphi(w)\right\rangle$$

$$= \exp\left(-\|w\|^2\right)\left\langle e(w), e\left(\exp\left(f_T H\right)w\right)\right\rangle$$

$$= \exp\left(\left\langle w, \exp\left(f_T H\right)w\right\rangle - \|w\|^2\right)$$

$$= \exp\left(\int_0^T\left(\overline{w(t)}\left(\exp\left(f\ H\right)w\right)(t) - |w(t)|^2\right)dt\right)$$

$$= \log\left\{\left\langle \varphi(w), \exp\left(\int_0^T f(t)\,d\Lambda_t(H)\right)\varphi(w)\right\rangle\right\}$$

$$= \int_0^T\left(\overline{w(t)}\exp\left(f(t)H\right)w(t) - |w(t)|^2\right)dt$$

$$= \int_{\mathbb{R}}\int_0^T\left(\exp f(t)x\right)d_x\left(E(x)w\right)(t)\cdot\overline{w(t)}\,dt \quad \int_0^T |w(t)|^2\,dt$$

$$= \int_{\mathbb{R}\times[0,\,T]} d_x\left(E(x)w\right)(t)\cdot\bar{w}(t)\,dt$$

Problem: Compute $\frac{d}{dt}\exp(A(t))$ and $\frac{d^2}{dt^2}\exp(A(t))$ where $A(t)$ is a matrix. Use those formulae to obtain $d\exp(A(t))$ where $A(t)$ is a matrix valued diffusion process.

$$d\exp(A) = \exp(A)\cdot\left(\frac{I - \exp(-adA)}{ad\ A}\right)(dA)$$

so $$\frac{d}{dt}\exp(A(t)) = \exp\left(A(t)g\left(ad\ A(t)\right)\frac{dA(t)}{dt}\right)$$

where $$g(x) = \left(\frac{1 - \exp(-x)}{x}\right)$$

$$\frac{d^2}{dt^2}\exp(A(t)) = \left(\frac{d}{dt}\exp A\right)g(ad\ A)\left(\frac{dA}{dt}\right) + \exp(A)\frac{d}{dt}g(ad\ A)\left(\frac{dA}{dt}\right)$$

$$g(ad\ A)(A'(t)) = \left(\frac{I-\exp(-ad\ A)}{ad\ A}\right)(A'(t))$$

$$= (I-\exp(-ad\ A))(ad\ A)^{-1}(A')$$

$$= (ad\ A)^{-1}(A') - A^{-1}(ad\ A)^{-1}(A')A$$

$$\frac{d}{dt}g(ad\ A)(A') = (ad\ A)^{-1}(A'') - (ad\ A)^{-1}(ad\ A')(ad\ A)^{-1}(A')$$

$$+ A^{-1}A'A^{-1}(ad\ A)^{-1}(A')A$$

$$+ A^{-1}(ad\ A)^{-1}(ad\ A')(ad\ A)^{-1}(A')A$$

$$- A^{-1}(ad\ A)^{-1}(A'')A - A^{-1}(ad\ A)^{-1}(A')\times A'$$

Transferencies per CMI Lecture 17th June 2015

❖❖❖❖❖

[184] Generalized quantum filtering for non-demolition measurements in the series of Balavkin.

[1] $\mathcal{H} = \mathbb{C}^d \otimes L^2(\mathbb{R}_+)$

Noise Hilbert space (Fock space) :$\Gamma s(\mathcal{H})$

$$u \in \mathcal{H} \Rightarrow u : \mathbb{R}_+ \to \mathbb{C}^d,\ \int_0^\infty \|u(t)\|^2\, dt < \infty$$

[2] $\hbar$ = system Hilbert space.

[3] Exponential vector in $\Gamma s(\mathcal{H})$: $e(u)\ u \in \mathcal{H}$

$$e(u) = \bigoplus_{n=0}^{\infty} \frac{u\otimes^n}{\sqrt{\lfloor n}}$$

$$u\otimes^n(t_1,t_2,..,t_n) = \underline{u}(t_1)\otimes\underline{u}(t_2)\otimes\ldots\otimes\underline{u}(t_n)$$

$u,\ u \in \mathcal{H}$

$$\Rightarrow \quad \langle e(u), e(v)\rangle = \exp(\langle u, v\rangle) = \exp\left(\int_0^\infty \langle \underline{u}(t), \underline{v}(t)\rangle\, dt\right)$$

[4] $\mathcal{L}(\mathcal{H}) \Rightarrow \lambda(H)\, e(u) = \underset{n=0}{\overset{\infty}{\oplus}} \underset{h=0}{\overset{\infty}{\oplus}} \left(u^{\otimes k} \otimes H u \otimes u^{\otimes n-k} / \sqrt{\lfloor n}\right)$

$\lambda(H)$ = conservation operator field

$$a(u)\, e(v) = \langle u, v\rangle\, e(v)$$

$$a(u)\frac{v\otimes^{n}}{\sqrt{\lfloor n}} = \langle u, v\rangle \frac{v\otimes^{n-1}}{\sqrt{\lfloor n-1}}$$

$$\Rightarrow \qquad a(u)v^{\otimes n} \ \langle u, v\rangle \sqrt{n}\, v\otimes^{n-1}$$

$a(u) \simeq$ annehilation operator field.

Creation operator feild

$a^{+}(x) \simeq$ adjoint of $a(u)$.

$$a^{+}(u)\, e(v) = \frac{d}{dt} e(v+tu)\Big|_{t=0}$$

$$= \sum_{n=1}^{\infty} \frac{1}{\sqrt{\lfloor n}} \overset{n-1}{\underset{k=0}{\oplus}} v^{k} \otimes u \otimes v^{n-k-1}$$

$$\text{So, } a^{+}(u)v^{\otimes n-1} = \frac{1}{\sqrt{\lfloor n}} \overset{n-1}{\underset{k=0}{\oplus}} \Sigma v^{k} \otimes u \otimes v^{n-k-1}$$

Check:

$$\left\langle a^{+}(u)v^{\otimes n-1}, w^{\otimes n}\right\rangle = \frac{1}{\sqrt{n}} \sum_{k=0}^{n-1} \left\langle \overset{n-1}{\underset{k=0}{\oplus}} \Sigma^{v^{k}} \otimes u \otimes v^{n-k-1}, w^{\otimes n}\right\rangle$$

$$= \frac{1}{\sqrt{n}} \sum_{k=0}^{n-1} \langle v, w\rangle^{n-1} \langle n, w\rangle$$

$$= \sqrt{n}\langle v, w\rangle^{n-1} \langle n, w\rangle$$

$$= \langle u, w\rangle \sqrt{n} \left\langle v^{\otimes n-1}, w^{\otimes n-1}\right\rangle$$

$$= \left\langle v^{\otimes n-1}, a(u)w^{\otimes n}\right\rangle$$

[5] Let $\{|e_k\rangle\}_{k=1}^{d}$ be the standard basis for $\mathbb{C}^{d}$.

Let $\Lambda_j^k(t) = \lambda\left(|e_j\rangle\langle e_k|\chi_{[0,t]}\right)$, $1 \le j, k \le d$.

Joint moment generally function of $a_j^{(t)} = a\left(|e_j\rangle \chi_{(t)}\right)$

$a_j^{(+)}(t) = a^+\left(|e_j\rangle \chi_{[0,t]}\right)$ and $\lambda\left(|e_j\rangle\langle e_k| \chi_{[0,t]}\right) = \Lambda_j^+(t)$

Let $\varphi_j(t)$, $\psi_s(t)$, $\eta_{jk}(t)$ be complex valued functions of time, Then,

$$X \triangleq \sum_{j=1}^{d} \int_0^T \varphi_j(t) da_j(t) + \psi_s(t) da_s^+(t) + \eta_k^j(t) d\Lambda_j^k(t)$$

$$= a\left(\bar{\varphi}\chi_{[0,T]}\right) + a^+\left(\psi\chi_{[0,T]}\right) + \Lambda\left(\eta\chi_{[0,T]}\right)$$

where $$\varphi = \sum_{j=1}^{d} \varphi_j |e_j\rangle \in \mathcal{H},\ \psi = \sum_{j=1}^{d} \psi_j |e_j\rangle \in \mathcal{H}$$

$$\eta = \sum_{j,k=1}^{d} \eta_k^j |e_j\rangle\langle e_k| \in \mathcal{B}(\mathcal{H}) = \mathbb{B}\left(L^2\left(\mathbb{R}_+ 1\right)\right) \otimes \mathbb{B}\left(\mathbb{C}^d\right)$$

Case I: $\eta = 0$. Then

$$X = \sum_{j=1}^{d} \int_0^T \left(\varphi_j(t) da_j + \psi j(t)\right)$$

$$= a\left(\bar{\varphi}_T\right) + a^+\left(\psi_T\right)\ \phi_T = \varphi\chi_{[0,T]}$$

$$\psi_T = \Psi\chi_{[0,T]}$$

$\left[a(u), a^+(u)\right] = \langle u, v\rangle$ is well know, since

$$a(u)a^+(v)w^{\otimes n} = a(u)\frac{1}{\sqrt{n+1}} \underset{n+1}{\overset{1}{\oplus}} \Sigma w^k \otimes u \otimes w^{n-k}$$

$$= \left\langle s^{\otimes n}, a(u)a^+(v)w^{\otimes n}\right\rangle = \left\langle a^+(u)s^{\otimes n}, a(v)w^{\otimes n}\right\rangle$$

$$= \frac{1}{(n+1)}\left\langle \sum_{k=0}^{n} s^k \otimes u \otimes^{n-k}, \sum_{k=0}^{n} w^k \otimes v \otimes w^{n-k}\right\rangle$$

$$= \frac{1}{(n+1)}\langle u, v\rangle \sum_{k=0}^{n} \langle s, w\rangle^n + \frac{1}{(n+1)} \sum_{0\le k\ne j\le n} \langle u, w\rangle\langle s, v\rangle\langle s, w\rangle^{n-1}$$

$$= \langle u, v\rangle\langle s, w\rangle^n + n\langle u, w\rangle\langle s, v\rangle\langle s, w\rangle^{n-1}$$

On the other hand,

$$\left\langle s^{\otimes n}, a_+(v)a(u)w^{\otimes n}\right\rangle = \left\langle a(v)s^{\otimes n}, a(u)w^{\otimes n}\right\rangle$$

$$= \langle s,v\rangle\langle u,w\rangle\langle s,w\rangle^{n-1}$$

[5] So,

$$\left\langle s^{\otimes n}, \left[a(u), a^+(u)\right]w^{\otimes n}\right\rangle = \langle u,v\rangle\left\langle s^{\otimes n}, w^{\otimes n}\right\rangle$$

[*b*] Let *A, B* be two operator in a hilbert space $\mathcal{H}$ such that [*A*, *B*] commutes with both A and *B*. Let

$$F(t) \triangleq e^{t(A+B)} = e^{tA}\, G(t)$$

Then, $$F'(t) = e^{tA}(AG(t) + G'(t)) = (A + B)\, e^{tA}$$

$$G'(t) = e^{-tA}Be^{tA}G(t)$$

$$= \exp(-t\, ad\, A)\,(B) \times G(t)$$

$$= (B - t[A, B])\, G(t)$$

Since $(\text{ad } A)^n(B) = 0$, $n \geq 2$. Since $[B, [A, B]] = 0$, it follows that

$$G(t) = e^{tB - \frac{t^2}{2}[A,B]} = e^{tB}e^{-\frac{t^2}{2}[A,B]}$$

So, $$e^{t(A+B)} = e^{tA}\cdot e^{tB}\cdot e^{-\frac{t^2}{2}[A,B]}$$

$\therefore$ $$e^{a(u)+a^+(v)} = e^{a^+(v)}\,.\, ea(u)\,.\, e^{\frac{1}{2}\left[a(u),\, a^+(v)\right]}$$

$$= e^{\langle u,v\rangle/2}\cdot e^{a^+(v)}\cdot e^{a(u)}$$

$$e^{-\|e(w)\|^2}\left\langle e(w), e^{a(u)+a^+(v)}e(w)\right\rangle = e^{\langle u,u\rangle/2}\left\langle e^{a(v)}e(w), e^{a(u)}e(w)\right\rangle$$

$$= e^{\langle u,u\rangle/2}e^{\langle w,v\rangle+\langle u,w\rangle}$$

This gives the joint moment generally function of the creation and annihilation processes $a_j(t), a_j^+, 0 \leq t \leq T$.

$\therefore$ $$\mathbb{E}_w \exp\left(\sum_{j=1}^{d}\int_0^T \varphi_j(t)da_j(t) + \psi_j(t)da_j^T(t)\right)$$

$$= \mathbb{E}_w\left[\exp\left(a\left(\bar{\Phi}_T\right) + a^+\left(\psi_T\right)\right)\right]$$

$$= \exp\left(\left\langle \bar{\Phi}_T, \psi_T \right\rangle / 2 + \left\langle w, \psi_T \right\rangle + \left\langle \bar{\varphi}_T, w \right\rangle\right)$$

Thus, $$\int_0^T \varphi_j(t) da_j(t) + \psi_j(t) da_j^t(t)$$

is a complex Gaussian random variable in the coherent state $|w\rangle \exp\left(-\|w\|^2/z\right)$.

Moment generating functional $\Lambda^i_j(t), 1 \le i, j \le d$

$$\Lambda_\eta(T) \triangleq \int_0^T \eta_i^j(t) d\Lambda^i_j(t) = \lambda\left(\int_0^T \eta_i^j(t) |e_j\rangle\langle e_i| d\chi_{[0,t]}\right)$$

$$= \lambda\left(\underline{\eta}_T\right), \text{ where } \underline{\eta}_T = \eta\chi_{[0,T]}$$

$$\underline{\eta}_T(t) = \left\{\eta_i^j(t)|e_j\rangle\langle e_j|, 0 \le t \le T0, w\right.$$

$$= \mathbb{E}_w \exp\left(\int_0^T \eta_i^j(t) d\Lambda^i_j(t) = \mathbb{E}_w \exp\left(\lambda\left(\eta_T\right)\right)\right)$$

$$= \exp\left(-\|w\|^2\right)\left\langle e(w), \exp\left(\lambda\left(\eta_T\right)\right)e(w)\right\rangle$$

$$= \exp\left(\left\langle w, \exp\left(\eta_T\right)w\right\rangle \exp\left(-\|w\|^2\right)\right)$$

$$= \exp\left(\int_0^\infty \left\langle w(t), \exp\left(\eta_T(t)\right)w(t)\right\rangle dt\right)\exp\left(-\|w\|^2\right)$$

Suppose $\eta_i^j(t) = \eta_i^j$ is independent of time.

$$\exp\left(\eta_T(t)\right) = \exp\left(\eta\chi_{[0,T]}(t)\right)$$

$$= I + \sum_{n=1}^\infty \frac{\eta^n}{\underline{|n}}\chi_{[0,T]}(t)$$

$$= I + \left(\exp(\eta) - 1\right)\chi_{[0,T]}(t)$$

$$= \int_0^\infty \left\langle w(t), \exp\left(\eta_T(t)\right)w(t)\right\rangle dt - \|w\|^2$$

$$= \int_0^T \left\langle w(t), \left(\exp(\eta) - 1\right)w(t)\right\rangle dt$$

Let $\eta = \int_{\mathbb{R}} \lambda \, d P(\lambda)$ be the spectral representation of η

η is assumed to be Hermitian. Then,

$$\int_0^T \langle w(t), (\exp(\eta)-1) w(t) \rangle \, dt = \int_{[0, T] \times \mathbb{R}} dt \, d_\lambda \langle w(t), P(\lambda) w(t) \rangle (e^\lambda - 1)$$

If $$w(t) = \begin{cases} w_0 & 0 \le t \le T \\ 0 & 0. w \end{cases}$$

Then this equal $T \int_{\mathbb{R}} d_\lambda \langle w_0, P(\lambda) w_0 \rangle (e^\lambda - 1)$

So, $$\mathbb{E}_w \exp\left(\int_0^T \eta_i^j \, d\Lambda_j^t(t) \right) - \mathbb{E}_w \exp\left(\lambda\left(\eta \chi_{[0, T]}\right)\right)$$

$$= \exp\left(\int_{\mathbb{R}}^T (e^\lambda - 1) d_\lambda \langle w_0, P(\lambda) w_0 \rangle \right)$$

Let $F(\lambda) = \langle w_0, P(\lambda) w_0 \rangle$. Then the above equals $\exp\left(T \int_{\mathbb{R}} (e^\lambda - 1) d F(\lambda) \right)$

which shows that in the state $e\left(w_0 \chi_{[0, T]}\right) \exp\left(-\|w_0\|^2 \frac{T}{2} \right)$

$\lambda(\eta \chi_{[0, T]})$ has the Poisson distributor with mean

$$T F(\infty) = T \|w_0\|^2$$

[7] Quantum Ito's formula:

$$\left\langle e(w_1), da\left(u \chi_{[0, t]}\right) da^+\left(v \chi_{[0, t]}\right) e(w_2) \right\rangle$$

$$= \frac{d}{d\varepsilon} \left\langle e(w_1), da\left(u \chi_{[0, t]}\right) e\left(w_2 + \varepsilon v d_t \chi_{[0, t]}\right) \right\rangle \Big|_{\varepsilon=0}$$

$$= \frac{d}{d\varepsilon} \left\langle e(w_1), \left\langle u d\chi_{[0, t]}, w_2 + \varepsilon v d\chi_{[0, t]} \right\rangle e\left(w_2 + \varepsilon v d\chi_{[0, t]}\right) \right\rangle \Big|_{\varepsilon=0}$$

$$= \frac{d}{d\varepsilon} \left\{ \left(\left\langle u, d\chi_{[0, t]} w_2 \right\rangle + \varepsilon \left\langle u, \left(d\chi_{[0, t]}\right)^2 v \right\rangle \right) \right.$$

$$\left. \times \exp\left(\langle w_1, w_2 \rangle + \varepsilon \left\langle w_1, d\chi_{[0, t]} v \right\rangle \right) \right\} \Big|_{\varepsilon=0}$$

$$= \left[\left\langle w_1, d\chi_{[0, t]} v \right\rangle \left\langle u, d\chi_{[0, t]} w_2 \right\rangle + \left\langle u, d\chi_{[0, t]} v \right\rangle \right] \exp(\langle w_1, w_2 \rangle)$$

$$= \langle u(t), u(t)\rangle dt \langle e(w_1), e(w_2)\rangle$$

So, $$da_i(t)da_j^t(t) = \langle e_i, e_j\rangle dt = \delta_{ij}\, dt$$

This implies

$$da\left(u\chi_{[0,t]}\right)da^+\left(v\chi_{[0,t]}\right) = \langle u(t), v(t)\rangle dt$$

Also writing $$\Lambda_H(t) = \lambda\left(H\chi_{[0,t]}\right) = \lambda\left(\langle e_i\,|H|\,e_j\rangle |e_i\rangle\langle e_j|\chi_{[0,t]}\right)$$

$$= \langle e_i\,|H|\,e_j\rangle \lambda\left(|e_i\rangle\langle e_j|\chi_{[0,t]}\right)$$

$$= \langle e_i\,|H|\,e_j\rangle \Lambda_i^j(t)$$

We have $$\left\langle e(w_2), da\left(u\chi_{[0,t]}\right)d\Lambda_H(t)e(w_1)\right\rangle$$

$$= \frac{d}{d\varepsilon}\left\langle e(w_2), da\left(u\chi_{[0,t]}\right)e\left(e^{eH\,d\chi_{[0,t]}}w_1\right)\right\rangle\Bigg|_{\varepsilon=0}$$

$$= \frac{d}{d\varepsilon}\left\langle e(w_2), \left\langle ud\chi_{[0,t]}, e^{\varepsilon H d\chi_{[0,t]}}w_1\right\rangle e\left(e^{\varepsilon H d\chi_{[0,t]}}w_1\right)\right\rangle\Bigg|_{\varepsilon=0}$$

$$= \frac{d}{d\varepsilon}\left\{\left(\left\langle ud\chi_{[0,t]}w_1\right\rangle + \varepsilon\left\langle u, Hd\chi_{[0,t]}w_1\right\rangle\right)\right.$$

$$\left.\times\left\langle e(w_2), e\left(w_1 + \varepsilon H w_1 d\chi_{[0,t]}\right)\right\rangle\right|_{\varepsilon=0}$$

$$= \frac{d}{d\varepsilon}\left\{\left(\left\langle u, d\chi_{[0,t]}w_1\right\rangle + \varepsilon\left\langle u, Hd\chi_{[0,t]}w_1\right\rangle\right)\right.$$

$$\left.\left.\times\left(\exp\left(\langle w_2, w_1\rangle + \varepsilon\left\langle w_2 H w_1 d\chi_{[0,t]}\right\rangle\right)\right)\right\}\right|_{\varepsilon=0}$$

$$= \left\{\left\langle u, Hd\chi_{[0,t]}w_1\right\rangle + \left\langle u, d\chi_{[0,t]}w_1\right\rangle\left\langle w_2 H w_1 d\chi_{[0,t]}\right\rangle\right\}\langle e(w_2), e(w_1)\rangle$$

$$= \left\langle u, Hd\chi_{[0,t]}w_1\right\rangle\langle e(w_2), e(w_1)\rangle$$

$$\left\langle u, H\chi_{[0,t]}Hw_1\right\rangle\langle e(w_2), e(w_1)\rangle$$

$$\left\langle e(w_2), da\left(H^* u\chi_{[0,t]}\right)e(w_1)\right\rangle$$

Thus, $$da\left(u\chi_{[0,t]}\right)d\Lambda_H(t)$$

$$= da\left(H^* u\chi_{[0,t]}\right)$$

In particular $da_i(t)d\Lambda_k^j(t)$

$= \delta_{ki}da_j(t)$. Taking adjoints, $d\Lambda_H(t)da^+\left(u\chi_{[0,t]}\right)$

$= da^+\left(Hu\chi_{[0,t]}\right)$ $sc\; d\Lambda_j^k(t)da_i^+(t)$

$= \delta_{ki}da_i^+(t)$

Finally $\left\langle e(w_2), d\Lambda_{H_1}(t)d\Lambda_{H_2}(t)e(w_1)\right\rangle$

$$= \frac{\partial^2}{\partial\varepsilon_1\partial\varepsilon_2}\left\langle e\left(\exp\left(\varepsilon_1 H_1^* d\chi_{[0,t]}\right)w_2\right), e\left(\exp\left(\varepsilon_2 H_2 d\chi_{[0,t]}\right)w_1\right)\right\rangle\Big|_{\varepsilon_1=\varepsilon_2}$$

$$= \frac{\partial^2}{\partial\varepsilon_1\partial\varepsilon_2}\exp\left\langle w_2 + \varepsilon_1 H_1^* w_2 d\chi_{[0,t]}, w_1 + \varepsilon_2 H_2 w_1 d\chi_{[0,t]}\right\rangle\Big|_{\varepsilon_1=\varepsilon_2}$$

$$= \left\langle H_1^* w_2, d\chi_{[0,t]}H_2 w_1\right\rangle\left\langle e(w_2), e(w_1)\right\rangle$$

$$= \left\langle w_2, d\chi_{[0,t]}H_1 H_2 w_1\right\rangle\left\langle e(w_2), e(w_1)\right\rangle$$

$$= \left\langle e(w_2), d\lambda\left(H_1 H_2 \chi_{[0,t]}\right)e(w_1)\right\rangle$$

$$= \left\langle e(w_2), d\Lambda_{H_1H_2}(t)e(w_1)\right\rangle$$

So, $\delta\Lambda_{H_1}\,\delta\Lambda_{\mathrm{H}_2} = \delta\Lambda_{H_1H_2}$

Summary of the quantum Ito formulae :

$$d\Lambda_b^a(t)d\Lambda_d^c(t) = \varepsilon_d^a d\Lambda_b^c(t) \quad a, b, c, d = 0, 1, 2, ..., d$$

where $\Lambda_0^0(t) = t$, $\Lambda_0^i(t) = 0_j(t)$,

$$\Lambda_i^0(t) = a_i^+(t),$$

$$i \geq 1,\ \varepsilon_d^a = \delta_d^c, a, d \geq 1, \varepsilon_0^a = 0, \varepsilon_d^0 = 0$$

Note: $\left(d\chi_{[0,t]}\right)^2 = d\chi_{[0,t]}$

$$\left\langle w_2, d\chi_{[0,t]}H\, w_1\right\rangle = d\left\langle w_2, d\chi_{[0,t]}H\, w_1\right\rangle$$

so $\left\langle w_2, \left(d\chi_{[0,t]}\right)^2 H\, w_1\right\rangle = d\left\langle w_1, \chi_{[0,t]}H\, w_1\right\rangle$

$$\left\langle w_2, d\chi_{[0,t]} H w_1 \right\rangle \left\langle w_3, d\chi_{[0,t]} \tilde{H} w_4 \right\rangle = 0$$

*Unitary evolution in system + bath space.

*Noisy Schrödinger eqn.

$$dU(t) = \varepsilon i H_0 + P(t)dt + L_j(t)da_j(t) + \tilde{L}_j(t)da_j^{+}(t) + S_k^j(t)d\Lambda_j^k(t)U(t)$$

$H_0, P, L_j, \tilde{L}_j, S_k^j$ are system operator *i.e.* $\mathcal{L}(\hbar)$ and $H_0^* = H_0$.

Find the relations between $P, L_j, \tilde{L}_j, S_k^j$ so that $U(t)$ is a uintary operator in $\hbar \otimes \Gamma s(\mathcal{H})$

$$\mathcal{H} = \mathbb{C}^d \otimes L^2(\mathbb{R}_+)$$

Ans: We write

$$\Lambda_0^k = a_k, \quad \Lambda_k^0 = a_k^t$$

$$L_k = S_k^0, \tilde{L}_k = S_0^k, \Lambda_0^0 = t$$

Then, $$dU = \left((-iH_0 + P)dt + S_\nu^\mu d\Lambda_\mu^\nu\right)U(t)$$

$$dU^* = U^*\left(\left(iH_0 + P^*\right)dt + S_\nu^{\nu *} d\Lambda_\nu^\mu\right)$$

$$0 = d(U^* U) = dU^* \cdot U + U^* \cdot dU + dU^* \cdot dU$$

$$= U^*\left\{(\rho + \rho^*)dt + \left(S_\nu^{\mu *} + S_\mu^\nu\right)d\Lambda_\nu^\mu + S_\nu^{\mu *} S_\sigma^\rho d\Lambda_\nu^\mu d\Lambda_\rho^\sigma\right\}U$$

$$d\Lambda_\nu^\mu \, d\Lambda_\rho^\sigma = \varepsilon_\nu^\mu \, d\Lambda_\rho^\sigma$$

So for unitary of U, we require $\left(P + P^*\right)dt + \left(S_\nu^{\mu *} + S_\mu^\nu\right)d\Lambda_\nu^\mu + S_\nu^{\mu *} S_\sigma^\rho \varepsilon_\rho^\mu d\chi_\nu^\sigma = 0$

or equivalently $P + P^* + S_0^{0*} + S_0^0 + S_\nu^{\mu *} S_\sigma^\rho \varepsilon_\rho^\mu = 0$

$$S_k^{j*} + S_j^k + S_k^{\mu *} S_\sigma^\rho \varepsilon_\rho^\mu = 0$$

Einstein Summation convention is used here: Roman indices *j k l m n* etc. vary over 1 to *d* and greek indices μ ν ρ σ etc, vary over 0 to *d*.

The above eqn. are the same as

$$P + P^* + S_0^{0*} + S_0^0 + S_\nu^{j*} S_0^j = 0$$

$$S_k^{j*} + S_j^k + S_k^{m*} S_j^m = 0$$

We can absorb P into S_0^0 etc. and write the hudson Partharasathy eqn. as

$$dU(t) = \left[-iH_0 dt + S_\nu^\mu(t) d\Lambda_\mu^\nu(t)\right] U(t)$$

Where the unitary condition becomes.

$$S_\nu^{\mu^*} + S_\mu^\nu + S_\nu^{\sigma^*} S_\mu^\rho \varepsilon_\rho^\sigma = 0$$

or $$S_\nu^{\mu^*} + S_\mu^\nu + \sum_{j=1}^{d} S_\nu^{j*} S_\mu^j = 0,\ 0 \le \mu, \nu \le d.$$

Evolution of a Heisenberg observable in H. P. Theory Let $X \in \mathcal{L}(\hbar)$, $X^* = X$. Pro

$$j_t(X) = U(t)^* X U(t) \equiv U(t)^* (X \otimes I)\, U(t)$$

Then $$dj_t(X) = dU^*(t) X U(t) + U^*(t) X\, dU(t) + dU^*(t) X\, dU(t)$$

$$= U^*(t)\left\{i[H_0, X] + \left(S_\nu^{\mu^*} X + X S_\mu^\nu\right) d\Lambda_\nu^\mu + S_\nu^{\mu^*} X S_\sigma^\rho d\Lambda_\nu^\mu d\Lambda_\rho^\sigma\right\} U(t)$$

$$= j_t\left\{i[H_0, X] dt + \left(S_\nu^{\mu^*} X + X S_\mu^\nu + S_\nu^{\sigma^*} X\, S_\mu^\rho\, \varepsilon_\rho^\sigma\right) d\Lambda_\nu^\mu\right\}$$

$$= j_t\left\{i[H_0, X] dt + j_t\left(\theta_\mu^\nu(X)\right) d\Lambda_\nu^\mu(t)\right\}$$

❖❖❖❖❖

[185] Evolution of the system state for H. P equn.

Let $$\rho(0) = \rho_s(0) \otimes |\varphi(u)\rangle\langle\varphi(u)| \rho_s(0) \in \mathbb{B}(\hbar) Tr(\rho_s(0)) = 1$$

$\rho_s(0) \simeq$ Initial state of the system.

$$|\varphi(u)\rangle \simeq \exp\left(-\frac{\|u\|^2}{2}\right)|e(u)\rangle = \text{Initial state of the bath}$$

$\rho(0)$ = Initial state of system plus bath.

Let $\rho_s(t)$ be system state of time t. Then,

$$Tr(\rho_s(t) X) = Tr(\rho(0) j_t(X)) = Tr\left[\rho_s(0) Tr\left(|\varphi(u)\rangle\langle\varphi(u)| j_t(X)\right)\right]$$

by Heisenberg Schrodinger duality.

So,

$$d\,Tr(\rho_s(t) X) = Tr(\rho_s'(t) X) dt$$

$$= Tr\left(\left(\rho_s(0)\otimes|\varphi(u)\rangle\langle\varphi(u)|\right)dj_t(X)\right)$$

$$= Tr\left(\left(\rho_s(0)\otimes|\varphi(u)\rangle\langle\varphi(u)|\right)\left(j_t(i[H_0, X])dt + j_t\left(\theta_\mu^\nu(X)\right)d\Lambda_\nu^\mu\right)\right)$$

$$= Tr\left(\rho_s(t)i[H_0, X]\right)dt + \overline{u_\mu(t)}u_\nu(t)dtTr\left(\rho_s(t)\theta_\mu^\nu(X)\right)$$

$$-i\,Tr\left(H_0, \rho_s(t)X\right)dt + \overline{u_\mu(t)}\,u_\nu(t)Tr\left(\theta_\mu^{\nu*}(\rho_s(t))X\right)$$

or $$\rho_s'(t) = -i\left[H_0, \rho_s(t)\right] + \overline{u_\mu(t)}u_\nu(t)\theta_\mu^{\nu*}(\rho_s(t))$$

This is the generalized Sudarhan Lindblad eqn. on the quantum Liouville Fokker Planck eqn.

Here $\theta_\mu^{\nu*}(\cdot)$ is the dual of $\theta_\mu^{\nu*}(\cdot)$.

Evaluation of $\theta_\mu^\nu(\cdot)$

$$Tr\left(\theta_\mu^{\nu*}(\rho_s)X\right) = Tr\left(\rho_s\theta_\mu^\nu(X)\right)$$

$$= Tr\left(\rho_s.\left(S_\nu^{\mu^*}X + X\,S_\mu^\nu + S_\nu^{\sigma*}X\,S_\mu^\rho\,\varepsilon_\sigma^\rho\right)\right)$$

$$= Tr\left(\left(\rho_s\,S_\nu^{\mu^*} + S_\mu^\nu\rho_s + \varepsilon_\sigma^\rho S_\nu^\sigma\rho_s S_\nu^{\sigma*}\right)X\right)$$

So, $$\theta_\mu^{\nu*}(\rho_s) = \rho_s\,S_\nu^{\mu*} + S_\mu^\nu\rho_s + \varepsilon_\sigma^\rho S_\mu^\rho\,\rho_s S_\nu^{\sigma*}$$

$$= \rho_s\,S_\nu^{\mu*} + S_\mu^\nu\,\rho_s + \sum_{j\ge 1} S_\mu^j\rho_s + S_\nu^{j*}$$

Let

$$\overline{u_\mu(t)}u_\nu(t)S_\mu^\nu = L_0(t)$$

Then,

$$\overline{u_\mu(t)}u_\nu(t)S_\nu^{\mu*} = L_0(t)^*$$

Let

$$\overline{u_\mu(t)}S_\mu^j(t) = L_j(t)$$

So, $u_\mu(t)S_\mu^j(t)^* = L^*_j(t)$

and the generalized Sudarshan Lindblad eqn. becomes

$$\rho_s'(t) = -i\left[H_0, \rho_s(t)\right] + L_0(t)\rho_s(t) + \rho_s(t)L_0^*(t) + \sum_{1\le j\le d} L_j(t)\rho_s(t)J_j^*(t)$$

❖❖❖❖❖

[186] Quantum filtering (Belavkin) State process

$$dj_t(X) = j_t\left(\theta^{\nu}_{\mu}(X)\right)d\Lambda^{\mu}_{\nu}(t)$$

(Assume that the terms $i\,[H_0, X]$ has been absorbed into $\theta^0_0(X)$)

Here $\quad j_t(X) = U^*(t)XU(t)$

$U(t)$ being the unitary evolution of the H. P. theory

Let

$$\int_0^t C^a_b(t',k)\,d\Lambda^b_a(t') = Y_{in,k}(t)\, k = 1,2,\ldots,d$$

where $C^a_b(t,k) \in \mathbb{C}$ are such that

$$\left[Y_{in,k}(t), Y_{in,j}(s)\right] = 0 \,\forall\, t, s, k, j$$

This can be guaranteed if

$$\begin{aligned} &C^a_b(t,k)C^p_q(t,j)\,d\Lambda^b_a(t)\,d\Lambda^q_p(t) \\ &= C^p_q(t,j)C^a_b(t,k)\,d\Lambda^q_p(t)\,d\Lambda^b_a(t)\,\forall, k, j, t \end{aligned}$$

or

$$\begin{aligned} &C^a_b(t,k)C^p_q(t,j)\,\varepsilon^q_p\,d\Lambda^q_a(t) \\ &= C^p_q(t,j)C^a_b(t,k)\,\varepsilon^q_p\,d\Lambda^q_a(t) \end{aligned}$$

or

$$\begin{aligned} &C^p_q(t,k)C^a_b(t,j)\,\varepsilon^q_a \\ &= C^p_q(t,j)C^a_b(t,k)\,\varepsilon^q_a \end{aligned}$$

or $$\underline{\underline{C}}(t,j)\,\underline{\underline{\varepsilon}}\,\underline{\underline{C}}(k) = \underline{\underline{C}}(t,k)\,\underline{\underline{\varepsilon}}\,\underline{\underline{C}}(t,j) \quad 1 \le j,k \le d\; \forall t \ge 0$$

i.e. the matrices $\underline{\underline{C}}(t,k)$, $k = 1, 2, \ldots, d$ are $\underline{\underline{\varepsilon}}$-cumutative $\forall\, t \ge 0$.

Let $$\eta^{in}_{t]} = \sigma\left\{Y_{in,k}(s), s \le t, 1 \le k \le d\right\}$$

$\eta^{in}_{t]} \simeq$ Input measurement algebra.

* $\eta^{in}_{t]}$ is an Abelian algebra

$\left[d(h), \eta^{in}_{t]}\right] = 0. \simeq$ System operators commute with measurement algebra, because

$$\eta^{in}_{t]} \subset \mathcal{L}(\Gamma s(\mathcal{H}))$$

By $\mathcal{L}(\hbar)$ we mean $\mathcal{L}(\hbar) \otimes I$ and by $\mathcal{L}(\Gamma s(\mathcal{H}))$ we mean $I \otimes \mathcal{L}(\Gamma s(\mathcal{H}))$. So,

$[\mathcal{L}(\hbar), \mathcal{L}(\Gamma s(\mathcal{H}))] = 0$

$U(t)$ lives in $\hbar \otimes \Gamma s(\mathcal{H})$, *i.e.*

$$U(t) \in \mathcal{L}(h\, \Gamma s(\mathcal{H}))$$
$$= \mathcal{L}(\hbar) \otimes \mathcal{L}(\Gamma s(\mathcal{H}))$$

$\eta_{t]}^{out} \triangleq U^*(t)\eta_{t]}^{in} U(t) \simeq$ Output measurement algebra.

The unitarity condition on $U(t)$ depends only on system operators, Hence $d_t U^*(t)YU(t) = 0$

$$\forall Y \in \mathcal{L}(\Gamma s(\mathcal{H}_{s]}))s \leq t$$

This happen becuse $d\Lambda_\nu^\mu(t)$ commutes with $\mathcal{L}(\Gamma s(\mathcal{H}_{s]}))$ for $t \geq s$

This is particular,

$$d_t U^*(t) Y_{in,\,k}(s) U(t) = 0,\ t \geq s$$

Hence

$$U^*(t)Y_{in,k}(s)U(t) = U^*(s)Y_{in,\,k}(s)U(s) \equiv Y_{out,\,k}(s),\ t \geq s$$

So,

$$\eta_{t]}^{out} = \sigma\{Y_{out,\,k}(s), s \leq t, 1 \leq k \leq d\}$$
$$= U^*(t)\eta_{t]}^{in} U(t)$$

$\eta_{t]}^{out}$ is also atelian since for $t \geq s$

$$= [Y_{out,\,k}(t), Y_{out,\,j}(s)]$$
$$= [U^*(t)Y_{in,\,k}(t)U(t), U^*(s)Y_{in,\,j}(s)U(s)]$$
$$= U^*(t)[Y_{in,\,k}(t), Y_{in,\,j}(s)]U(t) = 0$$

Further, for $t \geq s$,

$$[j_t(X), Y_{out,\,k}(s)] = [U^*(t)XU(t), U^*(t)Y_{in,\,k}(s)U(t)]$$
$$= U^*(t)[X, Y_{in,\,k}(s)]U(t) = 0$$

Thus the measurement algebra $\eta_{t]}^{out}$, $t \geq 0$ satisfy the non demolition condition in addition to being Abelian. In particular in any state on $\hbar \otimes \Gamma s(\mathcal{H})$, the conditional expectation

$$\Pi_t(X) = \mathbb{E}\left[j_t(X) \mid \eta_{t]}^{out}\right], X \in \mathcal{L}(\hbar)X^* \text{ is well defined}$$

In any state we can write down the joint probability distribution of the observable.

$\left(j_t(X), Y_{out,k}(s), s \leq t, 1 \leq k \leq d\right)$ and have define the above conditional expectation.

Let $\mathbb{E}$ denote expectation in the coherent state

$$|f\varphi(u)\rangle = \exp\left(-\frac{\|u\|^2}{2}\right)|f e(u)\rangle, \text{ where } f \in \hbar, \|f\| = 1.$$

Define

$$\mathbb{E}_t(Z) = \mathbb{E}\{U^*(t)\, ZU(t)\}, Z \in \mathcal{L}(\hbar \otimes \Gamma s(\mathcal{H}))$$

Then

Theorem 1:

$$\mathbb{E}\left\{U^*(t)ZU(t)\middle|\eta_{t]}^{out}\right\} = U^*(t)\cdot\mathbb{E}_t\left\{z \mid \eta_{t]}^{in}\right\}\cdot U(t) \qquad (a)$$

(This is the quantum theorem)

Proof: Obviously rhs of (a) is in $\eta_{t]}^{out}$. Future for any $\xi \in \eta_{t]}^{in}$ we have

$$= \mathbb{E}\left\{U^*(t)\mathbb{E}_t\left\{Z \mid \eta_{t]}^{in}\right\}U(t)U^*(t)\xi U(t)\right\}$$

$$= \mathbb{E}\left\{U^*(t)\mathbb{E}_t\left\{Z \mid \eta_{t]}^{in}\right\}\xi U(t)\right\}$$

$$= \mathbb{E}_t\mathbb{E}_t\left\{Z\xi \mid \eta_{t]}^{in}\right\} = \mathbb{E}_t\{Z\,\xi\}$$

$$= \mathbb{E}\left\{U^*(t)ZU(t)U^*(t)\xi U(t)\right\}$$

$$= \mathbb{E}\left\{\mathbb{E}\left\{U^*(t)ZU(t) \mid \eta_{t]}^{out}\right\}U^*(t)\xi\, U(t)\right\}$$

qed.

Theorem 2: Let $F(t) \in \eta_{t]}^{in'}$ *i.e.*

$\left[F(t), \eta_{t]}^{in}\right] = 0$. Suppose

$$\mathbb{E}_t(Z) = \mathbb{E}\left\{U^*(t)ZU(t)\right\} = \mathbb{E}\left\{F^*(t)ZF(t)\right\}$$

Then, $\mathbb{E}_t\left\{Z \mid \eta_{t]}^{in}\right\} = \dfrac{\mathbb{E}\left\{F^*(t)ZF(t) \mid \eta_{t]}^{in}\right\}}{\mathbb{E}\left\{F^*(t)F(t) \mid \eta_{t]}^{in}\right\}}$

Proof: Let $\xi \in \eta_{t]}^{in}$. Thus, $\mathbb{E}_t\left\{\left[\dfrac{\mathbb{E}\left\{F^*(t)ZF(t) \mid \eta_{t]}^{in}\right\}}{\mathbb{E}\left\{F^*(t)F(t) \mid \eta_{t]}^{in}\right\}}\xi\right]\right\}$

$$= \mathbb{E}\left\{\left[\frac{\mathbb{E}\left\{F^*(t)ZF(t) \mid \eta_{t]}^{in}\right\}}{\mathbb{E}\left\{F^*(t)F(t) \mid \eta_{t]}^{in}\right\}}F^*(t)\xi F(t)\right]\right\}$$

$$\left(\because \left[F(t), \eta_{t]}^{in}\right] = 0\right)$$

$$= \mathbb{E}\left\{\left[\frac{\mathbb{E}\left\{F^*(t)ZF(t) \mid \eta_{t]}^{in}\right\}}{\mathbb{E}\left\{F^*(t)F(t) \mid \eta_{t]}^{in}\right\}}\mathbb{E}\left\{F^*(t)\xi F(t) \mid \eta_{t]}^{in}\right\}\right]\right\}$$

$$= \mathbb{E}\left\{F^*(t)ZF(t)\right\} = \mathbb{E}_t\{Z\}$$

* Computation of $F(t)$ for the filtering problem:

$$d\left(U^*(t)XU(t)\right)$$

$$dj_t(X) = j_t\left(\theta_\nu^\mu(X)\right)d\Lambda_\mu^\nu(t)$$

$$= \left(-i[H_0, X]\text{has absorbed into}\,\theta_0^0(X)\right)$$

$$d\mathbb{E}\left(U^*(t)XU(t)\right) = d\left\langle f\varphi(u), j_t(X)f\varphi(u)\right\rangle$$

$$= u_\nu(t)\overline{u_\mu(t)}\,dt\left\langle f\varphi(u), j_t\left(\theta_\nu^\mu(X)\right)f\varphi(u)\right\rangle$$

We want that this be equal to $d\,\mathbb{E}\left(F^*(t)XF(t)\right)$ where $F(t) \in \eta_{t]}^{in'}$. Let therefore $F(t)$ satisfy

$$dF(t) = \left(B_0^{(t)} + \sum_{k=1}^{d} B_k^{(t)}dY_{in,k}(t)\right)F(t)$$

where $B_\mu(t) \in \mathcal{L}(h)$. Since $F(t)$ can be expressed in terms of $B_\mu(s)$, $Y_{in,k}(s)$ $s \le t$, $1 \le k \le d$ it follows that $F(t) \in \eta_{t]}^{in'}$. Now,

$$d\,\mathbb{E}\left(U^*(t)XU(t)\right) = u_\nu(t)\overline{u_\mu(t)}dt\left\langle f\varphi(u), F^*(t)\theta_\nu^\mu(X)F(t)f\varphi(u)\right\rangle$$

and $$dF^*(t)XF(t) = dF^*(t)XF(t)+F^*(t)XdF(t)+dF^*(t)XdF(t)$$

$$= F^*(t)\left\{B_\mu^* dY_{in,\mu}^* X + XB_\mu dY_{in,\mu} + B_k^* XB_j\, dY_{in,k}^* dY_{in,j}\right\}F(t)$$

(where $Y_{in,0}(t) = t$)

Now,

$$dY_{in,k}(t) = C_\nu^\mu(t,k)\,d\Lambda_\mu^\nu(t)$$

$$\left\langle f\varphi(u)F^* B_\mu^*(t)dY_{in,\mu}^*(t)XFf\varphi(u)\right\rangle$$

$$\left\langle f\varphi(u)F^* B_\mu^*(t)XFf\varphi(u)\right\rangle\overline{C_\beta^\alpha(t,\mu)}\,u_\beta(t)\overline{u_\alpha(t)}dt$$

$$\left\langle f\varphi(u)F^* X B_\mu(t)Ff\varphi(u)\right\rangle C_\beta^\alpha(t,\mu)u_\alpha(t)\overline{u_\beta(t)}dt$$

$$+\left\langle f\varphi(u)F^*(t), B_k^* X B_j F(t)f\varphi(u)\right\rangle\overline{C_\beta^\alpha(t,k)}C_\sigma^\rho(t,j)\,\varepsilon_\rho^\alpha u_\sigma(t)\overline{u_\beta(t)}dt$$

So we require that

$$u_\nu(t)\overline{u_\mu(t)}\,\theta_\nu^\mu(X)$$

$$= B_\mu^* X\overline{C_\beta^\alpha(t,\mu)}\,u_\beta(t)\overline{u_\alpha(t)} + XB_\mu C_\beta^\alpha(t,\mu)u_\alpha(t)\overline{u_\beta(t)}$$

$$+B_\mu^* X B_j\overline{C_\beta^\alpha(t,k)}\,C_\sigma^\rho(t,j)\,\varepsilon_\rho^\alpha u_\sigma(t)\,\overline{u_\beta(t)}$$

Define $$g_\mu(t) = C_\beta^\alpha(t,\mu)\,u_\alpha(t)\overline{u_\beta(t)}$$

so, $$\overline{g_\mu(t)} = \overline{C_\beta^\alpha(t,\mu)}\overline{u_\alpha(t)}u_\beta(t)$$

Also define $l_{kj}(t) = \overline{C_\beta^\alpha(t,k)}C_\sigma^\rho(t,j)\,\varepsilon_\rho^\alpha\, u_\sigma(t)\overline{u_\beta(t)}$ then the condition for $F(t)$ is

$$\overline{g_\mu(t)}B_\mu^*(t)X + Xg_\mu(t)B_\mu(t)+l_{kj}(t)B_k^*(t)XB_j(t) = u_\nu(t)\overline{u_\mu(t)}\theta_\nu^\mu(X)$$

Recall that

$$\theta_\mu^\nu(X) = i[H_0,X]\delta_0^\nu\delta_\mu^0 + S_\nu^{\mu*}X + XS_\mu^\nu + S_\nu^{k*}XS_\mu^k$$

$$= \left(i\delta_0^\nu\delta_\mu^0 H_0 + S_\nu^{\mu*}\right)X + X\left(-i\delta_0^\nu\delta_\mu^0 H_0 + S_\mu^\nu\right) + S_\nu^{k*}XS_\mu^k$$

So, the condition $F(t)$ is that

$$\overline{g_\mu(t)}B_\mu^*(t) = \left(i\delta_0^\nu\delta_\mu^0 H_0 + S_\nu^{\mu*}\right)\bar{u}_\nu u_\mu$$

or equivalently,

$$g_\mu(t)B_\mu(t) = u_\nu(t)\overline{u_\mu(t)}S_\nu^\mu(t) - iH_0$$

and $l_{kj}(t)B_k^*(t)XB_j(t) = u_v(t)\overline{u_\mu(t)}S_v^{k*}XS_\mu^k$

Derivation of kushuris eqn. from Kallianpur Striebel formula. $\varphi : \mathbb{R}^n \to \mathbb{R}$

$$\pi_t(\varphi) = \frac{\int \varphi(X_t)L_t(X)ZP[dX]}{\int L_t(X \mid Z)P[dX]}$$

where state model is

$$d\underline{X}(t) = \underline{f}(t, \underline{X}(t))dt + \underline{\underline{g}}(t, \underline{X}(t))d\underline{B}(t)$$

and measurement model is

$$d\underline{Z}(t) = \underline{h}(t, \underline{X}(t))dt + \sigma_v d\underline{V}(t)$$

$$L_t(X \mid Z) = \exp\left(\frac{1}{\sigma_v^2}\int_0^t \underline{h}(s, X(s))^T d\underline{Z}(s) - \frac{1}{2}\underline{h}(s, X(s))^T \underline{h}(s, \underline{X}(s))\right)$$

$$dL_t(X \mid Z) = \frac{1}{\sigma v^2}h^T(t, X(t))L_t(X \mid Z)dZ(t)$$

by Ito's formula.

$$\int \varphi(X_{t+dt})L_{t+dt}(X \mid Z)P[dx]$$

$$= \int(\varphi(X_t) + dt\,\mathcal{L}\varphi(X_t))$$

$$\left(L_t(X \mid Z) + \frac{1}{\sigma v^2}h^T(t, X(t)dZ(t)L_t(X \mid Z))\right)P[dX]$$

Write $\quad \sigma_t(\varphi) = \int \varphi(X_t)L_t(X \mid Z)P[dX]$

Thus,

$$d\sigma_t(\phi) = \left\{\frac{1}{\sigma_v^2}\int \varphi(X_t)L_t(X \mid Z)\underline{h}^T(t, \underline{X}(t))P[dX]\right\}dZ$$

$$+\left\{\int \mathcal{L}\varphi(X_t)L_t(X \mid Z)P[dX]\right\}dt$$

$$= \frac{1}{\sigma_{V^2}}\sigma_t\left(\varphi h^T\right)dZ + \sigma_t(\mathcal{L}\varphi)dt$$

$$d\sigma_t(1) = \frac{1}{\sigma_{V^2}}\sigma_t\left(\varphi h^T\right)dZ$$

$$\pi_t(\varphi) = \frac{\sigma_t(\varphi)}{\sigma_+(t)}$$ By Ito's formula,

$$\pi_t(\varphi) = \frac{\sigma_t(\varphi)}{\sigma_+(t)} \text{ By Ito's formula,}$$

$$d\pi_t(\varphi) = \frac{\sigma_t(\varphi)}{\sigma_t(t)} - \frac{\sigma_t(\varphi)d\sigma_t(t)}{\sigma_t(1)^2} + \frac{\sigma_t(\varphi)}{\sigma_t(1)^3}(d\sigma_t(11))^2 - \frac{d\sigma_t(\varphi)d\sigma_t(t)}{\sigma_t(1)^2}$$

$$= \frac{1}{\sigma_{V^2}}\frac{\sigma_t(\varphi h^T)}{\sigma_t(1)}dZ + \frac{\sigma_t(\mathcal{L}\varphi)dt}{\sigma_t(1)}$$

$$-\frac{1}{\sigma_{V^2}}\frac{\sigma_t(h^T)\sigma_t(\varphi)}{\sigma_t(1)^2}dZ + \frac{\sigma_t(\varphi)}{\sigma_t(t)^3}\frac{1}{\sigma_{V^2}}\sigma_t(h^T)\sigma_t(h)dt$$

$$-\frac{1}{\sigma_{V^2}}\frac{\sigma_t(\varphi h^T)\sigma_t(h)}{\sigma_t(1)^2}dt$$

$$= \pi_t(\mathcal{L}\varphi)dt + \frac{1}{\sigma_{V^2}}\left\{\left(\pi_t(\varphi h^T) - \pi_t(\varphi)\pi_t(h^T)\right)dz\right.$$

$$\left.-\left(\pi_t(\varphi h^T) - \pi_t(\varphi)\pi_t(h^T)\right)\pi_t(h)dt\right\}$$

$$= \pi_t(\mathcal{L}\varphi)dt + \frac{1}{\sigma_{V^2}}\left(\pi_t(\varphi h^T) - \pi_t(\varphi)\pi_t(h^T)\right)\times\left(dZ - \pi_t(h)dt\right)$$

$\pi_t(\mathcal{L}\varphi)dt$ which is Kushner's equation.

Assume that $F(t) \in \eta_{t]}^{in'}$ has been construction on the solution to the qsd

$$dF(t) = \left(B_0(t) + \sum_{k=1}^{d} B_h(t)dY_{in,k}(t)\right)F(t)$$

$$\equiv B_\mu(t)dY_{in,\mu}(t)F(t)$$

$$= B_\mu(t)F(t)dY_{in,\mu}(t) \;\; Y_{in,0}(t) = t$$

so that

$$\mathbb{E}_{|f\varphi(u)\rangle}\left(U^*(t)ZU(t)\right) = \mathbb{E}_{|f\varphi(u)\rangle}\left(F^*(t)ZF(t)\right)$$

Then by the quantum gaussian theorem and the $Z \in h\mathcal{L}(\hbar\Gamma_s(\mathcal{H}_{t]}))$ other result proved, we have the quantum Kallianpur Striebel formula:

$$\pi_t(X) = \mathbb{E}\left\{j_t(X)\,|\,\eta_{t]}^{out}\right\} = U^*(t)\left[\frac{\mathbb{E}\left\{F^*(t)XF(t)\,|\,\eta_{t]}^{in}\right\}}{\mathbb{E}\left\{F^*(t)XF(t)\,|\,\eta_{t]}^{in}\right\}}\right]U(t)$$

Let $\qquad \sigma_t(X) = \mathbb{E}\left\{F^*(t)X\,F(t)\,|\,\eta_{t]}^{in}\right\}$

Then

$$\pi_t(X) = U^*(t)\left\{\frac{\sigma_t(X)}{\sigma_t(1)}\right\}U(t)$$

$$\pi_{t+dt}(X) = U^*(t+dt)\left\{\frac{\sigma_t(X)}{\sigma_t(1)} + d\left\{\frac{\sigma_t(N)}{\sigma_t(1)}\right\}\right\}U(t+dt)$$

$$= U^*(t)\left\{\frac{\sigma_t(X)}{\sigma_t(1)}\right\}U(t) + U^*(t+dt)\,d\left\{\frac{\sigma_t(X)}{\sigma_t(1)}\right\}U(t+dt)$$

$$= \pi_t(X) + U^*(t+dt)\,d\left\{\frac{\sigma_t(X)}{\sigma_t(1)}\right\}U(t+dt)$$

we've used $U^*(t)\,\xi\,U(t) = U^*(t)\,\xi\,U(s)\;\forall\; t \geq S$

for all. $\xi \in \eta_{s]}^{in}$. Hence,

$$d\pi_t(X) = U^*(t+dt)\,d\left\{\frac{\sigma_t(X)}{\sigma_t(1)}\right\}U(t+dt)$$

Now

$$d\sigma_t(X) = \sigma_t\left(B_\mu^* X\right)dY_{in,\mu}^* + \sigma_t\left(X\,B_\mu\right)dY_{in,\mu}$$

$$+\,\sigma_t\left(B_\mu^* X\,B_j\right)dY_{in,k}^* \cdot dY_{in,j}$$

So $\quad d\left\{\frac{\sigma_t(X)}{\sigma_t(1)}\right\} = \frac{\sigma_t(X)}{\sigma_t(1)} + \sigma_t(X)\sum_{r=1}^{\infty}\frac{(-1)^r\left(d\sigma_t(1)\right)^r}{\sigma_t(1)^{r+1}}$

$$= \sum_{r=1}^{\infty} \frac{(-1)^r}{\sigma_t(1)^{r+1}} d\sigma_t(X)(d\sigma_t(1))^r$$

$$= \frac{\sigma_t(B_\mu^* X)}{\sigma_t(1)} dY_{in,\mu}^* + \frac{\sigma_t(XB_\mu)}{\alpha_t(1)} dY_{in,\mu}^*$$

$$+ \frac{\sigma_t\left(B_\mu^* X\, B_0\right)}{\sigma_t(1)} dY_{in,\,k}^* dY_{in,\,j}$$

$$= + \frac{\sigma_t(X)}{\sigma_t(1)} \sum_{r \geq 1} (-1)^r \left(\frac{\sigma_t\left(B_\mu^*\right)}{\sigma_t(1)} \right) dY_{in,\,\mu}^*$$

$$\left. + \frac{\sigma_t\left(B_\mu\right)}{\sigma_t(1)} dY_{in,\,\mu}^* + \frac{\sigma_t\left(B_\mu^*\, B_j\right)}{\sigma_t(1)} dY_{in,\,\mu}^* dY_{in,\,j} \right)^r$$

$$+ \left(\frac{\sigma_t\left(B_\mu^* X\right)}{\sigma_t(t)} dY_{in,\,\mu}^* + \frac{\sigma_t\left(X\, B_\mu\right)}{\sigma_t(1)} dY_{in,\,\mu} + \frac{\sigma t\left(B_k^* X\, B_j\right)}{\sigma_t(1)} dY_{in,\,k}^* dY_{in,\,j} \right)$$

$$\times \sum_{r \geq 1} (-1)^r \left(\frac{\sigma_t\left(B_\mu^*\right)}{\sigma_t(1)} dY_{in,\,\mu}^* + \frac{\sigma_t\left(B_\mu\right)}{\sigma_t(1)} dY_{in,\,\mu} \right.$$

$$\left. + \frac{\sigma t\left(B_k^* B_j\right)}{\sigma_t(1)} dY_{in,\,k}^* dYin,\, j \right)^r$$

So we finally get the quantum Kushner equation:

$$\begin{aligned} d\pi_t(X) = {} & \pi_t\left(B_\mu^* X\right) dY_{out,\,\mu}^* + \pi_t\left(X\, B_\mu\right) dY_{out,\,\mu} \\ & + \pi_t\left(B_k^* X\, B_j\right) dY_{out,\,k}^* dY_{out,\,j} \\ & + \left(\pi_t(X) + \pi_t\left(B_\mu^* X\right) dY_{out,\,in}^* + \pi_t\left(X\, B_\mu\right) dY_{out,\,\mu} \right. \\ & \left. + \pi_t\left(B_k^* X\, B_j\right) dY_{out,k}^* d\, Y_{out,\,j}\right) \\ & \times \sum_{r \geq 1} (-1)^r \left(\pi_t\left(B_\mu^*\right) dY_{out,\,in}^* + \pi_t\left(B_\mu\right) dY_{out,\,\mu} \right. \\ & \left. + \pi_t\left(B_k^* B_j\right) d\, Y_{out,\,k}^* dY_{out,\,j}\right)^r \end{aligned}$$

An example

Let $\mathcal{H} = L^2(\mathbb{R}_+)$ and $a(t)$, $a^+(t)$ the annihilation and creation process respectively:

$da(t)\, da^+(t) = dt$

let $$\hbar = \mathbb{C}^2 \text{ and } H_0 = \alpha\sigma_Z = \begin{pmatrix} \alpha & 0 \\ 0 & -\alpha \end{pmatrix}$$

Let $$L = \lambda\sigma_X = \begin{pmatrix} 0 & \lambda \\ \lambda & 0 \end{pmatrix}$$

$$L^+ = \bar{\lambda}\sigma_x = \begin{pmatrix} 0 & \bar{\lambda} \\ \bar{\lambda} & 0 \end{pmatrix}$$

$$d\,U(t) = w\left(-\left(iH_0 - \frac{1}{2}LL^+\right)dt + Lda(a) - L^+a^+(t)\right)U(t)$$

Then $dU^* \cdot U + U^* \cdot dU + dU^* \cdot dU = 0$

and hence $U(t)$ is a unitary operator $\forall\, t \geq 0$

We take $Y_{in}(t) = \mu a(t) + \bar{\mu} a^+(t) \quad \mu \in \mathbb{D}$

Now that

$$-iH_0 + \frac{1}{2}LL^+ = \begin{pmatrix} -i\alpha & 0 \\ 0 & i\alpha \end{pmatrix} + \frac{1}{2}\begin{pmatrix} |\lambda|^2 & 0 \\ 0 & |\lambda|^2 \end{pmatrix}$$

$$= \begin{pmatrix} \frac{1}{2}|\lambda|^2 - i\alpha & 0 \\ 0 & \frac{1}{2}|\lambda|^2 + i\alpha \end{pmatrix}$$

So our evolution eqn. is

$$dU(t) = \left[\begin{pmatrix} \frac{1}{2}|\lambda|^2 - i\alpha & 0 \\ 0 & \frac{1}{2}|\lambda|^2 + i\alpha \end{pmatrix} dt\right.$$

$$\left.\begin{pmatrix} 0 & \lambda da(t) - \bar{\lambda} da^+(t) \\ \lambda da(t) - \bar{\lambda} da^+(t) & 0 \end{pmatrix}\right] U(t)$$

$$dU(t)|fe(u)\rangle = \left(\left(-iH_0 + \frac{1}{2}LL^+\right)dt + Lda - L^+da^+\right)U(t)|fe(u)\rangle$$

$$= \left[\left(-iH_0 + \frac{1}{2}LL^+\right)dt + Lu(t)dt - \frac{L^+\left(\mu da + \bar{\mu}\, da^+\right)}{\mu} + \frac{\mu L^+}{\mu}da\right]$$

$$U(t)\left|fe(u)\right\rangle$$

$$= \left[\left(-iH_0 + \frac{1}{2}LL^+ + u(t)\left(L + \frac{\mu}{\bar{\mu}}L^+\right)\right)dt - \frac{L^+}{\bar{\mu}}dY_{in}(t)U(t)\right]\left|fe(u)\right\rangle$$

So we can take

$$F'(t) = \left(\left(-iH_0 + \frac{1}{2}LL^+\right)\left(L + \frac{\mu}{\bar{\mu}}L^+\right)dt - \frac{L^+}{\mu}dY_{in}(t)\right)F(t)$$

$$= \begin{pmatrix} \frac{1}{2}|\lambda|^2 - i\alpha & 0 \\ 0 & \frac{1}{2}|\lambda|^2 + i\alpha \end{pmatrix} + \begin{pmatrix} 0 & \lambda \\ \lambda & 0 \end{pmatrix} u(t)dt$$

$$+ e^{i\phi t}\begin{pmatrix} 0 & \bar{\lambda} \\ \bar{\lambda} & 0 \end{pmatrix} u(t)dt - \frac{1}{\mu}\begin{pmatrix} 0 & \bar{\lambda} \\ \bar{\lambda} & 0 \end{pmatrix} dY_{in}(t)\Bigg] F(t) \quad (\phi = 2arg\,(\mu))$$

$$d\,F(t) = \begin{pmatrix} \left(\frac{1}{2}|\lambda|^2 - i\alpha\right)dt & \left(\lambda + \bar{\lambda}e^{i\phi}\right)u(t)dt - \frac{\bar{\lambda}}{\mu}dY_{in}(t) \\ \left(\lambda + \bar{\lambda}e^{i\phi}\right)u(t)_{dt} - \frac{\bar{\lambda}}{\mu}dY_{in}(t) & \left(\frac{1}{2}|\lambda|^2 - i\alpha\right)dt \end{pmatrix}$$

Equivalently,

$$dF(t) = \left(\underline{\underline{F}}_0(t)dt + \underline{\underline{F}}_1 dY_{in}(t)\right)F(t)$$

where $\underline{\underline{F}}_0(t) = \begin{pmatrix} \frac{1}{2}|\lambda|^2 - i\alpha & \left(\lambda + \bar{\lambda}e^{i\phi}\right)u(t) \\ \left(\lambda + \bar{\lambda}e^{i\phi}\right)u(t) & \frac{1}{2}|\lambda|^2 - i\alpha \end{pmatrix}$ and

$$\underline{\underline{F}}_1 = \begin{pmatrix} 0 & 1 \\ 1 & 0 \end{pmatrix}\left(\frac{\bar{\lambda}}{\mu}\right)$$

$$d\left(F^*(t)XF(t)\right) = dF^*(t)XF(t) + F^*(t)X\,dF(t) + dF^*(t)X\,dF(t)$$

$$= F^*(t)\Big\{\left(F_0^* dt + F_1^* dY_{in}\right)X$$

$$+X\left(F_0 dt + F_1 dY_{in}\right) + F_1^* X F_1 \left(dY_{in}\right)^2 F(t)\Big\}$$

Since

$$(dY_{in})^2 = |\mu|^2\, dt, \text{ we get}$$

$$d\sigma_t(X) = \sigma_t\left(F_0^* X + XF_0 + F_1^* X F_1\right)dt + \sigma_t\left(F_1^* X + XF_1\right)dY_{in}$$

So,

$$d\left\{\frac{\sigma_t(X)}{\sigma_t(1)}\right\} = \frac{\sigma_t(X)}{\sigma_t(1)} - \frac{\sigma_t(X)d\sigma_t(1)}{\sigma_t(1)^2} + \frac{\sigma_t(X)}{\sigma_t(1)^3}(d\sigma_t(1))^2 - \frac{d\sigma_t(X)d\sigma_t(1)}{\sigma_t(1)^2}$$

So,

$$\begin{aligned} d\pi_t(X) &= \pi_t\left(F_0^* X + X F_0 + F_1^* X F_1\right)dt \\ &\quad -\pi_t(X)\pi_t\left(F_0^* + F_0 + F_1^* F_1\right)dt + \pi_t(X)\pi_t\left(F_1^* F_1\right)^2 |\mu|^2\, dt \\ &\quad -\pi_t\left(F_1^* X + XF_1\right)\pi_t\left(F_1^* + F_1\right)|\mu|^2\, dt \\ &\quad -\pi_t(X)\pi_t\left(F_1^* + F_1\right)dY_{out} \\ &\qquad +\pi_t\left(F_1^* X + XF_1\right)dY_{out} \\ &= \left(\pi_t\left(F_1^* X + XF_1\right) - \pi_t(X)\pi_t\left(F_1^* + F_1\right)\right) \\ &\quad \left(dY_{out}(t) - \pi_t\left(F_1^* + F_1\right)\right)|\mu|^2\, dt \\ &\qquad +\left(\pi_t\left(F_0^* X + XF_0 + F_1^* X F_1\right) - \pi_t(X)\pi_t\left(F_0^* + F_0 + F_1^* F_1\right)\right)dt \end{aligned}$$

Another example: Photon counting

Here we consider

$$H_0 = a\sigma_Z = \begin{pmatrix} \alpha & 0 \\ 0 & \alpha \end{pmatrix}, \alpha \in \mathbb{R}$$

$$L = \lambda\sigma_X = \begin{pmatrix} 0 & \lambda \\ \lambda & 0 \end{pmatrix}$$

$$dU(t) = \left[\left(-iH_0 + \frac{1}{2}LL^+\right)dt + Lda(t) - L^+ da^+(t)\right]U(t)$$

$$Y_{in}(t) = \Lambda(t) = \lambda\left(\chi_{[0,t]}\right)$$

$$\mathcal{H} = L^2(\mathbb{R}_+),\ \hbar = \mathbb{D}^2$$

❖❖❖❖❖

[187] Distribution of L_{Ta}

$$L_{Ta} = \int_0^{Ta} \delta(B_s)ds, a \geq 0$$

$$\mathbb{E}\left[\exp(\lambda L_{Ta}) | B_0 = X\right] = u_{\lambda,a}(X) \text{ say}$$

Then
$$u_{\lambda,a}(X) = \left(1 + \lambda\delta(X)h\right)\mathbb{E}\left[u_{\lambda,a}(B_h) | B_0 = X\right] + 0(h)$$

$$= \left(1 + \lambda h\delta(X)\right)\left(u_{\lambda,a}(X) + \frac{1}{2}u''_{\lambda,a}(X)h\right) + 0(h)$$

Thus,

$$\frac{1}{2}u''_{\lambda,a}(X) + \lambda u_{\lambda,a}(0)\delta(X) = 0 \quad (1)$$

$$u''_{\lambda,a}(X) = 0, X \neq 0$$

$$u_{\lambda,a}(X) = \begin{cases} A_1X + B_1, & X > 0 \\ A_2X + B_2, & X < 0 \end{cases}$$

(1) ⇒
$$\frac{1}{2}u'_{\lambda,e}(X) + \lambda u_{\lambda,a}(0)\theta(X) = 0$$

⇒
$$u_{\lambda,a}(0+) = u_{\lambda,a}(0-)\, u_{\lambda,a}(0) \equiv u_{ba}(0)$$

$$= B_1 = B_2 \equiv B \text{ say}$$

Then,
$$u_{\lambda,a}(X) = \begin{cases} A_1X + B, & X > 0 \\ A_2X + B, & X < 0 \end{cases}$$

(1) ⇒
$$\frac{1}{2}u'_{\lambda,a}(0+) - \frac{1}{2}u'_{\lambda,a}(0-) + \lambda u_{\lambda,a}(0) = 0$$

⇒
$$\frac{1}{2}(A_1 - A_2) + \lambda B = 0$$

⇒
$$A_2 = A_1 + 2\lambda B$$

$$u_{\lambda,a}(X) = \begin{cases} A_1X + B, & X > 0 \\ (A_1 + 2\lambda B)X + B, & X < 0 \end{cases}$$

$$\lim_{x \to a} u_{\lambda,a}(X) = 1$$

⇒
$$A_1 a + B = 1$$

$$B = 1 - A_1 a$$

$$u_{\lambda,a}(X) = \begin{cases} A_1(X - a) + 1, & X > 0 \\ A_1X + (1 + 2\lambda X)(1 - A_1 a), & X < 0 \end{cases}$$

$$\lim_{x \to -\infty} u_{\lambda, a}(X)$$

$\Rightarrow A_1 + 2\lambda(-A_1 a + 1 = 0)$

$$\Rightarrow \qquad A_1 = \frac{-2\lambda}{1-2\lambda a}$$

So, $$B = 1+\frac{2\lambda a}{1-2\lambda a} = \frac{1}{1-2\lambda a}$$

Finally

$$\mathbb{E}\left[\exp\left(-sL_{T_a}\right) \mid B(0)=0\right]$$

$$u_{-s, a}(0) = 1 - A, a|_{\lambda = -s} = B|_{\lambda=-s} \frac{1}{1+2sa}$$

Thus, $$P\{L_{Ta} \in dX\} = \frac{1}{2a}\exp\left(\frac{-X}{2a}\right), X \geq 0$$

$$P\{L_{Ta} \in dX\} = \exp\left(\frac{-X}{2a}\right), X \geq 0$$

$$P\{L_{Ta}^{-} > X\} = \exp\left(\frac{-X}{2a}\right), X \geq 0$$

Note $a \to L_{Ta}$ has universe given by

$t \to S_{\tau t}$

Where S is the maximum process B. M. and

$$\tau_t = \min\{s \geq 0 \mid L_s \geq t\}$$

Thus,

for $\quad X \geq 0, P\{S_{\tau_t} \leq \text{X}\} = P\{t \leq L_{TX}\} = \exp\left(\frac{-t}{2X}\right)$

❖❖❖❖❖

[188] Quantization of Linear Gauss-Markov processes

Classical sde : $d\underline{X}_t = \underline{\underline{A}}_t \underline{X}_t dt + \underline{\underline{G}}_t d\,\underline{B}_t$

generation

$$df(\underline{X}_t) = \left[\left(\underline{\underline{A}}_t \underline{X}_t\right)^T f'(\underline{X}_t) + \frac{1}{2}Tr\left(G_t G_t^T f''(\underline{X}_t)\right)\right]dt + d\,\underline{B}_t^T\, \underline{\underline{G}}_t^T f'(\underline{X}_t)$$

Quantization $f(X_t) = j_t(f)$

$$df_t(f) = j_t\left(\theta_{0t}(f)\right)dt + \sum_k j_t\left(\theta_{kt}(f)\right)dB_k$$

$$\text{Where } \theta_{0t}(f)(X) = X^T \underline{\underline{A}}_t^T \underline{\nabla}_X f(\underline{X}) + \frac{1}{2}Tr\left(G_t G_t^T \nabla_X \nabla_X^T f(X)\right)$$

$$\theta_{kt}(f) = \sum_j G_{jkt} \frac{\partial f(X)}{\partial X_j}$$

Evans Hudson flows.

$p_j = -i\dfrac{\partial}{\partial X_j}$, $X_j = q_j$. Assume A_t, G_t one time independent

$$\theta_0(f) = \left[i\underline{q}^T \underline{\underline{A}}^T \underline{p} - \frac{1}{2}Tr\left(GG^T p\, p^T\right)\right](f)$$

$$\theta_k(f) = i\sum_j G_{jk} p_j f \equiv i\left(G^T p\right)_k (f)$$

i.e. $\underline{\theta} = i\underline{\underline{G}}^T \underline{p} \quad \underline{\theta} = ((g_k))$

Other schemes:

$$d\underline{q}_t = \underline{p}_t\, dt,\ d\underline{p}_t = -\underline{\underline{A}}\,\underline{q}_t\, dt + \underline{\underline{G}}d\,\underline{B}_t\, \underline{\underline{A}}^T = \underline{\underline{A}}$$

$$H\left(\underline{q}, \underline{p}, t\right) = \frac{1}{2}\underline{p}^T \underline{p} + \frac{1}{2}\underline{q}^T \underline{\underline{A}}\,\underline{q} - \underline{q}^T \underline{\underline{G}}\,\underline{u}_t$$

$$\underline{u}_t = \frac{d\underline{B}_t}{dt}$$

$$\dot{\underline{q}} = \frac{\partial H}{\partial \underline{p}} = \underline{p},\quad \dot{\underline{p}} = \frac{-\partial H}{\partial \underline{q}} = -\underline{\underline{A}}\,\underline{q} + \underline{\underline{G}}\,\underline{u}_t$$

Quantum evolution:

$$dU(t) = \Big[-iH_t dt + (\alpha q + \beta p)da(t) - \left(\bar{\alpha} q + \bar{\beta} p\right)da^*(t)$$
$$-\frac{1}{2}(\alpha q + \beta p)\left(\bar{\alpha} q + \bar{\beta} p\right)dt\Big]U(t)$$

This is the H.P. equation for a harmonic oscillator with noise and damping. Consider for simplicity a 1- D harmonic oscillator with damping and noise:

$H_0 = a^+ a$, $A(t) \simeq$ annihilation process

$A^+(t) \simeq$ creation process H.P. eqn.

$$dU(t) = \left[-ia^+ a\, dt + \left(\alpha a + \beta a^+\right)dA(t) - \left(\bar{\alpha} a^+ \bar{\beta} a\right)dA^+(t)\right.$$

$$-\frac{1}{2}\left(\alpha a+\beta a^{+}\right)\left(\bar{\alpha}a^{+}+\bar{\beta}a\right)dt\Big]U(t)$$

$$dA\,dA^{+} = dt,\ [a, a^{+}] = 1,\ [a, A(t)] = 0$$

$$[a^{+}, A(t)] = 0$$

Filtering (Belavkin)

$$dU(t)|f(u)\rangle = \Big[-\left(ia^{+}a+\frac{1}{2}\left(\alpha a+\beta a^{+}\right)\left(\bar{\alpha}a^{+}+\bar{\beta}a\right)\right)dt$$

$$+\left(\alpha a+\beta a^{+}\right)u(t)dt$$

$$+\left(\bar{\alpha}a^{+}+\bar{\beta}a\right)u(t)dt-\left(\bar{\alpha}a^{+}+\bar{\beta}a\right)dB(t)\Big]U(t)\,|f\varphi(u)\rangle$$

where $f \in L^2(\mathbb{R})$, $\|f\| \le 1$,

$$\varphi(u) = \exp\left(-\frac{\|u\|^2}{2}\right)|e(u)\rangle,\ u \in \mathcal{H} = L^2(\mathbb{R}_+)$$

and

$$B(t) = A(t) + A^{+}(t).$$

We takes as measurement

$$Y_{out}(t) = U^{*}(t)Y_{in}(t)U(t),\ Y_{in}(t) = B(t)$$

Let $F(t)$ be such that $F(0) = I$

$$dF(t) = \left[\mathcal{F}_1(t)dt+\mathcal{F}_2(t)dB(t)\right]\mathcal{F}(t)$$

where

$$\mathcal{F}_1(t) = -\Big(ia^{+}a+\frac{1}{2}\left(\alpha a+\beta a^{+}\right)\left(\bar{\alpha}a^{+}+\bar{\beta}a\right)$$

$$+\left(\alpha a+\beta a^{+}+\bar{\alpha}a^{+}+\bar{\beta}a\right)u(t)\Big)$$

$$= -\left(ia^{+}a+\frac{1}{2}\left(|\alpha|^2 aa^{+}+|\beta|^2 a^{+}a+\alpha\bar{\beta}a^2+\bar{\alpha}\beta a^{+2}\right)\right)$$

$$+\left((\alpha+\bar{\beta})a+(\bar{\alpha}+\beta)a^{+}\right)u(t)$$

$$= -\left(i+\frac{1}{2}\left(|\alpha|^2+|\beta|^2\right)\right)a^{+}a-\frac{1}{2}|\alpha|^2$$

$$-\frac{1}{2}\left(\alpha\bar{\beta}a^2+\bar{\alpha}\beta a^{+2}\right)+u(t)\left(\left(\alpha+\bar{\beta}\right)a+(\bar{\alpha}+\beta)a^+\right)$$

and

$$\mathcal{F}_2(t) = -\left(\bar{\alpha}a^+ + \bar{\beta}a\right)$$

We have

$$dj_t(X) = d\left(U^*(t)X(t)\right)$$

$$= dU^*(t)XU(t)+U^*(t)X\,dU(t)+dU^*(t)X\,dU(t)$$

$$= U^*(t)\Big\{-i\left[a^+a,X\right]dt+\left(\left(\bar{\alpha}a^++\bar{\beta}a\right)X-X\left(\bar{\alpha}a^++\bar{\beta}a\right)\right)dA^+$$

$$+\left(-\left(\alpha a+\beta a^+\right)X+X\left(\alpha a+\beta a^+\right)\right)dA$$

$$-\frac{1}{2}\left(\left(\alpha a+\beta a^+\right)\left(\bar{\alpha}a^++\bar{\beta}a\right)X+X\left(\alpha a+\beta a^+\right)\left(\bar{\alpha}a^++\bar{\beta}a\right)\right)dt$$

$$= +\left(\alpha a+\beta a^+\right)X\left(\bar{\alpha}a^++\bar{\beta}a\right)dt\Big\}U(t)$$

$$= j_t\Big\{-i\left[a^+a,X\right]dt+\left[\bar{\alpha}a^++\bar{\beta}a,X\right]dA^+-\left[\alpha a+\beta a^+,X\right]dA$$

$$-\frac{1}{2}\left\{\left(\alpha a+\beta a^+\right)\left(\bar{\alpha}a^++\bar{\beta}a\right),X\right\}dt$$

$$+\left(\alpha a+\beta a^+\right)X\left(\bar{\alpha}a^++\bar{\beta}a\right)dt\Big\}$$

In particular

$$dj_t(a) = j_t\left\{ia+\left(\alpha a+\beta a^+\right)a\left(\bar{\alpha}a^++\bar{\beta}a\right)\right\}dt+\left(-\bar{\alpha}\,dA^++\beta\,dA\right)$$

$$+j_t\Big\{-\frac{1}{2}\left\{\alpha a+\beta a^+\left(\bar{\alpha}a^++\bar{\beta}a\right),a\right\}$$

$$+\left(\alpha a+\beta a^+\right)a\left(\bar{\alpha}a^++\bar{\beta}a\right)\Big\}dt$$

Now,

$$\left(\alpha a+\beta a^+\right)a\left(\bar{\alpha}a^++\bar{\beta}a\right)-\frac{1}{2}\left\{\left(\alpha a+\beta a^+\right)\left(\bar{\alpha}a^++\bar{\beta}a\right),a\right\}$$

$$= \frac{1}{2}\left(\alpha a+\beta a^+\right)\left[a,\bar{\alpha}a^++\bar{\beta}a\right]$$

$$+\left[\alpha a+\beta a^+,a\right]\left(\bar{\alpha}a^++\bar{\beta}a\right)$$

$$= \frac{1}{2}\left(\alpha a + \not{\beta} a^{+}\right)(\bar{\alpha}) - \beta\left(\not{\bar{\alpha}} a^{+} + \bar{\beta} a\right)$$

$$= \frac{1}{2}\left(|\alpha|^2 - |\beta|^2\right)a$$

Thus,

$$dj_t(a) = j_t\left(\left(i + \frac{1}{2}\left(|\alpha|^2 - |\beta|^2\right)\right)a\right)dt + \beta dA(t) - \bar{\alpha}\, d\, A^{+}(t)$$

Taking the adjoint gives

$$dj_t(a^{+}) = j_t\left(\left(-i + \frac{1}{2}\left(|\alpha|^2 - |\beta|^2\right)\right)a\right)dt + \bar{\beta} dA^{+}(t) - \alpha\, dA(t)$$

Thus,

$$dj_t(a) = j_t\left(i + \frac{1}{2}\left(|\alpha|^2 - |\beta|^2\right)\right)j_t(a)dt + \beta dA(t) - \bar{\alpha} dA^{+}(t)$$

and $$dj_t(a^{+}) = \left(-i + \frac{1}{2}\left(|\alpha|^2 - |\beta|^2\right)\right)j_t\left(a^{+}\right)dt + \bar{\beta} dA^{+}(t) - \alpha dA(t)$$

There are the quantum Longevir equations for a harmonic oscillator. Thus,

$$dj_t(a + a^{+}) = j_t\left(i\left(a - a^{+}\right)\right)dt + \frac{1}{2}\left(|\alpha|^2 - |\beta|^2\right)j_t\left(a + a^{+}\right)dt$$

$$+(\beta - \alpha)dA(t) + \left(\bar{\beta} - \bar{\alpha}\right)dA^{+}(t)$$

and

$$dj_t\left(a - a^{+}\right) = j_t\left(i\left(a + a^{+}\right)\right)dt + \frac{1}{2}\left(|\alpha|^2 - |\beta|^2\right)j_t\left(a - a^{+}\right)dt$$

Writing $a + a^{+} = q$, i $(a - a^{+}) = p$ gives

$$dj_t(q) = j_t(p)dt + \frac{1}{2}\left(|\alpha|^2 - |\beta|^2\right)j_t(q) + (\beta - \alpha)dA(t) + \left(\bar{\beta} - \bar{\alpha}\right)dA^{+}(t)$$

$$dj_t(p) = -j_t(q)dt + \frac{1}{2}\left(|\alpha|^2 - |\beta|^2\right)j_t(p)dt$$

$$+i(\beta + \alpha)dA(t) - i\left(\bar{\alpha} + \bar{\beta}\right)dA^{+}(t)$$

❖❖❖❖❖

[189] Quantization of a free robot having two link with noise.

$$dU(t) = \left(-iH_0 dt + \frac{1}{2}LL^+ dt + Lda(t) - L^+ da^+(t)\right)U(t)$$

$$da \,.\, da^+ = dt.\ H_0 = \frac{1}{2}\underline{p}^T \underline{\underline{M}}(q)^{-1}\underline{p}$$

Linearize w.r.t. $\underline{q} = \underline{q}_0$. $\underline{\underline{M}}(\underline{q}) \approx \underline{\underline{M}}(\underline{q}) + M'(q_0)(\delta q \otimes I)$

$\equiv \underline{\underline{M}}_0 + \underline{\underline{M}}_1(\delta q \otimes I)$ Write q is place of $\delta q = \underline{q} = \underline{q}_0$.

Thus, $$H_0 = \frac{1}{2}\underline{p}^T(\underline{\underline{M}}_0 + \varepsilon \underline{\underline{M}}_1(\delta \underline{q} \otimes \underline{\underline{I}}))^{-1}\underline{p}$$

$$\approx \frac{1}{2}\underline{p}^T(\underline{\underline{M}}_0^{-1} - \varepsilon M_0^{-1}M_1(\underline{q}\otimes I)M_0^{-1})p$$

$$= \frac{1}{2}\underline{p}^T \underline{\underline{M}}_0^{-1}\underline{p} - \frac{\varepsilon}{2}\underline{p}^T M_0^{-1}M_1(\underline{q}\otimes I)M_0^{-1}\underline{p}$$

After a canonical transformation, this becomes

$$H_0 = \frac{p_1^2}{2m_1} + \frac{p_2^2}{2m_2} + \varepsilon \sum_{i\,j\,k=1,2} a(i\,j\,k)q_i p_j p_k + O(\varepsilon^2)$$

$$= \underline{\underline{H}}_{00} + \varepsilon \underline{\underline{H}}_1 + O(\varepsilon^2)$$

$$H_{00} = -\frac{1}{2m_1}\frac{\partial^2}{\partial q_1^2} - \frac{1}{2m_2}\frac{\partial^2}{\partial q_2^2}.$$

$$u_{n_1 n_2}^{(q_1, q_2)} = \frac{A_1}{\pi}\cos(n_1 q_1)\cos(n_2 q_2) + \frac{A_2}{\pi}\cos(n_1 q_1)\sin(n_2 q_2)$$
$$+\frac{A_3}{\pi}\sin(n_1 q_1)\cos(n_2 q_2) + \frac{A_4}{\pi}\sin(n_1 q_1)\sin(n_2 q_2)$$

where $|A_1|^2 + |A_2|^2 + |A_3|^2 + |A_4|^2 = 1.$

$n_1, n_2, n_3, n_4 = 0, 1, 2, \ldots$

Secular matrix

$$u_1 = \frac{1}{\pi}\cos(n_1 q_1)\cos(n_2 q_2)$$

$$u_2 = \frac{1}{\pi}\cos(n_1 q_1)\sin(n_2 q_2)$$

$$u_3 = \frac{1}{\pi}\sin(n_1 q_1)\cos(n_2 q_2)$$

$$u_4 = \frac{1}{\pi}\sin(n_1 q_1)\sin(n_2 q_2)$$

$$\langle u_1 | \tilde{H}_1 | u_1 \rangle = \frac{1}{\pi^2} \int_{[0,2\pi]^2} \cos(n_1q_1)\cos(n_2q_2)\Bigg(-(a(111)q_1 + a(211)q_2)\frac{\partial^2}{\partial q_1^2}$$

$$- (a(122)q_1 + a(222)q_2)\frac{\partial^2}{\partial q_2^2} - 2(a(112)q_1 + a(212)q_2)\frac{\partial^2}{\partial q_1 \partial q_2}\Bigg)$$

$$\cos(n_1q_1)\cos(n_2q_2)\, dq_1dq_2$$

$$= -\frac{1}{\pi^2}\int \cos(nq_1)\cos(nq_2)[(a(111)q_1 + a(211)q_2)(-n_1^2)$$

$$+ (a(122)q_1 + a(222)q_2)(-n_2^2))\cos(n_1q_1)\cos(n_2q_2)$$

$$+ 2(a(112)q_1 + a(212)q_2)(n_1n_2)\sin(n_1q_1)\sin(n_2q_2)\, dq_1dq_2]$$

$$\tilde{H}_1 = a(ijk)q_i p_j p_k \quad \tilde{H}_1^* = a(ijk)p_j p_k q_i$$

$$H_1 = \frac{(\tilde{H}_1 + \tilde{H}_1^*)}{2} = a(ijk)q_i p_j p_k + \frac{1}{2}a(ijk)[p_j p_k, q_i]$$

$$= \tilde{H}_1 - \frac{i}{2}a(ijk)p_k + (-i\delta_{ij}p_k - i\delta_{ik}p_i)$$

$$= \tilde{H}_1 - \frac{i}{2}(a(ijk)p_k + a(ii)p_j)$$

$$= \tilde{H}_1 - \frac{i}{2}(a(iik) + a(iki))p_k$$

$$\langle u_1 | ip_k | u_1 \rangle = \left\langle u_1 \left| \frac{\partial}{\partial q_k} \right| u_1 \right\rangle$$

$$= \frac{1}{\pi^2}\int_{[0,2\pi]^2} \cos(n_1q_1)\cos(n_2q_2)\frac{\partial}{\partial q_k}(\cos(n_1q_1)\cos(n_2q_2))\, dq_1dq_2$$

$$= 0.$$

So,

$$\langle u_1 | H_1 | u_1 \rangle = \langle u_1 | \tilde{H}_1 | u_1 \rangle$$

$$\langle u_1 | \tilde{H}_1 | u_2 \rangle = \frac{1}{\pi^2}\int \cos(n_1q_1)\cos(n_2q_2)a(ijk)$$

$$(q_i p_j p_k \cos(n_1q_1)\sin(n_2q_2))\, dq_1dq_2$$

$$= -\frac{1}{\pi^2}\int \cos(n_1q_1)\cos(n_2q_2)\Bigg((a(111)q_1 + a(211)q_2)\frac{\partial^2}{\partial q_1^2}$$

$$+ (a(122)q_1 + a(222)q_2)\frac{\partial^2}{\partial q_2^2}$$

$$+2(a(112)q_1 + a(212)q_2)\frac{\partial^2}{\partial q_1 \partial q_2}\Bigg)(\cos(n_1 q_2)\sin(n_2 q_2))\,dq_1 dq_2$$

Likewise for the other elements of the secular matrix.

Calculation of the eigen function and eigenvalue of a perturbed Hamiltonian when the unperturbed Hamiltonian is degenerate.

$$H = H_0 + \varepsilon V$$

$$H_0|E_n, \beta\rangle^{(0)} = E_k^{(0)}|E_k, \beta\rangle^{(0)}, 1 \le \beta \le n_k, k \ge 1.$$

Let $\varepsilon|E_k\rangle^{(1)}$ be a first order perturbation to the any state of the form $\sum_{\beta=1}^{n_k} c(\beta)|E_k,\beta\rangle^{(0)}$ and $\varepsilon\, E_k^{(1)}$ the corresponding perturbation to the energy value $E_k^{(0)}$. They

Then

$$H_0|E_k\rangle^{(1)} + V|E_k\rangle^{(0)} = E_k^{(0)}|E_k\rangle^{(1)} + E_k^{(1)}|E_k\rangle^{(0)}$$

so

$$^{(0)}\Big\langle E_{km}, \beta' \,|\, (H_0 - E_k^{(0)}) \,|\, E_k\Big\rangle^{(1)} = E_k^{(1)} \sum_{m\beta} c(\beta)\,\delta_{km}\,\delta_{\beta\beta'}$$

or

$$(E_m^{(0)} - E_k^{(0)})\,^{(0)}\langle E_{km}, \beta' \,|\, E_k\rangle^{(1)} + \sum_{\beta} c(\beta)^{(0)}\langle E_m, \beta' \,|\, V \,|\, E_k, \beta\rangle = E_k^{(1)} c(\beta')\,\delta_{km}$$

Taking $m = k$ gives

$$\sum_{\beta} {}^{(0)}\langle E_k, \beta' \,|\, V \,|\, E_k, \beta\rangle^{(0)}\, c(\beta) = E_{(k)}^{(1)} c(\beta')$$

So writing $\underline{c} = ((c(\beta)))_{\beta=1}^{n_k}$ gives

$$\det\left\{\left(\left({}^{(0)}\langle E_k, \beta' \,|\, V \,|\, E_k, \beta\rangle^{(0)}\right)\right)_{\beta',\beta} - E_k^{(1)} I_{n_k}\right\} = 0$$

and thus, we get the possible value or $E_k^{(1)}$ as the eigenvalues of the $n_k \times n_k$ matrix $\left(\left({}^{(0)}\langle E_k, \beta' \,|\, V \,|\, E_k, \beta\rangle^{(0)}\right)\right)_{\beta',\beta}$.

Let us denote these eigenvalues by

$E_k^{(1)}(\beta), 1 \le \beta \le n_k$ and the corresponding eigenvalues as $n_k\left(\left(c_{k\beta}(\beta')\right)\right)_{\beta'=1}^{n_k}$. Then the state $\sum_{\beta'=1}^{n_k} c_{k\beta}^{(\beta')}|E_k\,\beta'\rangle$ gets perturbations to an $|E_k\rangle^{(1)}$ which is determined by

$$ {}^{(0)}\langle E_m, \beta'' \mid E_k \rangle^{(1)} = \left(E_m^{(0)} - E_m^{(0)}\right)^{-1} \sum_{\beta'=1}^{n_k} c_{k\beta}(\beta')\, {}^{(0)}\langle E_m, \beta'' \mid V \mid E_k, \beta' \rangle \quad m \neq k $$

so

$$ |E_k\rangle^{(1)} = \sum_{\substack{m\neq k \\ 1\leq\beta'\leq n_m}} |E_m, \beta''\rangle^{(0)}\, {}^{(0)}\langle E_m, \beta'' \mid E_k \rangle^{(1)} $$

$$ = \sum_{\substack{1\leq\beta''\leq n_m, \\ m\neq k}} \frac{|E_m, \beta''\rangle^{(0)}\, {}^{(0)}\langle E_m, \beta'' \mid V \mid E_k, \beta' \rangle c_{k\beta}(\beta')}{E_k^{(0)} - E_m^{(0)}} $$

The eigenstates of the perturbed Hamiltonian are upto first order in given by

$$ \sum_{\beta'} c_{k\beta}(\beta') |E_k, \beta'\rangle^{(0)} + \varepsilon \sum_{\substack{1\leq\beta''\leq n_m, \\ m\neq k}} \frac{c_{k\beta}(\beta') |E_m, \beta''\rangle^{(0)}\, {}^{(0)}\langle E_m, \beta'' \mid V \mid E_k, \beta' \rangle}{E_k^{(0)} - E_m^{(0)}} $$

Here $((c_{k\beta}(\beta)))_{\beta'=1}^{n_\beta}$ is the eigenvalue of $\left(\left(\langle E_k, \beta' \mid V \mid E_k\, \beta'' \rangle\right)\right)_{1\leq\beta,\, \beta''\leq n_k}$ corresponding to the eigenvalue $E_k^{(1)}(\beta)$. The energy of the above perturbed state is $E_k^{(0)} + E_k^{(1)}\beta + O(\varepsilon^2)$.

Quantum Robots

Quantum noise in a quantum Robot:

$$ H_0(\underline{q}, \underline{p}) = \frac{1}{2} \underline{p}^T \underline{\underline{M}}(\underline{q})^{-1} \underline{p} + V(\underline{q}) $$

$$ \underline{q} = \begin{pmatrix} q_1 \\ q_2 \end{pmatrix}, \ \underline{p} = \begin{pmatrix} p_1 \\ p_2 \end{pmatrix}, \ [q_\alpha, p_\beta] = i\delta_{\alpha\beta}\,. $$

$$ \underline{\underline{M}}(\underline{q}) = \underline{\underline{M}}_0 + \cos(q_2) \underline{\underline{M}}_1 $$

$$ \underline{\underline{M}}(q)^{-1} = (M_0 + \cos(q_2) M_1)^{-1} $$

$$ = M_0^{-1}(I + \cos(q_2) M_1 M_0^{-1})^{-1} $$

Let $M_1 M_0^{-1} = A$. Then $\underline{\underline{M}}(q)^{-1} = M_0^{-1}\left(I + \sum_{r=1}^{\infty} (-1)^r (\cos q_2)^r A^r\right)$

$$ = M_0^{-1} + \sum_{r=1}^{\infty} (-1)^r (\cos q_2)^r M_0^{-1} A^r $$

$$ M_0^{-1} A^r = M_0^{-1} M_1 M_0^{-1} M_1 \ldots M_1 M_0^{-1} $$

$$ V(\underline{q}) = a_1 \sin(q_1) + a_2 \sin(q_1 + q_2) $$

$$ H_0 = \frac{1}{2} p^T M_0^{-1} p + a_1 q_1 + a_2(q_1 + q_2) + \varepsilon H_1(\underline{q}, \underline{p}) $$

where

$$\varepsilon H_1(\underline{q}, \underline{p}) = a_1(\sin(q_1) - q_1) + a_2(\sin(q_1 + q_2) - (q_1 + q_2))$$

$$+\frac{1}{2}\sum_{r=1}^{\infty}(-1)^r(\cos q_2)^r \underline{p}^T \underline{\underline{M}}_0^{-1} \underline{\underline{A}}^r \underline{p}$$

$$M_0^{-1} = O^{-1}O^T \quad OO^T = I, N = \text{diag}[\lambda_1, \lambda_2].$$

$$\underline{\underline{O}}^T \underline{p} = \tilde{\underline{p}}, \underline{\underline{Q}}^T \underline{q} = \tilde{\underline{q}}$$

Canonical transformations. $[\tilde{q}_\alpha, \tilde{p}_\beta] = i\delta_{\alpha\beta}$

or $$O^T = \begin{bmatrix} O_{11} & O_{21} \\ O_{12} & O_{22} \end{bmatrix}, Q^T = \begin{bmatrix} Q_{11} & Q_{21} \\ Q_{12} & Q_{22} \end{bmatrix}$$

$$\tilde{p}_\alpha = O_{\beta\alpha} p_\beta, \tilde{q}_\alpha = O_{\beta\alpha} q_\beta. \text{ Then,}$$

$$[O_{\beta\alpha} q_\beta, O_{\rho\sigma} p_\rho] = i\delta_{\beta\sigma}$$

or $$Q_{\beta\alpha} O_{\rho\sigma} \delta_{\beta\rho} = \delta_{\alpha\sigma}$$

$$Q_{\beta\alpha} \mathrm{O}_{\beta\sigma} = \delta_{\alpha\sigma}$$

$$Q^T O = I. \text{ So, } Q = O.$$

$$\tilde{\underline{p}} = \underline{\underline{O}}^T \underline{p}, \tilde{\underline{q}} = \underline{\underline{Q}}^T \underline{q}$$

$$H_0 = \frac{1}{2}\tilde{\underline{p}}^T \underline{\underline{\Lambda}}^{-1} \tilde{\underline{p}} + c_1\tilde{q}_1 + c_2\tilde{q}_2 + \varepsilon \tilde{H}_1(\tilde{q}_1, \tilde{\underline{p}})$$

$$= \frac{1}{2}\tilde{\underline{p}}^T \underline{\underline{\Lambda}}^{-1} \tilde{\underline{p}} + \underline{c}^T \tilde{\underline{q}} + \varepsilon \tilde{H}_1(\tilde{\underline{q}}, \tilde{\underline{p}})$$

where

$$[a_1 + a_2, a_2]\underline{q} = c^T \tilde{q} = \underline{c}^T \underline{\underline{Q}}^T \underline{q},$$

so, $$\underline{\underline{Q}}\underline{c} = \begin{bmatrix} a_1 + a_2 \\ a_2 \end{bmatrix} = \underline{\underline{O}}\underline{c}$$

$$\underline{c} = O^T \begin{bmatrix} a_1 + a_2 \\ a_2 \end{bmatrix}$$

$$c_1 = O_{11}(a_1 + a_2) + O_{21}, a_2, c_2 = O_2(a_1 + a_2) + O_{22}a_2.$$

We denote the canonicals transformation variables $\tilde{\underline{q}}$ and $\tilde{\underline{p}}$ by $\underline{q}, \underline{p}$, and $\tilde{H}_1$ by

H_1 for the sake of notational simplicity.

Then,

$$H_0 = \frac{1}{2\lambda_1} p_1^2 + \frac{1}{2\lambda_2} p_2^2 + c_1 q_1 + c_2 q_2 + \varepsilon H_1(q_1, q_2, p_1, p_2)$$

To calculate the approximate eigenvalue and eigenfunction of H_0 using perturbation theory, We must first calculate the same for the 1 - D Hamiltonian

$$H_{01} = \frac{1}{2m_1} p_i^2 + cq \quad (q, p) = i.$$

This is the Hamiltonian of a particle in a gravitational field.

Stationary state Schrödinger equation for particle in a gravitational field.

$$\left[-\frac{1}{2m} \frac{d^2}{dx^2} + cx \right] u(x)s = E u(x)$$

Solved using areas function.

Let $x - \frac{E}{c} = Y$. Then Let $u(x) = u\left(\frac{E}{c} + Y\right) = v(Y)$.

We have $-\frac{1}{2m} v''(Y) + Yv(Y) = 0.$

Let $\quad Y = \alpha\xi,\ v(Y) = v(\alpha\xi) = w(\xi)$. Then

$$w''(\xi) = \alpha^2 v''(\alpha\xi) = 2m\alpha^2 Yv(y) = 2m\alpha^3 \xi w(\xi)$$

Let $\quad 2m\,\alpha^3 = 1$, *i.e*, $\alpha = (2m)^{-1/3}$. Then,

$$w''(\xi) = \xi w(\xi).$$

Let $$w(\xi) = \sum_{n=0}^{\infty} c_n \xi^{n+s}$$

Then

$$\sum_n (n+s)(n+s-1) c_n \xi^{n+s-2} = \sum_n c_n \xi^{n+s+1}$$

$$S(S-1)\,c_0 = 0,$$

$$\Rightarrow \quad (S+2)(S+1)\,c_2 = 0,$$

$$S(S+1)c_1 = 0,$$

$$(n+S)(n+S-1)\,c_n = c_{n-3},\ n \geq 3.$$

$$c_1 \neq 0$$

$S = 0$ or $1 \Rightarrow c_0 \neq 0$, $c_1 \neq 0$ can be assumed, but $c_2 = 0$.

$S = -1 \Rightarrow c_1 \neq 0$, $c_2 \neq 0$ can be assumed but $c_0 = 0$.

$S = -2 \Rightarrow c_2 \neq 0$ can be assumed but $c_0 = c_1 = 0$.

Let $S = 0$. Then

$$w(\xi) = \sum_{n\geq 0} c_{3n}\,\xi^{3n} + \sum_{n\geq 0} c_{3n+1}\,\xi^{3n+1}$$

$$c_{3n} = \frac{c_{3n-3}}{3n(3n-1)} = \frac{c_{3n-6}}{3n(3n-1)(3n-3)(3n-4)} = \cdots$$

$$= \frac{c_0}{(3n(3n-0)\cdots 3)\,((3n-1)(3n-4)\cdots(2))}$$

$$= \frac{c_0}{3^n \lfloor n \prod_{r=0}^{n-3}(3r+2)}$$

$$c_{3n+1} = \frac{c_{3n-2}}{(3n+1)(3n)} = \frac{c_{3n-5}}{(3n+1)(3n)(3n-2)(3n-3)} = \cdots$$

$$= \frac{c_1}{3^n \lfloor n \prod_{r=1}^{n}(3r+1)}$$

$$= \frac{c_1}{3^n \lfloor n \prod_{r=1}^{n+1}(3r-2)}$$

$$\text{As } n \to \infty,\ c_{3n} \approx \frac{c_0}{(3^n \lfloor n)^2},\ c_{3n+1} \approx \frac{c_1}{(3^n \lfloor n)^2}$$

So we had to investigate properties of

$$f(X) = \sum_{n\geq 0} \frac{X^{3n}}{(3^n \lfloor n)^2}$$

This will dictate the asymptotic behaviour of $u(X)$ as $|X| \to \infty$.

$$U(t+dt)\left\{\frac{\sigma_{t+dt}(X)}{\sigma_{t+dt}(1)}\right\}U(t+dt)$$

$$= U^*(t+dt)\left\{\frac{\sigma_t(X)}{\sigma_t(1)}+d\left(\frac{\sigma_t(X)}{\sigma_t(1)}\right)\right\}U(t+dt)$$

Quantum robot with Levy noise.

$$d\psi(t) = \left[-iH\,dt + id\sum_{\alpha=1}^{N(t)} Y_\alpha V_\alpha\right]\psi(t)$$

where $V_1, V_2, \ldots$ are operator and $Y_1, Y_2, \ldots$ are iid r.v's. $N(t)$ is a Poisson.

Thus,

$$d\psi(t) = \left[-iHdt + iY_{N(t)+1}V_{N(t)+1}dN(t)\right]\psi(t)$$

For $d\langle\psi,\psi\rangle = 0$, we require

$$i(H^*-H)dt + Y_{N(t)+1}^2 V_{N(t)+1}^2\,dN(t) = 0$$

$$+\,2Y_{N(t)+1}V_{N(t)+1}\,dN(t) = 0$$

So $$H^* = H,\ Y_{N+1}^2 V_{N+1}^2 + 2Y_{N+1}V_{N+1} = 0$$

This cannot be satisfied in general.

Suppose $V_N^{\prime s}$ are orthogonal projection. Then $V_N^2 = V_N$ and the above condition for unitainty reduces to

$$Y_{N+1}^2 + 2Y_{N+1} = 0$$

Let $Y_{N+1} = 2\xi_{N+1}$ where ξ_{N+1} are Bernoulli r.v's, assuming values 0 & – 1.

Then the unitarity condition can be satisfied.

So

$$i(H^*-H)dt + iY_{N(t)+1}dN(t)(V_{N(t)+1} - V_{N(t)+1}^*)$$

$$+\,Y_{N(t)+1}^2\,dN(t)V_{N(t)+1}^*V_{N(t)+1} = 0$$

Thus, $H^* = H,$

$$iY_N(V_N - V_N^*) + Y_N^2 V_N^* V_N = 0\ \ \forall N \geq 1.$$

Let $Y_N, N \geq 1$ be iid Bernoulli r.v's. with $p\{Y_N = 1\} = p, p\{Y_N = 0\} = q = 1 - p$. The we require

$$H^* = H \text{ and } i(V_N - V_N^*) + V_N^* V_N = 0 \; \forall N \geq 1.$$

Assuming this, unitarity is of the stochastic evolution is guaranteed.

❖❖❖❖❖

[190] Robot in a box approximation:

$$H_0 = \frac{p_1^2}{2\lambda_1} + \frac{p_2^2}{2\lambda_2} + \delta V(q_1, q_2, p_1, p_2)$$

$$= H_{00} + \delta . V$$

Consider the 1-D Hamiltonian

$\frac{p_1^2}{2\lambda_1}$ with boundary consider $\psi(q_1) = \psi(q_1 + 2\pi)$.

$$-\frac{1}{2\lambda_1}\frac{d^2\psi(q_1)}{dq_1^2} = E\psi(q_1)$$

$$\Rightarrow \qquad \psi(q_1) = A_1 \cos\left(\sqrt{2\lambda_1 E}\, q_1\right) + A_2 \sin\left(\sqrt{2\lambda_1 E}\, q_1\right)$$

$$\psi(q_1 + 2\pi) = \psi(q_1) \Rightarrow \sqrt{2\lambda_1 E} = n, n = 0, 1, 2 \ldots$$

or
$$E = \frac{n^2}{2\lambda_1}, n = 0, 1, 2, \ldots$$

So eigenvalue of $H_{00} = \frac{p_1^2}{2\lambda_1} + \frac{p_2^2}{2\lambda_2}$ are (unnormalised)

$$u_n(q_1, q_2) = (a_1 \cos(n_1 q_1) + a_2 \sin(n_1 q_1))$$
$$(b_1 \cos(n_2 q_2) + b_2 \sin(n_2 q_2))$$

$n_1, n_2 = 0, 1, 2 \ldots$ with energy eigenvalues

$$E_n = \frac{n_1^2}{2\lambda_1} + \frac{n_2^2}{2\lambda_2}, n = (n_1, n_2) \in \{0, 1, 2, \ldots\}^2$$

Calculate the shift in the energy levels
4 fold degeneracy of $(n_1, n_2)^{th}$ level:

$$|n_1, n_2\rangle \in \{\cos(n_1 q_1)\cos(n_2 q_2), \cos(n_1 q_1)\sin(n_2 q_2)$$

$$\sin(n_1 q_1)\cos(n_2 q_2), \sin(n_1 q_1)\sin(n_2 q_2)\}$$

Robots is Levy noise (continuation)

$$d|\psi(t)\rangle = (-iHdt + Y_{N(t)+1} V_{N(t)+1}\, dN(t))|\psi(t)\rangle.$$

$$i(V_N - V_N^*) + V_N^* V_N = 0. \ \forall N \geq 1.$$

Let $$|\psi(t)\rangle = \left|\psi^{(0)}(t)\right\rangle + \left|\psi^{(1)}(t)\right\rangle + \left|\psi^{(2)}(t)\right\rangle + \ldots$$

where $\left|\psi^{(k)}(t)\right\rangle$ is $O(\|V\|^k)$. Then

$$\frac{d}{dt}\left|\psi^{(0)}(t)\right\rangle = -iH\left|\psi^{(0)}(t)\right\rangle,$$

$$d\left|\psi^{(k+1)}(t)\right\rangle = -iH\left|\psi^{(k+1)}(t)\right\rangle dt + Y_{N(t)+1} V_{N(t)+1}\left|\psi^{(k)}(t)\right\rangle dN(t)$$

so $$\left|\psi^{(0)}(t)\right\rangle = \exp(-itH)|\psi(0)\rangle = U_0(t)\left|\psi_{(0)}\right\rangle,$$

$$\left|\psi^{(k+1)}(t)\right\rangle = \int_0^t U_0(t-\tau) Y_{N(\tau)+1} V_{N(\tau)+1}\left|\psi^{(k)}(\tau)\right\rangle dN(\tau)$$

$$= \int Y_{N(\tau_k)+1} Y_{N(\tau_{k-1})+1} \ldots Y_{N(\tau_0)+1}$$

$$0 < \tau_0 < \tau_1 < \ldots < \tau_k < t$$

$$U_0(t-\tau_k) V_{N(\tau_k)+1} U_0(t-\tau_{k-1}) V_{N(\tau_{k-1})+1} \cdots$$

$$U_0(t-\tau_0) V_{N(\tau_0)+1} U_0(\tau_0)$$

$$dN(\tau_k)\, dN(\tau_{k-1}) \ldots dN(\tau_0))\,|\psi(0)\rangle$$

❖❖❖❖❖

[191] Quantum relative entropy evolution in the Sudarshan-Lindblad picture. Same Hamiltonian with different noise processes. How does the relative entropy evolve.

$$\frac{d\rho_t}{dt} = -i[H_0, \rho_0] - \frac{\varepsilon}{2}(L^* L\rho_t + \rho_t L^* L - 2L\rho_t L^*)$$

$$\frac{d\sigma_t}{dt} = -i[H_0, \sigma_t] - \frac{\varepsilon}{2}(M^* M\sigma_t + \sigma_t M^* M - 2M\sigma_t M^*)$$

$$S_t = S(\rho_t | \sigma_t) = Tr(\rho_t \log \rho_t) - Tr(\rho_t \log \sigma_t)$$

$$\frac{d}{dt} Tr(\rho_t \log \rho_t) = Tr\left(\frac{d\rho_t}{dt} \log \rho_t\right)$$

$$= -i\,Tr([H_0,\rho_t]\log\rho_t) - \frac{\varepsilon}{2}Tr(\theta_L(\rho_t)\log\rho_t)$$

$$\frac{d}{dt}Tr(\rho_t\log\sigma_t) = -i\,Tr([H_0,\rho_t]\log\sigma_t) - \frac{\varepsilon}{2}Tr(\theta_L(\rho_t)\log\sigma_t) + Tr\left(\rho_t\frac{d}{dt}\log\sigma_t\right)$$

$$\log\sigma_t = z_t\,\sigma_t = e^{z_t}$$

$$\frac{d\sigma_t}{dt} = \sigma_t\frac{(I-\exp(ad\,z_t))}{(ad\,z_t)}\frac{(dz_t)}{dt}$$

$$\frac{d\log\sigma_t}{dt} = \frac{ad\,z_t}{1-\exp(ad\,z_t)}\left(\sigma_t^{-1}\frac{d\sigma_t}{dt}\right)$$

$$\frac{ad\,z}{1-\exp(ad\,z)} = \sum_{n=0}^{\infty}C_n(ad\,z)^n \quad \frac{ad\,z}{-ad\,z-\frac{(ad\,z)^2}{2}-\cdot-\cdot}$$

$$\frac{d\log\sigma_t}{dt} = \sum_{n=0}^{\infty}C_n(ad\log\sigma_t)^n\left(\sigma_t^{-1}\frac{d\sigma_t}{dt}\right)$$

$$Tr\left(\rho_t\frac{d\log\sigma_t}{dt}\right) = C_0\,Tr\left(\rho_t\sigma_t^{-1}\frac{d\sigma_t}{dt}\right)$$

$$+\sum_{n\geq 1}C_n\,Tr\left(\rho_t(ad\log\sigma_t)^n\left(\sigma_t^{-1}\frac{d\sigma_t}{dt}\right)\right)$$

$$(ad\log\sigma_t)\left(\sigma_t^{-1}\frac{d\sigma_t}{dt}\right) = \sigma_t^{-1}\,ad\log\sigma_t\left(\frac{d\sigma_t}{dt}\right)$$

$$(ad\log\sigma_t)^n\left(\sigma_t^{-1}\frac{d\sigma_t}{dt}\right) = \sigma_t^{-1}(ad\log\sigma_t)^n\left(\frac{d\sigma_t}{dt}\right)$$

$$Tr\left(\rho_t(ad\log\sigma_t)^n\left(\sigma_t^{-1}\frac{d\sigma_t}{dt}\right)\right) = Tr\left(\rho_t\sigma_t^{-1}(ad\log\sigma_t)^n\left(\frac{d\sigma_t}{dt}\right)\right)$$

$$= -i\,Tr\left(\rho_t\sigma_t^{-1}(ad\log\sigma_t)^n[H_0,\sigma_t]\right)$$

$$-\frac{\varepsilon}{2}Tr\left(\rho_t\sigma_t^{-1}(ad\log\sigma_t)^n\,\theta_M(\sigma_t)\right)$$

$$\frac{d}{dt}Tr(\rho_t\log\sigma_t) = -i\,Tr([H_0,\rho_t]\log\sigma_t)$$

$$-\frac{\varepsilon}{2}Tr(\theta_L(\rho_t)\log\sigma_t)$$

$$+\sum_{n\geq 1} C_n Tr\left(\rho_t(ad\log\sigma_t)^n\left(\sigma_t^{-1}\left(-i[H_0,\sigma_t]-\frac{\varepsilon}{2}\theta_M(\sigma_t)\right)\right)\right)$$

$$\rho_t = \rho_t^{(0)}+\varepsilon\,\rho_t^{(1)}+O(\varepsilon^2)$$

$$\sigma_t = \sigma_t^{(0)}+\varepsilon\,\sigma_t^{(1)}+O(\varepsilon^2)$$

$$i\frac{d\rho_t^{(0)}}{dt} = [H_0,\rho_t^{(0)}]$$

$$i\frac{d\sigma_t^{(0)}}{dt} = [H_0,\sigma_t^{(0)}]$$

Assume $\rho_0 = \sigma_0$.

Thus $\rho_t^{(0)} = U_0(t)\rho_0 U_0^*(t)$

$$\sigma_t^{(0)} = U_0(t)\rho_0\,U_0^*(t) = \rho_t^{(0)}$$

where $U_0(t) = \exp(-itH_0)$.

$$i\frac{d\rho_t^{(1)}}{dt} = [H_0,\rho_t^{(1)}]-\frac{1}{2}\theta_L\left(\rho_t^{(0)}\right)$$

$$i\frac{d\sigma_t^{(1)}}{dt} = [H_0,\sigma_t^{(1)}]-\frac{1}{2}\theta_M\left(\rho_t^{(0)}\right)$$

$$\rho_t^{(1)} = \frac{i}{2}\int_0^t U_0^*(t-\tau)\theta_L\left(\rho_\tau^{(0)}\right)U_0^*(t-\tau)\,d\tau$$

$$\sigma_t^{-(1)} = \frac{i}{2}\int_0^t U_0(t-\tau)\theta_M(\rho_\tau^{(0)})U_0^*(t-\tau)\,d\tau$$

$$\log(A+\varepsilon B) = Z.A+\varepsilon B = e^z = e^{Z_t\,\varepsilon Z_1}+O(\varepsilon^2)$$

$$= \varepsilon^{Z_0}\frac{(I-\exp(-ad\,Z_0))}{(ad\,Z_0)}(Z_1)+e^{Z_0}$$

$$e^{Z_0} = A,\ A\frac{(I-\exp(-ad\,Z_0))}{ad\,Z_0}(Z_1) = B$$

$$\frac{A(I-ad(A^{-1}))}{ad\log A}(Z_1) = B$$

$$ad\log A = \log Ad(A)$$

$$Ad(A) = e^{ad \log A}$$

$$Z_1 = \frac{ad \log A}{I - Ad(A^{-1})}(A^{-1}B)$$

$$= \sum_{n=0}^{\infty} C_n (ad \log A)^n (A^{-1} B)$$

$$= \sum_{n=0}^{\infty} C_n A^{-1} (ad \log A)^n (B)$$

$$= \sum_{n=0}^{\infty} C_n A^{-1} (\log Ad(A))^n (B)$$

$$\frac{d}{dt} Tr(\rho_t \log \rho_t) - \frac{d}{dt} Tr(\rho_t \log \sigma_t)$$

$$Tr\left(\frac{d\rho_t}{dt} \log \rho_t\right) - \frac{d}{dt} Tr(\rho_t \log \sigma_t)$$

$$= -iTr([H_0, \rho_t] \log \rho_t) - \frac{\varepsilon}{2} Tr(\theta_L(\rho_t) \log \rho_t) - \frac{d}{dt} Tr(\rho_t \log \sigma_t)$$

$$= -\frac{\varepsilon}{2} Tr\left(\theta_L\left(\rho_t^{(0)} \log \rho_t^{(0)}\right)\right)$$

$$+ i\varepsilon\, Tr([H_0, \rho_t^{(1)}] \log \rho_t^{(0)}) + i\varepsilon\, Tr([H_0, \rho_t^{(0)}])$$

$$\sum_{n \geq 0} C_n \rho_t^{(0)^{-1}} ((\log Ad(\rho_t^{(0)}))^n (\sigma_t^{(1)}))$$

$$+ \frac{\varepsilon}{2} Tr(\theta_L(\rho_t^{(0)}) \log \rho_t^{(0)})$$

$$+ i\varepsilon \sum_{n \geq 1} C_n [Tr \rho_t (ad \log \sigma_t)^n \sigma_t^{-1} [H_0, \sigma_t]]_1$$

$$+ \frac{\varepsilon}{2} \sum_{n \geq 1} C_n\, Tr(\rho_t^{(0)} (ad \log \rho_t^{(0)})^n \theta_M (\rho_t^{(0)})) + O(\varepsilon^2)$$

Here $[X]_1$ denotes two coefficient of $X_1\varepsilon$ in the expansion

$$X = \sum_{m=0}^{\infty} X_m \varepsilon^m$$

$$\frac{d\rho_t}{dt} = -i[H_0, \rho] - \frac{\varepsilon}{2} \theta_L(\rho_t)$$

$$\frac{d\sigma_t}{dt} = -i[H_0, \sigma_t] - \frac{\varepsilon}{2} \theta_M(\rho_t \sigma_t)$$

$$\rho_0 = \sigma_0.$$

$$\rho_t = \rho_t^{(0)} + \varepsilon\,\rho_t^{(1)}, \sigma_t = \rho_t^{(0)} + \varepsilon\,\sigma_t^{(1)}$$

$$S(\rho_t \,|\, \sigma_t) = Tr(\rho_t \log \rho_t - \rho_t \log \sigma_t)$$

$$\log(A + \varepsilon B) = Z_0 + \varepsilon Z_1 \quad \exp(Z_0) = A.$$

$$A + \varepsilon B = \exp(Z_0 + \varepsilon Z_1) = \varepsilon \exp(Z_0) g(adZ_0)(Z_1)$$

$$Z_1 = g(ad \log A)^{-1}(A^{-1}B)$$

$$= g(\log Ad(A))^{-1}(A^{-1}B)$$

$$\log \rho_t = \log(\rho_t^{(0)} + \varepsilon\,\rho_t^{(1)})$$

$$= \log \rho_t^{(0)} + \varepsilon\, g(\log Ad(\rho_t^{(0)}))^{-1}\left(\rho_t^{(0)^{-1}} \rho_t^{(1)}\right)$$

$$\log \sigma_t = \log \rho_t^{(0)} + \varepsilon\, g(\log Ad(\rho_t^{(0)}))^{-1}(\rho_t^{(0)^{-1}} \sigma_t^{(1)})$$

(with neglect of $O(\varepsilon^2)$)

$$Tr(\rho_t \log \rho_t - \rho_t \log \sigma_t)$$

$$= \varepsilon\left\{Tr\left(\rho_t^{(0)} g\left(\log Ad\,\rho_t^{(0)}\right)^{-1}\right)\left(\rho_t^{(0)^{-1}} \rho_t^{(1)}\right)\right.$$

$$\left. -Tr\left(\rho_t^{(0)} g\left(\log Ad\left(\rho_t^{(0)}\right)\right)^{-1}\left(\rho_t^{(0)^{-1}} \sigma_t^{(1)}\right)\right)\right\}$$

$$S(\rho_t|\sigma_t) = \varepsilon \cdot Tr\left\{\rho_t^{(0)} g\left(\log Ad\,\rho_t^{(0)}\right)^{-1} \rho_t^{(0)}\left(\rho_t^{(1)} - \sigma_t^{(1)}\right)\right\} + O(\varepsilon^2)$$

$$\rho_t^{(0)} = U_0(t)\rho(0)U_0^*(t), U_0(t) = \exp(-itH_0)$$

$$Ad\left(\rho_t^{(0)}\right) = Ad(U_0(t))\,Ad(\rho(0))\,Ad(U_0(t))^{-1}$$

$$g(\log Ad\,\rho_t^{(0)})^{-1} = Ad(U_0(t))\,g(Ad(\rho(0)))^{-1}\,Ad(U_0(t))^{-1}$$

$$\rho_t^{(0)} g(\log Ad\,\rho_t^{(0)})^{-1} \rho_t^{(0)^{-1}} = Ad(\rho_t^{(0)})\,g(\log Ad\,\rho_t^{(0)})^{-1}$$

$$= Ad\,U_0(t) \cdot Ad\,\rho(0) \cdot \cancel{Ad U_0(t)}\ \cancel{Ad(U_0(t))^{-1}}\ g(Ad\,\rho(0))^{-1}\,AdU_0(t)^{-1}$$

$$= AdU_0(t)\,f(Ad\,\rho(0))\,AdU_0(t)^{-1}$$

where $f(x) = x\,g(x)^{-1}$

So,

$$S(\rho_t|\sigma_t) = \varepsilon\cdot Tr\left\{Ad\,U_0(t)\,f(Ad\,\rho(0))\,AdU_0(t)^{-1}\left(\rho_t^{(1)}-\sigma_t^{(1)}\right)\right\}+O(\varepsilon^2)$$

$$\rho_t^{(1)}-\sigma_t^{(1)} = -\frac{1}{2}\int_0^t Ad\,(U_0(t-\tau))(\theta_L-\theta_M)\left(\rho_\tau^{(0)}\right)d\tau$$

So,

$$S(\rho_t|\sigma_t) = -\frac{\varepsilon}{2}\int_0^t Tr\left\{Ad\,U_0(t)\,f(Ad\,\rho(0))\,Ad(U_0(-\tau))(\theta_L-\theta_M)\left(\rho_\tau^{(0)}\right)\right\}d\tau$$

$$= -\frac{\varepsilon}{2}Tr\left\{AdU_0(t)\,f(Ad\,\rho(0))\int_0^t Ad(U_0(-\tau))((\theta_L-\theta_M)(\rho_\tau^{(0)}))\,d\tau\right\}$$

$$= -\frac{\varepsilon}{2}Tr\left\{Ad\,U_0(t)\,f(Ad\,\rho(0))\int_0^t (Ad\,U_0(-\tau))(\theta_L-\theta_M)(Ad\,U_0(\tau))(\rho(0))\,d\tau\right\}$$

$$+\,O(\varepsilon^2)$$

Note that if θ operates linearly on density matrices, then

$$Ad(U_0(-\tau))\cdot\theta\cdot Ad\,U_0(\tau)(\rho)$$

$$= Ad(U_0(-\tau))\cdot\theta\cdot(U_0(\tau)\rho U_0(-\tau))$$

$$= U_0(-\tau)\theta(U_0(\tau)\rho U_0(-\tau))U_0(\tau)$$

Perturbation with random inputs field and random nonlinear perturbing $\mathcal{L}(\varphi)+\varepsilon\,N(\varphi) = S$ force. $\mathcal{L}$ = Linear operator invertible

$\varphi = \varphi_0+\varepsilon\,\varphi_1+O(\varepsilon^2)$. $\mathcal{N}$ = Nonlinear operator.

$$\varphi_0 = \mathcal{L}^{-1}(S),$$

$$\mathcal{L}(\varphi_1) = -\mathcal{N}(\varphi_0)\,S \simeq \text{ random field.}$$

$$\varphi_1 = -\mathcal{L}^{-1}\mathcal{N}(\varphi_0).$$

Random vector in the nonlinearity:

$$\mathcal{L}(\varphi)+\varepsilon\sum_{k=1}^{p}\theta_k\mathcal{N}_k(\varphi) = S$$

S, $\simeq$ random field, $(\theta_k)_{k=1}^{p}\varepsilon\,\mathbb{R}^p$ = random vector.

$$\varphi = \varphi_0+\varepsilon\,\varphi_1+O(\varepsilon^2).$$

$$\mathcal{L}(\varphi_0) = S,\ \mathcal{L}(\varphi_1)+\sum_{k=1}^{p}\theta_k\mathcal{N}_k(\varphi_0) = 0.$$

$$\varphi_1 = -\sum_{k=1}^{p}\theta_k\mathcal{L}^{-1}\mathcal{N}_k(\varphi_0)$$

$$\mathbb{E}\exp(\langle\lambda, S\rangle) = M_S(\lambda)\ \lambda \simeq \text{a non random field.}$$

→ Moment generating functional of *S*.

$$M_{\varphi_0}(\lambda) = \mathbb{E}\exp\left(\left\langle\lambda, \mathcal{L}^{-1}S\right\rangle\right)$$

$$= \mathbb{E}\exp\left(\left\langle\mathcal{L}^{-*}\lambda, S\right\rangle\right)$$

$$= \mathbb{E}\exp\left(\left\langle\mathcal{L}^{*-1}\lambda, S\right\rangle\right)$$

$$= M_S(\mathcal{L}^{*-1}\lambda)$$

$$\mathbb{E}\left[\exp(\langle\lambda, \varphi\rangle)\middle|S\right] = \mathbb{E}\left[\exp\left(-\sum_{k=1}^{p}\theta_k\left\langle\lambda, \mathcal{L}^{-1}\mathcal{N}_k(\varphi_0)\right\rangle\right)\middle|S\right]$$

$$= \mathbb{E}\left[\left[M_\theta\left(-\left\langle\lambda, \mathcal{L}^{-1}\mathcal{N}_k(\varphi_0)\right\rangle, k=1,2,..,p\right)\right]\middle|S\right]$$

$$\mathbb{E}\exp\left(\langle\lambda, \varphi_1\rangle\right) = M_{\varphi_1}(\lambda) = \mathbb{E}\left[M_\theta\left(-\left\langle\lambda, \mathcal{L}^{-1}\mathcal{N}_k(\varphi_0)\right\rangle, k=1,2,..,p\right)\right]$$

$$= \mathbb{E}M_\theta\left(-\left\{\left\langle\mathcal{L}^{*-1}\lambda, \mathcal{N}_k(\varphi_0)\right\rangle\right\}_k\right)$$

$$= \mathbb{E}M_\theta\left(-\left\{\left\langle\mathcal{L}^{*-1}\lambda, \mathcal{N}_k\,\pounds^{-1}(S)\right\rangle\right\}_k\right)$$

$$= \int M_\theta\left(-\left\{\left\langle\mathcal{L}^{*-1}\lambda, \mathcal{N}_k\,\mathcal{L}^{-1}(S)\right\rangle\right\}_k\right)dP(S)$$

Higher order perturbation is:

$$\varphi = \varphi_0+\varepsilon\,\varphi_1+\varepsilon^2\varphi_2+\ldots+\varepsilon^n\varphi_n+\ldots$$

$$\mathcal{N}(\varphi)-\mathcal{N}(\varphi_0) = \sum_{k=1}^{\infty}\mathcal{N}^{(k)}(\varphi_0)(\varepsilon\varphi_1+\varepsilon^2\varphi_2+\ldots)\otimes k/\underline{\lfloor k}$$

$$= \sum_{k=1}^{\infty}\varepsilon^n\sum_{k\geq 1}\frac{\mathcal{N}^{(k)}(\varphi_0)}{\underline{\lfloor k}}\sum S\left\{\varphi_1^{m_1}\otimes\varphi_2^{m_2}\otimes\ldots\varphi_k^{m_k}\right\}$$

$$m_1+2m_2+\ldots+km_k = n_j\ m_j\geq 1\forall\ j$$

S ≃ Symmetrizer.

So purturbation theory gives

$$= \mathcal{L}(\varphi_n)+\sum_{k\geq 1}\frac{\mathcal{N}^{(k)}(\varphi_0)}{\lfloor k}\sum S\left\{\varphi_1^{\ m_1}\otimes\ldots\otimes\varphi_k^{\ m_k}\right\} m_1+2m_2+\ldots+km_k \ = n-1$$

$$\varphi_n = -\sum_{k\geq 1}\frac{1}{\lfloor k}\mathcal{L}^{-1}\sum\mathcal{N}^{(k)}(\varphi_0)S\left\{\varphi_1^{\ m_1}\otimes\ldots\otimes\varphi_n^{\ m_n}\right\}m_1+2m_2+\ldots+km_k = n-1$$

❖❖❖❖❖

[192] Evolution of quantum relative entropy (Exercise)

$$dU(t) = -i(H_0\,dt+P_t dt+L_{1t}dA_t+L_{2t}dA_t^{+}+S_t d\Lambda_t)U(t)$$

$$dA_t\,dA_t^{+} = dt, d\Lambda_t\,dA_t^{+} = dA_t^{+}, (d\Lambda_t)^2 = d\Lambda_t$$

For unitarity of $U(t), d(U^*U) = 0\ (-)$

$$i(\rho_t^* - \rho_t)+L_{2t}^* L_{2t}^* = 0,$$

$$S_t^* S_t + i(S_t^* - S_t) = 0,$$

$$i(L_{1t}^{+} - L_{2t}) + S_t^* L_{2t} = 0$$

Thus, we may take $P_t = -\frac{i}{2}L_{2t}^* L_{2t}$,

$$L_{1t}^{+} = L_{2t} - iS_t^* L_{2t} = (I - iS_t^*)L_{2t}$$

Let $Q_t = I - iS_t^*$. Then the condition on S_t returns

$$(I-Q_t)(I-Q_t^*)+Q_t+Q_t^* = 0$$

or $\quad Q_tQ_t^* = I,\ L_{1t}^* = Q_tL_{2t}$

i.e. Q_t is a unitary operator.

So $$dU(t) = \left(-iH_0dt-\frac{1}{2}L_t^*L_t - i\,L_t^*Q_t^*dA_t - iL_t dA_t^* - i(I-Q_t^*)d\Lambda_t\right)U(t)$$

Compute $$\frac{d}{dt}STr_2\left(U(t)\left(\rho_s(0)\otimes|\varphi(u)\rangle\langle\varphi(u)|U^*(t)\right)\right)$$

Remaining Q_t on Q_t^* and L_t^* as iL_t, we get

$Q_t^* Q_t = I$, L_t arbitrary and

$$dU(t) = \left[-iH_0 dt - \frac{1}{2} L_t^* L_t \, dt + L_t Q_t dA_t - L_t^* dA_t^* - i(I - Q_t) d\Lambda_t \right] U(t)$$

❖❖❖❖❖

[193] Particle in a classical and quantum noise field.

$$H_0(t) = \frac{(\underline{p} + eA(t, \underline{q}))^2}{2m} - e\Phi(t, \underline{q}) + \Phi_0(\underline{q})$$

$$\approx \frac{p^2}{2m} + \frac{e}{2m}((\underline{p}, \underline{A}) + (\underline{A}, \underline{p})) - e\Phi(t, \underline{q}) + \Phi_0(q)$$

$\underline{A}(t, \underline{q}), \Phi(t, \underline{q})$ are classical random fields.

$$\underline{A}(t, \underline{q}) = \sum_{\sigma=1}^{N} w_\alpha(t) \underline{\psi}_\sigma(\underline{q})$$

$$\Phi(t, \underline{q}) = -\int_0^t \text{div}\, A\, dt$$

$$= -\sum_{\alpha=1}^{N} \left(\int_0^t w_\alpha(t')\, dt' \right) \text{div}\, \underline{\psi}_\alpha(\underline{q})$$

$\{w_\alpha(t)\}_{\alpha=1}^{N} \simeq$ zero mean Gaussian noise processes.

Let $|n\rangle, E_n, n = 0, 1, 2, \ldots$ be the eigenfunction and eigenvalue of $H_{00} = \underline{p}^2 / 2m + \Phi_0(\underline{q})$

Define

$$V_t = \frac{\tau}{2m}((\underline{p}, \underline{A}_t) + (\underline{A}_t, \underline{p})) - e\Phi_t$$

$$= \sum_\alpha w_\alpha(t) \frac{e}{2m} \left((\underline{p}, \underline{\psi}_\alpha(\underline{q})) + (\underline{\psi}_\alpha(\underline{q}), \underline{p}) \right)$$

$$- e \sum_\alpha \left(\int_0^t w_\alpha(t')\, dt' \right) (\text{div}\, \underline{\psi}_\alpha(\underline{q}))$$

Let

$$V_\alpha(\underline{q}, \underline{p}) = \frac{e}{2m}((\underline{p}, \underline{\psi}_\alpha(\underline{q})) + (\underline{\psi}_\alpha(\underline{q}), \underline{p}))$$

$$W_\alpha(\underline{q}) = -\tau \operatorname{div} \underline{\psi}_\alpha(\underline{q})$$

and $$v_\alpha(t) = \int_0^t w_\alpha(t')dt'$$

Then, $$V_t = \sum_{\alpha=1}^{N} (V_\alpha(\underline{q}, \underline{p}) w_\alpha(t) + W_\alpha(\underline{q}) v_\alpha(t))$$

where V_α, W_α are now non-random system operator and $w_\alpha(t)$, $v_\alpha(t)$ are jointly zero mean Gaussian random processes.

We absorb the term $v_\alpha(t)$ and $w_\alpha(q)$ into $w_\alpha(t)$ and $V_\alpha(\underline{q}, \underline{p})$.

So finally, we get

$$V_t = \sum_{\alpha=1}^{N} V_\alpha(\underline{q}, \underline{p}) w_\alpha(t)$$

We assume that classical noise terms are $O(\varepsilon^2)$ and quantum noise terms are $O(\varepsilon)$. Then, that Schrödinger unitary evolution becomes

$$dU(t) = \Big[-iH_{00}\,dt - i\varepsilon^2 \sum_\alpha V_\alpha w_\alpha(t)\,dt - \frac{1}{2}\varepsilon^2 L_t^* L_t dt + \varepsilon L_t \exp(i\varepsilon Z_t) dA_t - \varepsilon L_t^* dA_t^* - i(I - \exp(i\varepsilon Z_t)) d\Lambda_t\Big] U(t)$$

where $Z_t^* = Z_t$ is a system operator.

Writing $\exp(i\varepsilon Z_t) \approx I + i\varepsilon Z_t - \frac{\varepsilon^2}{2} Z_t^2$ gives

$$dU(t) = \Big[-iH_{00}\,dt + \varepsilon(L_t\,dA_t - L_t^* dA_t) + \varepsilon^2 \Big\{-i\sum_\alpha V_\alpha w_\alpha(t) - \frac{1}{2} L_t^* L_t + iL_t Z_t dA_t + \frac{i}{2} Z_t^2 d\Lambda_t\Big\}\Big] U(t)$$

❖❖❖❖❖

[194] Galactine evolution in gtr.

$$n\sum_{\alpha=1}^{p} a_\alpha^n \geq \left(\sum_{\alpha=1}^{p} a_\alpha\right)^n \qquad \frac{a_\alpha^n}{\sum a_\beta^n}$$

$$\sum_{\alpha} n\left(\frac{a\alpha}{\sum a\alpha}\right)^n \geq 1 \; \sum_{\alpha} n x_{\alpha}^n \geq 1 \; \sum x\alpha = 1$$

$$\delta\Gamma^{\alpha}_{\mu\nu,\alpha} = \frac{1}{2}\delta\left(g^{\alpha\beta}g_{\alpha\beta,\mu}\right)_{,\nu}$$

$$\begin{aligned}
\delta\Gamma^{\alpha}_{\mu\nu,\alpha} &= \delta\left(g^{\alpha\beta}\Gamma_{\beta\mu\nu}\right)_{,\alpha} = \frac{1}{2}\delta\left(g^{\alpha\beta}\left(g_{\beta\mu,\nu}+g_{\beta\nu,\mu}-g_{\mu\nu,\beta}\right)\right)_{,\alpha} \\
&= \frac{1}{2}\Big(-g^{\alpha\rho}g^{\beta\sigma}\delta g_{\rho\sigma}\left(g_{\beta\mu,\nu}+g_{\beta\nu,\mu}-g_{\mu\nu,\beta}\right) \\
&\quad + g^{\alpha\beta}\left(\delta g_{\beta\mu,\nu}+\delta g_{\beta\nu,\mu}-\delta g_{\mu\nu,\beta}\right)\Big)_{,\alpha} \\
&= -\frac{1}{2}\left(g^{\alpha\rho}g^{\beta\sigma}\Gamma_{\beta\mu\nu}\right)_{,\nu}\delta g_{\rho\sigma} - \frac{1}{2}g^{\alpha\rho}g^{\beta\sigma}\Gamma_{\beta\mu\nu}\,\delta g_{\rho\sigma,\alpha} \\
&\quad + \frac{1}{2}g^{\alpha\beta}_{,\alpha}\left(\delta g_{\beta\mu,\nu}+\delta g_{\beta\nu,\mu}-\delta g_{\mu\nu,\beta}\right) \\
&\quad + \frac{1}{2}g^{\alpha\beta}\left(\delta g_{\beta\mu,\nu\alpha}+\delta g_{\beta\nu,\mu\alpha}-\delta g_{\mu\nu,\alpha\beta}\right)
\end{aligned}$$

$$\begin{aligned}
&= -\frac{1}{2}\left(g^{\alpha\rho}\Gamma^{\sigma}_{\mu\nu}\right)_{,\alpha}\delta g_{\rho\sigma} - \frac{1}{2}g^{\alpha\rho}\Gamma^{\sigma}_{\mu\nu}\delta g_{\rho\sigma,\alpha} \\
&\quad + \frac{1}{2}g^{\alpha\beta}_{,\alpha}\left(\delta g_{\beta\mu,\nu}+\delta g_{\beta\nu,\mu}-\delta g_{\mu\nu,\beta}\right) \\
&\quad + \frac{1}{2}g^{\alpha\beta}\left(\delta g_{\beta\mu,\nu\alpha}+\delta g_{\beta\nu,\mu\alpha}-\delta g_{\mu\nu,\alpha\beta}\right) \\
&= -\frac{1}{2}\Big[\left(g^{11}\Gamma^{1}_{\mu\nu}\right)_{,1}\delta g_{11} + \left(g^{22}\Gamma^{2}_{\mu\nu}\right)_{,2}\delta g_{22} \\
&\quad + \left(g^{33}\Gamma^{3}_{\mu\nu}\right)_{,3}\delta g_{33} + \Gamma^{0}_{\mu\nu,0}\,\delta g_{00}\Big] \\
&\quad - \frac{1}{2}\left[g^{11}\Gamma^{1}_{\mu\nu}\,\delta g_{11,1} + g^{22}\Gamma^{2}_{\mu\nu}\,\delta g_{22,2} + g^{33}\Gamma^{3}_{\mu\nu}\delta g_{33,3} + \Gamma^{0}_{\mu\nu}\delta g_{00,0}\right] \\
&\quad + \frac{1}{2}\sum_{r=1}^{3} g^{\nu\nu}_{,r}\left(\delta g_{\nu\mu,\nu}+\delta g_{\nu\nu,\mu}-\delta g_{\mu\nu,\nu}\right)
\end{aligned}$$

$$\begin{aligned}
\delta\Gamma^{\alpha}_{\mu\alpha,\nu} &= \frac{1}{2}\delta\left(g^{\alpha\beta}g_{\alpha\beta,\mu}\right)_{,\nu} \\
&= \frac{1}{2}\left(-g^{\alpha\rho}g^{\beta\sigma}g_{\alpha\beta,\mu}\delta g_{\rho\sigma} + g^{\alpha\beta}\delta g_{\alpha\beta,\mu}\right)_{,\nu} \\
&= -\frac{1}{2}\left(g^{\alpha\rho}g^{\beta\sigma}g_{\alpha\beta,\mu}\right)_{,\nu}\delta g_{\rho\sigma} - \frac{1}{2}g^{\alpha\rho}g^{\beta\sigma}g_{\alpha\beta,\mu}\delta g_{\rho\sigma,\nu}
\end{aligned}$$

$$+ g^{\alpha\beta}_{,\nu}\,\delta g_{\alpha\beta,\mu} + g^{\alpha\beta}\,\delta g_{\alpha\beta,\mu\nu}$$

$$= \frac{1}{2}g^{\rho\sigma}_{,\nu\mu}\,\delta g_{\rho\sigma} + \frac{1}{2}g^{\rho\sigma}_{,\mu}\,\delta g_{\rho\sigma,\nu} + \frac{1}{2}g^{\rho\sigma}_{,\nu}\,\delta g_{\rho\sigma,\mu} + \frac{1}{2}g^{\rho\sigma}\,\delta g_{\rho\sigma,\mu\nu}$$

$$= \frac{1}{2}\sum_{\rho=0}^{3} g^{\rho\rho}\delta g_{\rho\rho,\mu\nu} + \frac{1}{2}\sum_{k=1}^{3} g^{kk}_{,\nu}\,\delta g_{kk,\mu}$$

$$+ \frac{1}{2}\sum_{k=1}^{3} g^{kk}_{,\mu}\,\delta g_{kk,\nu} + \frac{1}{2}\sum_{k=1}^{3} g^{kk}_{,\mu\nu}\,\delta g_{kk}$$

$$\left(\delta\Gamma^{\alpha}_{\mu\nu}\right)\Gamma^{\beta}_{\alpha\beta} = \left(\delta\Gamma^{0}_{\mu\nu}\right)\Gamma^{k}_{0k} + \left(\delta\Gamma^{1}_{\mu\nu}\right)\cdot\Gamma^{k}_{1k} + \left(\delta\Gamma^{2}_{\mu\nu}\right)\cdot\left(\Gamma^{3}_{23}\right)$$

$$\delta\Gamma^{0}_{\mu\nu} = \delta\left(g^{0\alpha}\,\Gamma_{\alpha\mu\nu}\right) = g^{0\alpha}\cdot\delta\Gamma_{\alpha\mu\nu} + (\delta g^{0\alpha})\cdot\Gamma_{\alpha\mu\nu}$$

$$= \delta\Gamma_{0\mu\nu} + \frac{1}{2}\delta g^{0\alpha}\left(g_{\alpha\mu,\nu} + g_{\alpha\nu,\mu} - g_{\mu\nu,\alpha}\right)$$

❖❖❖❖❖

[195] $d(X_t - a)^+ = \theta(X_t - a)dX_t + dL_t^a$

X_t = B.H.

So, $(X_t - a)^t - L_t^a$ = Martingale.

Let $L_t^a = \left(L_t^{a_1},..,L_t^{a_n}\right)$, $a = (a, ..., a_n)$.

Then
$$d\,f\left(L_t^a\right) = \sum_j f_{,j}\left(L_t^a\right)dL_t^{a_j}$$

$$= \sum_j f_{,j}\left(L_t^a\right)d\left(X_t - a_j\right)^t + d \text{ (Martingale).}$$

$$= d\sum_j f_{,j}\left(L_t^a\right)\left(X_t - a_j\right)^t - \sum_j (X_t - a_j)^t\, d\, f_{,j}\left(L_t^a\right)^t + d \text{ (Martingale)}$$

$$= d\sum_j f_{,j}\left(L_t^a\right)\left(X_t - a_j\right)^t - \sum_j (X_t - a_j)^t\left(\sum_k f_{,jk}\left(L_t^a\right)d\,L_t^{ak}\right) + d \text{ (Martingale)}$$

$$= d\sum_j f_{,j}\left(L_t^a\right)\left(X_t - a_j\right)^t - \sum_j (X_t - a_j)^t$$

$$\sum_k f_{,ik}(L_t^a)\,\delta(X_t - a_k)dt \times d\text{(Martingale)}$$

$$= \left(\sum_{j,k} f_{,jk}\left(L_t^{\underline{a}}\right) d(a_k - a_j)^t\, d(X_t - a_k)^t\, L_t^{ak} \right) + d \text{ (Martingale)}$$

$$= d\sum_j f_{,j}\left(L_t^{\underline{a}}\right)(X_t - a_j)^t - \sum_{j,k} f_{,jk}\left(L_t^{\underline{a}}\right)(a_k - a_j)^t\, d(X_t - a_k)^t + d$$

(**Note:** $\varphi(X_t)\,dL_t^x = \varphi(X_t)\delta(X_t - x)\,dt = \varphi(x)\delta(X_t - x)\,dt = \varphi(x)\,dL_t^x$

$= \varphi(x)\,d(X_t - x)^t + d$ (Martingale))

With the Einstien summation convention (*i.e.* over repeated induces), we they have,

$$d\,f(L_t^{\underline{a}}) = d(f_{,j}(L_t^{\underline{a}})(X_t - a_j)^t) - f_{,jk}(L_t^{\underline{a}})(a_k - a_j)^t\, d(X_t - a_k)^t + d \text{ (Martingale)}$$

$$= d(f_{,j}(L_t^{\underline{a}})(X_t - a_j)^t)$$

$$-(a_k - a_j)^t[d(f_{,jk}(L_t^{\underline{a}})(X_t - a_k)^t - f_{,jkl}(L_t^{\underline{a}})(X_t - a_k)^t\, dL_t^{\underline{al}})] + d \text{ (Martingale)}$$

$$= d(f_{,j}(L_t^{\underline{a}})(X_t - a_j)^t)$$

$$-(a_k - a_j)^t\, d(f_{,jk}(L_t^{\underline{a}})(X_t - a_k)^t)$$

$$+(a_k - a_j)^t\,(a_l - a_k)^t\, f_{,jkl}\, d(X_t - a_l)^t + d \text{ (Martingale)}$$

Combining, we get assuming $a_1 > a_2 > \cdots > a_n$,

$$f(L_t^{\underline{a}}) = \sum_{j_1} f_{,j_1}(L_t^{\underline{a}})(X_t - a_{j_1})^t - \sum_{j_1 > j_2} (a_j - a_{j_2}) f_{,j_1 j_2}(L_t^{\underline{a}})(X_t - a_{j_2})^t$$

$$+ \sum_{j_1 > j_2 > j_3} (a_{j1} - a_{j2})(a_{j2} - a_{j3}) f_{,j_1 j_2 j_3}(L_t^{\underline{a}})(X_t - a_{j_3})^t$$

$$\cdots + (-1)^{n-1}(a_1 - a_2)(a_2 - a_3)\ldots(a_{n-1} - a_n)$$

$$f_{,12\ldots,}(L_t^{\underline{a}})(X_t - a_n)^t + \text{Martingale.}$$

❖❖❖❖❖

[196] Group representation and robot configuration estimation.

Optical sources are located at points $\xi_1, \xi_2, \ldots, \xi_k$ on the first link and at $\eta_1, \eta_2, \ldots, \eta_m$ on the second link.

Here ξ_α is the distance of the αth source on the first link from the origin and η_α is the distance of the αth source on the second link from its joints with the first link.

The position of ξ_α is ($\xi_\alpha \cos\theta_1 \cos\phi_1\, \xi_\alpha \cos\theta_1 \sin\phi,\, \xi_\alpha \sin\theta_1$)

and the position of η_α on the second link is

$$\left((l_1\cos\theta_1 + \eta_\alpha\cos(\theta_1+\theta_2)\right)\cos\phi,\ (l_1\cos\theta_1$$

$$+\eta_\alpha\cos(\theta_1+\theta_2))\sin\phi,\ l_1\sin\theta + \eta_\alpha\sin(\theta_1+\theta_2)\big)$$

The rotation applied to the first link is

$$R_1 = R_Z(\phi)R_Y(\theta_1)$$

and to the second link is

$$R_2 = R_Z(\phi)R_Y(\theta_1+\theta_2)$$

and the second link is

$$R_2 = R_Z(\phi)R_Y(\theta_1+\theta_2)$$

$$\begin{aligned}
R'(t) &= \phi' R_Z(\phi)X_3R_Y(\theta_1) + \theta_1' R_1 X_2 \\
&= \phi' R_Z(\phi)R_Y(\theta_1)\exp(-ad(X_2)\theta_1)(X_3) \\
&= \phi' R_1(X_3\cos\theta_1 - X_1\sin\theta_1) + \theta_1' R_1 X_2 \\
&= R_1(\theta_1' X_2 - \phi'\sin\theta_1 X_1 + \phi'\cos\theta_1 X_3) \\
R_2'(t) &= R_2((\theta_1'+\theta_2')X_2 - \phi'\sin(\theta_1+\theta_2)X_1 + \phi'\cos(\theta_1+\theta_2)X_3)
\end{aligned}$$

The total signal field emitted by the optical lowers is

$$S(\{\xi_j\},\psi_1(R_1\underline{\xi}_1\, R_1\underline{\xi}_2,\ldots,R_1\xi_k))$$

$$+\psi_2(R_1\underline{l}_1 + R_2\underline{\eta}_1,\ R_1\underline{l}_1 + R_2\eta_2,\ldots,R_1\underline{l}_m + R_2\underline{\eta}_m)$$

$$\equiv \sum_{j=1}^{k}\psi_{ij}(R_1\underline{\xi}_j) + \sum_{j=1}^{m}\psi_{2j}(R_1\underline{l}_1 + R_2\underline{\eta}_j)$$

Suppose instead, we place a while continuous of optical sources along the lengths the first & second links.

When $R_1 = I$, $R_2 = I$, let the signal received be So $(\{\xi_j\},\{\eta_j\})$. Then

$$S(\{\xi_j\},\{\eta_j\}) = So(\{\underline{\underline{R}}_1\underline{\xi}_j\},\{\underline{\underline{R}}_2\underline{\eta}_j\}) + W(\{\xi_j\},\{\eta_j\})$$

Assume $\{\xi_j\}$, $\{\eta_j\}$ can be varied (*i.e.* the positional) Noise of the optical sources are variable.

Let $$\begin{bmatrix}\underline{\xi}_1\\ \vdots \\ \underline{\xi}_k\end{bmatrix} = \underline{\xi},\quad \begin{bmatrix}\underline{\eta}_1\\ \vdots \\ \underline{\eta}_m\end{bmatrix} = \underline{\eta}$$

Then

$$\begin{bmatrix} \underline{\underline{R}}_1 \underline{\xi}_1 \\ \vdots \\ \underline{\underline{R}}_1 \underline{\xi}_k \end{bmatrix} = (\underline{\underline{I}}_k \otimes \underline{\underline{R}}_1)\underline{\xi}$$

$$\begin{bmatrix} \underline{\underline{R}}_2 \underline{\eta}_1 \\ \vdots \\ \underline{\underline{R}}_2 \underline{\eta}_n \end{bmatrix} = (\underline{\underline{I}}_m \otimes \underline{\underline{R}}_2)\underline{\eta}$$

So, $$S(\underline{\xi}, \underline{\eta}) = S_0((\underline{\underline{I}}_k \otimes \underline{\underline{R}}_1)\underline{\xi}, (\underline{\underline{I}}_m \otimes \underline{\underline{R}}_2)\underline{\eta}) + W(\underline{\xi}, \underline{\eta})$$

For $R, S \in SO(3)$, let

$$\psi_0(R, S) = S_0((I_k \otimes R)\underline{\xi}_0, (I_m \otimes S)\underline{\eta}_0)$$

where $\underline{\xi}_0, \underline{\eta}_0$ are given and fixed. Writing

$$(I_k \otimes R_0)\underline{\xi}_0 = \underline{\xi},$$

$$(I_m \otimes S_0)\underline{\eta}_0 = \underline{\eta},$$

We get

$$S_0(\xi, \eta) = \psi_0(R_0, S_0),$$

$$\begin{aligned} S(\xi, \eta) &= \psi_0(R_1 R_0, R_2 S_0) + W(R_0, S_0) \\ &= \psi(R_0, S_0) \end{aligned}$$

By varrying $\underline{\xi}, \underline{\eta}$, we measure $\psi(R_0, S_0)$ for all $R_0, S_0 \in SO(3)$.

Let $\pi_l,\ l = 1, 2, \ldots$ be the crops of $SO(3)$.

Let $$\int (\pi_{l_1}(R_0) \otimes \pi_{l_2}(S_0)) \psi_0(R_0, S_0)\, dR_0\, dS_0$$

$$SO(3) \times SO(3) = \widehat{\psi}_0[l_1, l_2].$$

Then, also let

$$\widehat{\psi}_0[l_1, l_2] = \int (\pi_{l_1}(R_0) \otimes \pi_{l_2}(S_0)) \psi(R_0, S_0)\, dR_0\, dS_0$$

Then, we get

$$\widehat{\psi}[l_1, l_2] \approx \left[\pi_{l_1}(R_1^{-1}) \otimes \pi_{l_2}(R_2^{-1})\right] \widehat{\psi}_0[l_1, l_2]$$

So R_1, R_2 can be estimated as

$$\left(\hat{R}_1, \hat{R}_2\right) = \underset{R_1, R_2}{\arg\min} \sum_{l_1, l_2} \left\| \hat{\psi}[l_1, l_2] - \left(\pi_{l_1}^*(R_1) \otimes \pi_{l_2}^*(R_2)\right) \hat{\psi}_0(l_1, l_2) \right\|^2$$

Let $$R = R_Z(\phi) R_Y(\theta) = e^{\phi X_3} e^{\theta X_1}$$

$$= e^{i\phi L_3} e^{i\theta L_1}$$

$$\langle l, m' | \pi_l(R) | l, m \rangle = e^{im\phi} \delta_{m, m'} \left\langle l, m' \middle| e^{i\theta L_1}, d\pi_l \middle| l, m \right\rangle$$

$d\pi_l(L_1) = \sum_{r=-l}^{+l} e^{ir\theta} P_{l,r}$ Spectral rep'n O the Hamiltonian matrix $d\pi_l(L_1)$.

$$\langle l, m' | \pi_l(R) | l, m \rangle = e^{im\phi} \delta_{m,m'} \sum_{|r| \le l} e^{ir\theta} \langle l, m' | P_{l,r} | l, m \rangle$$

$\underline{P} \equiv \underline{\xi}$ = Initial position of optical source on First link.

$\theta \equiv \underline{\eta}$ = Initial position of optical source on second link.

Position of $\underline{P}$ after link t is

$$\underline{P}(t) = \underline{\underline{R}}_1(t)\underline{\xi}$$

are position of θ after time t is

$$\underline{\theta}(t) = \underline{\underline{R}}_1(t)\underline{p}_0 + R_2^{(t)} \underline{\underline{R}}_2(t)(\underline{\eta} - \underline{p}_0)$$

Let $R_1(t) = R(t), R_2(t) R_1(t) = S(t)$. Then

$$\underline{P}(t) = \underline{\underline{R}}(t)\underline{\xi}, \underline{\theta}(t) = \underline{\underline{R}}(t)\underline{p}_0 + \underline{\underline{S}}(t)(\underline{\eta} - \underline{p}_0)$$

Signal received at the origin is

$$S(t) = f_1(\underline{P}(t)) + f_2(\underline{Q}(t))$$

$$= f_1(\underline{\underline{R}}(t)\underline{\xi}) + f_2(\underline{\underline{R}}(t)\underline{p}_0 + \underline{\underline{S}}(t)(\underline{\eta} - \underline{p}_0))$$

$$= S(t, \underline{\xi}, \underline{\eta}) . \xi \in B_1,$$

$\underline{\eta} \in B_2$,

Where B_1 is initial volume occupied by the first link and B_2 is the initial volume occupied by the second link.

Let $$\xi = R_0 \xi_0 \qquad R_0 \in SO(3)$$

$$\eta - p_0 = S_0 \eta_0 \qquad S_0 \in SO(3)$$

$\xi_0 \in B_1, \eta_0 \in B_2$ are fixed. Then,

$$S(t, \xi, \eta) = S(t, R_0, S_0) = f_1(R\, R_0\, \xi_0) + f_2(R\, p_0 + SS_0\eta_0)$$

$$\int_{SO(3)} S(t, R_0, S_0)\pi_l(R_0)dR_0 = \pi_l(R^{-1})\hat{f}_1[l], l \geq 1 \tag{1}$$

since $\int_{SO(3)} \pi_l(R_0)d\,R_0 = 0,\ l \geq 1$.

From (1) therefore R can be determined (estimated). Hence the function $S_0 \to f_2(Rp_0 + SS_0\eta_0) = h(t, S_0)$ is known.

Call this function as $g_t(SS_0\,\eta_0)$

$$\int_{SO(3)} h(t, S_o)\pi_l(S_0)dS_0 = \pi_l(S^{-1})\hat{g}_t[l] \tag{2}$$

From (2) S can be estimated.

3-D Fourier analysis:

$$f_1(\underline{\xi}) = \int \hat{f}_1(\underline{k})\exp(i\underline{k}\cdot\underline{\eta})d^3k$$

$$f_2(\underline{\eta}) = \int \hat{f}_2(\underline{k})\exp(i\underline{k}\cdot\underline{\eta})d^3k$$

Let $f_1(\underline{\xi}) = A\cdot\exp(i\underline{k}_1\cdot\underline{\xi})$

$f_2(\underline{\xi}) = B\cdot\exp(i\underline{k}_2\cdot\underline{\xi})$.

Then, $S(t, \underline{\xi}, \underline{\eta}) = S(t, R_0, S_0)$

$$= A\cdot\exp(i(\underline{k}_1, RR_0\,\xi_0)) + B\cdot\exp(i(\underline{k}_2, Rp_0))\cdot\exp(i(\underline{k}_2, SS\underline{\eta}_0))$$

$$S(t, \xi, \eta) = f_1(R(t)\xi) + f_2(R(t)\,p_0 + S(t)(\eta - p_0))$$

$$f_1(\xi) = \sum_{l,m} f_1[l, m, \|\xi\|]Y_{lm}(\hat{\xi})\quad \eta - p_0 \to \eta.$$

$$f_2(R(t)\,p_0 + S(t)\eta) = \sum_{l,m} f_2[R(t)l, m\,\|\eta\|]Y_{lm}(\hat{\eta})$$

$$f_1(R(t)\xi) = \sum_{l,m} f_1[l, m, \|\underline{\xi}\|]Y_{lm}(R(t)\hat{\xi})$$

$$= \sum_{l,m,m'} f_1[l, m\,\|\underline{\xi}\|](\pi_l(R(t))^{-1})_{m'm}Y_{lm'}(\hat{\xi})$$

$$\int S(t, \xi, \eta)\overline{Y_{lm}(\hat{\xi})}d\Omega(\hat{\xi}) = \sum_{m'}[\pi_l(R^{-1}(t))]_{mm'},\ f_1[l, m', \|\xi\|], l \geq 1$$

This eqn. can be used to determine $R(t)$ and hence

$$\psi(t, \underline{\eta}) = S(t, \underline{\xi}, \underline{\eta} + p_0) - f_1(R(t)\underline{\xi})$$

$$= f_2(R(t)\underline{p}_0 + S(t)\underline{\eta})$$

is known $\forall \underline{\eta}$. Hence writing

$$f_2(R(t)p_0 + \underline{\eta}) = \sum_{l,m} f_2[t, l, m, \|\underline{\eta}\|] Y_{lm}(\eta)$$

we get

$$f_2(R(t)p + S(t)\underline{\eta}) = \sum_{l,mm'} f_2[t, l, m\|\underline{\eta}\|][\pi_l(S^{-1}(t))]_{m'm} Y_{lm'}(\hat{\eta})$$

$$\equiv \psi(t, \underline{\eta})$$

so

$$\int_{S^2} \psi(t, \underline{\eta}) \overline{Y_{lm}(\hat{\eta})} d\Omega(\hat{\eta}) = \sum_{m'} [\pi_l(S(t))^{-1}]_{mm'} f_2[t, l, m'\|\underline{\eta}\|]$$

using which $S(t)$ can be obtained.

Let $\{Y_{lm}(\hat{n}) || m| \leq l\} \hat{n} = (\theta, \varphi) \in S^2$ be the spherical harmonics of degree and $(l = 0, 1, 2, ...)$.

For $R \in SO(3)$,

$$Y_{lm}(R^{-1}\hat{n}) = \sum_{|m'| \leq l} [\pi_l(R)]_{m'm} Y_{lm'}(\hat{n})$$

where $\pi_l : SO(3) \to U_{2l+1}(\mathbb{C})$ is an variables representation of $SO(3)$. We can express $R \in SO(3)$ in terms of Euler angles. $R = R(\varphi, \theta, \psi) = R_Z(\varphi) R_X(\theta | R_Z(\psi))$.

Let $\{X_1, X_2, X_3\}$ be the standard generation of $SO(3)$, $X_1, X_2, X_3 \in g(SO(3))$ (Lie algebra of $SO(3)$).

$$R_X(\theta) = \exp(\theta X_1), R_Y(\theta) = \exp(\theta X_2), R_Z(\theta) = \exp(\theta X_3)$$

$$[X_1, X_2] = X_3, [X_2, X_3] = X_1, [X_3, X_1] = X_2$$

We wish to express $\pi_l(R) = \pi_l(R_Z(\varphi) R_X(\theta) R_Z(\psi))$ as a fourier series in (φ, θ, ψ).

Now, $\pi_l(R_Z(\varphi) R_X(\theta) R_Z(\psi))$

$$= \pi_l(R_Z(\varphi)) \cdot \exp(\theta d\pi_l(X_1)) \cdot \pi_l(R_Z(\psi))$$

Let $\quad Y_{lm} = |0, m\rangle, Y_{lm}(\theta, \varphi) = \langle \theta, \varphi | l, m\rangle$

Then,

$$\langle l, m' | \pi_l(R(\varphi, \theta, \psi)) | l, m\rangle = \exp(i(m'\varphi + m\psi))\langle l, m' | \exp(\theta d\pi_l(X_1)) | l, m\rangle$$

Let $X_k = iL_k, k = 1, 2, 3$. So $d\pi_l(X_1) = i d\pi_l(L_1)$.

$d\pi_l(L_1)$ is a hermitian matrix since

$\exp(i\theta d\pi_l(L_1)) = \pi_l(e^{i\theta L_1})$ is a unitary matrix.

Let $\tilde{L}_k = d\pi_l(L_k), 1 \le k \le 3$. Then $\{\tilde{L}_k\}_{k=1}^3$ are the standard angular momentum operator for a spin l-particle. Let $\tilde{L}^2 = \sum_{k=1}^{3} \tilde{L}_k^2$. Then

$\tilde{L}^2 |l, m\rangle = l(l+1), \ \tilde{L}_3 |l, m\rangle = m|l, m\rangle$. Let

$\tilde{L}_1 = \tilde{L}_1 + i\tilde{L}_2, \tilde{L} = \tilde{L}_1 - i\tilde{L}_2$. Then

$$\begin{aligned}
[\tilde{L}_1, \tilde{L}_3] &= [\tilde{L}_1, \tilde{L}_3] + i[\tilde{L}_2, \tilde{L}_3] \\
&= -i\tilde{L}_2 - \tilde{L}_1 = -\tilde{L}_+ \\
[\tilde{L}_1, \tilde{L}_3] &= [\tilde{L}_1, \tilde{L}_3] - i[\tilde{L}_2, \tilde{L}_3] \\
&= -i\tilde{L}_2 + \tilde{L}_1 = \tilde{L}_-
\end{aligned}$$

So
$$\begin{aligned}
\tilde{L}_3\tilde{L}_+|l, m\rangle &= ([\tilde{L}_3, \tilde{L}_+] + \tilde{L}_+\tilde{L}_3)|l, m\rangle \\
&= (\tilde{L}_+ + \tilde{L}_+\tilde{L}_3)|l, m\rangle \\
&= (m+1)\tilde{L}_+|l, m\rangle
\end{aligned}$$

$$\begin{aligned}
\tilde{L}_3\tilde{L}-|l, m\rangle &= \left([\tilde{L}_3, \tilde{L}_-] + \tilde{L}_-\tilde{L}_3\right)|l, m\rangle \\
&= \left(-\tilde{L}_- + \tilde{L}_-\tilde{L}_3\right)|l_1 m\rangle = (m-1)\tilde{L}_-|l, m\rangle.
\end{aligned}$$

So,
$$\begin{aligned}
\tilde{L}_+|l, m\rangle &= (l, m)|l, m+1\rangle \\
\tilde{L}_-|l, m\rangle &= d(l, m)|l, m-1\rangle
\end{aligned}$$

So, $\quad |c(l, m)|^2 = \langle l, m | \tilde{L}_- \tilde{L}_+ | l, m\rangle$

$$= \left\langle l,m \mid \tilde{L}_1^2 + \tilde{L}_2^2 - L_3 \mid l,m \right\rangle$$

$$= l(l+1) - m^2 - m = l(l+1) - m(m+1)$$

$$|d(l,m)|^2 = \left\langle l,m \mid \tilde{L}_+ \tilde{L}_- \mid l,m \right\rangle$$

$$= \left\langle l,m \mid \tilde{L}_1^2 + \tilde{L}_2^2 + L_3 \mid l,m \right\rangle$$

$$= l(l+1) - m(m-1)$$

So, $\langle l,m' \mid d\pi_l(L_1) \mid l,m \rangle = \langle l,m' \mid \tilde{L}_1 \mid l,m \rangle$

$$= \left\langle l,m' \left| \left(\frac{\tilde{L}_+ + \tilde{L}_-}{2} \right) \right| l,m \right\rangle$$

$$= \frac{1}{2}(c(l,m)\delta[m'-m-1] + d(l,m)\delta[m'-m+1])$$

$$= \frac{1}{2}\left((l(l+1)-m(m+1))^{\frac{1}{2}}\delta[m'-m-1] + (l(l+1)-m(m-1))^{\frac{1}{2}}\delta[m'-m+1] \right)$$

So, $[\pi_l(R)]_{m'm} = \sum_{|r|\le l} \exp\{i(m'\varphi + m\psi + r\theta)\} P_r[l,m',m]$

where $\sum_{|r|\le l} P_r[l,m',m] = \langle l,m' \mid \pi_l(L_1) \mid l,m \rangle$

$$= \frac{1}{2}(c(l,m)\delta[m'-m-1] + d(l,m)\delta[m'-m+1])$$

$$\underline{\underline{P}}_r[l] = ((P_r[l,m',m]))_{|m'|,|m|\le l}$$

$$r = -l, -l+1, \ldots, l-1, l$$

Form a revolution of the identity:

$$P_r[l]P_s[l] = \delta_{rs} P_r[l],$$

$$P_r[l]^* = P_r[l],$$

$$\sum_{|r|\le l} P_r[l] = I_{2l+1}$$

❖❖❖❖❖

[197] For Gautom: Ion trap gate design:

$$H_0 = \sum_{k=1}^{N} w_k a_k^t a_k + \frac{egB_0}{2m}\sigma_z \quad ...(1)$$

$$H_I(t) = \sum_{k=1}^{N} (f_{xk}(t)\sigma_x + f_{yk}(t)\sigma_y + f_{zk}(t)\sigma_z)(a_k + a_k^t) \quad ...(2)$$

Note $A(t, \underline{r}) = \sum_{K\sigma}\left(a_k e_{k\sigma} e^{-ikX} + a_{k\sigma}^t e_{k\sigma}^t e^{ikX}\right)\lambda_k$

$$k.X = |K|t - K.r = k_\mu X^\mu.$$

Coulomb gauge div $A = 0 \Rightarrow (K, e_{K\sigma}) = 0$.

$$\underline{B}^{(q)}|_{\underline{r}-0} = \nabla \times A|_{\underline{r}=\underline{0}} = \sum_{K,\sigma}\left[i a_{k\sigma}\underline{K} \times \underline{e}_{k\sigma} e^{-i|k|t} - i a_{k\sigma}^t \underline{K} \times \underline{e}_{k\sigma}^t e^{i|k|t}\right]\lambda_k$$

Pan the quantum em field through (quantum magnetic field) a filter to filter depends also on the wave propagation direction K. $e^{-i|k|t} \rightarrow h(t, \underline{K})$

$$\text{Then } \underline{B}^{(q)}|_{\underline{r}-0} = \sum_{K,\sigma}\{i\lambda_K h(t, \underline{k})[a_{K\sigma}\underline{K} \times \underline{e}_{K\sigma}] - i\lambda_K \bar{h}(t, \underline{K})[a_{K\sigma}^t \underline{K} \times \underline{e}_{K\sigma}^t]\}$$

Interaction energy between quantum magnetic field and spin of the atom:

$$H_I(t) = \frac{eg}{2m}\left(\sigma^{-1}, B^{(q)}(t)\right)$$

$$= \frac{ieg}{2m}\sum_{\underline{K},\sigma}\lambda_K\left[a_{K\sigma}\left(\sigma^{-1}, \underline{K} \times \underline{e}_{K\sigma}\right)h(t, \underline{k}) + a_{k\sigma}^t\left(\bar{\sigma}, \underline{K} \times \underline{e}_{K\sigma}^{+}\right)\overline{h(t, k)}\right]$$

This is of the form (2) with $f_{Xk}(t)$, $f_{Yk}(t)$, $f_{Zk}(t)$ expressed im terms of $h(t, \underline{K})$

Unperturbed eigen states of the field plus atom:

$|r_1 r_2 ... r_N s\rangle$ $r_k = 0, 1, 2, ..., s = 0, 1$.

$s = 0 = -\frac{1}{2}$ separate $s = 1 = -\frac{1}{2}$ spains

$$H_0|r_1 r_2 ... r_N s\rangle = \sum_{j=1}^{N} r_j w_j + \frac{eg B_0}{2m}(2s-1), s = 0, 1.$$

$$= E(r_1, ..., r_n, s)(r_1, ..., r_N, s)$$

Matrix elements of $H_I(t)$:

$$\langle r_1' r_2' ... r_N' s' | H_I(t) | r_1 r_2 ... r_N s\rangle = \sum_{k=1}^{N}\left[f_{Xk}(t)\langle s'|\sigma_X|s|\rangle + f_{Yk}(t)\langle s'|\sigma_y|s\rangle\right.$$

$$+ f_{Zk}(t)\langle s' | \sigma_z | s\rangle]\langle r_k' | a_k + a_k^t | r_k\rangle$$

$$\prod_{j=j,\hat{j}\neq k}^{N} \delta[r_j' - r_j]$$

❖❖❖❖❖

[198] Quantum filtering in Coherent states.

Summary of Gough's paper:

$$\eta_{t]}^{in} = \sigma\{Y_s^{in} \ s \leq t\}.$$

$$\eta_{t]}^{out} = U^*(t)\eta_{t]}^{in} U(t)$$

$$dU(t) = \left(L_b^a(t) d\Lambda_a^b(t)\right)U(t)$$

$$U(t) \simeq \text{unitary } X \in \hbar.$$

$$\mathbb{E}\left[j_t(X) | \eta_{t]}^{out}\right] = \mathbb{E}\left[U^*(t) X U(t) | \eta_{t]}^{out}\right]$$

$$= U^*(t)\mathbb{E}_t(X | \eta_{t]}^{in})U(t).$$

Proof: Let $Z \in \eta_{t]}^{in}$. Then $Z(t) = U^*(t)\, Z\, U(t) \in \eta_{t]}^{out}$.

$$\mathbb{E}[(j_t(X) - U^*(t)\mathbb{E}(X | \eta_{t]}^{in})U(t))Z(t)]$$

$$= \mathbb{E}[U^*(t)\, XZ U(t)]$$

$$- \mathbb{E}[U^*(t)\mathbb{E}_t[XZ | \eta_{t]}^{in}]U(t)]$$

$$= \mathbb{E}_t[XZ] - \mathbb{E}_t\,\mathbb{E}_t[X z | \eta_{t]}^{in}] = 0$$

$$= \mathbb{E}_t(\xi) \underset{def}{\triangleq} \mathbb{E}[U^*(t)\xi U(t)]$$

Suppose

$$\mathbb{E}_t(X) = \mathbb{E}[F^*(t) X F(t)] \equiv \mathbb{E}_{F(t)}(X)$$

where $F(t) \in \eta_{t]}^{in}$ *i.e.* $[F(t), \eta_{t]}^{in}] = 0$.

Then $\mathbb{E}_t(X | \eta_{t]}^{in}) = \dfrac{\mathbb{E}[F^*(t) X F(t) | \eta_{t]}^{in}]}{\mathbb{E}[F^*(t) F(t) | \eta_{t]}^{in}]}$

(Quantum Kallianpan-Streibel formula)

Pt: Let $Z \in \eta_{t]}^{in}$. Then since $[F(t), Z] = [F^*(t), Z] = 0$,

we have

$$\begin{aligned}
\mathbb{E}_t[\mathbb{E}_t(X \mid \eta_{t]}^{in}) Z] &= \mathbb{E}_t(XZ) \\
&= \mathbb{E}[F^*(t) XZ\, F(t)] \\
&\quad \mathbb{E}_t\left[\frac{\mathbb{E}[F^*(t) X\, F(t) \mid \eta_{t]}^{in}]}{\mathbb{E}[F^*(t) F(t) \mid \eta_{t]}^{in}]} Z\right] \\
&= \mathbb{E}_t\left[\frac{\mathbb{E}[F^*(t) XZ\, F(t) \mid \eta_{t]}^{in}]}{\mathbb{E}[F^*(t) F(t) \mid \eta_{t]}^{in}]}\right] \\
&= \mathbb{E}\left[\frac{\mathbb{E}[F^*(t) XZ\, F(t) \mid \eta_{t]}^{in}] F^*(t) F(t)}{\mathbb{E}[F^*(t) F(t) \mid \eta_{t]}^{in}]}\right] \\
&= \mathbb{E}[F^*(t) XZ\, F(t)]
\end{aligned}$$

(first condition on $\eta_{t]}^{in}$, then cancel out $\mathbb{E}[F^*(t) \mid \eta_{t]}^{in}]$)

Let

$$\sigma_t(X) = \mathbb{E}[F^*(t) X\, F(t) \mid \eta_{t]}^{in}]$$

Then

$$\mathbb{E}_t(X \mid \eta_{t]}^{in}) = \frac{\sigma_t(X)}{\sigma_t(1)}$$

and so,

$$\begin{aligned}
\pi_t(X) &\underset{def}{\overset{\Delta}{=}} E(j_t(X) \mid \eta_{t]}^{out}) \\
&= U^*(t) \frac{\sigma_t(X)}{\sigma_t(1)} U(t)
\end{aligned}$$

$$d\sigma_t(X) = d\,\mathbb{E}[F^*(t) X\, F(t) \mid \eta_{t]}^{in}]$$

Note: $F^*(t) X\, F(t) \in \eta_{t]}^{in}$

Example:

$$dU(t) = \left(L_b^a(t)\, d\Lambda_a^b(t)\right) U(t)$$

We want

$$\mathbb{E}[U^*(t)XU(t)] = \langle f\varphi(w)|U^*(t)XU(t)|f\varphi(u)\rangle$$

$$|\varphi(u)\rangle = \exp\left(-\|u\|^2/2\right)|e(u)\rangle$$

$$= \langle f\varphi(u)|F^*(t)XF(t)|f\varphi(u)\rangle$$

where $F(t) \in \eta_{t]}^{in}$

Let
$$|\psi(t)\rangle = U(t)|f\varphi(u)\rangle$$

$$d|\psi(t)\rangle = dU(t)|f\varphi(u)\rangle$$

$$= L_b^a(t)\,d\Lambda_a^b(t)|\psi(t)\rangle$$

$$d\langle f\varphi(u)|U^*(t)XU(t)|f\varphi(u)\rangle$$

$$= \langle f\varphi(u)|U^*(t)X\,dU(t)|f\varphi(u)\rangle$$

$$+\langle f\varphi(u)|dU^*(t)XU(t)|f\varphi(u)\rangle$$

$$+\langle f\varphi(u)|dU^*(t)X\,dU(t)|f\varphi(u)\rangle$$

$$= \left\langle f\varphi(u)|U^*(t)XL_b^aU(t)|f\varphi(u)\right\rangle$$

$$= u_b(t)\overline{u_a(t)}\,dt$$

$$+\left\langle f\varphi(u)|U^*(t)L_b^{a*}XU(t)|f\varphi(u)\right\rangle$$

$$u_a(t)\overline{u_b(t)}\,dt$$

$$+\left\langle f\varphi(u)|U^*(t)L_b^{a*}XL_d^cU(t)|f\varphi(u)\right\rangle$$

$$\varepsilon_c^a u_a(t)\overline{u_p(t)}\,dt$$

So we require

$$\frac{d}{dt}\langle f\varphi(u)|F^*(t)XF(t)|f\varphi(u)\rangle$$

$$= \Big\langle f\varphi(u)|F^*(t)\Big(XL_b^a u_b(t)\overline{u_a(t)} + L_b^a{}^*Xu_a(t)\overline{u_b(t)}$$

$$+ L_b^{a*}XL_d^c\varepsilon_c^a u_d(t)\overline{u_b(t)}\Big)F(t)|f\varphi(u)\Big\rangle$$

Let $\quad dF(t) = A(t)F(t)\,dt + B(t)F(t)\,dY_{t]}^{in}$

Then $F(t) \in \eta_{t]}^{in}$ assuming $A(t), B(t) \in \mathcal{L}(h)$, since $\{\eta_{t]}^{in}, t \geq 0\}$ is an Abelian algebra and commutes with $\mathcal{L}(h)$.

Then $d\langle f\varphi(u) | F^*(t) X F(t) | f\varphi(u)\rangle$

$$= \langle f\varphi(u) | F^* XdF + dF^* XF + dF^* X\, dF | f\varphi(u)\rangle$$

$$= \langle f\varphi(u) | F^*(XA + A^* X)F | f\varphi(u)\rangle dt$$

$$= \langle f\varphi(u) | F^* X B F | f\varphi(u)\rangle c_b^a u_b \bar{u}_a\, dt$$

$$= +\langle f\varphi(u) | F^* B^* X F | f\varphi(u)\rangle \bar{c}_b^a u_a \bar{u}_b\, dt$$

$$= +\langle f\varphi(u) | F^* B^* X B F | f\varphi(u)\rangle \bar{c}_b^a c_\beta^\alpha \varepsilon_\alpha^a u_\beta \bar{u}_b\, dt$$

where we've assuming that

$$Y_t^{in} = c_b^a \Lambda_a^b(t), \quad c_b^a \,\varepsilon\, \mathbb{C}$$

So we require

$$= A + c_b^a u_b \bar{u}_a B$$

$$= L_b^a u_b \bar{u}_a$$

and $B^* X B \bar{c}_b^a c_\beta^\alpha \varepsilon_\alpha^a u_\beta \bar{u}_b$

$$= L_b^a X L_d^c \varepsilon_c^a u_d \bar{u}_b$$

$$= \sum_{j\geq 1}\left(\sum_b \bar{u}_b L_b^j\right)\times\left(\sum_d ud\, L_d^j\right) \forall X$$

May not be generally possible

Suppose however we can final processes $c_b^a(j)\Lambda_a^b(t)$

j = 1, 2, ... which are in $\eta_{t]}^{in}$ *i.e.* which commute with $c_b^a \Lambda_a^b(s), s \leq t$ for some $c_b^a(t) \in \mathbb{C}$:

Example of Abelian measurements.

Let $$Y_{t]}^{in} = c_b^a(j)\Lambda_a^b(t), \quad t \geq 0, \quad c_b^a(j) \in \mathbb{C}.$$

For $\left\{Y_t^{j\,in}, t \geq 0\right\}_{j=1, 2, N}$ *t* form a commutative family of operators, we require that

$$dY_t^{j\,in}.dY_t^{k\,in} = dY_t^{k\,in}.dY_t^{j\,in}\ \forall j, k, t$$

i.e. $c_b^a(j)c_q^p(k)d\Lambda_a^b(t).d\Lambda_p^q(t)$

$$= c_q^p(k)c_b^a(j)d\Lambda_p^q(t).d\Lambda_a^b(t)$$

or $$c_b^a(j)c_q^p(k)\varepsilon_p^b\, d\Lambda_a^q(t)$$

$$= c_q^p(k)c_b^a(j)\varepsilon_a^q\, d\Lambda_p^b(t)$$

or $$c_q^p(j)c_b^a(k)\varepsilon_a^q = c_q^p(k)c_b^a(j)\varepsilon_a^q$$

or $$\sum_{q\geq 1} c_q^p(j)c_b^q(k) = \sum_{q\geq 1} c_q^p(k)c_p^q(j) \ \forall\, j, k$$

i.e. the matrix $\left(\left(c_q^p(j)\right)\right)_{1\leq p,q\leq N} = \widetilde{C}(j)$

Commute $\forall j$ (*i.e.* necessary condition).

The general necessary and sufficient condition are

$$\underline{\underline{C}}(j)\cdot\underline{\underline{\varepsilon}}\cdot\underline{\underline{C}}(k) = \underline{\underline{C}}(k)\cdot\underline{\underline{\varepsilon}}\cdot\underline{\underline{C}}(j) \quad \forall\, k, j$$

❖❖❖❖❖

[199] Design of quantum gates with Levy noise.

Let $\{W(t), t\geq 0\}$ be a Levy process:

$$W(t) = \sum_{\alpha=1}^{N(t)} X_\alpha$$

$\{N(t), t\geq 0\}$ Poisson, $\{X_\alpha\}_{\alpha=1}$ iid in Ex. $F'_X(x) = f_X(x)$

Ito's formula $(dW(t))^2 = X_{N(t)+1}^2\, dN(t)$.

$$H(t) = H_0 + P(t) + \frac{dW(t)}{dt}V(t)$$

$$dU(t) = -iH(t)U(t)dt$$

$$= -i[(H_0 + P(t))dt + dW(t)V(t)]U(t)$$

Unitarity of $U(t)$:

$$d(U^*U) = dU^*.U + U^*.dU + dU^*.dU$$

$$= U^*(i(P^* - P)dt + (dw)^2V^2)U$$

So $P^* - P = i\frac{(dw)^2}{dt}V^2 = iX^2_{N(t)+1}\frac{dN(t)}{dt}V^2(t)$

So we can take $P(t) = i\,Q(t), Q^* = Q$

$$-z\,i\,Q = i\,X^2_{N(t)+1}\frac{dN(t)}{dt}V^2(t)$$

or $$Qdt = -\frac{1}{2}X^2_{N(t)+1}V^2(t)\,dN(t)$$

So the Schrödinger can with jerk noise is

$$dU(t) = \left[-iH_0dt - \frac{1}{2}X^2_{N(t)+1}V^2(t)\,dN(t) - i\,X_{N(t)+1}V(t)\,dN(t)\right]U(t)$$

$$= \left[-i\,H_0dt - \left(i\,X_{N(t)+1}V(t) + \frac{1}{2}X^2_{N(t)+1}V^2(t)\right)dN(t)\right]U(t)$$

Solution upto $O(V^2)$:

$$U(t) = U_0(t) - \int_0^t U_0(t-\tau)\left(i\,X_{N(\tau)+1}V(\tau) + \frac{1}{2}X^2_{N(\tau)+1}V^2(\tau)\right)U_0(\tau)\,dN(\tau)$$

$$- \int_{0<\tau_2<\tau_1<t} U_0(t-\tau_1)V(\tau_1)\,U_0(\tau_1-\tau_2)\,V(\tau_2)\,U_0(\tau_2)$$

$$X_{N(\tau_1)+1}X_{N(\tau_2)+1}\,dN(\tau_1)\,dN(\tau_2)$$

where $U_0(t) = \exp(-itH_0)$

$$\mathbb{E}\{U(t)\} = U_0(t) - i\int_0^t \lambda\mu_X U_0(t-\tau)V(\tau)U_0(\tau)\,d\tau$$

$$-\frac{1}{2}\int_0^t \lambda(\mu_X^2 + \sigma_X^2)U_0(t-\tau)V^2(\tau)U_0(\tau)\,d\tau$$

$$- \int_{0<\tau_2<\tau_1<t} U_0(t-\tau_1)\,V(\tau_1)\,U_0(\tau_1-\tau_2)\,V(\tau_2)\,U_0(\tau_2)$$

$$\mathbb{E}\{X_{N(\tau_1)+1}X_{N(\tau_2)+1}\,dN(\tau_1)\,dN(\tau_2)\}$$

Now, For $\tau_2 < \tau_1$,

$$\mathbb{E}\{X_{N(\tau_1)+1}X_{N(\tau_2)+1}\,dN(\tau_2)\} = \lambda d\tau_1\,\mathbb{E}\{X_{N(\tau_1)+1}X_{N(\tau_2)+1}\,dN(\tau_2)\}$$

Now, $\mathbb{E}\{X_{N(\tau_1)+1}X_{N(\tau_2)+1}\,dN(\tau_2)\}$

$$= \mathbb{E}\{X_{N(\tau_1)-N(\tau_2)+d\tau_2+N(\tau_2)+d\tau_2+1}X_{N(\tau_2)+1}dN(\tau_2)\}$$

$$= \sum_{r\geq 0} \frac{e^{-\lambda(\tau_1-\tau_2-d\tau_2)}(\lambda(\tau_1-\tau_2-d\tau_2))^n}{\lfloor r}$$

$$\mathbb{E}\{X_{r+N(\tau_2+d\tau_2)+1}X_{N(\tau_2)+1}dN(\tau_2)\}$$

$$= e^{-\lambda(\tau_1-\tau_2)}\mathbb{E}\Big[X_{N(\tau_2)+1}+dN(\tau_2)X_{N(\tau_2)+1}+dN(\tau_2)\Big]$$

$$+\mathbb{E}(X)\sum_{r\geq 1}\frac{e^{-\lambda(\tau_1-\tau_2)}(\lambda(\tau_1-\tau_2))^r}{\lfloor r}$$

$$\mathbb{E}(X)\lambda d\tau_2$$

$$= (\mathbb{E}(X))^2\lambda d\tau_2 \quad (1-e^{-\lambda(\tau_1-\tau_2)})$$

$$+e^{-\lambda(\tau_1-\tau_2)}(\lambda d\tau_2\,\mathbb{E}(X)^2)$$

So finally, for $\tau_1 > \tau_2$,

$$\mathbb{E}\{X_{N(\tau_1)+1}X_{N(\tau_2)+1}dN(\tau_1)dN(\tau_2)\} = \lambda^2 d\tau_1 d\tau_2 (EX)^2$$

So $$\mathbb{E}\{U(t)\} = U_0(t)-i\lambda\mu\chi\int_0^t U_0(t-\tau)V(\tau)U_0(\tau)d\tau$$

$$-\frac{1}{2}\lambda(\mu_X^2+\sigma_X^2)\int_0^t U_0(t-\tau)V^2(\tau)U_0(\tau)d\tau$$

$$-\int_{0<\tau_2<\tau_1<t}\lambda^2\mu\chi^2U_0(t-\tau_1)V(\tau_1)U_0(\tau_1-\tau_2)V(\tau_2)U_0(\tau_2)d\tau_1 d\tau_2$$

Let $\underline{\underline{U}}_0$ be a derived gate. We with to component $\mathbb{E}\{\|\underline{\underline{U}}_d-U(\tau)\|^2\}$ upto $O(V^2)$ terms.

Now, $\|\underline{\underline{U}}_d-U(\tau)\|^2$

$$= \mathbb{E}\Bigg\|\underline{\underline{U}}_d-U_0(\tau)+\int U_0(\tau-\tau)\left(iX_{N(\tau)+1}V(\tau)+\frac{1}{2}X^2_{N(\tau)+1}V^2(\tau)\right)U_0(\tau)d\tau$$

$$-\int U_0(\tau-\tau_1)V(\tau_1)U_0(\tau_1-\tau_2)V(\tau_2)U_0(\tau_2)X_{N(\tau_1)+1}dN(\tau_1)dN(\tau_2)X_{N(\tau_2)+1}\Bigg\|^2$$

$$= \mathbb{E}\Bigg[\Bigg\|\underline{\underline{W}}_d+\int_0^t\left(iX_{N(\tau)+1}\tilde{V}(\tau)+\frac{1}{2}X^2_{N(\tau)+1}\tilde{V}^2(\tau)\right)d\tau$$

$$-\int\tilde{V}(\tau_1)\tilde{V}(\tau_2)X_{N(\tau_2)+1}X_{N(\tau_2)+1}dN(\tau_1)dN(\tau_2)\Bigg\|^2\Bigg]$$

$$+O(\|V\|^3)$$

where $\underline{\underline{W}}_d = \underline{\underline{U}}_0(-\tau)\underline{\underline{U}}_d - I$

and $\tilde{V}(t) = U_0(-t)V(t)U_0(t)$

❖❖❖❖❖

[200] Problems in electrostatics

A point charge q is placed at $(0, 0, d)$.

A conducting sphere of radius R maintained at potential V_0 has its centre at $(0, 0, 0)$. Assume $R > d$.

Find the potential $V(r, \theta)$, $0 \leq r \leq R$, $0 < \theta < \pi$.

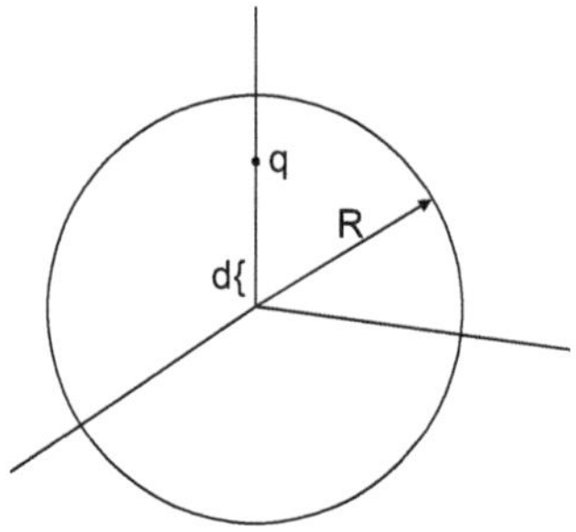

Solution:

$$\nabla^2 V = -\frac{q}{2\pi\,\varepsilon_0}\frac{\delta(r-d)}{r^2}\delta(\cos\theta)$$

$$\delta(\cos\theta) = \sum_{l=0}^{\infty} c_l P_l(\cos\theta)$$

Then, $$c_l = \int_{-1}^{1} P_l(X)\delta(X)\,dX = P_l(0)$$

So $$\delta(\cos\theta) = \sum_{l=0}^{\infty} P_l(0).P_l(\cos\theta)$$

$$= \frac{1}{r^2}\frac{\partial}{\partial r}r^2\frac{\partial V(r,\theta)}{\partial r} + \frac{1}{r^2\sin\theta}\frac{\partial}{\partial\theta}\sin\frac{\partial V(r,\theta)}{\partial\theta}$$

$$= -\frac{q}{2\pi\,\varepsilon_0 r^2}\delta(r-d)\sum_{l=0}^{\infty} P_l(0)P_l(\cos\theta)$$

Let

$$V(r,\theta) = \sum_{l=0}^{\infty} V_l(r)P_l(\cos\theta)$$

Since $$\frac{-1}{\sin\theta}\frac{d}{d\theta}\sin\theta\frac{d}{d\theta}P_l(\cos\theta) = l(l+1)P_l(\cos\theta),$$

we get

$$\frac{1}{r^2}\frac{d}{dr}(r^2 V_l'(r)) - \frac{l(l+1)V_l(r)}{r^2} = \frac{-q\,P_l(0)\delta(r-d)}{2\pi\,\varepsilon_0 r^2}$$

or $$r^2 V_l''(r) + 2rV_l'(r) - l(l+1)V_l(r) = \frac{-qP_l(0)}{2\pi}\delta(r-d), l \geq 0$$

$$V_l(r) = a(l)r^l, \quad r < d$$

$$V_l(r) = b(l)r^l + \frac{c(l)}{r^{l+1}} \qquad R > r > d$$

$$d^2(V_l'(d+) - V_l'(d-1)) = \frac{-qP_l(0)}{2\pi\,\varepsilon_0}$$

$$V_l(d+) = V_l(d-)$$

So,

$$d^2\left(l\,a(l)d^{l-1} - l\,b(l)d^{l-1} - \frac{(l+b(\cos))}{d^{l+2}}\right) = \frac{-qP_l(0)}{2\pi\,\varepsilon_0}$$

$$V(R,\theta) = V_0$$

$$\Rightarrow \quad \sum_{l=0}^{\infty}\left(b(l)R^l + \frac{c(l)}{R^{l+1}}\right)P_l(\cos\theta) = V_0$$

$$\Rightarrow \quad b(l)R^l + \frac{c(l)}{R^{l+1}} = V_0\int_{-1}^{1} P_l(X)\,dx$$

or $$b(0) + \frac{c(0)}{R} = V_0,$$

$$b(l)R^l + \frac{c(l)}{R^l} = 0, l \geq 1.$$

$$c(0) = R(V_0 - b(0)), c(l) = -R^{2l}\,b(l), l \geq 1$$

*Solving Laplacian equation with random boundary.

Let $S(\underline{\theta})$ be the boundary (2-D surface in $\mathbb{R}^3$) paramatized by the random parameter $\underline{\theta} \in \mathbb{R}^p$.

$\tau_\theta^{(X)}$ is the first hitting time of 3-D Brownian motion $B|.|$ at the boundary sterity at X. Then

$$u(X,\underline{\theta}) = \mathbb{E}_B\{\varphi(B(\tau_\theta(X)))\}, X \in B(\underline{\theta})\,\partial B(\underline{\theta}) = S(\underline{\theta})$$

satisfies

$$\begin{aligned}
\nabla^2 u(X, \underline{\theta}) &= 0,\ X \,\varepsilon\, B(\underline{\theta}), \\
u(X, \underline{\theta}) &= \varphi(X), X \in \partial B(\underline{u}) = S(\underline{\theta}) \\
\mathbb{E}u(X, \underline{\theta}) &= \int u(X, \underline{\theta}) dF(\underline{\theta}) \\
&= \int \mathbb{E}_B \{\varphi(B(\tau_0(X)))\} dF(\underline{\theta}) \\
&= \mathbb{E}\{\varphi(B(\tau_0(X)))\}
\end{aligned}$$

satisfies $\mathbb{E}\{\varphi(B(\tau_0(X)))\} = 0$

❖❖❖❖❖

Wave Equation with Random Boundary, Robots in a Fluid, Cosmology with Stochastic Perturbations, Interaction Between Graviton, Photon and Electron-Positron Fields, Evans-Hudson Flows for Quantization of Stochastic Evolutions

[201] Solving the wave eqn. with random boundary. Let $B(\underline{\theta})$ be the region in $\mathbb{R}^3$ (open connection set) with boundary $S(\underline{\theta}) = \partial B(\underline{\theta})$. We wish to solve

$$\nabla^2 u(X, \underline{\theta}) + k^2 u(X, \underline{\theta}) = -\delta(X), X \in B(\underline{\theta})$$

$$u(\underline{X}, \underline{\theta}) = 0, \ \underline{X} \in S(\underline{\theta})$$

Denote the solution by $G(\underline{X}_0, \underline{\theta}, k)$

Consider now the boundary value problem

$$\nabla^2 V(\underline{X}, \underline{\theta}) + k^2 V(\underline{X}, \underline{\theta}) = -\frac{\rho(\underline{X}, \underline{\theta})}{\varepsilon}, \ \underline{X} \,\varepsilon\, B(\underline{\theta})$$

$$V(\underline{X}, \underline{\theta}) = \varphi(\underline{X}, \underline{\theta}), \ \underline{X} \in S(\underline{\theta})$$

$$\int [V(\underline{X}, \underline{\theta}) \nabla^2 G(\underline{X}, \underline{X}_0, \underline{\theta}, k)$$

$$B(\underline{\theta}) - G(\underline{X}, \underline{X}_0, \underline{\theta}, k) \nabla^2 V(\underline{X}, \underline{\theta})] \, d^3X$$

$$= -V(\underline{X}_0, \underline{\theta}) - k^2 \int_{B(\underline{v})} V(\underline{X}, \underline{\theta}) G(\underline{X}, \underline{X}_0, \underline{\theta}, k) d^3X$$

$$= \int_{S(\underline{\theta})} \left[V(X, \theta) \frac{\partial G(\underline{X}, X_0, \theta, k)}{\partial \hat{n}} - G(X, X_0, \underline{\theta}, k) \frac{\partial V(\underline{X}, \underline{\theta})}{\partial \hat{n}} \right] dS$$

so

$$= -V(\underline{X}_0, \underline{\theta}) - k^2 \int_{B(\underline{v})} V(\underline{X}, \underline{\theta}) G(\underline{X}, \underline{X}_0, \underline{\theta}, k) d^3X$$

$$= -\int_{B(\underline{\theta})} G(\underline{X}, \underline{X}_0, \underline{\theta}, k) \left(-\frac{\rho(\underline{X}, \underline{\theta})}{\varepsilon} - k^2 V(\underline{X}, \underline{\theta}) \right) d^3X$$

$$= \int_{S(\underline{\theta})} \varphi(\underline{X}, \underline{\theta}) \frac{\partial G}{\partial \hat{n}}(\underline{X}, \underline{X}_0, \underline{\theta}, k)\, dS(X)$$

$$V(\underline{X}_0, \underline{\theta}) = \frac{1}{\varepsilon} \int_{B(\underline{\theta})} \rho(\underline{X}, \underline{\theta}) G(\underline{X}, \underline{X}_0, \underline{\theta}, k)\, d^3X - \int_{S(\underline{\theta})} \varphi(\underline{X}, \underline{\theta}) \frac{\partial G}{\partial \hat{n}}(\underline{X}, \underline{X}_0, \underline{\theta}, k)\, dS(X)$$

Calculate $\mathbb{E}\{V(\underline{X}_0, \underline{\theta})\}$ and more generally

$$\mathbb{E}\left\{\exp\left(\int_{B(\underline{\theta})} \psi(\underline{X}) V(\underline{X}, \underline{\theta})\, d^3X\right)\right\}$$

❖❖❖❖❖

[202] Lectures on Electromagnetics

Transmission line with Brownian local time input

$$v(t) = \int_0^t \delta(X - B(s))\, ds$$

$$V(Z, \omega) = V_+(\omega) e^{-\gamma(\omega)z} + V_-(\omega) e^{\gamma(\omega)Z}$$

$$I(Z, \omega) = Z_0(\omega)^{-1}(V_+(\omega) e^{-\gamma(\omega)Z} - V_-(\omega) e^{\gamma(\omega)Z})$$

$$V_+(\omega) = \frac{V(\omega)}{1 + \Gamma(\omega)}$$

$$\Gamma(\omega) = \frac{V_-(\omega)}{V_+(\omega)} = \left\{\frac{Z_L(\omega) - Z_0(\omega)}{Z_L(\omega) + Z_0(\omega)}\right\} e^{-2\gamma(\omega)d}$$

$$V(\omega) = \int_{0<s<t<\infty} \delta(X - B(s)) e^{-j\omega t}\, ds\, dt$$

$$V_T(\omega) = \int_{0<s<t<T} \delta(X - B(s)) e^{-j\omega t}\, ds\, dt$$

$$= \int_0^T \frac{1}{j\omega}\left(e^{-j\omega s} - e^{-j\omega t}\right) \delta(X - B(s))\, ds$$

$$= \int_0^T \frac{2e^{-j\omega(s+T)/2}}{\omega} \sin\left\{\frac{\omega(T-s)}{2}\right\} \delta(X - B(s))\, ds = V_T^X(\omega)$$

$$\int_{\mathbb{R}} f(X) V_T^X(\omega)\, dX = 2\int_0^T \frac{f(B(s)) \sin\left\{\frac{\omega(T-s)}{2}\right\}}{\omega} e^{-\frac{j\omega(s+T)}{2}}\, ds$$

$$V(\omega) = \int_0^\infty \delta(X - B(s))\, ds \int_s^\infty e^{-j\omega t}\, dt$$

$$= \int_0^\infty \delta(X - B(s))\, ds\, e^{-j\omega s}\left(\frac{1}{j\omega} + \pi\delta(\omega)\right)$$

$$= \int_0^\infty \frac{e^{-j\omega s}}{j\omega}\delta(X - B(s))\, ds + \pi\int_0^\infty \delta(X - B(s))\, ds\, \delta(\omega)$$

Filter out the de component to get source voltage as

$$V_0(\omega) = \int_0^\infty \frac{e^{-j\omega s}}{j\omega}\delta(X - B(s))\, ds$$

$$j\omega\mathbb{E}\{V_0(\omega)\} = \int_0^\infty e^{-j\omega s}\, \mathbb{E}\{\delta(X - B(s))\}\, ds$$

$$= \int_0^\infty e^{-j\omega s}\, \mathbb{E}\{\delta(X - B(s))\}\, ds$$

$$= \int_0^\infty \frac{e^{-j\omega s}}{\sqrt{2\pi s}} e^{-X^{2/25}}\, ds$$

$$\omega^2\, \mathbb{E}\left\{V_0(\omega_1)\overline{V_0(\omega_2)}\right\} = \int_{[0,\infty]^2} e^{-j(\omega_1 s_1 - \omega_2 s_2)}\, \mathbb{E}\{\delta(X - B(s_1))\delta(X - B(s_2))\}\, ds_1\, ds_2$$

$$= \frac{1}{2\pi}\int_{\substack{s_1 > s_2 > 0 \\ \xi \in \mathbb{R}}} \exp\left\{-\frac{(x-\xi)^2}{2(s_1 - s_2)}\right\}\frac{\delta(X-\xi)}{\sqrt{s_1 - s_2}\sqrt{s_2}} e^{-\frac{\xi^2}{2}s_2}\, d\xi\, ds_1\, ds_2$$

$$\times \exp\{-j(\omega_1 s_1 - \omega_2 s_2)\}$$

$$+ \exp\{-j(\omega_1 s_2 - \omega_2 s_1)\}$$

$$= \frac{1}{2\pi}\int_{s_1 > s_2 > 0} \exp\left(-\frac{X^2}{2s_2}\right)\exp\{-j(\omega_2 - \omega_1)s_1 + (\omega_2 - \omega_1)s_2\}\, ds_1\, ds_2$$

$$= \frac{1}{2\pi}\int_0^\infty \exp\left(-\frac{x^2}{2s_2}\right)\cdot \exp\{j(\omega_2 - \omega_1)s_2\}$$

$$\times \exp\{j(\omega_2 - \omega_1)s_2\}$$

$$\times\left(\frac{1}{j(\omega_2 - \omega_1)} + \pi\delta(\omega_2 - \omega_1)\right) dS_2$$

$$= \frac{1}{2\pi}\delta(\omega_2 - \omega_1)\int_0^\infty e^{-x^2/2s_2}\, ds_2$$

$$+ \frac{1}{2\pi j}(\omega_2 - \omega_1)\int_0^\infty \exp\left(-\frac{X^2}{2s_2}\right)\exp\{j_2(\omega_2 - \omega_1)s_2\}ds_2$$

Continuous distribution of voltage and current sources along the line

$$-\frac{\partial V(Z,\omega)}{\partial Z} = (R(Z) + j\omega L(Z))I(Z,\omega) + F_V(Z,\omega)$$

$$-\frac{\partial I(Z,\omega)}{\partial Z} = (G(Z) + j\omega C(Z))\,V(Z,\omega) + F_I(Z,\omega)$$

or
$$-\frac{d}{dZ}\begin{bmatrix} V(Z,\omega) \\ I(Z,\omega) \end{bmatrix} = \begin{bmatrix} 0 & R(Z) + j\omega L(Z) \\ G(Z) + j\omega C(Z) & 0 \end{bmatrix}\begin{bmatrix} V(Z,\omega) \\ I(Z,\omega) \end{bmatrix} + \begin{bmatrix} F_V(Z,\omega) \\ F_I(Z,\omega) \end{bmatrix}$$

Distributed parameter fluctuations:

Let
$$A(Z) = -\begin{bmatrix} 0 & R(Z) + j\omega L(Z) \\ G(Z) + j\omega C(Z) & 0 \end{bmatrix}$$

$$\delta A(Z) = \begin{bmatrix} 0 & R(Z) + j\omega L(Z) \\ G(Z) + j\omega C(Z) & 0 \end{bmatrix}$$

$$\frac{\partial \Phi(Z,Z')}{\partial Z} = A(Z)\Phi(Z,Z'), Z \geq Z'$$

$$\Phi(Z',Z') = I$$

$\Phi \simeq$ state transmission matrix

$$\begin{bmatrix} V(Z,\omega) \\ I(Z,\omega) \end{bmatrix} = -\int_0^z \Phi(Z,\xi)\begin{bmatrix} F_V(\xi,\omega) \\ F_I(\xi,\omega) \end{bmatrix}d\xi$$

$$\frac{\partial}{\partial Z}\delta\Phi(Z,Z') = \delta A(Z)\cdot\Phi(Z,Z') + A(Z)\cdot\delta\Phi(Z,Z')$$

So
$$\delta\Phi(Z,Z') = \int_0^z \Phi(Z,\xi)\delta A(\xi)\Phi(\xi,Z')d\xi$$

$$\begin{bmatrix} \delta V(Z,\omega) \\ \delta I(Z,\omega) \end{bmatrix} = -\int_0^z \delta\Phi(Z,\xi)\begin{bmatrix} F_V(\xi,\omega) \\ F_I(\xi,\omega) \end{bmatrix}d\xi$$

$$= -\int_{0<\xi,\, Z'<Z} \Phi(Z,\xi)\delta A(\xi)\Phi(\xi, Z')\begin{bmatrix} F_V(Z',\omega) \\ F_I(Z',\omega) \end{bmatrix} d\xi\, dZ'$$

Voltage and current correlations:

$$= \mathbb{E}\left\{\begin{bmatrix} \delta V(Z,\omega) \\ \delta I(Z,\omega) \end{bmatrix} \otimes \begin{bmatrix} \delta \bar{V}(Z_2,\omega) \\ \delta \bar{I}(Z_2,\omega) \end{bmatrix}\right\}$$

$$- \int_{\substack{0<\xi_1,\, Z_1'<Z_1, \\ 0<\xi_2, Z_2'<Z_2}} (\Phi(Z,\xi)\otimes\Phi(Z,\xi_2))\, \mathbb{E}\{\delta \underline{A}(\xi_1)\otimes \delta\bar{\underline{A}}(\xi_2)\}$$

$$(\Phi(\xi_1, Z_1')\otimes\Phi(\xi_2, Z_2))$$

$$\begin{bmatrix} F_V(Z_1',\omega) \\ F_I(Z_1',\omega) \end{bmatrix} \otimes \begin{bmatrix} \bar{F}_V(Z_2',\omega) \\ \bar{F}_I(Z_2',\omega) \end{bmatrix}$$

$$d\xi_1, d\xi_2\, dZ_1'\, dZ_2'$$

Note that $\mathbb{E}\{\delta \underline{A}(\xi_1)\otimes \delta\bar{\underline{A}}(\xi_2)\}$

can be completed from $\mathbb{E}\left\{\begin{bmatrix} \delta R(\xi_1) \\ \delta L(\xi_1) \\ \delta G(\xi_1) \\ \delta C(\xi_1) \end{bmatrix} [\delta R(\xi_2), \delta L(\xi_2), \delta G(\xi_2), \delta C(\xi_2)]\right\}$

❖❖❖❖❖

[203] Ppt. for Nanorobotics.

1. Nanorobot acts as a small chemical senior near a cell. Its size is of the order of a cell size.
2. Biological nanorobots swim in the body flux in the form of protien things and can same as chemical deductors and also chemical injectors.
3. nanoscale tongs can be operated using SEM (scanning electron microscope) and TEM (transmission electron microscope) for manipulating cells and causing collidar reactor and mutations.
4. The dynamics of a nano-robot involves analyzing the joint motion of nanorobots and biological molecules.

Let M be the man of a bio-molecule and M the man of the nanorobot. Molecular collision on the robots take place at poission times $\tau_1 < \tau_2 < \ldots$ where $\tau_{\alpha+1} - \tau_\alpha$, $\alpha \geq 1$ are iid exponential r.v's with density $\lambda e^{-\lambda x}$, $x \geq 0$. Let $N(t)$ be the number of collision on the nanorobot in the $[0, t]$. Let $\underline{V}(t)$ be the velocity of the colliding robot at time t and $\underline{v}_n$. be the velocity of the colliding biomolecule just before n^{th} collision. Then momentum conservation implies

$$m\underline{v}_n + M\underline{V}(\tau_n-) = m\underline{v}_n + M\underline{V}(\tau_n+)$$

where $\underline{v}'_n$ is the velocity of the colliding biomolecule after the nth collision. Also energy conservation implies

$$\frac{1}{2}mv_n^2 + \frac{1}{2}M\underline{V}(\tau_n)^2 = \frac{1}{2}mv_n'^2 + \frac{1}{2}M\underline{V}(\tau_{n+})^2$$

Also $V(\tau_{n+1-}) = V(\tau_{n+})$ if the robot does not move in an external potential. If however, there is an external potential $U(\underline{r})$, then

$$\frac{d\underline{r}(t)}{dt} = \underline{V}(t), \tau_{n+} < t < \tau_{n+1}-,$$

$$\frac{d\underline{V}(t)}{dt} = -U'(\underline{r}(t)),\ \tau_{n+} < t < \tau_{n+1}$$

$$r(\tau_{n+}) = r(\tau_{n-}),\ V(\tau_{n+}) \text{ is determined by}$$

(1) and (2) $\{\underline{v}_n : n \geq 1\}$ are iid $N(\underline{0}, \frac{3KI}{2}\underline{\underline{I}}_3)$ random variables.

❖❖❖❖❖

[204] The dynamics of an extended nano robot body a fluid environment studied using joint fluid dynamics and particle mechanics. Let $\underline{v}(t, \underline{r})$ be then fluid velocity field. Let B be the body of the nanorobot with surface S. Then the ith component of the viscous force on the body is given by

$$f_i^{visc} = \int_S \tau_{ij} n_j dS = \int_B \frac{\partial \tau_{ij}}{\partial X^j} dV$$

$$f_{ij}^{visc} = \eta\left(\frac{\partial v_j}{\partial X_i} + \frac{\partial v_i}{\partial X_j}\right)$$

Thus, $$f_i^{visc} = \eta\int_S \left(\frac{\partial v_j}{\partial X_i} + \frac{\partial v_i}{\partial X_j}\right)\eta_j\, dS$$

f_i^{visc} describes the contribution to the translators motion of the nano-robot. Rotation is designed by viscous torque about a fixed point.

$$\frac{dL}{dt} = \int_S \underline{r} \times \tau_{ij}\, \hat{e}_i\, \eta_j\, dS$$

$$\underline{r}\times\hat{e}_i = X_k\hat{e}_k\times\hat{e}_i = X_k\varepsilon(kim)\hat{e}_m$$

So,
$$\frac{d\underline{L}}{dt} = \left(\int_S \varepsilon\,(kim)\,X_k\,\tau_{ij}\,\eta_j\,dS\right)\hat{e}_m$$
$$= \int_S \underline{r}\times(\underline{\tau}.\hat{n})\,dS$$

The other (kinematical) expression for $\underline{L}$ can be computed using the Euler angle representation of rotations.

$$\underline{L} = \rho\int_B (\underline{r}_0^{(t)} + R(\varphi(t),\theta(t),\psi(t))\underline{\xi})$$
$$\times(\dot{r}_0(t) + R(\varphi,\theta,\psi))\,d^3\xi$$

where $\underline{r}_0(t)$ is the C.M. position

Since $\int_B \underline{\xi}d^3\xi = 0$, we get

$$\underline{L}(t) = \rho\underline{r}_0(t)\times\dot{\underline{r}}_0(t)$$
$$= +\rho\int_B R(\varphi,\theta,\psi)\underline{\xi}\times\dot{R}(\varphi,\theta,\psi)\underline{\xi}\,d^3\xi$$

Thus we get

$$\frac{d\underline{L}}{dt} = \dot{\rho}\underline{r}_0\times\ddot{\underline{r}}_0 + \rho\int_B R\xi\times\ddot{R}\xi\,d^3\xi$$

❖❖❖❖❖

[205] The effect of the nano-robot on the fluid motion within the body msut also be studied.

This is standard by choosing a reference frame relative to which the body is at rest and the fluid is flowing part it. Let ϕ be the velocity potential in steady state. Then

$$\nabla^2\phi(\underline{r}) = 0 \quad \underline{r}\notin B$$
$$\underline{v}(\underline{r}) = -\nabla\phi(\underline{r}),$$
$$-\underline{v}(\underline{r})\cdot\hat{r} = \frac{\partial\phi}{\partial r} = 0 \text{ for } \underline{r}\in S = \partial B$$

Thus steady flow can be analyzed by using the Newmann problem in elecrostatic.

❖❖❖❖❖

[206] Quantum effects.

Let $\underline{f}(t)$ be the force on a rised body and $\underline{\tau}(t)$ the torque. Then if $\underline{r}_0(t)$ is the position of the CM of the body and (φ, θ, ψ) are the Euler angles of rotation the interaction Lagrangian of the rigid body with the forces and torques is

$$= (\tau_\varphi \dot{\varphi} + \tau_\theta \dot{\theta} + \tau_\psi \dot{\psi}) + (\underline{f}, \dot{\underline{r}}_0)$$

The kinetic energy is

$$\frac{1}{2} m \dot{r}_0^2 + \frac{1}{2} (\dot{\rho}, \dot{\theta}, \dot{\psi}) J(\varphi, \theta, \psi) \begin{pmatrix} \dot{\varphi} \\ \dot{\theta} \\ \dot{\psi} \end{pmatrix}$$

and the gravitational potential energy is

$$V(\underline{r}_0, \varphi, \theta, \psi)$$

❖❖❖❖❖

[207] Quantum stochastic filtering control

$$dU(t) = L_b^a(t) d\Lambda_a^b U(t)$$

$$0 = d(U^*U) = dU^* \cdot U + U^* \cdot dU + dU^* \cdot dU$$

$$= U^* \left(L_b^a {}^* d\Lambda_b^a + L_b^a d\Lambda_a^b + L_b^a {}^* L_d^c \, d\Lambda_a^b \cdot d\Lambda_c^d \right) U$$

$$\Rightarrow \quad = U^* \left(\left(L_b^a {}^* + L_a^b \right) d\Lambda_b^a + \varepsilon_c^b \, L_b^a {}^* L_d^c \, d\Lambda_a^\alpha \right) U$$

$$\Leftrightarrow \quad L_b^a {}^* + L_a^b + \varepsilon_c^d L_a^b {}^* L_a^c = 0$$

i.e.

$$L_b^a {}^* + L_a^b + \sum_{j \geq 1} L_j^b {}^* L_a^j = 0$$

$L_b^a(t)$ are operator in $h \otimes \Gamma_s(\mathcal{H}_{t]})$

Let $$Y_i(t) = C_b^a \Lambda_a^b(t) \quad C_b^a \in \mathbb{C}$$

Then $\{Y_i(t) : t \geq 0\}$ is a commuting family of operator in $\Gamma s(\mathcal{H})$. Let

$$Y_0(t) = U^*(t) Y_i(t) U(t), t \geq 0$$

Then $$U^*(T) Y_i(t) U(T) = Y_0(t), T \geq t$$

provided that $$[Y_i(t), L_b^a(s)] = 0 \ \forall S \geq t, \forall a, b$$

This will be so if

$$L_b^a(t) = F_b^a(S_b^a(t), u(t))$$

where $S_b^a(t)$ are operator in $\underline{\mathcal{H}}$ and $u(t) = \psi(t, Y_i(t))$

y is an ordinary function and F_b^a are also ordinary function representable as Taylor series.

Let X be an operator in $\hbar$

Consider $j_t(X) = U^*(t)\, X\, U(t)$

Then $d_{jt}(X) = dU^*(t) X U(t) + U^*(t)\, X\, dU(t) + dU^*(t) X\, dU(t)$

$$= U^*(t)\left(L_b^a(t)^* X\, d\Lambda_a^b(t) + X L_b^a(t)\, d\Lambda_a^b(t) + L_b^a(t)^* X\, L_d^c(t)\, d\Lambda_b^a(t)\, d\Lambda_c^d(t)\right) U(t)$$

$$= U^*(t)\, (L_b^a(t)^* X + X\, L_a^b(t) + \varepsilon_c^d\, L_d^b(t)^* X\, L_a^c(t))\, U(t)\, d\Lambda_b^a(t)$$

$$= j_t\left(\theta_a^b(t, X)\right) d\Lambda_b^a(t)$$

where $\theta_a^b(t, X) = L_b^a(t)^* X + X\, L_a^b(t) + \varepsilon_c^d\, L_a^b(t)^* X\, L_a^c(t)$

We note that

$$\theta_a^b(t, X) = \theta_a^b(S(t), u(t), X)$$

where $S(t) = ((S_b^a(t)))$

$$j_t(\theta_a^b(t, X)) = \theta_a^b(j_t(S(t)), j_t(u(t)), j_t(X))$$

Note that

$$j_t(u(t)) = j_t(\psi(t, Y_i(t))) = \psi(t, j_t(Y_i(t)))$$

$$= \psi(t, Y_0(t))$$

The qsde for $j_t(X)$ reads

$$dX(t) = dj_t(X) = \theta_a^b(j_t(S(t)), j_t(u(t)), j_t(X))\, d\Lambda_b^a$$

$$= \theta_a^b(\tilde{S}(t), \psi(t, Y_0(t)), X(t))\, d\Lambda_b^a(t)$$

The function ψ of t and the output process $Y_0(t)$ is to be chosen so that

$$\mathbb{E}\int_0^T F(X(t), \psi(t, Y_0(t)))\,dt$$

is a minimum

Note that $\{Y_0(t), t \geq 0\}$ is a commuting family of operators (sine $Y_0(t) = U^*(T)Y_i(t)U(T), T \geq t$) and $Y_0(t)$ commutes with $X(s), s \geq t$ and also with

$$\tilde{S}(s), s \geq t \quad (\tilde{S}_b^a(t) = U^*(t)S_a^b(t)U(t)$$

so
$$\tilde{S}_b^a(s) = U^*(s)S_c^b(s)U(s)$$
$$Y_0(t) = U^*(s)Y_i(t)U(s), s \geq t$$

and

$$[Y_i(t), S_a^b(s)] = 0 \;\forall\, t, s$$

since $Y_i(t) \in \mathcal{L}(\Gamma s(\mathcal{H}))$,

$S_a^b(s) \in \mathcal{L}(\hbar)$.

Let
$$V(t) = \min_{\psi(s,.)} \mathbb{E}\left[\int_t^T F(X(s), \psi(s, Y_0(s)))\,ds\,\mathbb{B}_{t]}\right] \qquad (\alpha)$$

$t \leq s \leq T$ where $\mathbb{B}_{t]} = \sigma\{Y_0(s), s \leq t\}$.

Note that since $Y_0(t),\ t \geq 0$ form a commuting family and since $[Y_0(t), X(s)] = 0$, $s \geq t$, it follows that $\mathbb{B}_{t]}$ commutes with $\int_t^T F(X(s), \psi(s, Y_0(s)))\,ds$

$\forall$ function ψ and hence the conditional expectation in (α) is well defined.

We have

$$V(t) = \min_{\psi(t_j.)} \{\mathbb{E}[F(X(t), \psi(t, Y_0(t))) \mid \mathbb{B}_{t]}]dt$$
$$+ \mathbb{E}[V(t+dt) \mid \mathbb{B}_{t]}]\}$$

We note that $V(t)$ is a function of $\{Y_0(s), s \leq t\}$

Suppose $dV(t) = F_1(t)dt + F_2(t)dY_0(t)$

where $F_1(t), F_2(t)$ are function of $\{Y_0(s), s \leq t\}$

Then we get

$$\min_{\psi(t_j.)}\left\{F_1(t)+F_2(t)\mathbb{E}[dY_0(t)\,|\,\mathbb{B}_{t]}]/dt+\mathbb{E}[F(X(t),\psi(t,Y_0(t))\,|\,\mathbb{B}_{t]}]\right\}=0$$

Now,

$$\begin{aligned}
Y_0(t) &= U^*(t)\,c_b^a\Lambda_a^b(t)U(t)\\
&= c_b^a\,U^*(t)\Lambda_a^b(t)U(t)\\
dY_0(t) &= c_b^a[dU^*(t)\Lambda_a^b(t)U(t)\\
&\quad +U^*(t)\Lambda_a^b(t)dU(t)+dU^*(t)\Lambda_a^b(t)dU(t)\\
&\quad +d\Lambda_a^b(t)+dU^*(t)d\Lambda_a^b(t)U(t)\\
&\quad +U^*(t)d\Lambda_a^b(t)dU(t)]\\
&= dU^*(t)Y_i(t)U(t)+U^*(t)Y_i(t)dU(t)+dU^*(t)Y_i(t)dU(t)\\
&\quad +dU^*(t)dY_i(t)U(t)+U^*(t)dY_i(t)dU(t)+dY_i(t)\\
&= dU^*(t)dY_i(t)U(t)+U^*(t)dY_i(t)dU(t)+dY_i(t)\\
&= dY_i(t)+j_t\left(L_b^a(t)^*d\Lambda_b^a(t)c_q^p\,d\Lambda_p^q(t)c_q^p d\Lambda_p^q(t)L_b^a(t)d\Lambda_a^b(t)\right)\\
&= dY_i(t)+c_q^p\,j_t\left(L_b^a(t)^*\varepsilon_p^a\,d\Lambda_b^q(t)+L_b^a(t)\varepsilon_a^q\,d\Lambda_p^b(t)\right)\\
&= dY_i(t)+c_q^p\;\varepsilon_p^a\,j_t(L_b^a(t)^*)d\Lambda_b^q(t)\\
&\quad +c_q^p\;\varepsilon_a^q\,j_t\left(L_b^a(t)\right)d\Lambda_p^b(t)\\
&= dY_i(t)+[c_a^p\;\varepsilon_p^q\,j_t\left(L_b^q(t)^*\right)\\
&\quad +c_q^b\;\varepsilon_b^q\,j_t(L_a^b(t))]d\Lambda_b^a(t)\\
&= dY_i(t)+j_t\left(c_a^p\;\varepsilon_b^q L_b^q(t)^*+c_q^b\;\varepsilon_p^q L_a^p(t)\right)d\Lambda_b^a(t)
\end{aligned}$$

Let

$$R_a^b(t) = c_a^p\;\varepsilon_q^p L_b^q(t)^*+c_q^b\;\varepsilon_p^q L_a^p(t)$$

Then,

$$dY_0(t) = dY_i(t)+j_t(R_a^b(t))d\Lambda_b^a(t)$$

Assume that the expectation $\mathbb{E}$ is defined w.r.t the state $|f\varphi(v)\rangle\;\; f\in\mathscr{h},\|f\|=1,$

$$v\in\mathcal{H},|\varphi(v)\rangle = \exp\left(-\frac{\|v\|^2}{2}\right)|e(v)\rangle$$

Then

$$\begin{aligned}
\mathbb{E}[dY_0(t)\,|\,\mathbb{B}_{t]}] &= \mathbb{E}[dY_i(t)]+\mathbb{E}[j_t(R_a^b(t))\,\mathbb{E}[d\Lambda_b^a(t)]\,|\,\mathbb{B}_{t]}]\\
&= c_b^a\left\langle f\varphi(v)\,|\,d\Lambda_a^b(t)\,|\,f\varphi(v)\right\rangle\\
&\quad +\mathbb{E}\left[\,j_t(R_a^b(t))\left\langle f\varphi(v)\,|\,d\Lambda_b^a(t)\,|\,f\varphi(v)\right\rangle\middle|\mathbb{B}_{t]}\right]
\end{aligned}$$

$$= \exp(-\|v\|^2)\Big[c_b^a v_b(t)\overline{v_b(t)}dt + \mathbb{E}\Big[j_t(R_a^b(t))v_a(t)\overline{v_b(t)dt}\Big]\Big|\mathbb{B}_{t]}\Big]$$

so

$$\mathbb{E}[dY_0(t)|\mathbb{B}_{t]}]/dt = \exp(-\|v\|^2)\Big[\mathbb{E}[j_t(R_b^a(t)) + c_b^a]\overline{v_a(t)}v_b(t)|\mathbb{B}_{t]}\Big]$$

So $\min_{\psi(t_j.)}\Big\{F_1(t) + F_2(t)\exp(-\|v\|^2)[\mathbb{E}[j_t(R_b^a(t))|\mathbb{B}_{t]}] + c_b^a]\overline{v}_a(t)v_b(t)$

$$+\mathbb{E}[F(X(t), \psi(t, Y_0(t)))/\mathbb{B}_{t]}]\} = 0$$

Note that

$R_b^a(t)$ is a linear combination of $\{L_q^p(t), L_q^p(t)^*\}_{p,q\geq 0}$

and hence is expressible as a function of $\left(S_b^a(t)\right), Y_i(t)$

Thus $j_t(R_b^a(t))$ is a function of $(\tilde{S}_b^a(t)), Y_0(t)$

or more precisely as a function of $(\tilde{S}_b^a(t)), u(t) = \psi(t, Y_0(A))$

We write

$$V(t) = V(t, \mathbb{B}_{t]}) \quad \mathbb{B}_{t]} = \sigma\{Y_0(s), s \leq t\}$$

Then $V(t+dt) = V(t+dt, \mathbb{B}_{t]}, dY_0(t))$

$$= V(t, \mathbb{B}_{t]}) + \frac{\partial V}{\partial t}(t, \mathbb{B}_{t]})dt + \sum_{k=1}^{\infty} V_k(t)(dY_0(t))^k$$

where $V_k(t) = \frac{1}{\lfloor k}\frac{\partial^k V(t, \mathbb{B}_{t]}, \xi)}{\partial \xi^k}\Big|_{\xi=0} \equiv V_k(t, \mathbb{B}_{t]})$

So, $\mathbb{E}[V(t+dt)|\mathbb{B}_{t]}] = V(t, \mathbb{B}_{t]}) + \frac{\partial V}{\partial t}(t, \mathbb{B}_{t]})dt$

$$+\sum_{k=1}^{\infty} V_k(t)\mathbb{E}[dY_0(t)^k|\mathbb{B}_{t]}]$$

$$\mathbb{E}[dY_0(t)^k|\mathbb{B}_{t]}] = \mathbb{E}\Big\{U^*(t)[(dY_i(t) + R_a^b(t)d\Lambda_b^c(t))^k]U(t)|\mathbb{B}_{t]}\Big\}$$

$$= \mathbb{E}\left\{U^*(t)\,\mathbb{E}\left[\left(\left(c_a^b + R_a^b(t)\right)d\Lambda_b^a(t)\right)^k\right]U(t)|\mathbb{B}_{t]}\right\}$$

$$= \mathbb{E}\left[j_t\left(\prod_{mk=1}^{k}\left(c_{am}^{bm} + R_{am}^{bm}(t)\right)\left[\prod_{m=1}^{k} d\Lambda_{bm}^{am}(t)\right]\right)\Big|\mathbb{B}_{t]}\right]$$

Now $\prod_{m=1}^{k} d\Lambda_{bm}^{am} = d\Lambda_{b_1}^{a_1}d\Lambda_{b_2}^{c_2}\ldots d\Lambda_{b_k}^{a_k}$

$$= \varepsilon_{b_2}^{a_1} \varepsilon_{b_3}^{a_2} \dots \varepsilon_{b_k}^{a_{k-1}} \, d\Lambda_{b_1}^{a_k}$$

and hence,

$$\mathbb{E}\left(\prod_{m=1}^{k} d\Lambda_{bm}^{am}(t)\right) = \left(\prod_{j=1}^{k-1} \varepsilon_{b_{j+1}}^{a_j}\right) v_{a_k}(t)\overline{v_{b_1}(t)}\, dt$$

so
$$\mathbb{E}\left[V(t+dt) \mid \mathbb{B}_{t]}\right] = V(t,\mathbb{B}_{t]}) + \frac{\partial V}{\partial t}(t,\mathbb{B}_{t]})dt + \sum_{k=1}^{\infty} V_k(t,\mathbb{B}_{t]}) \prod_{j=1}^{k-1} \varepsilon_{b_{j+1}}^{a_j}$$

$$\times \mathbb{E}\left[j_t\left(\prod_{m=1}^{k}\left(c_{am}^{bm} + R_{am}^{bm}(t)\right)\right) d\Lambda_{b_1}^{a_k} \mid \mathbb{B}_{t]}\right]$$

$$= V(t,\mathbb{B}_{t]}) + \frac{\partial V}{\partial t}(t,\mathbb{B}_{t]})dt$$

$$+ \sum_{k=1}^{\infty}\left\{ V_k(t,\mathbb{B}_{t]}) \prod_{j=1}^{k-1} \varepsilon_{b_{j+1}}^{a_j}\right.$$

$$\left. \times \mathbb{E}\left[j_t\left(\prod_{m=1}^{k}\left(c_{am}^{bm} + R_{am}^{bm}(t)\right)\right) \mid \mathbb{B}_{t]}\right] v_{a_k}(t)\overline{v_{b_1}(t)}\right\}$$

So finally, our quantum stochastic BHJ eqn. is

$$-\frac{\partial V}{\partial t}(t,\mathbb{B}_{t]}) = \min_{\psi(t_{j.})}\left\{\mathbb{E}\left[F(X(t),\psi(t,Y_0(t))) \mid \mathbb{B}_{t]}\right]\right.$$

$$+ \sum_{k=1}^{\infty}\left\{ V_k(t,\mathbb{B}_{t]})\, \mathbb{E}\left(j_t\left(\prod_{m=1}^{k}\left(c_{am}^{bm} + R_{am}^{bm}(t)\right)\right)\right)\right.$$

$$\left. \mid \mathbb{B}_{t]}] v_{a_k}(t)\overline{v_{b_1}(t)}\right\}$$

❖❖❖❖❖

[208] Quantum gravity metric

Dirac field in curved stochastic tetrad (approximate)

Let $e_a^{\mu}(x)$ be a tetrad *i.e.*

$$\eta^{ab} e_a^{\mu} e_b^{\nu} = g^{\mu\nu}$$

where $((\eta^{ab})) = diag\,[1,-1,-1,-1]$

Consider $\gamma^a e_a^{\mu} p_{\mu} - m$

where $\left(\begin{smallmatrix} p \\ \mu \end{smallmatrix}\right)_{\mu=0}^{3}$ are real numbers and $\left(\gamma^a\right)_{a=0}^{3}$ are the

Dirac Gamma matrices. They satisfy

$$\gamma^a\gamma^b + \gamma^b\gamma^a = 2\eta^{ab}$$

We have

$$\begin{aligned}&(\gamma^a e_a^\mu p_\mu + m)(\gamma^a e_a^\mu p_\mu - m) \\ &= \gamma^a\gamma^b e_a^\mu e_b^\nu p_\mu p_\nu - m^2 \\ &= n^{ab} e_a^\mu e_b^\nu p_\mu p_\nu - m^2 \\ &= g^{\mu\nu} p_\mu p_\nu - m^2\end{aligned}$$

which is the Klin Goarden operator with $i\partial_\mu$ replaced by the scalar p_μ.

Hence, a candidate for the approximate Dirac equation in the metric $g^{\mu\nu}(x)$ is

$$\left[i\gamma^a e_a^\mu(x)\partial_\mu - m\right]\psi(x) = 0$$

Note that we can take

$$\gamma^0 = \begin{pmatrix} I_2 & 0 \\ 0 & -I_2 \end{pmatrix}, \gamma^r = \begin{pmatrix} 0 & \sigma_r \\ -\sigma_r & 0 \end{pmatrix}, \ 1 \le r \le 3$$

In flat space time, the Lagrangian density for the free Dirac field is

$$\mathcal{L}(\psi, \psi^+, \partial_\mu\psi) = \psi^+ i\partial_0\psi - \psi^+(\alpha, -i\nabla)\psi - m\psi^+\beta\psi$$

where

$$\alpha^r = \gamma^0\gamma^r = \begin{pmatrix} 0 & \sigma_r \\ \sigma_r & 0 \end{pmatrix}, \ 1 \le r \le 3, \ \beta = \gamma^0$$

In curved space-time, this must be replaced by

$$\psi^+ i\partial_0\psi - \frac{1}{2}\psi^+\left((\tilde{\alpha}, -i\nabla) + (-i\nabla, \tilde{\alpha})\right)\psi - m\psi^+\tilde{\beta}\psi$$

where

$$\tilde{\alpha}^r = \left(\tilde{\gamma}^0\right)^{-1}\tilde{\gamma}^r, \tilde{\beta} = \tilde{\gamma}^{0^{-1}}$$

where

$$\tilde{\gamma}^\mu = \gamma^a e_a^\mu(x)$$

$$\begin{aligned}\tilde{\gamma}^0_{(x)} &= \gamma^a e_a^0(x) = \gamma^0 e_0^0(x) + \gamma^r e_r^0(x) \\ &= e_0^0\begin{pmatrix} I & 0 \\ 0 & -I \end{pmatrix} + e_r^0\begin{pmatrix} 0 & \sigma_r \\ -\sigma_r & 0 \end{pmatrix} \\ &= \begin{pmatrix} e_0^0 & \sum_1^3 e_r^0\sigma_r \\ -\sum_1^3 e_r^0\sigma_r & -e_0^0 \end{pmatrix}\end{aligned}$$

The action for the interaction of matter with an enfield (Lagrangian density)

$$\mathcal{L}_{\text{int}} = -\psi^{+}_{(x)}\left(\tilde{\alpha}^{\mu}_{(x)} + \tilde{\alpha}^{\mu^{+}}_{(x)}\right)\psi_{(x)}\Lambda^{(x)}_{\mu}$$

$$\tilde{\alpha}^{\mu} = \left(I, \left(\tilde{\alpha}^{r}\right)^{3}_{r=1}\right)$$

Note $$\sum_{a=0}^{3} e^{\mu}_{a}\sigma_{a} = e^{\mu}_{0}I_{2} + \sum_{r=1}^{3} e^{\mu}_{r}\sigma_{r}$$

$$= \begin{pmatrix} e^{\mu}_{0} + e^{\mu}_{3} & e^{\mu}_{1} + ie^{\mu}_{2} \\ e^{\mu}_{1} - ie^{\mu}_{2} & e^{\mu}_{0} - e^{\mu}_{3} \end{pmatrix}$$

Action for the gravitational field (Lagrangian density)

$$\mathcal{L}_{grov} = g^{\mu\nu}\left(\Gamma^{\alpha}_{\mu\nu}\Gamma^{\beta}_{\alpha\beta} - \Gamma^{\alpha}_{\mu\beta}\Gamma^{\beta}_{\nu\alpha}\right)\sqrt{-g}$$

We evaluate this in terms of e^{a}_{μ} for small perturbation from flat space-time:

$$e^{\mu}_{a} = \delta^{\mu}_{a} + f^{\mu}_{a} \quad f^{\mu}_{a} \text{ small}$$

$$e^{a}_{\mu} \approx \delta^{a}_{\mu} + h^{a}_{\mu}$$

$$e^{\mu}_{a}e^{a}_{\nu} = \delta^{\mu}_{\nu}$$

$$\Rightarrow \quad \left(\left(e^{a}_{\mu}\right)\right) = \left(\left(e^{\mu}_{a}\right)\right)^{-1}$$

$$\approx \left(\left(\delta^{\mu}_{a} - f^{\mu}_{a} + f^{\mu}_{b}f^{b}_{a}\right)\right)$$

We write $$f^{\mu}_{a} = h^{\mu}_{a} \equiv g^{\mu\alpha}\eta_{ab}h^{b}_{\alpha}$$

$$g_{\mu\nu} = e^{a}_{\mu}e^{b}_{\nu}\eta_{ab} = \eta_{ab}\left(\delta^{a}_{\mu} + h^{a}_{\mu}\right)\left(\delta^{b}_{\nu} + h^{b}_{\nu}\right)$$

$$= \eta_{\mu\nu} + \eta_{\mu b}h^{b}_{\nu} + \eta_{\nu a}h^{a}_{\mu} + \eta_{ab}h^{a}_{\mu}h^{b}_{\nu}$$

$$\equiv \eta_{\mu\nu} - h_{\mu\nu} + h_{\nu\mu} + h_{a\mu}h^{a}_{\nu}$$

$$g^{\mu\nu} \approx \eta_{\mu\nu} + h_{\mu\nu} - h_{\nu\mu} - h_{a\mu}h^{a}_{\nu} + \left(h_{\mu\alpha} - h_{\alpha\mu}\right)\left(h_{\alpha\nu} - h_{\nu\alpha}\right)$$

Upto 4th degree terms in $\left\{h^{a}_{\mu}\right\}$, $\mathcal{L}_{grov}$ has the form

$(abcd;\ \mu\nu\alpha\beta\gamma\rho)\ h^{a}_{\mu,\nu}\ h^{b}_{\alpha,\beta}\ h^{c}_{\gamma}\ h^{d}_{\rho}$

where $(abcd\ \mu\nu\alpha\beta\gamma\rho)$ are constants. The interaction part of the Dirac field with the gravitational field in approximately (Upto first degree in $h^{a}_{\mu}, h^{a}_{\mu\nu}$)

$$\text{Re}\left\{2i\bar{\psi}_a \psi_{b,\alpha} h_c^\alpha \alpha_{ab}^C + i\bar{\psi}_a \psi_b h_{c,\alpha}^\alpha \alpha_{ab}^c\right\}$$

The interaction part of the Dirac field with the em and gravitational fields in approximately

$$\text{Re}\left\{2\,\bar{\psi}_a A_\alpha \psi_b h_c^\alpha \alpha_{ab}^C + \bar{\psi}_a \psi_b h_c^\alpha A_\alpha \alpha_{ab}^c\right\}$$

This is obtained by replacing $i\partial_\alpha$ by A_α in the previous expression

The component of the Lagrangian density that describe induction between the Dirac field and the electromagnetic field is

$$\mathcal{L}_{Dem} = \psi^+ \alpha^\mu \psi A_\mu = \psi^+ \psi A_0 + \sum_{r=1}^{3} \psi^+ \alpha^r \psi A_r$$

Finally the component that describe interaction between the gravitational field and the electromagnetic field is

$$\mathcal{L}_{grem} = F_{\alpha\beta} F_{\mu\nu}\left(\eta^{\mu\alpha}\delta g^{\nu\beta} + \eta^{\nu\beta}\delta g^{\mu\alpha}\right)$$

where
$$\begin{aligned}\delta g^{\nu\beta} &= g^{\nu\beta} - \eta^{\nu\beta} \\ &= \eta^{ab}\left(h_a^\nu \delta_b^\beta + h_b^\beta \delta_a^\nu + h_a^\nu h_b^\beta\right)\end{aligned}$$

Thus
$$\begin{aligned}\mathcal{L}_{grem} &= F_{\alpha\beta} F_{\mu\nu}\Big(\eta^{\mu\alpha}\eta^{ab}\left(h_a^\nu \delta_b^\beta + h_b^\beta \delta_a^\nu + h_a^\nu h_b^\beta\right) \\ &\quad + \eta^{\nu\beta}\eta^{ab}\left(h_a^\mu \delta_b^\alpha + h_b^\alpha \delta_a^\mu + h_a^\mu h_b^\beta\right)\Big) \\ &= F_{\alpha\beta} F_{\mu\nu}\Big(\eta^{\mu\alpha}\left(h_a^\nu \eta^{a\beta} + h_b^\beta \eta^{\nu b} + \eta^{ab} h_a^\nu h_b^\beta\right) \\ &\quad + \eta^{\nu\beta}\left(h_a^\mu \eta^{a\alpha} + h_b^\alpha \eta^{\mu b} + \eta^{ab} h_a^\mu h_b^\beta\right)\Big)\end{aligned}$$

❖❖❖❖❖

[209] ρ_t is a density matrix. Then

$$Tr\left(\rho_t \frac{d}{dt}\log\rho_t\right) = 0$$

Proof: Let $\log\rho_t = X_t$. Then

$$\rho_t = \exp(X_t)$$

$$\begin{aligned}\frac{d\rho_t}{dt} &= \rho_t \cdot \frac{(I - \exp(-ad\,X_t))}{ad\,X_t}\left(\frac{dX_t}{dt}\right) \\ &= \rho_t \int_0^1 \exp(-s\,ad\,X_t)\,ds\left(\frac{dX_t}{dt}\right)\end{aligned}$$

$$= \rho_t \int_0^1 \rho_t^{-s}\left(\frac{dX_t}{dt}\right)\rho_t^s\, ds$$

$$= \int_0^1 \rho_t^{1-s}\cdot\frac{dX_t}{dt}\cdot\rho_t^s\, ds$$

So, $$0 = Tr\left(\frac{d\rho_t}{dt}\right) = \int_0^1 Tr\left(\rho_t^{1-s}\frac{dX_t}{dt}\rho_t^s\right)ds$$

$$= \int_0^1 Tr\left(\frac{dX_t}{dt}\right)ds = Tr\left(\rho_t\frac{dX_t}{dt}\right)$$

$$= Tr\left(\rho_t\frac{d\log\rho_t}{dt}\right)$$

❖❖❖❖❖

[210] Study projects

1. Stochastic problems in cosmology
2. Calculation of photon propagator
3. Calculation of electron propagator
4. Approximate quantum general relativity.
5. Quantum teleportation

1. $$R_{\mu\nu} = \frac{1}{2}Rg_{\mu\nu} = G_{\mu\nu} = KT_{\mu\nu}$$

$$g_{\mu\nu}(x) = g_{\mu\nu}^{(0)}(x) + \delta g_{\mu\nu}(x)$$

$$T_{\mu\nu}(x) = T_{\mu\nu}^{(0)}(x) + \delta T_{\mu\nu}(x)$$

$$G_{\mu\nu}^{(0)} = KT_{\mu\nu}^{(0)}$$

$$\delta G_{\mu\nu} = K\delta T_{\mu\nu}$$

$\delta g_{\mu\nu} \simeq$ stochastic fluctuation in background metrics $g_{\mu\nu}^{(0)}$

$\delta G_{\mu\nu} \simeq$ stochastic fluctuation in background Energy tenor $G_{\mu\nu}^{(0)}$

$\delta T_{\mu\nu} \simeq$ stochastic fluctuation in background Energy-momentum tenor $T_{\mu\nu}^{(0)}$

$$T^{\mu\nu} = g^{\mu\nu}g^{\nu\beta}T_{\alpha\beta} = \left(g^{(0)\mu\alpha} + \delta g^{\mu\alpha}\right)\left(g^{(0)\nu\beta} + \delta g^{\nu\beta}\right)\left(T_{\alpha\beta}^{(0)} + \delta T_{\alpha\beta}\right)$$

$$= g^{(0)\mu\alpha}g^{(0)\nu\beta}T_{\alpha\beta}^{(0)} + g^{(0)\mu\alpha}T_{\alpha\beta}^{(0)}\delta g^{\nu\beta}$$

$$+ g^{(0)\nu\beta}T_{\alpha\beta}^{(0)}\delta g^{\mu\alpha} + g^{(0)\mu\alpha}g^{(0)\nu\beta}\delta T_{\alpha\beta}$$

$$\left(\left(g^{(0)\alpha\beta}\right)\right) = \left(\left(g^{(0)}_{\alpha\beta}\right)\right)^{-1}$$

$$\delta g^{\mu\nu} = -g^{(0)\mu\alpha}g^{(0)\nu\beta}\delta g_{\alpha\beta}$$

The Einstien field equation can also be expressed as

$$R_{\mu\nu} = K\left(T_{\mu\nu} - \frac{1}{2}Tg_{\mu\nu}\right)$$

where $\quad T = g^{\mu\nu}T_{\mu\nu}$

So

$$R^{(0)}_{\mu\nu} = K\left(T^{(0)}_{\mu\nu} - \frac{1}{2}T^{(0)}g^{(0)}_{\mu\nu}\right)$$

$$\delta R_{\mu\nu} = K\left(\delta T_{\mu\nu} - \frac{1}{2}\delta T\cdot g^{(0)}_{\mu\nu} - \frac{1}{2}T^{(0)}\delta g_{\mu\nu}\right)$$

The second equation determines stochastic fluctuations $\delta g_{\mu\nu}$ in the metric produced by stochastic fluctuations $\delta T_{\mu\nu}$ in the energy momentum term

$$T^{(0)} = g^{(0)\mu\nu}T^{(0)}_{\mu\nu},$$

$$\delta T = \delta g^{\mu\nu}\cdot T^{(0)}_{\mu\nu} + g^{(0)\mu\nu}\delta T_{\mu\nu}$$

$$= -g^{(0)\mu\alpha}g^{(0)\nu\beta}\cdot\delta g_{\alpha\beta}\cdot T^{(0)}_{\mu\nu} + g^{(0)\mu\nu}\delta T_{\mu\nu}$$

$$= -T^{(0)\alpha\beta}\delta g_{\alpha\beta} + g^{(0)\mu\nu}\delta T_{\mu\nu}$$

$$\delta R_{\mu\nu} = \delta\left\{\Gamma^{\alpha}_{\mu\alpha,\nu} - \Gamma^{\alpha}_{\mu\nu,\alpha} - \Gamma^{\alpha}_{\mu\nu}\Gamma^{\beta}_{\alpha\beta} + \Gamma^{\alpha}_{\mu\beta}\Gamma^{\beta}_{\nu\alpha}\right\}$$

$$\delta\Gamma^{\alpha}_{\mu\alpha,\nu} = \left(\delta\Gamma^{\alpha}_{\mu\alpha}\right)_{,\nu}$$

$$\Gamma^{\alpha}_{\mu\alpha} = g^{\alpha\beta}\Gamma_{\beta\mu\alpha} = \frac{1}{2}g^{\alpha\beta}\left(g_{\beta\mu,\alpha} + g_{\beta\alpha,\mu} - g_{\alpha\mu,\beta}\right) = \frac{1}{2}g^{\alpha\beta}g_{\alpha\beta,\mu}$$

$$\therefore \quad \delta\Gamma^{\alpha}_{\mu\alpha} = -\frac{1}{2}g^{(0)\alpha\beta}\delta g_{\alpha\beta,\mu} - \frac{1}{2}g^{(0)}_{\alpha\beta,\mu}\left(g^{(0)}_{\alpha\rho}g^{(0)\beta\sigma}\delta g_{\rho\sigma}\right)$$

$$\delta\Gamma^{\alpha}_{\mu\alpha,\nu} = -\frac{1}{2}g^{(0)\alpha\beta}_{,\nu}\delta g_{\alpha\beta,\mu} - \frac{1}{2}g^{(0)\alpha\beta}\delta g_{\alpha\beta,\mu\nu}$$

$$-\frac{1}{2}g^{(0)}_{\alpha\beta,\mu}g^{(0)\alpha\rho}g^{(0)\beta\sigma}\delta g_{\rho\sigma,\nu}$$

$$-\frac{1}{2}\left(g^{(0)}_{\alpha\beta,\mu}g^{(0)\alpha\rho}g^{(0)\beta\sigma}\right)_{,\nu}\delta g_{\rho\sigma}$$

$$\delta\Gamma^{\alpha}_{\mu\nu,\alpha} = \left(\delta\Gamma^{\alpha}_{\mu\nu}\right)_{,\alpha}$$

$$\begin{aligned}
\delta\Gamma^{\alpha}_{\mu\nu} &= \delta\left(g^{\alpha\beta}\Gamma_{\beta\mu\nu}\right) = \delta g^{\alpha\beta}\Gamma^{(0)}_{\beta\mu\nu} + g^{(0)\alpha\beta}\delta\Gamma_{\beta\mu\nu} \\
&\quad - g^{(0)\alpha\rho}g^{(0)\beta\sigma}\Gamma^{(0)}_{\beta\mu\nu}\delta g_{\rho\sigma} \\
&\quad + \frac{1}{2}g^{(0)\alpha\beta}\left(\delta g_{\beta\mu,\nu} + \delta g_{\beta\nu,\mu} - \delta g_{\mu\nu,\beta}\right) \qquad (a) \\
\left(\delta\Gamma^{\alpha}_{\mu\nu}\right)_{,\alpha} &= -\left(g^{(0)\alpha\rho}\Gamma^{(0)\sigma}_{\mu\nu}\right)_{,\alpha}\delta g_{\rho\sigma} \\
&\quad - g^{(0)\alpha\rho}\Gamma^{(0)\sigma}_{\mu\nu}\delta g_{\rho\sigma,\alpha} + \frac{1}{2}g^{(0)\alpha\beta}_{,\alpha}\left(\delta g_{\beta\mu,\nu} + \delta g_{\beta\nu,\mu} - \delta g_{\mu\nu,\beta}\right) \\
&\quad + \frac{1}{2}g^{(0)\alpha\beta}\left(\delta g_{\beta\mu,\nu\alpha} + \delta g_{\beta\nu,\mu\alpha} - \delta g_{\mu\nu,\alpha\beta}\right) \\
\delta\left(\Gamma^{\alpha}_{\mu\nu}\Gamma^{\beta}_{\alpha\rho}\right) &= \Gamma^{(0)\alpha}_{\mu\nu}\delta\Gamma^{\beta}_{\alpha\rho} + \Gamma^{(0)\beta}_{\alpha\beta}\delta\Gamma^{\alpha}_{\mu\nu} \\
\delta\left(\Gamma^{\alpha}_{\mu\beta}\Gamma^{\beta}_{\nu\alpha}\right) &= \Gamma^{(0)\alpha}_{\mu\beta}\delta\Gamma^{\beta}_{\nu\alpha} + \Gamma^{(0)\beta}_{\nu\alpha}\delta\Gamma^{\alpha}_{\mu\beta}
\end{aligned}$$

where $\delta\Gamma^{\alpha}_{\mu\nu}$ is given by (a)

Thus the equation

$$\delta R_{\mu\nu} = K\left(\delta T_{\mu\nu} - \frac{1}{2}\delta T\cdot g^{(0)}_{\mu\nu} - \frac{1}{2}T^{(0)}\delta g_{\mu\nu}\right)$$

is a linear second order partial differential equation for $\delta g_{\mu\nu}(x)$ driven by the random source

$\delta T_{\mu\nu}(x)$. The coefficients in this pde are functions of $X = (t, \underline{r}) = (X^{\mu})$ determined by the background metrics $g^{(0)}_{\mu\nu}(x)$.

The equation of continuity

$T^{\mu\nu}_{:\nu} = 0$ follows from the bianchi identity

$G^{\mu\nu}_{:\nu} = 0$. Thus

$$\left(T^{(0)\mu\nu} + \delta T^{\mu\nu}\right)_{:\nu} = 0$$

Now

$$0 = T^{\mu\nu}_{:\nu} = T^{\mu\nu}_{,\nu} + T^{\mu\alpha}\Gamma^{\nu}_{\alpha\nu} + T^{\alpha}\Gamma^{\mu}_{\alpha\nu}$$

so

$$T^{(0)\mu\nu}_{,\nu} + T^{(0)\mu\alpha}\Gamma^{(0)\nu}_{\alpha\nu} + T^{(0)\nu\alpha}\Gamma^{(0)\mu}_{\alpha\nu} = 0$$

$$\delta T^{\mu\nu}_{,\nu} + \delta T^{\mu\alpha}\Gamma^{(0)\nu}_{\alpha\nu} + T^{(0)\mu\alpha}\delta\Gamma^{\nu}_{\alpha\nu} + \delta T^{\nu\alpha}\Gamma^{(0)\mu}_{\alpha\nu} + T^{(0)\nu\alpha}\Gamma^{\mu}_{\alpha\nu} = 0 \quad (b)$$

$$\begin{aligned}
\delta T^{\mu\nu} &= (\delta T^{\mu\nu}) = (\delta g^{\mu\alpha}g^{\nu\beta}T_{\alpha\beta}) \\
&= g^{(0)\mu\alpha}g^{(0)\nu\beta}\delta T_{\alpha\beta} - g^{(0)\mu\rho}g^{(0)\alpha\sigma}g^{(0)\nu\beta}T^{(0)}_{\alpha\beta}\delta g_{\rho\sigma} \\
&\quad - g^{(0)\mu\alpha}g^{(0)\nu\rho}g^{(0)\beta\sigma}T^{(0)}_{\alpha\beta}\delta g_{\rho\sigma}
\end{aligned}$$

So, (*b*) gives as linear pde for $\delta T_{\alpha\beta}$ in terms of $\delta g_{\mu\nu}(x)$. Thus, we get linear pde's for $\delta T_{\mu\nu}(x)$ and $\delta g_{\mu\nu}(x)$ of second order with coefficients depending on $T^{(0)}_{\mu\nu}(x)$ and $g^{(0)}_{\mu\nu}(x)$.

❖❖❖❖❖

[211] Quantum teleportation

Alice and Bob share the entangled state

$$\frac{1}{\sqrt{d}}\sum_{\alpha=1}^{d}\left|u_\alpha^A \otimes u_\alpha^B\right\rangle$$

where $\left\{\left|u_\alpha^A\right\rangle\right\}_{\alpha=1}^{d}$ and $\left\{\left|u_\beta^B\right\rangle\right\}_{\beta=1}^{d}$ are

ONB's for $\mathcal{H}_A$ and $\mathcal{H}_B$ respectively.

Alice wishes to transmit the state

$$\left|x^C\right\rangle = \sum_{\alpha=1}^{d} x_\alpha\left|u_\alpha^C\right\rangle$$

to bob where $\left\{\left|u_\alpha^C\right\rangle\right\}_{\alpha=1}^{d}$ is an ONB for $\mathcal{H}_C$.

$\mathcal{H}_A \otimes \mathcal{H}_C$ is possenel by Alice and $\mathcal{H}_B$ by Bob.

Assume that there exist operator (unitary) $S_{\alpha\beta}^{AC}$

such that $\left|e_{\alpha\beta}^{AC}\right\rangle = S_{\alpha\beta}^{AC}\sum_{\alpha'=1}^{d}\left|u_{\alpha'}^A \otimes u_{\alpha'}^C\right\rangle\sqrt{d}$, $1 \le \alpha, \beta \le d$

is an ONB for $\mathcal{H}_A \otimes \mathcal{H}_C$. Let $S = S_{\alpha\beta}^{AC}$.

The initial state of Alice and Bob with C attached at Alice is enol is

$$\left|\psi\right\rangle = \left|x^C\right\rangle \otimes \frac{1}{\sqrt{d}}\sum_{\alpha=1}^{d}\left|u_\alpha^A \otimes u_\alpha^B\right\rangle$$

$$\equiv \frac{1}{\sqrt{d}}\sum_{\alpha=1}^{d}\left|u_\alpha^A \otimes x^C \otimes u_\alpha^B\right\rangle$$

Alice takes a measurement of her side in the PVM $\left\{\left|e_{\alpha\beta}^{AC}\right\rangle\left\langle e_{\alpha\beta}^{AC}\right| \{\alpha, \beta = 1, 2, ..., d\}\right\}$ and notes the outcome (α, β). Then the state $\left|\psi\right\rangle$ collapses to

$$\rho_{\alpha\beta}^{ABC} = \left(I_B \otimes \left|e_{\alpha\beta}^{AC}\right\rangle\left\langle e_{\alpha\beta}^{AC}\right|\right)\left|\psi\right\rangle\left\langle\psi\right|\left(I_B \otimes \left|e_{\alpha\beta}^{AC}\right\rangle\left\langle e_{\alpha\beta}^{AC}\right|\right)$$

Bob's state because

$$\rho^B = Tr_{A,C}\left(\rho_{\alpha\beta}^{ABC}\right) = \frac{1}{\sqrt{d}}Tr_{A,C}\sum_{\alpha'=1}^{d}\sum_{\beta'=1}^{d}\left|e_{\alpha\beta}^{AC}\right\rangle\left\langle e_{\alpha\beta}^{AC}\right|\left|u_{\alpha'}^A \otimes x^C \otimes u_\alpha^B\right\rangle$$

$$\left\langle u_{\beta'}^{A}\otimes x^{C}\otimes u_{\beta'}^{B}\right\|\left|e_{\alpha\beta}^{AC}\right\rangle$$

$$= \frac{1}{\sqrt{d}}\sum_{\alpha',\beta'=1}^{d}\left\langle e_{\alpha\beta}^{AC}\mid u_{\alpha'}^{A}\otimes x^{C}\right\rangle\left\langle u_{\beta'}^{A}\otimes x^{C}\mid e_{\alpha\beta}^{AC}\right\rangle$$

$$\left|u_{\alpha'}^{B}\right\rangle\left\langle u_{\beta'}^{B}\right|$$

$$= \frac{1}{(\sqrt{d})^{3}}\sum_{\alpha'\beta'\rho\sigma}\left\langle S(u_{\rho}^{A}\otimes u_{\rho}^{C})\mid u_{\alpha'}^{A}\otimes x^{C}\right\rangle\left\langle u_{\beta'}^{A}\otimes x^{C}\mid S u_{\sigma}^{A}\otimes u_{\sigma}^{C}\right\rangle$$

$$\left|u_{\alpha'}^{\beta}\right\rangle\left\langle u_{\beta'}^{B}\right|$$

$$= \frac{1}{d^{3/2}}\sum_{\mu\nu\rho\sigma'\beta'} x_{\nu}x_{\mu}\left\langle S\left(u_{\rho}^{A}\otimes u_{\rho}^{C}\right)\mid u_{\alpha'}^{A}\otimes u_{\mu}^{C}\right\rangle$$

$$\left\langle u_{\beta'}^{A}\otimes u_{\nu}^{C}\mid S\mid u_{\sigma}^{A}\otimes u_{\sigma}^{C}\right\rangle\left|u_{\alpha'}^{B}\right\rangle\left\langle u_{\beta'}^{B}\right|$$

Write

$$S\left|u_{\alpha'}^{A}\otimes u_{\beta'}^{C}\right\rangle = \sum_{\alpha''\beta''} S(\alpha''\beta''\mid\alpha'\beta')\left|u_{\alpha''}^{A'}\otimes u_{\beta''}^{C}\right\rangle$$

Then

$$\rho_{B} = \frac{1}{d^{3/2}}\sum_{\substack{\mu\nu\rho\sigma\alpha'\beta'\\ \alpha_1\beta_1\alpha_2\beta_2}}\bar{x}_{\nu}x_{\mu}\bar{S}(\alpha_1\,\beta_1\mid\rho,\rho)\,\delta(\alpha',\alpha_1)\,\delta(\mu,\beta_1)$$

$$S(\alpha_2,\beta_2\mid\sigma,\sigma)\,\delta(\beta',\alpha_2)\,\delta(\nu,\beta_2)\left|u_{\alpha'}^{B}\right\rangle\left\langle u_{\beta'}^{B}\right|$$

$$= \frac{1}{d^{3/2}}\sum_{\mu\nu\rho\sigma\alpha'\beta'}\bar{x}_{\nu}\,x_{\mu}\,\bar{S}(\alpha'\mu\mid\rho\rho)\,S(\beta'\nu\mid\sigma\sigma)\left|u_{\alpha'}^{B}\right\rangle\left\langle u_{\beta}^{B}\right|$$

We then assume that the operators $S_{\alpha\beta}^{AC}$ have the form

$$S_{\alpha\beta}^{AC} = I_{A}\otimes S_{\alpha\beta}^{C}$$

where $S_{\alpha\beta}^{C}$ acts only on $\mathcal{H}_{C}$.

Then, writing S^{C} in place of $S_{\alpha\beta}^{C}$, we get

$$S(\alpha\mu\mid\rho\rho) = \left\langle u_{\rho}^{A}\otimes u_{\rho}^{C}\mid I_{A}\otimes S^{C}\mid u_{\alpha}^{A}\otimes u_{\mu}^{C}\right\rangle$$

$$= \delta_{\rho\alpha}\left\langle u_{\rho}^{C}\mid S^{C}\mid u_{\mu}^{C}\right\rangle \equiv \delta_{\rho\sigma}S^{C}(\rho,\mu)\left\langle u_{\rho}^{C}\mid S^{C}\mid u_{\mu}^{C}\right\rangle$$

Thus,

$$\rho^{B} = \frac{1}{d^{3/2}}\sum_{\mu\nu\rho\sigma\alpha'\beta'} x_{\nu}x_{\mu}\,\delta_{\rho\alpha'}\,\bar{S}^{C}(\rho,\mu)\,S(\sigma,\nu)\,\delta_{\sigma\beta'}\left|u_{\alpha'}^{B}\right\rangle\left\langle u_{\beta'}^{B}\right|$$

$$= \frac{1}{d^{3/2}} \sum_{\mu\nu\rho\sigma} \bar{x}_\nu x_\mu \bar{S}^C(\rho,\mu) S^C(\sigma,\nu) \left|u_\rho^B\right\rangle \left\langle u_\sigma^B\right|$$

Alice conveys the result (α, β) of her measurement to Bob and hence Bob knows the operator

$S^C = S^C_{\alpha\beta}$, or equivalently the matrix $\left(\left(S^C(\rho,\sigma)\right)\right)_{1\le\rho,\sigma\le d}$.

Define

$$\left|\left(\bar{S}^B u^B\right)_\mu\right\rangle = \sum_\rho \bar{S}^C(\rho,\mu)\left|u_\rho^B\right\rangle$$

$$\equiv \bar{S}^B\left|u_\mu^B\right\rangle$$

Then $$\rho^B = \sum_{\mu,\nu} \bar{X}_\nu X_\mu \bar{S}^B \left|u_\mu^B\right\rangle\left\langle u_\nu^B\right| S^B$$

$$= \bar{S}^B\left(\sum_\mu X_\mu\left|u_\mu^B\right\rangle\right)\left(\sum_\nu \bar{X}_\nu\left\langle u_\nu^B\right|\right)S^{B^T}$$

Since Bob knows the operator S^B (or equivalently $\bar{S}^B$ and S^{B^T}), he applies its inverse to his (prove)

state to $\bar{S}^B\left(\sum_\mu X_\mu\left|u_\mu^B\right\rangle\right)$ obtain

$\sum_\mu X_\mu\left|u_\mu^B\right\rangle$. Hence he receives Alice's $\left|x^C\right\rangle$.

For Navnet

❖❖❖❖❖

[212] A quantum atomic receiver is excited by a quantum e-mfield. The Hamiltonian of the field is

$$H_f = \sum_{\alpha=1,2}^{\infty} a_\alpha^+ a_\alpha \qquad (1)$$

where a_α, a_α^+ are annichultation and creation operators respectively. The magnetic vector potential of the quantum e-m field has the form

$$A^m(x) = A^m(t,r) = \sum_\alpha\left(a_\alpha e_\alpha^m \exp(ik_\alpha X) + a_\alpha^+ e_\alpha^{m^+} \exp(-ik_\alpha X)\right)\Big/\sqrt{k_\alpha^0}$$

where e_α^μ an polarization vectors

(See for example Weinberg, " The quantum theory of fields Vol .1" or Sakurai Advanced quantum mechanics")

Here $k_\alpha = \left(k_\alpha^\mu\right)_{\mu=0}^3$ and $k_{\alpha,x} = k_{\alpha\mu}x^\mu$

where $k_{\alpha 0} = k_\alpha^0,\ k_{\alpha r} = -k_\alpha^r,\ r = 1, 2, 3$.

The field energy calculated using

$$H_f = \int \left[\sum_{m=1}^{3} \left(\frac{\partial A^m}{\partial t} \right)^2 + |\nabla \times A|^2 \right] d^3X$$

gives the expression (1). The standard commutator relatives are $\left[a_\alpha, a_\beta^+\right] = \delta_{\alpha\beta}$.

The unperturbed atom Hamiltonian of the atom plus field is

$$H_0 = \frac{p^2}{2m} + V(\underline{r}) + \sum_{\alpha=1}^{N} a_\alpha^+ a_\alpha$$

where the dynamical variables of the atom $\underline{p}$, $\underline{r}$ commute with the field dynamical variables $\left\{a_\alpha, a_\alpha^+\right\}$. After interaction assuming the atom is at the origin, the Hamiltonian is

$$\begin{aligned} H(t) &= \frac{(p+eA)^2}{2m} + V(\underline{r}) + \sum_{\alpha=1}^{N} a_\alpha^+ a_\alpha \\ &= \frac{p^2}{2m} + V(\underline{r}) + \frac{e}{2m}((\underline{p}, \underline{A}) + (\underline{A}, \underline{p})) + \frac{e^2 A^2}{2m} + \sum_{\alpha=1}^{N} a_\alpha^+ a_\alpha \\ &= H_0 + H_I(t) \end{aligned}$$

where $H_I(t) = \dfrac{e}{2m}\left((\underline{p}, \underline{A}) + (\underline{A}, \underline{p})\right) + \dfrac{e^2}{2m}\underline{A}^2$

Now if we superpose currents of different frequencies we can generate a quantum e-m field with a_α replaced by $f_\alpha(t)\, a_\alpha$ and a_α^+ by $\overline{f_\alpha(t)}\ a_\alpha$ where $f_\alpha(t)$ is a complex scalar function of time. In this case the interaction Hamiltonian gets replaced by

$$\begin{aligned} H_I(t) = {} & \frac{e}{2m} \sum_{\alpha=1}^{N} \left((\underline{p}, \underline{e}_\alpha)\, f_\alpha(t) a_\alpha + (\underline{p}, \overline{e}_\alpha)\, \overline{f_\alpha(t)} a_\alpha^+ \right) \\ & + \frac{e^2}{2m} \sum_{\alpha=1}^{N} \left(f_\alpha(t) a_\alpha \underline{e}_\alpha + \overline{f_\alpha(t)} a_\alpha^+ \overline{e}_\alpha \right)^2 \end{aligned} \qquad (2)$$

The current source is generated by modulating PAM pulses with an information bearing sequence $\{I_n\}_{n=1}^{M}$. Here $f_\alpha(t)$ has the form $f_\alpha(t) = \sum_{n=1}^{M} I_n\, p_{n\alpha}(t)$ where $p_{n\alpha}$ are PAM pulses.

(Note that the coulomb gauge condition on A is div $A = 0$ which gives

$$0 = \left(\underline{k}_\alpha, e_\alpha\right) = \sum_{r=1}^{3} k_\alpha^r e_\alpha^r\ .)$$

The interaction Hamiltonian (2) can be expressed as

$$H_I(t) = e\sum_{n=I}^{M} I_n V_n^{(1)}(t) + e^2 \sum_{n,\,m=1}^{M} I_n I_m N_{nm}^{(2)}(t)$$

where

$$V_n^{(1)}(t) = \frac{1}{2m}\sum_{\alpha=1}^{N} a_\alpha(\underline{p}, \underline{e}_\alpha)\, p_{n\alpha}(t) + a_\alpha^{+}(\underline{p}, \underline{e}_\alpha)\, \overline{p_{n\alpha}(t)}$$

and

$$V_{nm}^{(2)}(t) = \frac{1}{2m}\sum_{\alpha=1}^{N}\left(p_{n\alpha}(t) a_\alpha \underline{e}_\alpha + \bar{p}_{n\alpha}(t) a_\alpha^{+} \underline{\bar{e}}_\alpha,\ p_{m\alpha}(t) a_\alpha \underline{e}_\alpha + \bar{p}_{m\alpha}(t) a_\alpha^{+} \bar{e}_\alpha\right)$$

$$= \frac{1}{2m}\sum_{\alpha} p_{n\alpha}(t)\, p_{m\alpha}(t) (\underline{e}_\alpha, \underline{e}_\alpha) a_\alpha^2 + \overline{p_{n\alpha}(t)}\,\overline{p_{m\alpha}(t)} (\underline{\bar{e}}_\alpha, \underline{\bar{e}}_\alpha) a_\alpha^{+2}$$

$$+ \overline{p_{n\alpha}(t)}\, p_{m\alpha}(t) (\underline{\bar{e}}_\alpha, \underline{e}_\alpha) a_\alpha^{+} a_\alpha + p_{n\alpha}(t)\, \overline{p_{m\alpha}(t)} (\underline{e}_\alpha, \underline{\bar{e}}_\alpha)\, a_\alpha\, a_\alpha^{+}$$

❖❖❖❖❖

[213] Fluid dynamics in the Schwarzschild metrics in state variable form.

Incompressible fluid ρ = constant

$$\left((\rho + p)v^\mu v^\nu - pg^{\mu\nu}\right)_{:\nu} = f^\mu$$

$$g_{00} = \alpha(r) = 1 - \frac{2m}{r},\ m = \frac{GM}{c^2},$$

$$g_{11} = -\alpha(r)^{-1} = \left(1 - \frac{2m}{r}\right),$$

$$g_{22} = -r^2,\ g_{33} = -r^2 \sin^2\theta$$

$$d\tau^2 = \left(1 - \frac{2m}{r}\right)dt^2 - \left(1 - \frac{2m}{r}\right)^{-1} dr^2 - r^2(d\theta^2 + \sin^2\theta\, d\varphi^2)$$

$$\gamma = \frac{dX^0}{d\tau} \quad X^0 = t, X^1 = r, X^2 = \theta, X^3 = \varphi$$

$$v^\mu = \frac{dX^\mu}{d\tau} \quad v^0 = \gamma$$

$$g_{\mu\nu}v^\mu v^\nu = 1 \quad v_\mu v^\mu = 1$$

$$g_{00}\gamma^2 + g_{11}v^{1^2} + S_{22}v^{2^2} + g_{33}v^{3^2} = 1$$

$$\Rightarrow \quad \alpha\gamma^2 - \alpha^{-1}v^{1^2} - r^2v^{2^2} - r^2\sin^2\theta v^{3^2} = 1$$

$$\gamma = \alpha^{-\frac{1}{2}}\left[1 + \alpha^{-1}v^{-1^2} + r^2v^{2^2} + r^2\sin^2\theta v^{3^2}\right]^{\frac{1}{2}}$$

$$= \left[\alpha^{-1} + \alpha^{-2}v^{1^2} + \alpha^{-1}r^2v^{2^2} + \alpha^{-1}r^2\sin^2\theta v^{3^2}\right]^{\frac{1}{2}}$$

$$(v^k) = (v^1, v^2, v^3), (v^\mu) = (v^0, v^1, v^2, v^3)$$

$$((\rho + p)v^\nu)_{:\nu}v^\mu + (\rho + p)v^\nu v^\mu_{:\nu} = p^{,\mu} + f^\mu \tag{1}$$

$$\Rightarrow \quad ((\rho + p)v^\nu)_{:\nu} = v^\alpha(p_{,\alpha} + f_\alpha) \tag{2}$$

$$\Rightarrow \quad \left((\rho + p)\sqrt{-g}\, v^\nu\right)_{,\nu} = v^\alpha\left(p_{,\alpha} + f_\alpha\right)\sqrt{-g} \tag{3}$$

$$g = -r^4\sin^2\theta \quad \sqrt{-g} = r^2\sin\theta$$

$$((\rho + p)r^2\sin\theta v^0)_{,0} + ((\rho + p)r^2\sin\theta v^r)_{,r} = r^2\sin\theta(v^0p_{,0} + (v, \nabla p) + v^\alpha f_\alpha)$$

$$(v, \nabla p) \underset{=}{\Delta} v^k p_{,k}$$

So $(\rho + p)r^2\sin\theta v^0_{,0} + ((\rho + p)r^2\sin\theta v^r)_{,r} = r^2\sin\theta(p_{,0}v^0 + v^r p_{,r} + v^\alpha f_\alpha)$ (4)

Substituting (2) into (1) gives

$$v^\mu v^\alpha(p_{,\alpha} + f_\alpha) + (\rho + p)v^\nu v^\mu_{:\nu} = p^{,\mu} + f^\mu \tag{5}$$

(5) $\Rightarrow$ (with $\mu = 0$)

$$(g^{00} - v^{0^2})\, p_{,0} + f^0 = v^0(v^r p_{,r} + v^\alpha f_\alpha) + (\rho + p)v^\nu v^0_{:\nu} \tag{6}$$

Combining (4) and (6) gives

$$\begin{bmatrix} (\rho + p)r^2\sin\theta & -r^2\sin\theta v^0 \\ (\rho + p)v^0 & -(g^{00} - v^{0^2}) \end{bmatrix}\begin{bmatrix} v^0_{,0} \\ p_{,0} \end{bmatrix}$$

$$= \begin{bmatrix} -\left((\rho+p)r^2 \sin\theta v^k\right)_{,k} + r^2 \sin\theta (v^k p_{,k} + v^\alpha f_\alpha) \\ f^0 - v^0 (v^k p_{,k} + v^\alpha f_\alpha) - (\rho+p) v^\nu v^\alpha \Gamma^0_{\nu\alpha} \end{bmatrix}$$

This eqn. can be solved for $v^0_{,0}$ and $p_{,0}$

Now, (5) can be expressed as

$$(\rho+p)\left[v^0\left(v^k_{,0} + \Gamma^k_{0m} v^m\right) + v^l\left(v^k_{,l} + \Gamma^k_{lm} v^m\right)\right]$$
$$= g^{kk} p_{,k} + f^k - v^k\left(v^0 p_{,0} + v^m p_{,m} + v^\alpha f_\alpha\right)$$

Solving for $p_{,0}$ from (7) gives

$$p_{,0} = \Big\{(\rho+p)v^0\left[\left((\rho+p)r^2 \sin\theta v^k\right)_{,k} - r^2 \sin\theta\left(v^k p_{,k} + v^\alpha\right)\right]$$
$$+ (\rho+p)r^2 \sin\theta\left[f^0 - v^0(v^k p_{,k} + v^\alpha f_\alpha) - (\rho+p)v^\nu v^\alpha \Gamma^0_{\nu\alpha}\right]\Big\}$$
$$= \left\{(\rho+p)r^2 \sin\theta v^{0^2} - (g^{00} - v^{0^2})(\rho+p)r^2 \sin\theta\right\} \tag{9}$$

(9) express $p_{,0}$ are function of (X^μ) p, $p_{,k}$, v^k, $v^k_{,n}$ (k, m correspond all to spatial derivatives). Substituting for $p_{,0}$ from (9) into (8) gives $v^k_{,0}$ in terms of X^μ, p, $p_{,k}$, v^k, v^k_m (all spatial derivatives). This (8) and (9) can be expressed in the form

$$\frac{d}{dt}\begin{bmatrix} p(r, \underline{r}) \\ \underline{v}(t, \underline{r}) \end{bmatrix} = \begin{bmatrix} F_1(p, \underline{\nabla} p, \underline{v}, v_{,k}, X) \\ F_2(p, \underline{\nabla} p, \underline{v}, \underline{v}_{,k}, X) \end{bmatrix} \tag{10}$$

❖❖❖❖❖

[214] Introducing quantum Stochastic process in quantum field theory.

Dirac equation

$$(\gamma^\mu(i\partial_\mu + eA_\mu) - m)\psi = 0 \tag{1}$$

$$A_\mu = A_\mu(X), X = (t, \underline{r})$$

$$\gamma^0 = \begin{pmatrix} I & 0 \\ 0 & -I \end{pmatrix}, \gamma^r = \begin{pmatrix} 0 & \sigma_r \\ -\sigma_r & 0 \end{pmatrix}, 1 \le r \le 3$$

(1) to the same as

$$[-(i\partial_0 + eV) + (\alpha, -i\nabla + cA) + \beta m]\psi = 0$$

where $A = (A^r)^3_{r=1}, V = A^0, A_r = -A^r$

Thus $$i\partial_0\psi = ((\alpha, -i\nabla + cA) + \beta m - eV)\psi$$

$$\sigma_r = \gamma^0\gamma^r = \begin{pmatrix} 0 & \sigma_r \\ \sigma_r & 0 \end{pmatrix}, \ 1 \le r \le 3,$$

Notation $(\underline{\alpha}, \underline{p}) = \sum_{r=1}^{3} \alpha_r p_r, \ \beta = \gamma^0$

$$H_D^{(t)} = H_{0D} + eH_{1D}(t)$$

is the Dirac Hamiltonian

$H_{0D} = (\alpha, -i\nabla) + \beta m$ is the Free Dirac Hamiltonian

$H_{1D}(t) = (\alpha, A) - V$ is the interaction Hamiltonian.

Eigenstates of $\mathbf{H_{0D}}$:

$$((\alpha, p) + \beta m)u_\sigma(p) = E(p)u_\sigma(p), \sigma = 1, 2$$

$$((\sigma, p) + \beta m)v_\sigma(-p) = -E(p)v_\sigma(-p), \sigma = 1, 2$$

Second quantized wave function of H_{0D}:

$$\psi(X) = \int\left(\frac{a_\sigma(p)}{\sqrt{E(p)}}u_\sigma(p)e^{-ip.X} + \frac{b_\sigma^+(b)v_\sigma(p)}{\sqrt{\Gamma s(p)}}e^{ip.X}\right)d^3p$$

$$p_X = p_\mu X^\mu = p^0 t - \sum_{r=1}^{3} p^r X^r = E(p)t - (p, r)$$

For ψ to satisfy the Free Dirac equation

$$(i\gamma^\mu\partial_\mu - m)\psi = 0$$

Thus, $(\gamma^\mu p_\mu - m)u_\sigma(p) = 0$

and $(\gamma^\mu p_\mu + m)v_\sigma(p) = 0, \sigma = 1, 2$

Equivalently, $$p^0 u_\sigma(p) = ((\alpha, p) + m)u_\sigma(p)$$

$$-p^0 v_\sigma(-p) = ((\alpha, p) + m)v_\sigma(-p)$$

$$p^0 = +\sqrt{m^2 + \beta^2}$$

** Now the wave function $v_\sigma(-p)$ has – ve energy

Free Dirac Hamiltonian after 2nd quantization:

$$\tilde{H}_{D0} = \int \psi^T(X) H_{D0} \psi(X) d^3X$$
$$= \int E(P)(a_\sigma^+(P) a_\sigma(P) - b_\sigma^+(P) b_\sigma(P)) d^3P$$

since

$$\frac{1}{(2\pi)^3} \int e^{i(p-p')\cdot X} d^3X = \delta^3(P - P')$$

$$p^0 = \sqrt{m^2 + p^2}$$

$$p^{01} = \sqrt{m^2 + p'^2}$$

and

$$u_{\sigma'}^+(P) u_\sigma(P) = E(P) \delta\sigma'\sigma$$
$$v_{\sigma'}^+(P) v_\sigma(P) = E(P) \delta\sigma'\sigma$$
$$v_{\sigma'}^+(-P) u_\sigma(P) = 0$$

Note that we have chosen the normalization so that $\frac{u_\sigma(P)}{\sqrt{E(P)}}$ and $\frac{v_\sigma(P)}{\sqrt{E(P)}}$ have unit length.

Intraction Hamiltonian of Boson field $A_\mu(X)$ with electron field in the second quantization.

$$\tilde{H}_{1D}(t) = \int \psi^+(X)((\alpha A(X)) - V(X)) \psi(X) d^3X$$

$$X = (t, r) \in \mathbb{R}^4 \; d^3X = d^3r$$

$$= \int \frac{a_\sigma^+(P) a_{\sigma'}(P')}{\sqrt{EE'}} u_\sigma^+(P')((\alpha, A) - V) u_{\sigma'}(P) \exp(i(P - P')\cdot X) d^3X \, d^3P \, d^3P'$$

$$= + \int \frac{b_\sigma(P) b_{\sigma'}^+(P')}{\sqrt{EE'}} v_\sigma^+(P)((\alpha, A) - V) v_{\sigma'}(P') \exp(-i(P - P')\cdot X) d^3X \, d^3P \, d^3P'$$

$$= + \int \frac{b_\sigma(P) a_{\sigma'}(P')}{\sqrt{EE'}} v_\sigma^+(P)((\alpha, A) - V) u_{\sigma'}(P') \exp(-i(P + P')\cdot X) d^3X \, d^3P \, d^3P'$$

$$= + \int \frac{a_\sigma^+(P) b_{\sigma'}^+(P')}{\sqrt{EE'}} u_\sigma^+(P)((\alpha, A) - V) v_\sigma(P') \exp(i(P + P')\cdot X) d^3X \, d^3P \, d^3P'$$

Let $\int A(X) e^{-(P-P')\cdot X} d^3r = \int A(t, r) e^{-i(P-P')\cdot r} d^3r$

$$= \hat{A}(t, P - P')$$

$$\int V(t, r) e^{-i(P-P').r} d^3r = \hat{V}(t, P - P')$$

Note

$$\left\{a_\sigma(P), a_{\sigma'}^T(P')\right\} = \delta_{\sigma\sigma'}\delta^3(P-P')$$
$$\left\{b_\sigma(P), b_{\sigma'}^+(P')\right\} = \delta_{\sigma\sigma'}\delta^3(P-P')$$

All other operations antiommute.

So

$$\tilde{H}_{1D}(t) = \int\frac{a_\sigma^+(P)a_{\sigma'}(P')}{\sqrt{EE'}}u_\sigma^+(P)((\alpha,\hat{A}(t,P-P'))-\hat{V}(t,P-P'))u_{\sigma'}(P')$$
$$= \exp(i(E-E')t)d^3Pd^3P'$$
$$= -\int\frac{b_{\sigma'}^+(P')a_{\sigma'}(P)}{\sqrt{EE'}}v_\sigma^+(P)((\alpha,\hat{A}(t,P'-P))-\hat{V}(t,P'-P))v_{\sigma'}(P')$$
$$= \exp(-i(E-E')t)d^3Pd^3P'$$
$$= -\int\frac{a_{\sigma'}(P')b_\sigma(P)}{\sqrt{EE'}}v_\sigma^+(P)((\alpha,\hat{A}(t,-P-P'))-\hat{V}(t,-P-P'))u_{\sigma'}(P')$$
$$\exp(-i(E+E')t)d^3Pd^3P'$$
$$+\int\frac{a_\sigma^+(P)b_{\sigma'}(P')}{\sqrt{EE'}}u_\sigma^+(P)((\alpha,\hat{A}(t,P+P'))-\hat{V}(t,P+P'))v_{\sigma'}(P')$$
$$\exp(i(E+E')t)d^3Pd^3P'$$

Take $\hat{A}(t,P),\hat{V}(t,P)$ as the sum of a quantum Stochastic process plus a quantized em-field.

$$A_0^\mu(t,\underline{r}) = A_0^\mu(X) = \int\left(\frac{a_5(K)e_s^\mu(K)}{\sqrt{|K|}}e^{-ik\cdot X}+\frac{a_s^+(K)e_s^\mu(K)^+}{\sqrt{|\underline{K}|}}e^{ik\cdot X}\right)d^3K$$

$$k = (k^0,\underline{K}) = (|\underline{K}|,\underline{K})$$

Note the Lorentz gauge condition gives

$$\partial_\mu A^\mu = 0$$

i.e. $\quad k_\mu e_S^\mu(K) = 0$

or $\quad |\underline{K}|e_S^0(K) = \sum_{r=1}^{3}K^r e_S^r(K)$

Thus S = 1, 2, 3 *i.e.* the photon field has 3 degrees of polarization.

$$[a_S(K), a_{S'}^+(K')] = \delta_{ss'}\delta^3(\underline{K}-\underline{K}')$$

Then the energy of the quantized em-field is

$$H_{ph}^{(0)} = \frac{1}{2}\int\left[\left|\nabla V + \frac{\partial A}{\partial t}\right|^2 + |\underline{\nabla} \times \underline{A}|^2\right] d^3X$$

$$= \int |\underline{K}| a_s^+(\underline{K}) a_s(\underline{K}) d^3K$$

assuming

$$\sum_{\mu=0}^{3} e_s^{\mu}(K)\overline{e_{s'}^{\mu}(K)} = \delta_{ss'}$$

In addition to $H_{ph}^{(0)}$, we have quantum Stochastic photon field:

$$A_1^{\mu}(X) = A_1^{\mu}(t, \underline{r})$$

$$A_1^{\mu}(t, r)dt = \sum_{\rho, \sigma} C^{\mu}(\rho, \sigma, \underline{r}) d\Lambda_{\sigma}^{\rho}(t)$$

$$\hat{A}_1^{\mu}(t, \underline{K})dt = \sum_{\rho, \sigma} \hat{C}^{\mu}(\rho, \sigma, \underline{K}) d\Lambda_{\sigma}^{\rho}(t)$$

Take for example the first interaction term in $\tilde{H}_{1D}(t)$ with $A^{\mu} = A_0^{\mu} + A_1^{\mu}$.

Then calling this term by $\tilde{H}_{1D}(t)|_1$, we get

$$\tilde{H}_{1D}(t)|_1\, dt = \int \frac{a_{\sigma}^+(P) a_{\sigma'}(P')}{\sqrt{EE'}} u_{\sigma}^+(P)((\alpha, \hat{A}(t, P-P')) - \hat{V}(t, P-P'))u_{\sigma'}(P')$$

$$\exp(i(E-E')t) d^3P d^3P'$$

Let $\alpha^0 = 1 = \alpha_0$, $\alpha^r = -\alpha_r$, $r = 1, 2, 3$. Then

$$\tilde{H}_{1D}(t)|_1\, dt = \int \frac{a_{\sigma}^+(P) a_{\sigma'}(P')}{\sqrt{EE'}} u_{\sigma}^+(P)(\alpha_{\mu} \hat{A}^{\mu}(t, P-P'))u_{\sigma'}(P')$$

$$\exp(i(E-E')t) d^3P d^3P'$$

$$\hat{A}_0^{\mu}(t, P) = \int \frac{a_s(\underline{K}) e_s^{\mu}(\underline{K})}{\sqrt{|\underline{K}|}} e^{-i|K|t} \delta(\underline{K} - \underline{P}) d^3K$$

$$+ \int \frac{a_s^+(\underline{K}) e_s^{\mu}(\underline{K})}{\sqrt{|\underline{K}|}} e^{-|K|t} \delta(\underline{K} + \underline{P}) d^3k$$

$$= a_s(\underline{P}) e_s^{\mu}(\underline{P}) e^{-i|\underline{P}|t} \Big/ \sqrt{|\underline{P}|} + a_s^+(\underline{P}) e_s^{\mu}(\underline{P}) e^{i|P|t} \Big/ \sqrt{|\underline{P}|}$$

So $\tilde{H}_{1D}(t)|_1\, dt$

$$= \int \frac{a_\sigma^+(P)a_{\sigma'}(P')}{\sqrt{EE'|\underline{P}|}} u_\sigma^+(P)(\alpha_\mu e_s^\mu(\underline{P}'))u_{\sigma'}(P')(a_s(\underline{P}-\underline{P}')\exp(-i|\underline{P}-\underline{P}')t)$$

$$+ a_s^+(\underline{P}-\underline{P}')\exp(-i|\underline{P}-\underline{P}'|t)\exp(i(E-E')t)\, d^3Pd^3P'$$

$$+ \int \frac{a_\sigma^+(P)a_{\sigma'}(P')}{\sqrt{EE'}} u_\sigma^+(P)(\alpha_\mu \widehat{C}^\mu(P,\sigma,P-P'))u_{\sigma'}(P')$$

$$\exp(i(E-E')t)d^3Pd^3P'd\Lambda_\sigma^\rho(t)$$

$$= \int \frac{a_\sigma^+(P)a_{\sigma'}(P')a_s(P-P')}{\sqrt{EE'|\underline{\rho}|}} u_\sigma^+(P)(\alpha_\mu e_s^\mu(P-P'))u_{\sigma'}(P')$$

$$\exp\{i(E-E'-|\underline{P}-\underline{P}'|)t\}d^3Pd^3P'$$

$$+ \int \frac{a_\sigma^+(P)a_{\sigma'}(P')a_s^+(P-P')}{\sqrt{EE'|P|}} u_\sigma^+(P)(\alpha_\mu e_s^\mu(P-P'))u_{\sigma'}(P')$$

$$\exp\{i(E-E'+|\underline{P}-\underline{P}'|)t\}d^3Pd^3P'$$

$$dX_\sigma^P(t)\int \frac{a_\sigma^+(P)a_{\sigma'}(P')}{\sqrt{EE'}} u_\sigma^+(P)(\alpha_\mu \widehat{C}^\mu(P\sigma PP'))u_{\sigma'}(P')$$

$$\exp(i(E-E')t)d^3Pd^3P$$

❖❖❖❖❖

[215] Left invariant measures on a general real Lie group.

Let $\{X_1, ..., X_n\}$ be a basis for g, the Lie algebra of G with structure constants $C\in(\alpha\beta\gamma)$:

$$[X_\alpha, X_\beta] = C\,\varepsilon(\alpha\beta\gamma)\,X_\gamma$$

Parameterize g ε G as

$$g = g(x_1, ..., x_n) = e^{x_1X_1}... e^{x_n X_n}, x_1, ..., x_n \in \mathbb{R}$$

$$\frac{\partial g}{\partial x_n} = gX_n$$

$$\frac{\partial g}{\partial x_{n-1}} = e^{x_1X_1},... e^{x_{n-1}}X_{n-1}e^{x_nX_n}$$

$$= g\exp(-x_n\, ad\, X_n)(X_{n-1})$$

$$\frac{\partial g}{\partial x_{n-2}} = g\exp(-x_n\, ad\, X_n)\exp(-x_{n-1}\, ad\, X_{n-1})(X_{n-2})$$

$$\frac{\partial g}{\partial x_{n-k}}$$

$$= g\exp(-x_n\, ad\, X_n)\exp(-x_{n-1}\, ad\, X_{n-1})\ldots$$

$$\exp(-x_{n-k+1}\, ad\, X_{n-k+1})(X_{n-k})$$

$$0 \le k \le n-1$$

Let $\exp(-x_n\, ad\, X_n)\ldots\exp(-x_{n-k+1}\, ad\, X_{n-k+1})(X_{n-k})$

$$= \sum_{r=1}^{n} f_{rk}(x_{n-k+1}, x_{n-k+2}, x_n) X_r$$

Then

$$\frac{\partial}{\partial x_{n-k}} = \sum_{r=1}^{n} f_{rk}(x_{n-k+1},.., x_n)\tilde{X}_r\,,\ 0 \le k \le n-1$$

Let

$$((f_{rk}(x_{n-k+1},.., x_n)))_{\substack{1\le r\le n\\ 0\le k\le n-1}} = \underline{\underline{F}}(x_1,.., x_n) \in \mathbb{R}^{n\times n}$$

Then with $\underline{\underline{H}}(x_1,.., x_n) = \underline{\underline{F}}(x_1,.., x_n)^{-1}$, we get

assuming $$\underline{\underline{H}} = (h_{rk}(x_1,.., x_n))_{\substack{1\le r\le n,\\ 0\le k\le n\le 1}}$$

that $$\tilde{X}_r = \sum_{k=0}^{n-1} h_{kr}(\underline{x})\frac{\partial}{\partial x_{n-k}}$$

and hence writing

$$\det(\underline{\underline{H}}(\underline{x})) - \det((h_{kr})) = (\det \underline{\underline{F}}(\underline{x}))^{-1},$$

We have that

$(\det \underline{\underline{F}}(\underline{x}))d^n x$

is the Haar measure on G.

Hint generator for Gaussian semigroups:

$$\mathcal{L} = \frac{1}{2}\sum_{r,s=1}^{n} \lambda_{rs}\tilde{X}_r\tilde{X}_s$$

$$= \frac{1}{2}\sum_{r,s,k_1,k_2} \lambda_{rs} h_{k_1 r}(\underline{x})\frac{\partial}{\partial x_{n-k_1}} h_{k_2 s}(\underline{x})\frac{\partial}{\partial x_{n+k_2}}$$

Haar measure on $SO(3)$ in terms of Euler angles.

Let $\{X_1, X_2, X_3\}$ be the standard generation of $SO(3)$:

$$
\begin{aligned}
[X_1, X_2] &= X_3, [X_2, X_3] = X_1, [X_3, X_1] = X_2 \\
R_Z(\phi) &= e^{\phi X_3}, R_Y(\phi) = e^{\phi X_2}, R_X(\phi) = e^{\phi X_1} \\
R(\phi, \theta, \psi) &= g = R_Z(\phi) R_X(\theta) R_Z(\psi) \\
&= \exp(\phi X_3)\cdot\exp(\theta X_1)\cdot\exp(\psi X_3)
\end{aligned}
$$

$\frac{\partial g}{\partial \psi} = X_3$. Let $f(g) = f_1(\varphi, \theta, \psi)$. Thus

$$
\begin{aligned}
\tilde{X}_3 f(g) &= \left.\frac{d}{dt} f(ge^{tX_3})\right|_{t=0} \\
&= \frac{\partial}{\partial \psi} f(g) = \frac{\partial}{\partial \psi} f_1(\phi, \theta, \psi) \\
\frac{\partial g}{\partial \phi} &= X_3 g = \exp(\phi X_3) X_3 \exp(\theta X_1)\exp(\psi X_3) \\
&= \exp(\phi X_3)\exp(\theta X_1)\exp(-\theta ad\, X_1)(X_3)\exp(\psi X_3) \\
&= \exp(\phi X_3)\exp(\theta X_1)(X_3 \cos\theta + X_2 \sin\theta)\exp(\psi X_3) \\
&= \cos\theta\, g X_3 + \sin\theta\, g \exp(-\psi\, ad\, X_3)(X_2) \\
&= \cos\theta\, g X_3 + \sin\theta\, g (X_2 \cos\psi + X_1 \sin\psi)
\end{aligned}
$$

So,

$$
\begin{aligned}
\frac{\partial f_1}{\partial \phi} &= \sin\theta \sin\psi\, \tilde{X}_1 f + \sin\theta \cos\psi\, \tilde{X}_2 f + \cos\theta \tilde{X}_3 f \\
\frac{\partial f_1}{\partial \theta} &= \frac{\partial f(g)}{\partial \theta} \\
\frac{\partial g}{\partial \theta} &= \exp(\phi X_3)\exp(\theta X_1) X_1 \exp(\psi X_3) \\
&= g \exp(-\psi\, ad\, X_3)(X_1) \\
&= g(X_1 \cos\psi - X_2 \sin\psi)
\end{aligned}
$$

Thus,

$$
\frac{\partial f_1}{\partial \theta} = \cos\psi \tilde{X}_1 f - \sin\psi \tilde{X}_2 f
$$

So

$$\begin{bmatrix} \sin\theta\sin\psi & \sin\theta\cos\psi & \cos\theta \\ 0 & 0 & 1 \\ \cos\psi & -\sin\psi & 0 \end{bmatrix} \begin{bmatrix} \tilde{X}_1 \\ \tilde{X}_2 \\ \tilde{X}_3 \end{bmatrix} = \begin{bmatrix} \frac{\partial}{\partial\phi} \\ \frac{\partial}{\partial\psi} \\ \frac{\partial}{\partial\theta} \end{bmatrix}$$

Note that $\tilde{X}_\alpha,\ \alpha = 1, 2, 3$ are left invariant vector fields on $SO(3)$.

$$\tilde{X}_1 \wedge \tilde{X}_2 \wedge \tilde{X}_3 =$$

Invariance of the Dirac equation under the rotation group.

$$= R \in SO(3)$$

$$\left[E - \sum_{k=1}^{3} \alpha_k \sum_{n=1}^{3} R_{kn}\, p_n' - \beta m\right] S\tilde{\psi} = 0$$

or
$$\left[E - \sum_{n=1}^{3} p_n' \sum_{k=1}^{3} R_{kn}\, \alpha_k - \beta m\right] S\tilde{\psi} = 0$$

For invariance, we require

$$S^{-1}\left(\sum_{k=1}^{3} R_{kn}\, \alpha_k\right) S = \alpha_n, 1 \le n \le 3$$

and $S^{-1}\beta S = \beta$

Let
$$S = \sum C^{km}\, \alpha_k\, \alpha_m \equiv C_{km}\alpha_k\alpha_m$$

Summation over the repeated implies k, m being implies

Then $\beta S = S\beta$ holds since $\alpha_k\beta = -\beta\alpha_k$

We thus require that

$$R_{kn}\, \alpha_k\, C_{pq}\alpha_p\alpha_q = C_{pq}\, \alpha_p\alpha_q\alpha_n$$

or
$$R_{kn}\, C_{pq}\, \alpha_k\, \alpha_p\, \alpha_q = C_{pq}\, \alpha_p\alpha_q\alpha_n$$

Now
$$\begin{aligned} \alpha_k\, \alpha_p\, \alpha_q &= (2\delta_{kp} - \alpha_p\alpha_k)\alpha_q \\ &= 2\delta_{kp}\alpha_q - \alpha_p(2\delta_{kq} - \alpha_q\alpha_k) \\ &= 2\delta_{kp}\alpha_q - 2\delta_{kq}\alpha_p + \alpha_p\alpha_q\alpha_k \end{aligned}$$

So the requirement is

$$R_{kn}C_{pq}(2\delta_{kp}\alpha_q - 2\delta_{kq}\alpha_p + \alpha_p\alpha_q\alpha_k) = C_{pq}\alpha_p\alpha_q\alpha_n$$

or

$$2R_{pn}C_{pq}\alpha_q - 2R_{qn}C_{pq}\alpha_p = C_{pq}(\delta_{kn} - R_{kn})\alpha_p\alpha_q\alpha_k$$

or

$$2(R^TC)_{nq}\alpha_q - 2(R^TC^T)_{np}\alpha_p = C_{pq}(\delta_{kn} - R_{kn})\alpha_p\alpha_q\alpha_k$$

Let

$$\frac{C - C^T}{2} = F, \ \frac{C + C^T}{2} = G$$

Then $F^T = -F$, $G^T = G$ and the requirement becomes

$$4(R^TF)_{np}\alpha_p = F_{pq}(\delta_{kn} - R_{kn})\alpha_p\alpha_q\alpha_k + a(\delta_{kn} - R_{kn})\alpha_k$$

where

$$a = G_{pq}\alpha_p\alpha_q = G_{pp}\alpha_p^2 = G_{pp}^I = Tr(G)I = Tr(C)\cdot I$$

Now

$$\alpha_p = \begin{pmatrix} 0 & \sigma_p \\ \sigma_p & 0 \end{pmatrix}$$

So

$$\alpha_p\alpha_q = \begin{pmatrix} 0 & \sigma_p \\ \sigma_p & 0 \end{pmatrix}\begin{pmatrix} 0 & \sigma_q \\ \sigma_q & 0 \end{pmatrix} = \begin{pmatrix} \sigma_p\sigma_q & 0 \\ 0 & \sigma_p\sigma_q \end{pmatrix}$$

$$\alpha_p\alpha_q\alpha_k = \begin{pmatrix} \alpha_p\alpha_q & 0 \\ 0 & \sigma_p\sigma_q \end{pmatrix}\begin{pmatrix} 0 & \sigma_k \\ \sigma_k & 0 \end{pmatrix}$$

$$= \begin{pmatrix} 0 & \sigma_p\sigma_q\sigma_k \\ \sigma_p\sigma_q\sigma_k & 0 \end{pmatrix}$$

$$= \begin{pmatrix} 0 & i\varepsilon(pqr)\sigma_r\sigma_k \\ i\varepsilon(pqr)\sigma_r\sigma_k & 0 \end{pmatrix}$$

$$= -\varepsilon(pqr)\varepsilon(rkl)\alpha_l$$

The requirement is this

$$4(R^TF)_{np}\alpha_p = -\varepsilon(pqr)\varepsilon(rkl)F_{pq}(\delta_{kn} - R_{kn})\alpha_l + a(\delta_{kn} - R_{kn})\alpha_k$$

or

$$4(R^TF)_{nk} = a(\delta_{kn} - R_{kn}) + \varepsilon(pqr)\varepsilon(rkl)F_{pq}(\delta_{ln} - R_{ln})$$

Writing $(X_p)_{qr} = \varepsilon(pqr)$

we have that $\{X_p\}_{p=1}^3$ are the $SO(3)$ generators.

and the requirement is

$$4(R^TF)_{nk} = a(\delta_{kn} - R_{kn}) - Tr(FX_r)((X_r)_{kn} - (X_rR)_{kn})$$

or $$-4FR = a(I-R) - Tr(FX_r)X_r(I-R)$$

Thus,

$$F = \frac{a}{4}(I-R^T) - \frac{1}{4}Tr(FX_r)X_r(I-R^T)$$

The requirement $F^T = -F$ gives

$$\frac{a}{4}(I-R) + \frac{1}{4}Tr(FX_r)(I-R)X_r = \frac{a}{4}(R^T-I) + \frac{1}{4}Tr(FX_r)X_r(I-R^T)$$

since $$X_r^T = -X_r$$

equivalently,

Now, $$X_1 = \begin{pmatrix} 0 & 0 & 0 \\ 0 & 0 & 1 \\ 0 & -1 & 0 \end{pmatrix},$$

$$X_2 = \begin{pmatrix} 0 & 0 & -1 \\ 0 & 0 & 0 \\ 1 & 0 & 0 \end{pmatrix},$$

$$X_3 = \begin{pmatrix} 0 & 1 & 0 \\ -1 & 0 & 0 \\ 1 & 0 & 0 \end{pmatrix},$$

Since $F^T = -F$, we can write $F = f_r X_r$ where $f_r \in \mathbb{R},\ 1 \le r \le 3$. Then,

$$Tr(FX_r) = f_s Tr(X_s X_r) = -2f_r$$

since $TR(X_s X_r) = -2\delta_{rs}$. The expression for F become

$$f_r X_r = \frac{a}{4}(I-R^T) + \frac{1}{2}f_r X_r(I-R^T)$$

$$= \left(\frac{a}{4}I + \frac{1}{2}f_r X_r\right)(I-R^T)$$

or,

$$\left(\sum_r f_r X_r\right)(I+R^T) = \frac{a}{4}(I-R^T)$$

so $$\sum_r f_r X_r = \frac{a}{4}(I-R^T)(I+R^T)^{-1}$$

$$= \frac{a}{4}(R-I)(R+I)^{-1}$$

Now $(R-I)(R+I)^{-1}$ is anti symmetric because,

$$((R-I)(R+I)^{-1})^T = (R^T+I)^{-1}(R^T-I)$$

$$(R+I)^{-1}(I-R) = -(R-I)(R+I)^{-1}$$

Hence there exists a solution $(f_r)_{r=1}^3$ and the proof of rotation invariance of the Dirac eqn. is proved.

Lorentz invariance of the Dirac equation

$$(E-\alpha_X p_X - \alpha_Y p_Y - \alpha_Z p_Z - \beta m)\psi = 0$$

$$\Rightarrow \quad [\gamma(E'+vp'_X) - \alpha_X\gamma(p'_X+vE') - \alpha_Y p_Y - \alpha_Z p_Z - \beta m]\psi = 0$$

$$\Rightarrow \quad [\gamma'(1-v\alpha_X)E' - \gamma(\alpha_X - v)p'_X - \alpha_Y p_Y - \alpha_Z p_Z - \beta m]\psi = 0$$

$$\gamma = (1-v^2)^{-\frac{1}{2}}$$

Infinitesimal Lorentz boosts: $\gamma = 1$, $v = \delta v$

$$[(1-\delta v\alpha_X)E' - (\alpha_X - \delta v)p'_X - \alpha_Y p_Y - \alpha_Z p_Z - \beta m]$$

$$[I+i\delta X]\tilde{\psi} = 0$$

$$\delta X^* = \delta X$$

We require that

$$(I-i\delta X)(1-\delta v\alpha_X)(I+i\delta X) = I+\delta\rho$$

$$(I-i\delta X)(\alpha_X-\delta v)(I+i\delta X) = \tilde{\alpha}_X = \alpha_X + \delta\alpha_X$$

$$(I-i\delta X)(\alpha_Y)(I+i\delta X) = \tilde{\alpha}_Y = \alpha_Y + \delta\alpha_Y$$

$$(I-i\delta X)(\alpha_Z)(I+i\delta X) = \tilde{\alpha}_Z = \alpha_Z + \delta\alpha_Z$$

$$(I-i\delta X)\beta(I+i\delta X) = \tilde{\beta} = \beta + \delta\beta$$

So

$$-\delta v\,\alpha_X = \delta\rho$$

$$-\delta v + i\,[\alpha_X, \delta X] = \delta\alpha_X$$

$$i[\alpha_Y, \delta_X] = \delta\alpha_Y$$

$$i[\alpha_Z, \delta X] = \delta\alpha_Z$$

$$i[\beta, \delta X] = \delta\beta$$

$$
\begin{aligned}
(1+\delta v\alpha_X)(\alpha_X+\delta\alpha_X) &= \alpha_X+\delta v+\delta\alpha_X=\alpha_X+\Delta\alpha_X \\
(1+\delta v\alpha_X)(\alpha_Y+\delta\alpha_Y) &= \alpha_Y+\delta v\alpha_X\alpha_Y+\delta\alpha_Y=\alpha_Y+\Delta\alpha_Y \\
(1+\delta v\alpha_X)(\alpha_Z+\delta\alpha_Z) &= \alpha_Z+\delta v\alpha_X\alpha_Z+\delta\alpha_Z=\alpha_Z+\Delta\alpha_Z \\
(1+\delta v\alpha_X)(\beta+\delta_\beta) &= \beta+\delta v\alpha_X\beta+\delta\beta=\beta+\Delta\beta \\
\Delta\alpha_X &= i[\alpha_X,\delta X] \\
\Delta\alpha_Y &= \delta v\alpha_X\alpha_Y+i[\alpha_Y,\delta X] \\
\Delta\alpha_Z &= \delta v\alpha_X\alpha_Z+i[\alpha_Z,\delta X] \\
\Delta\beta &= \delta v\alpha_X\beta+i[\beta,\delta X] \\
i\delta X &= \lambda\alpha_X \\
\Delta\alpha_Y &= 0
\end{aligned}
$$

$$\Rightarrow \qquad \Delta\alpha_X=0,\ \delta v\alpha_X\alpha_Y+\lambda[\alpha_Y,\alpha_X]=0$$

$$\Rightarrow \qquad (\delta v-\lambda)\alpha_X\alpha_Y+\lambda\alpha_Y\alpha_X=0$$

$$\Rightarrow \qquad \lambda=\frac{\delta v}{2}$$

$$\Delta\alpha_Z=0\Rightarrow\ \delta v\alpha_X\alpha_Z+\lambda[\alpha_Z,\alpha_X]=0$$

$$\Rightarrow \qquad \delta v\alpha_X\alpha_Z+\frac{\delta v}{2}(\alpha_Z\alpha_X-\alpha_X\alpha_Z)=0$$

$\Rightarrow \dfrac{\delta v}{2}(\alpha_X\alpha_Z+\alpha_Z\alpha_X)=0$ which is true

$\Delta\beta=0\Rightarrow\ \delta v\alpha_X\beta+\dfrac{\delta v}{2}[\beta,\sigma_X]=0$ which is true

Finite boosts

$$[\gamma(1-v\alpha_X)E'-\gamma(\alpha_X-v)p'_X-\alpha_Y p_Y-\alpha_Z p_Z-\beta m]S\tilde{\psi}=0$$

Premultiply by $\gamma^{-1}S^{-1}(1-v\alpha_X)^{-1}$ to get

$$
\begin{aligned}
S^{-1}(1-v\alpha_X)^{-1}(\alpha_X-v)S &= \alpha_X \\
\gamma^{-1}S^{-1}(1-v\alpha_X)^{-1}\alpha_Y S &= \alpha_Y
\end{aligned}
$$

$$\gamma^{-1}S^{-1}(1-v\alpha_X)^{-1}\alpha_Z S = \alpha_Z$$

$$\gamma^{-1}S^{-1}(1-v\alpha_X)^{-1}\beta S = \beta$$

Let

$$(1-v\alpha_X)^{-1} = \lambda+\mu\alpha_X$$

Then $$1 = (1-v\alpha_X)(\lambda+\mu\alpha_X)$$

$$= \lambda-\mu v+(\mu-\lambda v)\alpha_X = 0$$

$$\Rightarrow \quad \lambda-\mu v = 1,\ \mu-\lambda v = 0$$

$$\Rightarrow \quad \mu = \lambda v,\ \lambda(1-v^2) = 1,\ \lambda = \gamma^2 = (1-v^2)^{-1},$$

$$\mu = \gamma^2 v$$

$$(1-v\alpha_X)^{-1}\alpha_Y = (\lambda+\mu\alpha_X)\alpha_Y = \lambda\alpha_Y+\mu\alpha_X\alpha_Y$$

$$\alpha_Y = \gamma^{-1}S^{-1}(1-v\alpha_X)^{-1}\alpha_Y S$$

$$= \gamma S^{-1}(\alpha_Y+v\alpha_X\alpha_Y)S$$

Let $$S = a+b\alpha_X$$

Let $$S^{-1} = c+d\alpha_X$$

Then $$(a+b\alpha_X)(c+d\alpha_X) = 1$$

$$\Rightarrow \quad ac+bd = 1,\ ad+bc = 0$$

$$S^{-1}(\alpha_Y+v\alpha_X\alpha_Y)S = (c+d\alpha_X)(\alpha_Y+v\alpha_X\alpha_Y)(a+b\alpha_X)$$

$$= (c\alpha_Y+(cv+d)\alpha_X\alpha_Y+dv)(a+b\alpha_X)$$

$$= a(c+dv)\alpha_Y+b(c+dv)\alpha_Y\alpha_X$$

$$+(cv+d)a\alpha_X\alpha_Y-b(cv+d)\alpha_Y$$

$$= (ac-bd+(ad-bc)v)\alpha_Y+\alpha_X\alpha_Y(acv+ad-bc-bdv)$$

We require that

$$\gamma(ac-bd+v(ad-bc) = 1,$$

$$(ac-bd)^{v+a\alpha-bc} = 0$$

So, $$ad-bc = (bd-ac)v$$

and hence,

$$\gamma(ac-bd+v^2(bd-ac)) = 1$$

or
$$(ac-bd)(1-v^2) = \frac{1}{\gamma} = \sqrt{1-v^2}$$

$$ac-bd = \frac{1}{\sqrt{1-v^2}} = \gamma$$

Now
$$ac+bd = 1,\ ad+bc = 0$$

$$\Rightarrow \quad d = -\frac{bc}{a}$$

$$ac-\frac{b^2c}{a} = 1$$

$$\Rightarrow \quad c(a^2-b^2) = a$$

$$c = \frac{a}{a^2-b^2},\ d = \frac{-b}{a^2-b^2}$$

$$\gamma = ac-bd = \frac{a^2}{a^2-b^2}+\frac{b^2}{a^2-b^2} = \frac{a^2+b^2}{a^2-b^2}$$

$$= \frac{1+\left(\frac{b}{a}\right)^2}{1-\left(\frac{b}{a}\right)^2}$$

$$\left(\frac{b}{a}\right)^2[1+\gamma] = \gamma-1 = \frac{b}{a} = \sqrt{\frac{x-1}{x+1}}.$$

❖❖❖❖❖

[216] Let $\underline{B}(t)\in\mathbb{R}^n$ be n-dimensional Brownian motion and $\underline{\underline{F}}(Y)\in\mathbb{R}^{p\times n}, Y\in\mathbb{R}^p$, *i.e.*

$\underline{\underline{F}}:\mathbb{R}^p\to\mathbb{R}^{p\times n}$ is a smooth function, consider the Stratanovich side

$$\partial\underline{y}(t) = \underline{\underline{F}}(\underline{y}(t))\partial\underline{B}(t) \tag{1}$$

Find the generator of the process $\{\underline{y}(t):t\geq 0\}$. When (1) is converted into an Ito sde, it becomes

$$d\underline{y}(t) = \underline{\underline{F}}(\underline{y}(t))d\underline{B}(t)+\underline{\underline{F}}_{,k}(\underline{y}(t))F_{kj}\,dB_j d\underline{B}$$

$$= \underline{\underline{F}}(\underline{y})d\underline{B} + \underline{\underline{F}}_{,k}(\underline{y})F_{kj}\,\underline{e}_j dt$$

$$= \underline{\underline{F}}(\underline{t})d\underline{B} + F_{j,k}(\underline{y})F_{kj}(\underline{y})dt$$

Summation over the repeated indices j, k is implied. Here $\underline{F}_j(\underline{y})$ is the jth column of $\underline{F}(\underline{y})$

Now $\quad \underline{F}_{j,k}F_{kj} = F_{kj}F_{j,k}$

$= \underline{\underline{F}}_{,k}(\underline{\underline{F}}^T)_k$ where $(\underline{\underline{F}}^T)_k$ is the kth column $\underline{\underline{F}}^T$. The generator of $\{\underline{y}(t)\}$ is therefore given by

$$\mathcal{L}\psi(\underline{y}) = \tilde{F}_k^T\,\underline{\underline{F}}_{,k}\underline{\nabla}_y\psi(\underline{y})\underline{\underline{F}}\,\underline{\underline{F}}^T\underline{\nabla}_y\underline{\nabla}_y^T + \frac{1}{2}Tr(\psi(y))$$

where $\tilde{F}_k^T$ is the kth row of $\underline{\underline{F}}$

Note that $\tilde{F}_k^T\,\underline{\underline{F}}_{,k} = (\tilde{F}_k^T\,\underline{\underline{F}})_{,k} - F_{k,k}^T\,\underline{\underline{F}}$

Example:

Let $\quad \underline{F}(\underline{y}) = \begin{bmatrix} \underline{y} \\ \underline{H}(\underline{y}) \end{bmatrix} \in \mathbb{R}^n, \underline{y} \in \mathbb{R}^p, H(\underline{y}) \in \mathbb{R}^{n-p}$

Consider the Stratanovich sde

$$\partial\begin{pmatrix} \underline{y}(t) \\ \underline{H}(\underline{y}(t)) \end{pmatrix} = \underline{\underline{P}}(\underline{y}(t))\partial\underline{B}(t) \quad \underline{B}(t) \in \mathbb{R}^n \text{ is n-dimensional Brownian motion.}$$

where $\underline{\underline{P}}(\underline{y})$ is the projection of $\mathbb{R}^n$ onto the tangent mainfold defined by the p dimensional surface $\underline{y} \to \underline{F}(\underline{y}) \in \mathbb{R}^n$, *i.e.*

if $\quad \underline{\underline{A}}(\underline{y}) = \underline{F}'(\underline{y}) = \begin{bmatrix} \underline{\underline{I}}_p \\ \underline{\underline{H}}'(\underline{y}) \end{bmatrix} \in \mathbb{R}^{n\times p}$

Thus, $\quad \underline{\underline{P}}(\underline{y}) = \underline{\underline{A}}(\underline{y})(\underline{\underline{A}}^T(\underline{y})\,\underline{\underline{A}}(\underline{y}))^{-1}\underline{\underline{A}}^T(\underline{y})$

$$= (I + H'^T H')^{-1}[\underline{\underline{I}}_p, \underline{\underline{H}}'^T] = \begin{bmatrix} \underline{\underline{I}}_p \\ \underline{\underline{H}}'(q) \end{bmatrix}$$

$$= \begin{bmatrix} (I + H'^T H')^{-1} & (I + H'^T H')H'^T \\ H'(I + H'^T H')^{-1} & H'(I + H'^T H')^{-1}H'^T \end{bmatrix}$$

Let $\underline{B}(t) = \begin{pmatrix} \underline{B}_1(t) \\ \underline{B}_2(t) \end{pmatrix}$ where $\underline{B}_1(\cdot) \in \mathbb{R}^p \;\; \underline{B}_2(\cdot) \in \mathbb{R}^{n-p}$

are independent Brownian motions

Then the equation

$$\partial\begin{pmatrix} y \\ H(y) \end{pmatrix} = p(y)\partial B$$

is equivalent to

$$\partial y = (I + H'(y)^T H'(y))^{-1}(\partial \underline{B}_1 + H'(y)^T B_2)$$

or writing

$$Q_1(\underline{y}) = (I + H'(y)^T H'(y))^{-1},$$

$$Q_2(\underline{y}) = (I + H'(y)^T H'(y))^{-1} H'(y)^T$$

$$\partial \underline{y} = \underline{\underline{Q}}_1(\underline{y})\partial \underline{B}_1 + \underline{\underline{Q}}_2(\underline{y})\partial \underline{B}_2$$

or equivalently in the Ito sense

$$dy = Q_1(y)\,d\underline{B}_1 + Q_2(y)\,d\underline{B}_2 + \frac{1}{2}(Q_{1,k}(\underline{y})dy_k d\underline{B}_1 + Q_{2,k}(\underline{y})\,dy_k d\underline{B}_2)$$

or

$$dy = \underline{\underline{Q}}_1(\underline{y})\,d\underline{B}_1 + \underline{\underline{Q}}_2(\underline{y})\,d\underline{B}_2 + \frac{1}{2}\left(Q_{1,k}(\underline{y})Q_{1km}^{(y)}\underline{e}_m + Q_{2,k}(\underline{y})Q_{2km}(\underline{y})\underline{f}_m\right)dt$$

where $\underline{e}_m$ is the $p \times 1$ vector with 1 in the mth position and O at all other position. $\underline{f}_m$ is the $(n - p) \times 1$ vector with 1 in the mth position and O at all the other position.

Let $\tilde{\underline{Q}}_{1k}$ denote the kth column of $\underline{\underline{Q}}_1$ and $\tilde{\underline{Q}}_{2k}$ the kth column of $\underline{\underline{Q}}_2$. Then the above Ito sde is the same as

$$dy = \underline{\underline{Q}}_1 d\underline{B}_1 + \underline{\underline{Q}}_2 d\underline{B}_2 + \frac{1}{2}(Q_{1,k}\tilde{Q}_{1k} + Q_{2,k}\tilde{Q}_{2k})dt$$

summation over the repeated index $k = 1, 2, \ldots, p$ being implied.

The generator of $\{y(t)\}$ is given by

$$\mathcal{L}\psi(y) = \frac{1}{2}\left(Q_{1,k}\tilde{Q}_{1k} + Q_{2,k}\tilde{Q}_{2k}\right)^T \underline{\nabla}_y \psi(y) + \frac{1}{2}Tr\left\{\left(Q_1 Q_1^T + Q_2 Q_2^T\right)\nabla_y \nabla_y^T \psi(y)\right\}$$

Laplace Beltrami operator on μ.

Metrics

$$d\begin{pmatrix} y \\ \underline{H}(\underline{y}) \end{pmatrix}^T d\begin{pmatrix} \underline{y} \\ H(\underline{y}) \end{pmatrix} = d\underline{y}^T(\underline{\underline{I}}_p + \underline{\underline{H'}}(\underline{y})^T \underline{\underline{H'}}(\underline{y}))\,d\underline{y}$$

So the metrics on μ is

$$\underline{\underline{G}}(\underline{y}) = \underline{\underline{I}}_p + H'(y)^T H'(y) = Q_1(\underline{y})^{-1}$$

Laplace-Beltrami operator on μ is

$$\Delta_\mu = \frac{1}{2}\left|\underline{\underline{G}}^{(y)}\right|^{-\frac{1}{2}} \underline{\nabla}_y^T\left(\left|G(\underline{y})\right|^{\frac{1}{2}} \underline{\underline{G}}(\underline{y})^{-1}\nabla_y\right)$$

$$\Delta_\mu\,\psi = \frac{1}{2}|G|^{-\frac{1}{2}} \underline{\nabla}_y^T\left(|G|^{\frac{1}{2}} G^{-1}\nabla_y\psi\right)$$

$$= \frac{1}{2}\underline{\nabla}_y^T\left(\underline{\underline{G}}^{-1}\underline{\nabla}_y\psi\right) + \frac{1}{4}\left(\underline{\nabla}_y \log|G|\right)^T \underline{\underline{G}}^{-1}\underline{\nabla}_y\psi$$

$$= \frac{1}{2}Tr\left(\underline{\underline{G}}^{-1}\underline{\nabla}_y\underline{\nabla}_y^T\psi\right) + \frac{1}{2}(\underline{\underline{G}}^{-1})_{km,k}\,\psi_{,m} + \frac{1}{4}(\underline{\nabla}_y \log|G|)^T \underline{\underline{G}}^{-1}\underline{\nabla}_y\psi$$

Note that $\quad \dfrac{\partial}{\partial y_k}|G| = (G^{-1})_{rs}\,|G|\,G_{rs,k}$

So, $\quad \nabla_y \log|G| = (G^{-1})_{rs}\nabla_y G_{rs,k}$

$$Q_1Q_1^T + Q_2Q_2^T = (I + H'^T H')^{-2} + (I + H'^T H')^{-1} H'^T H'(I + H'^T H')^{-1}$$

$$= (I + H'^T H')^{-1} = G^{-1}.$$

❖❖❖❖❖

[217] Path integral approach to Dirac relativstic wave equation

$$H = (\alpha, p) + \beta m + V,\ n\delta t = t$$

$$e^{-i\delta t} H_0\, e^{-i\delta t V}$$

$$\lim_{n\to\infty}\left(e^{-i\delta t} H_0\, e^{-i\delta t V}\right)^n = e^{-it(H_0+V)} = e^{-itH}$$

$$\left\langle \sigma'', q'' \,|\, e^{-i\delta t H_0}\, e^{-i\delta t V} \,|\, \sigma', q' \right\rangle = \left\langle \sigma'', q'' \,|\, e^{-i\delta t((\alpha,\, p)+\beta m)} \,|\, \sigma', q' \right\rangle$$

$$= -i\delta t\,\delta_{\sigma'',\sigma'} V(q')\,\delta(q'' - q')$$

$$= \int [e^{-i\delta t}((\alpha,\, p') + \beta m)]_{\sigma'',\sigma'}\, e^{ip'.(q'-q'')}\,dp'$$

$$= -i\delta t\,\delta_{\sigma'',\sigma'} V(q')\,\delta(q'' - q')$$

$$\int_0^\infty \exp(-it((\alpha, p') + \beta m))\exp(-st)\,dt$$

$$= (s+i((\alpha, p')+\beta m))^{-1}$$

$$= \frac{s-i((\alpha, p')+\beta m)}{S^2+E^2}$$

$$\mathcal{L}^{-1} \to \cos(E(p')t)\underline{\underline{I}}_4 - \frac{i}{E(p)}H_0(p')\sin \mathbb{E}(p')$$

$$\left(\left(\langle\sigma'', q'' \mid \exp(-i\delta t H_0)\cdot\exp(-i\delta t V) \mid \sigma', q'\rangle\right)\right)_{1\le\sigma', \sigma''\le\alpha}$$

$$= \int\left[\cos(E(p')\delta t)\underline{\underline{I}}_4 - \frac{i}{E(p')}H_0(p')\sin(E(p')\delta t)\right]$$

$$\exp(ip'\cdot(q'-q''))d^3p'$$

$$-i\delta t\cdot I_4 V(q')\delta(q'-q'')$$

$$= \int_{\mathbb{R}^3} \cos(E(p)\delta t)e^{ip.q}d^3p$$

$$= \frac{1}{2}\int \exp(i(p.q+E(p)\delta t))d^3p$$

$$+\frac{1}{2}\exp(i(p.q+E(p)\delta t))d^3p$$

$$\pi \int_{\substack{0<p<\infty\\0<\theta<\pi}} \exp\left(i\left(p\,|\,q\,|\cos\theta+\delta t\sqrt{p^2+m^2}\right)\right)p^2\sin\theta\, dp\, d\theta$$

$$+\pi \int_{\substack{0<p<\infty\\0<\theta<\pi}} \exp\left(i\left(p\,|\,q\,|\cos\theta-\delta t\sqrt{p^2+m^2}\right)\right)p^2\sin\theta\, dp\, d\theta$$

$$= \int_0^\infty \frac{i\pi p}{|\underline{q}|}\times(2i)\sin(p\,|\,\underline{q}\,|)\exp\left(i\delta t\sqrt{p^2+m^2}\right)dp$$

$$+\int_0^\infty \frac{i\pi p}{|\underline{q}|}(-2i)\sin(p\,|\,\underline{q}\,|)\exp\left(-i\delta t\sqrt{p^2+m^2}\right)dp$$

$$= \frac{4\pi}{|\underline{q}|}\left\{\int_0^\infty p\sin(p\,|\,\underline{q}\,|)\cos\left(\delta t\sqrt{p^2+m^2}\right)dp\right\}$$

(Let $p = m\tan x\ dp = m\sec^2 x\, dx$)

$$= \frac{4\pi}{|\underline{q}|}\int_0^{\frac{\pi}{2}} m\sec^2(x)\sin(m\,|\,\underline{q}\,|\tan x)\cdot\cos(\delta t\, m\sec x)\, m\tan x\, dx$$

$$= \frac{4\pi m^2}{|\underline{q}|} \int_0^{\frac{\pi}{2}} \sin x \sec^2(x) \sin(m|\underline{q}|\tan x) \cdot \cos(\delta t\, m \sec x) dx$$

$$I(\underline{v}) = \int \frac{H_0(\underline{p})}{E(p)} \sin(E(p)\delta t) \exp(i\underline{p} \cdot \underline{q}) d^3 p$$

$$= \int_{\substack{0<p<\infty,\\ 0<\theta<\pi,\\ 0<\varphi<2\pi}} \frac{H_0(p\cos\varphi\sin\theta,\ p\sin\varphi\sin\theta,\ p\cos\theta)}{E(p)} \sin(E(p)\delta t)$$

$$\times \exp(ip|\underline{q}|\cos\theta) p^2 \sin\theta\, dp\, d\theta\, d\varphi$$

$$H(\underline{p}) = \begin{pmatrix} mI_2 & (\underline{\sigma}, \underline{p}) \\ (\underline{\sigma}, \underline{p}) & -mI_2 \end{pmatrix}$$

$$\int_0^{2\pi} H_0(p\cos\varphi\sin\theta,\ p\sin\varphi\sin\theta,\ p\cos\theta) d\varphi$$

$$\int_0^{2\pi} \begin{pmatrix} mI_2 & p(\sigma_1\cos\varphi\sin\theta + \sigma_2\sin\varphi\sin\theta + \sigma_3\cos\theta) \\ p(\sigma_1\cos\varphi\sin\theta + \sigma_2\sin\varphi\sin\theta + \sigma_3\cos\theta) & -mI_2 \end{pmatrix} d\varphi$$

$$= 2\pi \begin{pmatrix} mI_2 & \sigma_3 p\cos\theta \\ \sigma_3 p\cos\theta & -mI_2 \end{pmatrix}$$

So.

$$I(\underline{q}) = 2\pi \int_{\substack{0<p<\infty\\ 0<\theta<\pi}} \begin{pmatrix} mI_2 & \sigma_3 p\cos\theta \\ \sigma_3 p\cos\theta & -mI_2 \end{pmatrix} \frac{\sin\{E(p)\delta t\}}{E(p)} p\exp(ip|\underline{q}|\cos\theta)$$

$$p^2 \sin\theta\, dp d\theta$$

$$= \begin{pmatrix} m \cdot J_1(|\underline{q}|, \delta t) \cdot \underline{\underline{I}}_2 & \underline{\underline{\sigma}}_3 J_2(|\underline{q}|, \delta t) \\ \underline{\underline{\sigma}}_3 J_2(|\underline{q}|, \delta t) & -mJ_1(|\underline{q}|, \delta t) \underline{\underline{I}}_2 \end{pmatrix}$$

where $J_1(|\underline{q}|, \delta t) = 2\pi \int_{\substack{0<p<\infty\\ 0<\theta<\pi}} \frac{\sin\{E(p)\delta t\}}{E(p)} \exp(ip|\underline{q}|\cos\theta) p^2 \sin\theta\, dp\, d\theta$

$$= \int_0^\infty \frac{4\pi p}{|\underline{q}|} \sin(p|\underline{q}|) \sin\frac{\{E(p)\delta t\}}{E(p)} dp$$

and

$$J_2(|\underline{q}|, \delta t) = 2\pi \int_{\substack{0<p<\infty\\ 0<\theta<\pi}} \frac{\sin\{E(p)\delta t\}}{E(p)} \exp(ip|\underline{q}|\cos\theta) p^3 \sin\theta\cos\theta\, dp\, d\theta$$

$$= 2\pi\int_0^\infty dp \cdot \frac{\sin(E(p)\delta t)}{E(p)} p \int_{-p|\underline{q}|}^{p|\underline{q}|} \frac{\exp(ix)}{|\underline{q}|^2} x\,dx$$

$$\int_{-p|\underline{q}|}^{p|\underline{q}|} xe^{ix}dx = \frac{xe^{ix}}{i}\Bigg|_{-p|\underline{q}|}^{p|\underline{q}|} - \int_{-p|\underline{q}|}^{p|\underline{q}|} \frac{e^{ix}}{i}dx$$

$$= \frac{p|\underline{q}|x^2\cos(p|\underline{q}|)}{i} + e^{ix}\Big|_{-p|\underline{q}|}^{p|\underline{q}|}$$

$$= -2ip|\underline{q}|\cos(p|\underline{q}|) + 2i\sin(p|\underline{q}|)$$

So,

$$J_2(|\underline{q}|, \delta t) = \int_0^\infty \left[\frac{-4\pi i p^2|\underline{q}|\cos(p|\underline{q}|)}{E(p)|\underline{q}|^2} + \frac{4\pi i p\sin(p|\underline{q}|)}{E(p)|\underline{q}|^2}\right] \times \sin\{E(p)\delta t\}dp$$

$$= \frac{4\pi i}{|\underline{q}|^2}\int_0^\infty \frac{p}{E(p)}\sin\{p|\underline{q}|\}\sin\{E(p)\delta t\}dp$$

$$= -\frac{4\pi i}{|\underline{q}|}\int_0^\infty \frac{p^2}{E(p)}\cos\{p|\underline{q}|\}\sin\{E(p)\delta t\}dp$$

The Dirac path integral Kernal is then

$$\lim_{\substack{\delta t\to 0\\ N\to\infty\\ N\delta t = T}} \int \prod_{\substack{n=0\\ \underline{q}_0=\underline{\xi}\\ \underline{q}_N=\underline{\eta}}}^{N-1} \begin{pmatrix} mJ_1(|\underline{q}_{n+1}-\underline{q}_n|, \delta t).\underline{\underline{I}}_2 & \underline{\underline{\sigma}}_3 J_2(|\underline{q}_{n+1}-\underline{q}|, \delta t) \\ \underline{\underline{\sigma}}_3 J_2(|\underline{q}_{n+1}-\underline{q}_n|, \delta t) & -mJ_1(|\underline{q}_{n+1}-\underline{q}|, \delta t)\underline{\underline{J}}_2 \end{pmatrix}$$

$$\exp(-iV(\underline{q}_n)\delta t)$$

$$d^3\underline{q}_1 \ldots d^3\underline{q}_{N-1}$$

$$= \underline{\underline{K}}_D(\underline{\eta}, \underline{\xi}, T)$$

Changing the integration variable from p to $p' = p\delta t$ gives

$$J_1(|\underline{q}|, \delta t) = 4\pi\int_0^\infty \frac{\sin\left(\sqrt{p'^2 + m^2\delta t^2}\right)}{\sqrt{p'^2 + m^2\delta t^2}} \frac{p'}{|\underline{q}|\delta t}\sin\left(p'\frac{|\underline{q}|}{\delta t}\right)\frac{dp'}{\delta t}$$

$$\delta t \overset{\approx}{\to} 0 \approx 4\pi\int_0^\infty \frac{\sin(p')}{(p')} \frac{p'}{\left|\frac{\underline{q}}{\delta t}\right|}\sin\left(p'\left|\frac{\underline{q}}{\delta t}\right|\right)\frac{dp'}{(\delta t)^3}$$

So writing $\dot{q} = (q' - q'')/\delta t$ gives

$$J_1(|\underline{q}' - \underline{q}''|, \delta t) \approx \frac{4\pi}{(\delta t)^3}\int_0^{\infty} \frac{\sin(p)}{|\dot{q}|} \sin(p|\dot{q}|)\,dp$$

$$J_2(|\underline{q}|, \delta t) = \frac{4\pi i}{|\underline{q}|^2}\int_0^{\infty} \frac{p'}{\sqrt{p'^2 + m^2\delta t^2}} \sin\left(p'\frac{|\underline{q}|}{\delta t}\right)$$

$$\times \sin\left(\sqrt{p'^2 + m^2\delta t^2}\right) dp' / \delta t$$

$$- \frac{4\pi i}{|\underline{q}|^2}\int_0^{\infty} \frac{p'}{\delta t\sqrt{p'^2 + m^2\delta t^2}} \cos\left(p'\frac{|\underline{q}|}{\delta t}\right) \sin\left(\sqrt{p'^2 + m^2\delta t^2}\right)\frac{dp'}{\delta t}$$

$$\delta t \xrightarrow{\approx} 0 \approx \frac{4\pi i}{|\underline{q}|^2}\int_0^{\infty} \sin(p|\underline{q}|/\delta t)\sin(p)\,\delta p/\delta t$$

$$- \frac{4\pi i}{|\underline{q}|\delta t^2}\int_0^{\infty} \cos\left(-p\frac{|\underline{q}|}{\delta t}\right)\sin(p)\,dp$$

or $J_2(|\underline{q}' - \underline{q}''|, \delta t)$

$$\approx \frac{4\pi i}{(\delta t)^3}\left\{\frac{1}{|\dot{\underline{q}}|^2}\int_0^{\infty} \sin(p|\dot{\underline{q}}|)\sin(p)\,dp - \frac{1}{|\dot{\underline{q}}|}\int_0^{\infty} \cos(p|\dot{\underline{q}}|)\sin(p)\,dp\right\}$$

$$H_0 = (\alpha, p) + \beta m$$

$$\alpha_r = \begin{pmatrix} 0 & \sigma_r \\ \sigma_r & 0 \end{pmatrix}$$

$$\beta = \begin{pmatrix} I & 0 \\ 0 & -I \end{pmatrix}$$

$$dU(t) = [-(iH_0 + \varepsilon^2 LL_2^*)\,dt + \varepsilon L\,dA_t - \varepsilon L^*\,dA_t^*]U(t)$$

$$U_0(t) = \exp\{-it((\alpha, p) + \beta m)\}$$

$$U_1(t) = \int_0^t U_0(t-\tau)(L\,dA_\tau - L^*\,dA_\tau^*)U_0(\tau)$$

$$= \int_0^{\infty} \exp(-itH_0)\exp(-st)\,dt = (S + iH_0)^{-1}$$

$$= (S + i((\alpha, p) + \beta m))^{-1}$$

$$= \frac{S - i((\alpha, p) + \beta m)}{S^2 + E^2}$$

$$E^2 = p^2 + m^2$$

$$U_0(t) = \exp(-itH_0)$$

$$= E(p) = \left(\cos(E_t) - \frac{iH_0}{E} \sin(E_t) \right) \theta(t)$$

$$\underline{f}(\underline{p}), \underline{g}(\underline{p}) \in L^2(\mathbb{R}^3) \otimes \mathbb{C}^4$$

$$\int \| \underline{f}(\underline{p}) \|^2 d^3 p = 1$$

$$\int \| \underline{g}(\underline{p}) \|^2 d^3 p = 1$$

$$\langle g\varphi(v) | U_1(t) | f\varphi(u) \rangle$$

$$\left(\varphi(u) = \exp\left(-\frac{\| u \|^2}{2} \right) | e(u) \rangle \right)$$

$$= \int_0^t \langle g | U_0(t-\tau) L U_0(\tau) | f \rangle \langle \varphi(v) | dA_\tau | \varphi(x) \rangle$$

$$+ \int_0^t \langle g | U_0(t-\tau) L^* U_0(\tau) | f \rangle$$

$$\langle \varphi(v) | dA_\tau^* | \varphi(v) \rangle$$

$$\frac{\langle g\varphi(v) | U_1(t) | f\varphi(u) \rangle}{\langle \varphi(v) | \varphi(u) \rangle}$$

$$= \int_0^t [\langle g | U_0(t-\tau) L U_0(\tau) | f \rangle u(\tau) + \langle g | U_0(t-\tau) L^* U_0(\tau) | f \rangle v(\tau)] d\tau$$

$$\langle g | U_0(t-\tau) L U_0(\tau) | f \rangle = \int \underline{g}(p)^* \left[\cos(E(p)(t-\tau)) - \frac{iH_0^{(p)}}{E(p)} \sin(E(p)(t-\tau)) \right]$$

$$L(p) \left[\cos(E(p)\tau) - \frac{iH_0(p)}{E(p)} \sin(E(p)\tau) \right] \underline{f}(p) d^3 p$$

$$= \int \underline{g}(\underline{p})^* L(p) \underline{f}(\underline{p}) \cos(E(p)(t-\tau)) \cos(E(p)\tau) d^3 p$$

$$-i \int \frac{\underline{g}(\underline{p})^* L(p) H_0(p)}{E(p)} \underline{f}(\underline{p}) \cos(E(p)(t-\tau)) \sin(E(p)^2) d^3 p$$

$$-i \int \frac{\underline{g}(p)^* H_0(p) L(p)}{E(p)} \underline{f}(p) \sin(E(p)(t-\tau_1)) \cos(E(p)\tau) d^3 p$$

$$-\int \frac{\underline{g}(\underline{p})^* H_0(\underline{p}) L(\underline{p}) H_0(\underline{p})}{E^2(\underline{p})} \underline{f}(\underline{p}) \sin(E(\underline{p})(t-\tau)) \sin(E(\underline{p})\tau) d^3p$$

+ Similar terms with $L(\underline{p}) \to L^*(\underline{p}), u(\tau) \to \overline{v}(\tau)$

$$\int_0^t u(\tau)\cos(E(t-\tau))\cos(E\tau)d\tau = \frac{1}{2}\int_0^t u(\tau)(\cos(Et)+\cos(E(t-2\tau)))d\tau$$

$$\xrightarrow[t\to\infty]{} \frac{1}{2}\left(\cos(Et)\hat{u}(0)+\frac{1}{2}\exp(iEt)\hat{u}(2E)+\frac{1}{2}\exp(-iEt)\hat{u}(-2E)\right)$$

$$= \frac{1}{4}\left[\exp(iEt)\left(\hat{u}(0)+\hat{u}(2E)\right)+\exp(-iEt)(\hat{u}(0)+\hat{u}(-2E))\right]$$

$$\int_0^t u(\tau)\cos(E(t-\tau))\sin(E\tau)d\tau = \frac{1}{2}\int_0^t u(\tau)(\sin(Et)+\sin(E(2\tau-t)))d\tau$$

$$\xrightarrow[t\to\infty]{} \frac{1}{2i}\left[e^{iET}-e^{-iEt}\right]\hat{u}(0)+\frac{1}{2i}\left[e^{-iEt}\hat{u}(-2E)+e^{iEt}\hat{u}(2E)\right]$$

$$= -\frac{i}{2}e^{iEt}\left[\hat{u}(0)+\hat{u}(2E)\right]+\frac{i}{2}e^{-iEt}\left[\hat{u}(0)-\hat{u}(-2E)\right]$$

For real functions $u(t)$, $v(t)$,

$$\int_0^\infty u(\tau)\cos(E(t-\tau))\cos(E\tau)d\tau = \frac{1}{2}\left[+\mathrm{Re}(e^{-iEt}\hat{u}(2E))\right]\hat{u}(0)\cos(Et)$$

and

$$\int_0^\infty u(\tau)\cos(E(t-\tau))\sin(E\tau)d\tau = \frac{1}{2}\left[\hat{u}(0)\sin(Et)-\mathrm{Im}\left(\hat{u}(2E)e^{iEt}\right)\right]$$

$$\int_0^\infty u(\tau)\sin(E(t-\tau))\cos(E\tau)d\tau = \frac{1}{2}\int_0^\infty u(\tau)(\sin(Et)+\sin(E(t-2\tau)))d\tau$$

$$= \frac{1}{2}\left[\sin(Et)\hat{u}(0)+\mathrm{Im}\left\{\hat{u}(2E)e^{iEt}\right\}\right]$$

$$\int_0^\infty u(\tau)\sin(E(t-\tau))\sin E\tau,d\tau = \frac{1}{2}\int u(\tau)(\cos(E(t-2\tau))-\cos(Et))d\tau$$

$$= \frac{1}{2}\left[\mathrm{Re}\left(\hat{u}(2E)e^{iEt}\right)-\hat{u}(0)\cos(Et)\right]$$

Let $\quad \frac{1}{2}\left(\hat{u}(0)+\hat{u}(2E)\right) = \alpha_u(E)$

$$\frac{1}{2}\left(\hat{u}(0)-\hat{u}(2E)\right) = \beta_u(E)$$

Then, $\dfrac{\langle g\varphi(v)|U_1(t)|f\varphi(u)\rangle}{\langle\varphi(v)|\varphi(u)\rangle}$

$$t \overset{\approx}{\rightarrow} \infty \approx T_u + \tilde{T}_v$$

where
$$T_u = \int \underline{g}(\underline{p})^* L(\underline{p}) \underline{f}(\underline{p}) \operatorname{Re}\left(\alpha_u(E(p))e^{iE(p)t}\right)d^3p$$
$$-\int g(\underline{p})^* H_0(p) L(\underline{p}) H_0(\underline{p}) \underline{f}(\underline{p}) \operatorname{Re}\left(\beta_u(E(p))e^{iE(p)t}\right)d^3p$$
$$-i\int \frac{\underline{g}(\underline{p})^* L(\underline{p}) H_0(\underline{p}) \underline{f}(\underline{p})}{E(\underline{p})} \operatorname{Im}\left(\beta_u(E(p))e^{iE(p)t}\right)d^3p$$
$$-i\int \frac{\underline{g}(p)^* H_0(p) \underline{L}(\underline{p}) \underline{f}(\underline{p})}{E(p)} \operatorname{Im}\left(\alpha_u\left(E(p)e^{iE(p)t}\right)d^3p\right)$$

and $\tilde{T}_v$ is the same as T_u but with u replaced

Note by v and $L(p)$ by $L^*(p)$.

$$H_0(\underline{p}) = (\alpha, p) + \beta m = \begin{pmatrix} mI_2 & (\sigma, p) \\ (\sigma, p) & -mI_2 \end{pmatrix}$$

and $L(\underline{p})$ is a 4×4 matrix valued find $\underline{p}$.

The Vector Langevin equation

$$\ddot{q}(t) + \beta q \sum(t) + Kq \sum(t) = \sigma\frac{d\beta(t)}{dt}$$

gives on discretization (Here $\Sigma -1$ denotes the units delay operator) the following expression for the los-likelihood functions.

$$\log p(\underline{q}\,|\{w_k, A_k\}) = \frac{1}{2}\sigma^2\left(I_N + \underline{\underline{\beta}}\Delta + \underline{\underline{K}}\Delta^2 + Z^2 - (2 + \underline{\underline{\beta}}\Delta Z^{-1})\right)\underline{q}$$

$$= \Delta^2 \sum_k \underline{e}(\sigma_k) \otimes \underline{\underline{J}}^{-1} \underline{\underline{C}}_k - \Delta^2 \left(\sum_k \underline{e}(w_k) \otimes \underline{\underline{J}}^{-1} \underline{A}_k\right)^* \left(\underline{\underline{I}} \otimes \underline{\underline{J}}^2\right)$$

$$= \left(I_N + \underline{\underline{\beta}}\Delta + K\Delta^2 + Z^2 - (2 + \underline{\underline{\beta}}\Delta Z^{-1})\right)\underline{q}$$

$$= -\Delta^2 \sum_k \underline{e}(\alpha_k) \otimes \underline{J}^{-1} \underline{C}_k - \Delta^2 \sum_k e(w_k) \otimes \underline{\underline{J}}^{-1} \underline{A}_k \Big)$$

The *m*/*e* of $\{w_k, A_k\}$ can be obtained from this expression and further the CRLB of $\{w_k, A_k\}$ can also be obtained from this. The CRLB for $\{w_k, A_k\}$ can also be derived from this expression.

❖❖❖❖❖

[218] Quantum fidelity between two states.

Let $|u\rangle^{RA}$ be a purification of ρ_A and $|v\rangle^{RAB}$ a purification of ρ_B

Thus,

$$Tr_R\left[|u\rangle^{RA}\langle u|^{RA}\right] = \rho_A,$$

$$Tr_R\left[|u\rangle^{RB}\langle u|^{RB}\right] = \rho_B$$

Let
$$|u\rangle^{RA} = \sum_{\alpha}\sqrt{p_\alpha}\,|e_\alpha\rangle^R|e_\alpha\rangle^A$$

$$|v\rangle^{RA} = \sum_{\alpha}\sqrt{p_\alpha}\,|f_\alpha\rangle^R|f_\alpha\rangle^B$$

$\left\{|e_\alpha\rangle^R\right\}$ and $\left\{|f_\alpha\rangle^R\right\}$ are ONB's for $\mathcal{H}_R$,

$\left\{|e_\alpha\rangle^A\right\}$ and $\left\{|f_\alpha\rangle^B\right\}$ are ONB's for $\mathcal{H}_A = \mathcal{H}_B$.

$$^{RA}\langle u|v\rangle^{RB} = \sum_{\alpha,\beta}\sqrt{p_\alpha}\sqrt{q_\beta}\;{}^R\langle e_\alpha|f_\beta\rangle^R\;{}^A\langle e_\alpha|f_\beta\rangle^B$$

$$\rho_A = \sum_{\alpha}p_\alpha|e_\alpha\rangle^A\,{}^A\langle e_\alpha|$$

$$\rho_B = \sum_{\alpha}q_\alpha|f_\alpha\rangle^B\,{}^B\langle f_\alpha|$$

$$\sqrt{\rho_A}\sqrt{\rho_B} = \sum_{\alpha,\beta}\sqrt{p_\alpha q_\beta}\;{}^A\langle e_\alpha|f_\beta\rangle^B|e_\alpha\rangle^A\,{}^B\langle f_\beta|$$

Let
$$U|e_\alpha\rangle = |f_\alpha\rangle$$

$$\sqrt{\rho_A}\sqrt{\rho_B} = \sum_{\alpha,\beta}\sqrt{p_\alpha q_\beta}\,\langle e_\alpha|U|e_\beta\rangle|e_\alpha\rangle\langle e_\beta|U$$

$$Tr\left[\sqrt{\rho_A}\sqrt{\rho_B}\,U\right] = \sum_{\alpha,\beta}\sqrt{p_\alpha q_\beta}\,\langle e_\alpha|U|e_\beta\rangle Tr\left[|e_\alpha\rangle\langle e_\beta|\right]$$

$$Tr\left(\sqrt{\rho_A}\sqrt{\rho_B}\,U\right) = \sum_{\alpha}\sqrt{p_\alpha q_\alpha}\,\langle e_\alpha|U|e_\alpha\rangle$$

$$Tr\left(\sqrt{\rho_A}\sqrt{\rho_B}\,W\right) = \sum_{\alpha,\beta}\sqrt{p_\alpha q_\beta}\,\langle e_\alpha|U|e_\beta\rangle\langle e_\beta|U^*W|e_\alpha\rangle$$

$$\left\|\sqrt{\rho_A}\sqrt{\rho_B}\right\|_1 = \sup\left|Tr\left(\sqrt{\rho_A}\sqrt{\rho_B}\,W\right)\right|\; W^*W = I$$

$$|u\rangle = \sum_{\alpha}\sqrt{p_\alpha}\,|u_\alpha\rangle|e_\alpha\rangle$$

$$|v\rangle = \sum_{\alpha} \sqrt{q_\alpha} |v_\alpha\rangle |f_\alpha\rangle$$

$$\langle v | u\rangle = \sum_{\alpha,\beta} \sqrt{p_\alpha q_\beta} \langle f_\beta | e_\alpha\rangle \langle v_\beta | u_\alpha\rangle$$

$$\max_{\{u_\alpha, v_\beta\}} \langle v | u\rangle$$

$$\left(\left(\sqrt{p_\alpha q_\beta}\langle f_\beta | e_\alpha\rangle\right)\right) = A$$

$$\left(\left(\langle V_\beta | u_\alpha\rangle\right)\right) = W$$

$$\sqrt{\rho_A}\sqrt{\rho_B} = \sum \sqrt{p_\alpha q_\beta}\langle e_\alpha | f_\beta\rangle$$

❖❖❖❖❖

[219] Estimating the random rotation and translation applied to an object (3-D) using the Maximum A posteriori algorithm

$f(\underline{r}) \approx$ original an object

$$g(\underline{r}) = f(\underline{\underline{R}}^{-1}(\underline{r} - \underline{a})) + w(\underline{r})$$

$R \in SO(3), a \in \mathbb{R}^3$ are random with a pdf

$p(\underline{\underline{R}}, \underline{a})\, dR\, d\underline{a}$ where $d\underline{a}$ is the Lebergue measure on $\mathbb{R}^3$ ad $d\mathbb{R}$ is the Haar measure on $SO(3)$. Note that the rotation translation group is $SO(3)\ S\ \mathbb{R}^3$ (semidirect product)

which is a locally compact non-Abelian group Radon transform

$$(\mathbb{R}g)(\hat{n}, p) = \int g(\underline{r})\delta(p - (\hat{n}, \underline{r}))d^3r$$

$$= \int f(R^{-1}(\underline{r} - \underline{a}))\delta(p - (\hat{n}, \underline{r}))d^3r + (\mathbb{R}w)(\hat{n}, p)$$

$$= \int f(\underline{r})\delta(p - (\hat{n}, R\underline{r} + \underline{a}))d^3r + (\mathbb{R}w)(\hat{n}, p)$$

$$= \int f(\underline{r})\delta(p - (\hat{n}, a) - (R^T\hat{n}, \underline{r}))d^3r + (\mathbb{R}w)(\hat{n}, p)$$

$$= (\mathbb{R}f)(R^T\hat{n}, p - (\hat{n}, a)) + \mathbb{R}w(\hat{n}, p)$$

$$(\mathcal{F}_p\mathbb{R}g)(\hat{n}, w) = \int_{\mathbb{R}} \exp(-jwp)(\mathbb{R}g)(\hat{n}, p)dp$$

$$= \exp(-jw(\hat{n}, a))(\mathcal{F}_p\mathbb{R}f)(R^T\hat{n}, w) + (f_p\,\mathbb{R}\,w)(\hat{n}, w)$$

Let $\underline{a} \sim N(\underline{\mu a}, \underline{\underline{\Sigma a}})$ be independent of R. Then

$$\mathbb{E}\{(\mathcal{F}_p \mathbb{R} g)(\hat{n}, w)\} = \exp\left\{-jw(\hat{n}, \underline{\mu a}) - \frac{1}{2} w^2 \hat{n} \Sigma_a \hat{n}\right\}$$

$$\int (\mathcal{F}_p \mathbb{R} f)(R^T \hat{n}, w)\, p(R)\, dR$$

assuming $\mathbb{E}\, w(\underline{r}) = 0$

Let $\quad R = R_0 + \delta R \quad R_0 = \mathbb{E}\{R\}$

Then $(\mathcal{F}_p Rf)(R^T \hat{n}, w)$

$$\begin{aligned}
&= (\mathcal{F}_p \mathbb{R} f)(R_0^T \hat{n} + \delta R^T . \hat{n}, w) \\
&\approx (\mathcal{F}_p Rf)(R_0^T \hat{n}, w) \\
&\quad + \delta \cdot \hat{n}^T \cdot \delta \underline{\underline{R}} \cdot \nabla_{\hat{n}} (\mathcal{F}_p Rf)(R_0^T \hat{n}, w) \\
&\quad + \frac{1}{2} \hat{n}^T \delta R \nabla_{\hat{n}} \nabla_{\hat{n}}^T (\mathcal{F}_p Rf)(R_0^T \hat{n}, w) \delta R^T \hat{n}
\end{aligned}$$

So,

$$\begin{aligned}
&\approx \mathbb{E}\{(\mathcal{F}_p Rf)(R^T \hat{n}, w)\} \\
&\approx (\mathcal{F}_p Rf)(R_0^T \hat{n}, w) \\
&\quad + \frac{1}{2} Tr\left\{\left(\nabla \nabla^T \mathcal{F}_p Rf\right)\left(R_0^T \hat{n}, w\right) \mathbb{E}\left\{\delta R^T \hat{n} \hat{n}^T \delta R\right\}\right\}
\end{aligned}$$

$$\begin{aligned}
\left(\mathbb{E}\left\{\delta R^T \hat{n} \hat{n}^T \delta R\right\}\right)_{\alpha\beta} &= \mathbb{E}\left\{(\delta R)_{\rho\alpha} n_\rho n_\sigma (\delta R)_{\sigma\beta}\right\} \\
&= n_\rho n_\sigma \mathbb{E}\left\{(\delta R)_{\rho\alpha} (\delta R)_{\alpha\beta}\right\}
\end{aligned}$$

Let $\{L_1, L_2, L_3\}$ be generator of $SO(3)$.

Thus $\delta R = \sum_{\alpha=1}^{3} \xi_\alpha L_\alpha \quad \xi_\alpha = \text{small } r.v\text{'s}$

They

$$\begin{aligned}
(\mathbb{E}\{\delta R^T \hat{n} \hat{n}^T \delta R\})_{\alpha\beta} &= n_\rho n_\sigma \mathbb{E}(\xi_k \xi_m)(L_k)_{\rho\alpha} (L_m)_{\sigma\beta} \\
&= \sum_{k,m=1}^{3} \left(L_k^T \hat{n}\right)_\alpha \left(L_m^T \hat{n}\right)_{\beta'} \mathbb{E}(\xi_k \xi_m)
\end{aligned}$$

$$\underline{\rho} = ((\rho_{km}))$$

Let $\rho_{km} = \mathbb{E}(\xi_k \xi_m)$. Then,

$$\mathbb{E}\{(\mathcal{F}_p Rg)(\hat{n}, w)\} \approx (\mathcal{F}_p Rf)(R_0^T \hat{n}, w) + \frac{1}{2} Tr \left\{ \nabla\nabla^T \mathcal{F}_p Rf \left(R_0^T \hat{n}, w \right) \left(\sum_{k,m=1}^{3} \rho_{km} L_k^T \hat{n} \hat{n}^T L_m \right) \right\}$$

For Rohit Rana

$$\sum_{k=1}^{p} \underline{\underline{B}}_k h_k[n] * \underline{q}[n] = \hat{\tau}_T[n]$$

$$H_k(Z) = \frac{2(1 - r_k \cos(w_k' \Delta) Z^{-1})}{(1 + r_k^2 Z^{-2} - 2 r_k \cos(w_k \Delta) Z^{-1})}$$

$$h_k(n) * \underline{q}[n] = \underline{v}_k[n]$$

$$h_k[n] = \left(r_k^n e^{jn\Delta w_k'} + e_k^n e^{-jn\Delta w_k'} \right) u[v]$$

$$\theta_k[n] = 2\,\mathrm{Re} \sum_{m=0}^{n} r_k^{n-m} e^{j(n-m)\Delta w_k'} \underline{q}[m]$$

$$\approx 2\,\mathrm{Re}\left[r_k^n e^{jn\Delta w_k'} \theta(r_k e^{j\Delta w_k'}) \right]$$

$$= \mathbb{E} \sum_{h=0}^{N} \left\| \sum_{k=1}^{p} \underline{\underline{B}}_k \underline{\theta}_k[n] - \underline{\tau}[n] \right\|^2$$

$$\theta_k[n] = h_k[n] + \underline{\underline{g}}[n] * \left(\underline{\tau}[n] + \sigma \underline{W}[n] \right)$$

$$\mathbb{E} \left\| \sum_{k=1}^{p} \underline{\underline{B}}_k h_k[n] * \underline{\underline{g}}_k[n] * (\underline{\tau}_m[n] + \sigma \underline{W}[n]) - \underline{\tau}_m[n] - \sigma \underline{W}(n) \right\|^2$$

❖❖❖❖❖

[220] Quantum Stochastic filtering

$$dX(t) = \mu(X(t))\,dt + \sigma(X(t))\,dB(t)$$

$$df(X(t)) = \mathcal{L}f(X(t))\,dt + \sigma(X(t)) f'(X(t))\,dB(t)$$

$$\mathcal{L} = \mu\frac{d}{dx}+\frac{1}{2}\sigma^2\frac{d^2}{dx^2}$$

State model $dj_t(X) = j_t(\mathcal{L}_0X)dt + j_t(\mathcal{L}_1X)dA_t + j_t(\mathcal{L}_2X)dA_t^*$

$$X \equiv f\mathcal{L}_0 = \mathcal{L}, \mathcal{L}_1 \equiv \sigma\frac{d}{dx} = \mathcal{L}_2$$

measurement

$$dY_t = Y_t dZ_t$$

$$z_t \in \eta_{t]}^{out} \quad \pi_t(X) = \mathbb{E}\left[j_t(X) \mid \eta_{t]}^{out}\right]$$

$$\mathbb{E}\{(j_t(X)-\pi_t(X))Y_t\} = 0$$

$$d\xi_t = f_\beta^\alpha(t)d\Lambda_\alpha^\beta(t) \in \eta_{t]}^{in}$$

$$\xi_t = \int_0^t f_\beta^\alpha(s)d\Lambda_\alpha^\beta(s)$$

$$U_t^*(I\otimes\xi_t)U_t = Z_t$$

$$\begin{aligned} dZ_t &= dU_t^*\xi_t U_t + U_t^*\xi_t dU_t + d\xi_t \\ &\quad + dU_t^*\xi_t dU_t + dU_t^* d\xi_t U_t + U_t^* d\xi_t dU_t \end{aligned}$$

Let $\quad dU_t = \left(L_\beta^\alpha(t)d\Lambda_\alpha^\beta(t)\right)U(t)$

so that U_t is unitary. Then,

$$\begin{aligned} dZ_t &= d\xi_t + U_t^* L_\beta^\alpha(t)^* d\Lambda_\beta^\alpha(t) d\xi_t U_t + U_t^* L_\beta^\alpha(t) d\xi(t) d\Lambda_\alpha^\beta(t) U(t) \\ &= f_\beta^\alpha(t)d\Lambda_\alpha^\beta(t) + j_t\left(L_\beta^\alpha(t)^* d\Lambda_\beta^\alpha(t) d\Lambda_\sigma^\rho(t) f_\rho^\sigma(t)\right) \\ &\quad + j_t\left(L_\beta^\alpha(t) f_\rho^\sigma(t) d\Lambda_\sigma^\rho(t) d\Lambda_\alpha^\beta(t)\right) \\ &= f_\beta^\alpha(t)d\Lambda_\alpha^\beta(t) + f_\rho^\sigma(t) j_t\left(L_\beta^\alpha(t)^* \uparrow \varepsilon_\sigma^\alpha\, d\Lambda_\beta^\rho(t)\right) \\ &\quad + f_\rho^\sigma(t)\, j_t\left(L_\beta^\alpha(t)\, \varepsilon_\alpha^\rho d\Lambda_\sigma^\beta(t)\right) \\ &= f_\beta^\alpha(t)d\Lambda_\alpha^\beta(t) + f_\rho^\sigma(t)\, \varepsilon_\sigma^\alpha j_t\left(L_\beta^\alpha(t)^*\right)d\Lambda_\beta^\rho(t) \end{aligned}$$

$$+ f_\rho^\sigma(t)\varepsilon_\alpha^\rho j_t\left(L_\beta^\alpha(t)\right)d\Lambda_\sigma^\beta(t)$$

$$= f_\beta^\alpha d\Lambda_\alpha^\beta + \left[f_\rho^\sigma\, \varepsilon_\sigma^\alpha j_t\left(L_\beta^{\alpha*}\right) + f_\sigma^\beta\, \varepsilon_\alpha^\sigma j_t\left(L_\rho^\alpha\right)\right]d\Lambda^\rho$$

❖❖❖❖❖

[221] $\psi_t(\alpha) = \mathbb{E}e^{i\alpha z(t)} = \exp\left(-\frac{\alpha^2 t\sigma^2}{2} + \lambda t\int_{\mathbb{R}}(e^{i\alpha x}-1)\,dF(x)\right)$

$$= \exp\left(-\frac{\sigma^2\alpha^2 t}{2}\right)\sum_0^\infty \frac{e^{-\lambda t}(\lambda t)^n}{\underline{|n}}\left(\int_{\mathbb{R}} e^{i\alpha x}dF(x)\right)^n$$

$$f_Z(x,t) = \frac{1}{\sigma\sqrt{2\pi t}}\exp\left(-\frac{x^2}{2\sigma^2 t}\right)\left[e^{-\lambda t}\,\delta(x) + \sum_{n=1}^\infty \frac{e^{-\lambda t}(\lambda t)^n}{\underline{|n}} f_Y^{*n}(x)\right]$$

Let

$Y \sim N\left(0,\sigma_Y^2\right)$.

$$f_Z(x,t) = \frac{1}{\sigma\sqrt{2\pi t}} e^{-\frac{x^2}{2\sigma^2 t}} e^{-\lambda t}$$

$$+\sum_{n=1}^\infty \frac{e^{-\lambda t}(\lambda t)^n}{\underline{|n}} \frac{1}{\left(\sigma^2 t + n\sigma_Y^2\right)}\exp\left(\frac{-x^2}{2\left(\sigma^2 t + n\sigma_Y^2\right)}\right)$$

$$= \sum_{n=0}^\infty \frac{e^{-\lambda t}(\lambda t)^n}{\underline{|n}} N\left(0,\sigma^2 t + n\sigma_Y^2\right)$$

≈ Tremor plus jerk noise probability density.

pdf of noise

$$\exp\left(-\frac{1}{2}\|\underline{W}\|^2 + \varepsilon\psi(\underline{W})\right)$$

Example:

$$\mathbb{E}e^{\alpha\delta Wt} = e^{\delta t\psi_1(\alpha)}$$

$$\psi_1(\alpha) = \frac{1}{2}\|\alpha\|^2 + \varepsilon\int\left(e^{\alpha x}-1\right)dF(x)$$

Independent increment process ≈ W_H = Brownian motion + Compound Poisson process.

$$e^{\delta t\psi_1(\alpha)} \approx e^{\frac{\delta t}{2}\|\alpha\|^2}\left(1 + \varepsilon\,\delta t\int\left(e^{\alpha x}-1\right)dF(x)\right)$$

pdf or $$\delta W_t \approx \frac{1}{\sqrt{2\pi}} e^{-\frac{W^2}{2\delta t}} * \left(\delta(t) + \varepsilon\, \delta t \int \left(\delta(W-x) - \delta(W)\right) dF(x)\right)$$

$$= \frac{1}{\sqrt{2\pi}} e^{-\frac{W^2}{2\delta t}} + \frac{\varepsilon\, \delta t}{\sqrt{2\pi}} \left(\int \left(e^{-\frac{(W-x)^2}{2\delta t}} - e^{-\frac{W^2}{2\delta t}} \right) dF(x) \right)$$

Let $\int dF(x) = F(\infty) = \lambda$. Then,

$$p_{\delta W_t}(W) \approx \frac{1}{\sqrt{2\pi}} e^{-\frac{W^2}{2\delta t}} + \frac{\varepsilon\, \delta t}{\sqrt{2\pi}} \left(\int e - \frac{(W-x)^2}{2\delta t} dF(x) - \lambda e^{-\frac{W^2}{2}} \delta t \right)$$

= Superposition of Gaussian densities.

$$\log p_{\delta W_t}(W) \approx -\frac{1}{2}\log(2\pi) - \frac{W^2}{2\delta t} + \varepsilon\, \delta t \left[\int \exp\left\{ \frac{\left(W^2 - (W-x)^2\right)}{2\delta t} \right\} dF(x) - \lambda \right]$$

$$= -\frac{1}{2}\log(2\pi) - \frac{W^2}{2\delta t} + \varepsilon\, \delta t \left[\int \exp\left\{ \frac{x(2W-x)}{2\delta t} \right\} dF(x) - \lambda \right]$$

$$-\log p_{\delta W_t}(W) \approx \frac{W^2}{2\delta t} + \varepsilon\, \delta t \left[\int \exp\left(\frac{Wx}{\delta t}\right) \exp\left(-\frac{x^2}{2\delta t}\right) dF(x) - \lambda \right]$$

$$= \frac{W^2}{2\delta t} + \varepsilon\, \delta t \left[\int \exp\left(\frac{Wx}{\delta t}\right) \times \left(\sum_{n=0}^{\infty} \frac{(-1)^n}{\underline{\lfloor n}} \frac{x^{2n}}{(2\delta t)^n} \right) dF(x) - \lambda \right]$$

$$\triangleq \frac{W^2}{2\delta t} + \varepsilon\, \psi(W, \delta t)$$

(Approximation)

❖❖❖❖❖

[222] Proof of convergence of CRLB for Gaussian White noise.

$$\dot{\underline{\xi}}(t) = \underline{F}\left(t, \underline{\xi}(t), \underline{\theta}\right) + \underline{\underline{G}}\left(t, \underline{\xi}(t), \underline{\theta}\right) \underline{W}(t)$$

$\underline{\theta}$ = parameter to be estimated. Let $\underline{\theta} = \theta_0 + \delta\underline{\theta}$

with $\|\delta\underline{\theta}\|\ \|\underline{W}\|$ negligible. Then we have approximately

$$\dot{\underline{\xi}}(t) = \underline{F}_0\left(t, \underline{\xi}(t)\right) + \frac{\partial F}{\partial \underline{\theta}}\left(t, \underline{\xi}(t), \underline{\theta}_0\right) \delta\underline{\theta} + G_0\left(t, \underline{\xi}(t)\right) \underline{W}(t)$$

$$\underline{W}(t) = \frac{d\underline{B}(t)}{d_0}$$

$$d\underline{\xi}(t) = \left(\underline{F}_0\left(t,\underline{\xi}(t)\right) + F_1\left(t,\underline{\xi}(t)\delta\underline{\theta}\right)dt + \underline{\underline{G}}_0\left(t,\underline{\xi}(t)\right)d\underline{B}(t)\right)$$

Log-likelihood for $\delta\underline{\theta}$ based on observation collected over the time interval $[0, T]$:

$$-p\left(\underline{\xi}(t):0\leq t\leq T\mid\delta\underline{\theta}\right)$$

$$= \frac{1}{2}\int_0^T \left\| G_0\left(t,\underline{\xi}(t)\right)^{-1}\left(\underline{\dot{\xi}}(t) - \underline{F}_0\left(t,\underline{\xi}(t)\right) - \underline{\underline{F}}_1\left(t,\underline{\xi}(t)\right)\delta\underline{\theta}\right)\right\|^2 dt$$

Fisher info-matrix

$$\underline{\underline{J}}_T\left(\delta\underline{\theta}\right) = -\mathbb{E}\left\{\frac{\partial^2}{\partial\delta\underline{\theta}\,\partial\delta\underline{\theta}^T}\log p\left(\underline{\xi}(t):0\leq t\leq T|\delta\underline{\theta}\right)\right\}$$

$$= -\mathbb{E}\left\{\int_0^T\left(\underline{\underline{F}}_1\left(t,\underline{\xi}(t)\right)^T\left(\underline{\underline{G}}_0\underline{\underline{G}}_0^T\right)^{-1}\underline{\underline{F}}_1^T\right)dt\right\}$$

In the absence of noisy let the solution be $\underline{\xi}_0(t)$:

$$\underline{\dot{\xi}}_0(t) = \underline{F}_0\left(t,\underline{\xi}_0(t)\right) + \frac{\partial F}{\partial\underline{\theta}}\left(t,\underline{\xi}_0(t),\underline{\theta}_0\right)\delta\underline{\theta}$$

Let $\quad \underline{\xi}_0(t). = \underline{\xi}_0(t) + \delta\underline{\xi}(t)$. Then,

$$\delta\underline{\dot{\xi}}(t) \approx \frac{\partial F_0}{\partial\underline{\xi}}\left(t,\underline{\xi}_0(t)\right)\delta\underline{\xi}(t) + \frac{\partial F}{\partial\underline{\theta}}\left(t,\underline{\xi}_0(t),\underline{\theta}_0\right)\delta\underline{\theta} + \underline{\underline{G}}_0\left(t,\underline{\xi}_0(t)\right)\underline{W}(t)$$

$$= \underline{\underline{A}}(t)\delta\underline{\xi}(t) + \underline{\underline{\theta}}(t)\delta\underline{\theta} + \underline{\underline{S}}(t)\underline{W}(t)$$

$$\underline{\underline{A}}(t) = \frac{\partial F_0}{\partial\underline{\xi}}\left(t,\underline{\xi}_0(t)\right),\ \underline{\underline{\theta}}(t) = \frac{\partial F}{\partial\underline{\theta}}\left(t,\underline{\xi}_0(t)\underline{\theta}_0\right)$$

$$\underline{\underline{S}}(t) = G_0\left(t,\underline{\xi}_0(t)\right)$$

Linearized observation based CRLB

$$-\log p\left(\delta\underline{\xi}(t):0\leq t\leq T|\delta\underline{\theta}\right)$$

$$= \frac{1}{2}\int_0^T\left\|\underline{\underline{S}}(t)^{-1}\left(\delta\underline{\dot{\xi}}(t) - \underline{\underline{A}}(t)\delta\underline{\xi}(t) - \underline{\underline{Q}}(t)\delta\underline{\theta}\right)\right\|^2 dt$$

For general Levy process

$$-\log p = \frac{1}{2}\int_0^T \left\| \underline{\underline{S}}(t)^{-1}\left(\delta\dot{\underline{\xi}}(t) - \underline{\underline{A}}(t)\delta\underline{\xi}(t) - \underline{\underline{Q}}(t)\delta\underline{\theta}\right)\right\|^2 dt$$

$$+\delta\times\int_0^T \psi\left(\underline{\underline{S}}(t)^{-1}\left(\delta\dot{\underline{\xi}}(t) - \underline{\underline{A}}(t)\delta\underline{\xi}(t) - \underline{\underline{Q}}(t)\delta\underline{\theta}\right)\right)dt$$

❖❖❖❖❖

[223] Cosmology with stochastic perturbation.

$$R_{00} = \frac{3S''(t)}{S(t)}$$

where $g_{00} = 1,\ S_{11} = \dfrac{-S^2}{1-kr^2},\ S_{22} = -S^2r^2,$

$$g_{33} = -S^2r^2\sin^2\theta$$

$$R_{\mu\nu} - \frac{1}{2}Rg_{\mu\nu} = K\left((\rho+p)v_\mu v_\nu - pg_{\mu\nu} + \mathcal{F}_{\mu\nu}\right) \quad ...(1)$$

$\mathcal{F}_{\mu\nu}$ = Stochastic perturbation in the energy-momentum tensor of matter.

$(1) \Rightarrow -R = K(\rho + p - 4p + \mathcal{F})$

$$= K(\rho - 3p + \mathcal{F}) \quad ...(2)$$

So (1) becomes

$$R_{\mu\nu} = K\left((\rho+p)v_\mu v_\nu - \frac{1}{2}(\rho-3p+\mathcal{F})g_{\mu\nu} - pg_{\mu\nu} + \mathcal{F}_{\mu\nu}\right)$$

$$= K\left((\rho+p)v_\mu v_\nu + \frac{(p-\rho)}{2}g_{\mu\nu} + \mathcal{F}_{\mu\nu} - \frac{1}{2}\mathcal{F}g_{\mu\nu}\right) \quad ...(3)$$

$$v_\mu = V_\mu + \delta v_\mu \quad \rho = \rho_0 + \delta\rho$$

$$p = p_0 + \delta p$$

Assume metric perturbation are negligible. Then

$$R_{00} = \frac{3S''(t)}{S(t)} = K\left(\rho + p + \frac{(p-\rho)}{2}\right) = \frac{1}{2}K(3p+\rho)$$

$$(V^\mu) = (1, 0, 0, 0)$$

$$V_\mu = (1, 0, 0, 0)$$

$$\left((\rho+p)v^{\mu}v^{\nu}-pg^{\mu\nu}+\mathcal{F}^{\mu\nu}\right)_{:\nu}=0.$$

$$\left((\rho+p)v^{\nu}\right)_{:\nu}v^{\mu}+(\rho+p)v^{\nu}v^{\mu}_{:\nu}-p^{\mu}+\mathcal{F}^{\mu}=0$$

$\mathcal{F}^{\mu}=\mathcal{F}^{\mu\nu}_{:\nu}$. So,

$$\left((\rho+p)v^{\nu}\right)_{:\nu}+v_{\mu}\left(\mathcal{F}^{\mu}-p'^{\mu}\right)=0$$

$$\left((\rho+p)v^{\nu}\sqrt{-g}\right)_{,\nu}+v_{\mu}\left(\mathcal{F}^{\mu}-p'^{\mu}\right)\sqrt{-g}=0.$$

$$\left(\left((\rho_0+p_0)\delta v^{\nu}+(\delta\rho+\delta p)V^{\nu}\right)\sqrt{-g}\right)_{,\nu}$$

$$-\delta v^{\mu}p_{0,\mu}\sqrt{-g}+V^{\mu}\mathcal{F}_{\mu}\sqrt{-g}=0$$

$$\left((\delta\rho+\delta p)\sqrt{-g}\right)_{,0}+\left((\rho_0+p_0)\delta v^{\mu}\sqrt{-g}\right)_{,\mu}$$

$$-\delta v^{0}p_{0,0}\sqrt{-g}+\mathcal{F}_0\sqrt{-g}=0$$

$$\left((\delta\rho+\delta p)\frac{S^3r^2\sin\theta}{\sqrt{1-kr^2}}\right)_{,0}+\left((\rho_0+p_0)\frac{\delta v^0S^3r^2\sin\theta}{\sqrt{1-kr^2}}\right)_{,0}$$

$$+(\rho_0+p_0)\left(\frac{\delta v^kS^3r^2\sin\theta}{\sqrt{1-kr^2}}\right)_{,k}-\delta v^0\frac{p_0'S^3r^2\sin\theta}{\sqrt{1-kr^2}}+\frac{\mathcal{F}_0S^3r^2\sin\theta}{\sqrt{1-kr^2}}=0$$

$\delta\rho=\delta\rho(t,r,\theta,\varphi)$,

$$\delta p=\delta p(t,r,\theta,\varphi)$$

$$\delta v^{\mu}=\delta v^{\mu}(t,r,\theta,\varphi)$$

$$g_{\mu\nu}(V^{\mu}+\delta v^{\mu})(V^{\nu}+\delta v^{\nu})=1$$

$$\Rightarrow\quad g_{\mu\nu}V^{\mu}V^{\nu}=g_{00}=1 \qquad \text{(True anyway)},$$

$$g_{\mu\nu}(V^{\mu}\delta v^{\nu}+V^{\nu}\delta v^{\mu})=0$$

$$\Rightarrow\quad \delta v^0=0.$$

So

$$\left((\delta\rho+\delta p)S^3\right)_{,0}\frac{r^2\sin\theta}{\sqrt{1-kr^2}}+(\rho_0+p_0)S^3\left[\left(\frac{\delta v^1r^2\sin\theta}{\sqrt{1-kr^2}}\right)\right.$$

$$\left.+\frac{r^2}{\sqrt{1-kr^2}}\left(\delta v^2\sin\theta\right)_{,0}+\frac{r^2\sin\theta}{\sqrt{1-kr^2}}\delta v^2_{,\theta}\right]+\frac{\mathcal{F}_0S^3r^2\sin\theta}{\sqrt{1-kr^2}}=0$$

$$\Rightarrow \qquad \left((\delta\rho+\delta p)S^3\right)_{,0}+(\rho_0+p_0)S^3\left[\delta v^1{}_{,1}+\delta v^1\times\left[\frac{2}{r}+\frac{kr}{\left(1-kr^2\right)}\right]\right.$$

$$\left.+\,\delta v^2_{,\theta}+\delta v^2{}_{,\theta}+\delta v^2{}_{,\theta}\right]+\mathcal{F}_0 S^3=0$$

This is the perturbed equation of continuity in the presence of external random forces with the Robertson Walker background metric.

The Navier-Stokes eqn's

$$(\rho+p)v^\nu v^\mu_{:\nu}-v^\mu v^\alpha\left(\mathcal{F}_k-p_{,k}\right)-p'^\mu+\mathcal{F}^\mu=0$$

and its perturbed form is

$$(\delta\rho+\delta p)V^r_{:0}+(\rho_0+p_0)\left(\delta v^r_{:0}+\delta v^k V^r_{:k}\right)-V^rV^\alpha\mathcal{F}_\alpha$$

$$+V^rV^\alpha\delta p_{,\alpha}+\delta v^rV^\alpha p_{0,\alpha}+V^r\delta v^\alpha p_{0,\alpha}-\delta p'^r+\mathcal{F}^r=0$$

$$V^r_{:0} = V^r_{,0}+\Gamma^r_{00}V^0=0$$

$$\delta v^r_{:0} = \delta v^r_{,0}+\Gamma^r_{0k}\delta vk$$

$$= \delta v^r_{,0}+\Gamma^r_{0r}\delta v^r \qquad \text{(no summation)}$$

$$\Gamma^r_{0r} = \frac{1}{2}g^{rr}g_{rr,0} = \frac{1}{2}\left(\log g_{rr,0}\right) = \frac{S'(t)}{S(t)}$$

$$V^r_{:k} = V^r_{,k}+\Gamma^r_{\mu k}V^\mu = \Gamma^r_{0k}$$

$\Gamma^r_{0k}=0$ for $r\neq k$ and $\Gamma^r_{0r} = \dfrac{S'}{S}$, $1\le r\le 3$.

$(V^r=0,\ 1\le r\le 3)$. So,

$$(\rho_0+p_0)\left(\delta v^r_{,0}+\frac{S'}{S}\delta v^r\right)+p_0'\delta v^r-\delta p'^r+\mathcal{F}^r=0$$

Note $\quad \delta p'^r = g^{rr}\delta p_{,r}$

$$\Rightarrow \qquad \delta p'^1 = g^{11}\delta p_{,1} = -\frac{\left(1-kr^2\right)}{S^2}\delta p_{,1},$$

$$\delta p'^2 = g^{22}\delta p_{,2} = -\frac{1}{S^2r^2}\delta p_{,2},$$

$$\delta p'^3 = g^{33}\delta p_{,3} = -\frac{1}{S^2 r^2 \sin^2\theta}\delta p_{,3},$$

For Rohit Rana

❖❖❖❖❖

[224] $\left(S^2 \underline{\underline{J}} + S\underline{\underline{\beta}} + \underline{\underline{K}}\right)\delta \underline{q}(t) = \underline{\tau}_I(t) + \underline{\tau}_T(t) + \underline{W}(t)$

$$\mathcal{L}^{-1}\left\{\left(S^2 \underline{\underline{J}} + S\underline{\underline{\beta}} + \underline{\underline{K}}\right)^{-1}\right\} = \underline{\underline{g}}(t)$$

$$\delta \underline{q}(t) = \underline{\underline{g}}(t) * \left(\underline{\tau}_I(t) + \underline{\tau}_T(t) + \underline{W}(t)\right)$$

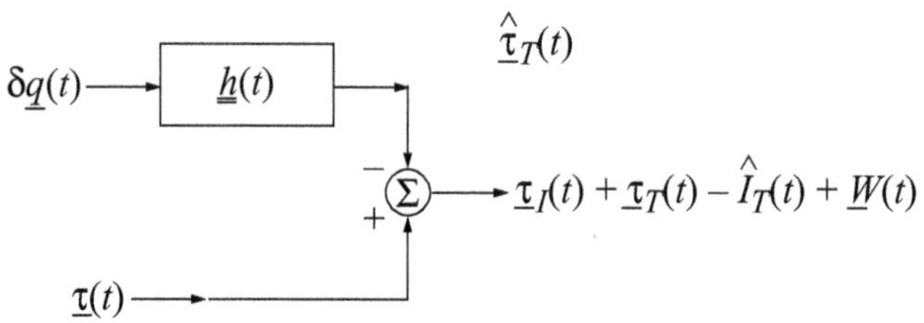

$$= \underline{\tau}_I(t) + \underline{\tau}_T(t) + \underline{W}(t)$$

$$\hat{\tau}_T(t) = \sum_k B_k h_k(t) * \underline{\underline{g}}(t) * \left(\underline{\tau}_I(t) + \underline{\tau}_T(t) + \underline{W}(t)\right)$$

$$\underline{\tau}_T(t) = \sum_{|k|\le p} \underline{A}_k e^{jw_k t}$$

$$\underline{A} - k = \bar{\underline{A}}_k,$$

$$w - k = -w_k.$$

$$\underline{A}_0 = 0.$$

$$h_k(t) = \frac{S}{S^2 + 2\zeta_k w_k' S + w_k'^2}$$

$$= \frac{S}{\left(S + \zeta_k w_k'\right)^2 + w_k'^2\left(1 - \zeta_k^2\right)}$$

$$\left|H_k(jw)\right|^2 = \frac{w^2}{\left(w^2 - w_k'^2\right)^2 + 4\zeta_k^2 w_k'^2 w^2}$$

$$\frac{d}{dw}\left|H_k(jw)\right|^2 = 0$$

$$\Rightarrow \left(\left(w^2 - w_k'^2\right)^2 + 4\zeta^2 w_k'^2 w^2\right)$$

$$= 2w^2\left(w^2 - w_k'^2\right) + 4\zeta^2 w^2$$

$$\Rightarrow \quad w^2 - w_k'^2 = 0 \text{ or } w^2 - w_k'^2 = 2w^2.$$

$$w = \pm w_k'.$$

$$\underline{\tau}_I(t) = \int_{\mathbb{R}} \hat{\underline{\tau}}_I(w) e^{jwt} \frac{dw}{2\pi}$$

$$\underline{\underline{g}}(t) * \left(\underline{\tau}_I(t) + \underline{\tau}_T(t) + \underline{W}(t)\right)$$

$$= \int \underline{\underline{G}}(jw) \hat{\underline{\tau}}_I(w) e^{jwt} \frac{dw}{2\pi}$$

$$+ \sum_k \underline{\underline{G}}(jw_k) \underline{A}_k e^{jw_k t} + \underline{\underline{g}}(t) + \underline{W}(t)$$

$$h_k(t) * \underline{\underline{g}}(t) * \left(\underline{\tau}_I(t) + \underline{\tau}_T(t) + \underline{W}(t)\right)$$

$$= \int H_k(jw) \underline{\underline{G}}(jw) \hat{\underline{\tau}}_I(w) e^{jwt} \frac{dw}{2\pi}$$

$$+ \sum_m H_k(jw_m) \underline{\underline{G}}(jw_m) \underline{A}me^{jw_m t} + h_k(t) * \underline{\underline{g}}(t) * \underline{W}(t)$$

$$\underline{\tau}(t) - \hat{\underline{\tau}}_T(t)$$

$$= \underline{\tau}_I(t) + \underline{\tau}_T(t) + \underline{W}(t) - \sum_k B_k \int H_k(jw) \underline{\underline{G}}(jw) \hat{\underline{\tau}}_I(w) e^{jwt} \frac{dw}{2\pi}$$

$$- \sum_{k,m} B_k H_k(jw_m) \underline{\underline{G}}(jw_m) \underline{A}_m e^{jw_m t}$$

$$- \sum_k B_k h_k(t) + \underline{\underline{g}}(t) * \underline{W}(t)$$

$$\lim_{T \to \infty} \frac{1}{2T} \int_{-T}^{T} \mathbb{E}\left\{\left\|\underline{\tau}(t) - \hat{\underline{\tau}}_T(t)\right\|^2\right\} dt$$

$$= Tr\left\{\int \left(\underline{\underline{I}} - \sum_k B_k H_k(w) \underline{\underline{G}}(w)\right) \underline{\underline{S}}_W(w)\right.$$

$$\left(\underline{\underline{I}}-\sum_k \bar{B}_k \overline{H_k(w)}\,\underline{\underline{G}}^*(w)\right)\frac{dw}{2\pi}\Bigg\}$$

$$+\sum_m \left\|\left(\underline{\underline{I}}-\sum_k B_k H_k(j\alpha_m)\underline{\underline{G}}(\cdot)\alpha_m\right)\hat{\underline{\tau}}_{Id}[m]\right\|^2$$

$$+\sum_m \left\|\left(\underline{\underline{I}}-\sum_k B_k H_k(jw_m)\underline{\underline{G}}(jw_m)\right)\underline{A}_m\right\|^2$$

$$\{\alpha_m\}_m \cap \{w_r\}_r = \phi.$$

Forfinite T,

$$\frac{1}{2T}\int_{-T}^{T}\mathbb{E}\left\{\left\|\underline{\tau}(t)-\hat{\underline{\tau}}_T(t)\right\|^2\right\}dt$$

$$= Tr\left\{\int\left(\underline{\underline{I}}-\sum_k B_k H_k(w)\underline{\underline{G}}(w)\right)\cdot \underline{\underline{S}}_w(w)\right\}\cdot$$

$$\left\{\cdot\left(\underline{\underline{I}}-\sum_k \bar{B}_k \overrightarrow{H_k(w)}\,\underline{\underline{G}}^*(w)\right)\frac{dw}{2\pi}\right\}$$

$$+\frac{1}{(2\pi)^2}\int\left(\left(\underline{\underline{I}}-\sum_k B_k H_k(w)\underline{\underline{G}}(w)\right)\hat{\underline{\tau}}_I(w)\right)^*$$

$$\left(\left(\underline{\underline{I}}-\sum_k B_k H_k(w')\underline{\underline{G}}(w')\right)\hat{\underline{\tau}}_I(w')\right)\left\{\frac{\sin(T(w-w'))}{T(w-w')T}\right\}dwdw$$

$$+\sum_{m,m'}\left[\left(\underline{\underline{I}}-\sum_k B_k H_k(w_m)\underline{\underline{G}}(w_m)\right)A_m\right]^*$$

$$\left[\left(\underline{\underline{I}}-\sum_k B_k H_k(w'_m)\underline{\underline{G}}(w'_m)\right)\underline{A}_m\right]\left\{\frac{\sin\{(w_m-w'_m)T\}}{\pi T(w_m-w'_m)}\right\}$$

Radon transform of the Maxwell equations.

$$\int div\underline{E}(t,\underline{r})\delta\left(p-(\underline{r},\hat{n})\right)d^3r$$

$$= \int \sum_{\alpha} \frac{\partial E_\alpha}{\partial X_\alpha}(t, \underline{r}) \delta\left(p - \left(\underline{r}, \hat{n}\right)\right) d^3 r$$

$$= -\sum_{\alpha} \int E_\alpha(t, \underline{r}) \frac{\partial}{\partial X_\alpha} \delta\left(p - \left(\underline{r}, \hat{n}\right)\right) d^3 r$$

$$= \sum_{\alpha} \int n_\alpha E_\alpha(t, \underline{r}) \delta'\left(p - \left(\underline{r}, \hat{n}\right)\right) d^3 r$$

$$= \frac{\partial}{\partial p} \mathbb{R}\left(\hat{n}, E\right)\left(t, p, \hat{n}\right)$$

$$\int \underline{\nabla} \times \underline{E}(t, \underline{r})\, \delta\left(p - \left(\underline{r}, \hat{n}\right)\right) d^3 r = \int \underline{\nabla} \delta\left(p - \left(\underline{r}, \hat{n}\right)\right) \times \underline{E}(t, \underline{r})\, d^3 r$$

$$= \int -\hat{n}\, \delta'\left(p - \left(\underline{r}, \hat{n}\right)\right) \times \underline{E}(t, \underline{r})\, d^3 \underline{r}$$

$$= -\hat{n} \times \frac{\partial}{\partial p} \mathbb{R}E\left(t, p, \hat{n}\right)$$

$$\int \frac{\partial \underline{H}(t, \underline{r})}{\partial t} \delta\left(p - \left(\hat{n}, \underline{r}\right)\right) d^3 r = \frac{\partial}{\partial t} \mathbb{R}H\left(t, p, \hat{n}\right)$$

$$\underline{\nabla} \times \underline{E} = -\mu \frac{\partial H}{\partial t}$$

$$\Rightarrow \qquad -\hat{n} \times \frac{\partial}{\partial p} \mathbb{R}E\left(t, p, \hat{n}\right) - \mu \frac{\partial}{\partial t} \mathbb{R}H\left(t, p, \hat{n}\right)$$

or
$$\hat{n} \times \frac{\partial}{\partial p} \mathbb{R}E\left(t, p, \hat{n}\right) = \mu \frac{\partial}{\partial t} \mathbb{R}H\left(t, p, \hat{n}\right)$$

$$\nabla \times H = J + \varepsilon \frac{\partial E}{\partial t}$$

$$\Rightarrow \qquad \hat{n} \times \frac{\partial}{\partial p} \mathbb{R}\underline{E}\left(t, p, \hat{n}\right) = -\mathbb{R}\underline{J}\left(t, p, \hat{n}\right) - \mu\, \varepsilon \frac{\partial}{\partial t} \mathbb{R}\underline{E}\left(t, p, \hat{n}\right)$$

Wave equation

$$\frac{\partial^2 \psi(t, \underline{r})}{\partial t^2} - c^2 \nabla^2 \psi(t, \underline{r}) = 0.$$

$$\Rightarrow \qquad \frac{\partial^2}{\partial t^2} (\mathbb{R}\psi)\left(t, p, \hat{n}\right) - c^2 \int \left(\nabla^2 \psi(t, \underline{r})\right) \delta\left(p - \left(\hat{n}, \underline{r}\right)\right) d^3 r = 0$$

$$\Rightarrow \qquad \frac{\partial^2 \mathbb{R}\psi\left(t,p,\hat{n}\right)}{\partial t^2} - c^2 \frac{\partial^2}{\partial p^2}\mathbb{R}\psi\left(t,p,\hat{n}\right) = 0.$$

❖❖❖❖❖

[225] Quantum Boltzmann Equation

$$i\rho' = \left[\sum H_\alpha, \rho\right] + \left[\sum_{\alpha<\beta} V_{\alpha\beta}, \rho\right] + i\theta(\rho)$$

$$i\rho_1' = \left[H_1, \rho_1\right] + (N-1)Tr_2\left[V_{12}, \rho_{12}\right] + i\, Tr_{23..N}\theta(\rho)$$

$$i\rho_{12}' = \left[H_1 + H_2 + V_{12}, \rho_{12}\right] + (N-2)Tr_3\left[V_{13}, \rho_{123}\right]$$

$$+(N-2)Tr_3\left[V_{23}, \rho_{123}\right] + i\, Tr_{3..N}\theta(\rho)$$

$$\rho_{12}(t) = \exp\left(-itad\left(H_1 + H_2\right)\right)\tilde{\rho}_{12}(t)$$

$$= \rho_1(t) \otimes \rho_2(t) + g_{12}(t)$$

$$i\left(\rho_1' \otimes \rho_2 + \rho_1 \otimes \rho_2' + g_{12}'\right)$$

$$\approx \left[H_1 + H_2 + V_{12}, \rho_1 \otimes \rho_2 + g_{12}\right]$$

$$ig_{12}' \approx \left[V_{12}, \rho_1 \otimes \rho_2\right] + \left[V_{12}, g_{12}\right]$$

$$g_{12}(t) \approx -i\int_0^t \left[V_{12}(\tau), \rho_1(\tau) \otimes \rho_2(\tau)\right] d\tau$$

$$\approx -i\int_0^t \left[V_{12}, \exp\left(-iad\left(H_1 + H_2\right)(t-\tau)\right)\right] d\tau\left(\rho_1(0) \otimes \rho_2(0)\right)$$

$$g_{12}(t) \approx -i\int_0^t \exp\left(-iad\left(V_{12}\right)(t-\tau)\right)\left[V_{12}, \rho_1(\tau) \otimes \rho_2(\tau)\right] d\tau$$

$$\approx -i\int_0^t \exp\left(-iad\left(V_{12}\right)(t-\tau)\right) ad\left(V_{12}\right) \exp\left(-iad\left(H_1 + H_2\right)\tau\right)$$

$$\left(\rho_1(0) \otimes \rho_2(0)\right) d\tau$$

$$= -i\exp\left(-iad\left(V_{12}\right)t\right) ad\left(V_{12}\right)$$

$$\int_0^t \exp\left(iad\left(V_{12}\right)\tau\right) \cdot \exp\left(-iad\left(H_1 + H_2\right)\tau\right)\left(\rho_1(0) \otimes \rho_2(0)\right) d\tau$$

$$i\rho_1' \approx \left[H_1, \rho_1\right] + (N-1)Tr_2\left[V_{12}, \rho_1(t) \otimes \rho_2(t)\right]$$

$$\rho_1(t) \approx -i(N-1)\int_0^t \exp\left(-iadH_1(t-\tau)\right)Tr_2\left[V_{12},\rho_1(\tau)\otimes\rho_2(\tau)\right]d\tau$$

$$\approx -i(N-1)Tr_2\int_0^t \exp\left(-iadH_1(t-\tau)\right)$$

$$\left[V_{12},\exp\left(-iad\left(H_1+H_2\right)\tau\right)\rho_1(0)\otimes\rho_2(0)\right]d\tau$$

$$= -i(N-1)Tr_\alpha\int_0^t\left[\exp\left(-i(t-\tau)adH_1\right)\right](V_{12}),\exp\left(-iad\right)$$

$$\nabla\times E = -\mu\frac{\partial H}{\partial t}$$

$$\frac{1}{\varepsilon}\nabla\times H = \frac{J}{\varepsilon}+\frac{\partial \underline{E}}{\partial t}$$

$$\underline{J}(t,\underline{r}) = \sum_n \psi_n(\underline{r})\frac{dB_n(t)}{dt}$$

$$\underline{E}(t,\underline{r}) = \sum_n \psi_n(\underline{r})\underline{E}_n(t)$$

$$\underline{H}(t,\underline{\gamma}) = \sum_n \psi_n(\underline{r})\underline{H}_n(t) - \mu\sum_n \psi_n(\underline{r})d\underline{H}_n(t)$$

$$= \sum_n \underline{\nabla}\psi_n(\underline{r})\times\underline{E}_n(t)$$

$$\sum_n \psi_n(\underline{r})d\underline{E}_n(t)$$

$$= \frac{1}{\varepsilon}\sum_n \nabla\psi_n\times H_n dt - \frac{1}{\varepsilon}\sum\psi_n dB_n(t)$$

$$d\underline{E}_n(t) = \frac{1}{\varepsilon}\sum\left(\underline{\psi}_m(\underline{r})\underline{\nabla}\psi_n(\underline{r})d^3r\right)\times H_n(t)dt - \frac{1}{\varepsilon}dB'_n(t)$$

❖❖❖❖❖

[226] Design of unitary gates by perturbing the free Dirac equation.

$(i\gamma^\mu\partial_\mu - m)\psi = 0$ = Free Dirac equation.

$$\gamma^0 = \begin{pmatrix} I & 0 \\ 0 & -I \end{pmatrix}\quad \gamma^r = \begin{pmatrix} 0 & \sigma_r \\ -\sigma_r & 0 \end{pmatrix}$$

$$\gamma^r\gamma^s = \begin{pmatrix} 0 & \sigma_r \\ -\sigma_r & 0 \end{pmatrix}\begin{pmatrix} 0 & \sigma_s \\ -\sigma_s & 0 \end{pmatrix} = \begin{pmatrix} -\sigma_r\sigma_s & 0 \\ 0 & -\sigma_r\sigma_s \end{pmatrix}.$$

$$\gamma^r\gamma^s + \gamma^s\gamma^r = 2\delta^{rs}.$$

$$\gamma^{r+} = -\gamma^r.\ \gamma^0\gamma^r = \begin{pmatrix} 0 & \sigma_r \\ \sigma_r & 0 \end{pmatrix}$$

$$\gamma^r\gamma^0 = \begin{pmatrix} 0 & -\sigma_r \\ -\sigma_r & 0 \end{pmatrix}\ \gamma^0\gamma^r + \gamma^r\gamma^0 = 0.$$

$$\gamma^{02} = I,\ \gamma^0\gamma^r = \alpha_r = \begin{pmatrix} 0 & \sigma_r \\ \sigma_r & 0 \end{pmatrix}.$$

So Dirac's equation is the same as

$$\left(i\partial_0 + i\gamma^0\gamma^r\partial_r - m\gamma^0\right)\psi = 0.$$

or
$$i\partial_0\psi = \left(-i\gamma^0\gamma^r\partial_r + m\gamma^0\right)\psi$$

$$= \left(\sum_{r=1}^{3}\alpha_r p_r + m\beta\right)\psi$$

$$\beta = \gamma^0, p_r = -i\partial_r\ \alpha_r = \begin{pmatrix} 0 & \sigma_r \\ \sigma_r & 0 \end{pmatrix}$$

$$\alpha_r^T = \alpha_r.$$

Free Dirac Hamiltonian

$$H_0 = (\alpha, p) + \beta m.$$

Perturbed Dirac Hamiltonian

$$H(t) = (\alpha, p + eA) + \beta m - eV$$
$$= H_0 + e((\underline{\alpha}, A) - V)$$
$$\underline{A} = \underline{A}(t, r),\ V = V(t, \underline{r})$$

Computation of $\exp(-itH_0)$:

$$H_0\begin{pmatrix} \underline{\chi} \\ \underline{\varphi} \end{pmatrix} = E\begin{pmatrix} \underline{\chi} \\ \underline{\varphi} \end{pmatrix}$$

$$\Rightarrow\quad ((\alpha, p) + \beta m)\begin{pmatrix} \underline{\chi} \\ \underline{\varphi} \end{pmatrix} = E\begin{pmatrix} \underline{\chi} \\ \underline{\varphi} \end{pmatrix}$$

$$\Rightarrow\begin{pmatrix} m\underline{\underline{I}} & (\underline{\underline{\sigma}}, p) \\ (\underline{\underline{\sigma}}, p) & -m\underline{\underline{I}} \end{pmatrix}\begin{pmatrix} \underline{\chi} \\ \underline{\varphi} \end{pmatrix} = E\begin{pmatrix} \underline{\chi} \\ \underline{\varphi} \end{pmatrix}$$

$$(\sigma, p)\underline{\varphi} = (E - m)\underline{\chi}$$

$$(\sigma, p)\underline{\chi} = (E + m)\underline{\chi}$$

$$\Rightarrow \quad (\sigma, p)^2\underline{\varphi} = \left(E^2 - m^2\right)\underline{\varphi}$$

$$\Rightarrow \quad E^2 = p^2 + m^2 E = \pm\sqrt{p^2 + m^2}$$

$$\begin{pmatrix} \underline{\chi} \\ \underline{\varphi} \end{pmatrix} = \begin{pmatrix} (\sigma, p)\underline{\varphi} \\ (E - m)\underline{\varphi} \end{pmatrix}$$

$$\|\underline{\chi}\|^2 + \|\underline{\varphi}\|^2 = 1$$

$$\Rightarrow \quad \left(\frac{p^2}{(E - m)^2} + 1\right)\|\underline{\varphi}\|^2 = 1$$

$$\Rightarrow \quad \|\underline{\varphi}\|^2 = \frac{(E - m)^2}{E^2 + m^2 + p^2 - 2mE}$$

$$= \frac{(E - m)^2}{2E^2 - 2mE} = \frac{(E - m)}{2E}$$

$$\|\underline{\varphi}\| = \sqrt{\frac{E - m}{2E}}.$$

Let

$$\varphi_A^+ = \sqrt{\frac{E - m}{2E}}\begin{pmatrix} 1 \\ 0 \end{pmatrix} \quad E = \sqrt{p^2 + m^2}$$

$$\varphi_A^- = \sqrt{\frac{E + m}{2E}}\begin{pmatrix} 1 \\ 0 \end{pmatrix} \quad \varphi_B^- = \sqrt{\frac{E + m}{2E}}\begin{pmatrix} 0 \\ 1 \end{pmatrix}$$

$$\varphi_B^+ = \sqrt{\frac{E - m}{2E}}\begin{pmatrix} 0 \\ 1 \end{pmatrix}$$

An ONB of 4 eigenvectors of H_0 and

(1) $$\left|\psi_A^+\right\rangle = \begin{pmatrix} \dfrac{(\sigma, p)}{(E - m)}\varphi_A^+ \\ \varphi_A^+ \end{pmatrix}$$

(2) $$\left|\psi_B^+\right\rangle = \begin{pmatrix} \dfrac{(\sigma, p)}{(E - m)}\varphi_B^+ \\ \varphi_B^+ \end{pmatrix}$$

(3)
$$\left|\psi_A^-\right\rangle = \begin{pmatrix} \dfrac{-(\sigma, p)}{(E+m)}\varphi_A^- \\ \varphi_A^- \end{pmatrix}$$

(4)
$$\left|\psi_B^-\right\rangle = \begin{pmatrix} \dfrac{-(\sigma, p)}{(E+m)}\varphi_B^- \\ \varphi_B^- \end{pmatrix}$$

Check for orthogonality.

$$\left\langle\psi_A^+\middle|\psi_A^-\right\rangle = -\left\langle \varphi_A^+ \middle| \frac{(\sigma,p)^2}{\left(E^2-m^2\right)} \middle| \varphi_A^- \right\rangle + \left\langle\varphi_A^+\middle|\varphi_A^-\right\rangle$$

$$= \left(-\frac{p^2}{\left(E^2-m^2\right)}+1\right)\left\langle\varphi_A^+\middle|\varphi_A^-\right\rangle$$

$$= \frac{\left(E^2-m^2-p^2\right)}{\left(E^2-m^2\right)}\left\langle\varphi_A^+\middle|\varphi_A^-\right\rangle = 0$$

$\because \quad E^2 = m^2 + p$

$$\left\langle\psi_A^+\middle|\psi_B^+\right\rangle = \left\langle \varphi_A^+ \middle| \frac{(\sigma,p)^2}{(E-m)^2} \middle| \varphi_B^+ \right\rangle + \left\langle\varphi_A^+\middle|\varphi_B^+\right\rangle$$

$$\left(\frac{p^2}{(E-m)^2}+1\right)\left\langle\varphi_A^+\middle|\varphi_B^+\right\rangle = 0$$

$\because$
$$\left\langle\varphi_A^+\middle|\varphi_B^+\right\rangle = 0$$

Likewise for the other cases.
Free Dirac unitary evolution operator.

Let $\qquad E = +\sqrt{m^2+p^2}$

$$U_0(t) = \exp(-itH_0) = \exp(-iEt)\left(\left|\psi_A^+\right\rangle\left\langle\psi_A^+\right| + \left|\psi_B^+\right\rangle\left\langle\psi_B^+\right|\right)$$

$$+\exp(iEt)\left(\left|\psi_A^-\right\rangle\left\langle\psi_A^-\right| + \left|\psi_B^-\right\rangle\left\langle\psi_B^-\right|\right)$$

$$\left|\psi_A^+\right\rangle\left\langle\psi_A^+\right| + \left|\psi_B^+\right\rangle\left\langle\psi_B^+\right|$$

$$\begin{pmatrix} \frac{(\sigma,p)}{(E-m)} & \varphi_A^+ \\ \varphi_A^+ \end{pmatrix}\left(\varphi_A^{+*}\frac{(\sigma,p)}{(E-m)},\varphi_A^{+*}\right)+\begin{pmatrix} \frac{(\sigma,p)}{(E-m)} & \varphi_B^+ \\ \varphi_B^+ \end{pmatrix}\left(\varphi_B^{+*}\frac{(\sigma,p)}{(E-m)}\varphi_B^{+*}\right)$$

$$= \begin{pmatrix} \frac{(\sigma,p)}{(E-m)}\left(\varphi_A^+\varphi_A^{+*}+\varphi_B^+\varphi_B^{+*}\right)\frac{(\sigma,p)}{(E-m)}, & \frac{(\sigma,p)}{(E-m)}\left(\varphi_A^+\varphi_A^{+*}+\varphi_B^+\varphi_B^{+*}\right) \\ \left(\varphi_A^+\varphi_A^{+*}+\varphi_B^+\varphi_B^{+*}\right)\frac{(\sigma,p)}{(E-m)}, & \varphi_A^+\varphi_A^{+*}+\varphi_B^+\varphi_B^{+*} \end{pmatrix}$$

$$= \begin{pmatrix} \frac{p^2}{2E(E-m)}I_2 & \frac{(\sigma,p)}{2E} \\ \frac{(\sigma,p)}{2E} & \frac{(E-m)}{2E}I_2 \end{pmatrix}$$

$$\left|\psi_A^-\right\rangle\left\langle\psi_A^-\right|+\left|\psi_B^-\right\rangle\left\langle\psi_B^-\right|$$

is obtained by replacing E with $-E$ in the above expression.

Thus finally,

$$U_0(t) = \exp(-iET)\begin{pmatrix} \frac{p^2}{2E(E-m)}I_2 & \frac{(\sigma,p)}{2E} \\ \frac{(\sigma,p)}{2E} & \frac{(E-m)}{2E}I_2 \end{pmatrix}$$

$$+\exp(-iET)\begin{pmatrix} \frac{p^2}{2E(E+m)}I_2 & -\frac{(\sigma,p)}{2E} \\ -\frac{(\sigma,p)}{2E} & \frac{(E+m)}{2E}I_2 \end{pmatrix}$$

General solution to the Dirac equation.

$$H_0|\,\psi > \; = E|\,\psi >$$

gives for $E = +\sqrt{m^2+p^2} = E(p)$.

$$<\underline{r}|\psi> \; = \underline{\psi}(\underline{r})$$

$$\left[C_1\begin{pmatrix} \frac{(\sigma,p)}{(E-m)} & \varphi_A^+ \\ \varphi_A^+ \end{pmatrix}+C_2\begin{pmatrix} \frac{(\sigma,p)}{(E-m)} & \varphi_B^+ \\ \varphi_B^+ \end{pmatrix}\right]\exp(ip.r)$$

$$|\,C_1\,|^2+|\,C_2\,|^2 = 1.$$

and likewise for $-E(p)$.

The general solution is therefore the super position.

$$\underline{\psi}(\underline{r},t) = \int \exp\{-i(E(p)t - p\cdot r)\}$$

$$\left\{C_1(p)\begin{pmatrix} \dfrac{(\sigma,p)}{\sqrt{2E(p)(E(p)-m)}} & \underline{e}_1 \\ \sqrt{\dfrac{E(p)-m}{2E(p)}} & \underline{e}_1 \end{pmatrix} + C_2(p)\begin{pmatrix} \dfrac{(\sigma,p)}{\sqrt{2E(p)(E(p)-m)}} & \underline{e}_2 \\ \sqrt{\dfrac{E(p)-m}{2E(p)}} & \underline{e}_2 \end{pmatrix}\right\} d^3p$$

$$+\int \exp\{i(E(p)t + p\cdot r)\}$$

$$\left\{C_3(p)\begin{pmatrix} \dfrac{-(\sigma,p)}{\sqrt{2E(p)(E(p)+m)}} & \underline{e}_1 \\ \sqrt{\dfrac{E(p)+m}{2E(p)}} & \underline{e}_1 \end{pmatrix} + C_4(p)\begin{pmatrix} \dfrac{-(\sigma,p)}{\sqrt{2E(p)(E(p)+m)}} & \underline{e}_2 \\ \sqrt{\dfrac{E(p)+m}{2E(p)}} & \underline{e}_2 \end{pmatrix}\right\} d^3p$$

where $\underline{e}_1 = \begin{pmatrix} 1 \\ 0 \end{pmatrix}, \underline{e}_2 = \begin{pmatrix} 0 \\ 1 \end{pmatrix}.$

Define the 4 × 4 matrices

$$\underline{Q}_+(p) = \begin{pmatrix} \dfrac{(\sigma,p)}{\sqrt{2E(p)(E(p)-m)}} & \underline{\underline{0}} \\ \underline{\underline{0}} & \sqrt{\dfrac{E(p)-m}{2E(p)}}\underline{\underline{I}}_2 \end{pmatrix}$$

and

$$\underline{Q}_-(p) = \begin{pmatrix} \dfrac{-(\sigma,p)}{\sqrt{2E(p)(E(p)-m)}} & \underline{\underline{0}} \\ \underline{\underline{0}} & \sqrt{\dfrac{E(p)+m}{2E(p)}}\underline{\underline{I}}_2 \end{pmatrix}$$

and let, $\underline{u} = \begin{pmatrix} 1 \\ 1 \end{pmatrix}$. Then

$$\psi(\underline{r}, t) = \int \exp\{-i(E(p)t - p\cdot r)\}$$

$$\{C_1(p)\underline{Q}_+(p)(\underline{u}\otimes\underline{e}_1) + C_2(p)\underline{Q}_+(p)(\underline{u}\otimes\underline{e}_2)\}d^3p$$

$$+\int \exp\{i(E(p)t + p\cdot r)\}$$

$$\{C_3(p)Q_-(p)(\underline{u}\otimes \underline{e}_1)+C_4(p)Q_-(p)(\underline{u}\otimes \underline{e}_2)\}d^3p$$

Perturbed Dirac equation operator

$$iU'(t) = (H_0 + e((\alpha, A) - V))\,U(t)$$

$$U(t) \approx U_0(t) - ie\int_0^t U_0(t-\tau)\big((\alpha, A(\tau,\underline{r})) - V(\tau,\underline{r})\big)U_0(\tau)d\tau$$

$$U(t) = U_0(t)\,W(t),$$

$$W(t) = I - ie\int_0^t U_0(-\tau)\big((\alpha, A_\tau) - V_\tau\big)U_0(\tau)d\tau$$

$$= I - ie\int_0^t \exp\{-i\tau((\alpha,p)+\beta m)\}\big((\alpha, A_\tau) - V_\tau\big)\exp\{i\tau((\alpha,p)+\beta m)\}dt$$

In momentum space, $\underline{r} = i\underline{\nabla}p = i\dfrac{\partial}{\partial \underline{p}}$ and we get

$$W(t) = I - ie\int_0^t \exp\{-i\tau((\alpha,p)+\beta m)\}\left(\left(\alpha, A\left(\tau, i\frac{\partial}{\partial p}\right)\right) - V\left(\tau, i\frac{\partial}{\partial p}\right)\right)$$

$$\exp\{i\tau((\alpha,p)+\beta m)\}d\tau$$

Let $\quad \underline{A}(t,\underline{r}) = \dfrac{1}{2}\underline{B}(t)\times \underline{r}$

so $\quad \underline{\nabla}\times\underline{A} = \underline{B}(t)$

and $\quad V(t,\underline{r}) = -(E(t),\underline{r})$

Then

$$W(t) = I - ie\int_0^t \exp\{-i\tau((\alpha, p)+\beta m)\}$$

$$\times\left[\frac{1}{2}(\alpha, \underline{B}(\tau)\times i\underline{\nabla}p) + (E(\tau), i\underline{\nabla}p)\right]\exp\{i\tau((\alpha, p)+\beta m)\}d\tau$$

To evaluate this, we need to evaluate

$$\underline{\nabla}p\{\exp\{i\tau((\alpha, p)+\beta m)\}\underline{\psi}(\underline{p})\}$$

Where $\underline{\psi}(\underline{p})$ is a 4 component momentum space wave function

Now $\exp\left\{it\left((\alpha, p)+\beta m\right)\right\}$

$$\begin{pmatrix} \frac{p^2}{2E(p)}\left\{\frac{\exp(iE(p)t)}{(E(p)-m)}+\frac{\exp(-iE(p)t)}{(E(p)-m)}\right\} & \frac{(\sigma_j p)}{2E(p)}Zi\sin(E(p)t) \\ -\frac{(\sigma_j p)}{2E(p)}Zi\sin(E(p)t) & \frac{(E(p)-m)}{2E(p)}e^{iE(p)t} \\ & +\frac{E(p)+m}{2E(p)}e^{iE(p)t} \end{pmatrix}$$

$$= \begin{pmatrix} \frac{p^2}{2E}\frac{(2E\cos(Et)+2mi\sin(Et))}{(E^2-m^2)} & \frac{i(\sigma, p)}{E}\sin(Et) \\ -\frac{i(\sigma, p)\sin(Et)}{E} & \cos(Et)\frac{-im\sin(Et)}{E} \end{pmatrix}$$

$$\begin{pmatrix} \left(\cos(Et)+\frac{im}{E}\sin(Et)\right)I_2 & \frac{i(\sigma, p)}{E}\sin(Et) \\ -\frac{i(\sigma, p)\sin(E\underline{t})}{E} & \left(\cos(Et)-\frac{im}{E}\sin(Et)\right)I_2 \end{pmatrix}$$

Check for unitarity

$$\frac{\partial E}{\partial p_k} = \frac{p_k}{E}$$

So,

$$\frac{\partial U_0(t)}{\partial p_k} = \begin{pmatrix} \frac{p_k}{E}\left(-\sin(Et)-\frac{im}{E^2}\sin(Et)+\frac{imtp_k}{E^2}\cos(Et)I_2,\right) \\ \frac{p_k}{E}\left\{i\frac{\sigma, p}{E}\sin(Et)-i\frac{(\sigma, p)}{E}t\cos(Et)-i\frac{\sigma_k}{E}\cos(Et),\right\} \end{pmatrix}$$

$$\begin{pmatrix} \frac{p_k}{E}\left\{-i\frac{(\sigma, p)}{E}\sin(Et)+i\frac{(\sigma, p)}{E}t\cos(Et)\right\}+i\frac{\sigma_k}{E}\sin(Et) \\ \frac{p_k}{E}\left(-t\sin(Et)+\frac{im}{E^2}\sin(E,)-\frac{imt\cos(Et)\cos(Et)}{E}\right) \end{pmatrix}$$

Alternate way to compute exp $(-itH_0)$

Laplace transform of exp ($-itH_0$):

$$\int_0^\infty e^{-itH_0}e^{-st}dt = (SI+iH_0)^{-1}$$

$$= (SI_4+i(\alpha, p)+i\beta m)^{-1}$$

$$= \frac{(SI_4+i(\alpha, p)+i\beta m)}{S^2+((\alpha, p)+\beta m)^2}$$

$$= \frac{SI_4-i((\alpha, p)+\beta m)}{S^2+E^2} \quad E^2 = p^2+m^2$$

Inverse Laplace tempering gives

$$\exp(-itH_0) = \cos(Et)I_4 - i\frac{(\alpha, p+\beta m)}{E}\sin(Et) = U(t)$$

$$= \begin{pmatrix} \cos(Et)I_2 & 0 \\ 0 & \cos(Et)I_2 \end{pmatrix} - \frac{i\sin(Et)}{E}\begin{pmatrix} mI_2 & (\sigma, p) \\ (\sigma, p) & -mI_2 \end{pmatrix}$$

$$= \begin{pmatrix} \left(\cos(Et)-\frac{im}{E}\sin(Et)\right)I_2 & -i\frac{(\sigma, p)}{E}\sin(Et) \\ -i\frac{(\sigma, p)}{E}\sin(Et) & \left(\cos(Et)+\frac{im}{E}\sin(Et)\right)I_2 \end{pmatrix}$$

Dirac Hamiltonian is quantum noise

$$H_0 = (\alpha, p) + \beta m \quad p = -i\underline{\nabla}$$

$$H(t) = (\alpha, p+eA_t)+\beta m - eV_t$$

$$dU(t) = \Big[-i((\alpha, p)+\beta m)dt + e\left(L(p)dA(t) - L^*(t)dA^*(t)\right)$$

$$-\frac{e^2}{2}(p)L^*(p)dt\Big]U(t)$$

or

$$U(t) = \left[-\left(iH_0+\frac{e^2}{2}LL^*\right)dt + e\left(LdA - L^*dA^*\right)\right]U(t)$$

$$U(t) = U_0(t)+eU_1(t)+e^2U_2(t)+O(e^3)$$

$$dU_0(t) = iH_0U_0(t)dt$$

$$dU_1(t) = -(iH_0)U_1(t)dt + LU_0(t)dA(t) - L^*U_0(t)dA^*(t)$$

$$dU_2(t) = -iH_0U_2(t)dt - \frac{1}{2}LL^*U_0(t)dt + LU_1^{(t)}dA(t)L^*U_1(t)dA^*(t)$$

$$Q_1 = Q_1(p) = \begin{pmatrix} \frac{p^2 I_2}{2E(E-m)} & \frac{(\sigma, p)}{2E} \\ \frac{(\sigma, p)}{2E} & \frac{(E-m)}{2E} I_2 \end{pmatrix}$$

and $$Q_2 = Q_2(p) = \begin{pmatrix} \frac{p^2 I_2}{2E(E+m)} & \frac{-(\sigma, p)}{2E} \\ \frac{-(\sigma, p)}{2E} & \frac{(E+m)}{2E} I_2 \end{pmatrix}$$

So, $$U_1(t) = \int_0^t U_0(t-\tau)LU_0(\tau)dA(\tau) - \int_0^t U_0(t-\tau)L^*U_0(\tau)dA^*(\tau)$$

$$\text{Bath shate} = |\varphi(u)\rangle = \exp\left(-\frac{1}{2}\|u\|^2 e(u)\right)$$

$u \in L^2(\mathbb{R}_+)$.

❖❖❖❖❖

[227] Let consider $|\varphi_n(u)\rangle = U^{\otimes n}$, $\|u\| = 1$

This is a state in which the bath has n photon all in the state| $|u\rangle \in L^2(\mathbb{R}_+)$ More generally let $u_1,..,u_p \in L^2(\mathbb{R}_+)$ be mutually orthogonal *i.e.* $\langle u_\alpha, u_\beta \rangle = \delta_{\alpha\beta}$. consider

$$\left|\varphi_{n_1 \dots n_p}(u_1,..,u_p)\right\rangle = \left|n_1^{n_1} \dots n_p^{n_p}\right\rangle$$

$$= \left(\sum_{\sigma \in \zeta_n} |v_{\sigma 1} \otimes \dots \otimes_{\sigma n}\rangle\right) \times \left(\lfloor n_1 .. \lfloor n_p\right)^{-1/2}$$

$$= \left(\lfloor n_1\right)^{1/2} \cdot \sum_{\sigma \in \zeta_n} |v_{\sigma 1} \otimes \dots \otimes_{\sigma n}\rangle \left(\frac{\lfloor n}{\lfloor n_1 .. \lfloor np}\right)^{-1/2}$$

where $v_1 = v_2 = v_{n_1} = u_1, v_{n_1+1} = .. = v_{n_1+n_2}, .., v_{n_1+n_2} .. + n_{p-1} + 1 = v_{n_1+n} + .. +$

$\left|\varphi_{n_1..n_p}\left(u_1..u_p\right)\right\rangle$ is the state in which n_1 photoin in the bath are in the state $|u_1\rangle$ n_2 photon are in the state $|u_2\rangle$,..., n_p photn are in the state $|u_p\rangle$

Let $|f\rangle, |g\rangle$ be in $L^2(\mathbb{R}^3) \otimes \mathbb{C}^4$

$$\langle g\varphi(v), U_1(t) f\varphi(u)\rangle \; f(p') = \langle p' | F\rangle$$

$$g(p') = \langle p' | g\rangle$$

$$= \int_0^t \langle g|U_0(t-\tau) L U_0(\tau) f|\rangle \langle \varphi(v)|dA(\tau)|\varphi(u)\rangle$$

$$+\int_0^t \langle g|U_0(t-\tau) L^* U_0(\tau) f|\rangle \langle \varphi(v)|dA^*(\tau)|\varphi(u)\rangle$$

$$= \left(\int \overline{\underline{g}(p')}^T \exp\left(-i(t-\tau)(\alpha, p') + \beta m\right)\right)$$

$$L(p')\exp\left(-i\tau\left((\alpha, p) + \beta m\right)\right) \underline{f}(p') d^3 p'$$

$$\mathbb{R}^3 \times [0, t] \times u(\tau) d\tau \times \langle \varphi(v) | \varphi(u)\rangle$$

$$+\left(\int_{\mathbb{R}^3 \times [0, t]} \overline{\underline{g}(p')}^T \exp\left(-i(t-\tau)(\alpha, p') + \beta m\right)\right)$$

$$L^*(p')\exp\left(-i\tau\left((\alpha, p) + \beta m\right) \underline{f}(p') \bar{v}(\tau) d\tau\right) \times \langle \varphi(v) | \varphi(u)\rangle$$

$$= \int_{\mathbb{R}^3 \times [0, t]} \underline{g}(p') Q_{,}(p') \exp\left\{-i(t-\tau) E(p')\right\}$$

$$+Q_2(p')\exp\left\{i(t-\tau) E(p')\right\} L(p')$$

$$\left(Q_1(p')\exp\left\{-i\tau E(p')\right\} + Q_2(p')\exp\left\{i\tau E(p')\right\}\right) \underline{f}(p') d^3 p' u(\tau) d\tau$$

$$+ \int_{R^3 \times [0, t]} \underline{g}(p')^* \left(Q_1(p')\exp\left\{-i(t-\tau) E(p')\right\}\right.$$

$$+Q_2(p')\exp\left\{i(t-\tau) E(p')\right\}\Big) \cdot L^*(p') \cdot$$

$$\cdot\left(Q_1(p')\exp\left(-i\tau E(p')\right) + Q_2(p')\exp\left(i\tau E(p')\right)\right)$$

$$\underline{f}(p') \bar{v}(\tau) d^3 p' d\tau \; \langle \varphi(v) | \varphi(u)\rangle$$

$$= \left(\int_0^t u(\tau)d\tau\right)\int_{\mathbb{R}^3} g^*(p')Q_1(p')L(p')Q_1(p')f(p')\exp\{-itE(p')\}d^3p'$$

$$+\int_{\mathbb{R}^3} g^*(p')Q_2(p')L(p')Q_2(p')f(p')\exp\{itE(p')\}d^3p'$$

$$+\left(\int_0^t u(\tau)d\tau\right)$$

$$\left\{\int_{\mathbb{R}^3} g^*(p')Q_1(p')L^*(p')Q_1(p')f(p')\exp\{-itE(p')\}d^3p'\right.$$

$$\left.+\int_{\mathbb{R}^3} g(p')^* Q_2(p')L^*(p')Q_2(p')f(p')\exp\{itE(p')\}d^3p'^2\right\}$$

$$+\int_{\mathbb{R}^3} \underline{g}(p')^* Q_1(p')L(p')Q_2(p')\underline{f}(p')d^3p'$$

$$\left(\int_0^t u(\tau)\exp\{i(2\tau-p)E(p')\}d\tau\right)$$

$$+\int_{\mathbb{R}^3} \underline{g}(p')^* Q_2(p')L(p')Q_1(p')\underline{f}(p')d^3p'$$

$$\left(\int_0^t u(\tau)\exp\{i(t-2\tau)E(p')d\tau\}\right)$$

❖❖❖❖❖

[228] Waves in a plasma

The pressure is eliminated from this equation by a curl operation ad the result is a pde for the stream function algorithms by Expressing the Navier. stokes equation as a partial differential equation in terms of the stream function of the fluid. Velocity field of the fluid at different space-time points. (By estimating these fields, we get a more accurate picture of the fluid field).

$$\frac{\partial f_1(t,\underline{r},\underline{v})}{\partial t}+(\underline{V},\underline{\nabla}r)f_1+\frac{q}{m}\left(E(t,\underline{r})+\underline{V}\times\underline{B}(t,\underline{r})\underline{\nabla}v\right)f_0$$

$$-\frac{q}{m}\left(\underline{\nabla}\Phi(\underline{r}),\underline{\nabla}v\right)f_1 = \frac{-f_1}{\tau}$$

$$f_0 = C\cdot\exp\left(-\frac{1}{kT}\left(mV^2+q\Phi(\underline{r})\right)\right)$$

$$f_1(t, \underline{r}, \underline{v}) = \int \hat{f}_1(w, \underline{k}, \underline{v}) \exp(j(wt - \underline{k}, \underline{r}))dwd\underline{k}$$

$$j(w - (\underline{k}, \underline{v}))\hat{f}_1(w, \underline{k}, \underline{v})$$

$$\int((\underline{k} - \underline{k}')\Phi(\underline{k}'), \underline{\nabla}v)\hat{f}_1(w, \underline{k}', \underline{v})d\underline{k}' = \hat{f}_1(w, \underline{k}, \underline{v})/\tau(v)$$

$$\hat{f}_1(w, \underline{k}, \underline{v}) + \frac{jq}{m}\int \frac{(\underline{k} - \underline{k}', \underline{\nabla}v)}{((w - (\underline{k}, \underline{v}) - \tau m))}\hat{f}_1(w, \underline{k}', \underline{v})d\underline{k}'$$

$$= \frac{q}{kT}\int(\underline{V}, \underline{E}(w, \underline{r}))f_0(\underline{r}, \underline{v})\exp(-j\underline{k}, \underline{r})d^3r$$

$$\hat{f}_1(w, \underline{k}, \underline{v}) + j\alpha\int \frac{(\underline{k} - \underline{k}', \underline{\nabla}v)\hat{f}_1(w, \underline{k}', \underline{v})}{\left(w - (\underline{k}, \underline{v}) - \frac{j}{\tau(v)}\right)}d^3k'$$

$$= \frac{q}{KT}\int(\underline{V}, \hat{\underline{E}}(w, \underline{r}))f_0(\underline{r}, \underline{v})\exp(-i\underline{k} \cdot \underline{r})d^3r$$

We shall develop algorithms for estimate parameters of the flind like fluid viscosity density [remove tje vetpoc;u fo;d voa tje strems, function from noisy measurements of the velocity field of the fluid difference space time points. By eshtity then fluids, we can get a more acavate piline of the find fild. The algorithms we shall develop shall be based on by exprining the Navier stoks equation as a particle differently equation in tram of the sheam function ordoer fild. (The premme is the limuided from this equation by a aerl opeation and the result is a pde for the shean fund) $\underline{\psi}(t, \underline{r})$ $\underline{r} = (a, y, z)$ which has the form

$$\frac{\partial\underline{\psi}(t, \underline{r})}{\partial t} + \Delta^{-1}(\Delta\underline{\psi} \times \underline{\nabla} \times \underline{\psi}) = v\Delta\underline{\psi} - \Delta^{-1}(\underline{\nabla} \times \underline{f})$$

where $\Delta = \nabla^2$ is the Laplacing f is the external force fluid (random) and $v = \frac{\eta}{\rho}$ η = viscosity ρ, density. We the represent $\underline{\psi}(t, \underline{r})$ by a vector $\psi(t) \in \mathbb{R}^{3N^3}$ where N^3 is the number of spatial pixels and the (1) can be put in the form

$$\frac{d\underline{\psi}(t)}{dt} = \underline{\underline{A}}\underline{\psi}(t) + \underline{\underline{B}}(\underline{\psi}(t) \otimes \underline{\psi}(t)) + \underline{g}(t)$$

Where $\underline{\underline{A}}$ is an $\mathbb{R}3N^3 \times 3N^3$ matrix and $\underline{\underline{B}}$ is an $\mathbb{R}^3N^3 \times 9N^6$ matrix. $\underline{g}(t) \in R^{3N^3}$ is a random process. $\underline{g}(t)$ can be modelled as a which Gaussian random process with an autocorrelation

$$\mathbb{E}\{\underline{g}(t_1) \cdot \underline{g}(t_2)^T\} = \underline{\underline{Rg}}\delta(t_1 - t_2)$$

and the (2) is represented by a stochastic diffracted equation

$$d\underline{\psi}(t) = (\underline{\underline{A}}\underline{\psi}(t) + \underline{\underline{B}}(\underline{\psi}(t) \oplus \underline{\psi}(t)))dt + \sqrt{\underline{\underline{Rg}}}d\underline{W}(t)$$

Where $\underline{W}(t) \in \mathbb{R}^{3N^3}$ is a Brownian motions process. we take measurements on the velocity fuils v = $\underline{\nabla} \times \underline{\psi}$ to different times at different spatial piexel and these noisey measurement are described by the measurement model

$$v(t)dt = d\underline{\xi}(t) = \underline{\underline{C}}\underline{\psi}(t)dt + d \in (t)$$

where $\varepsilon(t)$ is measurement noise. Then the problem is to construct

$$\mathbb{E}\left[\underline{\psi}(t) \mid \underline{\xi}(t), \tau \leq t\right] = \widehat{\underline{\psi}}(t \mid t)$$

and hence the velocity fluid estimate

$$v(t) = \underline{\underline{C}}\widehat{\underline{\psi}}(t \mid t)$$

This idea can be extended to study the motion of the galaxies find (*i.e.*) large scale fluid dynamic) Specially let $g^0_{\mu\nu}(X)$ denote the background homogeneous and isotropic Robertson-Walker matrix and $g_{\mu\nu}(X) = g^0_{\mu\nu}(X) + \delta g_{\mu\nu}(X)$ its perturbation caused by small random forces having an energy momentum tensor $\delta\mathcal{F}^{\mu\nu}$. The Einstein Fluid equations

$$R_{\mu\nu} - \frac{1}{2}Rg_{\mu\nu} = K(T_{\mu\nu} + \delta\mathcal{F}_{\mu\nu})$$

with $T_{\mu\nu} = (\rho + p)v_\mu v_\nu - pg_{\mu\nu}$

have an unperturbed solution

$$R^{(0)}_{\mu\nu} - \frac{1}{2}R^{(0)}g^{(0)}_{\mu\nu} = KT^{(0)}_{\mu\nu}$$

$$T^{(0)}_{\mu\nu} = (\rho_0(t) + p(t))V^\mu V^\nu - p_0(t)g^{(0)}_{\mu\nu}(X)$$

with $\left(V^\mu\right) = (1, 0, 0, 0)$ (commoving coordinate)

$g^0_{\mu\nu}(X)$ is deimted by a scale factor $S(t)$ of the universe and a curvature constant k, which in 0, 1 or – 1 corresponding to flat, spartial or hyperbolic universe constant $\{S(t), \rho_0(t), p_0(t)\}$ satisfy nonlinear differential equations and to solve then, we must supplements then with an equation of state $p_0(t) = f(\rho_0)(t)$. These solution give expanding universes contracting or cycloidally expanding and conically expanding and contracting universing. These are called the friedmann model

The perturbed equation are

$$\delta R^{\mu\nu} - \frac{1}{2}\left(\delta R \cdot g^{(0)\mu\nu} + R^{(0)}\delta g^{\mu\nu}\right) = K\left(\delta T^{\mu\nu} + \delta\mathcal{F}^{\mu\nu}\right)$$

where

$$\delta T^{\mu\nu} = (\delta\rho + \delta p)V^\mu V^\nu + (\rho_0 + p_0)\left(V^\mu \delta v^\nu + V^\nu \delta u^\mu\right)$$

$$-\delta p g^{(0)\mu\nu} - p_0 \delta g^{\mu\nu}$$

Now, $\delta\rho$, δp, $\delta g_{\mu\nu}$, $\delta g\mu\nu$, δv^μ are function of space time $X = (t, \underline{r}) = (t, r, \theta, \varphi)$.

The above equation are linear partial differented equations for $\delta g_{\mu\nu}$, δv^{μ},. $\delta\rho$, δp with coefficients depend on $X = (t, \underline{r})$. We shall be developing program to solve these pde's. The fluid dynamical equation follow from the Einsten field equation. If the have ground metric $g^{(0)}_{\mu\nu}(X)$ is assumed to be unpertubed, then the fluid dynamical equation which describe the perturbed galaction are

$$\left((\rho + p)v^{\mu}v^{\nu} - pg^{\mu\nu}\right):\nu = \delta\mathcal{F}^{\mu}$$

where $\delta\mathcal{F}^{\mu} = -\delta\mathcal{F}^{\mu\nu}_{:\nu}$. We choose has $g_{\mu\nu} = \delta\mathcal{F}^{(0)}_{\mu\nu}$.
These yield

$$\left((\rho + p)v^{\nu}\right)_{:\nu} - v^{\nu}p_{,\nu} = v^{\nu}\delta\mathcal{F}_{\nu} \quad (\beta)$$

or

$$\left((\rho + p)\theta^{\nu}\sqrt{-g}\right)_{,\nu} = v^{\nu}\left(p_{,\nu} + \delta\mathcal{F}_{\nu}\right)\sqrt{-g}$$

These equation describe matter conservation. Its perturbed version is

$$= \left((\rho_0 + p_0)\sqrt{-g}\delta v^{\nu}\right)_{,\nu} + \left((\delta\rho + \delta p)V^{\nu}\sqrt{-g}\right)_{,\nu}$$

$$= V^{\nu}\left(\delta p_{,\nu} + \delta\mathcal{F}_{\nu}\right)\sqrt{-g}$$

$$+\delta v^{\nu}p_{0,\nu}\sqrt{-g} + V^{\nu}\sqrt{-g}\,\delta p_{,\nu} + V^{\nu}\sqrt{-g}\,\delta\mathcal{F}_{\nu}$$

or

$$\left((\rho_0 + p_0)\delta v^{\nu}\sqrt{-g}\right)_{,\nu}$$

$$+\left((\delta\rho + \delta p)\sqrt{-g}\right)_{,0}$$

$$+\left(\delta p_{,0} + \delta\mathcal{F}_0\right)\sqrt{-g}$$

$$+\delta v^{0}p_{0,0}\sqrt{-g}$$

$$+\sqrt{-g}\,\delta p_{,0} + \sqrt{-g}\,\delta\mathcal{F}_0 = 0 \quad (1)$$

Substituting (β) into (α) gives

$$(\rho + p)V^{\nu}V^{\mu}_{:\nu} + V^{\mu}V^{\nu}\left(p_{,\nu} + \delta\mathcal{F}_{\nu}\right)$$

$$\delta\mathcal{F}^{\mu} + P^{,\mu}$$

and its perturbed version is

$$(\rho_0 + p_0)\left(V^{\nu}\,\delta u^{\mu}_{:\nu} + \delta v^{\nu}V^{\mu}_{:\nu}\right)$$

$$+ V^{\mu}V^{\nu}\left(\delta p_{,\nu} + \delta\mathcal{F}_{\nu}\right)$$

$$+p_{0,\nu}\left(V^{\nu}\delta v^{\mu} + V^{\mu}\delta V^{\nu}\right) = \delta\mathcal{F}^{\mu} + \delta p^{,\mu}$$

on for $\mu = k = 1, 2, 3,$

$$(\rho_0 + p_0)\left(\delta v^k_{,0} + 2\Gamma^k_{0k}\delta v^k\right) + p_{0,0}\delta v^k$$

$$= \delta\mathcal{F}^k + g^{kk}\delta p_{,k}$$

there perturbed equation can be expressed on

$$(\rho_0(t) + p_0(t))\left[\frac{\partial \delta v^k(t,\underline{r})}{\partial t} + (\log g_{kk})_{,0}\,\delta v^k(t,\underline{r})\right]$$

$$+ p_0'(t)\delta v^k(t,\underline{r}) = \delta\mathcal{F}^k + g^{kk}\delta p_{,k} \tag{2}$$

From the statistics of $\delta\mathcal{F}^k(t, \underline{r})$, we can solve (1) and (2) for $\delta v^k(t, \underline{r})$ $\rho_0(t)$ $1 \le k \le 3$ assuming an equation of state $p = f(\rho)$ so that

$$\delta p(t,\underline{r}) = f'(\rho_0(t))\delta\rho(t,\underline{r})$$

and calculated the density and velocity correlation.

Newtonian cosmology with stochastic perturbation

$$S(t) = \text{radius of the universe}$$

$$\rho(t) = \text{man density}$$

$$p(t) = \text{pressure}$$

$$\underline{u}(t, \underline{r}) = H(t)\underline{r}\ H(t) = \frac{S'(t)}{S(t)} \simeq \text{Hubbis constant}$$

Energy momentum tensor

$$T^{\mu\nu} = (\rho + p)v^\mu v^\nu - pg^{\mu\nu}$$

$$\underline{v^0} \approx 1, v = V(t,r)\hat{r}$$

$T^{\mu\nu}_{:\nu} = f_\mu \simeq$ external random force

$$\Rightarrow \qquad \left((\rho + p)V^\nu\right)_{:\nu} u^\mu + (\rho + p)U^\nu V^\mu_{:\nu} - p'^\mu = f^\mu \tag{α}$$

$$v_\mu v^\mu = 1 \Rightarrow v^\nu v^\mu_{:\nu}$$

$$\Rightarrow \quad \left((\rho + p)V^\nu\right)_{:V} = v^\mu p_{,\mu} + v^\mu f_\mu$$

$$\Rightarrow \left((\rho + p)v^\nu\sqrt{-g}\right)_{,\nu} = v^\mu p_{,\mu}\sqrt{-g} + v^\mu f_\mu\sqrt{-g} \tag{1}$$

Robertson Walker metric

$$g_{00} = 1, g_{11} = \frac{-S^2(t)}{1 - kr^2}, g_{22}\ g_{22} = -S^2(t)r^2,$$

$$g_{33} = -S^2(t)r^2\sin^2\theta$$

$$g = -S^6(t)v^4 \sin^2\theta$$

$$\sqrt{-g} = S^3(t)r^2 \sin\theta$$

So (1) gives will $v^0 = 1$, $v^r = 0$, $1 \le r \le 3$

Comoving co-ordinates

$$\left((\rho + p)S^3\right)_{,0} = p_{,0}S^3 + f_0 S^3 \tag{2}$$

$$\left(\rho S^3\right)_{,0} + 3pS^2 S_{,0} = f_0 S^3 \tag{2'}$$

Interpretation

$\frac{4}{3}\pi\rho_{(t)}S^3(t)$ = total mater even in the sphere of radius $S(t)$. According to (2),

$\frac{d}{dt}\left(\frac{4\pi}{3}\rho S^3\right)$ = rate of inverse of even the

$$\text{Sphere} = -4\pi p S^2 S'(t) + \frac{4\pi}{3} f_0(t) S^3(t)$$

= Total work done per units time due to internal pressure acting on the sphere boundary + total power generated with the

Volume due to external force.

($f_0(t)$ = Work done per units time) per units volume due to external force)

Thus, (α) can be expressed on

$$(\rho + p)v^\nu v^\mu_{:\nu} + v^\mu v^\alpha\left(p_{,\alpha} + f_\alpha\right) = f^\mu + p^{,\mu} \tag{2}$$

If we make the Newtonian approximation to arrange only radial forces, the since $p = p(t)$ is independent of r, the redial component of (2) gives approximately

$$= \left(v(t, \underline{r}) = H(t)\underline{r}, H(t) = \frac{S'(t)}{S(t)}\right)$$

$$(\rho + p)\left(H'(t)r + H^2(t)r\right) \; + H(t)r\left(p_{,0} + f_0\right)$$

$$= f_1(t)r - \frac{4\pi}{3}G\rho^2(t)r$$

where $f_1(t)r$ = Radial component of external force per unit volume

$$\text{Thus, } (\rho + p)\frac{S''(t)}{S(t)} + \left(\frac{S'(t)}{S(t)}\right)\left(p'(t) + f_0(t)\right)$$

$$= f_1(t) - \frac{4}{3}\pi G\rho^2(t) \tag{3}$$

So, our basis equation in the presence of rant on farces with time compound $f_0(t)$ and radial component $f_1(t)r$

$$\left(\rho S^{3'}\right)+3pS^2S' = f_0S^3, \tag{4a}$$

$$(\rho+p)S''+(p'+f_0)S' = f_1S-\frac{4\pi G}{3}\rho^2 S \tag{4b}$$

In the alesence of external faces $f_0 = 0, f_1 = 0$ (4) redues to

$$\left(\rho S^3\right)'+3pS^2S' = 0, \tag{5a}$$

$$(\rho+p)S''+S'P'+\frac{4\pi G}{3}\rho^2 S = 0 \tag{5b}$$

If further price is zero *i.e.p* = 0, then (5) reduces to

$$\rho S^3 = K\,(\text{constant}) \tag{6a}$$

$$S''+\frac{4\pi G\rho}{3}S = 0 \tag{6b}$$

(6*b*) can be explained

$$\frac{3S''}{S} = -4pGr \tag{6c}$$

and combining (6*c*) with (6*a*) gives

$$\frac{3S''}{S} = -\frac{4\pi GK}{S^3}$$

or
$$3S'' = -\frac{4\pi GK}{S^2}$$

or
$$\frac{3S'^2}{2} = \frac{4\pi GK}{S}+\alpha$$

$$S' = \pm\left(\frac{8\pi GK}{3S}+\beta\right)^{1/2}$$

β is a constant $f_0(t), f_1(t)$

for small stochsatic perturbation, let $\delta S(t)$, $\delta\rho(t)$, $\delta p(t)$ be the perturbation in $S(t)$, $\rho(t)$ and $p(t)$ relatively.

Excel general reactivity description

$$R^{\mu\nu}-\frac{1}{2}Rg^{\mu\nu} = K\left(\rho_{+p)}v^{\mu}v^{\nu}-pg^{\mu\nu}\right)$$

$$\Rightarrow \quad -R = K(\rho+p-4p) = K(\rho-3p)$$

$$= K\left\{(\rho + p)v^{\mu}v^{\nu} - pg^{\mu\nu} - \frac{1}{2}(\rho - p)g^{\mu\nu}\right\}$$

$$\Rightarrow \qquad R^{\mu\nu} = K\left\{(\rho + p)v^{\mu}v^{\nu} - pg^{\mu\nu} - \frac{1}{2}(\rho - p)g^{\mu\nu}\right\}$$

$$= K\left\{(\rho + p)v^{\mu}v^{\nu} + \frac{1}{2}(p - \rho)g^{\mu\nu}\right\}$$

$$\Rightarrow \qquad R_{00} = K\left\{p + \rho + \frac{1}{2}(p - \rho)\right\} = \frac{1}{2}K(3p + \rho)$$

$$R_{11} = -\frac{1}{2}K(p - \rho)S^2/\left(1 - kr^2\right)$$

$$R_{22} = -\frac{1}{2}K(p - \rho)r^2S^2$$

$$R_{33} = -\frac{1}{2}K(p - \rho)r^2S^2\sin^2\theta$$

$$\frac{\left(1 - kr^2\right)R_{11}}{S^2R_{00}} = \frac{\rho - p}{\rho + 3p}$$

$$R_{00} = \Gamma^{\alpha}_{0\alpha,0} - \Gamma^{\alpha}_{00,\alpha} - \Gamma^{\alpha}_{00}\Gamma^{\beta}_{\alpha\beta} + \Gamma^{\alpha}_{0\beta}\Gamma^{\beta}_{0\alpha}$$

$$= \left(\Gamma^{1}_{01} + \Gamma^{2}_{02} + \Gamma^{3}_{03}\right)_{,0} + \left(\Gamma^{12}_{01} + \Gamma^{22}_{2} + \Gamma^{33}_{03}\right)$$

$$\Gamma^{1}_{01} = \frac{1}{2}g^{11}\left(g_{10,1} + g_{11,0}\right) = \frac{1}{2}\left(\log\delta_{11,10}\right) = \frac{S'}{S}$$

$$\Gamma^{2}_{02} = \frac{1}{2}\left(\log S_{22}\right)_{,0}$$

$$\frac{S'}{S} = \Gamma^{3}_{03}$$

$$R_{00} = 3\left(\frac{S'}{S}\right)^{1} + 3\left(\frac{S'}{S}\right)^{2} = \frac{3S''}{S}$$

So, $$\frac{3S''}{S} = \frac{1}{2}K(3p + \rho)$$

In the pressure of external force

$$R_{\mu\nu} - \frac{1}{2}Rg_{\mu\nu} = K\left(T_{\mu\nu} + \delta.\mathcal{F}_{\mu\nu}\right)$$

$$\Rightarrow -R = K(T + \mathcal{F})$$

$$R_{\mu\nu} = K\left(T_{\mu\nu} - \frac{1}{2}(T + \delta\cdot\mathcal{F})g_{\mu\nu} + \delta\cdot\mathcal{F}_{\mu\nu}\right)$$

$$T_{\mu\nu} = (\rho + p)v_\mu v_\nu - pg_{\mu\nu}$$

(oo) equation

$$\delta R_{00} + \frac{3S''}{S} = K\left(\rho + p + \frac{(p-\rho)}{2} g_{00} + \delta \cdot \frac{\mathcal{F}}{00} - \frac{\delta}{2} \mathcal{F} g_{0\delta}\right)$$

$$\left((\rho + p)v^\mu v^\nu - pg^{\mu\nu}\right)_{:\nu} + \mathcal{F}^\mu, \mathcal{F}^\mu = \mathcal{F}^{\mu\nu}_{:\nu}$$

$$\Rightarrow \quad \left((\rho + p)v^\nu\right)v^\mu + (\rho + p)v^\nu v^\mu_{:\nu} - p^{,\mu} + \mathcal{F}^\mu = 0$$

$$\Rightarrow \quad \left((\rho + p)v^\nu\right)_{:\nu} - v^\nu v^\mu + v_\mu + \mathcal{F}^\mu = 0$$

$$\Rightarrow \quad \left((\rho + p)v^\nu\sqrt{-g}\right)_{,\mu} = v^\mu\left(p_{,\mu} - \mathcal{F}_\mu\right)\sqrt{-g}$$

$$\Rightarrow \quad (\rho + p)\left(v^\mu\sqrt{-g}\right)_{,\mu} + \rho_{,\mu}v^\mu\sqrt{-g} = -v^\mu{}_{\delta \cdot}\mathcal{F}_\mu\sqrt{-g}$$

$$v^\mu = V^\mu + \delta . v^\mu \ (1, 0, 0, 0)$$

$$\rho = \rho_0 + \delta\rho$$

$$g_{\mu\nu} = g^{(0)}_{\mu\nu} + \delta \cdot h\mu\nu$$

$$g_{\mu\nu} = g^{(0)}(1 + \delta \cdot h)\ Tr\left[\left(\left(g^{(0)}_{\mu\nu}\right)\right)^1 \left(\left(h_{\mu\nu}\right)\right)\right]$$

$$\sqrt{-g} = \sqrt{-g^{(0)}}\left(1 + \frac{\delta \cdot h}{2}\right)$$

$$(\delta\rho + \delta p)\left(V^\mu\sqrt{-g^{(0)}}\right)_{,\mu}$$

$$+(\rho_0 + p_0)\left(u^\mu\sqrt{-g^{(0)}} + V^\mu\frac{\sqrt{-g^{(0)}}\,h}{2}\right)$$

$$= +\delta\rho_{,\mu}V^\mu\sqrt{-g^{(0)}}$$

$$= +\rho_{0,0}\left(u^0\sqrt{-g^{(0)}} + \sqrt{-g^{(0)}}\frac{h}{2}\right)$$

$$= -\mathcal{F}_0\sqrt{-g^{(0)}}$$

$$\Rightarrow \quad (\delta\rho + \delta p)\left(\sqrt{-g^{(0)}}\right)_{,\mathcal{F}_0} + (\rho_0 + p_0)\left(\sqrt{-g^{(0)}}u^\mu\right)_{,\mu}$$

$$+\frac{1}{2}\left((\rho_0 + p_0)\sqrt{-g^{(0)}}h\right)_{,0} + \delta\rho_{,0}\sqrt{-g^{(0)}}$$

$$+\rho_{0,0}\left(u^0\sqrt{-g^{(0)}}+\frac{h}{2}\sqrt{-g^{(0)}}\right)+\mathcal{F}_0\sqrt{-g^{(0)}}=0$$

$$\Rightarrow\qquad (\delta\rho+\delta p)\left(\log\sqrt{-g^{(0)}}\right)_{,0}$$

$$+(\rho_0+p_0)\left(\sqrt{-g^{(0)}}u^{\mu}\right)_{,\mu}\Big/\sqrt{-g^{(0)}}$$

$$+\frac{1}{2}\left((\rho_0+p_0)\sqrt{-g^{(0)}}h\right)_{,0}\Big/\sqrt{-g^{(0)}}$$

$$+\delta\rho_{,0}+\rho_{0,0}\left(u^0+\frac{h}{2}\right)+\mathcal{F}_0=0$$

$$\delta R_{00}=K\left(\delta\rho+\delta p+\left(\frac{p_0-\rho_0}{2}\right)h_{00}+\left(\frac{\delta p-\delta\rho}{2}\right)+\mathcal{F}_{00}-\frac{1}{2}\mathcal{F}\right)$$

$$=K\left(\frac{1}{2}(3\delta p-\delta\rho)+\left(\frac{p_0-\rho_0}{2}\right)h_{00}+\mathcal{F}_{00}=\frac{1}{2}\mathcal{F}\right)$$

$$(\rho+p)v^{\nu}v^{\mu}_{:\nu}+v^{\mu}v^{\alpha}\left(p_{,\alpha}-\mathcal{F}'_{\alpha}\right)-p^{,\mu}+\mathcal{F}^{\mu}=0$$

Ignore matrix, perturbation

$$v^{\mu}=V^{\mu}+u^{\mu}$$

$$(\rho_0+p_0)\left(u^{\nu}V^{\mu}_{:\nu}+V^{\nu}u^{\mu}_{:\nu}\right)+(\delta\rho+\delta p)V^{\mu}_{:0}$$

$$+p_{0,\alpha}\left(V^{\alpha}u^{\mu}+V^{\mu}u^{\alpha}\right)+\delta p_{,\alpha}V^{\mu}V^{\alpha}$$

$$-\delta p^{,\mu}+\mathcal{F}^{\mu}-V^{\mu}V^{\alpha}\mathcal{F}_{\alpha}$$

$$V^{\mu}_{:\nu}=\Gamma^{\mu}_{\alpha\nu}V^{\alpha}=\Gamma^{\mu}_{0\nu}$$

$$(\rho_0+p_0)\left(u^{r}_{:0}+u^{\nu}\Gamma^{r}_{0\nu}\right)+p_{0,0}u^{r}-\delta p^{,r}+\mathcal{F}^{r}=0$$

$$(\rho_0+p_0)\left(u^{r}_{,0}+2\Gamma^{r}_{r0}u^{\nu}\right)+p_{0,0}u^{r}-\delta p^{,r}+\mathcal{F}^{r}$$

$$\Gamma^{r}_{r0}u^{\nu}=\Gamma^{r}_{r0}u^{r}=\frac{1}{2}g^{rr}g_{rr,0}u^{r}=\frac{S'(t)}{S(t)}u^{r}$$

$$(\rho_0+p_0)\left(\frac{\partial\underline{u}}{\partial t}+\frac{2S'(t)}{S(t)}\underline{u}\right)+p_0'(t)\underline{u}+\underline{\nabla}\delta p+\underline{\mathcal{F}}=0$$

$$\Gamma^{\alpha}_{\mu\alpha,r} = \left(g^{\alpha\beta}\Gamma_{\beta\mu\alpha}\right),_r g_{\mu\nu} = \eta\mu\nu + \in\eta_{\mu\nu}$$

Stochastic problem is cosmology.

$$d\tau^2 = dt^2 - \frac{S^2(t)}{1-kr^2}dr^2 - S^2(t)r^2\left(d\theta^2 + \sin^2\theta d\varphi^2\right)$$

Calculate the different equation satisfied by *S*,

When the equation momentum tensor of matter

$$\begin{bmatrix} \rho+p & & & \\ & -(\rho+p) & 0 & \\ & 0 & -(\rho+p) & \\ & & & -(\rho+p) \end{bmatrix} - p\left(\left(g^{\mu\nu}\right)\right)$$

i.e.

$$T^{00} = \rho + p - p = \rho,$$

$$T^{11} = -(\rho+p) + \frac{p\left(1-Kr^2\right)}{S^2(t)}$$

$$T^{22} = -(\rho+p) + p/S^2(t)r^{2'}$$

$$T^{33} = -(\rho+p) + /S^2(t)r^2\sin^2\theta$$

$$\begin{cases} \ddot{S} = F_1\left(\dot{S}, S, p, \rho\right) \\ S\cdot = F_2\left(S, p, \rho\right) \end{cases}$$

Now consider small random addition to the equation momentum tensor coning from em radiation

$$\ddot{S} = F_1\left(\dot{S}, S, p, \rho\right) + \varepsilon\, w_1(t)$$
$$\ddot{S} = F_2\left(\dot{S}, S, p, \rho\right) + \varepsilon\, w_2(t)$$

Calculate charge in *S*, *p*, *r* by linearization.

$$S(t) = \delta_0(t) + \varepsilon\,\delta S(t) + O\left(\varepsilon^2\right)$$

$$P = p_0(t) + \varepsilon\,\delta p(t) + O\left(\varepsilon^2\right)$$

$$\rho = \rho_0(t) + \varepsilon\,\delta\rho(t) + O\left(\varepsilon^2\right)$$

$$\delta\ddot{S} = F_{1,i}(t)\delta\dot{S} + F_{1,2}(t)\delta S(t) + F_{1,3}(t)\delta p(t) + F_{1,4}(t)\delta\rho(t) + w_1(t) \text{ etc.}$$

Generalization:

$$g^{(0)}_{\mu\nu}(x) \simeq \text{background}$$

$$g_{\mu\nu}(x) = g_{\mu\nu}^{(0)}(x) + \varepsilon\, \delta g_{\mu\nu}(x) \text{ perturbated matrix}$$

$$G_{\mu\nu} = R_{\mu\nu} - \frac{1}{2} R \delta_{\mu\nu}$$

$$G_{\mu\nu}^{(0)+\varepsilon\{C_1(\mu\nu\rho_1\,\sigma\alpha\beta,x)\delta g_{\alpha\beta}(x)} + C_2(\mu\nu\rho\sigma, x)\delta g_{\rho\sigma,\,x}(x)\} + O(\varepsilon)^2$$

$$= G_{\mu\nu}^{(0)} + \varepsilon\, G_{\mu\nu}^{(1)} + O(\varepsilon^2)$$

$$G_{\mu\nu}^{(0)} = K T_{\mu\nu}^{(0)}$$

$$G_{\mu\nu}^{(1)} = K\,\delta T_{\mu\nu} = K\left(T_{\mu\nu}^{(1)} + \mathcal{F}_{\mu\nu}\right)$$

$$T^{\mu\nu} = T^{\mu\nu(0)} + \varepsilon\, T^{\mu\nu(1)}$$

$$\left(\left(T^{\mu\nu} + \mathcal{F}^{\mu\nu}\right)_{:\nu}\right) = 0$$

$$T_{\nu}^{(1)\mu\nu} + T^{(0)\mu\alpha}\Gamma_{\alpha\nu}^{\nu} + T^{(0)\alpha\nu}\Gamma_{\alpha\nu}^{\mu} + \mathcal{F}_{,\nu}^{\mu\nu} = 0$$

$$T_{,\nu}^{(1)\mu\nu} = -T^{(0)\mu\alpha}\Gamma_{\alpha\nu}^{(1)\nu} - T^{(0)\alpha\nu}\Gamma_{\alpha\nu}^{(1)\mu} - \mathcal{F}^{\mu}$$

$$g_{\mu\nu} = g_{\mu\nu}^{(0)+\varepsilon h_{\mu\nu}}$$

$$T_{,\nu}^{(0)\mu\nu} = 0$$

$$\Gamma_{,\nu}^{(1)\mu\nu} = g^{\mu\nu(0)}\Gamma_{\nu\alpha\beta}^{(1)} + g^{\mu\nu(2)}\Gamma_{\nu\alpha\beta}^{(0)}$$

$$= \frac{1}{2} g^{\mu\nu(0)}\left(h_{\nu\alpha,\,\beta} + h_{r\beta,\,\alpha} - h_{\alpha\beta,\,\nu}\right)$$

$$-\frac{1}{2} h^{\mu\nu}\left(g_{\nu\alpha,\,\beta}^{(0)} + g_{\nu\beta,\,\alpha}^{(0)} - g_{\alpha\beta,\,\nu}^{(0)}\right)$$

$$= \frac{1}{2} g^{\mu\nu(0)}\left(h_{\nu\alpha,\,\beta} + h_{\nu\beta,\,\alpha} - h_{\alpha\beta,\,\nu}\right) - \frac{1}{2} h^{\mu\nu}\Gamma_{\nu\alpha\beta}^{(0)}$$

$$\left[(\delta\rho + \delta p)V^{\mu}V^{\nu} + (\rho_0 + p_0)\left(V^{\mu} + \delta v^{\nu} + V^{\nu}\delta v^{\mu}\right)\right.$$

$$\left. - p_0 h^{\mu\nu} - \delta p\, g^{\mu\nu(0)}\right]_{,\,r}$$

$$= -f^{\mu} + T^{(0)\mu\alpha} h_{,\,\alpha} - T^{(0)\alpha\nu}\left(g^{\mu\beta(0)}\Gamma_{\beta\alpha\nu}^{(1)} + h^{\mu}\Gamma_{\alpha\beta}^{(0)}\right)$$

$$C_1\left(\mu\nu\rho\sigma\alpha\beta,_{x}\right) h_{\rho\sigma,\,\alpha\beta}(x) + C_2(\mu\nu\rho\sigma\alpha, x) h_{\rho\sigma,\,\alpha}(x)$$

$$= K\left\{(\rho_0 + p_0)\left(V^{\mu}\delta v^{\nu} + V^{\nu}\delta v^{\mu}\right) - p_0 h^{\mu\nu} - \delta p g^{(0)\mu\nu} + \mathcal{F}^{\mu\nu}\right\}$$

$$g_{\mu\nu}^{(0)} V^{\mu} V^{\nu} = 1$$

$$g_{\mu\nu}^{(0)} \left(V^{\mu} \delta v^{\nu} + V^{\nu} \delta v^{\mu} \right) + h_{\mu\nu} V^{\mu} V^{\nu} = 0$$

$\mathcal{F}^{\mu\nu}_{(x)}$ = Random symmetry every momentum forcing tensor.

correlation of $\mathcal{F}^{\mu\nu}_{(x)}$ are known. We've then goo to find correlation of $\left\{\delta\rho(x), \delta p(x), h_{\mu\nu}(x)\right\}$ using the above equation. Assume that an equation of state $p = f(\rho)$ is who siven so that

$$\delta p(X) = f'\left(\rho_0(x)\right)\delta\rho(X)$$

Note: $g_{\mu\nu}(X) = g_{\mu\nu}^{(0)}(X) + \varepsilon\, h_{\mu\nu}(X)$

So, $g(X) = \text{dat}((g_{\mu\nu}))$

$$= g^{(0)}(X)\left(1 + \varepsilon h_{\mu}^{\mu}(X) + O\left(\varepsilon^2\right)\right)$$

where $h_{\mu}^{\mu} = g^{\mu\nu(0)} h_{\mu\nu}$

We define $h(x) = g^{(0)}(X) h_{\mu}^{\mu}(X)$. The final set of equation to be solved are:

$$\left[(\delta\rho + \delta p) V^{\mu} V^{\nu} + (\rho_0 + p_0)\left(V^{\mu} \delta V^{\nu} + V^{\nu} \delta v^{\mu}\right) - p_0 h^{\mu\nu} - \delta p g^{(0)\mu\nu}\right]_{,\,\nu}$$

$$= -\mathcal{F}^{\mu} + T^{(0)\mu\alpha} h_{,\,\alpha}$$

$$-T^{(0)\alpha\nu}\left(g^{(0)\mu\beta}\Gamma^{(1)}_{\beta\alpha\nu} + h^{\mu\rho}\Gamma^{(0)}_{\alpha\beta}\right) \tag{1}$$

$$C_1(\mu\nu\rho\sigma\alpha\beta, X) h\rho\sigma, \alpha\beta(X) + C_2(\mu\nu\rho\sigma\alpha, X) h\rho\sigma, \alpha(X)$$

$$= K\left\{(\rho_0 + p_0)\left(V^{\mu} \delta v^{\nu} + V^{\nu} \delta v^{\mu}\right) - p_0 h^{\mu\nu} - \delta p g^{(0)\mu\nu} + \mathcal{F}^{\mu\nu}\right\}$$

where $\mathcal{F}^{\mu\nu} = \mathcal{F}^{\mu\nu}_{,\nu}\ \delta p(X), = \mathcal{F}'\left(\rho 0(x)\delta\rho(x)\right)$

The uniperturbed equation:

$$G_{\mu\nu}^{(0)} = R_{\mu\nu}^{(0)} - \frac{1}{2} R^{(0)} g^{(0)\mu\nu} = K(\rho_0 + p_0) V^{\mu} V^{\nu} - p_0 g^{(0)\mu\nu}$$

are annual to have been solved (Robertson Walker solution)

The stochastic input to the perturbed equation in (($\mathcal{F}$<u>μν</u>(X))) and the Stochastic output variable to be solved for are

$$\left\{\delta\rho(X), \delta p(X), \delta v^{\mu}(X), h_{\mu\nu}(X)\right\}$$

According to (1) and (2), there satisfy linear second order partial differential equation with coefficient depending on the space time coordinating (*i.e* expressible in teams of $g_{\mu\nu}^{(0)}$ ≃ the Robertson or Walker solution. Thus the solution can be expressed as)

$$\delta\rho(X) = \int K_0^{\mu\nu}(X,Y)\mathcal{F}_{\mu\nu}(Y)d^4y$$

$$\delta\rho(X) = \int K_1^{\mu\nu}(X,Y)\mathcal{F}_{\mu\nu}(y)d^4y$$

$$= \mathcal{F}'(\rho_0(X))\delta\rho(X)$$

$$\delta v\underline{\mu}(X) = \int K_2^{\mu\nu\beta}(X,Y)\mathcal{F}_{\alpha\beta}(y)d^4y$$

$$h^{\mu\nu}(X) = \int K_{3\mu\nu}^{\alpha,\beta}(X,Y)\mathcal{F}_{\alpha\beta}(y)d^4y$$

The problem is to determine the kernels K_0, K_1, K_2, K_3 and hence the corrections in the output variable.

❖❖❖❖❖

[229] G = H Ⓢ N N = Abelian subgroup of G : $hNh^{-1} = N \forall h \in H$

Let $\chi \varepsilon \widehat{N}$ (character of N) N normal in G

$g = (h, n)$ $\qquad hNh^{-1} = \mathrm{N}\ \forall h \in H$

$$(h_1, n_1)\cdot(h_2, n_2) = (h, h_2, \alpha_h, (n_2)n_1)$$

$h \to \alpha_h$ from $H \to$ Ant (N) is a homomorphism *i.e.*

$\alpha_{h1h2} = \alpha_{h1} .\alpha_{h2}, h_1, h_2 \in H$

Let V be and irrep of H. Consider:

$(h, n) \to \chi(n)V_h$

$(h_1, n_1).(h_2, n_2) = (h_1\, h_2, \alpha_{h1}(n_2)n_1) = (h, n)$ say

$$\chi(n)V_h = \chi(\alpha_{h_1}(n_2)n_1)V_{h_1h_2}$$

$$= \chi(\alpha_{n_1}(n_2))\chi(n_1)V_{h_1}V_{h_2}$$

$$= \beta_{n1}^{-1}(\chi)(n_2)\chi(n_1)V_{h_1}V_{h_2}$$

$$= \chi_{h_1}(n_2)\chi(n_1)V_{h_1}V_{h_2}$$

where $\qquad \chi_h = \beta_h^{-1}(\chi)\ \beta_h^{-1}(\chi) = \chi.\alpha_h$

is a character of N *i.e.* $\beta_h(\chi) \in \widehat{N}\ \forall h \in H$

The above calculation knows that $(h, n) \to \chi(n)V_h$ is not necessity a rep'n of G

Let

$$H\chi = \{h \in H | \beta_h\chi = \chi\}$$

isotropy group of $\chi \in \widehat{N}$ in *H*.
Let *V* be an group of $H\chi$.

Define $U\chi((h, n)) = \chi(n)V_h \; n \in N_1 h \in H\chi$

$$U\chi((h_1, n_1)) \cdot U\chi((h_2, n_2)) = \chi(n_1)V_{h_1}\chi(n_2)V_{h_2}$$

$$= \chi(n_1 n_2)V_{h_1 h_2}, h_1, h_2 \in H\chi, n_1, n_2 \in N$$

$$U\chi((h_1, n_1)\cdot(h_2, n_2)) = U\chi((h_1 h_2, \alpha_{n_1}(n_2)n_1))$$

$$= \chi(\alpha_{n_1}(n_2)n_1)V_{h_1, h_2}$$

$$= \beta_{h_1^{-1}}\chi(n_2)\cdot\chi(n_1)V_{h_1 h_2} = \chi(n_2)\chi(n_1)V_{h_1 h_2}$$

$$= \chi(n_2 n_1)V_{h_1 h_2} = \chi(n_1 n_2)V_{h_1 h_2}$$

Thus,
$U\chi$ is a representation of $H\chi\; S\, lN = G_\chi$

Let $O_\chi = \{\beta_h \chi | h \in H\}$ = orbit of $\chi \in \widehat{N}$ under *H*.
Let *G* be a group and *H* a subgroup of *G*.
Let *V* be a rep'n of *H* in a vector space *Y*.
Let f: $G \to \gamma$ be such that
$f(gh) = V_h f(g) \forall h \in H$

Define $Ugf(x) = f(g^{-1}x)$, $g, x \in G$.

Clearly $(U_g f)(xh) = f(g^{-1}xh) = V_h f(g^{-1}x)$

$$= V_h(U_g f)(x) \forall h \in H, g, x \in G$$

This if

$$\mathcal{F} = \{f : G \to \gamma | f(gh) - V_h f(g) \forall g \in G, h \in H\}$$

then

$g \in G \to U_g : \mathcal{F} \to \mathcal{F}$ is a rep'n is $\mathcal{F}$.
U_g is called the representation indeed by *V* from *H* to *G* and knows by

$$U = Ind_H^G(V)$$

Coming back to the earlier them suppose *V* is an irrep of $H\chi$. Then, $U\chi$ is an group of $G\chi = H\chi \circledS N$. It can than be shown that $Ind_{G\chi}^G(U\chi)$ is an irep of *G*.

In the general case of in induced repairs from *HG* we note that $\mathcal{F}$ can also be usualized on $\underset{\alpha \in G/H}{\oplus} Y_\alpha$.

where $Y_\alpha = Y \,\forall\, \alpha$. For $g \in G$

$Ug\mathcal{F} \in Y_{g\alpha} \,\forall\, \mathcal{F} Y_\alpha$

(for the special case under considerator, $\chi \in \widehat{N}$)

$H_\chi = \{h \in H \mid \beta_h \chi = \chi\}$ and $O_\chi = \{\beta_h \chi : h \in H\}$

Then $O_\chi \cong H/H_\chi$,

and if $G_\chi = H_\chi \circledS N$, then $O_\chi \cong G/G_\chi$

$(H \circledS N)/(H_\chi \circledS N)$. Let $V\chi$ be a rep'n of $H\chi$ in Y and put $Y\chi = \underset{\alpha \in O\chi}{\oplus} Y_\alpha$, $Y_\alpha \cong Y$.

Then let $V = Ind_{H\chi}^{H}(V\chi)$

For $h \in H, f \in Y_\alpha$

$V(h)fY \in Y\beta_{h\alpha})$

❖❖❖❖❖

[230] Calculate using perturbation theory the statistics of a charged particle moving in a random electromagnetic field of the form

$$\underline{E}(t, \underline{r}) = \varepsilon \sum_{k=1}^{p} \xi_k(t) \underline{\psi}_k(\underline{r}) + \underline{E}_0(t)$$

$$\underline{B}(t, \underline{r}) = \varepsilon \sum_{k=1}^{p} \eta_k(t) \underline{\psi}_k(\underline{r}) + \underline{B}_0(t)$$

where $\{\xi_k(\cdot), \eta_k(\cdot)\}$ are while noise process

The equation of motion are

$$d\underline{r}_{(t)} = \underline{V}(t)dt$$

$$d\underline{V}(t) = \underline{E}_0(t)dt + \varepsilon \sum_{k=1}^{p} \underline{\psi}_k(\underline{r}(t))\sigma_k dB_k(t) + \underline{V}(t) \times \underline{B}_0(t)dt$$
$$+ \varepsilon \sum_{k=1}^{p} \underline{V}(t) \times \underline{\psi}_k(\underline{r}(t))\rho_k dB_k(t)$$

where $\{B_k(\cdot)\}_{k=1}^{p}$ are independent Brownian motion process. This can be cast is state variable form a

$$d\begin{pmatrix} \underline{r}(t) \\ \underline{r}(t) \end{pmatrix} = \underline{f_0}(t)dt + \underline{\underline{F}}(t)\begin{pmatrix} \underline{r}(t) \\ \underline{V}(t) \end{pmatrix}dt + \varepsilon \sum_{k=1}^{p} \Phi_k\left(\underline{r}(t)dB_k(t)\right)$$

$$+\varepsilon \sum_{k=1}^{p} \Phi_{2k}\left(\underline{r}(t)\right)\underline{V}(t)dB_k(t)$$

This is in term a special case of

$$d\underline{\xi}(t) = \underline{f_0}(t)dt + \underline{\underline{F}}(t)\underline{\xi}(t)dt$$

$$+\varepsilon \sum_{k=1}^{p} \underline{\varphi_k}\left(\underline{\xi}(t)\right)dB_k(t)$$

We identify the different vector and matrix valued function

$$\underline{f_0}(t) = \begin{bmatrix} \underline{0} \\ \underline{E_0}(t) \end{bmatrix}, \ \underline{\underline{F}}(t) = \begin{bmatrix} \underline{\underline{0}} & \underline{\underline{I}}_3 \\ \underline{\underline{0}} & \underline{\underline{\Omega}}_0(t) \end{bmatrix}$$

$$\underline{\underline{\Omega}}_0(t) = \begin{bmatrix} 0 & B_{03}(t) & -B_{02}(t) \\ B_{03}(t) & 0 & -B_{01}(t) \\ B_{02}(t) & -B_{01}(t) & 0 \end{bmatrix}$$

$$\underline{\underline{\Phi}}_{1k}(\underline{r}) = \sigma_k \underline{\psi_k}(\underline{r}),$$

$$\underline{\underline{\Phi}}_{2k}(\underline{r}) = \rho_k \begin{bmatrix} 0 & \psi_{k3}(\underline{r}) & -\psi_{k2}(\underline{r}) \\ \psi_{k3}(\underline{r}) & 0 & -\psi_{k1}(\underline{r}) \\ \psi_{k2}(\underline{r}) & -\psi_{k1}(\underline{r}) & 0 \end{bmatrix}$$

$$\underline{\varphi_k}(\underline{\xi}) = \underline{\underline{\Phi}}_{1k}(\underline{r}) + \underline{\underline{\Phi}}_{2k}(\underline{r})\underline{V}$$

$$\underline{\xi} = \begin{bmatrix} \underline{r} \\ \underline{v} \end{bmatrix}$$

❖❖❖❖❖

[231] Quantisation of a stochastic differential equation using the method of Evan Husdson flows

$$d\underline{\xi}(t) = \underline{\mu}\left(\underline{\xi}(t)\right)dt + \underline{\underline{\sigma}}\left(\underline{\xi}(t)\right)d\underline{B}(t) + \int \underline{f}\left(t, \underline{\xi}(t), \underline{x}\right)dN(t, dx)$$

Where $N(t,.)$ is a Poisson field with rate

$\mathbb{E}(N(t, E)) = \lambda(t)\, F(E)$ F being a measure

on (X, ξ). Then,

$$d\psi\left(\underline{\xi}(t)\right) = \left(\mathcal{L}_0\psi\right)\left(\underline{\xi}(t)\right)dt + \sum_{k=1}^{p}\left(\mathbb{A}_k\psi\right)\left(\underline{\xi}(t)\right)dB_k(t)$$

$$+\int\left(\underline{\psi}\left(\underline{\xi}(t)+\underline{f}\left(t,\underline{\xi}(t)x\right)\right)-\underline{\psi}\left(\underline{\xi}(t)\right)\right)dN(t,dx)$$

$$\mathcal{L}_0\psi\left(\underline{\xi}\right) = -\underline{\mu}\left(\underline{\xi}\right)^T\frac{\partial\psi\left(\underline{\xi}\right)}{\partial\underline{\xi}}+\frac{1}{2}Tr\left(\underline{\underline{\sigma}}\left(\underline{\xi}\right)\underline{\underline{\sigma}}^T\left(\underline{\xi}\right)\nabla_{\xi}\nabla_{\xi}^T\psi\right)$$

A special case is

$$A_k\psi\left(\underline{\xi}\right) = \sum_{m=1}^{n}\sigma_{mk}\left(\underline{\xi}\right)\frac{\partial}{\partial\xi_m},\ 1\le k\le p$$

$$d\underline{\xi}(t) = \underline{\mu}\left(\underline{\xi}(t)\right)dt+\underline{\underline{\sigma}}\left(\underline{\xi}(t)\right)d\underline{B}(t)\ +\sum_{k=1}^{p}\underline{f}_k\left(t_1\,\underline{\xi}(t)\right)dN_k(t)$$

where $N_1(.),\ldots,\ N_p(.)$ are independent Poisson process with rate $\lambda_1,,\ldots,\ \lambda_p$ respectively.

Then,

$$d\psi\left(\underline{\xi}(t)\right) = \mathcal{L}_0\psi\left(\underline{\xi}(t)\right)dt$$

$$+\sum_{k=1}^{p}\mathcal{L}_{0_k}\psi\left(\underline{\xi}(t)\right)db_k+\sum_{k=1}^{p}\left(\psi\left(\underline{\xi}+f_k\left(t,\underline{\xi}\right)\right)\right)-\psi\left(\underline{\xi}\right)dN_k$$

Now consider the quantum process $\Lambda_k(t), \mathrm{A}_k(t), A_k^+(t)$

where

$$d\Lambda_k\cdot d\Lambda_m = \delta_{km}d\Lambda_k,$$

$$dA_k.\ dA_j^+ = dt\delta_{kj},$$

$$dA_k\cdot dA_j^+ = dt\delta_{kj},\ dA_j^+dA_k$$

$$dA_j^+d\Lambda_k = d_{kj}\,dA_j$$

$$dA_k.\ dA_j^+ = dt\delta_{kj},$$

$$d\Lambda_k^m = \left(\Lambda_k+d\Lambda_k\right)^m-\Lambda_k^m$$

$$= \sum\Lambda_k^{r_1}d\Lambda_k\Lambda_k^{r_2}d\Lambda_k\cdots\Lambda_k^{rp-1}d\Lambda_k\,\Lambda_k^{r}$$

$$r_1+r_2+\cdot\cdot r_p = m-p,\ p\ge 1,\ r_1,.,r_p\ge 0$$

$$= \sum\Lambda_k^{r_1+r_2+\cdot r_p}d\Lambda_k$$

$$r_1+..+r_p = m-p$$

$$p\ge 1,\ r_1,\ldots,\ r_p\ge 0$$

$$= \left(\left(\Lambda_k + Id\right)^m - \Lambda_k^m\right) d\Lambda_k$$

Thus, is general

$$df\left(\Lambda_k\right) = \left(f\left(\Lambda_k + I\right) - f\left(\Lambda_k\right)\right) d\Lambda_k$$

More generally if $S_k(t)$, $k = 1, 2, \ldots$ are adapted process and

$$d\,M(t) = \sum_{k=1}^{p} Sk(t)\, d\Lambda_k(t), \text{ then}$$

$$df(M(t)) = \sum_{k=1}^{p} \left(f\left(M(t) + S_k(t)\right) - f\left(M(t)\right)\right) d\Lambda_k(t)$$

Let $\quad M(t) = A_k^m(t)$, Then

$$d\,M = m\, A_k^{m-1} dA_k \cdot F_k(t)$$

Let $dM = E_k(t)\, dA_k(t) + dA_k^+(t)$

Then

$$d(M^m) = \sum_{r_1 + r_2 = m-1, r_j \geq 0, j=1,2,3} M^{r_1} E_k M^{r_2} dA_k$$

$$+ \sum_{\substack{r_1 + r_2 = m-1, \\ r, \geq 0, j=1,2,3}} M^{r_1} F_k M^{r_2} dA_k^+$$

$$+ \sum_{\substack{r_1 + r_2 + r_3 = m-2, \\ r_j \geq 0, j=1,2,3}} M^{r_1} E_k M^{r_2} F_k M^{r_3} dt$$

Let $X(t) = e^{ipM(t)}$, where $M(t) = \int_0^t \left(E_k(s)\, dA_k(s) + F_k(s)\, dA_k^+(s)\right)$

Then

$$dX(t) = \sum_{m=0}^{\infty} \frac{(ip)^m}{\lfloor m} d\left(M^m(t)\right)$$

$$= \sum_{r_1, r_2 \geq 0} \frac{(ip)^{r_1 + r_2 + 1}}{\lfloor r_1 + r_2 + 1} M^{r_1} E_k M^{r_2} dA_k$$

$$+ \sum_{r_1, r_2 \geq 0} \frac{(ip)^{r_1 + r_2 + 1}}{\lfloor r_1 + r_2 + 1} M^{r_1} F_k M^{r_2} dA_k^+$$

$$+\sum_{r_1, r_2, r_3 \geq 0} \frac{(ip)^{r_1+r_2+r_3+2}}{\underline{|r_1+r_2+r_3+2}} M^{r_1} E_k M^{r_2} F_k M^{r_3} dt$$

$$= ip \sum \frac{\underline{|r_1}\,\underline{|r_2}}{\underline{|r_1+r_2+1}} \frac{(ir)^{r_1} M^{r_1}}{\underline{|r_1}} E_k \frac{(ip)^{r_2}}{\underline{|r_2}} M^{r_2} dA_k + \ldots$$

$$= ip \sum \frac{\Gamma(r_1+1)\Gamma(r_2+1)}{\Gamma(r_1+r_2+2)} \ldots$$

$$= ip \sum_{r_1, r_2, \geq 0} \beta(r_1+1, r_2+1) \frac{(ip)^{r_1} M^{r_1}}{\underline{|r_1}} \cdot E_k \cdot \frac{(ip)^{r_2}}{\underline{|r_2}} M^{r_2} dA_k + \ldots$$

$$= ip \sum_{r_1, r_2, \geq 0} \left(\int_0^1 x^{r_1} (1-x)^{r_2} dx \right) \frac{(ip)^{r_1} M^{r_1}}{\underline{|r_1}} \cdot E_k \cdot \frac{(ip)^{r_2} M^{r_2}}{\underline{|r_2}} dA_k + \ldots$$

$$= ip \left(\int_0^1 \exp(ip\,x M_z).E_k.\exp(ip(1-x)M)\,dx \right) dA$$

$$+ ip \left(\int_0^1 \exp(ip\,x M) F_k \exp(ip(1-x)M)\,dx \right) dA_k^{+}$$

$$+(i\,p)^2 \sum_{r_1, r_2 r_3 \geq 0} \frac{\underline{|r_1}\,\underline{|r_2}\,\underline{|r_3}}{\underline{|r_1+r_2+r_3+2}} \frac{(ip)^{r_1} M^{r_1}}{\underline{|r_1}}$$

$$E_k \frac{(ip)^{r_2} M^{r_2}}{\underline{|r_2}} F_k \frac{(ip)^{r_3} M^{r_3}}{\underline{|r_3}} dt$$

Last term

$$-p^2 \sum_{r_1+r_2+r_3 \geq 0} \frac{\Gamma(r_1+1)\Gamma(r_2+1)\Gamma(r_3+1)}{\Gamma(r_1+r_2+r_3+3)} \frac{(ip)^{r_1} M^{r_1}}{\underline{|r_1}} E_k$$

$$\frac{(ip)^{r_2} M^{r_2}}{\underline{|r_2}} F_k \frac{(ip)^{r_3} M^{r_3}}{\underline{|r_3}} dt$$

$$= -p^2 \sum_{r_1, r_2 r_3 \geq 0} \beta(r_1+1, r_2+1)\beta(r_1+r_2+2, r_3+1)$$

$$= \frac{(ip)^{r_1} M^{r_1}}{\underline{|r_1}} E_k \frac{(ip)^{r_2} M^{r_2}}{\underline{|r_2}} F_k \frac{(ip)^{r_3} M^{r_3}}{\underline{|r_3}} dt$$

$$= -p^2 \sum_{r_1, r_2, r_3 \geq 0} \int_0^1 \int_0^1 x^{r_1} (1-x)^{r_2} y^{r_1+r_2+1} (1-y)^{r_3} \, dxdy$$

$$\frac{(ip)^{r_1} M^{r_1}}{\lfloor r_1} E_k \frac{(ip)^{r_2} M^{r_2}}{\lfloor r_2} F_k \frac{(ip)^{r_3} M^{r_3}}{\lfloor r_3}$$

$$= \Bigg(-p^2 \int_{[0,1]^2} y \exp(ipxyM) E_k$$

$$\exp(ip(1-x)yM) F_k \cdot \exp(ip(1-y)M) dx\, dy \Bigg) dt$$

❖❖❖❖❖

[232] Filtering for independent increment measurement noise:

$\{\chi(t), t \geq 0\}$ is a Markov process with generator L_t:

$$\mathbb{E}\left[\varphi(X(t+\delta t)) \middle| X(f) = x\right] = \varphi(x) + \delta t \cdot L_t \varphi(x) + O\left(st^2\right)$$

Measurement model :

$$dZ(t) = \underline{h}_t(X(t)dt) + d\underline{V}(t)$$

where $V(\cdot)$ is a stationary independent process independent of $X(\cdot)$.

$$\mathbb{E}e^{i\underline{\alpha}^T \underline{V}(t)} = e^{t\psi(\underline{\alpha})}$$

Density of $\underline{V}(t)$:

$$p_v(v) = \frac{1}{(2\pi)^m} \int e^{t\psi(\underline{\alpha})} e^{-i\underline{\alpha}^T \underline{v}} d^m \alpha$$

$$= \frac{1}{(2\pi)^m} \int e^{t\psi(iD)} e^{-i\underline{\alpha}^T \underline{v}} d^m \alpha$$

where $D = \dfrac{\partial}{\partial \underline{v}}$

So, $p_v(v) = e^{t\psi(iD)} \{\delta(\underline{v})\}$

Let $Y_t = \{\underline{Z}(s) : S \leq t\}$.

$$p\left(X(t+\delta t) \middle| Y_{t+\delta t}\right) = \frac{p(X(t+\delta t), Yt, dZ(t))}{p(Y_t, dZ(t))}$$

$$= \frac{\int p\left(dZ(t)\middle|X(t+\delta t)\right)p\left(X(t+\delta t)\middle|X(t)\right)p\left(X(t)\middle|Y_t\right)dX(t)}{\int \text{numerator}(\cdot)dX(t+\delta t)}$$

$$= \frac{\int p\left(dZ(t)\middle|X(t)\right)p\left(X(t+\delta t)\middle|X(t)\right)p\left(X(t)\middle|Y_t\right)dX(t)}{\int p\left(dZ(t)\middle|X(t)\right)\cdot p\left(X(t)\middle|Y_t\right)dX(t)}$$

$\pi_t(\varphi(X)) \triangleq \mathbb{E}(\varphi(X(t))Y_t)$.

Then $\pi_{t+\delta t}(\varphi(X))$

$$= \frac{\int\left(\varphi(X(t))+(L_t\varphi)(X(t))dt\right)k_p dV_{(t)}\left(dZ(t)-h_t(X(t))dt\right)p\left(X(t)\middle|Y_t\right)dX(t)}{(\text{Numerator with } \varphi=1)}$$

$$\pi_t + dt(\varphi(X)) = \frac{\sigma_{t+dt}(\varphi(x))}{\sigma_{t+dt}(1)}$$

where $\sigma_{t+dt}(\varphi)(X)) = \int\left(\varphi(x(t))+L_t\varphi(x(t))dt\right)$

$$\left(1+dt\,\psi(iD)\right)\delta\left(dZ(t)-h_t(x(t))dt\right)\; p\left(X(t)\middle|Y_t\right)dX(t)$$

$$= \pi_t\left\{\varphi(X)\delta\left(dZ(t)-h_t(X)dt\right)\right\}$$

$$+\, dt.\pi_t\left\{L_t\varphi(X)\left(\delta\left(Z(t)-h_t(X)dt\right)\right)\right\}$$

$$+\, dt\cdot\pi_t\left\{\varphi(X)\cdot\left(\psi(iD)\delta\right)\left(dZ(t)-h_t(X)dt\right)\right\}$$

$$\sigma_t + dt\,(1) = \pi_t\left\{\varphi\left(dZ(t)-h_t(X)dt\right)\right\}$$

$$+\, dt\cdot\pi t\left\{\varphi(X)\cdot\left(\psi(iD)\delta\right)\left(dZ(t)-h_t(X)dt\right)\right\}$$

So,

$$\pi_{t+dt}(\varphi(X)) = \frac{\left[\left\{\pi_t\delta\left(dZ-h_t(X)dt\right)\varphi(X)\right\}+\pi_t\left\{Lt\varphi(X)\left(\delta\left(dZ-h_t(X)dt\right)\right)+\varphi(X)\cdot\left(\psi(i\,D)\delta\right)\left(dZ-h_t(X)dt\right)\right\}\right]}{\left[\pi_t\left\{\delta\left(dZ-h_t(X)dt\right)\right\}+dt\,\pi_t\left\{\left(\varphi(X)\cdot\left(\psi(i\,D)\delta\right)\left(dZh_t(X)\right)dt\right)\right\}\right]}$$

$$= \frac{\pi_t\left\{\delta\left(dZ-h_t(X)dt\right)\varphi(X)\right\}}{\pi_t\left\{\delta\left(dZ-h_t(X)dt\right)\right\}}$$

$$+\, dt\cdot\frac{\left\{\pi_t\left\{\left(L_t\varphi(X)\delta\left(dZ-h_t(X)dt\right)+\varphi(X)\cdot\left(\psi(i\,D)\delta\right)\right)\left(dZ-h_t(X)dt\right)\right\}\right.}{\pi_t\left\{\delta\left(dZ-h_t(X)dt\right)\right\}}$$

$$-\left(\frac{\pi_t\{\varphi(X)(\psi(iD)\delta)(dZ-h_t(X)dt)\}}{\pi_t\{\delta(dZ-h_t(X)dt)\}}\right)$$

$$\times\pi_t\{\delta(dZ-h_t(X)dt)\varphi(X)\}\}$$

❖❖❖❖❖

[233] Hamiltonian formulation of geodesic motion

$$\mathcal{L} = \left(g_{00}(t,\underline{r})+2g_{0k}(t,\underline{r})\overset{\circ}{X}^k+gk(t,\underline{r})\overset{\circ}{X}^k\overset{\circ}{X}^m\right)^{1/2} = \frac{d\tau}{dt}$$

$$\overset{\circ}{\xi} = \frac{d\xi}{dt}$$

$$\underline{r} = \left(\left(X^k\right)\right)_{k=1}^3$$

$$p_k = \frac{\partial\mathcal{L}}{\partial\overset{\circ}{X}^k} = g_{k\mu}\frac{dX^\mu}{d\tau}$$

$$H = p_k\overset{\circ}{X}^k-\alpha = g_{k\mu}\frac{dX^\mu}{d\tau}\frac{dX^k}{dt}-\mathcal{L}$$

$$= \frac{d\tau}{dt}g_{k\mu}\frac{dX^k}{d\tau}\frac{dX^\mu}{d\tau}-\frac{d\tau}{dt}$$

$$= \frac{d\tau}{dt}\left[g_{k\mu}\frac{dX^k}{d\tau}\frac{dX^\mu}{d\tau}-1\right]$$

$$= \frac{d\tau}{dt}\left[1-g_{0\mu}\frac{dt}{d\tau}\frac{dX^\mu}{d\tau}-1\right]$$

$$= -g_{0\mu}\frac{dX^\mu}{d\tau}$$

Let $$p_0 = g_{0\mu}\frac{dX^\mu}{d\tau} = -H$$

Then $$p_\mu = g_{\mu\nu}\frac{dX^\nu}{d\tau}$$

$$\frac{dX^\mu}{d\tau} = g^{\mu\nu}p_\nu = p^\mu$$

$\therefore \quad g_{\mu\nu}p^{\mu}p^{\nu} = 1$

$$g_{00}p^{02} + 2g_{0r}p^{r} + g_{rs}p^{r}p^{s} - 1 = 0$$

$$p^{0} = -\frac{g_{0r}p^{r}}{g_{00}} + \left[\left(\frac{g_{0r}g_{0s} - g_{rs}}{g_{00}^{2}}\right)p^{r}p^{s} + \frac{1}{g_{00}}\right]^{1/2}$$

So $\quad H(t, \underline{r}, \underline{p})\left(\underline{p} = (p_r)_{r=1}^{3}\right)$

$$= -p_0 = g_{0\mu}p^{\mu} = -g_{00}p^{0} - g_{0r}p^{r}$$

$$= -g_{00}\left[\gamma_{rs}p^{r}p^{s} + g_{00}\right]^{1/2}$$

$$= -g_{00}\left[\gamma_{rs}p^{r\mu}p^{s\nu}p_{\mu}p_{\nu} + g_{00}^{-1}\right]^{1/2} \quad \gamma_{rs} = \frac{\left(g_{0r}g_{0s} - g_{rs}\right)}{g_{00}^{2}}$$

So

$$p_0^2 = g_{00}^2\left[g_{00}^{-1} + g^{r0}g^{s0}\gamma_{rs}p_0^2 + 2g^{r0}g^{sk}\gamma_{rs}p_0p_k + g^{rm}g^{sk}\gamma_{rs}p_mp_k\right]$$

This is a quantum equation for $p_0 = -H$ and can bve solved in term of g $\{p_k\}_{k=1}^{3}$
Alternatively

$g^{\mu\nu}p_{\mu}p_{\nu} = 1$

gives $\quad g^{00}p_0^2 + 2g^{0r}p_0p_r + g^{rs}p_rp_s - 1 = 0,$

So

$$p_0 = -\frac{g^{0r}}{g^{00}} + \left[\frac{\left(g^{0r}g^{0s} - g^{rs}\right)}{g^{00^2}}p_rp_s + \frac{1}{g^{00}}\right]^{1/2}$$

$$= -h^{r} + \left[\gamma^{rs}p_rp_s + \frac{1}{g_{00}}\right]^{1/2}$$

where $\quad h^{r} = \frac{-g^{0r}}{g^{00}}, \gamma^{rs} = \frac{\left(g^{0r}g^{0s} - g^{rs}\right)}{g^{00^2}}$

*Boltzmann kinetic tramport equation in gtr.

$$G_{\mu\nu} = R_{\mu\nu} - \frac{1}{2}Rg_{\mu\nu}$$

$$G_{\mu\nu} = K_{m0}\int f\left(t, \underline{r}, \underline{p}\right)\cdot d^{3}p \cdot V^{\mu}V^{\nu}$$

$$\underline{p} = \left((p_r)\right)_{r=1}^{3}$$

$$V_r = \frac{dX^{r}}{d\tau}, V_0 = \frac{dt}{d\tau}$$

$$p_r = g_{r\mu}\frac{dX^{\mu}}{d\tau} = g_{r\mu}V^{\mu}\, p_0 = g_{0\mu}V^{\mu}$$

$$\Rightarrow \quad p_{\mu} = g_{\mu\nu}V^{\nu}$$

$$V^{\mu} = g^{\mu\nu}p_{\nu} = g^{\mu\nu}p_r + g^{\mu 0}p_0$$

$$p_0 = -H\left(t, \underline{r}, \underline{p}\right) \qquad \underline{p} = (p_r)$$

So,

Let $$g_{\mu\nu} = \eta_{mv} + h_{mv}(t, \underline{r})$$

Then
$$\begin{aligned} G_{\mu\nu} = &+ C_1(\mu\nu\alpha\beta\rho\sigma)h_{\alpha\beta,\rho\sigma} + C_2(\mu\nu\alpha\beta\rho)h_{\alpha\beta,\rho} \\ &+ C_3(\mu\nu\alpha_1\beta_1\rho_1\alpha_2\beta_2\rho_2)h_{\alpha_1\beta_1,\rho_1}h_{\alpha_2\beta_2,\rho_2} \\ &+ C_4(\mu\nu\alpha_1\beta_1\rho\alpha_2\beta_2)h_{\alpha_1\beta_1,\rho}h_{\alpha_2\beta_2} \\ &+ C_5(\mu\nu\alpha_1\beta_1\rho_1\alpha_2\beta_2)h_{\alpha_1\beta_1,\rho_1\sigma_1}h_{\alpha_2\beta_2} + O\left(\|h\|^3\right) \end{aligned}$$

where the $C_j^{,s}$ are constants.

❖❖❖❖❖

Appendix

Syllabus for short course on electromagnetic control:

[1] Basics of electromagnetic: The Maxwell equations and their retarded potential solutions in constant dielectric and permeability media.

[2] Motion of charged particles in electromagnetic fields: Perturbative solutions to the Lorentz equations of motion.

[3] Motion of charged particles in random electromagnetic fields: The stochastic differential equation approach to computing the statistics of the motion.

[4] Jerks in external electromagnetic fields caused by Markov chain jumps in the input current density fields.

[5] Basic optimal control theory for continuous time dynamical systems: The variational approach and the Bellman-Hamilton-Jacobi dynamic programming approach.

[6] Optimal control of charged particle trajectories in electromagnetic fields.

[7] Optimal control of electromagnetic fields by input current sources.

[8] Optimal control of charged particle motion and electromagnetic fields with constraints on the input fields and input currents.

[9] Stochastic optimal control theory.

[10] Optimal control of the wave function and density matrix of a quantum system by modulation of the perturbing potential.

[11] Optimal control of the average values of a set of observables in a quantum system.

[12] Optimal control of the wave function of a quantum system using external electromagnetic fields.

[13] Optimal control of the electromagnetic fields produced by and antenna using the input current.

[14] Optimal control of the electromagnetic field at the output of a waveguide using input wire current probes.

[15] Optimal control of the motion of a conducting fluid by input electromagnetic fields.

[16] Optimal control of the velocity fields and electromagnetic field in a conducting fluid by the input current density field.

[17] Optimal control of scattering amplitudes in quantum field theory by external current sources.

[18] Optimal control of a *d* link Robot arm carrying current using an external electromagnetic field.

[19] Optimal control of a spinning top carrying current (single link Robot arm) with an external electromagnetic field.

[20] Basics of Lagrangian and Hamiltonian mechanics.

[21] Basics of fluid dynamics.

References

[1] Karishma Sharma, D.K. Upadhyay and H. Parthasarathy, Analysis of waveguides, resonators and microstrip patch antennas with arbitrary cross section using analytical functions of a complex variable", Technical Report, NSIT.

[2] Rohit Singla, Vijayant Agrawal and Harish Parthasarathy, " Studied in control of master-slave Robot systems", Technical Report, NSIT, 2014.

[3] Lokit Sharma, Harish Parthasarathy and D.K.Upadhyay," Computation of perturbed eigenmode solution in non-uniform transmission lines, IEEE conference, Communicated.

[4] Rohit Singh, Harish Parthaarathy and D.K. Upadhyay," Studies in transmission lines having distributed parameters that fluctuate in space time", Technical Report, NSIT, 2014.

[5] Ashwini, Harish Parthasarathy and D.K. Upadhyay," Adaptive LMS algorithm based estimation of an LTI filter based on quantum mechnical measurement of the wave function.

[6] Kumar Gautam, Navneet Sharma, T.K. Rawat and H.Parthasarathy," Computation of the information transmitted over a quantum channel described by Schrodinger's equation with the input being a classical source that modulates the system Hamiltonian.

[7] Neelu Nagpul, Vijayant Agrawal and Harish Parthasarathy, Maximum likelihood estimator of the pd control coefficient of a master slave robot with application to master tremor and environment noise estimation, Technical report, NSIT, 2014.

[8] Naman Garg, H.Parthasarathy,D.K.Upadhyay and Vijayant Agrawal," Estimating the configuration of a two link 3-D Robot carrying current from the electromagnetic filed pattern", Technical report, NSIT, 2014.

[9] Harish Parthasarathy and J.R.P. Gupta, K-variate generalization of Bernoulli filters with application to target signal filtering, Technical Report, NSIT, 2014.

[10] Navneet Sharma, T.K. Rawat and H.Parthasarathy," Initial state and channel estimation for quantum sytems", Technical Report, NSIT, 2014.

[11] Kumar Gautam, H.Parthasarathy and T.K.Rawat," Simulink model for 3-D harmonic oscillator unitary evolution tracking in a spatially uniform electromagnetic field, Technical report, NSIT, 2014.

[12] Naman Garg, Rohit Rana and Harish Parthasarathy," Noisy image of an object by a robot camera on a screen", Technical report, NSIT, 2014.

[13] Pragya Shilpi, Harish Parthasarathy and D.K. Upadhyay, Studies in em wave propagation in inhomogeneous and anisotropic media within wave-guides and cavity resonators, Technical Report, NSIT, 2014.

[14] Naman Garg, D.K. Upadhyay, Vijayant Agrawal and Harish Parthasarathy," Estimating parameters of nonlinear system with+statistical performance analysis," Technical Report, NSIT, 2014.

[15] Rohit Rana, Vijayant Agrawal and Harish Parthasarathy,: Wavelet based parameter estimation in nonlinear dynamical system with data comression", Technical Report, NSIT, 2014.

[16] Pradeep, J.K. Misra and Harish Parthasarathy," Fluid flow in Nonuniform cross section pipes with nonlinear viscosity, Technical report, NSIT, 2014.

[17] Navneet Sharma, Tarun K.Rawat, Kumar Gautam and Harish Parthasarathy, Gravitational swarm optimization for the design of quantum unitary gates, Technical report, NSIT, 2014.

[18] Vijayant Agrawal and Harish Parthasarathy, Studies in quantum mechanical models for nano-robotics, Technical report, NSIT, 2015.

[19] Naman Garg, D.K. Upadhyay, Vijayant Agrawal and Harish Parhtasarathy," Approximately nonlinear filtering in discrete time LIP models with applications to Robotics, Technical report, NSIT, 2015.

[20] Shilpa Dua, Jyotsna Singh and Harish Parthasarathy," Radon transform based rotation estimation of two and three dimensional images corrupted by noise," Technical report, NSIT, 2015.

[21] Manjeet Kumar, Tarun K.Rawat and Harish Parthasarathy, Gravitational search algorithm (GSA) for optimal estimation of filter coefficient and fractional delays of an LTI system, Technical Report, NSIT, 2015.

[22] Kumar Gautam, Garv Chauhan, Tarun Kumar Rawat, Harish Parthasarathy and Navneet Sharma," Realization of quantum gates using a 3-D quantum harmonic oscillator perturbed by an electromagnetic field", Technical report, NSIT, 2015.

[23] Kumar Gautam, Ph.D thesis," Design of quantum gates using physical systems with Schrodinger evolution, NSIT, 2015.

[24] Navneet Sharma, Ph.D thesis," Implementing quantum receivers and designing quantum gates using the GSA algorithm, NSIT.

[25] Rohit Singla, Harish Parthasarathy and Vijayant Agrawal," Levy noise in classical and quantum robot", paper in preparation, 2015.

[26] Rohit Rana, Vijayant Agrawal and Harish Parthasarathy," Lie group theoretic algorithms for constructing the extended Kanman filter for d-link robots," Paper in preparation, 2015.

[27] Vijayant Agrawal and Harish Parthasarathy," Combined state observer and state tracker for stochastic systems"m Technical report, NIT, 2015.

[28] Manjeet Kumar, Tarun K.Rawat and Harish Parthasarathy," Design of fractional delay Voltera filters in discrete time. Technical report, NSIT, 2015.

[29] Vishal Pandey, Jyotsna Singh and Harish Parthasarathy, "Estimation the rotation angle and velocity for two dimensional moving image fields,", Technical report, NSIT, 2015.

[30] S. Helgason, "Differential geometry, Lie groups and Symmetric spaces", American Mathematical Society,

[31] Vijayant Agrawal and Harish Parthasarathy, "Design of real time parameter estimator and disturbance observer for d link robots", Technical report, NSIT, 2015.

[32] Vijayant Agrawal and Harish Parthasarathy, "A generalized Lyapunov matrix for faster rate convergence of distrubance observer error in roboticcs" Tecnical report, NSIT, 2015.

[33] Vijayant Agrawal and Harish Parthasarathy, "EKF based filtering and control of a robot with disturance modeled as the sum of a dc and an ac component", Technial reporet, NSIT, 2015.

[34] Rohit Singla, Vijayant Agrawal and Harish Parthasarathy, "Stochastic Lyapunov energy", Nonlinear dynamice Journal , Springer, 2015.

[35] Rohit Rana, A. P. Mittal, Vijayant Agrawal and Harish Parthasarathy, "EKF for robot tremor estimation based on Lie group and Lie algebras, Technical report, NSIF, 2015.

[36] Rohit Rana, Vijayant Agrawal and Harish Parthasarathy, "Wavelet based tremor estimation in mechanical systems", IEEE conference, Indore, 2015.

[37] Naman Garg, Harish Parthasarathy, Vijayant Agrawal and D.K.Upadhyay "Sicuranza's LIP model applied to EKF based parameter estimation in a d-link robot", Technical report, NSIT, 2015.

Milton Keynes UK
Ingram Content Group UK Ltd.
UKHW031108180324
439528UK00004B/33

9 781032 384337